ENVIRONMENTAL SCIENCE

second edition

ENVIRONMENTAL SCIENCE

A GLOBAL CONCERN

William P. Cunningham

University of Minnesota

Barbara Woodworth Saigo

Southeastern Louisiana University

 Wm. C. Brown Publishers

Book Team

Editor *Kevin Kane*
Developmental Editor *Carol Mills*
Production Editor *Michelle M. Campbell*
Designer *Christopher E. Reese*
Art Editor *Margaret Rose Buhr*
Photo Editor *Lori Gockel*
Permissions Editor *Vicki Krug*
Visuals Processor *Amy L. Saffran*

 Wm. C. Brown Publishers

President *G. Franklin Lewis*
Vice President, Publisher *George Wm. Bergquist*
Vice President, Operations and Production *Beverly Kolz*
National Sales Manager *Virginia S. Moffat*
Group Sales Manager *Vincent R. Di Blasi*
Vice President, Editor in Chief *Edward G. Jaffe*
Marketing Manager *Paul Ducham*
Advertising Manager *Amy Schmitz*
Managing Editor, Production *Colleen A. Yonda*
Manager of Visuals and Design *Faye M. Schilling*
Production Editorial Manager *Julie A. Kennedy*
Production Editorial Manager *Ann Fuerste*
Publishing Services Manager *Karen J. Slaght*

WCB Group

President and Chief Executive Officer *Mark C. Falb*
Chairman of the Board *Wm. C. Brown*

Cover photo © IFA/Peter Arnold, Inc.

Unless otherwise credited, photographs © William P. Cunningham.

The credits section for this book begins on page 603, and is considered an extension of the copyright page.

Printed in the United States of America by Wm. C. Brown Publishers, 2460 Kerper Boulevard, Dubuque, IA 52001

10 9 8 7 6 5 4 3 2 1

My effort in this book is dedicated to the biosphere—a unique and dynamic system of which we all are a part—and to my children, who enjoy its beauty and its bounty. I have hope for the biosphere and for them. They represent its future . . . and they are good and intelligent people.

Barbara Woodworth Saigo

To Mary, Peggy, Mary Ann, and John, without whom this would not have been possible.

William P. Cunningham

C O N T E N T S

Preface xv

PART ONE Principles and Concepts

PART TWO Populations

PART THREE

Resources and Resource Economics

PART FOUR

Environmental Pollution and Other Dilemmas

PART FIVE Global Future

P R E F A C E

Environmental Science is a dynamic, rapidly-changing field that is becoming much more global in its outlook and concerns. Twenty years ago, the issues of greatest interest to most of us were local problems of air or water pollution, freeways cutting through our neighborhoods, or development in a nearby woods or rural area. Today, we live in an increasingly interconnected world. Modern communications and transportation make us aware of day-to-day conditions in places our parents never heard of. We are learning that what happens on the other side of the world can have a profound impact on us. Such problems as global climate change, loss of stratospheric ozone, ocean pollution, famine and war in faraway countries, and loss of tropical rain forests have become everyday subjects of discussion in newspapers and on television.

As citizens of this global community, we need a basis in scientific principles, as well as the social, political, and economic systems in which people live. Our primary purpose in writing this book is to bring together information from the natural sciences and human ecology to provide a foundation both for understanding the problems we face and finding ways to work to build a better future. We hope this book will be interesting, informative, and inspiring.

Our aim has been to achieve a balanced, realistic view of environmental conditions that neither wallows in "doom and gloom" pessimism, nor glosses over serious problems with unwarranted optimism. We hope you will reach the end of this book with a conviction that the need for environmental protection is urgent and essential. We can make our environment safer and more habitable, and there is time enough to do so if we begin to work now.

There are many opinions about why problems have arisen and what we should do about them. Our goal is not to force a single orthodox point of view, but to provide enough information so that students can analyze and discuss these challenging issues and form their own opinions about what should be done. Most of all, we hope this book will inspire its readers to become involved—individually and with others—to do something positive to protect and restore our collective environment.

AUDIENCE

This book is intended for use in a one- or two-semester course in Environmental Studies or Human Ecology at the college level. The vocabulary and level of discussion are simple and self-evident enough to be accessible to freshmen or sophomores with little or no science background. At the same time, we have presented enough data and scientific theory for the book to remain valuable as a reference source after the course is finished, or to be used as a text for an upper-division science course.

ORGANIZATION

This book is organized in five major sections that can be used in any sequence or combination, although we recommend beginning with at least an overview of part one for basic concepts that will enable deeper understanding of later topics.

Part one defines environmental science and presents some basic physical and ecological principles as a foundation for understanding environmental interactions. It also includes a chapter to introduce major natural ecosystems, as a preface to understanding how they are altered by human actions and why these actions may have broad significance. Part two covers fundamentals of population biology, and then applies these concepts to human populations.

Part three gives an overview of natural resources, and some of the issues and techniques in resource management. Part four starts with a brief survey of environmental health and toxicology, and then applies this knowledge to problems of pollution, waste management, and urbanization. Part five presents some options for our global future.

Beginning with a chapter on building sustainable cities, this section discusses environmental ethics and philosophy, activism, politics, environmental organizations, and some things each of us can do to help protect and improve our environment.

Except for chapters 2–4, the chapters within these sections are self-contained units that can be combined in any sequence that suits the interests of instructor and students. A two-semester class would probably cover all the material, while a single-semester class could omit some chapters.

THEMES

Global Environment

Our concern for worldwide environmental issues makes it essential to be familiar with geography and to be aware of conditions in countries far from our own. We have not limited our scope to the United States, therefore, but have drawn examples of environmental problems and ways people have solved those problems from all over the world.

Holistic Approach

This text emphasizes a holistic view of Earth and of environmental science. An in-depth, functional understanding of Earth and the interactions of its nonliving and living components requires a "systems" approach. The body of an organism is a complex, interactive system and is, in turn, part of increasingly larger systems. In a system, actions provoke responses. So it is with the environment and our activities within it and as a part of it.

Complex problems require creative solutions from many disciplines. Understanding our current situation requires knowledge not only from natural sciences, such as ecology, biochemistry, genetics, and environmental health, but also from social sciences, such as economics, history, and political science. While we can't present all these fields in a single book, we have tried to weave together information from many different fields to give a more complete view of where we are, how we got here, and what our options are for the future.

Social Justice

The worst environmental conditions are almost always suffered by the poorest people. Poverty, oppression, lack of education, and inequitable distribution of resources are often causative agents in environmental problems. We need wisdom and compassion, as well as knowledge, to make wise choices. We need a vision that works toward an economically prosperous, socially fulfilling, and biologically rich future for everyone if we are to have a stable, sustainable future for anyone. In this book, we find questions of social justice, ethics, and morals to be both a reason for environmental protection, and a way of achieving our objectives.

Stewardship

Our intelligence and skills give us advantages in managing our environment, but they also create obligations and duties. Technology is neither an evil force to be feared, nor a magic genie that will save us from our mistakes; rather, it is a tool that must be used carefully. Our power to change and shape our environment gives us a responsibility to be caretakers, not only for ourselves and our children, but also for all other living creatures.

Sustainable Development

Some authors argue that we must have an immediate, drastic reduction in our population, our levels of consumption, and our standard of living if we are to maintain a habitable environment into the future. Others argue that continued growth in science, technology, education, economy, and even population in some cases, are essential to provide the energy, capital, knowledge, and social stability needed for environmental protection. How can we reconcile these two positions? A middle course that we advocate is sustainable development (real economic growth per person and an increased standard of living for all) based on renewable resources, equitable sharing of benefits, and living in harmony with nature.

EXPANDED COVERAGE IN THIS BOOK

In addition to traditional subjects in Environmental Studies, such as ecosystem structure and function, biomes, population dynamics, air and water pollution, and energy, we have included topics that give a broader understanding of our place in the world. Among the subjects covered with both greater breadth and depth than in many books are toxicology and environmental health, urbanization and urban design, world land-use patterns and land ownership, environmental and resource economics, weather and climate, biological diversity, and endangered species and habitats. Since our emphasis in this book is humans interacting with nature, we have drawn from many disciplines to show how social, political, economic, and historic factors interact with natural forces to shape our environment.

Learning Aids

This book is designed to be useful as a self-education tool for students. To facilitate studying and to clarify important points, each chapter begins with a **summary of major concepts.** Headings within the chapter act as signposts to show relationships and to keep track of the orderly flow of information and concepts. **Key terms,** indicated by boldface type, are defined in context where they are first used, and are also listed in the glossary for quick reference. The glossary also contains additional key terms that will benefit the student.

At the end of each chapter, a **summary** reviews the material just covered. Two sets of questions are included in each chapter: objective questions to help students in self-review, and open-ended questions to stimulate critical thinking and discussion. **Boxed essays** and **case studies** demonstrate application of the principles discussed in each chapter. These studies are not editorials by well-known advocates in a particular area, but attempts to find a balanced position that presents both sides of the issue. Since these essays and case studies are self-contained for flexibility, they can be used in different sequence, assigned independently, or omitted, as best suits a particular class schedule.

Finally, the **writing strategy** is itself designed to facilitate learning and promote scientific literacy. It is personal and accessible to engage the reader in the substance of the text. In-text questions help to focus student attention and promote a problem-solving mindset. Presentation of alternative points of view encourages critical reading and thinking as the student moves through each chapter, rather than waiting for the end-of-chapter questions. Underlying concepts and themes are reinforced in multiple contexts, to develop learner confidence in recognizing and applying them. Explicit attention is given to representing the nature of science as a dynamic, human activity, rather than as a static, dogmatic authority.

SUPPLEMENTARY MATERIALS

Instructor's Manual

The Instructor's Manual and Test Item File, prepared by Douglas A. Wikum and Richard H. Wilson and revised by William Cunningham, is designed to assist instructors as they plan and prepare for classes using *Environmental Science*. Each chapter begins with an outline of the textbook chapter as well as chapter objectives and a listing of key terms. Suggested answers to the "Review Questions" and "Questions for Critical Thinking" are included, followed by "Additional Learning Experiences," listing relevant films, videotapes, and computer programs that are available. The Test Item File offers an average of fifty questions per chapter (including multiple-choice, fill-in-the-blank, and essay type) to assist the instructor in preparing tests and quizzes.

wcb TestPak with GradePak

wcb TestPak, a computerized testing service, provides instructions with either a mail-in/call-in testing program or the complete test item file on diskette for use with the IBM PC, Apple, or Macintosh computer. **wcb** TestPak requires no programming experience.

wcb GradePak, also a part of TestPak, is a computerized grade management system for instructors. This program tracks student performance on examinations and assignments. It will compute each student's percentage and corresponding letter grade, as well as the class average. Printouts can be made utilizing both text and graphics.

Transparencies

The text is accompanied by fifty overhead transparencies featuring illustrations found in the text. Labels have been enlarged to enhance their use in large lecture rooms.

Environmental Issues and Analysis Workbook

How does a person learn to recognize various facets of an issue, to assess the credibility of assurances and assertions, to understand the scientific validity of information, and to analyze the ways information is used in televised news reports, newspaper articles, magazine essays, face-to-face conversation, and persuasive advertisements and mailings? These are skills that will be essential to the scientifically literate person of the twenty-first century.

The Environmental Issues and Analysis Workbook, prepared by the authors and Gary S. Phillips, provides forty additional examples of environmental issues correlated to material found in chapters seven through twenty-six. Some examples are case studies; other examples are essays on controversial issues. The workbook challenges students by presenting them with both information and opinion. A general analysis model is presented, and each example is followed by thought-provoking questions suitable for individual student response or classroom discussion.

More than ever it is important to develop the mindset and skills that blend open-mindedness with critical thinking. This issues workbook gives students practice at this type of analysis.

ACKNOWLEDGMENTS

Many people have contributed in a variety of ways to the accomplishment of this book. It has dominated the Cunningham household for most of a decade, and every member of the family has made valuable contributions to the project. Mary Ann did preliminary work and research on several chapters. John organized illustrations and references and made innumerable photocopies. Peggy provided organizational skills and kept the household running during months of crisis and impending deadlines. Mary served as my personal editor and chief consultant throughout the many drafts. Without their help and support, this book would never have been possible.

This book was a collaboration that touched both of our lives deeply. Being veterans in the book-birthing process, the Saigo family provided support, direct assistance with the manuscript, and help with domestic responsibilities during months of peak writing, editing, and proofing. Our thanks to Heather, Holly, and Dustin for their assistance and understanding. Roy H. Saigo (Southeastern Louisiana University) provided valuable consultation and contributed in several ways to the successful completion of the project. Special appreciation goes to Gary S. Phillips (Iowa Lakes Community College) for assistance with end-of-chapter questions and the Environmental Issues and Analysis Workbook and to Michael D. Spiegler (Providence College) and Bill Webber (Valley Landfills, Corvallis, Oregon) for their assistance with individual chapters and for their overall support.

Dr. H. E. Wright, Jr., read the entire manuscript and made many helpful suggestions. Dr. M. L. Heinselman provided inspiration and much useful information. Leslie A. Will, Annemieke Kiss, Jennifer Hengelfelt, and Susan Hengelfelt provided invaluable assistance in proofreading, typing, and assembly in the final stages of manuscript preparation.

We would like to express our appreciation to the entire WCB book team for their enthusiasm, professionalism, patience, and creativity. John Stout first encouraged us to do this book and supported the early exploratory stages. Kevin Kane, Mary Porter, Carol Mills, and Renee Menne guided us to successful completion of the first edition. We would also like to thank Carol Mills and Kevin Kane for their continued development and support of the second edition. Our special thanks to Michelle Campbell, a wonderfully competent, steady helmsperson during the production of the second edition. To these and many others who contributed to the book we give our thanks.

Obviously, little of the contents of a book as comprehensive as this can be the original research of the authors. We owe a debt of gratitude to the entire scientific community through whose work and inspiration we know about the natural world and our effects on it. If credit is due, we share it with all our colleagues, living and dead, who have added to our store of knowledge. If errors persist, we accept responsibility and ask the assistance of our readers in identifying and correcting any mistakes. We thank users of the first edition who helped us to improve the second edition.

We gratefully acknowledge the invaluable assistance of many reviewers, whose constructive criticism was invaluable in the development of this text:

John Belshe
Central Missouri State Univ.

Bruce Bennett
Community College of Rhode Island

Bernard L. Clausen
University of Northern Iowa

John D. Cunningham
Keene State College

John Green
Nicholls St. University

James H. Grosklags
Northern Illinois University

Neil A. Harriman
University of Wisconsin–Oshkosh

Jon L. Hawker
Meramec Community College

Marion T. Jackson
Indiana State University

Wes Jackson
The Land Institute

David E. Kidd
University of New Mexico

Timothy Lyon
Ball State University

Christopher P. Marsh
Coastal Carolina College

Dennis M. McNair
University of Pittsburgh–Johnstown

Dr. Roland Mildner
Macomb Community College

Brian Myres
Cypress College

Larry Oglesby
Romona College

C. L. Rockett
Bowling Green State University

Dennis K. Shiozawa
Brigham Young University

Joseph L. Simon
University of South Florida

Dr. Peter G. Sutterlin
Wichita State University

Douglas Wikum
University of Wisconsin–Stout

Jerrold H. Zar
Northern Illinois University

PART ONE

Principles and Concepts

CHAPTER *1*

What Is Environmental Science?

Third Planet Operating Instructions:

This planet has been delivered wholly assembled and in perfect working condition, and is intended for fully automatic and trouble-free operation in orbit around its star, the sun. However, to insure proper functioning, all passengers are requested to familiarize themselves fully with the following instructions.

Loss or even temporary misplacement of these instructions may result in calamity. Passengers who must proceed without the benefit of these rules are likely to cause considerable damage before they can learn the proper operating procedures for themselves.

David R. Brower, founder of Friends of the Earth, in *Progress as if Survival Mattered*

CONCEPTS

Humans have used the world's environment, including its resources, with little knowledge or understanding of the consequences. Now our impact is showing, and potentially irreversible changes in the global environment have been set into motion.

Environmental science is a composite of natural and social sciences, mathematics, technology, business, law, ethics, philosophy, morality, and aesthetics. Ecology is its main scientific basis.

As our human population increases, there is a greater demand on natural resources and an accelerated rate of environmental deterioration. The future of our planet depends on how we deal with these three factors—human population, resource use, and pollution. Individuals, as well as governments, have roles to play in choosing whether to do nothing or to act affirmatively toward a sustainable future for ourselves and other species.

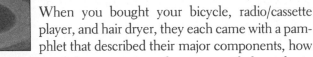

INTRODUCTION

When you bought your bicycle, radio/cassette player, and hair dryer, they each came with a pamphlet that described their major components, how they operated, and the precautions that you needed to take in using and servicing the equipment. Earth didn't come with an operating manual. In fact, we are still trying to understand its components and how they operate and interact not only in small, immediate systems, but also within the global system.

Since the Industrial Revolution, particularly, we've tended to use our planet's resources and biological systems without regard for the consequences of our actions. The best that can be said in our defense is that our technological capabilities grew so much more rapidly than our basic understanding of environmental interactions that we didn't anticipate the consequences. Also, there has been plenty of everything we needed and wanted somewhere in the world. We in the industrialized nations have developed a false sense of security about resource availability and a habit of being wasteful consumers.

The wear and tear on our planet is beginning to show. Some symptoms are subtle, like the absence of lichens growing on gravestones because of their sensitivity to atmospheric pollution that is not visible to us. Some symptoms are dramatic, like beaches littered with trash washed in from offshore disposal. Some of the most ominous symptoms, however, indicate fundamental, cumulative effects that are leading to permanent changes (Box 1.1).

BOX 1.1 The Global 2000 Report

During the administration of President Jimmy Carter, a massive document titled *The Global 2000 Report to the President* was generated. Hundreds of scientists and others participated in its preparation, involving three years and more than a dozen federal agencies. The purpose of the report was to identify the prevailing environmental trends up to 1980, then make projections to the year 2000 based on the premise that government policies and the operating policies of businesses and industries would *stay the same*. It also assumed there would be no major wars or other massive catastrophes. The projections were far from reassuring, and the potential social, economic, political, and environmental consequences inferred from the projections were rather grim.

The report sparked debate that continues, particularly among natural scientists, social scientists, and economists. Critics contend that things haven't become nearly as awful as the report indicated they would, so the report should be discredited. Defenders of the report point out that it contained "projections" based on the prevailing trends of the time and not "predictions" of inevitable outcomes. Furthermore, they say, it *has* had a positive impact in affecting public attitudes, education, and political awareness about the global future. Finally, because

of changes in attitudes and policies, we have simply extended the report's projections to sometime beyond the year 2000, but the projections remain valid on a longer time scale.

What did the report say? We suggest that you read the report or unbiased summaries of it. In the meantime, here are eight important projections it contained that you will see referred to throughout this textbook.

1. There will be continued, rapid human population growth, especially in the poorest countries. By the year 2000, more than four-fifths of the world's population will live in these countries. The income gap between poor and affluent nations will widen. More people will live in absolute poverty, without hope for improvement.
2. There will be a 90 percent increase in food production worldwide, but it will increase only marginally or actually decline in southern Asia, the Middle East, and the poorer African countries. As a consequence, food prices will double.
3. The combination of more people and rising needs will place great stresses on natural resources and natural systems. The greatest, most

immediate stresses will be on soils. There will be increasing desertification; greater soil deterioration due to erosion, compaction, waterlogging, and salinization; and greater loss of cropland to housing developments and highways.

4. Fuel sources will diminish. First, fuelwood will be depleted. Also, the readily accessible fossil fuel reserves (coal and oil) will decline substantially.
5. Forests will be depleted.
6. The tropical forests will be destroyed, with great loss of biological species through extinction—eliminating a tremendous genetic resource base that includes potentially valuable species. Lacking the forest cover, tropical soils will be degraded.
7. Water quality and water supplies will continue to decrease.
8. Air quality will continue to decrease, largely as a result of industrial growth. Some specific concerns are acid deposition, buildup of carbon dioxide and other gases that will enhance global warming (greenhouse effect), and depletion of the Earth's radiation shield (ozone layer).

We already may have set in motion a massive transformation, or metamorphosis, of our planet. For instance, the increasing level of carbon dioxide gas in the atmosphere is contributing to global warming. At the same time, pollutants called chlorinated fluorocarbons (chlorofluorocarbons, or CFCs) are accumulating in the upper atmosphere. When CFCs break down, they release chlorine, which participates in chemical reactions that destroy ozone, a gas in the atmosphere. The effect on our planet is significant because ozone forms an atmospheric layer that shields us from some of the most dangerous forms of solar radiation.

Why must we consider such changes to be irreversible? The actions and consequences of many things we set into motion outlive us. Chemical reactions involving CFC-originating chlorine, for instance, are like tiny video-game pursuers; ozone molecules are "gobbled up" one after another, and the chlorine is released to react again many times. Even if we all absolutely quit adding CFCs to the atmosphere, the full impact of CFCs that already have been liberated will not be realized for hundreds, perhaps thousands, of years. During this time of atmospheric change, living organisms and systems will be reacting to the new conditions. Some organisms react to negative conditions by dying. Some organisms are able to change in ways that allow them to survive. The point is that time cannot be reversed and neither can biological chains of events.

In this textbook we'll discuss the components of Earth, the third planet from our sun, and what we now know about how its components interact. We'll suggest operation and maintenance precautions. Unfortunately, we'll also have to discuss repair procedures—in some cases, emergency ones.

We'll begin, in this chapter, by introducing you to the broad concerns of environmental science. The topics mentioned in this chapter are dealt with in more detail later in the textbook; they are introduced here to provide a preview.

ENVIRONMENTAL SCIENCE: A SYNTHESIS

Environmental science is a composite entity, incorporating natural sciences—such as biology, chemistry, earth science, physics, and medicine—and social sciences—such as economics, political science, and sociology. It also involves history, mathematics and statistics, technology, business and management, law, ethics, philosophy, religion, morality, and aesthetics. Every aspect of human behavior has some relationship to the natural environment (figure 1.1). We rely on the natural environment to provide all of the physical aspects of our existence and, as our numbers grow, our collective demands on the environment increase.

Environmental science has its roots in natural history, the study of where and how organisms carry out their life cycles.

Natural history observations were the foundation for the science of ecology, the study of environmental factors and how organisms interact with them. Ecological research has given us hard, specific, analyzable information ranging from how specific individuals exchange molecules with their environments to how they affect and are affected by other organisms with which they live.

Resource management and resource technology also are significant features of environmental science. You probably learned in elementary school that natural resources are classified as either nonrenewable resources or renewable resources. Metal ores, coal, and oil are examples of nonrenewable resources; they can be retrieved from the environment once, but are not regenerated, at least not at a rate that does us any good in the foreseeable future. Biological resources, such as forests and fisheries, are classified as renewable resources. As you will see, we are finding that some biological resources are not truly renewable. When a forest is removed in such a way that the soil is left unprotected, for instance, erosion by wind and water may so degrade soil conditions that it is not possible for a new forest to become established.

Of all animal species that ever have inhabited Earth, humans have had the most rapid and drastic impact upon it. Our cleverness has given us incredible advantages for adjusting to and manipulating our environment. We can live anywhere because we have overcome adversities of weather and climate. For instance, we enclose our warm, vulnerable bodies in protective clothing and in temperature-controlled buildings and rely on food grown in favorable climates to sustain us. We have groomed wild lands and converted them to tended places of habitation and agricultural productivity. We have developed technologies that were only dreamed of by a few visionaries even a century ago. We have better health, less premature death from natural causes, and increased longevity.

Our impact on the environment is accelerating because our numbers are increasing at an accelerating rate (figure 1.2). An "exploding" human population places greater demands upon natural resources. It also causes an accelerated rate of environmental deterioration in three major ways: *through depletion of resources, disruption of natural environments, and pollution*. We are faced with what ecologist Paul Ehrlich calls the "population-resource-environment dilemma:" how can we deal with the needs of more people, while making the most of limited resources and minimizing environmental degradation (figure 1.3)?

This problem is the prime focus of environmental science. At its heart, it is a problem of human ecology, the interaction of humans with the environment. Understanding human ecology requires an understanding of demography, the study of human population dynamics. Most of this text incorporates demographic information, often linked to economic, political, sociological, religious, and behavioral factors.

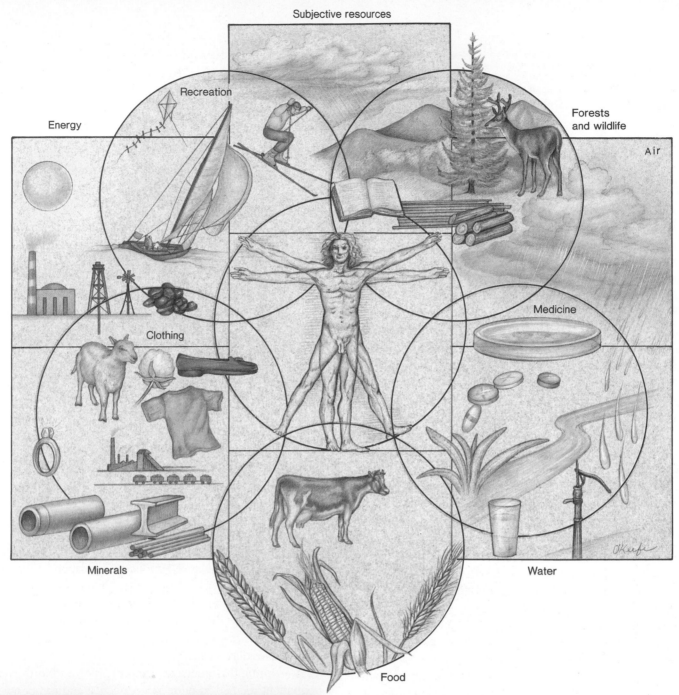

Subjective resources

Recreation

Energy

Forests and wildlife

Air

Medicine

Clothing

Minerals

Water

Food

FIGURE 1.1 Our interaction with the natural environment is complex and multidimensional. The natural environment is a source of both physical and aesthetic resources for us.

FIGURE 1.2 This crowd in Chapuotepec Park in Mexico City gives some impression of the population density in the world's largest urban area. With a natural increase rate of 2.4 percent, Mexico doubles its population every twenty-nine years. About 20 million people now live in the Mexico City metropolitan area, and the number grows by 750,000 every year.

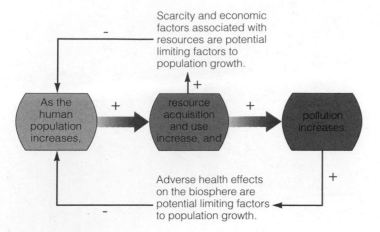

FIGURE 1.3 The population-resource-environment dilemma: more people use more resources, and increased resource processing creates more pollution. How are we to continue increasing our population without depleting our natural resources and destroying the life-support systems of the biosphere? The plus (+) signs show effects that are increased; the minus (−) signs show effects that are decreased.

Finally, environmental science deals with *priorities*. What conditions of the environment are acceptable and unacceptable? What ways of using the environment are best for people, for native vegetation and animals, and for maintaining usability of the land? How do you balance costs, economic benefits, and environmental impacts? How do you determine acceptable levels of risk? Where does armed conflict between countries, its threat or its deterrence, fit into the setting of priorities?

How, also, do you determine what is an acceptable *quality of life* for the world's people? Is fairness important? What do we mean by "quality of life"? The main focus of human population concerns often is at the most basic level: food. News accounts focus on dramatic famines and droughts. Debates about our exploding human population often are reduced to the question "How will we feed everyone?" Contrast the starkness of what it must be like to face each day worried mainly about obtaining enough calories to survive to the kind of life you have. Think also about what kind of life you hope to attain twenty years from now. Clearly, food is only one aspect of the quality of life. You also have expectations for education, health care, meaningful achievement, cultural and aesthetic development, and certain comforts and luxuries. How will all of these factors fit into the lives of your great-great-grandchildren? What will they say about how we managed Earth? Are we "buying" the present by "spending" the future? Many knowledgeable people believe that we are.

Human priorities are linked strongly, on the one hand, with economic and political expediencies and, on the other hand, with ethics. All of these questions are difficult, and their relevance is demonstrated daily on television news programs and in written news reports and analyses. In the end, *each* of us will decide, through our priorities, actions, and inactions, what the future of Earth will be.

PAST LESSONS AND FUTURE HOPES

History provides us with a context for understanding our current attitudes toward resource use and the environment.

Geographic and Resource Frontiers

Imagine what it must have been like for the European explorers who ventured to the New World. They encountered lands that had not been deforested or extensively cultivated. The plants and animals were unfamiliar, wild, and intimidating to the Europeans. They encountered native civilizations that were alien to what they were used to and that had different languages, values, technologies, arts, and social and religious practices. Native agricultural practices caused minimal disruption of the land.

History presents a picture of the political, social, and environmental changes that followed. No matter how awed and perhaps appreciative some visitors to the new lands may have been, to their governments the new lands represented domain to be claimed, conquered, and defended by military means, ensuring exclusive privileges to exploit natural resources (including slaves). France, Great Britain, Spain, and Portugal laid claim to parts of North, Central, and South America. As time went on, eastern North America became dominated by the British.

Even after the successful American Revolution, the North American continent served another, very important function for European countries. It was a place to which excess population could go. Settlers brought with them the domestic livestock, cultivated plants, weeds, and diseases of Europe. As time went on, the American populations of former immigrant Europeans grew, forcibly displacing Native American populations. Extensive forests were cut down and burned to "tame the wilderness" for settlement and agriculture. In about a century's time the great eastern forests were changed, drastically and forever. The prairies—the real agricultural mother lode—were exploited only for their wildlife, at least at first.

During the nineteenth century, the vast North American continent seemed to offer independence and prosperity to any who could claim it under the new laws of the land. In what would become the United States, the edge of settlement, the **frontier,** moved ever westward. Trading centers grew into great cities linked by trails, then dirt roads and waterways, then railroads, and finally highways. The prairies were opened to homesteading and ranching, the Native Americans were brutally disenfranchised of most of their lands and their traditional ways of life, and the sod was "busted" by iron plow blades pulled by oxen, horses, and mules. The extensive native prairies went the way of the native eastern forests. The lands were tamed and developed to more efficiently meet the needs of a growing human population.

One result has been progress and vast improvement in the quality of life for most Americans, but we have also paid a price and learned lessons. In about fifty years, large tracts of former prairie had been subjected to agricultural practices that were not appropriate for the environmental conditions that prevailed. Without massive networks of grass roots to hold it, topsoil was carried away by thunderstorm deluges and by roaring prairie winds. Like the virgin forests, the rich soil was thought to be a limitless natural resource. It wasn't and it isn't—it's still being eroded in massive amounts.

A natural legacy of our recent frontier era is the prevailing point of view that there are plenty of natural resources still to be exploited ("recovered" in economic jargon), that we can continue to refine technological means of removing them from the environment, and that the environment will take care of itself. This composite attitude is called the **frontier mentality.** Frontier mentality makes it difficult for such concepts as resource and energy conservation, recycling, and wilderness preservation to take hold in our society. No one is alive today who did not grow up benefiting from the riches of frontier exploitation. It is ingrained in us. Even though we have reached the geographical limits of the North American continent, the frontier mentality is still a dominating social and political force in the most recently settled regions—the western states, including Alaska. Its effect can be seen most strikingly in legislative debates over use of state and federal lands, oil and mineral exploration, and wildlife resources. Frontier mentality is not limited to the United States, however. You will be able to recognize it as a factor in many of the examples presented in this text.

Finite Planet, Global Systems

The Dust Bowl era during the 1930s was a lesson about how an environmental and human disaster can result from misuse of a resource (figure 1.4). It spurred the conservation movement in the United States and led to reformed farming practices, such as contour plowing, strip farming (figure 1.5), and abandonment of harrowing, which pulverizes the surface soil into fine particles.

In fact, the disastrous consequences of soil abuse are an age-old lesson that humans still haven't learned. The early centers of civilization in Asia Minor flourished as a result of rich agriculture on lands that now are unproductive. African deserts continue to spread, at least partly because of inappropriate land use. Complex tropical rain forest systems are being logged and burned to provide farming plots that the soil can sustain for less than a decade—frontier mentality at its most flagrant.

By contrast, some of the terraced fields of Pacific Asian and island countries have been husbanded carefully and remain productive after centuries of use. They provide examples of balanced use of a soil resource.

Soil may be our most critical global resource. It is obvious that, as long as the human population continues to increase, more

FIGURE 1.4 An abandoned farm with its drifts of windblown soil attests to the consequences of inappropriate cultivation techniques. Native plants—especially grasses—helped to form our rich prairie soils and protect them. Exposed and pulverized into fine dust by plowing and harrowing, however, topsoil was blown hundreds of miles in massive dust storms—a human and environmental disaster that was the dust bowl of the 1930s.

FIGURE 1.5 Contour plowing and strip planting on hilly terrain help preserve soil moisture and reduce erosion by slowing runoff and airflow across the soil surface.

food will need to be produced for us and for our domestic animals. Food production requires soil that has the proper texture, nutrient balance, and moisture content to grow crops. Technological advances have greatly increased crop yields, but with greater costs and environmental trade-offs. More of the world's land will need to be put into production to meet population growth, yet we are losing agricultural land yearly to degradation or development. Also, the best soil in the most favorable climates for agriculture is already under cultivation, buildings, or pavement (figure 1.6). We are expanding agriculture into increasingly marginal lands. It has been said that we are literally "mining" the soil to extract food, using it up the way we use up ores to extract minerals.

FIGURE 1.6 California's "Silicon Valley" now features research, development, and production facilities for the computer industry where once there were orchards and farmlands. Is this the best use of this region's rich, deep topsoil, considering increasing human needs for food?

On the other hand, is it possible that paving over and building on prime agricultural land might turn out to be a way of protecting this fertile resource until some future century when we might need to mine it or take down the buildings to put it back into food production?

Our atmosphere and water supplies also have been mistreated. Smoke from industrial processes, for instance, injects chemicals into the atmosphere at unprecedented rates. Some of these chemicals react with atmospheric gases. We already have mentioned ozone depletion. Also, sulfur-containing gases combine with water vapor molecules in the air to form sulfuric acid. Deposition of acidic rain, fog, or snow on vegetation and in bodies of water has negative effects on plant and animal life. Acid rain is an increasing problem in the moist, temperate Northern Hemisphere because of sulfurous gases released when coal is burned to generate electricity.

These are but a few dramatic examples of a lesson we have been slow in learning: *Everything in our global environment is connected to everything else. Changes in one component of the environment can affect many other components.*

Humans, as a significant cause of environmental change, are a relatively new development on our planet, yet we are now the most aggressively destabilizing factor in a complex web of interactive systems. One of the most volatile features of human-generated change is its rapid rate. The geological record shows that Earth has undergone continuous change for hundreds of millions of years, but the changes have been slow. Even "abrupt"

events, like the extinction of the dinosaurs, occurred over several *million* years. This rate of change pales beside what humankind has done in a few hundred years, even in decades. Another example of the force we exert on the environment is the incredibly accelerated rate of species extinction that is occurring as a result of human actions. Finally, lurking in the mind of each of us, is the reality of our potential power for total destruction of civilization and the world as we know it through nuclear war.

Sustainability

We have nowhere to go. There will be no Noah's Ark to move us to safe ground *if* Earth becomes uninhabitable. Space colonies have been hypothesized as a further frontier, but even if there were time and resources to create them, they still would not be as perfectly suited for human life as the planet on which we evolved. How would it be to live without forests, streams, seashores, and the multitude of natural species that presently share this world with us? Also, how would we react to the restrictive social and governmental structures that would be required to maintain order in an even more limited and self-contained environment than the one in which we now live?

PRINCIPLES AND CONCEPTS

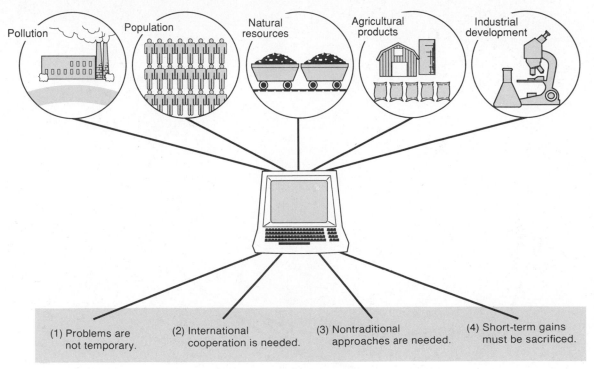

FIGURE 1.7 In *The Limits to Growth* (1974), members of the Club of Rome group presented computer simulation models involving the interrelationships of human population, demands for natural resources and agricultural products, industrial development, and pollution. Successful management of these relationships will require international cooperation, sacrifice of short-term gains in favor of long-term sustainability, and development of new approaches.

The future can be rich, rather than bleak, however, and we can make it happen. If we have nowhere to go, then our only viable choice is to "clean up our act" on Earth. This requires *planning for a sustainable future* based on ecological knowledge and clearly expressed priorities (figure 1.7). It will require *education*, individual *commitment*, and community, regional, state, and national *policies* translated into actions (Box 1.2). It will require global political and economic *cooperation*, the sooner the better. This course in environmental science will introduce you to several goals that must be parts of this plan. One of these goals is to identify the linkages of human actions within natural global systems. A second goal is to identify ways of predicting the impacts of our actions on natural systems. A third goal is to formulate meaningful responses that will offset or minimize these impacts.

A fourth goal—one that requires immediate action—is to decide how we will deal with our own population growth. Can we afford to deal with it by doing nothing? Most biologists and ecologists agree that it is foolish to think we can continue to increase our numbers at the current rate without eventually being unpleasantly checked by resource depletion and environmental degradation (figure 1.8). Furthermore, we must address the problem of equity. What minimal levels of expectation for health, comfort, and achievement must each person have to maintain economic and political stability within the world?

Why should *you*, with your life-style and level of education, be concerned about participating in these goals? People who are barely surviving from day to day do not have the luxury of worrying about global problems. Informed planning for the future rests squarely on the shoulders of those of us who, individually as citizens and collectively as nations, have the ability to formulate and implement policies.

There *is* hope for the future, but it requires present action. We must gather, as quickly as possible, enough scientific information about Earth's systems to write a sound "operating manual." Then we must follow the instructions. We will need to make appropriate choices about human population growth, resource utilization, and preservation of environmental quality. Both research and implementation require social and political commitment, expressed as policies that are backed by financial support to train scientists and to do research, monitoring, and enforcement. In a democratic society, responsibility for such national initiatives rests on each of us.

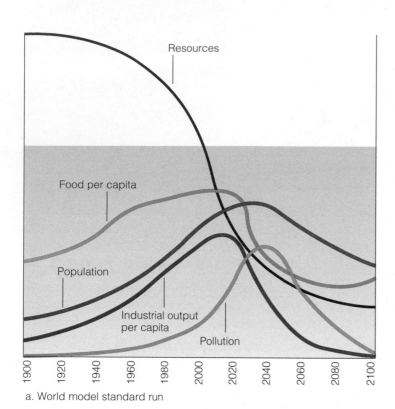

a. World model standard run

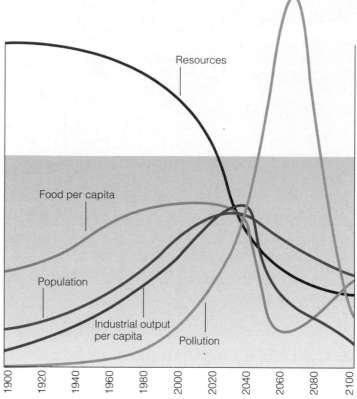

b. World model with natural resource reserves doubled

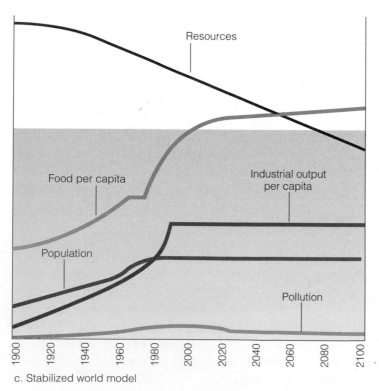

c. Stabilized world model

FIGURE 1.8 Computer models allow us to simulate potential effects of related variables. The modeled outcomes can be useful in planning personal financial budgets or a global resources "budget." These simulations relate world human population growth to availability of food and other resources, as well as to industrial output and pollution. In these three computer runs, resource and population parameters were varied. In *a*, declining resources and increasing degradation of the environment eventually result in declines in human aspects of the model. In *b*, the amount of world resources has been doubled, sustaining human population growth and industrialization for a somewhat longer time. However, the outcome is eventually the same as in *a* but with a more drastic rate of die-off at the end. In *c*, a stabilized, no-growth population variable slows resource depletion, stabilizes industrial output and pollution, and predicts a stable per person world food supply.

BOX 1.2 Why Should Business Care?

Are the interests of business and the integrity of the natural environment naturally opposing forces? In the past, it often seemed so. Exploitation of natural resources was directed more by financial balance sheets than by environmental concerns. More recently, however, businesses and industries have become more responsible for the environmental side effects of their operations, at least in the United States. The decades since the 1960s have seen much resistance and acrimony between the business sector and those who were seen as their opposition, often referred to as "those damned environmentalists." All of us, however—business leaders, workers, and citizens-at-large—have benefited from imposed controls over air and water pollution, toxic and hazardous waste production and disposal, mining and mineral extraction, and damage to wild species and habitats.

There is some evidence that the attitude of resistance and regulation skirting that sometimes prevailed in the United States is shifting to a "good-neighbor" attitude of compliance. Many companies pride themselves on their environmental awareness. The change has been influenced, no doubt, by the hard economic facts that go along with legal liabilities, as well as the expectations of society at large and "quality of life" aspects of planning for community economic development. Consider the following article printed in the December, 1987, issue of *The Rotarian*, published by Rotary International, an organization composed mainly of business leaders in communities

around the world. We have added italics for emphasis.

WHY SHOULD BUSINESS CARE?

"The World Resources Institute (WRI) is an internationally respected policy research center that helps governments, the private sector, environmental leaders, and other policymakers answer a fundamental question: How can societies meet human needs and nurture economic growth while preserving the world's natural resources and environmental integrity? James Gustave Speth, the president of WRI, lists four reasons why it is important for business to care about—and care for—the environment:

1. *National economies depend upon the wise use of natural resources.* The industries of agriculture, energy, forestry, pharmaceuticals, chemicals, fishing, real estate, recreation, and tourism could all be jeopardized by the poor management of the environment.

2. *The management of resources in developing countries has a direct impact on the economic well-being of developed countries.* The failure of a country to observe the policy of "sustainable development" could lead to economic restrictions, social turmoil, and even political instability. Any of these factors could affect access to both materials and markets, and result in long-term repercussions to the local and world economy.

3. *Global economic interdependence is affected by factors such as explosive population growth, which affects the consumption of natural resources.*

4. *It is in the best interests of corporations to resolve environmental problems before the government becomes involved and imposes regulations on industry.*

In Geneva, the International Chamber of Commerce (ICC) has announced an industry initiative to make available environmental information related to pollution-control technology. Its market includes industry, small to medium-sized companies, and governments interested in environmental quality. A new organization has been formed under ICC called the International Environmental Bureau (IEB). Founding members include a bevy of multinational companies such as Alcoa, Arco, Ciba-Geigy, Monsanto, Bosch, Ford, and 3M. According to David Roderick of IEB, 'The industries are convinced that timely investment in cost-effective pollution-control technology will encourage continuous economic development and address adverse effects on the environment that could prove unacceptable to society.'

The bottom line? Invest in the environment today before you have to pay a substantial penalty in the future."

From Drake McHugh, "Why Should Business Care?" in *The Rotarian*, December 1987. Copyright © 1987 Rotary International, Evanston, IL. Reprinted by permission.

SUMMARY

The complex interactions of Earth's physical and biological systems are, as yet, imperfectly understood. Until now, human impacts on these systems have been geographically localized. Now we are having a global impact.

Environmental science is a composite of natural and social sciences. The science of ecology is at its heart, a source of understanding how Earth's systems work and of predicting future scenarios based on this understanding.

Such natural resources as ores, coal, and oil are nonrenewable natural resources. Most biological resources are renewable—they can be reproduced and replenished.

Our rapidly increasing human population demands more natural resources, causing increased disruption of natural environments and increased pollution. These interrelationships are the prime focus of environmental science. Human ecology links demography, the study of human populations, to the principles of ecological science.

Environmental science also deals with priorities regarding acceptable environmental quality; preservation of natural species and habitats, especially when there is conflict with human wants and needs; freedom of nations to do as they wish within their own political boundaries; and issues of quality of life, fairness, and ethics.

The frontier mentality assumes that there is no shortage of resources to meet human needs and that all we need to do is explore for and exploit them. Increasingly, however, there is awareness that our resources are finite, and that we need to adopt an attitude of stewardship to make them last as long as possible. Stewardship means taking responsibility for our power over the world. Our role in the future can be positive, given intelligent and compassionate planning and enactment of such plans.

■ Questions for Critical Thinking

These questions are designed to get you thinking about your place in the world and the foundations for environmental science concerns that are explored in this text.

1. Compared to your friends and classmates, would you say that your life-style level is about the same, higher, or lower than theirs? What factors do you use for your comparison? How many of these factors are what you would consider to be "critical" for your survival?

2. Now imagine that you are a young woman or man in an underdeveloped country, poor, illiterate, and living with several family members in a one-room shelter without electricity or plumbing. Under those conditions, what factors would you consider "critical" for your survival? What would be necessary for you to attain a higher standard of living? How could you improve your life, or could you?

3. Of the "two yous" from questions 1 and 2 above, which is in a better position to control or improve her/his personal situation? What factors make the biggest difference in the issue of self-help and improvement? What is the relationship of these factors to being able to control or improve the lot of the collective social group?

4. Inventory the goods and services that affect your life each day. How are they obtained from the environment? What is the impact of how they are obtained and used upon the environment? What happens to things you use when you no longer want them? Could any of them be reused or recycled in such a way that would decrease the demand for extraction and processing of new resources?

5. In what specific ways do you, your friends, your family, and your community exhibit the "frontier mentality"?

6. In what specific ways do you affect Earth's environment, for better or for worse?

7. Do you believe that there is an environmental crisis in North America? In other parts of the world? In the world as a whole system? If you can accept the premise that there will be one eventually, when do you think we should begin to do something about it?

8. Do you believe that technology will find solutions to whatever problems we create? What is the basis of your opinion?

9. Are you confident that "the people in charge" of resource recovery, extraction, and utilization will be careful to manage resources wisely, with a minimum of negative environmental side effects? Who *are* "the people in charge" and how did they attain that status?

10. How do you plan to affect Earth's future environment? Lacking a plan, how do you think you'll affect it anyway?

■ Key Terms

demography (p. 4)
ecology (p. 4)
environmental science (p. 4)
frontier (p. 6)
frontier mentality (p. 7)
human ecology (p. 4)
natural history (p. 4)
nonrenewable resources (p. 4)
renewable resources (p. 4)

SUGGESTED READINGS

These references are applicable for several chapters of this book, so you should keep them in mind even if they are not cited subsequently.

■ American Society of Zoologists. 1985. *Science as a Way of Knowing II: Human Ecology.* American Society of Zoologists, 104 Sirius Circle, Thousand Oaks, CA 91360. This volume includes ten papers by prominent ecologists who explore, in nontechnical terms, specific issues of human population, resource use, and pollution. A comprehensive list of films and videotapes on human ecology is included.

■ Brown, Lester R., and Edward C. Wolf, et al. *State of the World 1991.* A Worldwatch Institute Report on Progress Toward a Sustainable Society. New York: W. W. Norton & Co., 1991. Thoughtful essays analyze human population, urbanization, energy issues, resource uses, agriculture pollution, and related issues. Published annually, *State of the World* is an excellent, readable reference.

■ Burch, Wm. R., Jr., ed. *Readings in Ecology, Energy, and Human Society: Contemporary Perspectives.* New York: Harper & Row, 1977–78.

■ Hawley, Amos H. *Human Ecology—A Theoretical Essay.* Chicago: The University of Chicago Press, 1986.

■ McRostie, C. N., ed. *Global Resources: Perspectives and Alternatives.* Baltimore: University Park Press, 1980. A collection of interesting and provocative essays.

■ World Resources Institute and the International Institute for Environment and Development. *World Resources 1991.* New York: Basic Books, 1991. This is an annual series, filled with facts, data, and analysis of the latest developments in population, human settlements, food and agriculture, forests and rangelands, wildlife and habitat, energy, fresh water, oceans and coasts, atmosphere and climate, and policies and institutions. It is a wonderful reference, filled with easily accessible information.

CHAPTER 2

Life on Earth

We travel together, passengers in a little spaceship, dependent upon its vulnerable reserves of air and soil; all committed for our safety to its security and peace; preserved from annihilation only by the care, the work, and I will say, the love we give our fragile craft. We cannot maintain it half fortunate, half miserable; half confident, half despairing; half slave to the ancient enemies of [humanity], half free in a liberation of resources undreamed of until this day. No craft, no crew can travel safely with such vast contradictions.

On their resolution depends the survival of us all.

Adlai E. Stevenson II, United States Ambassador to the United Nations, Farewell Address, 1965

CONCEPTS

Relationships between matter and energy are the basis for the organization and interactions of the biosphere, from cellular to ecosystem levels. Living cells, the basic units of life, are active processors of matter and energy.

It is useful to look at how the science of ecology fits into a scheme of complexity leading from the atom to the biosphere.

The presence of life on Earth is the result of the planet's specific physical circumstances: its elements, a steady influx of solar energy, moderate temperatures, the presence of liquid water, and a sheltering, nontoxic atmosphere.

INTRODUCTION

The appearance of Earth from space reveals our planet's jewellike beauty and its vulnerability. Shimmering blue oceans surround terra-cotta landmasses that bear a mosaic of vegetational patterns. Swirling cloud patterns prove the presence of an atmosphere, a "vapor ball," surrounding Earth. The extraterrestrial view of Earth also emphasizes that life-forms occupy but a thin, life-supporting layer of air, soil, and water at the interface of the planet's surface and its atmosphere. This globe-surrounding layer, characterized by the presence of living things, is called the **biosphere,** a "sphere of life" that blankets Earth. Earth is a very special planet because it *does* support life, as part of systems in which physical and biological forces interact (Box 2.1).

The purpose of the next several chapters is to review some of the major principles and concepts about biosphere interactions that will be useful to you as a global citizen. We begin in this chapter by introducing you to (1) the concept of cells and organisms as processors of matter and energy; (2) levels of ecological study; (3) basic concepts about matter and energy; and (4) some of the physical characteristics of Earth that make life possible. This selective, basic overview will help you to better understand how organisms interact within the biosphere. Such understanding of natural processes is essential to planning and implementing a sustainable future for Earth and its inhabitants, problems dealt with in the main body of this text.

BOX 2.1 A Small Blue Planet

As the Apollo astronauts circled high above Earth in their spacecraft they saw a sight that affords all of us of a new understanding of the world on which we live (box figure 2.1). They saw Earth as a small, serene, beautiful sphere sailing through the cold, dark emptiness of space. The pictures they brought back have amazed all of us with the liveliness and elegance of our planet. The colors are vivid and refreshing. There are striking red deserts and green forests overlain with fluffy white clouds. The predominant color of Earth, however, is the blue of the oceans. We see ourselves as other inhabitants of space might, as dwellers in a lovely, cool, blue planet—the only one that we know of anywhere in the universe with oceans of liquid water. If it weren't for the wonderful properties of water, life would not be possible here.

Water has many unique, almost magical qualities. It is the most versatile solvent that we have found. It distributes minerals and nutrients throughout the ecosystems within which we live, as well as through our own bodies. Water has the highest surface tension of any ordinary liquid. The interface between water and air has so much tensile strength that insects can walk on it and it can be drawn by capillary action through such small spaces as blood vessels and plant transport cells.

Most important of all, however, is the heat storage capacity of water. Water absorbs amazing amounts of heat, considering its small molecular size. It distributes that heat around the globe and releases it slowly. As a result, we have relatively even temperatures nearly everywhere on the planet.

The organisms that coinhabit Earth come in an incredible variety of sizes and shapes. They are valuable to us as sources of food and as scavengers of our wastes. They clothe the land and give us pleasure in their bright colors, lively songs, and graceful movements. No one who has witnessed thousands of migratory birds, or been enfolded in the richness of a temperate or tropical forest can help but be awed by the life force that exists here.

The view of Earth from space also expresses the sense of a living community. One of the remarkable things that one notices about Earth is the freshness and newness that it displays. The other planets in our solar system show the ravages of their incredible age. They are pockmarked with meteor craters, cracks, and bulges from eons of geological ac-

BOX FIGURE 2.1 The planet Earth.

tivity, erosion scars, and the numerous signs of wear and tear from billions of years of history. Our planet shows the renewing effects of the plants and animals that make up the biosphere. Wastes are recycled and scars are healed over. Although an inevitable consequence of living is dying, the life force that passes from one generation to the next ensures that Earth, as a whole, remains vigorous, youthful, and fertile.

We are extremely fortunate to have inherited a comfortable, self-regulating, nourishing, sustaining world. It is the only such place that we know of in the whole universe. We must keep it that way.

MATTER, ENERGY, AND LIFE

Let's begin our examination of the biosphere by considering how life-forms exist within it. Organisms are made up of one or more cells, the basic structural and functional units of life. Each cell is an entity, separated from its surrounding environment and other cells by a living membrane. Cells maintain their internal composition and life processes by using matter and energy from their environments. *Cells, and thus organisms, are active processors of matter and energy that they obtain from their environment* (figure 2.1). Matter and energy must, however, be made available to cells in very specific ways.

Consider energy; it must be made available to organisms in the form of food. Even though solar radiation is the energy basis for life on Earth, it cannot be used directly to fuel cellular activities. Light first must be absorbed and converted to food by **photosynthesis,** the process by which green plants and some microorganisms convert carbon dioxide, water, and light energy to simple sugars that can be used as cellular food.

Organisms also are selective in their ability to obtain and process matter. Matter exists in pure forms called elements, which can be combined into mixed forms called compounds. Cells use certain elements in abundance, others in trace amounts, and others not at all. Some substances are hoarded within cells, others are actively excluded. Carbon is particularly important because chains and rings of carbon atoms form the skeletons of **organic compounds.** Organic compounds are the material of which biomolecules and, therefore, living organisms are made. The four major categories of organic compounds are (1) carbohydrates; (2) lipids; (3) proteins; and (4) nucleic acids.

The ability of cells to organize matter and energy is of great significance because living cells are a primary source of *order* on Earth. What is meant by that statement? Only cells are able to take raw energy and inorganic components of the environment and organize them into complex structures of great diversity and beauty. Furthermore, the structures they form are of a consistent, replicable nature because they are built according to blueprints that are encoded in the genetic information (DNA) passed from generation to generation of cells and organisms. When a cell's organization cannot be maintained, it dies, and its components are gradually returned to simpler, less-ordered forms of matter.

LEVELS OF ECOLOGICAL STUDY

How do the lives of individual organisms relate to the biosphere as a whole? Such a sweeping question can only be answered if it is broken down into smaller questions that deal with each separate interaction and that can be investigated using the inquiring methods of science. Scientists try to make precise, specific ob-

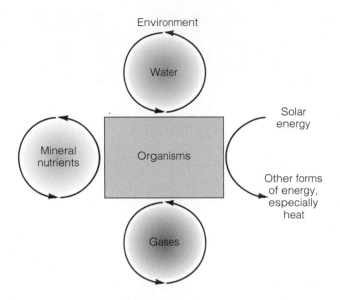

FIGURE 2.1 Organisms process matter and energy they obtain from the environment. Matter is recycled between organisms and the physical environment. Energy is transformed from high-level light energy to lower-level energy forms as it passes through the metabolic processes that sustain life.

servations, obtaining accurate data from which meaningful inferences and unifying theories can be derived. We look for *organization, patterns,* and *consistencies* in the natural world.

Ecologists and other biologists work at various levels of complexity. Figure 2.2 shows where the scope of ecology fits into an organizational hierarchy of biological study and also some of the quantitative aspects that have bearing on this organizational scheme.

The most basic ecological level of study deals with the interactions of individual organisms with their environments. In working with the life of a single organism, however, we're actually working on two levels—that of the individual organism and that of the **species.** The concept of a species is important to understand because it forms the basis for all our future discussions of specific life-forms. The word *species* literally means *kind.* In biology, species refers to all organisms that are genetically similar to the extent that they potentially breed in nature, producing live, fertile offspring. There are several qualifications and numerous important exceptions to this definition of species (especially among plants) but for our purposes here, it is a useful working definition.

An understanding of environmental interactions requires an understanding of populations. A **population** consists of all the members of the same species that live in the same area at the same time. Chapters 6 and 7 deal with population dynamics. All of the populations of organisms that live and interact in the same area comprise a **biological community.** What populations make up the biological community of which you are a part?

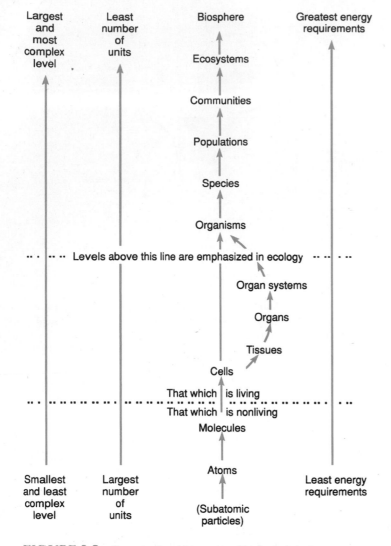

Largest and most complex level
Least number of units
Biosphere
Greatest energy requirements

↑
Ecosystems
↑
Communities
↑
Populations
↑
Species
↑
Organisms
↑

··· · ··· · Levels above this line are emphasized in ecology ··· ·· · ···

Organ systems
↑
Organs
↑
Tissues
↑
Cells
↑

That which | is living

·· ··· ··

That which | is nonliving

Molecules
↑
Atoms
↑

Smallest and least complex level
Largest number of units
(Subatomic particles)
Least energy requirements

FIGURE 2.2 Organizational hierarchy of biological study.

The population sign marking your city limits announces only the number of humans who live there, disregarding all the other populations of specific animals, plants, fungi, and microorganisms that are part of the biological community within the city's boundaries! Characteristics of biological communities are discussed in chapter 4.

Up to this point, we've referred only vaguely to "environmental interactions." Just what *is* an organism's **environment?** The environment includes all the factors that affect the life of an organism. **Physical (abiotic) factors** are nonliving components of the environment, such as climate, water, nutrients, sunlight, and soil. **Biological (biotic) factors** are the organisms and products of organisms (secretions, wastes, and remains) that are components of the environment.

A specific biological community and its physical environment, interacting to produce an exchange of matter and energy, comprise an ecological system—an **ecosystem.** An ecosystem may be very large, such as a forest, or very small, such as a pond or even the surface of your skin. Whatever its size, an ecosystem

is truly a *system*, characterized by specific interactions and exchanges of matter and energy. Ecosystem dynamics are discussed in chapters 3 and 4.

Our planet supports several broad regional types of ecosystems called **biomes,** which are discussed in detail in chapter 5. Each biome, wherever it occurs, is characterized by similar physical conditions and similar biological communities interacting in similar ways; however, the species composition varies geographically. Deserts, for instance, are characterized by low and unpredictable water availability and high rates of evaporation. In the New World, plants in the cactus family are significant components of desert communities. Desert cacti have evolved adaptations that allow them to flourish in the arid desert environment. In Old World deserts, many plants in an unrelated family, the euphorbs, show similar "cactuslike" adaptations.

At every level, ecological study, like other fields of scientific investigation, depends on instruments to help us make and interpret observations. Our ability to study global ecology has been greatly enhanced by data-gathering satellites. Not only do they enable us to step back and get a large, distant view of what our planet *looks* like, they also enable us to analyze what we see in ways our eyes cannot. Computers can be programmed to assign different values to the digitized input they receive, and these values can be correlated with specific environmental variables, giving us sophisticated and detailed analyses of such environmental components as landforms, weather, surface temperatures, and vegetation. Stunning visual images from space are one useful product of this technology (figures 2.3 and 2.4).

EARTH AS A SITE FOR LIFE

To our knowledge, Earth is the only planet in our solar system that provides a suitable environment for life, and *the composition and activities of Earth's organisms are a result of the chemical and physical properties of the planet.* The requirements for life include (1) availability of required chemical elements; (2) a steady influx of solar energy; (3) amenable temperatures; (4) the presence of liquid water; and (5) a sheltering, nontoxic atmosphere. Even our nearest planetary neighbors, Mars and Venus, do not meet these requirements.

Matter and the Energy of Atoms

Exchange of matter and energy is a fundamental theme of ecology that underlies much of what we talk about in this book. Each organism is a highly organized system of matter and energy interactions. Ecosystems and, indeed, the entire biosphere also are based on exchanges of matter and energy.

What, more precisely, is matter? It is the stuff of which the observable universe is composed. Matter has three interchangeable forms or phases: gas, liquid, and solid. The structure and behavior of matter is based on the fact that it is composed of discrete units called **atoms.** An atom is the smallest particle

FIGURE 2.3 Digitized, remote-sensed images can be programmed to yield geological information about Earth. Such information aids in ecological study and also has economic applications. Geological features are demonstrated in this Landsat image of the uranium-rich San Rafael Swell, which is 130 km (78 mi) southeast of Salt Lake City, Utah. The oval Swell (*center right*) is a plateau made of sedimentary rocks and traced by gullies and dry riverbeds. The golden area on the lower right is the San Rafael desert. A band of shale across the upper right of the Swell is blue, and the mountains of the Wasatch Plateau are red.

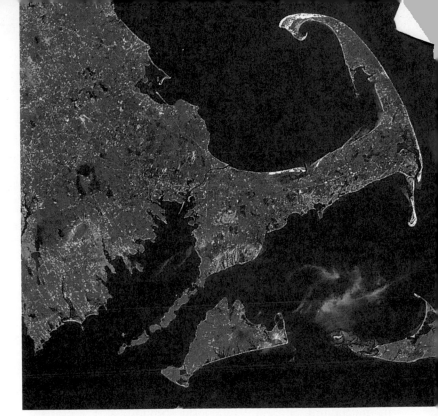

FIGURE 2.4 Satellite images also enable us to study vegetational and land-use patterns. Cape Cod, Massachusetts, juts into the Atlantic Ocean in this Landsat infrared image. The metropolitan areas of Boston, Massachusetts (*upper left*), Providence, Rhode Island (*lower left*), and numerous smaller cities are prominently blue. The red areas are vegetated.

that has the characteristics of an element and is, in turn, composed of several kinds of even smaller particles (figure 2.5). The three major kinds of subatomic particles are protons, neutrons, and electrons. **Neutrons** add mass to the central part, or nucleus, of an atom, but are electromagnetically neutral. **Protons** also add mass to the nucleus, but have a positive (+) charge. **Electrons** are negatively (−) charged particles that continually orbit the nucleus of the atom. Electrons are held in their paths, or **orbitals,** by attraction to the positive charge of the nucleus, which offsets their tendency to escape.

What has this to do with life? An atom is held together by energy, some of which is represented in the force that holds electrons in their orbitals. Some of this energy is released when electrons move from atom to atom. For this reason, many *energy* transfers in living cells are referred to as *electron* transfers. In fact, *the energy-binding and energy-releasing behavior of electrons is the basis for all the chemical reactions that occur in living cells.* Furthermore, the streams of electron transfers that maintain your metabolism are part of an energy flow that can be traced back through the food you ate and, eventually, to the sun. That sentence sums up a basic concept of life systems that will be elaborated on in the next two chapters: *the biosphere is maintained by a continuous influx of solar energy.*

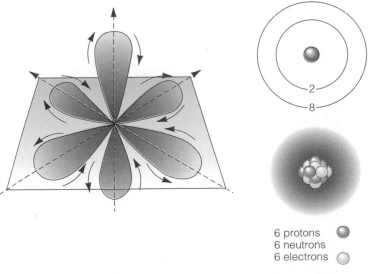

6 protons
6 neutrons
6 electrons

FIGURE 2.5 As difficult as it may be to imagine when you look at a solid object, all matter is composed of tiny, moving particles, separated by space and held together by energy. It is hard to capture these dynamic relationships in a drawing, so different models have been developed. Left and upper right models show two ways of depicting an atom having ten electrons. The inner two e⁻ are in a spherical orbital (not drawn in the first figure) and the other eight e⁻ are in three dumbbell-shaped orbitals at right angles to each other. The density of stippling in the first figure represents the relative probabilities of electrons being present in areas of the orbitals. The bottom model represents carbon[12], with a nucleus of six protons and six neutrons; the six electrons are represented as a fuzzy cloud of potential locations.

Two or more atoms can be combined to form molecules. The more atoms involved, the larger the molecule. Energy transfers are involved both in forming and breaking the chemical bonds that hold together the atoms in a molecule. Some kinds of bonds are easily broken (unstable bonds) and others are not (stable bonds). Both kinds of bonds are important to organisms.

What is the relative significance to organisms of *stable* versus *unstable* bonds? The body of an organism is composed of many kinds of molecules. Most of the molecules in a complex, multicellular organism are stable and long-lasting, such as those that make up the physical body structure, or are reused, such as enzymes, hemoglobin, and chlorophyll. The maintenance of a living body, however, requires a constant input of energy because life processes, collectively called **metabolism,** run on molecular energy transfers, or electron transfers. These transfers must occur *quickly* and with maximum *efficiency*; hence, molecules that are energy sources for metabolism are unstable or semistable.

Let's look at two of many possible examples of biologically important semistable and unstable molecules: glucose, a key sugar in metabolism, and ATP (adenosine triphosphate), a primary ready-energy substance found in cells. Glucose ($C_6H_{12}O_6$) is a fuel for cellular metabolism that is formed during photosynthesis (figure 2.6), then is converted to complex carbohydrates, including starch. It is made available to cells by subsequent digestion of carbohydrates, proteins, and lipids in food or internal storage tissues. Glucose is a *semistable* molecule, meaning that it can easily be destabilized and its bonds broken. Glucose is the primary fuel for metabolism via **cellular respiration,** in which the bonds that hold the atoms of the glucose molecule together are systematically broken, releasing energy for such cellular work as substance transport, synthesis of biomolecules, and movement (figure 2.7). In addition, some energy is always released as heat during metabolism. (Note that the concept of cellular respiration is *not* to be confused with use of the term respiration to refer to the act of breathing. In this text, we'll use the metabolic meaning exclusively.)

The role in metabolism of ATP, an *unstable* molecule, is related to two important factors of energy use: control and efficiency. If all of the bond energy in a glucose molecule were released suddenly and spontaneously, much energy would be released as heat (as occurs in combustion) rather than be converted to cellular work. The sudden burst of released energy could be destabilizing and harmful to other molecules in the cell. Instead, the glucose molecule is taken apart systematically, releasing bits of energy in a stepwise fashion. During respiration, some of the energy that is released from the chemical bonds of the glucose molecule is used to form ATP molecules (figure 2.8).

FIGURE 2.6 Photosynthesis is an active life process by which plants, protists, and bacteria that contain the green pigment, chlorophyll, capture solar energy and use part of that energy to build carbohydrate molecules. An immediate product of photosynthesis is the simple sugar, glucose. The carbon, hydrogen, and oxygen atoms of glucose, and the energy that holds them together, are the basis of metabolism for most organisms. The alphabetic symbols for the elements are used to communicate molecular composition; thus, $C_6H_{12}O_6$ is the chemical formula for glucose, which is composed of 6 carbon (C) atoms, 12 hydrogen (H) atoms, and 6 oxygen (O) atoms.

How then does ATP function in a cell? ATP readily breaks down to ADP, releasing energy that can be used for cellular work. An ATP molecule, thus, is a unit of "energy currency" that the cell can "spend" to do work.

What should you take from this brief discussion of energy and cells? For our subsequent discussions of energy interactions in the biosphere, remember this: (1) the cell is the basis for all life-forms on Earth, and (2) the energy economy of living systems, from individuals to ecosystems, is based on the energy transfers that occur at the cellular level.

PRINCIPLES AND CONCEPTS

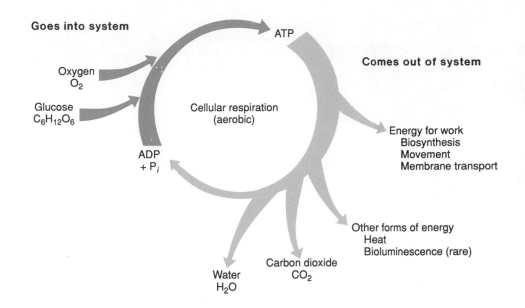

Goes into system

ATP

Comes out of system

Oxygen
O_2

Glucose
$C_6H_{12}O_6$

Cellular respiration
(aerobic)

Energy for work
  Biosynthesis
  Movement
  Membrane transport

ADP
+ P_i

Other forms of energy
  Heat
  Bioluminescence (rare)

Water
H_2O

Carbon dioxide
CO_2

FIGURE 2.7 Aerobic respiration releases energy from food for cellular use. Respiration consists of a series of enzymatically controlled reactions that systematically break the chemical bonds of glucose. As in combustion, oxygen is used and energy is released. Unlike combustion, bits of energy are captured in the chemical bonds of ATP molecules, instead of disappearing from the system in a flash of heat and light.

a. Chemical structure diagram of ATP

b. Word diagram of ATP components

Adenine + Ribose = Adenosine
  Adenosine + Phosphate = Adenosine monophosphate (AMP)
  Adenosine + Phosphate + Phosphate = Adenosine diphosphate (ADP)
  Adenosine + Phosphate + Phosphate + Phosphate = Adenosine triphosphate (ATP)

c. ATP formation during aerobic respiration

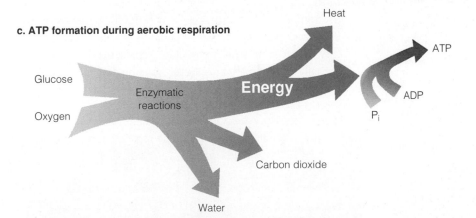

Heat

ATP

Glucose

Enzymatic
reactions

Energy

ADP

P_i

Oxygen

Carbon dioxide

Water

d. Destabilization of glucose by phosphorylation (phosphate transfer)

Glucose + ATP $\xrightarrow{enzyme}$ Glucose-P + ADP

Glucose phosphorylation can also be expressed as follows:

Glucose $\xrightarrow[\text{Mg}^{++}]{\text{hexokinase}}$ Glucose-6-phosphate + H^+
(ATP ADP)

FIGURE 2.8 (*a*) Like other organic molecules, ATP has a skeleton of linked carbon atoms to which nitrogen, hydrogen, and oxygen atoms are attached. Most of the chemical bonds in ATP are stable, enabling the molecule to retain its composition and shape. The main value of ATP to living systems, however, is the presence of two relatively unstable bonds that hold the two terminal phosphate groups to the rest of the molecule. These bonds are indicated with wavy lines. Compare the word diagram of ATP components (*b*) to the line diagram in (*a*). AMP, ADP, and ATP are identified by the number of phosphate units that are attached. All three compounds are important to the energy-transfer reactions of living cells. Energy transfers accompanied by phosphate transfers are called phosphorylation reactions. Diagram (*c*) shows the formation of ATP from ADP during cellular respiration, using energy liberated by the breakdown of glucose molecules. The energy trapped in this conversion is subsequently released to meet cellular energy demands. (*d*) Glucose is a fairly stable molecule that is destabilized by the transfer of energy from ATP in the earliest reaction series of respiration. This is another example of phosphorylation, an essential reaction of life. Note that the enzyme, hexokinase, and a mineral, magnesium, also must be present for the reaction to occur.

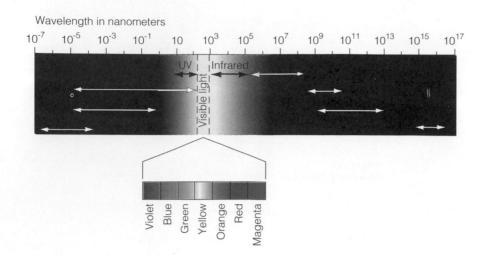

Wavelength in nanometers

Violet · Blue · Green · Yellow · Orange · Red · Magenta

FIGURE 2.9 The electromagnetic spectrum. From this large array of wavelengths, only a few bands within the visible spectrum are used to support life, via photosynthesis.

Solar Energy: Warmth and Light

Our sun is a star, a fiery ball of exploding hydrogen gas. Its thermonuclear reactions emit powerful forms of radiation, including potentially deadly ultraviolet and nuclear radiation (figure 2.9), yet life is here and is nurtured by and dependent upon this searing, energetic source. Solar energy is essential to life for two main reasons.

First, the sun provides warmth. Most organisms survive within a relatively narrow temperature range. In fact, each species tends to have its own range of temperatures within which it can function normally. At very high temperatures, biomolecules break or become distorted and, therefore, are nonfunctional. At very low temperatures, the chemical reactions of metabolism occur too slowly to enable organisms to grow and reproduce. Other planets in our solar system are either too hot or too cold to support life as we know it. Earth's water and atmosphere help to moderate, maintain, and distribute the sun's heat.

Second, organisms are dependent upon solar radiation for life-sustaining energy. No organisms are able to live on sunlight as a direct energy source; instead, photosynthesis converts sunlight to simple sugars, the primary energy source for metabolic work. Solar power generators provide a mechanical analogy to photosynthesis: they trap light energy, then convert it to electrical energy that can be used to do mechanical work.

How much potential solar energy is actually used by organisms? The amount of incoming, extraterrestrial solar radiation is incredible—about 2 gram-calories (gcal) per square centimeter (cm²) per minute. (One gcal is defined as the amount of heat needed to raise the temperature of one gram or milliliter of water 1°C at 15°C.) However, not all of this radiation reaches Earth's surface. As much as half of the incoming sunlight may be reflected or absorbed by atmospheric clouds, dust, and gases. In particular, harmful, short wavelengths are filtered out by gases (such as ozone) in the upper atmosphere; thus, the atmosphere is a significant guardian of the biosphere, protecting life-forms from harmful doses of ultraviolet and other forms of radiation.

Even given these energy reductions, however, there is much more solar energy than biological systems can harness, and more than enough for humans to harness for all of our projected energy needs if technology can enable us to tap it efficiently.

On a clear day at noon, only about 67 percent of direct solar radiation may reach Earth's surface, composed of about 10 percent ultraviolet, 45 percent visible, and 45 percent infrared radiation. Most of the solar radiation that does reach Earth's surface drives thermal patterns in the atmosphere and oceans or is reflected into space by water, snow, and land surfaces (table 2.1). (Seen from outer space, Earth shines about as brightly as Venus.)

Fortunately for life on Earth, some radiation is absorbed by photosynthetic organisms, powering photosynthesis. Even then, however, the amount of energy that can be used to build organic molecules is further reduced. Photosynthesis can only use certain wavelengths of solar energy that are within the visible light range of the electromagnetic spectrum. These wavelengths are in the ranges we perceive as red and blue light. Of the solar energy remaining after filtration and reflection, therefore, only about 1 to 2 percent of that which hits plants is captured for photosynthesis. This small percentage represents the energy base for virtually all life in the biosphere! Specific details of the energy economy of biological systems are stressed in the next two chapters.

TABLE 2.1 Dissipation of solar radiation as percentages of the total that reaches Earth's surface

Manner of energy dissipation	Percent
Reflected	30
Direct conversion to heat	46
Evaporation, precipitation	23
Wind, waves, and currents	0.2
Photosynthesis	0.8

Source: Data from M. King Hulbert, "The Energy Resources of the Earth" in *Scientific American*, 224(3):60–70, 1971.

PRINCIPLES AND CONCEPTS

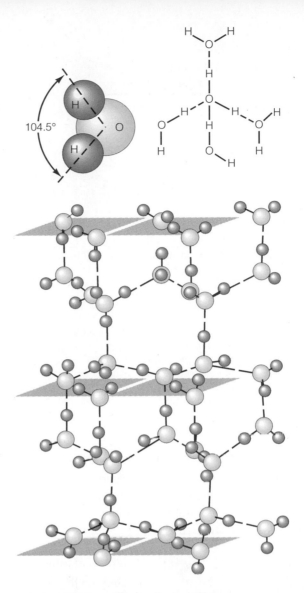

FIGURE 2.10 Compare this real-color Landsat view of England (largest body), Scotland (north of England), and Ireland (*left*) to the other remote-sensed images in this chapter. Vegetation appears green, as compared to red in the infrared-coded images. This image is a mosaic of forty-two separate scenes from data provided by the Thematic Mapper, an instrument aboard the Landsat satellite. Landsat orbits Earth at an altitude of about 900 km (540 mi).

A "Water Planet"

A prime distinction between Earth and other planets is that Earth is what naturalist and marine biologist Jacques Cousteau calls a "water planet," one that is blessed with an abundant supply of water in its liquid form. Earth's waters cover about three-fourths of its surface, with emergent landmasses occupying the rest (figure 2.10). Without water, there is no life. Not only is it essential for cell structure and metabolism, but water's unique physical and chemical properties have a direct impact on Earth's surface temperatures, its atmosphere, and the interactions of life-forms with their environments. It is a marvelous substance.

Why is water so important to life? There are several very specific reasons, most of which are rooted in the physical structure of the water molecule (figure 2.11).

FIGURE 2.11 The two hydrogen atoms of a water molecule are attached to the oxygen atom at a 104.5 degree angle. As a result, a water molecule is polar, having opposite charges on different ends of the molecule: the hydrogen atoms tend to be positive; whereas, the oxygen atoms tend to be negative. Consequently, the oxygen atom of one water molecule tends to attract hydrogen atoms from other molecules. Thus, water tends to be highly ordered, its molecules arranged in an open, three-dimensional lattice. When water freezes, the lattice expands, making frozen water less dense than liquid water, which is why ice floats. Most substances, by contrast, are more dense in their solid form.

(1) Water is the primary component of cells and makes up 60–70 percent (on average) of the weight of living organisms. It plumps out cells, thereby giving form and support to many tissues. It is not just "filler," however, but has vital biological roles. Generally speaking, a dry cell is a dead cell!

(2) Water is the medium in which all of life's chemical reactions occur, and it is an active participant in many of these reactions.

(3) Water is the only inorganic liquid that exists in nature and is the solvent in which most substances must be dissolved before cells can absorb, use, or eliminate them. These substances include food molecules, mineral nutrients, gases, hormones and

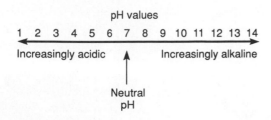

pH values

1 2 3 4 5 6 7 8 9 10 11 12 13 14

Increasingly acidic ← → Increasingly alkaline

Neutral
pH

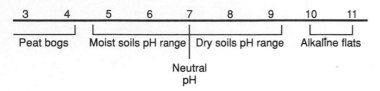

3   4   5   6   7   8   9   10   11

Peat bogs | Moist soils pH range | Dry soils pH range | Alkaline flats

Neutral
pH

FIGURE 2.12 pH values reflect the relative acidity (lower numbers) and alkalinity (higher numbers) of a solution, as determined by the relative concentration of hydrogen ions, H^+. Each number, from 1–14, differs from its neighbor on each side by tenfold on a negative logarithmic scale. pH 6 is ten times higher in H^+ concentration than pH 7, and pH 8 is ten times lower in H^+ concentration. The second diagram gives you an idea of the actual pH ranges of some natural environments. Solutions of the extremes of the pH scale are too caustic for organisms.

other chemical communicators, and waste byproducts of metabolism. When molecules dissolve in water, they often have a tendency to break into positively and negatively charged parts, called ions. Ions are significant participants in cellular reactions.

(4) Water molecules themselves can ionize, breaking into H^+ (hydrogen ions) and OH^- (hydroxyl ions). H^+ and OH^- ions assist in maintaining the acid/base balance in cells, helping to offset (buffer) fluctuations caused by the release of other ions during metabolism. **pH** is the term that refers to the relative abundance of H^+ ions in a solution. On a pH scale from 0–14, 7 is neutral, values lower than 7 are acidic, and those higher than 7 are basic (figure 2.12). A proper pH balance is critical to healthy cellular functioning.

(5) Water molecules are cohesive, tending to stick together because of the attraction of their polar hydrogen atoms to the oxygen atoms of adjacent water molecules (as shown in figure 2.12). Cohesion, for instance, holds together thin columns of water molecules as they are drawn upward through plant bodies. Because of its cohesiveness, water has a high surface tension, a condition in which the water surface, where it meets the air, acts like an elastic skin (figure 2.13).

(6) Water helps protect cells from temperature fluctuations because it has a high **heat capacity.** This means that water can absorb or release a large amount of heat energy before its own temperature changes. For this reason, large bodies of water have a moderating effect on local climates, as near our ocean coasts and the Great Lakes. Orchardists apply this principle when they spray water on fruit trees to limit frost damage to developing fruit. The ice layer that forms around the fruit moderates heat

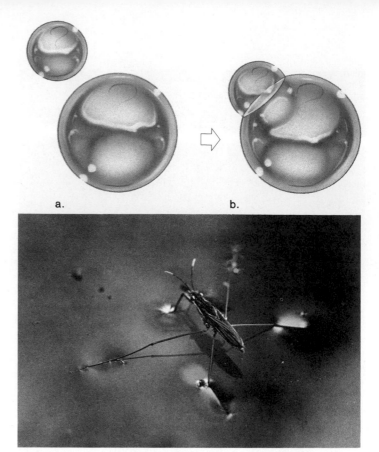

a.   b.

c.

FIGURE 2.13 Water molecules tend to hold on to one another strongly. One of the results of this attraction is surface tension, the tendency of liquid molecules to maintain a minimal surface area. (*a*) Bubbles form spheres—the shape with the least surface area—even if blown through an angular loop. (*b*) Where two bubbles touch, they form a flat surface. (*c*) Surface tension also is demonstrated by the resistance of a water surface to penetration, as when it is walked upon by a water strider.

loss to the air and may get the fruit through a cold night without formation of disruptive ice crystals within fruit tissues.

(7) Water exists as a liquid over a wide temperature range.

(8) Water has high melting and boiling points compared to other liquids. These truly remarkable characteristics make water unique among substances. The combination of high heat capacity, high boiling point, and high freezing point make water a stable, temperate environment for biomolecules, living cells, and aquatic organisms.

(9) A final, related characteristic is that water has a high **heat of vaporization.** Because of the amount of heat it absorbs in order to change from a liquid to a vapor state, water is an effective way for organisms to shed excess heat. Evaporative water loss from plant surfaces, called **transpiration,** helps offset high leaf temperatures on hot, sunny days. Terrestrial mammals pant or sweat to moisten evaporative cooling surfaces. Why do you feel less comfortable on a hot, humid day than on a hot, low-humidity day? It's because the water vapor-laden air inhibits the rate of evaporation from your skin, thereby impairing your ability to shed heat.

PRINCIPLES AND CONCEPTS

The Atmosphere

Primeval Earth was devoid of life-forms, a bleak surface of rocky landscape and shallow seas, bombarded harshly by solar radiation. The atmosphere was turbulent and charged with electrical activity. Scientists think that Earth's primitive atmosphere was radically different than it is now. The ancient atmosphere is thought to have been a thin band of water vapor (H_2O) and poisonous gases, mainly methane (CH_4), ammonia (NH_3), and hydrogen sulfide (H_2S). These gases probably were produced by such physical processes as volcanic activity, weathering, and solar irradiation of Earth's surface.

How did Earth's atmosphere evolve to its present non-toxic, nurturing composition—a mixture of about 78 percent nitrogen, 21 percent oxygen, 0.9 percent argon, 0.03 percent carbon dioxide, variable amounts of water vapor, and trace amounts (less than 0.001 percent) of more than a dozen other gases? We will not present a full discussion here, but it is certain that organisms played an important role in the development of the present atmosphere, just as they continue to have a significant influence on its present maintenance and evolution. Photosynthesis has been the source of nearly all the oxygen in Earth's atmosphere. Cellular respiration and combustion are major sources of carbon dioxide.

There is evidence that the atmosphere is still changing, mainly as a result of human activities. We have caused increased levels of carbon dioxide from combustion of organic fuels, increased levels of assorted gases released by industrial processes, and a decreased amount of ozone in the upper atmosphere. At present, Earth's atmosphere provides gases necessary for life, helps maintain surface temperatures, and filters out dangerous radiation. Are changes in its composition threatening these functions? Reasons for concern about present trends in the evolution of the atmosphere are summarized in chapter 23.

SUMMARY

The biosphere is the life-supporting zone of air, soil, and water that occurs at the interface of Earth's surface with its atmosphere. Life in the biosphere is based on highly organized exchanges of matter and energy between organisms and their environment.

The populations of different species that live and interact within a particular area are a biological community. The interactions of a community with the physical factors of its environment comprise the ecosystem level of organization and study. Large regional ecosystems around the world that possess similar physical conditions and biological communities are called biomes.

For life to exist on Earth, certain requirements must be met. These include the availability of required chemical elements, a steady influx of solar energy, amenable surface temperatures, the presence of liquid water, and a suitable atmosphere.

Matter is the observable material of which the universe is composed. It exists in three interchangeable phases: gas, liquid, and solid. Matter is made up of atoms, which are composed of particles called protons, neutrons, and electrons. The energy that holds atoms together forms the basis for energy transfers in the bodies of living organisms and, therefore, in the biosphere.

A steady influx of solar radiation provides the heat and light energy needed to support life in the biosphere. Water, which covers approximately three-fourths of Earth's surface, is readily available to life-forms. Because of its unique characteristics, it stabilizes the biosphere's temperature and provides the medium in which life processes occur. Earth's atmosphere provides gases necessary for life, helps maintain surface temperatures, and filters out dangerous radiation.

■ Review Questions

1. The biosphere or "sphere of life" is said to blanket Earth's surface. Describe those factors that define the upper and lower limits of the biosphere.
2. What is meant when we refer to cells as processors of matter and energy, and how does this description relate to the biosphere?
3. A species is a specific kind of organism. What general characteristics do individuals of a particular species share? Why is it so important for ecologists to differentiate among the various species in a biological community?
4. An organism's environment is described as all the things that affect the life of the organism, including biotic and abiotic factors. How is your environment similar to the environment of any other living organism? How does the human environment differ from that of other life-forms?
5. Earth is the only known planet to possess a temperature range that is suitable for life as we know it. Which physical factors of the biosphere contribute to this suitable temperature?
6. Daytime and nighttime temperature differences on Earth are seldom very extreme. What characteristics of the biosphere maintain these relatively constant temperatures?
7. Jacques Cousteau described our Earth as the "water planet." Why is it unlikely to find life as we know it existing on a planet lacking large quantities of water in the liquid phase as it occurs on Earth?
8. Molecular oxygen does not appear to have been a major component of Earth's primitive atmosphere. How can we explain the present composition of Earth's atmosphere, in which molecular oxygen is the second most abundant gas?

■ Questions for Critical Thinking

1. In 1965, Adlai Stevenson, the U.S. Ambassador to the United Nations, described Earth as a fragile spaceship. Explain why this is a valid analogy of our planet.

2. Schematic and mathematical models and hierarchies are commonly used by scientists to describe their observations. Why is it beneficial to scientists to arrange their data in this fashion? What modern technological developments have occurred that help scientists develop such models and hierarchies?

3. When you roast marshmallows, occasionally one will burst into flames if you are not careful. As the marshmallow burns, it releases energy. If you eat a marshmallow, energy is released in your body. How do the chemical reactions that take place in your body as the marshmallow's energy is released differ from the reactions involved when the marshmallow burns?

4. Most lakes support large populations of algae. In such lakes it is possible to observe a daily oxygen cycle. Dissolved oxygen levels begin increasing shortly after sunrise and continue to do so until the middle of the afternoon when they level off. As evening approaches, the dissolved oxygen levels begin to decline and continue to do so throughout the night. Why do you think this daily cycle occurs?

5. The efficiency of photosynthesis is estimated at approximately 1 percent. Despite this relatively poor efficiency, it is sufficient to support almost all life in the biosphere. Explain how this can be possible.

■ Key Terms

atoms (p. 16)
biological or biotic factors (p. 16)
biological community (p. 15)
biomes (p. 16)
biosphere (p. 14)
cellular respiration (p. 18)
ecosystem (p. 16)
electrons (p. 17)
environment (p. 16)
heat of vaporization (p. 22)
heat capacity (p. 22)
metabolism (p. 18)
neutrons (p. 17)
orbitals (p. 17)
organic compounds (p. 15)
pH (p. 22)
photosynthesis (p. 15)
physical or abiotic factors (p. 16)
population (p. 15)
protons (p. 17)
species (p. 15)
transpiration (p. 22)

SUGGESTED READINGS

■ Ehrlich, Paul R., and Jonathan Roughgarden. *The Science of Ecology*. New York: Macmillan Publishing Co., 1987. Synthesizes the several approaches of modern ecology, uses excellent examples based on current research, and gives balance to some of the interpretive controversies in ecology. Useful bibliography, consisting mainly of references cited.

■ Hanson, Robert W., ed. *Science and Creation— Geological, Theological and Educational Perspectives*. AAAS Series on Issues in Science and Technology. American Association for the Advancement of Science. New York: Macmillan Publishing Co., 1986. Explores the false dichotomy in the creation/evolution controversy through working examples of how different teachers, scientists, and theologians deal with the topic.

■ Johnson, Leland G. *Biology*. 2d ed. Dubuque: Wm C. Brown Publishers, 1987. A valuable reference source for additional information about the nature of matter, chemistry of life processes, characteristics of organisms, cells, metabolism, and other biological topics.

■ Kroschwitz, Jacqueline I., and Melvin Winokur. *Chemistry: General, Organic, Biological*. New York: McGraw-Hill, 1985. Gives clear explanations of basic chemistry principles, as well as practical examples.

■ Lovelock, James E. *Gaia: A New Look at Life*. Oxford: Oxford University Press, 1979. Presents the Gaia hypothesis, an intriguing but controversial hypothesis in which the biosphere is seen as a gigantic system with organism-like characteristics.

■ Van Matre, S., and B. Weiler. *The Earth Speaks*. Institute for Earth Education, Warrenville, Illinois, 1983. Interpretations of Earth as seen by scientists and poets.

PRINCIPLES AND CONCEPTS

CHAPTER 3

Matter and Energy in Ecosystems

'Statistically, the probability of any one of us being here is so small that you'd think the mere fact of existing would keep us all in a contented dazzlement of surprise. . . . Even more astounding is our statistical improbability in physical terms. The normal, predictable state of matter throughout the universe is randomness, a relaxed sort of equilibrium, with atoms and their particles scattered around in an amorphous muddle. We, in brilliant contrast, are completely organized structures, squirming with information at every covalent bond. We make our living by catching electrons at the moment of their excitement by solar photons, swiping the energy released at the instant of each jump and storing it up in intricate loops for ourselves. We violate probability, by our nature. To be able to do this systematically, and in such wild varieties of form, from viruses to whales, is extremely unlikely; to have sustained the effort without drifting back into randomness, was nearly a mathematical impossibility.'

Lewis Thomas, in *The Lives of a Cell*

C O N C E P T S

This chapter builds upon concepts developed in the preceding chapter, focusing them more specifically on the interactions between living systems and the physical environment.

Cells, organisms, and ecosystems are living systems, demonstrating influx, outflux, and utilization of specific atoms and energy. Living systems process matter and energy, assembling and disassembling biomolecules by the transfer of atoms and energy. Like physical systems, biological systems are governed by the principles of conservation of matter and thermodynamics. They also maintain homeostasis by active effort, resisting the tendencies toward disorder inherent in physical processes.

The metabolic processes of organisms are linked to global chemical cycling of water, carbon, oxygen, nitrogen, phosphorus, sulfur, and other substances. Matter tends to be cycled within ecosystems, but energy flows through, requiring constant replenishment to maintain the system.

Matter and energy processing in ecosystems is based on feeding interactions that begin with producer organisms and proceed through various categories of consumer organisms. Nutrients and energy are passed from population to population by linked feeding series called food chains, which typically are interlocked to form food webs. A population's position in a specific food chain determines its trophic level. Ecological pyramids of energy, biomass, and numbers provide quantitative demonstrations of how ecosystem dynamics reflect the physical principles of conservation of matter and thermodynamics.

INTRODUCTION

Living systems are *ordered*, and maintenance of this order is essential to their survival. They demonstrate **homeostasis,** a dynamic balance of many specific conditions that is actively maintained by opposing, compensating adjustments. In creating and maintaining order, cells are responsive to the conservation laws of the physical sciences. These principles of conservation of matter and thermodynamics provide a basis for understanding the dynamic interactions of matter and energy in living systems, from the cellular to ecosystem levels. In this chapter, we will emphasize the exchanges of matter and energy that occur in ecosystems and how these exchanges are related to homeostasis. Some examples of human actions are related to focus your understanding. Throughout this text, you will see how these principles are demonstrated.

SOME IMPORTANT CONCEPTS ABOUT MATTER AND ENERGY EXCHANGES

Why do you eat and drink? Where do the substances you ingest come from, and what does your body do with them? As you consider ecosystem dynamics, keep in mind yourself as an ecosystem component.

Conservation of Matter

It is probable that some of the molecules that make up your own body contain atoms that once made up the body of at least one dinosaur and, most certainly, the bodies of numerous smaller prehistoric organisms! Matter is used over and over again. It is transformed and combined in different ways, but it doesn't disappear. "Everything goes somewhere." These statements are a paraphrase of the physical principle of **conservation of matter.**

How does this principle apply to ecosystems and the biosphere as a whole? Ecosystem dynamics are based on matter that is physically present at a site. As components of a site are incorporated into the bodies of organisms, are fewer and fewer essential materials available to successive generations of organisms? Although this might seem to be so, it is not the case because materials are made available continuously by biological recycling. What are the most commonly recycled components in an ecosystem? To answer this question, think about the substances that make up biological molecules: carbon, hydrogen, oxygen, nitrogen, and other elements required by cells.

Conservation of matter also has a direct bearing on the human relationship with the biosphere. Particularly in affluent societies, we use natural resources to produce an incredible amount of "disposable" consumer goods. If everything goes somewhere, where do the things we dispose of go after the garbage truck leaves? As the sheer amount of "disposed-of stuff" increases, we are having greater problems finding places to put

FIGURE 3.1 Styrofoam cups, insulation, and floats do not biodegrade, but they do break apart into small pellets. These pellets are found far out at sea, where they are a hazard to marine life-forms that mistake them for food and ingest them. This shoreline is littered with plastic foam pellets and chips that have washed ashore.

it. A dramatic example of this concern is the New York City trash and garbage barge that was pushed down and back up the Atlantic coast in 1987 in search of a dump site. If ordinary garbage is becoming such a problem, multiply the concerns and problems manyfold for toxic and radioactive wastes.

These are large, public issues. What about our one-on-one human interactions with the environment, as related to conservation of matter? Styrofoam cups, plastic hamburger pods, and other items made of complex synthetics resist biological decomposition (figure 3.1). They are significant, virtually permanent forms of pollution. Some are dangerous, as well as unsightly— baby sea turtles and small fish ingest but cannot digest the styrofoam pellets; and fish, birds, and mammals become fatally entrapped by plastic six-pack beverage yokes. Remember the principle: everything goes somewhere!

Thermodynamics and Energy Transfers

How does the way organisms process energy differ from the way they process matter? Organisms use gases, water, and nutrients, then return them to the environment in altered forms as by-products of their metabolic processes. Year after year, century after century, the same atoms find endless reincarnation in new molecules synthesized by succeeding organisms as they feed, grow, and die. This exchange and continuity are made possible, however, by something that cannot be recycled—energy. Energy must be supplied from an external source to the ecosystem to keep it running. In nearly all of Earth's ecosystems, that energy comes from the sun.

Ecosystems are, for the most part, **open systems,** which means energy and matter may enter or leave. By comparison, matter is not exchanged between a closed system and its surroundings.

The study of **thermodynamics** deals with how energy is transferred in natural processes. More specifically, it deals with the relationships between heat, work, and energy. Heat is a transfer of energy as a result of temperature difference, and work is a transfer of energy that is not due to a temperature difference. Both heat and work are important kinds of energy transfers in organisms and in ecosystems.

The **first law of thermodynamics** is a principle of physics that recognizes that energy is *conserved*. It may be transferred into or out of a system, and as a result there is a change of energy within the system. It also may be transformed, changed from one form to another (e.g., from the energy in a chemical bond to heat energy). Like matter, energy can go somewhere, but is not lost. In a biological context, this principle often is stated thus: "Energy may be transferred or transformed, but it is not lost."

How does this principle apply to organisms? Cells are alive, using energy. Collectively, the energy-transforming processes of an organism are called its metabolism. The energy source for cellular metabolism is the chemical energy stored in the bonds that hold together food molecules. Most of this energy is used to do work, such as molecular synthesis, movement, or emission of light.

The **second law of thermodynamics** is a principle of physics that recognizes that, with each successive energy transfer or transformation, less energy is available to do work. It tends to be degraded or dissipated from a form that is useful to do mechanical work to a less useful form, usually heat. We can think of this as an energy "expenditure," the "cost" of doing work; thus, there is a tendency in natural processes to go from a state of order (e.g., high-quality energy, such as work energy) toward a state of increasing disorder (e.g., low-quality energy, such as heat energy). The second law recognizes the general trend toward disorder and death in the universe involving the degradation of matter and energy. Stated crudely, the second law of thermodynamics could be summarized as: "Left to themselves, things tend to degenerate."

How does the second law of thermodynamics apply to organisms and ecosystems? Organisms are made of cells. Cells are highly organized, both structurally and metabolically. The cell membrane, for instance, is an array of closely aligned molecules that regulates the transfer of substances into and out of the cell. As this work is conducted, some energy is transformed to heat energy and, therefore, cannot be used to do work. The membrane needs a continuous supply of chemical energy, which is provided by cellular respiration, mainly in the form of ATP molecules. If the energy supply is interrupted, however, membrane structure and function become disordered. Such disorganization can be either a cause or a consequence of cell death.

The energy relationships between producers and consumers in ecosystems also demonstrate the second law of thermodynamics, as we will elaborate further in a later section. In an ecosystem, **producers** are organisms that use energy and matter from the environment to synthesize food molecules. Most producers are green, photosynthetic organisms. Organisms that feed upon organic molecules are called **consumers,** such as animals, fungi, and most bacteria. The total **biomass** (the total amount of matter, by weight, tied up in the bodies of organisms) of producers in an ecosystem is greater than the total biomass of the consumers. Why is this so? Producers feed themselves *and* consumers. They use the food they produce for their own energy needs and store the excess. Much energy is tied up in the building of structural molecules, many of which are indigestible and, therefore, unavailable to animal consumers. Also, some energy is degraded and dissipated as heat during metabolism; thus, there is substantially less energy available to consumers in an ecosystem than was available to the producers.

Before we leave this topic, let's comment briefly on the heat released during metabolism. Just because it may be a "lower" form of energy does not mean that it is of no value to organisms. Think of your own body—it is warm, and you are able to perform quick movements while warm. Why do your muscle joints "stiffen up" when they become cold? Why do "cold-blooded" animals, such as frogs and insects, move sluggishly when they become dormant during winter? Chemical reactions are responsive to heat; in general, a higher temperature accelerates reaction rates, whereas colder temperatures reduce reaction rates. Metabolic heat has a direct relationship to survival of cells. It is an important factor of the cellular environment in which chemical reactions occur. It particularly helps to maintain a life-sustaining temperature level in cells when the outside environment is cold.

HOMEOSTASIS

Living systems, from the cellular to ecosystem levels, must maintain a dynamic internal balance to remain healthy. This need is demonstrated most evidently at the cellular level, where loss of stability can mean sudden death—consider the consequences of heart, liver, or kidney failure in a human body. The dynamic balance in living systems is called homeostasis (same status). A key word in this definition is "dynamic." Living systems are active, rather than static, so that conditions in a system change continuously in response to many environmental stimuli. Homeostasis is not, therefore, a static, unchanging condition, but a condition of fluctuating balance centered on some ideal state, or **optimum.** This dynamic balance is maintained by active and opposing adjustments and compensations, analogous to a tightrope walker maintaining balance with a sway bar. Understanding how organisms respond to environmental stresses is important to managing not only natural environments, but also agriculture (see Box 3.1).

Consider your own body as an example. It must maintain homeostasis in regard to numerous internal factors, including

BOX 3.1 Physiological Stress and Plant Distribution

Does a plant have a body? Does it have organs? Is a plant passive, or does it react actively to its environment? Does a plant experience stress? These may seem like simplistic questions, yet they ought to challenge your mental concept of plants in some way. We tend to think in animal terms, and you probably could answer each of the above questions quickly and affirmatively if animals were the subject. After all, plants seem so . . . inanimate! The facts are, plant bodies *are* sensitive to specific environmental conditions, which create physiological stresses, and they *do* respond to different kinds of stresses in order to maintain internal homeostasis.

Stress is any factor that causes a plant to grow or reproduce below its potential levels. Too much or too little heat, light, water, or specific nutrient, excess salt, or the combined effects (**synergism**) of any of these factors create stress on plant metabolic systems. Herbivory, parasitism, disease, and competition (chapter 4) create stress. Any kind of disturbance of a natural community or ecosystem also creates stress.

Why is research in stress physiology important? The bottom line is that physiological stress affects the health, survival, and reproductive potential of individual plants. Since plants make up the producer base of an ecosystem, plant responses to stress also have an impact on matter and energy flow within entire ecosystems. *The same concept applies both to natural ecosystems and to agricultural ecosystems.* The study of stress physiology in plants is, therefore, a significant research area to help us understand the natural distribution of plants and develop strains of cultivated plants that can produce well in less-than-ideal soil, climate, and nutrient conditions. Our challenge to understand stress physiology is given urgency by the increasing rate at which we are introducing stressful changes into the environment.

Adaptive responses to stressful environments can be categorized into two kinds of strategies: tolerance or avoidance. Tolerance mechanisms include metabolic, structural, and behavioral adjustments that enable plants to adjust to and remain normally active under mild stress and become less active during periods of severe stress. Closure of stomates to reduce water loss, movement of leaves to avoid direct solar exposure, lack of leafy parts in arid climates, and specialized photosynthetic mechanisms are examples of tolerance mechanisms. Avoidance mechanisms are innate metabolic adjustments that enable plants to suspend food production and growth by attaining a state of dormancy during periods of severe stress. Plants in temperate climates, for instance, have genetic patterns of seasonal growth, reproduction, and dormancy.

Looking at our planet, you can see a direct correlation between latitude and ecosystem productivity. The most productive ecosystems (both terrestrial and aquatic) are at or near the equator. Productivity is lower in the temperate zones and is very low in polar regions. Seasonal cold is the primary stress factor that affects plant distribution. It works in two ways: affecting plant tissues directly and affecting the availability of liquid water. The fact that we cannot control the seasonal tilt of our planet means that we must work within this overarching fact of life in planning for sustainable world ecosystems and agriculture.

Some vascular plants can survive temperatures in nature below $-60°C$, provided that they have an opportunity to go through a period of metabolic preparation (acclimation). Some vascular plants can survive heat up to about $+60°C$. At both extremes, survivability is better if the plants are in a state of lowered metabolic activity—winter or summer dormancy. Plants are more likely to be stressed by abrupt, unseasonal hot or cold episodes, and they are more susceptible to temperature extremes as seedlings and juveniles.

It may seem paradoxical that excesses of light and water can be stresses. If you've ever put a shade-adapted houseplant outside in direct sunlight, however, you may have seen how its leaves became sunburned, and it even may have died. In forest ecosystems, abrupt creation of a gap can have a drastic effect on the understory community, which suddenly is exposed to more intense light than its pigment systems and photosynthetic pathways are adapted for. The result can be a major change in species composition in the stressed locality. The vigor and survival of shade-loving species decline, while the germination and growth of light-loving species are encouraged. The result is change in the community composition. Likewise, many a houseplant dies of a suffocated root system as a result of well-intentioned overwatering! In natural ecosystems, plants that normally live "with wet feet," such as cultivated rice and wild wetlands plants, have specialized air passages to ensure adequate transport of oxygen for respiration and carbon dioxide for photosynthesis among internal tissues.

How is stress mediated at the cellular level? Exciting, continuing research is ongoing. Presently, much attention is directed toward internal cellular membrane systems, the diaphanous networks and layers of lipid/protein complexes that compartmentalize cytoplasm and form organelles. Membranes organize and carry out much of the cell's metabolic activity. Intricate enzymatic activities occur on membranes, and biochemical homeostasis is maintained by membrane-mediated activities. Because of their sensitivity to perturbation, membranes and membrane systems are thought to be the cell's stress sensors. We are just beginning to learn about such mechanisms. The implications for both understanding and managing natural plant species and for manipulating cultivated species are potentially fantastic.

concentration of water and salts, pH (relative acidity and alkalinity), food and mineral nutrients, hormones, temperature, muscle tonus, and so on. Your body is a complex of interacting systems, and the activity of each system affects the others. What is the alternative to homeostasis? It is a progressive tendency toward disorganization, in accordance with the second law of thermodynamics. On the organismal level, increasing disorder means decreasing vitality and, eventually, death.

The concept of homeostasis also applies to ecosystems. Usually it is expressed as ecosystem **stability,** a dynamic equilibrium among the physical and biological factors in an ecosystem (see also chapter 4).

Can the concept of homeostasis be applied at the biosphere level? Increasingly, we are becoming aware that even so vast a system as the whole biosphere is sensitive and responsive to changes that occur within it. During the past two centuries, changes in the biosphere have occurred at an accelerating pace, primarily as a result of the activities of one life-form—the human species. Some of these changes have been benign, even beneficial, while others have not. Persons who have expressed concern over destructive or potentially dangerous changes and our inability to predict their long-term consequences often have been criticized by others more involved in short-term economic projections. You, as a citizen, with your own attitudes, beliefs, and actions, are a participant in this debate, whether it involves local issues of solid waste disposal or global issues, such as nuclear testing, disposal of hazardous wastes, acid rain, and tropical deforestation.

Part of the problem of understanding homeostasis in the biosphere is that human lives are too short to lend perspective; biological and environmental processes and reactions require more than one human generation to reach a full response. Another part of the problem is that we have a planet rich in natural resources for which we are still finding uses. We've not had to worry about finiteness of resources or of the environment's capacity to absorb change without losing stability.

Part of the problem is that we have only recently entered an era in human technological and population development in which we are capable of causing fantastically rapid, destructive, and irrevocable changes in the biosphere. Does that sound like doom and gloom? It's easy to brush off such statements, but it's not wise. It's much more prudent to become informed and to guide our human actions based on emerging information and insights instead of tradition or wishful thinking.

MATTER AND ENERGY IN ECOSYSTEMS

Conservation of matter, the principles of energy transfer, and homeostasis are amply demonstrated in the actual workings of biological systems. As an example, consider photosynthesis and

Food manufacture: photosynthesis

$$6\ CO_2 + 6\ H_2O + \text{light energy} \quad \xrightarrow{\text{chlorophyll and enzymes}} \quad C_6H_{12}O_6 + 6\ O_2$$

carbon dioxide / water / / glucose / oxygen

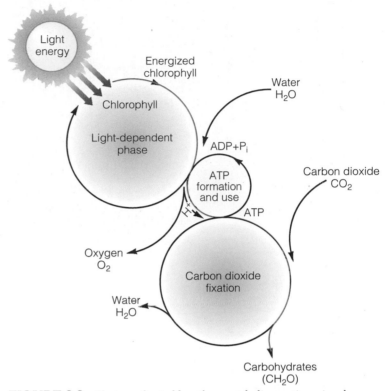

FIGURE 3.2 Photosynthesis, like other metabolic reactions, involves transfers of energy when chemical bonds are broken and formed. Also like other thermodynamic reactions, there is a net loss of concentrated energy during these transfers. The flow of energy through photosynthesis, from *light units to food units* in the form of carbohydrate molecules, is shown here. The diagram shows input of energy, water, and carbon dioxide, and output of carbohydrates and oxygen. The three circles represent reaction cycles within the living photosynthetic cells.

cellular respiration. They are complementary processes in maintaining life—what photosynthesis puts together, respiration takes apart, making components available again for photosynthesis:

Photosynthesis
$$6H_2O + 6CO_2 + \text{energy} \rightarrow C_6H_{12}O_6 + 6O_2$$

Respiration
$$6O_2 + C_6H_{12}O_6 \rightarrow 6H_2O + 6CO_2 + \text{energy}$$

Carbon, hydrogen, and oxygen atoms are recycled between organisms and the environment via these two processes.

What is the energy relationship between photosynthesis and respiration? Unlike atoms, energy is not recycled. Instead, according to the principles of thermodynamics, it *flows through* biological processes. With each successive transfer and transformation of energy, some energy is used to do work, some is bound up in stable molecules that are not usable by the next consumer, and some is released as heat. As a result, the amount of energy *available* to do work at each successive step decreases. Figure 3.2 depicts how energy flows through the two phases of

FIGURE 3.3 Each time an organism feeds, it becomes a link in a food chain. In an ecosystem, food chains become cross-connected when predators feed on more than one kind of prey, thus forming a food web. How many food chains make up the food web shown here? The arrows in this diagram and in figure 3.4 indicate the direction of predation. Which way would the arrows go if they indicated the flow of matter and energy instead?

photosynthesis, beginning with the maximum level that chloroplasts actually are capable of absorbing. Light energy from the sun sets off a reaction cycle that breaks water molecules, liberating oxygen. Energy is captured in the chemical bonds of ATP. The ATP formed in the first set of reactions is an energy source for the second set of reactions, which end with the formation of sugars = cellular food. A similar energy flow occurs during respiration.

These descending pathways of energy quality evoke an image of energy flowing downhill—an energy cascade—representing the gradual loss of energy that is available to do work. That is why, in order to maintain homeostasis in regard to available energy, an organism must continually acquire food molecules to stay alive.

Food Chains, Food Webs, and Trophic Levels

Photosynthesis is the base of the energy economy of all but a few special ecosystems, and ecosystem dynamics are based on how all organisms present share food resources. In fact, one of the major properties of an ecosystem is its **productivity,** the amount of biological matter (biomass) produced in a given area during a given period of time. For instance, a cornfield is a very

simple ecosystem, and we measure its agricultural productivity by the biomass of corn produced, expressed in the United States as bushels of corn per acre per year. Why is this assessment of cornfield productivity incomplete in biological terms? It doesn't take into account the biomass of other species, including grasshoppers, gophers, earthworms, and insectivorous birds that are also part of the ecosystem and, therefore, constitute a part of its ecological productivity.

Think about what you have eaten today and trace it back to its photosynthetic source. If you have eaten an egg, you can trace it back to a chicken, which ate corn. This is an example of a **food chain,** a linked feeding series. Now think about other possible feeding interactions involving you, the chicken, the corn plant, and a grasshopper. You could eat corn yourself, making the shortest possible food chain, or you could eat the grasshopper—some humans do.

In ecosystems, some consumers are adapted to feed on a single species, but most consumers utilize more than one food species; conversely, each species in an ecosystem tends to have several predators. In this way, individual food chains become interconnected to form a **food web** (figure 3.3). The matter of feeding becomes even more intricate when the energy needs

PRINCIPLES AND CONCEPTS

FIGURE 3.4 Harsh environments tend to have shorter food chains than environments with more favorable physical conditions. Compare the arctic food chains depicted here with the longer food chains in the food web in figure 3.3. The arrows indicate the direction of feeding.

and life histories of organisms that *parasitize* plants and animals in an ecosystem are considered. Perhaps you now can imagine the challenge ecologists face in trying to quantify and interpret the precise matter and energy transfers that occur in a natural ecosystem!

An organism's feeding status in an ecosystem can be expressed as its **trophic level.** In our first example, the corn plant is at the producer level; it transforms solar energy into chemical energy, producing food molecules. Other organisms in the ecosystem are consumers of the chemical energy harnessed by the producers. An organism that eats producers is a primary consumer. An organism that eats primary consumers is a secondary consumer, which may in turn be eaten by a tertiary consumer, and so on. Most terrestrial food chains are relatively short (seeds → mouse → owl), but aquatic food chains may be quite long (microscopic algae → copepod → minnow → crayfish → bass → osprey). The length of a food chain also may reflect the physical characteristics of an ecosystem (figure 3.4).

Organisms can be identified both by trophic level and by the *kinds* of food they eat (figure 3.5). **Herbivores** are plant eaters, **carnivores** are flesh eaters, and **omnivores** eat both plant and animal matter. What are humans? We are natural omnivores, by history and by habit. Tooth structure is an important clue to understanding animal food preferences, and humans are no exception. Our teeth are suited for an omnivorous diet, with a combination of cutting and crushing surfaces that are not highly adapted for one specific kind of food, as are the teeth of a wolf or a horse.

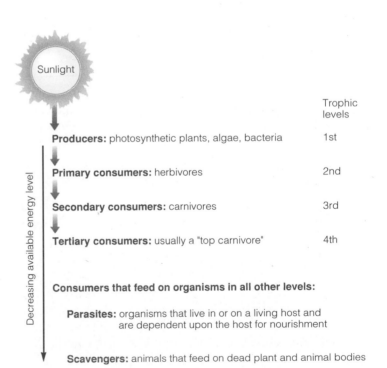

FIGURE 3.5 Organisms in an ecosystem may be identified by what they feed upon (produce, herbivore, carnivore, omnivore, scavenger, decomposer) or by consumer level (producer; 1°, 2°, 3°, consumer) or by trophic level (1st, 2nd, 3rd, 4th).

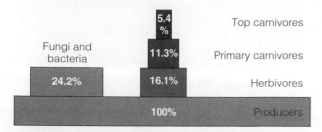

FIGURE 3.6 The classic example in this diagram is an aquatic ecosystem at Silver Springs, Florida, in which the producer energy base was calculated to be 20,810 kcal/m³/yr. The energy relationships between different trophic levels in an ecosystem can be depicted as a pyramid. Photosynthetic organisms are at the base of the pyramid and represent the maximal amount of primary food source for the ecosystem. A dramatically smaller amount of food energy is available at each successive trophic level: the percentage in each bar shows what percent of the energy represented in the *underlying* bar is actually incorporated at the next level. If we calculate the energy measurement for the top carnivores as a percentage of the producer base, the number is even more dramatic—it is only 0.1%. Fungi and bacteria, essential components of any ecosystem, feed at every level but are shown attached to the producer bar to represent the proportion of food energy they use.

Many kinds of organisms feed on the dead bodies of others. **Scavengers,** such as crows, ants, beetles, jackals, and many kinds of rodents, break apart dead plant and animal bodies into smaller pieces. This exposes inner tissues to small **decomposer** organisms—fungi and bacteria that grow on the tissues. A decomposer feeds by secreting digestive enzymes into its food source, then absorbing small molecules that are the products of digestion. This is called **absorptive nutrition.** It could be argued that decomposers are second in importance only to producers, because without their activity, nutrients would remain locked up in the organic compounds of dead organisms and discarded body wastes, rather than being made available to successive generations of organisms.

Ecological Pyramids

Visualize the final predator species in a food chain, whom other species do not kill to eat. This species occupies the top trophic level. What does this position mean in terms of matter and energy? In practical terms, it means there is "less room at the top." There are fewer organisms at this level in an ecosystem than at previous levels. True to the second principle of thermodynamics, there is less food energy available to the top trophic level than to preceding levels, which has an impact on other ecosystem characteristics. The energy and matter relationships of ecosystems suggest pyramidal models, which are not perfect, but are generally useful.

First, consider the **energy pyramid** (figure 3.6). The flow of energy through an ecosystem involves many transfers and transformations of energy in the metabolic pathways of organisms. Remember that for each link of the food chain, less energy is *available to do work* because much of it is undigested, used, or released as heat at each previous level. The losses often are

about 90 percent from one level to the next, so only about 10 percent of the energy in one consumer level may be usable by the next level.

The amount of usable energy that flows through an ecosystem has a direct bearing on productivity and, therefore, the composition of the biological community. Because of decreasing energy availability, a decreasing amount of biomass can be produced at each successive consumer level (figure 3.7). This idea can be expressed as a **biomass pyramid,** based on the total amount of matter that is tied up in the bodies of organisms at each level of the pyramid. The biomass pyramid usually is related to a **numbers pyramid,** in which populations of species at each trophic level tend to be smaller (figure 3.8).

Can you imagine discrepancies in the biomass and numbers models? Sometimes they can be turned upside down. The biomass pyramid, for instance, can be inverted by periodic fluctuations in producer populations (e.g., low plant and algal biomass present during winter in temperate aquatic ecosystems). The numbers pyramid also can be inverted. One coyote, for instance, can support numerous tapeworms. Numbers inversion also occurs at the lower trophic levels (e.g., one large tree can support numerous caterpillars).

It's tempting to think of ecosystems as being self-contained, enabling us to make neat conclusions about their operation. In fact, ecosystems influence one another and exchange materials with their neighbors, especially where they meet and intergrade. They affect each other's homeostasis. An obvious example is the range of effects a forest ecosystem has on the aquatic ecosystems with which it is associated. Streams are affected by soil, nutrients, and organic matter from the forest. Changes that affect the forest ecosystem, therefore, also affect the streams (figure 3.9).

This example also emphasizes the interrelatedness of ecosystems within the biosphere. As conservationist John Muir said, "When we try to pick out anything by itself, we find it hitched to everything else in the universe."

Material Cycles and Life Processes

Through metabolic processes, organisms participate in larger cycles that also have significant inorganic parts, including the **water cycle, carbon cycle, oxygen cycle,** and **nitrogen cycle.**

The Water Cycle

On a global scale, the role of living organisms in the water cycle (hydrologic cycle) is very small compared to the massive interaction of precipitation and evaporation from land and water surfaces (figure 3.10). (See chapter 16 for a more extensive discussion of the nonbiological details of the water cycle.) Collectively, plants and animals take in large quantities of water. Some of the water taken in by green plants is used in photosynthesis, liberating oxygen. Most water taken in by terrestrial organisms, however, is given back to the environment by

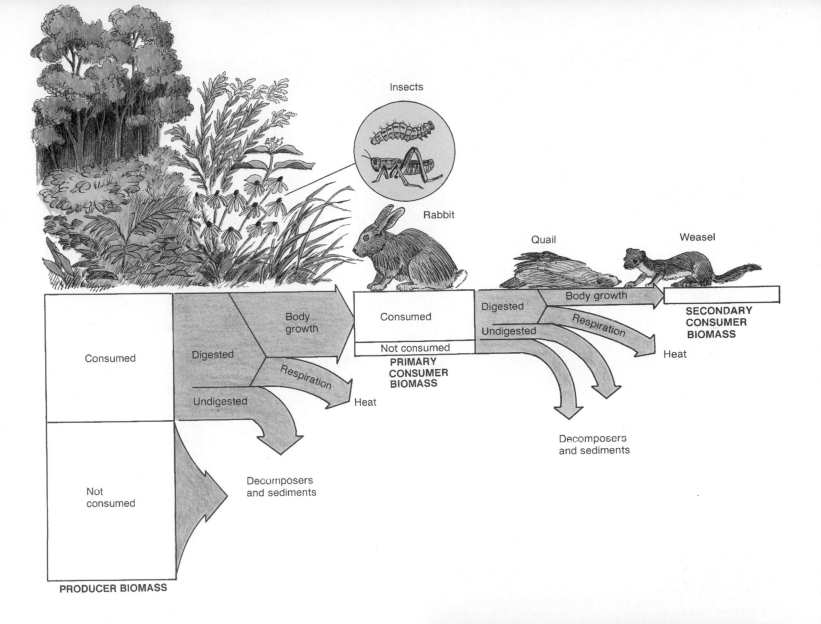

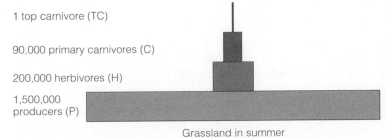

FIGURE 3.7 The energy pyramid is understood more clearly if it is related to a biomass pyramid, which represents the amount of biomass at each trophic level in a food chain. This figure illustrates how nutrients and energy become continuously less available to successive consumers.

1 top carnivore (TC)

90,000 primary carnivores (C)

200,000 herbivores (H)

1,500,000 producers (P)

Grassland in summer

FIGURE 3.8 Usually, smaller organisms are eaten by larger organisms, and it takes numerous small organisms to feed one large organism. This statement is demonstrated in studies of food chains in communities and can be represented visually as a numbers pyramid. The classic study represented in this pyramid shows numbers of individuals at each trophic level per 1,000 m² of grassland, and reads like this: to support one individual at the top carnivore (TC) level, there were 90,000 primary carnivores (C) feeding upon 200,000 herbivores (H) that in turn fed upon 1,500,000 producers (P).

FIGURE 3.9 Even mountain streams are only as clear as upstream activities allow them to be. The water on the right is fed by runoff from a wilderness area in Idaho, whereas the water on the left flows through a multiple-use forest where the ecosystem and soil have been disrupted by logging.

Matter and Energy in Ecosystems

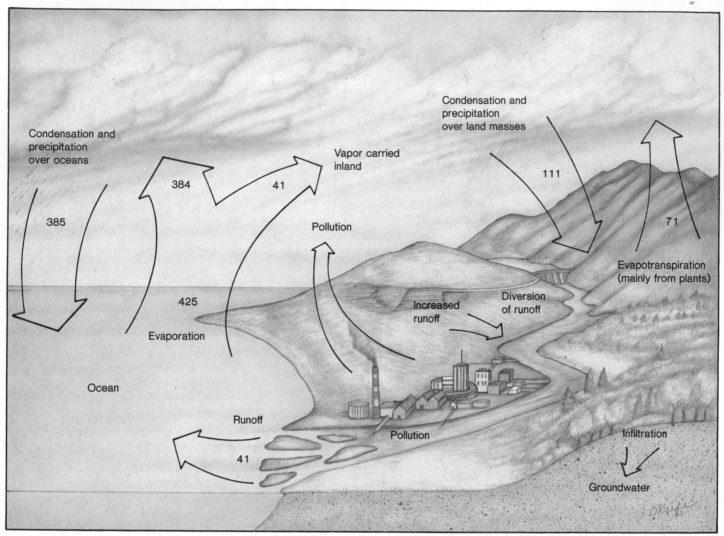

FIGURE 3.10 The water cycle (hydrologic cycle) involves movement of H_2O between the atmosphere, land and surface waters, and groundwater. The numbers in this illustration represent estimated quantities of water transported in units of 1,000 km³/yr.

evaporation from body surfaces, such as skin and leaves. Evaporation from forests, especially, has a significant effect on biome water cycling. Animals lose water vapor from their respiratory surfaces, as is dramatically illustrated when you can "see your breath" on a cold day! Animals also return water to the environment in their urine and excrement.

Plants participate in the water cycle in yet another way— as ground cover. A forest cover shades the soil surface and reduces air movement over it, thereby reducing evaporation of soil moisture. Vegetation also significantly reduces soil erosion due to runoff, especially on slopes and in areas of heavy rainfall (figure 3.11). In arid climates, tough woody shrubs and trees, such as mesquite, are moisture reservoirs for organisms and the soil itself. Their long roots draw moisture from deep water sources, making it available to soil and surface ecosystems (figure 3.12).

The Carbon Cycle

Carbon serves a dual purpose for organisms: (1) it is a structural component of organic molecules, and (2) the energy-holding chemical bonds it forms represent energy "storage." Carbon cycling in an ecosystem begins with the intake of carbon dioxide (CO_2) by photosynthetic organisms (figure 3.13). The carbon (and oxygen) atoms are incorporated into sugar molecules during photosynthesis. They are eventually released during respiration, closing the cycle.

The path followed by an individual carbon atom while it is cycling may be quite direct and rapid, depending on how it is used in an organism's body. Imagine for a moment what happens to a simple sugar molecule ($C_6H_{12}O_6$) you swallow in a glass of fruit juice. The sugar molecule is absorbed into your bloodstream, where it is made available to your cells for cellular

FIGURE 3.11 This is a clear-cut logging operation next to Redwood National Park in California. All of the forest strata have been removed— overstory, understory, and ground cover. The soil structure and contours also have been altered by the heavy equipment needed to create roadcuts and remove logs. This region is characterized by abundant rainfall, especially in winter. Given all of this information, what predictions might you make? Are all of your predictions negative?

respiration or for making more complex biomolecules. If it is used in respiration, then you may exhale the same carbon atoms as CO_2 the same day.

Those six carbon atoms instead may be used to repair recently damaged tissue or otherwise become a structural part of you. They could remain a part of your body until it dies and decomposes. Similarly, carbon in the wood of a thousand-year-old tree will only be released after the tree's death, when the wood is digested by fungi and bacteria that release carbon dioxide as a byproduct of their respiration.

Can you think of examples where carbon may not be recycled for even longer periods of time, if ever? Coal and oil are the compressed, chemically altered remains of plants that lived millions of years ago. Their carbon atoms (and hydrogen, oxygen, nitrogen, sulfur, etc.) are not released until the coal and oil are burned. Enormous amounts of carbon also are locked up as calcium carbonate ($CaCO_3$) used to build shells and skeletons of marine organisms, from tiny protozoans to corals. Most of these deposits are at the bottom of the oceans. The world's extensive

FIGURE 3.12 Long roots of the mesquite plant draw water from deep sources, thereby also making the water available to organisms that feed upon the mesquite. Shallower roots are able to absorb sporadic desert rainfall. They also represent a moisture reservoir for the upper soil level.

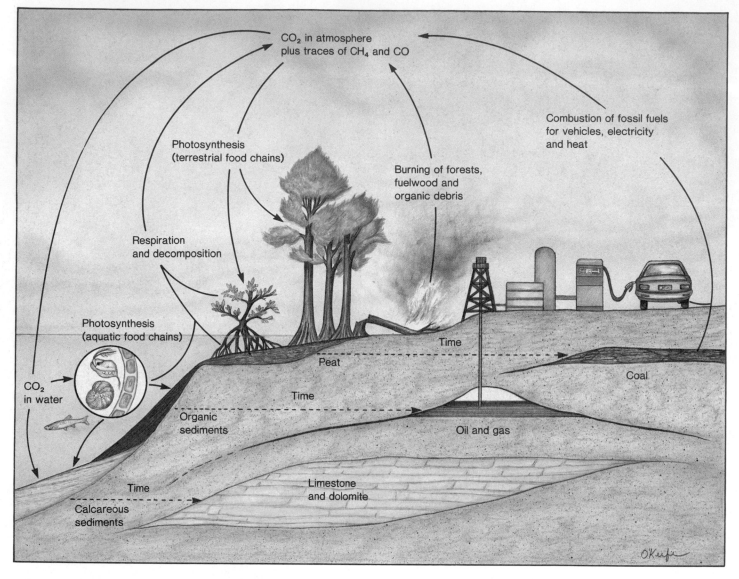

FIGURE 3.13 Atmospheric carbon dioxide is the "source" of carbon in the carbon cycle. It passes into ecosystems through photosynthesis and is captured in the bodies and products of living organisms. It is released to the atmosphere mainly by respiration and combustion. Carbon may be locked up for long periods in both organic and inorganic geological formations.

surface limestone deposits are biologically formed calcium carbonate from ancient oceans, exposed by geological events. The carbon in limestone has been locked away for millenia, which is probably the fate of carbon currently being deposited in ocean sediments. Eventually, even the deep ocean deposits are recycled as they are drawn down into deep molten layers and rereleased via volcanic activity. Geologists estimate that every carbon atom on Earth has made about thirty such round trips over the last four billion years.

How does tying up so much carbon in the bodies and by-products of organisms affect the biosphere? Favorably. It is part of homeostasis, balancing CO_2 generation and utilization. Carbon dioxide is one of the so-called "greenhouse gases" because it blocks radiation of heat from Earth's surface, retaining it instead

in the atmosphere. Photosynthesis and deposition of $CaCO_3$ remove atmospheric carbon dioxide; therefore, vegetation (especially large forested areas, such as the tropical rain forests) and the oceans are very important **carbon sinks.** Cellular respiration and combustion both release CO_2, so are referred to as **carbon sources** of the cycle.

Presently, natural fires and human-created combustion of organic fuels (mainly wood, coal, and petroleum products) release huge quantities of CO_2 at rates that seem to be surpassing the pace of CO_2 removal. Scientific concerns over the linked problems of increased atmospheric CO_2 concentrations, massive deforestation, and reduced productivity of the oceans due to pollution are included in later chapters.

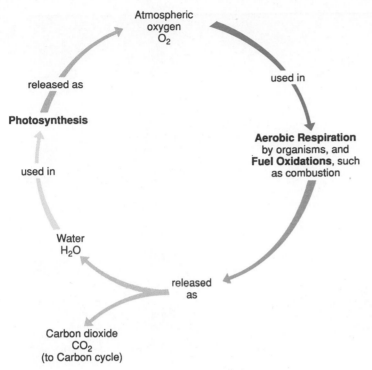

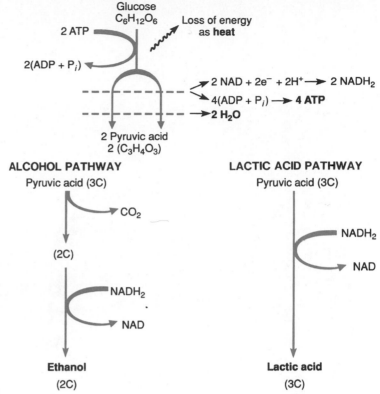

FIGURE 3.14 Oxygen is cycled through living systems mainly via two processes. It is released during photosynthesis, when hydrogen atoms are removed from water molecules and the released oxygen atoms combine to form oxygen molecules (O_2). Oxygen atoms are used during aerobic respiration, combining with electrons (e^-) and hydrogen ions (H^+) to form water (H_2O).

The Oxygen Cycle

Oxygen cycling is intimately linked with carbon cycling, as shown earlier, through photosynthesis and respiration (figure 3.14). It also is linked to combustion because oxygen must be present to burn organic matter. Why? What does oxygen do in both respiration and combustion?

Electrons (e^-) and H^+ ions are released when glucose molecules are taken apart during respiration. Excess electrons and H^+ ions are potentially disruptive to cellular processes. What does this have to do with oxygen? The negatively charged electrons quickly become associated with the positively charged hydrogen ions. Oxygen atoms attract the electrons, which bring along their attached H^+ ions, bonding hydrogen and oxygen atoms together in the specific ratio of 2:1, forming H_2O! Oxygen's role is to "clean up" the liberated electrons and H^+ ions—it is the final electron acceptor in the energy cascade of respiration. Now you also know how water is formed as a by-product of respiration. Similarly, water and carbon dioxide are released during combustion.

Actually, there is more than one kind of cellular respiration, as distinguished by the presence or absence of oxygen in the process. The type we have been discussing is called **aerobic respiration,** because it involves oxygen. By contrast, **anaerobic respiration** can occur in the absence of oxygen; however, without oxygen to accept electrons, the process does not go down

FIGURE 3.15 In anaerobic respiration, the initial by-products of glucose do not go into an oxygen-using cycle, but instead are converted to organic acids and alcohols. Carbon dioxide is released. Some energy is transferred to ATP for cellular use, but only about 2.5 percent of the total energy of each glucose molecule is released. Some microorganisms, such as yeasts, carry out the alcohol pathway of anaerobic respiration. Others, such as those that are used to make yogurt, carry out the lactic acid pathway. Lactic acid-forming respiration occurs in mammalian muscle tissue when not enough oxygen is available for aerobic respiration to occur, and it may cause muscle cramps. (Wondering about NAD and $NADH_2$? Like ATP, they are compounds that release and accept electrons and H^+ during cellular energy transfer reactions.)

all the way to CO_2 and H_2O. Instead, some energy and CO_2 are released, and smaller organic molecules are synthesized, usually alcohols and/or organic acids (figure 3.15). You may be familiar with **fermentation,** a type of anaerobic respiration carried out by yeast and some other organisms. Fermented beverages, such as beer and natural root beer, contain alcohol and are "carbonated"—contain carbon dioxide bubbles. Lactic acid-forming respiration occurs, for instance, in muscle tissues when the intensity of exercise (contraction) exceeds the tissue's oxygen supply, sometimes causing cramps. In this example, anaerobic respiration is temporary, until enough oxygen becomes present to complete respiration aerobically.

In addition to its participation in life processes, oxygen also reacts with a wide variety of elements to form such inorganic compounds as carbon monoxide (CO), carbon dioxide, sulfur dioxide (SO_2), nitrates (NO_3^-), phosphates (PO_3^-), and water. It is one of the most common elements in rocks, mainly as silicon

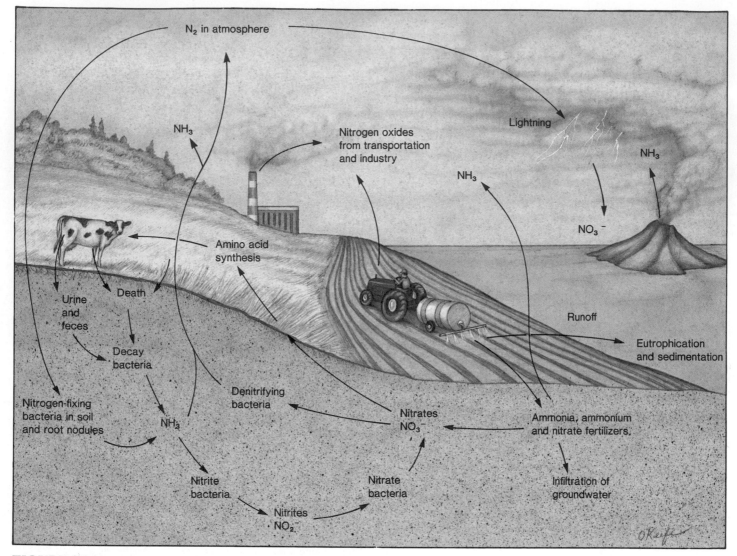

FIGURE 3.16 Nitrogen is incorporated into ecosystems when plants and bacteria use it to build their own amino acids and is released from ecosystems by bacterial decomposition. Natural and human interactions with the nitrogen cycle are depicted here also.

oxides, iron oxides, aluminum oxides, and carbonates—compounds that make up the main bulk of Earth's surface crust. In fact, the most common compound in Earth's crust is silicon dioxide (SiO_2), or quartz.

The Nitrogen Cycle

Organisms do not exist without amino acids, peptides, and **proteins**—organic molecules that contain nitrogen. The nitrogen atoms that form these important molecules are provided by producer organisms. Plants assimilate inorganic nitrogen from the environment and use it to build their own protein molecules, which are eaten by consumer organisms, digested and used to build their bodies, and so on. This sequence sounds very tidy, but there is one major problem. Even though N_2 is the most abundant gas (about 78 percent, by volume, of the atmosphere), plants cannot use N_2!

Where and how, then, do green plants get *their* nitrogen? The answer lies in the most complex of the gaseous cycles, the nitrogen cycle. Figure 3.16 summarizes the nitrogen cycle, emphasizing its biological aspects, even though some nitrogen is converted to usable form by nonbiological processes. The nitrogen cycle has "players" of all kinds, each with a specific role. In many ways, the most critical players of all are the several categories of bacteria that make nitrogen available to plants. Nitrogen gas (N_2) is stable, chemically unreactive, and found in the air, water, and soil. **Nitrogen-fixing bacteria** (including some blue-green bacteria) have a highly specialized ability to "fix" nitrogen, meaning they combine it with hydrogen to form ammonia (NH_3), some of which combines with H^+ in water that is present to become ammonium (NH_4^+). **Nitrite-forming bacteria** combine ammonia with oxygen, forming nitrites, which

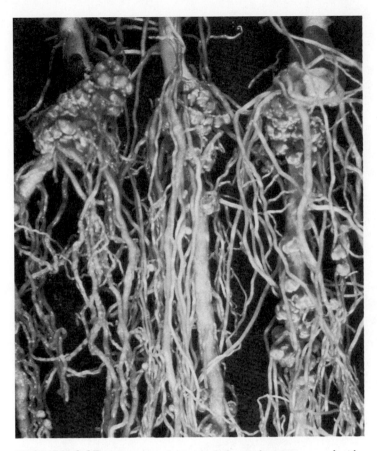

FIGURE 3.17 The roots of this adzuki bean plant are covered with bumps called nodules. Each nodule is a mass of root tissue containing many bacteria that help to convert nitrogen in the soil to a form the bean plants can assimilate and use to manufacture amino acids.

have the ionic form NO_2^-. **Nitrate-forming bacteria** then convert nitrites to nitrates, which have the ionic form NO_3^-. At this point nitrogen finally is in a form that can be absorbed and used by green plants—as nitrates! After nitrates have been absorbed into plant cells, they are broken down to ammonium, which is used to build amino acids, which become the building blocks for peptides and proteins.

You may have seen farmers injecting ammonium compounds into their fields to fertilize them. This treatment works because members of the bean family (legumes) and a few other kinds of plants have nitrogen-fixing bacteria actually living *within* their root tissues (figure 3.17). The bacteria are clustered in nodules, where they have a moist, nutrient-rich environment. This is an example of **mutualism,** an intimate, mutually beneficial "living-together" relationship between members of different species. Legumes and their mutualistic bacteria enrich the soil. Interplanting and rotating legumes with crops (such as corn) that use but cannot replace soil nitrates are beneficial farming practices that take practical advantage of the legume/bacteria relationship.

Nitrogen reenters the environment in several ways. The most obvious path is through the death of organisms. Their bodies

are decomposed by fungi and bacteria, releasing ammonia and ammonium ions, which then are available for nitrate formation. Organisms don't have to die to donate proteins to the environment, however. Plants shed their leaves, needles, flowers, fruits, and cones; animals shed hair, feathers, skin, exoskeletons, pupal cases, and silk. Animals also produce excrement and urinary wastes that contain nitrogenous compounds. Urinary wastes are especially high in nitrogen because they contain the detoxified wastes of protein metabolism. All of these by-products of living organisms decompose, replenishing soil fertility.

How does nitrogen reenter the atmosphere, completing the cycle? **Denitrifying bacteria** break down nitrates into N_2 and nitrous oxide (N_2O), gases that return to the atmosphere; thus, it would seem that denitrifying bacteria compete with plant roots for available nitrates. However, denitrification occurs mainly in waterlogged soils, which have low oxygen availability and a high amount of decomposable organic matter. These are suitable growing conditions for many wild plant species in swamps and marshes, but not for most cultivated crop species, except for rice, which is a domesticated wetlands grass.

Phosphorus and Sulfur Cycles

Minerals become available to organisms after they are released from rocks. Two mineral cycles of particular significance to organisms are those of phosphorus and sulfur. Why do you suppose phosphorus is a primary ingredient in fertilizers? At the cellular level, ATP and other energy-rich, phosphorus-containing compounds are primary participants in energy-transfer reactions, as we have discussed. The amount of available phosphorus in an environment can, therefore, have a dramatic effect on productivity. Abundant phosphorus stimulates lush plant and algal growth, which makes it a major contributor to water pollution.

The phosphorus cycle (figure 3.18) begins when phosphorus compounds are leached from rocks and minerals over long periods of time. Inorganic phosphorus is taken in by producer organisms, incorporated into organic molecules, and then passed on to consumers. It is returned to the environment by decomposition. An important aspect of the phosphorus cycle is the very long time it takes for phosphorus atoms to pass through it. Deep sediments of the oceans are significant phosphorus sinks of extreme longevity. Phosphate ores that now are mined to make detergents and inorganic fertilizers represent exposed ocean sediments that are millenia old. You could think of our present use of phosphates, which then are washed out into the river systems and eventually the oceans, as an accelerated "pouring" of phosphorus from source to sink. Aquatic ecosystems often are dramatically affected in the process because excess phosphates can stimulate explosive growth of algae and photosynthetic bacteria populations, upsetting ecosystem homeostasis (chapter 22). Notice also that in this cycle, as in the others, the role of organisms is only one part of a larger picture.

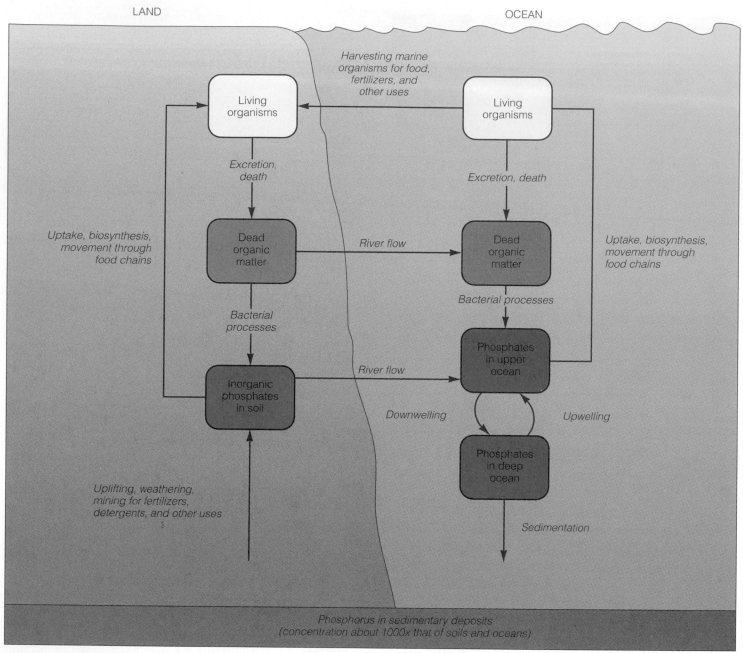

FIGURE 3.18 The phosphorus cycle includes relatively short-time biological components that involve the food chains of land and water. It also includes long-time geological processes of sedimentation, uplifting of landmasses, and weathering. Human activities, such as mining and use of agricultural fertilizers, now are prominent factors in the phosphorus cycle. Bacterial processes noted in the diagram refer mainly to decomposition and formation of phosphates.

Sulfur has important, specific roles in organisms. The biological cycling of sulfur is, again, only part of a larger cycle (figure 3.19). Human activities, however, have skewed the cycle, creating an impact on the homeostasis of many ecosystems. Industrial processes and combustion of coal to produce electrical power release large quantities of sulfur-containing gases, especially sulfur dioxide (SO_2), into the atmosphere. Atmospheric SO_2 reacts with H_2O to form sulfuric acid (H_2SO_4), causing acid precipitation. Acid precipitation has many deleterious environmental effects and is discussed in detail in chapter 23.

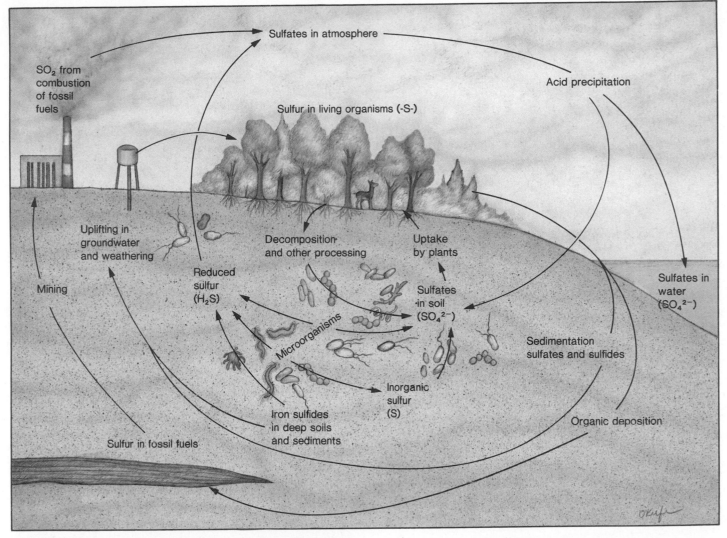

FIGURE 3.19 Sulfur is present mainly in rocks, soil, and water. It cycles through ecosystems when it is taken in by organisms. Combustion of fossil fuels is presently causing increased levels of atmospheric sulfur compounds, which is creating problems related to acid precipitation.

S U M M A R Y

Ecosystem dynamics are governed by physical laws, including the law of conservation of matter and the laws of thermodynamics. The recycling of matter is the basis of the cycles of elements that occur in ecosystems. Unlike matter, energy is not cycled. In ecosystems, solar energy enters the system and is converted to chemical energy by the process of photosynthesis. The chemical energy stored in the bonds that hold food molecules together is available for metabolism of organisms.

Matter and energy are processed through the trophic levels of an ecosystem via food chains and food webs. At each energy transfer point, less energy is available to do work, so energy must be supplied to an ecosystem continuously. The relationships between producers and consumers in an ecosystem, often depicted as pyramids, demonstrate this principle. Most of the energy that enters an ecosystem can be traced back to the sun.

The dynamic balance, or homeostasis, of organisms and ecosystems centers around an optimum state that is best suited for the healthy existence of the living system. To accomplish this, biological systems show an active condition of compensating adjustments.

The principles of conservation of matter, thermodynamics, and homeostasis are demonstrated in the ways organisms interact in material cycles in ecosystems. Water, carbon, oxygen, nitrogen, phosphorus, and sulfur, for instance, are recycled in ecosystems. The biosphere is a source of large quantities of essential elements, and in a given ecosystem, they are constantly reused by living organisms.

Review Questions

1. The number of board feet of lumber harvested per acre is a measure of the productivity of a forest. Why is this an incomplete measure of total biomass of a forest ecosystem?

2. Your body contains vast numbers of carbon atoms. How is it possible that some of these carbon atoms also may have been part of the body of a prehistoric creature?

3. In the biosphere, matter follows a circular pathway while energy follows a linear pathway. Explain.

4. Ecosystems require energy to function. Where does this energy come from? Where does it go? How does the flow of energy conform to the laws of thermodynamics?

5. Heat is released during metabolism. How is this heat useful to a cell and to a multicellular organism? How might it be detrimental, especially in a large, complex organism?

6. Photosynthesis and cellular respiration are complementary processes. Explain how they exemplify the laws of conservation of matter and thermodynamics, and homeostasis.

7. Explain the concept of an energy cascade in relation to the first law of thermodynamics. Explain it in relation to the second law of thermodynamics. How is the concept of energy cascades demonstrated in biomass production at successive levels of a food chain?

8. Water cycles at highly variable rates. Describe an example of a water cycle that occurs in a matter of several days and one that would take over a year to complete.

9. The population density of large carnivores is always very small compared to the population density of herbivores occupying the same ecosystem. Explain this in relation to the concept of an ecological pyramid.

Questions for Critical Thinking

1. Natural ecosystems tend to produce materials that are easily degraded or broken down. Humans, on the other hand, produce many materials that are not degradable. How does this affect the natural cycles of matter in the biosphere?

2. When you exercise, your body experiences an increase in metabolic activity. Discuss some of the ways your body maintains homeostasis. Do the same mechanisms work in an ecosystem?

3. Describe how humans are affecting the carbon cycle and the sulfur cycle. What effect might these human-induced changes have on natural ecosystems?

4. Nitrogen is cycled in the biosphere. Despite this fact, the raising of agricultural crops, such as corn, requires the application of nitrogen fertilizers. Why? Additional carbon, on the other hand, does not have to be applied to agricultural crops. Why?

5. In the industrialized nations of the world, a common food chain would be corn → beef animal → human, while in underdeveloped countries a typical food chain would be rice → human. Which food chain would feed the larger number of humans? Explain your answer in relation to the concept of energy transfers that occur at each trophic level.

6. Many environmentalists believe that it is totally unacceptable to discuss survivable nuclear warfare because of a predicted phenomenon they call the "nuclear winter." In essence, it means the screening of sunlight from reaching Earth's surface due to huge quantities of particulate material thrown into the atmosphere by nuclear blasts and resulting fires. What effect would the screening of sunlight have on Earth's food chains? Is it likely that homeostasis between photosynthesis and cellular respiration could be maintained under such conditions?

7. Humans are part of ecological pyramids. In recent years, the human population has been literally exploding. As the human population rapidly increases in size, what implications does this have on other naturally occurring food chains in the biosphere?

■ Key Terms

absorptive nutrition (p. 32)
aerobic respiration (p. 37)
anaerobic respiration (p. 37)
biomass (p. 27)
biomass pyramid (p. 32)
carbon cycle (p. 32)
carbon sinks (p. 36)
carbon sources (p. 36)
carnivores (p. 31)
conservation of matter
 (p. 26)
consumers (p. 27)
decomposers (p. 32)
denitrifying bacteria (p. 39)
energy pyramid (p. 32)
fermentation (p. 37)
first law of thermodynamics
 (p. 27)
food chain (p. 30)
food web (p. 30)
herbivores (p. 31)
homeostasis (p. 26)
mutualism (p. 39)

nitrate-forming bacteria
 (p. 39)
nitrite-forming bacteria
 (p. 38)
nitrogen cycle (p. 32)
nitrogen-fixing bacteria
 (p. 38)
numbers pyramid (p. 32)
omnivores (p. 31)
open systems (p. 26)
optimum (p. 27)
oxygen cycle (p. 32)
producers (p. 27)
productivity (p. 30)
proteins (p. 38)
scavengers (p. 32)
second law of
 thermodynamics (p. 27)
stability (p. 27)
synergism (p. 28)
thermodynamics (p. 27)
trophic level (p. 31)
water cycle (p. 32)

SUGGESTED READINGS

See also references suggested at end of chapter 1.

■ American Institute of Biological Sciences. *BioScience*. July/ August, 1986, "Ecology from Space" issue (vol. 36, no. 7), featuring several articles on using remote sensing satellite data to analyze biological communities. Includes computer-enhanced color photographs.

■ Bent, Henry A. 1977. "Entropy and the Energy Crisis." *Journal of Science Teaching* (44) 4:25–29. Implications of the second law of thermodynamics.

■ Bolin, B., and R. B. Cook. *The Major Biogeochemical Cycles and Their Interactions*. New York: John Wiley and Sons, 1983. Detailed treatment of nutrient cycling.

■ Chapin, F. Stuart, III, et al. 1987. "Plant Responses to Multiple Environmental Factors." *BioScience* (37) 1:49–57. Examination of resource allocations in plants, focusing on carbon and nitrogen, using a cost-benefit analysis approach.

■ Johnson, Leland G. *Biology*. 2d ed. Dubuque: Wm C. Brown Publishers, 1987. University textbook; a valuable reference source for additional information.

■ Kormondy, E. J. *Concepts of Ecology*. 3d ed. New York: Harper & Row, 1984. A good, readable ecology text.

■ Kroschwitz, Jacqueline I., and Melvin Winokur. *Chemistry: General, Organic, Biological*. New York: McGraw-Hill, 1985. College textbook that gives clear explanations of basic chemistry principles plus practical examples.

■ Miller, G. Tyler, Jr. *Energetics, Kinetics and Life: An Ecological Approach*. Belmont: Wadsworth, 1971. As the title implies, explains the relationship of thermodynamics to living systems.

■ Odum, Eugene P. *Basic Ecology*. Philadelphia: Saunders College Publishing, 1983. A premier ecology textbook with numerous, research-based examples and an extensive bibliography.

■ Odum, Howard T., and Elisabeth C. Odum. *Energy Basis for Man and Nature*. New York: McGraw-Hill, 1980. Energy analysis, principles, and applications of information.

■ Osmond, C. B., et al. 1987. "Stress Physiology and the Distribution of Plants." *BioScience* (37) 1:38–47. Good review of the subject.

C H A P T E R 4

Biological Communities

C O N C E P T S

All the populations inhabiting a given place at the same time make up a biological community. The nature of a community is determined by both the physical conditions present and the interactions among members of the community. Communities can be analyzed and compared on the basis of their productivity, diversity, and resilience. Communities demonstrate the principles of thermodynamics and homeostasis.

The presence of a species in a given place is the result of both its distribution history and its ability to tolerate local environmental conditions. The process of natural selection works within biological communities and may lead to resource partitioning and the determination of ecological niches.

Species within a community interact in many ways. The dominant plant species of a terrestrial community greatly influence the major habitat conditions, thus influencing community composition. Predation, in its broadest definition, is one of the most significant biotic forces in a community. Intraspecific and interspecific competition also affect community composition. Predation and competition may act as selective forces. Some interspecific relationships are symbiotic, often involving coadaptations of the partners.

Communities are dynamic, responding to daily and seasonal changes. They also exhibit long-term changes in composition called ecological succession. Communities often show patchiness, with natural or human-induced breaks in the predominant community type. Patches and edges create special opportunities for habitation within or between major community types.

The continual introduction of different species to a community is part of natural succession. Human introductions of species, however, have sometimes had disastrous effects on native communities.

Humans have unprecedented abilities to alter natural communities and the biosphere as a whole. Our greatest remaining challenge may be to learn to live according to the insights of ecology.

INTRODUCTION

Chapter 3 reviewed how matter and energy interact within ecosystems. Keeping these interactions in mind, let's look more closely at the composition and dynamics of biological communities. Each community consists of several to many populations. Each population is made up of all the members of a single species present, and therefore potentially interacting, in a given area. In an established community, the populations demonstrate homeostasis, staying more or less in equilibrium with each other and the ecosystem as a whole. In this way the community is homeostatic as well.

What determines the species composition of a community? Is it an accident of dispersal, or the result of change and adaptation over generations? How are populations affected by changes in the physical aspects of their environments? What effects do the diverse species in a community have on one another? Are some members of a community "more important" than others? Are communities "structured" or are they random sets of interactions? With all the variety of communities on our planet, are there any common principles or generalizations that can be made about community composition and intracommunity interactions?

These questions, and more, pose difficult challenges to ecologists. The science of ecology has progressed rapidly in recent years, enlightening and rendering obsolete many earlier generalities. In the past two decades, particularly, methods in ecological research have become increasingly refined. Statistical models, based on quantitative "input-output" analyses, have become important research tools. Also, there simply are more data available. Now it seems that the more we know, the more complex interpreting it becomes. We've begun to discover that each community may operate in some unique way, based on its own place and time, so that principles that govern certain kinds of communities don't apply to other kinds. We must, therefore, apply conclusions and generalizations carefully.

Community ecology, the study of biological communities, is now perhaps the most active research area in ecology. In this chapter we will present an overview of some major aspects of community ecology. As with the preceding chapters, we hope to emphasize ecological concepts and information that will prepare you to understand the concerns of environmental science.

SOME PROPERTIES OF COMMUNITIES

Three fundamental properties of biological communities and *ecosystems* are of primary concern to ecologists: productivity, diversity, and resilience. The productivity, diversity, and resil-

ience of communities are interrelated, but a problem facing community ecologists is to measure and attempt to understand the exact nature of those relationships in a given setting!

Productivity

A community's **productivity** is measured as rate of production of biomass, which also may be thought of as rate of energy conversion. Two major generalizations can be made about the relative productivity of global communities. (1) The average productivity of terrestrial ecosystems exceeds that of marine ecosystems, even though marine environments in coastal waters are *highly* productive. The oceans of the world cover about 71 percent of its surface, but average marine productivity is less than 25 percent of average terrestrial productivity. (2) The terrestrial communities of warm, moist regions are more productive than those occupying areas having cooler, drier climates. Tropical rain forests, for instance, are more productive than the temperate forests of eastern North America (figure 4.1).

Productivity is affected directly by such physical factors as climate, air and water currents, soils, landforms, and altitude or depth. Consider the impact of mountains on terrestrial environments. North America's three western mountain chains exert profound effects on the productivity of lands lying to their west and east because of their effect on precipitation (figure 4.2). Eastward-moving air masses pass over the Pacific Ocean, accumulating water vapor. As the moisture-laden air rises to cross the mountains, it cools, water condenses, and precipitation falls on the western slopes of first the Coast Range, then the Cascade-Sierras. This results in high productivity in the coastal forests, interior valleys, and west-slope forests. We capitalize on this productivity in our uses of this region for forest and agricultural resources. Drier air then flows over the east slopes of the Cascade-Sierra chain into the Great Basin and the southwestern plateaus, creating arid grassland and desert communities in the "rain shadow" of the mountains. The phenomenon is repeated on the eastern side of the Rocky Mountains.

What kinds of physical factors affect marine productivity? As on land, temperature is a factor. Warm tropical waters are more productive than the colder oceans. Light also is a factor, because the depth of light penetration limits photosynthesis. Nutrient availability is a third major factor. Compare the eventual resting place of a deceased marine organism to that of a deceased terrestrial organism. A car-killed raccoon lies by the side of the road and is feasted upon by scavengers and millions of microorganisms. Rain washes the products of decomposition to the soil, where they are available to plant roots. The nutrients of the raccoon's body are recycled quite quickly.

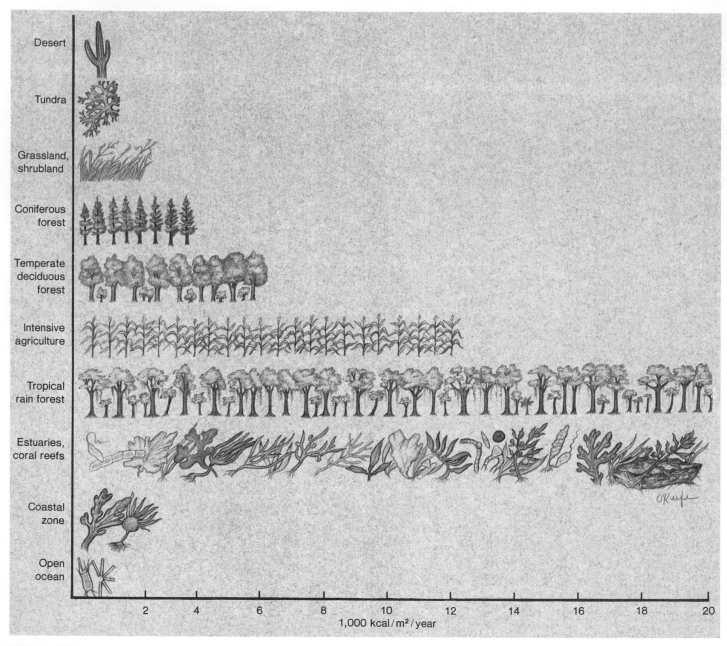

FIGURE 4.1 Gross primary productivity of major world ecosystems.

By contrast, a dead tuna sinks to the bottom of the ocean. It is feasted upon by assorted scavengers and decomposers along the way, but eventually the residual nutrients of its body become part of the bottom ooze, isolated from the upper ecosystem. Fisheries biologists consider this loss of nutrients a major factor limiting productivity of harvestable species. What would happen if some of the nutrient-rich ooze were brought to the surface? Would the result be a burst of increased productivity? In fact, this phenomenon does occur. It is called **upwelling,** in which convective currents within the body of water carry nutrients toward the surface. Sites where natural upwellings occur support rich fisheries (are highly productive).

Diversity

Measurements of **species diversity** are based on two parameters: the *number* of species present in a community (species richness) and their relative *abundance* (evenness). A high-diversity community, therefore, is one that has many species that have large populations.

Can we make generalizations about species diversity among communities? For one thing, species diversity is inversely related to latitude. In both marine and terrestrial communities, diversity is greater in the tropics than in temperate latitudes. Think of the multitudinous exotic fish that populate tropical coral reefs. These

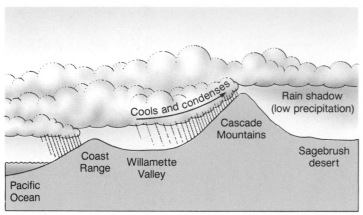

FIGURE 4.2 Mountain ranges exert an influence on climate in several ways, including effects on precipitation. A rain shadow occurs on the eastern side of the mountains of western North America, as shown here in Oregon.

reefs are home to 35–40 percent of all species of bony-skeleton fishes, and many more species are found in tropical lakes and river systems. Similarly, Panama has about six hundred bird species compared to Alaska's fifty.

Another diversity-related generalization has to do with the presence of **ecological equivalents** in communities that occupy similar habitats in different places. In these different locations, there is a convergence toward similar species types and similar community types (figure 4.3). Different species in these disparate communities perform similar roles, as defined by the ways they meet their needs for space, food, and other environmental essentials.

Resilience

A third major characteristic of communities is **resilience,** the capacity to recover from disturbances. The physical environment constantly challenges life-forms with disturbances. The weather, for instance, may offer daily changes in wind velocity and direction, precipitation, and temperature. Shoreline organisms must cope with constant wave action, sand abrasion, tides, and bashing by logs and driftwood. Humans cause both minor perturbations and major disruptions in communities. Consider, for instance, the impact of hiking trails, mowed parks, and "weedless" lakeshores on natural communities. On a larger scale, consider the impact of chainsaws and bulldozers on a woodlot, marsh, or rain forest community.

Change in community composition over time also is part of a community's resilience. What happens when a large forest tree falls? Soon the clearing develops characteristic plants that take advantage of the increased sunlight and open space, including new tree seedlings. Eventually the clearing disappears, becoming forest again.

Do you sense a familiar theme? Stated or unstated, the concept of community resilience includes homeostasis. Homeostasis is inherent in the larger concept of ecosystem stability, for

FIGURE 4.3 Ecological equivalents are unrelated species that fill similar niches in similar ecosystems in different places in the world. The Old World animals on the left (gerbil, jackal, gazelle) have North American equivalents (kangaroo rat, coyote, pronghorn).

communities tend to reach dynamic equilibria within their ecosystems. Furthermore, resilience in communities demonstrates vividly how the active, energy-using processes of organisms counter the tendencies toward disorder expressed in the principles of thermodynamics. To state it more dramatically, the members of a biological community create high levels of order from the physical environment by organizing matter and energy into complex systems.

The concept of **resistance** (**inertia**) also is related to community and ecosystem stability. It refers to the ability of a community to resist being changed by events, in a sense "absorbing" the changes without being disrupted.

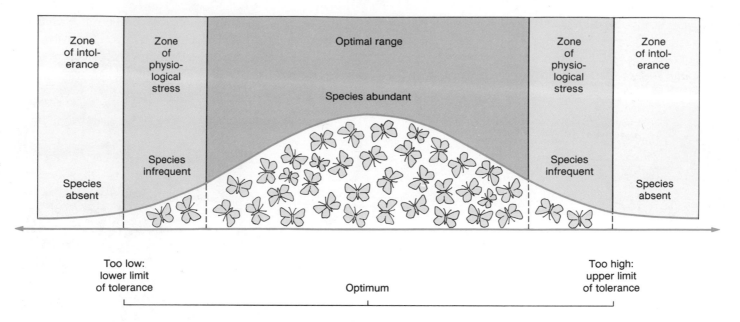

FIGURE 4.4   The concepts of limiting factors and tolerance ranges recognize that the presence of a species often is related to its ability to survive within a specific range of variation for each factor of the environment. The factor may be, for instance, the amount of shade, a particular mineral concentration, or the presence of a required food source. A species does best within an optimum range but is responsive to stresses created by higher or lower levels of the factor, as depicted by the hypothetical population curve in this drawing.

WHO LIVES WHERE, AND WHY?

"Why" questions often are the stimulus for scientific research, but the research itself centers on "how" questions. Why, we wonder, does a particular species live where it does? More to the point, how is it *able* to live there? How does it use the physical resources of its environment, and are some of its techniques unique? How does it interact with the other species present? What gives one species an edge over another species in a particular habitat?

In this section we will emphasize some specific ways species are limited by the physical aspects of their environment. We then will discuss some ways members of a biological community interact, pointing out some of the difficulties ecologists encounter when they attempt to discern patterns and make generalizations about community interactions and organization.

Overview

Each ecosystem consists of a biological community interacting with its physical environment. The physical factors determine, to a large extent, the nature of the biological factors. The physical environment provides basic limitations of climate, temperature, water, light, soil, and nutrient availability, but that is only part of the story. The biological community uses and modifies the environment. Trees and shrubs, for instance, protect soils from wind and water erosion, create shade, slow air movement, and preserve humidity within a forest ecosystem. A community also is somewhat self-determining in composition because specific members of the community create physical conditions and nutritional opportunities that are suitable for other members.

Limiting Factors and Tolerance Limits

Victor Shelford, a pioneer in American ecology, developed the principle of **limiting factors.** This principle states that for each physical factor in the environment, there are both minimum and maximum limits, called **tolerance limits,** beyond which a particular species cannot survive (figure 4.4). Furthermore, the one factor that is closest to a tolerance limit for a given species at a given time is the **critical factor.** The critical factor determines, more than any other physical factor, the abundance and distribution of a species in a given area. Water, light, and temperature are the most common critical factors for plants and animals, although something as specific as a particular pH or mineral requirement also can be a critical factor for a species.

Sometimes the requirements and tolerances of species can be useful indicators of specific environmental characteristics. The absence or presence of such species can tell us something about the community and the ecosystem as a whole. Locoweeds, for instance, are small legumes that grow where soil concentrations of selenium are high. Because selenium often is found with uranium deposits, locoweeds have an applied economic value as environmental indicators. Such indicator species also may demonstrate the effects of human activities. Lichens and eastern white pine are less restricted in habitat than locoweeds, but are indicators of air pollution, as they are extremely sensitive to acid precipitation. Bull thistle is a weed that grows on disturbed soil but is not eaten by cattle; therefore, an abundant population of bull thistle in a pasture is a good indicator of overgrazing. Anglers know that trout species require clean, well-oxygenated water, so the presence or absence of trout is an indicator of water quality.

Dispersal and Other Species

Keep in mind that the presence and abundance of a species in a community are determined by other than just physical factors. First, the presence of each species has *historical* roots. How were the founders of the population brought to the site? To what extent do the distribution and success of populations depend on serendipitous dispersal, linked to the ability to survive within the physical parameters of a site? These questions must be considered when we interpret community composition. The absence of a species might indicate some important fact about the physical environment, *but* it also might be a historical artifact.

Second, no species lives in a biological vacuum. The presence and vigor of a population are influenced by the populations of other species present, including food sources, predators, and those that help determine environmental characteristics.

Natural Selection and Ecology

What do genetics and evolution have to do with ecology? The answer is, "a great deal!" Each organism carries a specific genetic heritage, the legacy of generations of predecessors. Organisms inherit specific nutritional requirements and tolerances for environmental factors. Cultivated plants provide numerous good examples: citrus trees have a low tolerance for cold winters, whereas apple trees require winter chilling; most ornamental shrubs prefer soil that is neutral in pH, but azaleas and rhododendrons thrive in acidic soil.

How have species accumulated their particular structural and metabolic characteristics, their specific sets of genetic characteristics? Overwhelming scientific evidence supports the view that species have developed by a long series of past changes in genetic characteristics, and that they continue to change and adapt to their environments by natural selection. **Natural selection** is the complex process by which environmental pressures cause certain genetic combinations in a population to become more abundant than other combinations.

Natural selection acts upon the total genetic constitution of a population by influencing the fertility or survivorship of individuals within the population. Some environmental factors that can influence fertility or survivorship are (1) physiological stress due to inappropriate levels of some critical environmental factor, such as moisture, light, temperature, pH, or specific nutrient; (2) predation, including parasitism and disease; and (3) competition.

It's tempting to think of natural selection as a creative, molding process, but that view is too teleological, implying purposefulness. The selective process is subject to randomness, including—at the individual level—good luck or bad luck! Natural selection involves such day-to-day interactions as feeding and being fed upon; being stepped on or swatted; competition with other organisms for environmental needs; natural disasters, such as fires, floods, storms, earthquakes, landslides, volcanic eruptions; and even meteorite impacts. Natural selection involves plant seeds accidentally landing on soil instead of rocks, then receiving precipitation at the right times and in the right amounts to germinate and grow. It involves the chance that a particular grasshopper is caught in a spider web, instead of some other grasshopper in the population. As you can imagine, that kind of selection doesn't necessarily mean that the "best" genetic combination always survives. Over many generations, however, the combinations that do survive often are expressed as adaptive, genetically based characteristics that favor survival and reproduction of some organisms in a population over others.

Ecology and evolution are very closely related subjects. The ecological understanding of where and how an organism lives is rooted in an understanding of how it is adapted to its environment. Coming full circle, consideration of adaptations requires understanding the environmental forces that have shaped the genetic constitution of a population.

The Ecological Niche

Well-established communities often demonstrate **resource partitioning,** in which populations share limiting resources through specialization. By specializing in their choice of seeds, for instance, two populations of seed-eating birds can share the seed resources of a community with less competition.

Use of food resources is one part of an organism's **ecological niche,** the role a species plays in the community, including what, when, and how it uses its specific resources. A community can be thought of as a collection of interacting, niche-differentiated species adapted for the existing environmental conditions and complementing each other in the use of space, resources, and time.

Perhaps you haven't thought of time as an ecological factor, but in a community, niche specialization is a twenty-four hour phenomenon. Flowers of different species open at different times of day and night, attracting pollinators that are active at different times. Swallows and insectivorous bats both catch insects as they fly, but some insect species are active during the day and others at night, providing noncompetitive feeding opportunities for day-active swallows and night-active bats.

If the presence of a resource is predictable, a population may specialize to exploit it, reducing the level of interspecific competition. This phenomenon is demonstrated in the **vertical**

stratification of communities, whereby specific populations form subcommunities at different layers, or strata. Terrestrial forests and most aquatic ecosystems provide good examples of vertical stratification of niches within a community (figures 4.5 and 4.6).

Some species have narrow, well-defined niches, whereas others do not. Compare, for instance, the European starling and the golden-cheeked warbler. The starling, now a common resident in the United States, is a niche generalist. It has wide tolerance limits and is found in many habitats, from natural areas to inner cities. It is flexible in its choice of foods and nesting sites, choosing a tree nest hole or electrical transformer box with equal ease. By contrast, the golden-cheeked warbler is a niche specialist, with a very restricted habitat preference. The species is making its last stand on a juniper-covered limestone hillside in central Texas, threatened by development of its habitat for vacation homes. Species with such narrow tolerance limits are likely to be sensitive to small differences in microclimate, soil characteristics, food, nesting sites or materials, or some other critical factor. They lack flexibility to respond to changing environmental conditions. The giant panda, on the other side of the world, is another endangered niche specialist (figure 4.7).

SPECIES INTERACTIONS AND COMMUNITY DYNAMICS

Communities are organized groups of organisms that interact together with different species fulfilling different roles.

Dominant Plants: Defining the Conditions

Often a community has a group of producers that exerts a dominant influence on the community. Think of terrestrial communities with which you are familiar. A forest community is dominated by trees. A prairie community is dominated by grasses. The dominant plant species usually are the largest or most numerous forms, fixing the most energy and making up the largest proportion of biomass in a community. They provide a food base for most of the community, and also modify the environment, determining by their presence the other kinds of organisms that can live in the community. In a forest, for instance, the overstory trees modify the environmental conditions of a site, influencing all other plant and animal populations.

Predation

All organisms need food to live. Producers make their own food, and consumers get theirs by eating organic molecules produced by other organisms. In most communities, photosynthetic organisms are the producers. In the preceding chapter, you were introduced to several consumer categories: herbivore, carnivore, omnivore, scavenger, and decomposer. These may already have been familiar terms to you. With which of these categories do you associate the term "predator?" Ecologically, predator has a

much broader meaning than you might expect, and we will use this broader definition in our subsequent discussions.

A **predator,** in community ecology, is an organism that feeds directly upon another organism, whether or not it kills the prey to do so. Such feeding is called **predation.** This definition separates live-feeders from scavengers and decomposers. By this definition, herbivores, carnivores, and omnivores all are predators. Also in this broad sense, predation in a community includes feeding by **parasites,** organisms that live in or upon the body of a **host organism** and take nourishment from it, usually without killing it. **Pathogens,** disease-causing organisms, also may be included.

Predation is a potent and complex influence on the population balance of communities. It involves (1) all stages of the life cycles of predator and prey species, (2) many specialized food-obtaining mechanisms, and (3) specific prey-predator adaptations that either resist or encourage predation.

For instance, consider the impact of predation on various life cycle stages of a flowering plant. In temperate climates, such predation tends to have a seasonal character. The plant is susceptible as a young seedling, when ants, cutworms, or rabbits may clip it off and end its life. In some ecosystems, grazing by domestic cattle, sheep, and goats does, in fact, limit regeneration of trees and shrubs. At later stages in its lifetime, a plant is fed upon, especially by insects that are specialized to chew its leaves, pierce its tissues to suck its juices, tunnel its stems, leaves, and roots, lap its nectar, collect its pollen, or eat its flowers, fruits, and seeds. Such assaults are lifelong, perhaps for centuries in the case of long-lived plants.

Predation throughout the life cycle is very pronounced in marine intertidal animals. Eggs of a prey species attached to seaweeds and other surfaces are a ready food source for other animals. Many crustaceans, mollusks, and worms release eggs directly into the water, and the eggs and free-living larval and juvenile stages are part of the floating community, or **plankton** (figure 4.8). Planktonic animals feed upon each other and are food for successively larger carnivores, including small fish. As prey species mature, their predators change. For instance, barnacle larvae are planktonic, but the adults are sessile, living attached to a substrate. When barnacle larvae find places to become attached, they fall prey to limpets, single-shelled mollusks that graze by scraping rocky surfaces. Then, of the increasingly small proportions of eggs that survive to hatch, larvae that survive to juvenility, and juveniles that survive to adulthood, the population of a given invertebrate species is further reduced by predation upon the adults.

Can you visualize the problems in analyzing the trophic relationships of such a community? In the first place, the life cycle stages of a single species often occupy *different* trophic levels. Adult frogs, for instance, are carnivores, but the tadpoles of most species are grazing herbivores. Therefore, to sort out

PRINCIPLES AND CONCEPTS

100%

10%

Light intensity

1%

0.1%

Cerulean warbler

American redstart

Ovenbird

FIGURE 4.5 Stratification and light extinction in an Eastern deciduous forest. Vegetation is layered in response to graduated light intensity. The development of vegetational strata also affects the animal life in the community. Warblers, for instance, are adapted to a variety of foraging and nesting opportunities at different strata.

FIGURE 4.6 Giant kelp is a massive alga that forms dense "forests" off the Pacific coast of California. Like terrestrial forests, kelp forests are vertically stratified in terms of both physical factors and the biological communities. Gradients of light, temperature, nutrients, and water pressure provide many ecological niches for animal species.

FIGURE 4.7 The giant panda feeds exclusively on bamboos. Although its teeth and digestive system are those of a carnivore, it is not a good hunter, and has adapted to a vegetarian diet. To support its size, it must eat up to 40 lbs of bamboo leaves and stalks daily. This narrow specialization makes it extremely vulnerable to fluctuations in the bamboo population. In the 1970s huge acreages of bamboo flowered and died, and many pandas starved.

FIGURE 4.8 Adult forms of many marine invertebrates—such as crabs, snails, and starfish—have planktonic larval stages. Thus, adults and larvae do not compete for food resources. They participate in different food chains.

FIGURE 4.9 Communities on the rocks of ocean shorelines are dramatically stratified. Some species live where they are partially protected from the air during the low tides. They are the tidepool organisms familiar to beach explorers. Above the tidepool line are several other zones, where organisms are adapted to different degrees of exposure to wave action, desiccation, changes in salinity related to evaporation, and exposure to solar warming.

trophic relationships and relate them to community energy flow (e.g., energy pyramids), the different life cycle stages must be considered *independently*, rather than being lumped together as a single species. In the second place, how would you determine the impact of a single species on the community as a whole? Does the greatest impact come at the adult stage of the life cycle, where it may be more easily observed, or at some earlier stage? Pragmatically, marine ecologists have had to choose one or another stage to concentrate on in a single study, hoping to find linkages by comparing and coordinating the results of parallel and complementary investigations.

That predation plays a significant role in determining population composition and balance in biological communities is undisputed, although the mechanisms by which it works seem to vary among the communities that have been studied. This lack of common principles is extremely frustrating, defying our human desire for neat generalities. Keeping in mind that there are few communities that we know share exactly the same working "rules," let's look at two known ways predators can affect community homeostasis.

Obviously, direct feeding is one way predators affect a community; however, predators also have *indirect* effects on populations they do *not* feed upon. For instance, the rocky intertidal zones of coastal Washington are characterized by densely packed beds of mussels, barnacles, and seaweeds, all attached to the underlying rock or, in many cases, to each other (figure 4.9). Sometimes patches of rock are made available for new colonization, as when wave-driven logs knock loose existing residents. These patches are soon colonized by algae and larval barnacles, then by larval mussels. Eventually, the mussels tend to crowd out the barnacle populations. A carnivorous starfish that feeds on mussels, however, helps to limit their dominance of the community. If the starfish population is removed, the mussel population takes over; thus, in this community, starfish predation has both direct and indirect effects on the balance of populations.

Through its effects on populations, predation also affects evolution via natural selection. Do plants, for instance, have any defenses against creatures that would feed on them? They certainly can't escape or hide! One significant defense is the ability of plants to regenerate lost body parts, enabling them to survive repeated grazing or browsing. Many kinds of plants have physical and chemical defenses against predation, including tough, fuzzy, sticky, or spiny outer surfaces and many kinds of toxins that act upon contact or after ingestion. Furthermore, some plants have evolved ways to interact with predators in self-benefiting ways, such as for pollination and seed dispersal.

Competition

Competition is another kind of antagonistic relationship within a community. For what do organisms compete? To answer this question, think again about what all organisms need to survive: energy and matter in usable forms, space, and specific sites for life activities. These all are kinds of environmental resources. Plants compete for growing space for root and shoot systems so they can absorb and process sunlight, water, and nutrients. Animals compete for living, nesting, and feeding sites, food, water, and mates. Competition among members of the same species is called **intraspecific competition,** whereas competition between members of different species is called **interspecific competition.**

PRINCIPLES AND CONCEPTS

FIGURE 4.10 Notice how the creosote bushes in this view tend to be spaced apart from one another. Substances released by the roots of established plants retard the germination and growth of competitors, thus helping to share a critical resource—water—among existing plants.

Intraspecific competition can be especially intense because members of the same species have the same space and nutritional requirements; therefore, they compete directly for these environmental resources. How do plants cope with intraspecific competition? The inability of seedlings to germinate in the shady conditions created by parent plants acts to limit intraspecific competition by favoring the mature, reproductive plants. Many plants have adaptations for dispersing their seeds to other sites by air, water, or animals. You've seen dandelion plumes and probably have had sticky or burred seeds attach themselves to your clothing. Some plants secrete leaf or root exudates that inhibit the growth of seedlings near them, including their own and those of other species (figure 4.10). This strategy is particularly significant where water is a limiting factor.

Animals also have developed adaptive responses to intraspecific competition. Two major examples are varied life cycles and territoriality. The life cycles of many invertebrate species have juvenile stages that are very different from the adults in habitat and feeding. Compare, for instance, a leaf-munching caterpillar to a nectar-sipping adult butterfly, a herbivorous tadpole to an insectivorous frog, or a planktonic crab larva to its bottom-crawling adult form. In these examples, the adults and juveniles do not compete, even though they are members of the same species and live in the same habitats.

You may have observed robins chasing other robins during the mating and nesting season. Robins and many other vertebrate species demonstrate **territoriality,** in which they define an area surrounding their home site or nesting site and defend it, primarily against other members of their own species. Territoriality helps to allocate the resources of an area by spacing out the members of a population. It also promotes dispersal into adjacent areas by pushing grown offspring outward from the parental territory. In the case of robins, the territory is occupied by a mated pair. In the case of wolves and other social animals, the territory is occupied by a family-based unit.

Interspecific competition in a community can affect competing populations by influencing population sizes and sometimes by influencing population characteristics. As a result, competition may be a selective factor in the refinement of the ecological niche of a species. The role of competition as a selective force is expressed in the classic **principle of competitive exclusion.** This principle asserts that, as a result of natural selection through competition, two similar species in a community usually are not in direct competition for the same food because their niches differ in ways that reduce competition. Woodland warblers offer a classic example of competitive exclusion (figure 4.11).

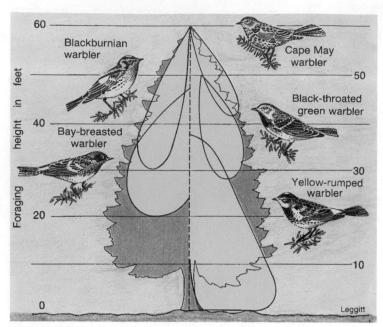

FIGURE 4.11 Resource partitioning and the concept of the ecological niche are demonstrated by the way several species of wood warblers use different strata of the same forest. This is a classic example of the principle of competitive exclusion.

Since elucidation of this principle as a primary cause of niche specialization, much more information has become available about specific communities. As a result, ecologists have moved away from giving competition too much credit for observable niche diversity. While it still seems to be the major influence in *some* communities that have been studied, competition among closely related forms clearly takes a back seat to predation and the influence of a third, unrelated species in many other communities. Once again, we find that the more we have learned, the greater our need to learn more!

Symbiosis

In contrast to predation and competition, symbiotic interactions between organisms are nonantagonistic. **Symbiosis** is the intimate living together of members of two or more species. Some of the most interesting examples of symbiosis are among marine organisms (Box 4.1). We've already mentioned the presence of nitrogen-fixing bacteria living in the roots of some legumes. Their association is a type of symbiosis called mutualism, because both members of the partnership benefit. **Commensalism** is a type of symbiosis in which one member clearly benefits and the other apparently is neither benefited nor harmed. Parasitism, described earlier as a form of predation, also may be considered a type of symbiosis, where one species benefits and the other is harmed. All of these relationships have a bearing on such ecological factors as resource utilization, niche specialization, diversity, predation, and competition. Symbiotic relationships often enhance the survival of one or both partners. Let's look at some examples.

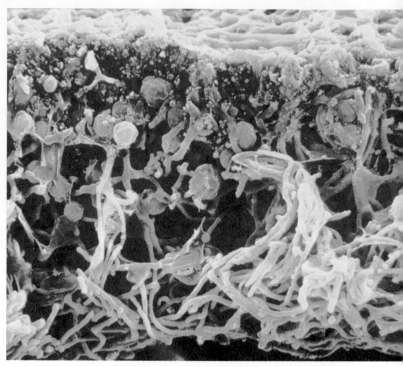

FIGURE 4.12 Mutualistic association of a fungus and photosynthetic partner in the body of a lichen. The fungus forms a filamentous matrix within which the spherical, single-celled photosynthetic partners live.

Fungi are mutualistic partners that help make up the bodies of lichens (figure 4.12). In a lichen, the fungal filaments (hyphae) provide a matrix in which photosynthetic bacteria or algae live, and the photosynthetic partner produces carbohydrates that presumably help feed the lichen body.

Many plant species have mycorrhizae, finely branched structures formed by an association of plant roots and the hyphae of specific fungi. Mycorrhizae assist with water and nutrient uptake, and many plant species require the relationship to thrive. Mycorrhizae are such significant structures that their presence or absence on citrus and pine roots and the germinating seeds of orchids can determine whether or not the plant survives.

Symbiotic relationships often show some degree of coadaptation of the partners, in which their structural and behavioral characteristics are shaped, at least in part, by the relationship. One of the most interesting cases of mutualistic coadaptation involves Central and South American swollen thorn acacias and acacia ants (figure 4.13). Acacia ant colonies live within the swollen thorns on the acacia tree branches and feed on two kinds of food provided by the acacias—nectar produced in nectaries at the leaf bases and special protein-rich structures (Beltian bodies) produced on leaflet tips. The acacias thus provide shelter and food for the ants. Although they make an energy expenditure to provide these services, they are not physically harmed by ant feeding.

BOX 4.1   Marine Symbiosis

BOX FIGURE 4.1 (*left*) Symbiotic relationships of all kinds are common among marine organisms. The large green sea anemone of the northern Pacific coast owes its color to the presence of photosynthetic protists that live within its tissues. How might this mutualistic relationship be related to the fact that these anemones occur in well-lit, intertidal waters?

BOX FIGURE 4.2 (*below*) "Cleaner fishes" are important members of reef communities and exemplify the complex, innate relationships between symbionts. The bright markings of cleaner fishes are instinctively recognized by fish of various species, which allow them to move freely over their bodies, into their gill chambers, and even into their mouths. The cleaner wrasses in this photograph are tending a moray eel on Australia's Great Barrier Reef.

Mutualistic, photosynthetic protists give green color to the bodies of a large sea anemone of the Pacific Northwest (box figure 4.1). They also live in the tissues of some corals, flatworms, and clams. The animals' tissues provide a stable environment and nutrients for the protists that presumably contribute oxygen and photosynthetic food.

Cleaning symbiosis is even more dramatic. The cleaning services of certain species of small fish and shrimp are sought by large fish. The cleaners move over the bodies of the larger fish and even into their gill covers and mouths to remove attached parasites, loose scales, wounded tissue, and other debris. The markings and specialized "advertising" behavior of cleaners cause their potential predators to respond by preparing to be cleaned, rather than to eat (box figures 4.2 and 4.3).

Marine environments also provide many examples of commensalism. The lugworm (*Urechis*) is a chubby, tube-dwelling marine worm that participates simultaneously in several commensal relationships with one or more members of three species (box figure 4.4). It provides home, haven, and even food for a pea crab, a scale worm, and a goby (fish). Because it hosts so many commensal guests, the lugworm also is called "the fat innkeeper."

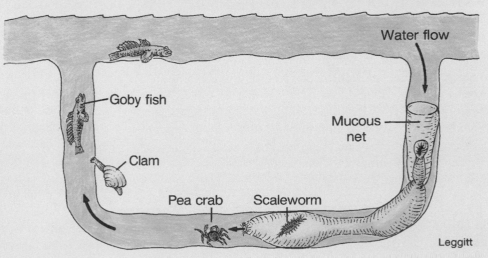

BOX FIGURE 4.3 The innkeeper worm (*Urechis caupo*) of the Pacific Coast of North America provides a classic example of symbiosis.

FIGURE 4.13 The swollen thorn acacia has a complex symbiotic relationship with certain ant species, shown here feeding on nutritious fluids produced in the row of nectaries on the leaf petiole.

What do the acacias get in return and how does the relationship relate to community dynamics? Ants tend to be aggressive defenders of their home areas, and acacia ants are no exception. They kill herbivorous insects that attempt to feed on their home acacia, thus reducing predation. They also trim away vegetation that grows around its base, thereby reducing competition. This is a fascinating example of how a symbiotic relationship fits into community interactions.

COMMUNITIES IN TRANSITION

So far our view of communities has focused on the day-to-day interactions of organisms with their environments, set in a context of survival and selection. In this section, we'll step back to look at some transitional aspects of communities, including where communities meet and how communities change over time.

Edges and Patches

A transitional zone develops where populations from two adjacent communities meet and overlap. This creates an **edge effect.** Where a forest meets a grassland, for instance, there is often a distinctive forest edge community, made up of grasses and perennial plants, as well as shrubs and young trees. This transitional zone provides habitat for a different set of animals or provides habitat that can be used by members of both adjacent communities. Edges, thus, tend to have great diversity.

Wildlife managers use the edge effect in developing habitat for wild game. Some gamebirds like to feed in grassy areas during the day and retreat to the trees for the night. Deer show a different pattern, staying in the woodland during the day and using the grassy perimeter for feeding in the early morning and early evening. Nongame species also benefit when this kind of variable habitat management is provided for game species.

Another diversity-promoting phenomenon is the natural **patchiness** of communities. In forests, for instance, open areas are created by the loss of overstory trees, rocky outcrops, streams or bodies of water, or areas where the soil is too wet, too dry, or in some other way unsuited for forest trees.

Ecological Succession

Biological communities have a beginning point in time. They begin simply and end in a state of complex homeostasis. The process by which a series of communities develops, replacing each other on a site over time, is called **ecological succession. Primary succession** occurs when a community begins to develop on a site that has not been occupied previously by a community, such as a new volcanic flow, island, sand or silt bed, or body of water. **Secondary succession** occurs when an existing community has been disrupted by some natural catastrophe—such as fire or flooding—or by a human activity—such as deforestation, plowing, or mining—and a series of communities subsequently develops at the site.

In primary succession on a terrestrial site, the new site first is colonized by a few hardy **pioneer species,** often microbes, mosses, and lichens. The pioneers alter the environment by their presence and their activities, creating different environmental conditions for additional organisms that arrive at the site. Their bodies create patches of organic matter in which protists and small animals can live. The organic acids they produce dissolve and etch rocky surfaces, contributing mineral matter to soil development. Organic debris accumulates in pockets and crevices, providing soil in which seeds can become lodged and grow. As the community of organisms continues to develop, it becomes more diverse and competition increases, but at the same time, new niche opportunities develop. The pioneer species disappear as the environment changes and species invasions progress, causing replacement of the preceding community. Figure 4.14 demonstrates primary succession in a terrestrial ecosystem.

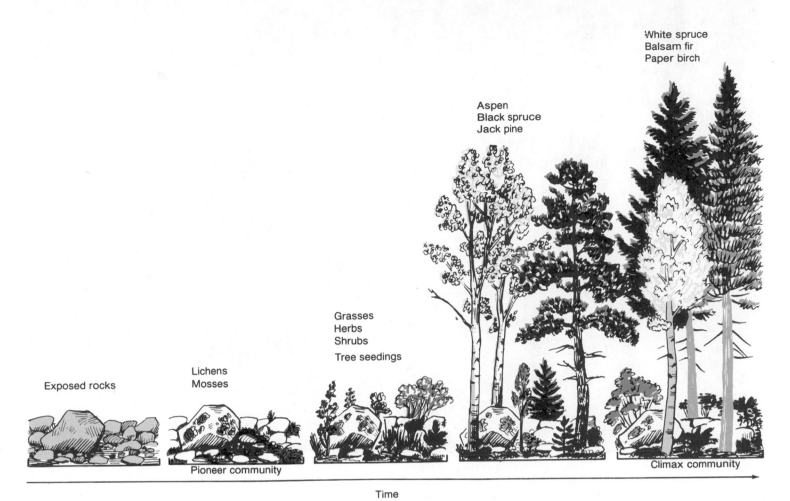

Aspen
Black spruce
Jack pine

White spruce
Balsam fir
Paper birch

Grasses
Herbs
Shrubs

Tree seedings

Lichens
Mosses

Exposed rocks

Pioneer community

Climax community

Time

FIGURE 4.14 Primary succession on a terrestrial site is shown in five stages, from left to right, beginning with rocks that are initially colonized by a pioneer community of lichens and mosses and ending with a climax forest community.

Figure 4.15 depicts succession in an aquatic ecosystem, showing how a developing community changes its own environment. The amount of open water in the lake or pond gradually decreases as vegetation encroaches from the margins, resulting in gradual community replacement progressing from the edges of the pond toward the center. Succession moves from open lake → shallow pond with highly vegetated edges → marshy area with rooted, emergent vegetation clear to the middle → wet soil with a few remaining cattails → grassland or forest.

Examples of secondary succession are easy to find. Observe an abandoned farm field or burned-over forest in a temperate climate. The bare soil first is colonized by rapidly growing annual plants (those that grow, flower, and die the same year) that have light, wind-blown seeds, and can tolerate full sunlight and exposed soil (figure 4.16). They are followed and replaced by perennial plants (those that live for several to many years), including grasses, various nonwoody flowering plants, shrubs, and trees. As in primary succession, plant species progressively change the environment on a site. Grasses have extensive root systems and many spread by rhizomes, reducing bare-soil site availability for the seeds of the fast-growing annuals. As a result, the site

FIGURE 4.15 Succession occurs in a pond as vegetation gradually encroaches from the margins toward the center. Eventually, the pond may fill in completely.

FIGURE 4.16 Fast-growing, prolific annuals such as this ragweed, thrive on disturbed ground and are typical early colonizers in succession.

becomes increasingly "grassy" and decreasingly "weedy" in appearance. As shrubs and young trees develop, they provide shade that is less suitable for the germination and growth of sun-loving grasses but is more suitable for plants that have different adaptations, favoring development of a shade-tolerant understory. Through this entire process, the suitability of the site for different kinds of animals also changes. The eventual community in this example usually is a mature prairie or forest, depending on the site.

Ecologists call such a mature biological community a **climax community.** The climax community is stable over a long period of time. It is characteristic of the soils and climate of a region and is in equilibrium with these physical conditions. It is more stable or resilient in its ability to resist and recover from external disturbances. It is characterized by high species diversity, narrow niche specialization, well-organized stratification, good nutrient conservation and recycling, and a large amount of total organic matter. Overall, the climax community exhibits a high degree of order.

In a climax community, the species composition and population sizes are relatively stable. Community functions, such as productivity and nutrient cycling, show homeostasis. No member of the community is causing dramatic changes or new conditions. Like other communities, the climax community is an open

system through which individual lifetimes, energy, nutrients, and organic matter pass; however, the nature of the climax community remains relatively constant through time.

The concept of the climax community is not as dogmatic as it once was because additional research indicates that what once were regarded as "final" climax communities still are changing. It's probably more accurate to say that the rate of succession is so slow in a climax community that, from the perspective of a single human lifetime, it is a stable community.

Table 4.1 summarizes five steps in terrestrial succession. These trends are generally true, but as we've mentioned before, every natural community is unique in its own way; therefore, there are other ways some of these steps can be interpreted in some cases.

Some landscapes never reach a stable climax in the traditional sense because they are characterized by and adapted to periodic disruption. They are called **equilibrium communities** or disclimax communities. Grasslands, the chapparal shrubland of California, and some kinds of coniferous forests, for instance, are shaped and maintained by periodic fires that have been a part of their history. They are, therefore, often referred to as **fire-climax communities.** Plants in these communities are adapted to resist fires and/or to reseed quickly after fires. In fact, many of the plant species we recognize as dominants in these com-

PRINCIPLES AND CONCEPTS

TABLE 4.1 Trends in succession leading to a terrestrial climax community

1. Productivity (rate of formation of organic material) increases with greater use of resources by the community.
2. Soil depth and organic content increase, as does the amount of organic matter held in the bodies of plants.
3. Species diversity often (but not always) increases.
4. Height, massiveness, and complexity of strata within the community expand, creating a wider variety of habitats and microhabitats (small areas within a larger environmental regime that vary in solar exposure, shade, temperature, and moisture conditions).
5. Populations burgeon and fade away in time as the environmental conditions that favor them are created, reach an optimum range, then become less favorable. As a result, communities replace one another on the site. The rate of replacement slows as larger, longer-lived species that maintain their populations substitute for smaller, shorter-lived species.

a.

b.

munities *require* fire to eliminate competition, to prepare seedbeds for germination of seedlings, or to open cones or thick seed coats (figure 4.17). Without fire, community structure may be quite different (figure 4.18).

Introduced Species and Community Change

Succession is the story of how communities are replaced (succeeded) by new communities. It is accomplished by the continual introduction of new members, some of which succeed in establishing populations, and the disappearance of old members. Each assemblage of organisms at each stage of succession alters the environment, thus affecting both its own members and newcomers to the community. Gradually, populations carve out their own ecological niches from the available opportunities. Often, organisms arrive together. Animal parasites, for instance, usually travel with their hosts. New predators are attracted into a community by the presence of suitable prey; specific plants attract certain herbivores, which attract certain carnivores. Scavenger and decomposer populations also change with changing feeding opportunities.

What happens to new species that are introduced after the community already has become established? There are several possible outcomes. Some species cannot compete within the established community, so they have little impact. Some species are able to fit into and become part of the community, defining new ecological niches. If, however, an introduced species preys upon or competes more successfully with one or more populations that are native to the community, the entire nature of the community can be altered.

FIGURE 4.17 Ponderosa pine grows in dense stands when it is young (*a*). Historically, fire thinned the saplings, creating the open stands that characterize the mature forest (*b*). With fire control measures, young stands now must be thinned by human labor in order to both reduce competition among young trees and reduce fire danger.

FIGURE 4.18 The open, sunny conditions of prairies favor growth of seedlings from adjacent woodland species. As a result, prairies are constantly subjected to invasion by woody shrubs and trees, as in this Iowa prairie. Natural wildfires helped to maintain prairies for centuries, but the remaining isolated tracts of natural prairie now are maintained by deliberate, controlled burning.

Some human introductions of European plants and animals to non-European communities have been disastrous to native species because of competition or overpredation. Oceanic islands offer classic examples of devastation caused by rats, goats, cats, and pigs liberated from sailing ships. All these animals are prolific, quickly developing large populations. Goats are efficient, nonspecific herbivores; they eat nearly everything vegetational, from grasses and herbs to seedlings and shrubs. In addition, their sharp hooves are hard on plants rooted in thin island soils. Rats and pigs are opportunistic omnivores, eating the eggs and nestlings of sea birds that tend to nest in large, densely packed colonies, and digging up sea turtle eggs. Cats prey upon nestlings of both ground- and tree-nesting birds. Native island species are particularly vulnerable because they have not evolved under circumstances that required them to have defensive adaptations to these predators.

Sometimes introduced species affect native species mainly by competing with them for limited resources. We already have mentioned the spread of starlings across the United States, from their introduction to the East Coast by homesick Europeans. The European weaver finch (English or house sparrow) has a similar history. The aggressive, adaptable birds quickly spread throughout the United States. They can nest in a wide variety of sites, including those required by such natives as bluebirds and purple martins, which have more specific nesting requirements.

Our own lovely Hawaiian Islands are a tragically continuing saga of extinction, not only of individual species but of whole taxonomic groups, as the result of introduced species (including humans). The human species has been the direct or indirect cause of most of the extinctions, especially through habitat destruction. When Europeans first arrived in the Hawaiian Islands, there were sixty-eight unique bird species. Of these, forty-one now are extinct, mainly because of deforestation to create pineapple and sugarcane plantations, resorts, and cities. Introduced cattle, goats, pigs, and rats have taken their toll on native vegetation and birds. A mosquito species accidentally introduced in 1826 has probably been the major factor causing extinction of native birds in otherwise undisturbed forests, because it carried an imported disease to which the native birds lacked resistance.

This last example parallels the introduction of European diseases, notably smallpox, to the Americas. The diseases had a horrendous impact on native human populations, weakening their ability to resist invading populations of Old World humans.

Introductions sometimes are made to solve problems created by previous introductions, but they can have adverse results. In Hawaii, mongooses were imported to help control rats, but instead became bird-nest predators like the rats. Our lessons from this and similar introductions have a new technological twist. Some of the ethical questions currently surrounding the release of genetically engineered organisms are based on concerns that they are novel organisms, and we might not be able to predict how they will interact with other species in natural ecosystems, let alone how they might respond to natural selective forces. We can't predict either their behavior or their evolution, it is argued.

PRINCIPLES AND CONCEPTS

THE HUMAN FACTOR

Several of the examples in preceding sections refer to the effects humans already have had on natural communities, but it's important to realize that our effects are not just in the past tense. This text is concerned with environmental problems we face as a result of past activities, to be sure, but it is more concerned with the future. What future problems will result from our present activities, and how might they be lessened by altering our actions according to the insights of ecology?

Consider the human species in historical context. Modern *Homo sapiens* populations lived for forty thousand years as relatively insignificant components of their natural communities. Cultural evolution has changed that irreversibly and has made us over from a natural animal to a technological animal. Technological human is the ultimate "introduced species" in terms of natural communities. We prey upon the communities of forests, fields, and oceans. We compete with natural members of communities by supplanting naturally occurring species with our own cultivated plants and domesticated livestock. We not only can alter the balance of natural communities, but we also are capable of obliterating them altogether, along with the physical conditions that characterize and support them.

Think, for a moment, about our collective effects on the biosphere in light of the matter and energy principles you have learned in the past three chapters. Compared to other species, human effects are immeasurably greater. We produce megatons of nonrecyclable wastes and use incredible amounts of energy in inefficient ways. In terms of matter and energy processing, we are squanderers. How long can the biosphere continue to maintain homeostasis under these conditions? Our greatest remaining challenge, as technological humans, may not be to find new ways to conquer and exploit the resources of the natural world, but to understand and live within it according to its ability to continue to support us.

S U M M A R Y

Biological communities are made up of populations of organisms occupying and interacting in a given area. Community ecologists attempt to determine the magnitude and importance of these interactions, which determine the productivity, diversity, and resilience of an ecosystem.

Productivity, measured in terms of biomass, is greatly affected by physical environmental factors. In general, terrestrial ecosystems tend to be more productive than marine ecosystems. Furthermore, communities in warm, moist tropical conditions tend to be more productive than those in cool, dry conditions, such as in temperate and polar latitudes.

The number and relative abundance of species present determine the species diversity. As with productivity, the greatest species diversity occurs in the tropics and is directly related to the tremendous diversity of ecological niches in these warm, moist regions of the world.

Resilience is the ability of an ecosystem to respond to disturbance. The rate at which an ecosystem returns to a state of homeostasis following a disruption can be taken as a measure of its resilience.

Physical factors set the minimum and maximum limits of the environment. For a particular species to survive, its tolerance ranges must fall within these limits. The one factor that is closest to the tolerance limits for a given species at a given time is the critical factor.

Natural selection is the process by which certain genetic combinations in a population become more abundant than other combinations. The process of natural selection affects the fertility or survivorship of individuals within a population. It can lead to adaptation.

The role a species plays in an ecosystem is the organism's ecological niche. It includes what resources the organism uses, when they are used, and how they are used. Niches allow usage of a limited amount of resources by a wide range of specialized species while minimizing competition.

Predation and competition affect population sizes and characteristics, so are factors in natural selection. Symbiotic relationships enhance the survival of one or both partners.

The biological community may be limited by the physical environment, but it also can cause change. Dominant plants are especially important modifiers of the physical environment. Changes in the physical environment result in changes in community composition. The process of gradual, long-term change is called succession and culminates with a long-lasting climax community. Climax communities represent a state of homeostasis between the biological community and the physical environment of a given area.

Natural and artificial introductions of species to an established community are a factor in community change. Many introductions of species by humans have been detrimental to native species. Humans are now the greatest factor of change in the biosphere—this is both a problem and a responsibility.

■ Review Questions

1. Productivity, diversity, and resilience are characteristics of all communities and ecosystems. Describe how these characteristics apply to the ecosystem in which you live.

2. Describe the general niche occupied by a bird of prey, such as a hawk or an owl. How can hawks and owls exist in the same ecosystem and not adversely affect each other?

3. All organisms within a biological community interact with each other. The greatest interaction, however, occurs between individuals of the same species. What concept discussed in this chapter can be used to explain this fact?

4. Predator/prey relationships play an important role in the energy transfers that occur in ecosystems. They also influence the process of natural selection. Explain how predators affect the adaptations of their prey. This relationship also works in reverse. How do prey species affect the adaptations of their predators?

5. Competition for a limited quantity of resources occurs in all ecosystems. This competition can be between members of different species (interspecific) or between members of the same species (intraspecific). Explain some of the ways an organism might deal with interspecific competition. How might it deal with intraspecific competition?

6. Each year fires burn large tracts of forest land. Describe the process of succession that occurs after a forest fire destroys the existing biological community. Is the composition of the final successional community likely to be the same as that which existed before the fire? What factors might alter the final outcome of the successional process?

7. Explain the concept of climax community. Why does the climax community exhibit a higher level of homeostasis than that found in other successional stages?

8. Discuss the dangers to existing community members of introduction of new species into ecosystems where they did not previously exist. What type of organism would be most likely to survive and cause problems in a new habitat?

■ Questions for Critical Thinking

1. To maintain agricultural productivity, farmers must expend energy to maintain their fields at a point of early productivity. Explain why there is a need to expend energy to accomplish this.

2. Throughout geologic time, events occurred that destroyed the biological communities existing in given areas. The eruption of a volcano is an example of such a disruptive event. Explain the process that occurs after such a geologic catastrophe. What types of organisms will eventually occur in these areas? What factors determine the final biological community?

3. An ecologist is planning to conduct a detailed study of the matter and energy relationships existing in a small coastal estuary. Discuss the problems that will be confronted by the ecologist as this community ecology study is conducted.

4. Each living organism has a specific tolerance range for environmental factors. Consider the temperature tolerance limits for organisms living in temperate portions of the world. What annual temperature range might such organisms expect to encounter? Now consider the annual temperature range for organisms living in the tropics. What would be the maximum and minimum temperature limits for such species? Which group of organisms would be most affected by temperature as a limiting factor? Why?

5. Imagine a species in which all the individuals are genetically identical. These organisms are currently well adapted to the conditions within the ecosystem where they exist. What could happen to this species if environmental conditions were to change within the ecosystem? What would have to happen to this species for the process of natural selection to work in its favor?

6. Because of human activities, the dominant plant species in many communities have been changed. Without prairie fires, for instance, trees invade grasslands and form forest communities. What consequences occur when there is a change in the dominant plant species in an ecosystem?

7. What would be the consequences in an ecosystem if all the decomposer organisms were killed?

8. Throughout the north central portion of the United States, forests and prairies have been replaced by corn and soybean fields. Despite these changes, whitetail deer and wild turkeys have experienced a dramatic increase in numbers. What environmental features of modern agricultural practices might have contributed to the rise in numbers of these two wildlife species?

■ Key Terms

climax community (p. 58)
commensalism (p. 54)
community ecology (p. 45)
critical factor (p. 48)
ecological equivalents (p. 47)
ecological niche (p. 49)
ecological succession (p. 56)
edge effect (p. 56)
equilibrium communities (p. 58)
fire-climax communities (p. 58)
host organism (p. 50)
interspecific competition (p. 52)
intraspecific competition (p. 52)
limiting factors (p. 48)
natural selection (p. 49)
parasites (p. 50)

patchiness (p. 56)
pathogens (p. 50)
pioneer species (p. 56)
plankton (p. 50)
predation (p. 50)
predator (p. 50)
primary succession (p. 56)
principle of competitive exclusion (p. 53)
productivity (p. 45)
resilience (p. 47)
resistance (inertia) (p. 47)
resource partitioning (p. 49)
secondary succession (p. 56)
species diversity (p. 46)
symbiosis (p. 54)
territoriality (p. 53)
tolerance limits (p. 48)
upwelling (p. 46)
vertical stratification (p. 49)

SUGGESTED READINGS

■ Ehrlich, Paul R., and Jonathan Roughgarden. *The Science of Ecology*. New York: Macmillan Publishing Co., 1987. The community ecology segments of this textbook are particularly appropriate for expanded reading in regard to this chapter.

■ Ehrlich, Paul R., and Anne H. Ehrlich. *Extinction—The Causes and Consequences of the Disappearance of Species*. New York: Random House, 1981. A fascinating treatment of extinction that emphasizes the short-term and potential long-term costs of extinction, how species are endangered by humans, the economic and political aspects of extinction, and what we can and should do about it. Loaded with examples.

■ May, Robert M. *Stability and Complexity in Model Ecosystems*. Princeton: Princeton University Press, 1973. Largely responsible for influencing our present understanding of the relationships between the number of species in a community and its stability in fluctuating or stable environments. May's work incorporates use of mathematical models for analysis and prediction.

■ Odum, Eugene P. *Basic Ecology*. Philadelphia: Saunders College Publishing, 1983. An excellent source of additional and more detailed information about biological communities and ecosystems.

C H A P T E R 5

Biomes: Kinds of Ecosystems

C O N C E P T S

Biomes consist of broad regional groups of related ecosystems. Their distribution is determined largely by climate, topography, and soils. The same or very similar ecological niches may be occupied by different species in geographically separated biomes.

It is important to understand the composition and dynamics of natural ecosystems for three basic reasons: (1) so we might better understand life on Earth, (2) so we can understand the contribution of each kind of ecosystem to the operation of the biosphere as a whole, and (3) so we can understand and predict the effects of human alteration of natural ecosystems.

North America supports several major biomes, including deserts, grasslands, temperate deciduous forests, coniferous forests, and both arctic and alpine tundra. Their special features and some vulnerabilities are presented.

Extensive tropical forests have major effects on the entire biosphere. Scientists are concerned about the loss of tropical rain forests because of their global ecological importance and the incredibly rapid rate at which they are being destroyed.

Freshwater ecosystems include running waters and stationary waters, all of which have distinctive physical and biological components that are vertically stratified.

Wetlands are characterized by the presence of standing water and include swamps, marshes, and bogs. Wetlands provide valuable ecological services, including water purification, flood control, desiltation, replenishment of groundwater, and habitat for many vertebrates, especially waterfowl and shorebirds.

Estuaries form where rivers and ocean waters meet and are highly productive, especially as nurseries for animals that move into the marine environment as they mature.

Marine environments occupy about 71 percent of Earth's surface (even more of its volume) and are distinctive and variable in regard to temperature, depth, and productivity. Vertical stratification is an important feature, and light penetration is a limiting factor for photosynthesis. Shorelines, in particular, are subject to direct human influences.

INTRODUCTION

Venerable yet vulnerable, the saguaro cactus is a good symbol of the desert community (figure 5.1). Just reading the quoted passage probably summoned up a visual image of the American Southwest, complete with hot, rocky hillsides, cacti, and rattlesnakes—a holistic view from the mind's window of one type of North American desert. The saguaro is uniquely American, but there are deserts all around the world. Deserts are a kind of biome, which is a particular type of biological community that develops in response to similar climate and landform conditions in different geographical regions. Most of you reading this book live in North America and have never visited another continent, yet you can identify with the terms desert, grassland, marsh, and forest.

A better understanding of the nature of specific biomes can help you be a better global citizen. News media reports inform us of encroaching deserts in Africa and destruction of tropical rain forests in South America, but these major ecological events often seem remote and of little significance to our own lives, because we don't live in either place. It's often difficult even to identify with something that happens in another region of our own country! Yet the concerns of environmental science are truly global, and our concerns as citizens of an increasingly interconnected world also must become broader.

It is our hope that this chapter will give you further understanding of the diversity of the biosphere and the conditions that maintain it. It emphasizes physical conditions, the biological community, and some specific vulnerabilities of each biome. In addition, this overview of Earth's biomes will set the stage for subsequent discussions of human interactions with and impacts upon them.

BIOME DISTRIBUTION

Many places on Earth share similar climatic, topographic, and soil conditions, and similar communities have developed in response to these conditions. These major ecosystem *types* are called biomes. Temperature and precipitation have synergistic effects on biome distribution (figure 5.2), and availability of water is usually the critical limiting factor for plant growth. Biome distribution also is influenced by the prevailing landforms of an area. Mountains, in particular, exert major influences on biological communities.

Figure 5.3 shows major biomes of the world. Because of its broad scope, this map ignores the many variations present within each major category. These variations are imposed mainly by watercourses, lakes, and differences in topography. Most of

FIGURE 5.1 Saguaro cacti are symbolic of southwestern deserts of North America.

the biomes have examples in North America. Our grasslands are similar to the veldts of Africa, steppes of Eurasia, and pampas of South America. Temperate deciduous forests occur in the middle latitudes of all major continents. Coniferous forests and woodlands are similar around the world. North American deserts have counterparts in Africa, Asia, and Australia.

Adaptation and niche specialization are nicely demonstrated in the biome concept. Organisms that fill similar niches in geographically separated but similar ecosystems usually are *different* species that have undergone similar adaptation independently, in response to similar environmental pressures. Most biomes are identified by the dominant plants of their communities (e.g., grassland or deciduous forest). The diversity of animal life and smaller plant forms characteristic of each biome is, in turn, influenced both by the physical conditions and the dominant vegetation.

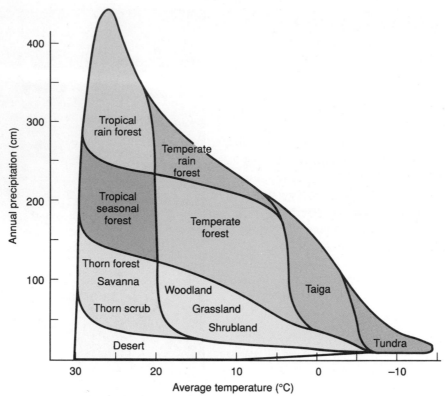

FIGURE 5.2 This schematic diagram illustrates the synergistic impact of two climatic factors—precipitation and temperature—on the development of biome types. Notice the two ways in which you can use the diagram, both consistent with the way scientists think and work. You can look at it as a *descriptive* presentation of three sets of data: annual precipitations, annual temperatures, and vegetation types. However, you also can use it as a basis of *prediction*. For instance, knowing the annual precipitation and temperature of a region, you can predict the expected vegetation type. If the existing vegetation type is counter to your prediction, you might want to explore the sources of the discrepancy between the prediction and the actual observation. Often, such discrepancies are the result of human actions, especially settlement and agriculture.

NORTH AMERICAN BIOMES

Figure 5.4 shows, with greater specificity, major kinds of biomes found in North America. We'll concentrate most of our discussion on North America because these community types are most likely to be familiar to you and are the ones that are affected most directly by your activities. As you read these descriptions, keep in mind that boundaries in nature usually are not as clear-cut as their printed descriptions may imply. If you live in a transitional region, you may recognize overlapping elements of two or more kinds of biomes.

Deserts

Desert climates have the least precipitation, and it is unpredictable from year to year. Daily and seasonal temperature ranges can be very great. Precipitation is so low that in some deserts, evaporation from soil and plant surfaces actually *exceeds* precip-

itation! How can this be? Moisture is brought to the surface from underground sources by deep plant roots.

North America has two major kinds of deserts, the **cool deserts** of the western mountain valleys and Great Basin (figure 5.5) and the **hot deserts** of the Southwest, extending into Mexico (figure 5.6). Most deserts in the United States are desert scrubland (steppe), because the 5–10 cm (2–4 in.) of annual precipitation supports a sparse, but often species-rich community that is dominated by tough, well-adapted shrubs and trees. Cool deserts are dominated by sagebrush, rabbitbrush, perennial grasses, and numerous smaller perennial plants. Hot desert communities are dominated by creosote bush, cacti, acacia, agave, and yucca species.

U.S. deserts are mainly mountainous or hilly with a gravelly or rocky surface, rather than sand. Desert soils are largely unprotected by vegetation, so they are subject to severe erosion from wind and occasional rains. They also lack or have just a

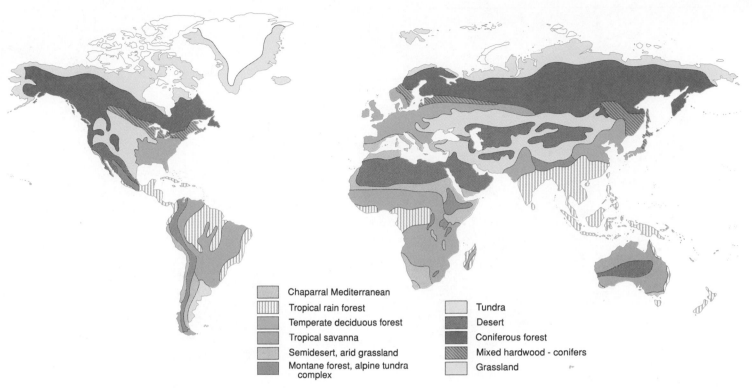

	Chaparral Mediterranean
	Tropical rain forest
	Temperate deciduous forest
	Tropical savanna
	Semidesert, arid grassland
	Montane forest, alpine tundra complex

	Tundra
	Desert
	Coniferous forest
	Mixed hardwood - conifers
	Grassland

FIGURE 5.3 This generalized map of world biomes will give you an overview of the vegetation types on different landmasses. At this scale, much detail cannot be shown, so such maps should be interpreted with that limitation in mind. See figure 5.6 for a more detailed view of North American ecosystems.

thin top layer of decaying plant material called **humus.** North American desert soils have been heavily influenced by volcanic ash; as a result, they are mineral-rich and potentially fertile.

How are desert organisms adapted to these harsh physical conditions? First consider their water needs. Our cool deserts generally have winter snow. During the growing season, however, desert plants must depend on subterranean water sources, runoff from adjacent slopes, and sporadic rainstorms. Animals also depend on these sources, plus moisture from plant tissues and dew. Because desert air has very low humidity, it loses heat rapidly at night. Nighttime temperatures, therefore, can be very chilly, causing even the low amount of atmospheric moisture to condense. (If you have camped on the desert overnight, you may have awakened damp!)

Desert precipitation varies. Deserts that have less than 2.5 cm (1 in.) of measurable precipitation support almost no vegetation. Deserts that have 2.5 to 5 cm (1 to 2 in.) annual precipitation have sparse vegetation (less than 10 percent of the ground is covered), and those plants have a variety of specializations to conserve water and resist exposure of moist inner tissues by predators. Seasonal leaf production, water-storage tissues, and thick epidermal layers help reduce water loss. Spines or serrated leaf edges discourage predators. Spines and ridges on cactus plants also help shade the main part of the shoot and disperse heat.

Animals of the desert have both structural and behavioral adaptations to meet their three most critical needs: food, water, and heat survival. Most desert animals live in burrows or rocky shelters where they can escape the main onslaught of daytime heat. As a result, they are most active between dusk and dawn. Pocket mice and kangaroo rats (and their Old World counterparts, gerbils) produce highly concentrated urine, losing very little water by elimination of body wastes. They are able to live mainly on water that is released when food is digested and utilized.

Deserts may seem formidable, but they are more vulnerable than you might imagine. Desert environments are easily disturbed by human activities and are slow to recover because the harshness of the desert climate severely reduces the resilience of desert communities. Here are five examples. Motorbikes and other recreational vehicles have created major erosion problems in some desert areas. Our southern deserts are being robbed of their biological diversity by profiteering "cactus rustlers." Entire populations of these slow-growing plants have been collected and, as a result, some cactus species are now extinct in their natural environments. Irrigated desert areas have great

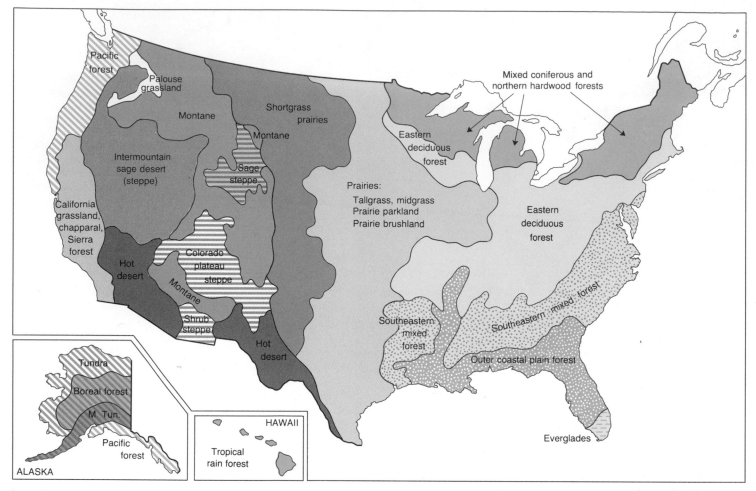

FIGURE 5.4 This map shows the distribution of major North American ecosystems within the United States. Within these general areas there are many distinctive plant associations and biological communities. In many of these regions, large areas of the native ecosystems have been replaced or otherwise altered by human settlements and managed, agricultural systems.

Source: Adapted from R. G. Bailey, *Ecoregions of the United States*, published by the U.S. Forest Service, prepared in cooperation with the U.S. Fish and Wildlife Service, 1976.

agricultural productivity, but at the expense of soil quality because minerals and salt from deeper soil levels are transported to the soil surface, then deposited there when the water evaporates. Many areas of the northern sagebrush deserts have been overgrazed, mainly by domestic livestock. Other areas are being converted to agricultural land in spite of uncertain future water availability to sustain intensive irrigation. The question is: When they are abandoned for human use, will they be able to recover their native vegetation, or will they become wasteland? Historical patterns of land use in the semidesert and desert areas of the Middle East and Africa suggest that mismanagement of our own desert lands could create permanent wastelands.

Grasslands: Prairies and Plains

The moderately dry continental climate of the central Great Plains, from Texas to Saskatchewan and from the Rocky Mountains to the Mississippi and Ohio Rivers, once supported diverse and species-rich grasslands (figure 5.7). Although they are dom-

FIGURE 5.5 Tough and aromatic sagebrush is the dominant plant of the cool deserts of North America. A part of our western lore, the sagebrush community includes smaller shrubs and other perennial flowering plants, bunchgrasses, mule deer, numerous burrowing rodents that are active mainly at night, and their predators.

PRINCIPLES AND CONCEPTS

FIGURE 5.6 The hot deserts of the American Southwest are dominated by large cacti, creosote bush, ocotillo, and paloverde trees—all perennial plants with adaptations to survive extreme temperatures and drought. Many species of annual plants also live in the hot desert. Their survival strategy involves germinating and producing seed during short periods of adequate rainfall, plus having seeds that can remain dormant for years.

FIGURE 5.7 Grasslands, such as this area in Badlands National Park, dominated much of the North American interior before they were converted to croplands and rangelands.

FIGURE 5.8   Prairie vegetation is not just grass, as this tall-grass prairie in North Dakota demonstrates. Many prairie blossoms rival cultivated species in beauty.

inated by grasses, grassland communities also include a rich array of deep-rooted, beautifully flowering, nongrass species (figure 5.8).

Native prairie soils are famous for their productivity, a collaboration of climate and vegetation. Seasonal cycles of temperature and precipitation contribute to abundant vegetative growth that both protects and enriches the soil. Prairie grasses produce massive systems of fine, relatively short-lived roots that decay rapidly, contributing to a deep upper soil layer that is rich in humus. Most North American prairie soils are neutral or somewhat alkaline.

Natural U.S. grasslands are of several major types, as defined by differences in precipitation, soil types, and the kinds of native grasses that dominated them before intensive settlement during the 1800s. Roughly from west to east, the prairies followed gradients of lesser to greater precipitation and rockier to finer soil. The Far West plains, or short-grass prairies, intergrade with cool desert, with the average 10 cm of precipitation supporting communities dominated by bunchgrasses, grasses that

grow in clumps (figure 5.9). Midgrass prairies once dominated the central United States. The eastern tall-grass prairies, dominated by Indian grass and the taller-than-a-person big bluestem, have precipitation exceeding 50 cm per year. The midgrass and tall-grass prairies are characterized by sod-forming grasses. Sod is a thick, dense complex of soil and grass roots. Lawn grasses, for instance, form sod.

Grasslands have few trees because of inadequate rainfall, large daily and seasonal temperature ranges, and frequent grass fires that kill woody seedlings. Exceptions are the narrow gallery forests that form corridors along rivers and streams through the grasslands.

Fire suppression and conversion of the rich prairie soil to farmland have greatly reduced our native grasslands. The most productive North American prairielands now grow wheat, corn, oats, sunflowers, flax, and other cultivated crops. Those lands that are less suited for grain crops because of water availability are used mainly as rangeland. Some of the dry western grasslands have been severely overgrazed, reducing the natural di-

FIGURE 5.9 The short-grass prairies are characterized by bunchgrasses—grasses that grow in clumps.

versity of the community and contributing to soil erosion. A few remnants of native prairie are protected by statute or by private ownership. This is fortunate because these places are havens for natural communities and are important study areas.

Before humans converted them to agriculture and rangeland, our grasslands were as diverse in impressive animals as Africa's Serengeti Plains. The density of large mammals was probably higher in the grasslands than in any other North American biome. Vast herds of bison were present, as were deer, elk, and pronghorns. The gray wolf, cougar, grizzly and black bears, gray fox, and badger were common predators, as well as eagles, several kinds of hawks, and some insectivorous birds. The large mammals have become extremely rare or extinct in most of their former ranges because of hunting, fencing, and habitat displacement. Many now are residents of adjacent forest communities. Significant smaller prairie herbivores are burrowing and tunneling rodents, rabbits, seed-eating birds, and—of course—insects. Shorebirds and migratory waterfowl occupied the millions of ponds, potholes, and marshy spots that once dotted the grasslands. These birds are now more restricted in their breeding sites and migration patterns because of wetland drainage for agriculture. Their populations are much diminished since the last century.

Temperate Deciduous Forests

Temperate climates around the world are characterized by abundant, year-round precipitation, even though the water is not available to plants evenly because of distinct winter and summer seasons. Winters are cold, usually snowy, and summers are hot, usually humid. As a result, species-rich forest communities can thrive. The climate supports lush summer plant growth when water is plentiful, but requires survival adaptations for the frozen season. One such adaptation is the ability to produce summer

leaves, then shed them at the end of the growing season—a **deciduous** pattern.

Soils usually are a mixture of weathered material from native bedrock and material deposited by ancient glaciers. Centuries of accumulated forest litter have been digested by a myriad of soil microorganisms, producing a rich, brown soil over much of this landscape. Compared to grassland soils, these soils have a thinner humus layer but still tend to be neutral and are quite fertile.

When the first European settlers came to North America, dense forests of broad-leaved deciduous trees covered most of the Eastern half of what is now the United States. This rich and varied biome includes associations of many tree species, including oaks, maples, birches, aspens, beech, elms, ashes, hickories, and black walnut. These large trees form a forest canopy over a diverse understory of smaller trees, shrubs, and herbaceous plants, including many spring flowers that grow, flower, set seed, and store carbohydrates before they are shaded by the canopy (figure 5.10). Nuts, berries, and other kinds of fruit are produced in abundance. The many kinds of food and shelter provided by plants afford many opportunities for niche specialization among animals. As a result, the fauna of this biome also is rich and varied.

Much of the original deciduous forest biome has been altered completely as a result of human occupation. Homesteads and settlements were carved out of the forest and later it was harvested for timber.

To the north, and at higher elevations in the south, the deciduous forest grades into coniferous forests, dominated by **conifers**—needle-bearing trees that produce seeds in cones. The northern transition is represented by extensive white pine forests that were largely exterminated by intensive logging in the late 1800s and early 1900s (figure 5.11).

Southern Evergreen Forests

To the south, eastern deciduous forests grade into various **evergreen** (nondeciduous) forest associations of the Atlantic and Gulf coasts, where the climate becomes increasingly warm and humid. Seasonal differences in temperature and humidity are less dramatic, resulting in southern pine forests (described under coniferous forests) and forests composed mainly of nondeciduous broad-leaved trees and shrubs, described here. Finally, at the southern tip of Florida, we have a small sample of true tropical forest biome.

Southern swamp forests (figure 5.12) include live oak, magnolias, and bald cypress, several kinds of "bay," several kinds of hollies, and other shrubs. Epiphytic Spanish "moss" (not really a moss, but a bromeliad of the pineapple family) decorates tree branches, especially those of the live oaks. (An **epiphyte** is a plant that grows attached to a substrate other than the ground, such as another plant.) The forests are penetrated and punctuated by numerous bayous, swamps, and marshes.

FIGURE 5.10   The deciduous forests of the eastern United States are stratified assemblages of tall trees, understory shrubs, and small, lovely, ground-covering species. In turn, a rich fauna is supported by the variety of niche opportunities that are present.

FIGURE 5.11   The northern white pine forests, as in Wisconsin, provided a booming industry in harvesting logs, in cutting, selling, and shipping lumber, and in making wool clothes and leather boots for hardy lumberjacks. Without forest replacement practices, however, the resource was rapidly depleted and logging companies moved to the lush forests of the West.

PRINCIPLES AND CONCEPTS

FIGURE 5.12 The Okefenokee Swamp exemplifies the southern swamp forest of North America. It includes many nondeciduous broad-leaved tree and shrub species and an abundance of aquatic plants and animals.

Southern Florida has a true tropical community, the hummock forest (figure 5.13), characterized by such nondeciduous, woody, broad-leaved plants as strangler fig, gumbo-limbo, and wild tamarind. Several species have toxic sap. Epiphytic bromeliads, orchids, ferns, and climbing vines are abundant. The Everglades National Park preserves examples of this unique community, home also to the American alligator, American crocodile, manatee, and many beautiful, large wading birds. The islands of the Florida Keys also support a distinctive tropical community rich in bird life, but they are not protected from recreational development. Even the Everglades, in spite of its legislated protection, is threatened by human land development, drainage projects, and introductions of exotic species.

Coniferous Forest Biomes

Several distinctive biomes are dominated by coniferous trees. Their distribution is determined mainly by latitude and altitude, which affect temperature, precipitation, and length of the growing season. In general, coniferous forest soils tend to be crumbly and sandy near the surface, with little humus, and somewhat acidic. Where the soil is waterlogged, it is distinctly acidic, as in the boggy northern forests. In mountain areas, particularly, fire has been an important and fairly regular factor in maintaining the coniferous forest. With forest fire control, many forest stands tend to become overly dense. As a result, foresters often must thin young stands of trees manually to insure well-spaced mature trees.

The **boreal forest** or **northern coniferous forest** stretches in a broad band of mixed coniferous and deciduous trees completely across northern North America. Dominant conifers are pines, tamarack, eastern hemlock, spruces, white cedar, and balsam fir. The common deciduous trees are birches, aspens, and maples. In this moist, cool biome, streams and wetlands abound, especially on recently glaciated landscapes. As a result, there are many lakes, potholes, and bogs (figure 5.14). Insects that have aquatic stages in the life cycle, such as mosquitoes and biting flies, are particularly abundant, to the consternation of human anglers and canoers.

Cold winters with heavy snowfall necessitate that animals migrate, hibernate, or develop heavy winter coverings; thus, the boreal forest is the home of many valuable fur-bearing species. Much of the history of European exploration and settlement of the "Great White North" was concentrated on harvesting furs and timber.

FIGURE 5.13 Florida's hummock forests are true tropical ecosystems. Nondeciduous broad-leaved trees and shrubs support climbing vines and epiphytic orchids and bromeliads. The forests are also rich in animal life, especially birds.

The northernmost edge of the boreal forest is a species-poor black spruce/sphagnum moss (peat moss) woodland, often called **taiga,** that intergrades with the treeless arctic tundra. The harsh climate limits both productivity and resilience of the taiga community. Cold temperatures, very wet soil during the growing season, and acids produced by fallen conifer needles and sphagnum inhibit full decay of organic matter. As a result, thick layers of semidecayed organic material called **peat** form. Boreal peat deposits are being explored as energy sources; however, the environmental disturbance of peat mining in this community could be severe and long-lasting, perhaps even permanent.

The very extensive **montane coniferous forests** of the western mountains are characterized by dramatic altitudinal stratification into distinct vegetation zones (figure 5.15). The higher the zone, the colder and harsher the climate becomes, with a shorter growing season. These conditions have a direct effect on productivity and species diversity at different altitudes.

The effect of this *altitudinal* zonation resembles the *latitudinal* distribution of biomes. The lowest zone is a transition from cool desert to juniper and shrub associations, blending into a zone dominated by such drought- and fire-resistant trees as ponderosa pine. The zone from about 2,500–3,500 m (8,200–11,500 ft) is similar in composition to the northern coniferous forest, having a biota adapted to cold, snowy winters. Sugar pine, mountain hemlock, several true firs, and other large conifers make up most of the montane forest. Trembling or quaking aspen is

FIGURE 5.14 The boreal forest of northern North America consists of mixed coniferous and deciduous trees interspersed with bogs and small lakes.

common in moist mountain valleys and on steep slopes where dominant conifers have been cleared by avalanches.

The uppermost montane forest zone is characterized by stunted, wind-sheared spruces and subalpine firs that are buried by deep, sometimes moving packs of snow during the winter. Exposed branches suffer severe wind chill and scouring by wind-driven ice particles. The ragged edge of this zone is called **timberline** (figure 5.16), and marks the transition to the treeless **alpine** zone.

The **southern pine forest,** another kind of U.S. coniferous forest ecosystem, is characterized by a warm, moist climate and, in the past, was subjected to frequent fires. Now it is managed extensively for timber and such resinous products as turpentine and rosin. The undergrowth includes saw palmetto and various thorny bushes (figure 5.17).

The coniferous forests of the Pacific coast represent yet another special set of environmental circumstances. Most of the other coniferous forests are characterized by seasonal extremes of cold or dryness, but **Pacific coast coniferous forests** are characterized by mild temperatures and abundant precipitation. Water vapor (clouds and fog) carried inland from the ocean condenses as it ascends the slopes of the Coast Range, unloading up to 200–250 cm (80–100 in.) of annual precipitation as air currents continue eastward to the interior regions. Trees grow rapidly, making the coastal forests an extremely valuable lumber resource.

Climax coastal forests that have not been logged are called virgin, or old-growth forests. (See chapter 4 discussion of climax community for a deeper understanding of the distinctive nature of an old-growth forest.) Old-growth forests support a rich community, including some species that live only in the canopy. These distinctive arboreal communities are as special and as vul-

FIGURE 5.15 Montane ecosystems are distributed in vertical zones because of the effects of altitude on physical factors such as temperature, length of growing season, availability of soil moisture, and wind exposure. This photograph shows the transition from valley floor to alpine tundra.

FIGURE 5.16 Timberline—the ragged transition zone that separates alpine and subalpine montane zones—often consists of islands of wind- and ice-stunted trees surrounded by expanses of ground-hugging species. Bright flowers attract pollinators during the short growing season, as here in the Cascade Mountains of Oregon.

FIGURE 5.17 This southern coniferous forest in Florida is dominated by slash pine and an understory of saw palmetto.

FIGURE 5.18 A photograph does not do justice to the massive size of mature coastal redwood trees. It does capture, however, the dappled, cathedral-like quality of the forest.

nerable as their counterparts in the tropical rain forest. Ancient trees in virgin forests are centuries old and often exceed 1.5 m (5 ft) in diameter.

The coastal forest extends from northern California to southeast Alaska and has several major, intergrading associations. The California end is characterized by giant redwood trees, the largest trees in the world and the largest organisms of any kind that ever have existed (figure 5.18). Individuals over 115 m (350 ft) tall and 3,500 years of age are preserved in Redwood National Park. The redwoods formerly were distributed over much of the Washington, Oregon, and California coast, but their distribution has been greatly reduced by logging pressures, including clear-cutting without regard to site preservation and restoration.

South of the redwoods are more arid environments that support several distinctive pine species, but the "typical" coastal forest extends northward from the redwoods. Its dominant tree species are Douglas fir, Sitka spruce, and western hemlock, all important lumber species. These moist forests characterize the coasts of Oregon, Washington, British Columbia, and Alaska. In addition to the tall canopy, a dense understory of berry-producing shrubs, rhododendrons, large ferns, and smaller plants provides ample food and shelter for a diverse fauna. Amphibians,

which require moist environments, are particularly well-represented. Because of its lushness and density, the coastal forest can feel overwhelming to a human venturing into it for the first time.

In its wettest parts, the coastal forest becomes a **temperate rain forest,** a cool, rainy forest that often is enshrouded in fog. Condensation from the canopy (leaf drip) is a major form of precipitation. Annual precipitation exceeds 250 cm (100 in.) in some places. Mosses, lichens, and ferns cover tree branches, old stumps, and the forest floor itself. The largest, best-preserved area of temperate rain forest is Olympic National Park in Washington.

Tundra

The climates of the arctic plains north of the boreal forest and the alpine communities above timberline are both too harsh for trees. This treeless landscape is called **tundra.** In the arctic, it is determined by latitude; in the mountains, it is determined by altitude. The tundra is characterized by a very short growing season, cold, harsh winters, and the potential for frost any month of the year.

The *arctic* tundra is a biome of low productivity, low diversity, and low resilience. Winters are long and dark. Only the top several centimeters of the soil thaw out in the summer, and the lower soil is permanently frozen **permafrost.** Permafrost prevents snowmelt water from being absorbed into the soil, so the surface soil is waterlogged during the summer. Try to imagine the difficulties encountered by plants in this kind of soil. Most of the year it is completely frozen, and even during the brief growing season, the permafrost is an impenetrable barrier to deep root growth. During the growing season, the soil is wet; furthermore, the thawed surface soil tends to flow downhill wherever there is a slope, tearing fine roots. In addition, the top layer buckles and heaves in response to cycles of freezing and thawing, further stressing root systems (figure 5.19).

The diversity and density of organisms in the arctic tundra are severely limited, especially during winter. Dominant tundra plants are dwarf shrubs, sedges, grasses, mosses, and lichens (figure 5.20). Its larger life-forms, such as the musk ox and caribou, are well adapted to life on the tundra. Many animals migrate or hibernate during winter. Flocks of migratory birds nest on the abundant summer wetlands, which also nurture hordes of blood-sucking insects that feed upon the summer flocks and herds (and tourists!).

Damage to the arctic tundra is slow to heal. At present, the greatest threat to this distinctive biome is human exploration for oil and natural gas. In the first place, the service roads and drillsite staging areas require special engineering to cope with the combination of wet summer soil and frost heaving. Massive amounts of gravel and concrete must be used to build up the platforms, which must be very broad. It is likely that present truck

FIGURE 5.19 Water expands when it freezes. When wet soil freezes, it buckles and heaves, forming ridges and domes. In the tundra, the result is formation of irregular polygons. The polygons trap water, forming breeding sites for waterfowl and insects.

FIGURE 5.20 A treeless landscape, the arctic tundra is vegetated by low-growing plants that withstand severely cold winters, a very short growing season, and other environmental stresses. In this biome, minor topographic differences provide niche opportunities for plants that can grow in the soggy soil over the permafrost and for plants that grow on better-drained small slopes or hummocks.

FIGURE 5.21 The harsh physical conditions of alpine tundra ecosystems are the result of altitude and slope. The growing season is short, ultraviolet radiation in the "thin" air is greater than at lower altitudes, and extreme day-night temperatures are possible even during the summer. Air currents and the thin, rocky soil contribute to the arid conditions of alpine tundra.

ruts and bulldozer tracks on the tundra landscape will not heal in our lifetimes. Second, some of the sites that seem to be the most promising for oil exploration are exactly the sites to which the caribou return for summer feeding and calving grounds.

The *alpine* tundra differs from the arctic tundra in several ways. Plants of the alpine tundra face different challenges than those of the arctic. The thin mountain air permits intense solar bombardment, especially by ultraviolet radiation; thus, many alpine plants have deep pigmentation that shields their inner cells. The glaring summer sun also causes very hot daytime ground temperatures, even though the night temperatures may return to freezing. Alpine soil is often gravelly or rocky and is wind-swept. Moisture drains quickly because of the sloping terrain (figure 5.21). Due to this combination of sun, soil, slope, and air currents, drought is a problem, as opposed to the wet conditions in the arctic tundra.

Mosses and lichens are abundant in this harsh environment. Flowering plants commonly grow as ground-hugging mats or compact cushions, and many of them are long-lived, dwarf perennials with deep, woody root systems. Smaller alpine animals live in sheltered burrows or rocky dens. Insects that are abundant during the summer enter dormant phases to survive the winter. Most birds and mammals are seasonal residents of the alpine tundra. Many more animals use the alpine feeding grounds during the summer but actually reside in the forests below timberline.

THE SIGNIFICANCE OF TROPICAL FOREST BIOMES

Except for the southeastern tip of the United States, North America doesn't support tropical biomes. Yet, the tropical rain forest and the tropical seasonal forests are a significant feature of biosphere climatic patterns. Vast areas of South America, particularly, are covered by a lush tropical rain forest, the ecosystem dynamics of which are still being explored. Because of their size and global impact, we'll devote some description here to the tropical forest biomes. Additional information is referred to in later chapters, including the impact of rain forest loss on biological diversity and extinction, potentially useful medicinal and food plants, climate and soil alteration, and human resources and populations.

Tropical Rain Forests

The humid tropical regions of South and Central America, Africa, Southeast Asia, and some of the Pacific Islands support one of the most complex and productive biome types in the

FIGURE 5.22 The layered communities of a tropical rain forest are directly related to the gradual extinction of light, from the brightness of the uppermost canopy to the deep shade of the forest floor.

PRINCIPLES AND CONCEPTS

world—the **tropical rain forest,** which you may associate with the word "jungle." Although there are several kinds of moist tropical forests, they have in common ample rainfall and temperatures that are warm to hot year-round. Rainfall is abundant in these biomes, more than 200 cm (80 in.) per year.

The soil of tropical rain forests is old, thin, acidic, and nutrient-poor, yet a lush, vertically stratified community is present (figure 5.22) (Box 5.1). The number of species present is mind-boggling. For instance, the Malay Peninsula has about 8,000 species of flowering plants. Great Britain, by comparison, has twice the land area but only 1,400 species. Costa Rica is smaller than West Virginia, but has 758 identified bird species, compared to 600 bird species in the entire United States. The number of insect species, alone, to be found in the canopy of tropical rain forests has been estimated to be in the millions! It is estimated that one-half to two-thirds of all species of terrestrial plants and insects live in tropical forests. This diversity is the result of the great variety of habitats and niche opportunities available and the long history of these ecosystems.

The nutrient cycles of these forests also are unique. Briefly, they differ from all other ecological systems in that *almost all (90 percent) of the nutrients in the system are contained in the bodies of the living organisms*. This is a vivid contrast to temperate forests, in which nutrients are held within the soil and made available for new plant growth. Plants and animals of the tropical rain forest grow rapidly and profusely only because of rapid decomposition of dead organic material. As nutrients become available, they are taken into the plants. The recycling rate is rapid.

When the forest is removed for logging, agriculture, and mineral extraction, the thin soil cannot support continued cropping and cannot resist erosion from the abundant rains. If the cleared area is too extensive, it cannot be repopulated by the rain forest community. Rapid deforestation in South America is occurring as part of attempts to resettle some of the population and promote economic growth, but the agricultural plots being carved out of the forest soon lose their fertility. During the Vietnam war of the 1970s, chemical defoliation with Agent Orange systematically destroyed large areas of Southeast Asian rain forest, as well as cultivated mangrove swamps.

Rapid, continuing loss of the rain forests promises to be a long-range disaster, both natural and human (figure 5.23). Specifically, what are the consequences? Identified problems are

1. potential conversion of rain forest to savanna;
2. unprecedented loss of species—perhaps *one million species* in the last two decades of the twentieth century;
3. loss of many small, isolated, indigenous human societies and their knowledge about rain forest species;
4. loss of a significant carbon sink, possibly leading to increased levels of atmospheric carbon dioxide;

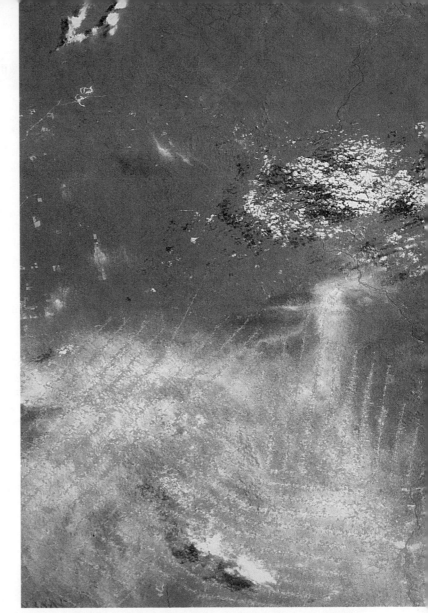

FIGURE 5.23 The Rondonia Development Project in western Brazil began with the building of a large, central road (running left to right in this Landsat image) deep into the tropical rain forests. Side roads extending from the main road were used to log and burn the forest, opening the land to subsistence farmers who were relocated there by the government because of severe poverty in other parts of Brazil. This computer-generated image—taken in September, 1981—provides infrared analysis. Areas of dense vegetation are red, whereas deforested areas are blue/green and white. The irregular white objects underlain by black are clouds and their shadows. Since the image was captured, millions more acres of forest have disappeared.

5. interference with regional water quality and hydrologic cycling because of increased runoff and evaporative loss;
6. economic drains on governments due to large investments in expensive projects that yield only short-term gains;
7. potential increase in political instability due to economic and social factors.

BOX 5.1   Tropical Rain Forests: Life in Layers

The richest and most productive ecosystems are in the tropical forests. These forests have been reduced to less than half of their former extent by human activities and now cover only about 7 percent of Earth's land area. In this limited area, however, is about two-thirds of the vegetation mass and about half of all living species in the world!

The largest, lushest, and most biologically diverse of the remaining tropical moist forests are in the Amazon River basin of South America, the Congo River basin of central Africa, and the large islands of southeast Asia (Sumatra, Borneo, and Papua, New Guinea). Whereas the forests of mainland Southeast Asia, western Africa, and Central America are strongly seasonal, with wet and dry seasons, the South American and central African forests are true rain forests. Rainfall is generally over 400 cm per year and falls more or less evenly throughout the year. It is said that such rain forests "make their own rain," because about half the rain that falls in the forests comes from condensation of water vapor released by transpiration from the trees themselves. Rain forests at lower elevations are hot and humid year-round. At higher elevations, tropical mountains intercept moisture-laden clouds, so the forests that blanket their slopes are cool, wet, and fog-shrouded. They are aptly and poetically called "cloud forests."

The ecology of a tropical moist forest is very different from that of temperate forests, with which you may be familiar. Tropical forests are mostly very old. Unlike temperate forests, they haven't been disturbed by glaciation or mountain building for hundreds of millions of years. This long period of evolution under conditions of ample moisture and stable temperatures has created an incredible

diversity of organisms of amazing shapes, colors, sizes, habits, and specialized adaptations. No other biome has even a fraction of the biological richness of the tropical forests.

Habitats in a tropical rain forest are stratified into three to five distinct layers from ground level to the tops of the tallest trees. Let's start at the top. Hundreds of tree species grow together in lush profusion, their crowns interlocking to form a dense, dappled canopy about 40 meters above the forest floor. These unusually tall trees are supported by relatively thin trunks that are reinforced by wedge-shaped buttresses, instead of having thick trunks and deep roots. A few emergent trees rise above the seemingly solid canopy into a world of sunlight, wind, and open space. Numerous species of birds, insects, reptiles, and small mammals live exclusively in the forest canopy, never descending below the crowns of the trees.

The forest understory is composed of small trees and shrubs growing between the trunks of the major trees, as well as climbing woody vines (lianas) and many epiphytes—mainly orchids, bromeliads, and arboreal ferns—that attach themselves to the trees. Some of the larger trees may support fifty to one hundred different species of epiphytes and an even larger population of animals that are specialized to live in the many habitats that are created. These understory layers are a world of bright but filtered light filled with animal activity.

By contrast, the forest floor is generally dark, humid, quiet, and rather open. Few herbaceous plants can survive in the deep shade created by the layered canopy of the forest trees, shrubs, and their epiphytes. The most numerous animals are ants and termites that scavenge on the detritus raining down from above. A few

rodent species gather fallen fruits and nuts. Rare predators such as leopards, jaguars, smaller cats, and large snakes hunt both on the ground and in the understory.

What happens at the soil level? The productivity of a tropical rain forest can be as high as 90 tons /ha/ year, and you might think that the soil that supports this incredible growth is rich and fertile. Instead, however, it is old, acidic, and nutrient poor. Ages of incessant tropical rains and high temperatures have depleted minerals, leaving an iron- and aluminum-rich podzol. Tropical forests have only about 10 percent of their organic material and nutrients in the soil, compared to boreal forests, which may have 90 percent of their organic material in litter and sediments.

The interactions of decomposers and living plant roots in the soil are, literally, the critical base that maintains the rain forest ecosystem. Tropical rain forests are able to maintain high productivity only through rapid recycling of nutrients. As you might suspect, the constant rain of detritus and litter that falls to the ground is quickly decomposed by populations of fungi and bacteria that flourish in the warm, moist environment. Some of these decomposers have symbiotic relationships with the roots of specific trees. Trees have broad, shallow root systems to capitalize on this surface nutrient source; an individual tree might create a dense mat of superficial roots 100 meters in diameter and 1 meter thick. In this way, nutrients are absorbed quickly and almost entirely and are reused almost immediately to build fresh plant growth, the necessary base to the trophic pyramid of this incredible ecosystem.

Tropical Seasonal Forests

Many areas in India, Southeast Asia, Australia, West Africa, the West Indies, and South America have tropical regions characterized by distinct wet and dry seasons, instead of uniform, heavy rainfall throughout the year. Temperatures are hot year-round. These areas have produced communities of **tropical seasonal forests**—semi-evergreen or partly deciduous forests tending toward open woodlands and grassy savannas dotted with scattered, drought-resistant tree species.

FRESHWATER ECOSYSTEMS

Fresh water, water other than seawater, is essential to life in all of the terrestrial biomes, and is present to varying degrees in all of them. Freshwater ecosystems are not themselves biomes, but are varied and distinctive parts of biomes. Their nature is determined more by individual site factors than by global climate or soil distribution. Beyond their introduction in this chapter, freshwater ecosystems also are discussed in chapter 16.

Freshwater ecosystems include the standing waters of ponds and lakes plus the flowing waters of rivers and streams. There also are some unique freshwater ecosystems, including underground rivers and subterranean caves. Freshwater ecosystems are as varied as their individual sites because they are influenced not only by characteristics of local climate, soil, and resident communities, but also by the surrounding terrestrial ecosystems and anything that happens uphill or upstream from them (figure 5.24). As with terrestrial ecosystems, the biological communities of freshwater ecosystems are limited and to a large extent determined by the physical characteristics of the environment, except that the surrounding medium is, significantly, water instead of air.

Aquatic organisms have the same basic needs as terrestrial organisms: carbon dioxide, water, and sunlight for photosynthesis; oxygen for respiration; and food and mineral nutrients for energy, growth, and maintenance. The availability of these necessities is influenced by such site characteristics as (1) substances that are dissolved in the water, such as oxygen, nitrates,

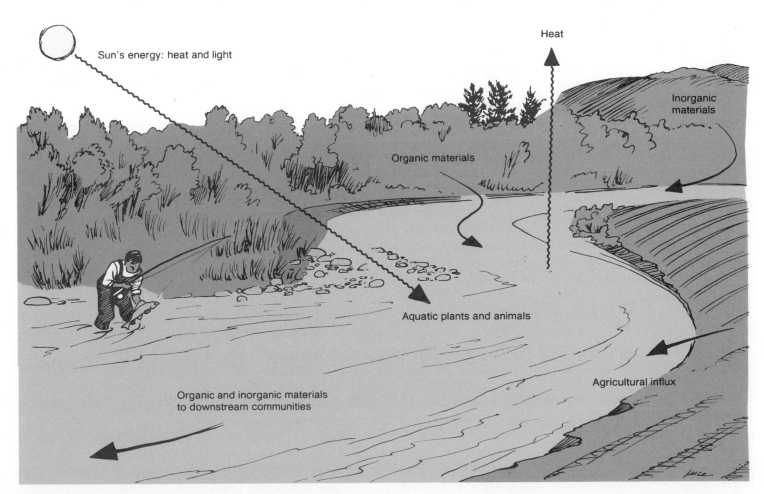

FIGURE 5.24 The character of freshwater ecosystems is greatly influenced by the immediately surrounding terrestrial ecosystem and even by ecosystems far upstream or far uphill from a particular site.

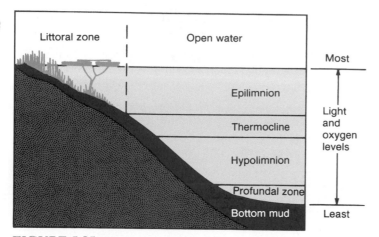

FIGURE 5.25 The layers of a deep lake are determined mainly by gradients of light, oxygen, and temperature. The epilimnion is affected by surface mixing from wind and thermal convections.

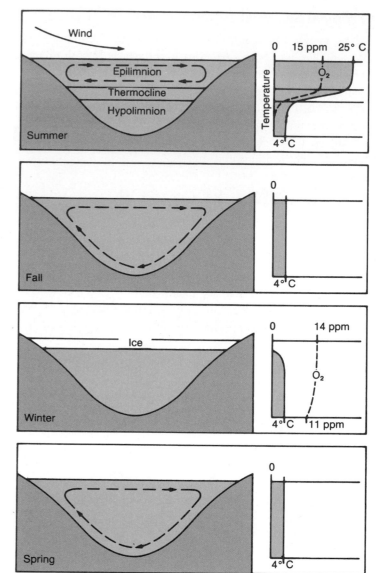

FIGURE 5.26 Lakes tend to have annual cycles based mainly on changes in water temperature and density. Many lakes experience a fall and spring overturn as cool, dense water (densest at 4° C) sinks. Nutrient upwelling, which enriches surface waters, results.

phosphates, potash (potassium compounds), and other byproducts of agriculture and industry; (2) suspended matter, such as silt and microscopic algae, that affect water clarity and, therefore, light penetration; (3) depth; (4) temperature; (5) rate of flow; (6) bottom characteristics (muddy, sandy, rocky); (7) internal convective currents; and (8) connection to or isolation from other aquatic ecosystems.

Vertical stratification is an important aspect of standing water ecosystems, especially in regard to gradients of light, temperature, nutrients, and oxygen. Organisms tend to form distinctive vertical subcommunities in response to this stratification of physical factors. The **plankton subcommunity** consists mainly of microscopic organisms that float freely within the water column. Although planktonic protists are able to adjust their position vertically, they are carried with the currents of the water body. Some nonplanktonic organisms are specialized to live at the air-water interface (e.g., water striders), and still others are able to swim freely in the open waters (e.g., fish). Finally, many animal species are bottom dwellers (e.g., snails, burrowing worms and insect larvae, bacteria), which make up the **benthos,** or bottom subcommunity. Oxygen levels are lowest in the benthic environment. Anaerobic bacteria can live in the very low-oxygen bottom sediments.

Deeper lakes are characterized by the presence of a warmer, upper layer that is mixed by the wind (the epilimnion) and a colder, deep layer that is not mixed (the hypolimnion) (figure 5.25). The two layers are separated by a distinctive transition zone called the thermocline. Most lakes in temperate zones go through an annual cycle that mixes upper and lower waters, redistributing oxygen, nutrients, and heat (figure 5.26).

What happens during the annual cycle of a lake? Autumn weather cools the lake surface. Since cold water is denser than warm water, surface water sinks and mixes with the lower waters. During winter the lake surface freezes, and a temperature gradient is established from the surface to the bottom. You might expect the deepest water to be the coldest. This is not true, however, because water is most dense at 4°C (4 degrees above freezing). As a result, the coldest water is near the surface, and the densest water is near the bottom. The nature of this winter temperature gradient leads to a **spring overturn.** In spring, the surface ice melts. As the oxygen-rich surface water warms to 4°C, it sinks, displacing nutrient-rich bottom waters upward. Responding to the spring availability of nutrients, the plankton community grows and reproduces rapidly, providing food for larger animals, especially fish. The annual cycle of a temperate zone lake is extremely important to its productivity.

Lakes have a tendency to undergo succession, which includes changes in the biological community as a response to increases in nutrient levels. Human sources of nutrient input can radically increase the rate at which a lake "ages," as discussed in chapter 22 (eutrophication).

Humans utilize freshwater communities as sources of food, for recreation, waste disposal, cooling nuclear power plants, and many other industrial uses. Waterways also are major transport lanes for barge and ship traffic, involving channel dredging and contamination from fuel and cargo leaks. Dams, canals, and other diversions have significant effects on freshwater resources. In these and other ways, humans have an enormous impact on the character of individual freshwater ecosystems and their biological communities.

Not all freshwater lakes contain what we would consider "fresh" water. The Caspian Sea, Dead Sea, Great Salt Lake, and numerous other salty lakes have been formed by evaporative shrinkage of large bodies of water. Deserts, particularly, may have alkaline or arsenic-containing lakes and potholes. Still other salty and mineral-rich ponds and small lakes are fed by mineral hot springs.

ESTUARIES AND WETLANDS: TRANSITIONAL ECOSYSTEMS

Estuaries are enclosed or semi-enclosed bodies of water that form where rivers enter the ocean, creating an area of mixed fresh water and ocean water. Estuaries usually contain rich sediments carried downriver, forming mudflats.

Estuaries are sheltered from the most drastic ocean action, but do experience tidal ebbs and flows. Daily tides may even cause the river levels to rise and fall up to several kilometers inland from the estuary. The combination of physical factors in estuaries makes them very productive and of high species diversity. They are significant "nurseries" for economically important fish, crustaceans (such as crabs and shrimp), and mollusks (such as clams, cockles, and oysters).

Estuaries are valued as ports and therefore are vulnerable to human use. Because they are fed by rivers, estuaries tend to accumulate silt. This may be a positive influence for the biological community, but hinders human navigation. As a result, estuaries are often dredged out, disrupting the bottom community. Estuarine waters also are often polluted by fuel oil and bilge wastes from boats and ships, seafood processing wastes, and whatever organic and inorganic substances are carried into the estuary by rivers and streams, including industrial and sewage effluents.

Where topography permits, an extensive fan-shaped sediment deposit called a **delta** may form at the river mouth. Deltas often are channeled by branches of the river, creating extensive coastal wetlands that are part of the larger estuarine zone.

Wetlands of several types form near bodies of water (table 5.1). Wetlands are ecosystems in which the rooted vegetation is surrounded by standing water at least part of the year. The water may or may not be flowing slowly through them. There are special names for specific kinds of wetlands, but we can lump them

TABLE 5.1 A wetlands classification	
Freshwater areas	
Inland fresh areas	
1. Seasonally flooded basins or flats	Soil is covered with water or is waterlogged during variable periods but is well drained during much of the growing season. In upland depressions and bottomlands. Bottomland hardwoods to herbaceous growth.
2. Fresh meadows	No standing water during growing season but is waterlogged to within a few inches of surface. Grasses, sedges, rushes, broad-leaved plants.
3. Shallow fresh marshes	Soil is waterlogged during growing season; often covered with 6 or more inches of water. Grasses, bulrushes, spike rushes, cattails, arrowhead, smartweed, pickerelweed. A major waterfowl area.
4. Deep fresh marshes	Soil is covered with 6 in. to 3 ft of water. Cattails, reeds, bulrushes, spike rushes, wild rice. Principal duck-breeding area.
5. Open freshwater	Water is less than 10 ft deep. Bordered by emergent vegetation. Pondweed, naiads, wild celery, water lily. Breeding, feeding, nesting area for ducks.
6. Shrub swamps	Soil is waterlogged, often covered with 6 or more inches of water. Alder, willow, buttonbush, dogwoods. Ducks nesting and feeding to limited extent.
7. Wooded swamps	Soil is waterlogged, often covered with 1 ft of water. Along sluggish streams, flat uplands, shallow lake basins. Tamarack, arbor vitae, spruce, red maple, silver maple in the North; water oak, overcup oak, tupelo, swamp black gum, cypress in the South.
8. Bogs/fens	Soil is waterlogged. Spongy covering of mosses; heath shrubs, sedges, sphagnum moss.
Coastal fresh areas	
9. Shallow fresh marshes	Soil is waterlogged during growing season by as much as 6 in. of water at high tides. Deep marshes along tidal rivers, sounds, and deltas on landward side. Grasses, sedges. Important areas for waterfowl.
10. Deep fresh marshes	Covered with 6 in. to 3 ft of water at high tide. Along tidal rivers and bays. Cattails, wild rice, giant cutgrass.
11. Open fresh water	Shallow areas of open water along fresh tidal rivers and sounds. Little or no vegetation. Important waterfowl areas.

(continued)

together into three major categories: swamps, marshes, and bogs/fens. Defined pragmatically, **swamps** are wetlands with trees; **marshes** are wetlands without trees; and **bogs/fens** are areas with waterlogged soils that tend to be peaty. Swamps and marshes tend to be associated with flowing water. Fens are characterized

TABLE 5.1 Continued

Saline water areas

Inland saline areas

12. Saline flats	Flooded after heavy precipitation. Soil is waterlogged within a few inches of surface during the growing season. Seablite, salt grass, saltbush. Fall waterfowl feeding areas.
13. Saline marshes	Soil is waterlogged during growing season; often covered with 2 to 3 ft of water. Shallow lake basins. Alkali hard-stemmed bulrush, widgeon grass, sago pondweed. Valuable waterfowl areas.
14. Open saline water	Permanent areas of shallow saline water. Depth variable. Sago pondweed, muskgrasses. Important waterfowl feeding areas.

Coastal saline areas

15. Salt flats	Soil is waterlogged during growing season. Occasionally to fairly regularly covered by high tide. Landward sides or islands within salt meadows and marshes. Salt grass, seablite, saltwort.
16. Salt meadows	Soil is waterlogged during growing season. Rarely covered with tide water. Landward side of salt marshes. Cord grass, salt grass, black rush. Waterfowl feeding areas.
17. Irregularly flooded salt marshes	Covered by tides at irregular intervals during the growing season. Along shores of nearly enclosed bays, sounds, etc. Needlerush. Waterfowl cover area.
18. Regularly flooded salt marshes	Covered at average high tide with 6 or more inches of water. Along open ocean and along sounds. Salt marsh cord grass along Atlantic. Alkali bulrush, glassworts along Pacific. Feeding area for ducks and geese.
19. Sounds and bays	Portions of saltwater sounds and bays shallow enough to be diked and filled. All water landward from average low tide line. Wintering areas for waterfowl.
20. Mangrove swamps	Soil covered at average high tide with 6 in. to 3 ft of water. Along coast of southern Florida. Red and black mangroves.

Source: Adapted from S. P. Shaw and C. G. Fredine, *Wetlands of the United States*, U. S. Fish and Wildlife Service Circular 39, 1956.

by the presence of upwelling waters, whereas bogs are fed mainly by precipitation. Bogs and fens have low productivity, but swamps and marshes generally have high productivity.

The water of marshes and swamps usually is shallow enough to allow full penetration of sunlight and seasonal warming. These mild conditions favor great photosynthetic ac-

tivity, resulting in high productivity at all trophic levels. In short, life is abundant and varied. Wetlands are major breeding, nesting, and migration staging areas for waterfowl and shorebirds.

Wetlands perform major ecosystem services, the importance of which cannot be overestimated. As mentioned previously, they support a great diversity of life-forms. What may be less obvious is their role in planetary water relationships. Wetlands act as traps and filters for water that moves through them. Runoff water is slowed as it passes through shallow, plant-filled areas, reducing flooding. As a result, sediments are deposited in the wetlands instead of traveling into rivers and, eventually, the oceans. In this way, wetlands both clarify surface waters and aid in the accumulation and formation of fertile land. Furthermore, chemical interactions in wetlands ecosystems neutralize and detoxify substances in the water. Finally, standing or slow-moving water seeps continually into the ground, helping to replenish underground water reservoirs called **aquifers.**

Wetlands continue to be lost at an accelerating pace, due mainly to human activities, as described in chapter 16. This destruction is of great concern because it means loss of the above services to the biosphere, as well as loss of essential habitats for a myriad of species.

MARINE ECOSYSTEMS

If the challenge of describing so many diverse wetlands in a few short paragraphs is great, that of describing briefly the many different ecosystems that make up about 71 percent of the Earth's surface is both challenging and extremely frustrating! Studying the physical and biological characteristics of oceans and seas is understandably fascinating, and the more we know the more fascinating it becomes. For now, however, we must give you just a brief overview.

The Oceans

The definitive factor of **marine** ecosystems is the presence of seawater. The average salinity of ocean water is about 3.6 percent. Another, perhaps startling, feature of the world's seas and oceans is that they are interconnected, forming a **World Ocean** (figure 5.27). The average depths of parts of the World Ocean range from about 54 m (180 ft) in the English Channel to about 4,030 m (13,220 ft) in the Pacific Ocean. The average World Ocean depth is estimated to be about 3,790 m (12,430 ft). Compare that to the average elevation of land above sea level, 840 m (2,760 ft). Taking into account all of the separate measurements of its parts, the World Ocean has a total volume of about 1,370,000,000 cubic km (329,000,000 cubic miles) of water, about eleven times the volume of land above sea level! These numbers cannot help but impress upon us the influence marine ecosystems have upon the biosphere.

Most marine communities are found in the shallower regions of the oceans and seas along continental shelves (figure

5.28), coral reefs, and oceanic islands. Life at greater depths is limited by darkness, cold temperatures, and pressure, and productivity is very low. Creatures of the great depths, mainly scavengers and predators that feed on the detritus rain and each other, are adapted to these conditions. Productivity in the bottom depths is very low.

Marine ecosystems show distinctive vertical stratification. With few exceptions, the food that supports large and diverse consumer populations is produced in well-lighted surface waters. In the open waters, **phytoplankton** (mainly planktonic algae) are the producers, the first link of most food chains. In the shallower waters, multicellular algae, or seaweeds, contribute to this role.

The ecosystems of marine shorelines and open waters are subject to and sometimes threatened by human activities, such as recreational use, real estate development, offshore trash disposal, oil and gasoline spills and leaks, radioactive waste disposal, and overexploitation of fish and shellfish. As with other communities, those of marine environments are subject to the pressures of habitat destruction, pollution, and human resource use.

Shorelines and Barrier Islands

Ocean shorelines, including rocky coasts, sandy beaches, and offshore barrier islands, are particularly rich in life-forms. Rocky shorelines, in particular, support an incredible density and diversity of organisms that grow attached to any solid substrate, including each other (figure 5.29). Sandy shorelines, on the other hand, provide homes for organisms that live among the sand grains and in burrows.

Sandy beaches are understandably popular to humans. Cities, resorts, and residences are built on beaches. Grasses and trees that hold the dunes are destroyed, destabilizing the soil system and thus increasing its susceptibility to wind and wave erosion. The constant battle to maintain such real estate is costly. Insurance is very expensive because of the hazards of natural erosion and storms. Building and maintaining sea walls and protecting and replenishing beaches are expensive, continuous processes. Protective structures often have their own effect on shore recontouring, and may even increase the rate of natural shoreline loss.

Barrier islands are low, narrow, sandy islands that form a rim offshore from a coastline. In North America, they are particularly characteristic of the Atlantic and Gulf coasts (figure 5.30). Barrier islands protect inland shores from the onslaught of the surf, especially during severe storms. Because they are so

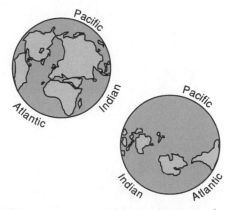

FIGURE 5.27 World Ocean. We're used to seeing the globe from an equatorial perspective. By contrast, these polar views show the extensiveness and distribution of the oceans. View (*a*) demonstrates that the continents dominate the Northern Hemisphere and that the oceans are interconnected there as well. View (*b*), over Antarctica, clearly shows the dominance of oceans versus dry land and shows the continuity of the oceans, forming a World Ocean.

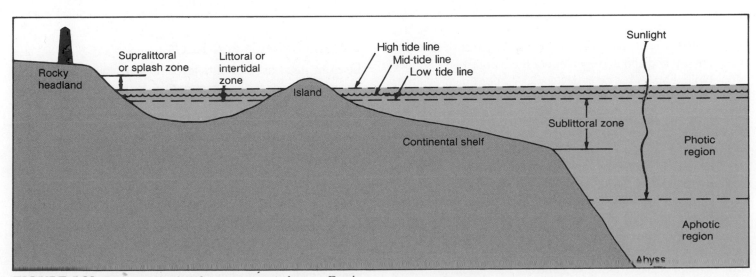

FIGURE 5.28 Shore profile identifying major vertical zones. Depth of light penetration is a limiting factor for photosynthetic organisms. As a result, surface waters are the most productive.

Biomes: Kinds of Ecosystems

FIGURE 5.29 The abundance and diversity of life in the intertidal zone is best appreciated by examining the tidepools that are left at low tide, as on this rocky shoreline.

FIGURE 5.30 Barrier islands protect inner coastlines from the full onslaught of surf and storms.

FIGURE 5.31 Corals grow mainly in shallow tropical seas, forming reefs just below the surface to form and protect oceanic islands. This aerial view of the Pacific island of Mooréa shows white coral sands and a line of breakers along the edge of the reef.

lovely, barrier islands have been tempting targets for real estate development, and about 20 percent of the barrier island surface in the United States has been developed. Unfortunately, barrier islands are *not* permanent. Their sands are constantly being redistributed by wind and water. Seaward shores of U.S. barrier islands erode at an average of about one-half m per year, but may lose up to 20 m per year. A single, major storm can do massive damage, destroying houses, and perhaps washing away most of an island. Because of these factors, buildings, roads, and bridges built on barrier islands are temporary structures, in spite of expensive measures to preserve the eroding shorelines.

Sooner or later, long-term environmental and economic factors will need to be given priority in coastal development planning. Perhaps it makes the most sense not to build on physically dynamic, unstable beaches and barrier islands.

Oceanic Islands and Reefs

Oceanic islands demonstrate particularly interesting ecosystems, because of the different ways they originate and the patterns by which they become colonized by organisms. Islands that originated by breaking away or being isolated from a continental landmass (e.g., British Isles, New Zealand) often have fairly complete, diverse flora and fauna that are clearly related to their continental sources. Volcanic and coral island communities, on the other hand, tend to show the results of chance colonization. Whatever their origins, island communities often have individual touches that result from chance colonization followed by natural selection among surviving immigrants.

Coral reefs form in clear, warm, tropical seas and are particularly well-developed in the south Pacific (figure 5.31). They are the accumulated calcareous skeletons of innumerable tiny colonial animals called corals. Each of the interconnected coral

PRINCIPLES AND CONCEPTS

animals builds a calcium carbonate chamber on the surface of the accumulated secretions of previous generations of animals. Coral reefs usually form along the edges of shallow, submerged banks or shelves. The depth at which they form is limited by the depth of light penetration, due at least in part to the presence of symbiotic photosynthetic protists in their tissues. Because of the photosynthetic relationship, the presence of dead versus living coral growths can be an index of either previous ocean levels or decreased light penetration due to increased turbidity. Coral reef communities rival tropical forest communities in species diversity, numbers of individuals, brilliance of color, and interesting forms of both plants and animals.

SUMMARY

Major ecosystem types called biomes are characterized by similar climates, soil conditions, and biological communities. Topography also is a major influence on the development and distribution of biomes. The major biomes found in North America are deserts; grasslands; temperate deciduous forests; southern broad-leaved evergreen forests; boreal, montane, southern, and coastal coniferous forests; and tundra.

Our deserts are characterized by low annual precipitation, unpredictable water availability, volcanic-influenced soil, and plants specialized to conserve water. There are distinct differences between "hot" and "cold" deserts. Grasslands, which occur in the central part of the continent, are dominated by grasses and other fire-adapted perennial plants. This biome, which was once the home of huge herds of grazing animals, has been converted almost entirely to agricultural usage.

Deciduous trees that are adapted to dramatic seasonal changes are the dominant plant forms in the temperate deciduous forests. The large trees provide a dense canopy over a diverse understory of smaller trees, shrubs, and herbaceous plants, creating a multitude of niches. The southern broad-leaved forests are dominated by nondeciduous trees.

Coniferous forest biomes vary greatly in temperature, precipitation, and species composition, depending mainly on altitude and latitude. Northern, southern, montane, and coastal coniferous forests are very distinct from one another.

North of the coniferous forests is a cold, harsh region called the arctic tundra. Alpine tundra exists above treeline in mountainous regions. The tundra is treeless, has a very short, cool growing season, and supports mosses, lichens, and dwarf plants. Many tundra animals are migrants found there only during the summer.

Tropical rain forests and tropical seasonal forests are of major concern to ecologists because of the large area of Earth's surface they cover. Tropical biomes are among the most diverse and productive of all biomes, containing most of Earth's plant and animal species. They are being destroyed at an alarming rate, creating irreversible changes that will affect the entire biosphere.

Freshwater ecosystems include mainly ponds, lakes, streams, and rivers. They are influenced greatly by the surrounding terrestrial ecosystems that contribute inorganic and organic matter to them. Physical factors and biological communities are stratified, especially in standing water. Spring overturn in lakes distributes nutrients vertically, enhancing productivity.

Estuaries and wetlands are special kinds of ecosystems that form where land and water meet. Estuaries are enclosed or semi-enclosed bodies of water that form where rivers enter oceans, and include rich biological communities. Wetlands are ecosystems where rooted vegetation is surrounded by standing water at least part of the year. The three major wetlands categories are swamps, marshes, and bogs/fens.

Marine ecosystems are defined by the presence of seawater. The most diverse and productive ecosystems are in the shallower water along the edges of continents, islands, and coral reefs. The penetration of light to support photosynthesis is a major factor in the productivity of marine ecosystems. Marine communities are dramatically stratified, which is especially visible on rocky shorelines.

Sandy beaches and barrier islands are characterized by their instability, yet are popular for real estate development. Oceanic islands are fascinating subjects of ecological study because of their histories and isolation from continental landmasses. Coral reefs form in tropical seas and include richly diverse, colorful biological communities.

■ Review Questions

1. Throughout the central portion of the North American continent is a large biome dominated by grasses. Describe how landforms and other physical factors control this biome.
2. Many species of desert animals rarely have access to standing water, yet they are able to obtain enough water to survive. How do you think these organisms are able to survive in an environment lacking water to drink?
3. During the day, deserts often appear barren of animal life. This is because most desert animals are nocturnal. Explain why this is a positive behavioral adaptation for desert animals and where and how these animals spend their days.
4. Many of the most productive agricultural areas of the world occur in grassland biomes. Explain why grasslands are so well suited for modern agricultural practices.
5. Plants that shed their leaves at the end of the growing season are referred to as deciduous plants. Why is this loss of leaves an important adaptation?
6. The greatest species diversity occurs in tropical rain forests. Describe some of the reasons why these ecosystems have such a rich flora and fauna.
7. What conditions determine the boundaries of the various communities found in freshwater ecosystems? How are aquatic ecosystems similar to terrestrial ecosystems? How do they differ?
8. Despite the fact that oceans cover approximately 71 percent of the Earth's surface, they are not as productive (on average) as terrestrial ecosystems. What factors might contribute to this data? Within marine ecosystems, where is the greatest productivity likely to occur?

Questions for Critical Thinking

1. No other species of living organism has had as dramatic an effect on the biosphere as we have. Why are we such an influential species and why have we caused such dramatic changes in Earth's ecosystems?

2. Certain biotic and abiotic factors characterize Earth's biomes. What physical and biological factors are characteristic of the biome in which you live? Where else in the world are similar characteristics found? Why are most of the species of plants and animals found in these other areas different from those in your biome?

3. In terms of biomass production, grasslands produce more large herbivores than any other biome. Why are grassland biomes capable of supporting such large population densities of grazing animals?

4. Discuss the differences in the biotic and abiotic factors of a deciduous broad-leaved forest and a nondeciduous broad-leaved forest, such as those found along the southern Atlantic and Gulf coast of the United States.

5. Coniferous forests occur over much of the North American continent. Why is this biome so widespread? What biotic and abiotic characteristics do the various kinds of coniferous forests share? In what ways do they differ?

6. Tundras occur near the polar regions as well as near the peaks of mountains above timberline. Discuss how both latitude and altitude can affect the development of these plant communities.

7. Tropical rain forests are being cleared at an increasingly rapid rate for use as agricultural lands. Why are these cleared lands relatively poor producers of agricultural products? The loss of these natural ecosystems is considered by many environmentalists to be the most serious problem of habitat destruction that faces us today. Why are these ecosystems so critical?

8. Islands exhibit some of the most unique assemblages of species found in the biosphere. Explain why island communities differ so significantly from continental communities.

9. Estuaries and wetlands serve many important functions within the biosphere. Besides providing nurseries for many species of aquatic organisms, they also play an important role in the water cycle. Explain the role of these ecosystems in Earth's water system. How has human technology made this role even more important?

Key Terms

alpine (p. 74)
aquifers (p. 84)
barrier islands (p. 85)
benthos (p. 82)
bogs (p. 83)
boreal forest (p. 73)
conifers (p. 71)
cool deserts (p. 66)
coral reefs (p. 86)
deciduous (p. 71)
delta (p. 83)
epiphyte (p. 71)
estuaries (p. 83)
evergreen (p. 71)
fens (p. 83)
fresh water (p. 81)
hot deserts (p. 66)
humus (p. 67)
marine (p. 84)
marshes (p. 83)
montane coniferous
    forests (p. 74)
northern coniferous
    forest (p. 73)
oceanic islands (p. 86)
ocean shorelines (p. 85)
Pacific coast coniferous
    forests (p. 74)
peat (p. 74)
permafrost (p. 76)
phytoplankton (p. 85)
plankton subcommunity
    (p. 82)
southern pine forest (p. 74)
spring overturn (p. 82)
swamps (p. 83)
taiga (p. 74)
temperate rain forest (p. 76)
timberline (p. 74)
tropical rain forest (p. 79)
tropical seasonal
    forests (p. 81)
tundra (p. 76)
wetlands (p. 83)
World Ocean (p. 84)

SUGGESTED READINGS

See also readings for chapters 3 and 4.

■ Attenborough, David. *The Living Planet*. Boston: Little, Brown & Co., 1984. Beautiful photographs and well-written text provide glimpses of Earth's natural communities.

■ Rodgers, C. L., and R. E. Kerstetter. *The Ecosphere: Organisms, Habitats, and Disturbances*. New York: Harper & Row, 1974. Treats the variety of biomes on Earth, as well as the effects of human disturbances.

■ Smith, Robert Leo. *Ecology and Field Biology*, 3d ed. New York: Harper & Row, 1980. A large, comprehensive textbook with detailed information that is particularly relevant to chapters 1–7 of this text. Includes a good treatment of biomes.

■ Whittaker, R. H. *Communities and Ecosystems*, 2d ed. New York: Macmillan Publishing Co., 1975. Excellent descriptions of natural communities and ecosystems maintain the value of this text even though more than a decade has passed, and we now have much more ecological data on which to base interpretations of community and ecosystem dynamics.

PART TWO

Populations

CHAPTER 6

Population Dynamics

Through the animal and vegetable kingdoms, nature has scattered the seeds of life abroad with the most profuse and liberal hand. . . . The germs of existence contained in this spot of earth, with ample food, and ample room to expand in, would fill millions of worlds in the course of a few thousand years. Necessity, that imperious all-pervading law of nature, restrains them within the prescribed bounds.

Thomas Malthus

CONCEPTS

Population dynamics is the study of changes in population size and density. The principles of population dynamics are fundamental to understanding how ecosystems function.

When resources are plentiful, most organisms reproduce at an exponential rate—a rate at which population size doubles at regular intervals. Some species will grow exponentially until they exceed the environment's carrying capacity, exhausting all available resources. When this happens, populations "crash," declining as quickly as they grew. Other species population growth is reg-

ulated so it reaches an equilibrium before exceeding the carrying capacity of the environment.

Many factors (both intrinsic and extrinsic) influence population size. Among these are natality, fecundity, life span, longevity, mortality, immigration, emigration, climate, weather, competition, predation, and disease. The sum of gains and losses due to these influences determines the net rate of growth. Stress and crowding due to competition for limited resources can be important population limiting factors both within a species and between species.

INTRODUCTION

In the spring, flowers bloom and set seed, birds raise nestlings, and mosquitoes hatch in teeming multitudes in marshes and swamps. All around is evidence of the amazing ability of living organisms to increase their kind. Populations of organisms—members of the same species that occupy a specific geographical area at a given time—tend to increase as far as their environment will allow. As a result, most populations are in a dynamic state of equilibrium. Their numbers wax and wane in a delicate balance that is influenced by limiting factors of the physical environment and interactions with other populations in the community.

Sometimes, however, a population will grow suddenly, in a **population explosion** that exhausts resources on which not only that particular species depends, but other members of its community as well. Migratory locusts, for instance, periodically erupt in huge clouds from arid lands to sweep across the countryside, stripping vegetation bare (figure 6.1). They not only die in great numbers themselves as they devour all available plant material, but they also destroy the food and disrupt the food chains on which other organisms, including humans, depend for survival.

A sudden population decline, called a **population crash**, can also cause serious problems for dependent populations. Anchovetta (anchovies) are small, sardinelike fish that once lived in vast numbers in the nutrient-rich, upwelling currents off the coast of Peru and Chile. They suffered a sudden population crash in the 1970s. This catastrophic decline was probably caused by a combination of overfishing by humans and changes in the ocean currents that had supplied nutrients necessary for phytoplankton and zooplankton on which the anchovetta depended. The fishing trade was devastated (figure 6.2) and millions of sea birds that had depended upon a diet of anchovetta starved to death or were so malnourished that reproduction failed.

What is it that determines whether a population will increase or decrease catastrophically, or whether it will maintain a delicate balance with its neighbors? Why is it that some populations grow to enormous numbers while others do not? In this chapter, we will look at some of the underlying factors that influence population dynamics. We will survey the components that lead to growth or decline of populations, and we will consider the mechanisms that regulate population size, distribution, and growth rate. One of the most important concepts to be gained from the study of population dynamics is that different strategies regulate population density, depending on the niche an organism occupies in its ecosystem and the stability of that ecosystem.

WOLVES AND MOOSE ON ISLE ROYALE: A CASE STUDY

The interaction of wolves and moose in Isle Royale National Park is an interesting case study that dramatically illustrates several important principles in population dynamics. Isle Royale National Park occupies the largest island in Lake Superior, the largest freshwater lake in the world. Isle Royale is a spectacular wilderness setting of high rocky ridges covered by a dense boreal forest. Cut off from the mainland of Minnesota and Ontario by 30 km (20 mi) of rough, deep, very cold water, the island is a mostly closed ecosystem that is a unique laboratory for studying large animal population dynamics.

Moose Population Growth and Decline

The original large herbivore of the island was the woodland caribou. Caribou were abundant until about 1900 but disappeared early in this century due to hunting and human-caused changes in the habitat. About the time that caribou became extinct on the island, moose first appeared. We suppose they swam from the mainland or crossed on ice that occasionally forms a bridge to the island. They must have found an ideal situation; the shrubs and aquatic plants on which they prefer to browse were plentiful, and there were no major predators to limit their population growth.

The number of moose increased slowly at first. In 1915, it is estimated that about two hundred moose were on the island, or one moose for each 2.6 square km (1 square mile). Then, in the 1920s, there was a moose population explosion on Isle Royale. As figure 6.3 shows, the number jumped from roughly three hundred in 1920 to about five thousand in 1928.

In the summer of 1929, famous wildlife biologist Adolph Murie went to the island to study the moose situation. He reported that all the tender branches on which the moose browse in the winter were eaten back as high as the moose could reach. Much of the summer food (aquatic plants and annuals) was also badly depleted. He predicted that disease and starvation would soon cause an extensive die-off. As you can see in figure 6.3, his prediction came true. By 1941, only 171 moose were found on the island—fewer than twenty years before. Clearly, unrestrained growth of the moose population surpassed the limits of the environment and resulted in a catastrophic population decline.

Moose-Wolf Equilibrium

In the early 1940s, shortly after the moose population on Isle Royale declined so precipitously, wolves appeared on the island in pairs and small groups, presumably having crossed on the ice during previous winters. There were no permanent human inhabitants on the island at the time, so we don't know exactly when the first wolves arrived or how many there were in those early years. By 1957, when the first systematic census was taken, twenty-one wolves were on the island (figure 6.3).

You might think that the wolves, arriving as they did when the moose were weakened by starvation, could have exterminated their prey completely. Instead, the wolves and moose established a dynamic balance, oscillating between six hundred to twelve hundred moose and fourteen to fifty wolves at any given time. Wolf predation prevented both excess population growth

FIGURE 6.1 Periodic population explosions of locusts are an example of irruptive population growth. Their sudden abundance disrupts food chains and destroys their own food supply.

and catastrophic decline of moose on the island (figure 6.4). For about thirty years, both species maintained an equilibrium in which the numbers of individuals that were born and died each year balanced each other (on average). There was enough vegetation for the moose to eat and stay healthy; and there were enough moose for the wolves to eat. We call such an equilibrium the **carrying capacity** of the environment. It is the maximum number of individuals of any species that can be supported on a long-term basis by a particular ecosystem.

Wolves in Trouble

In 1988, an alarming decline in the wolf population on Isle Royale was observed. The annual winter survey revealed only twelve wolves in three small packs, down from fifty wolves only six years earlier. Scientists proposed several possible causes for this population crash. It may have been the food supply. Although moose were plentiful, they were young and healthy—perhaps too difficult for the wolves to catch. Diseases might have been introduced by dogs or stray wolves from the mainland. Some scientists believe the problem may be genetic. When a population starts with only a few founding individuals and is highly inbred, as are the wolves of Isle Royale, deleterious mutations (gene defects) are likely to be expressed. A high rate of reproductive failures and infant mortality may result.

Fifty wolves may simply not be a viable population over the long-term, no matter how ideal the environment or how carefully the species is protected. This problem has important implications in managing other endangered species and in determining the size of protected habitat and wildlife preserves (chapter 15). As for Isle Royale, perhaps the wolf population now is stabilizing. Three pups were born in 1990.

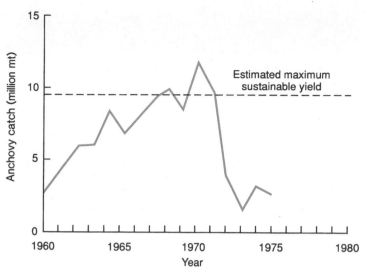

FIGURE 6.2   Anchovies on the coast of Peru and Chile suffered a population crash in 1973. Bird and human populations were hurt by the loss in the food chain.

FIGURE 6.4   Wolves eating a moose on Isle Royale. The wolves only succeed in killing about one moose in every ten they attack. Usually only the very young, old, and sick moose are vulnerable. A healthy adult can defend itself against a large pack.

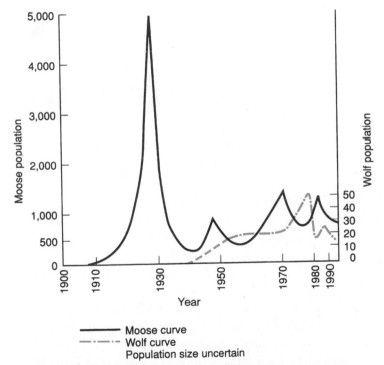

FIGURE 6.3   Growth of moose and wolf populations on the Isle Royale National Park, 1900 to 1985. Moose first appeared on the island shortly after 1900. By 1929, the moose population had grown to about 5,000, clearly overshooting the carrying capacity of the ecosystem. A catastrophic die-back occurred during the 1930s, when 97 percent of the moose died. The appearance of wolves after 1940 at first reduced the moose population, but then established a dynamic equilibrium around what appears to be the carrying capacity.

Source: David Mech, *The Wolves of Isle Royale*, 1966, published by National Park Service.

DYNAMICS OF POPULATION GROWTH

To explain the patterns of growth and interspecific interaction exhibited by the wolves and moose of Isle Royale, we need to know something about the principles of population biology. In this section, we will look at growth rates and patterns.

Exponential Growth and Doubling Times

The rapid increase of moose on Isle Royale in the early stages of colonization is an example of **exponential growth,** or growth at a constant *rate* of increase per unit of time. It is called exponential because the rate of increase can be expressed as a constant fraction, or exponent, by which the existing population is *multiplied*. This pattern also is called **geometric growth** because the sequence of growth follows a geometric pattern of increase, such as 2, 4, 8, 16, and so on. By contrast, a pattern of growth that increases at a constant *amount* per unit time is called an **arithmetic growth**. The sequence in this case might be 1, 2, 3, 4. or 1, 3, 5, 7. Notice in these examples that a constant amount is *added* to the population (figure 6.5).

The reason that moose populations (like those of most other organisms) have the ability to grow exponentially, given an appropriate environment, lies in the power of biological reproduction. Each cow moose reaches sexual maturity at age three or four. She then can produce one or two calves each year for the next eight to ten years if she remains healthy, has enough to eat, and survives that long. This means that each female moose can produce eight to twenty calves during its lifetime. If half of those calves are female and all survive to have as many offspring as their mother, the moose population will increase between four-

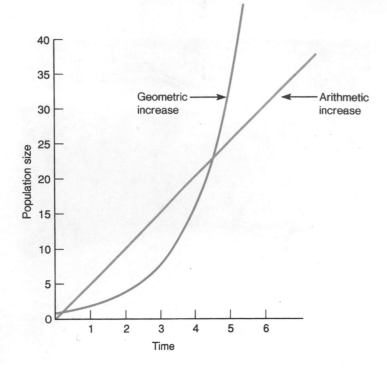

FIGURE 6.5 Arithmetic and geometric curves. Note that the geometric or exponential curve grows more slowly at first, but then accelerates past the arithmetic curve, which grows at a steady incremental pace throughout.

TABLE 6.1 Doubling times at various compound interest rates

Annual % increase	Doubling time (years)
0.1	700
0.5	140
1.0	70
2.0	35
5.0	14
7.5	9
10.0	7
100.0	0.7

TABLE 6.2 Biotic potential of houseflies (*Musca domestica*) in one year

Assuming that:

—a female lays 120 eggs per generation
—half of these eggs develop into females
—there are seven generations per year

	Total population	
Generation	If all females in each generation lay 120 eggs and then die	If all generations survive one year and all females reproduce maximally in each generation
1	120	120
2	7,200	7,320
3	432,000	446,520
4	25,920,000	27,237,720
5	1,555,200,000	1,661,500,920
6	93,312,000,000	101,351,520,120
7	5,598,720,000,000	6,182,442,727,320

Source: Data from Kormondy, *Elements of Ecology*.

population growing at 35 percent doubles every two years. You can also apply this rule to calculate doubling time of human populations. Countries growing at 4 percent per year will *double* their populations in 17.5 years. A country growing at a rate of 0.1 percent annually will double in seven hundred years. We will use this rule to discuss human population growth in chapter 7.

Biotic Potential

Neither moose nor humans are the fastest reproducing of all organisms. Many species have amazingly high reproductive rates that give them the potential to produce enormous populations very quickly, given unlimited resources and freedom from limiting factors. We call the maximum reproductive rate of an organism its **biotic potential.** Table 6.2 shows the potential number of offspring that a single female housefly (*Musca domestica*) and her offspring could produce in a year. The result is astounding. Each female fly lays an average of 120 eggs in each generation. The eggs hatch and mature into sexually active adults and lay their own eggs in fifty-six days. In one year (seven generations), if all its offspring survived long enough to reproduce, a single female could be the ancestor of *5.6 trillion flies.* If this rate of reproduction continued for ten years, the whole earth would be covered several meters deep with houseflies! Fortunately, this has not happened because of factors that limit the reproductive success of houseflies. This example, however, illustrates the potential for biological populations to increase rapidly.

to ten-fold in each ten-year generation. In other words, the population is increasing at a 14 to 35 percent annual growth rate.

As you can see in figure 6.3, the number of moose added to the population at the beginning of the growth curve is rather small. But within a very short time, the numbers begin to increase quickly. The growth curve produced by this constant rate of unfettered growth is called a **J curve** because of its shape. At 1 percent per year, a population or a bank account doubles in roughly seventy years (table 6.1). A useful rule of thumb is that if you divide seventy by the annual percentage growth, you will get the approximate doubling time in years. As a result, a moose

Catastrophic Declines and Population Oscillations

In the real world, there are limits to growth. When a population exceeds the carrying capacity of its environment or some other limiting factor comes into effect, death rates begin to surpass birth rates. The growth curve becomes negative rather than positive, and the population decreases as fast or faster than it grew. We call this the population crash or **dieback.** Looking back at figure 6.3, you can see catastrophic dieback of both moose and wolves on Isle Royale at different times.

The extent to which a population exceeds the carrying capacity of its environment is called **overshoot,** and the severity of the dieback is generally related to the extent of the overshoot. This pattern of population explosion followed by a population crash is called **irruptive** or **Malthusian growth.** It is named after the eighteenth-century economist Thomas Malthus, who concluded that human populations tend to grow until they exhaust their resources and become subject to famine, disease, or war. Malthus and his theories are discussed further in chapter 7.

Populations may go through repeated oscillating cycles of exponential growth and catastrophic crashes, as shown in figure 6.6. These cycles may be very regular if they depend on a few simple factors, such as the seasonal light and temperature-dependent bloom of algae in a lake. They also may be very irregular if they depend on complex environmental and biotic relationships that control cycles, such as the population explosions of migratory locusts in the desert or tent caterpillars and spruce budworms in northern forests.

Growth to a Stable Population

Not all biological populations go through these cycles of irruptive population growth and catastrophic decline. The growth rates of many species are regulated by both internal and external factors so that they come into equilibrium with their environmental resources. These species may grow exponentially when resources are unlimited, but their growth slows as they approach the carrying capacity of the environment. This pattern is called **logistic growth,** a mathematical description of its constantly changing rate.

Together, factors that tend to reduce population growth rates are called **environmental resistance.** In later sections of this chapter, we will look in more detail at these factors and how they limit growth and regulate population size. First, we will see how logistic growth compares to the Malthusian growth just discussed and what kinds of organisms we are talking about in these different patterns.

How does the growth curve of a stable population differ from the J curve of an exploding population? Figure 6.7 shows an idealized comparison between exponential and logistic growth. The J curve on the left in this figure represents the growth without restraint that we just discussed. It rises rapidly towards

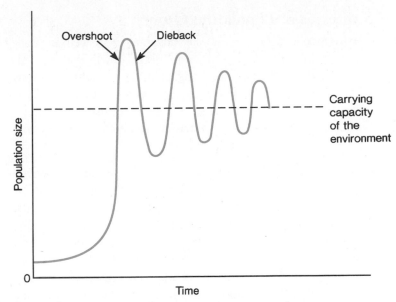

FIGURE 6.6 Population oscillations. Some species demonstrate a pattern of cyclic overshoot and dieback.

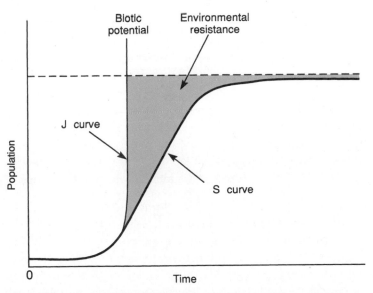

FIGURE 6.7 J and S population curves. The vertical J represents theoretical unlimited growth. The S represents population growth and stabilization in response to environmental resistance.

the maximum biotic potential of the species. The curve to the right represents logistic growth. We call this pattern an **S curve** because of its shape. It is also called a sigmoidal curve (for the Greek letter sigma). The area between these curves is the cumulative effect of environmental resistance. Note that the resistance becomes larger and the rate of logistic growth becomes smaller as the population approaches the carrying capacity of the environment.

Strategies of Population Growth

There appear to be evolutionary advantages, as well as disadvantages, in both Malthusian and logistic growth patterns. Although we should avoid implying intention in natural systems where the controlling forces may be entirely mechanistic, it sometimes helps us see these advantages in terms of "strategies" of adaptation and "logic" in different modes of reproduction.

Malthusian "Strategies"

Organisms with Malthusian growth patterns often tend to occupy low trophic levels in their ecosystems (chapter 3), or to be pioneers in succession (chapter 4). As generalists or opportunists, they move quickly into disturbed environments, grow rapidly, mature early, and produce many offspring. They usually can do little to care for their offspring or protect them from predation. They depend on sheer numbers and dispersal mechanisms to ensure that some offspring survive to adulthood (table 6.3). They have little investment in individual offspring, using their energy to produce vast numbers instead.

Many insects, rodents, marine invertebrates, parasites, and annual plants (especially the ones we consider weeds) follow this reproductive strategy. The moose of Isle Royale could also be considered members of this group, as are most species of the deer family. Their numbers generally are limited by predators or other controlling factors in the environment. They reproduce at the maximum rate possible, which offsets these losses. If the external factors that normally control their populations are inoperative, they tend to rapidly overshoot the carrying capacity of the environment and then dieback catastrophically, as we have just seen. Among this group are the weeds, pests, or other species we consider nuisances that reproduce profusely, adapt quickly to environmental change, and survive under a broad range of conditions.

Logistic "Strategies"

By contrast, species exhibiting logistic growth tend to occupy higher trophic levels in their ecosystems and reproduce at slower rates than do those with Malthusian patterns. These organisms are usually larger, live longer, mature more slowly, produce fewer offspring in each generation, and have fewer natural predators than the species below them in the ecological hierarchy. Some typical examples of this strategy are wolves, elephants, whales, and primates. Each of these species provides more care and protection for its offspring than do "lower" organisms.

Elephants, for instance, are not reproductively mature until they are eighteen to twenty years old. During youth and adolescence, a young elephant is part of a complex extended family that cares for it, protects it, and teaches it how to behave. A female elephant normally conceives only once every four or five years after she matures. The gestation period is about eighteen months; thus, an elephant herd doesn't produce many babies in a given year. Since they have few enemies (except humans) and live a long life (often sixty or seventy years), however, this low

TABLE 6.3 Characteristics of contrasting reproductive strategies	
Externally controlled growth	**Intrinsically controlled growth**
1. Short life	1. Long life
2. Rapid growth	2. Slower growth
3. Early maturity	3. Late maturity
4. Many small offspring	4. Fewer larger offspring
5. Little parental care or protection	5. High parental care and protection
6. Little investment in individual offspring	6. High investment in individual offspring
7. Adapted to unstable environment	7. Adapted to stable environment
8. Pioneers, colonizers	8. Later stages of succession
9. Niche generalists	9. Niche specialists
10. Prey	10. Predators
11. Regulated mainly by extrinsic factors	11. Regulated mainly by intrinsic factors
12. Low trophic level	12. High trophic level

reproductive rate produces enough elephants to keep the population stable, given appropriate environmental conditions.

An important underlying question to much of the discussion in this book is which of these strategies humans follow. Do we more closely resemble wolves and elephants in our population growth, or does our population growth pattern more closely resemble that of moose and rabbits? Will we overshoot the carrying capacity of our environment (or are we already doing so?), or will our population growth come into balance with our resources?

FACTORS THAT INCREASE OR DECREASE POPULATIONS

Now that you have seen population dynamics in action, let's focus on what happens *within* populations, which are, after all, made up of individuals. In this section, we will discuss how new members are added to and old members removed from populations. We also will examine the composition of populations in terms of age classes and introduce terminology that will apply in subsequent chapters.

Natality, Fecundity, and Fertility

Natality is the production of new individuals by birth, hatching, germination, or cloning. Natality is the main source of addition to most biological populations. Natality is usually sensitive to environmental conditions so that successful reproduction is tied strongly to nutritional levels, climate, soil or water conditions, and—in some species—social interactions between members of the species. The maximum rate of reproduction under ideal conditions varies widely among organisms and is a species-specific characteristic. We already have mentioned, for instance, the differences in natality between several different species.

Fecundity is the physical ability to reproduce, while **fertility** is a measure of the actual number of offspring produced. Because of lack of opportunity to mate and successfully produce offspring, many fecund individuals may not contribute to population growth. Human fertility often is determined by personal choice of fecund individuals.

Immigration

Organisms are introduced into new ecosystems by a variety of methods. Seeds, spores, and small animals may be floated on winds or water currents over long distances. This is a major route of colonization for islands, mountain lakes, and other remote locations. Sometimes organisms are carried as hitchhikers in the fur, feathers, or intestines of animals traveling from one place to another. They also may ride on a raft of drifting vegetation. Some animals travel as adults—flying, swimming, or walking—as did the moose and wolves that first colonized Isle Royale. In some ecosystems, a population is maintained only by a constant influx of immigrants. Schools of predatory fish, for example, are important members of ocean ecosystems; but they generally reproduce in shallows, shoals, or estuaries, and their numbers in the open ocean are maintained only by constant recruitment from such nursery areas.

Mortality and Survivorship

An organism is born and eventually it dies; it is mortal. Its death is a loss to the population. **Mortality,** in the population sense, can be expressed in two ways. The *mortality rate*, or probability of dying, is determined by dividing the initial population (the number alive at the beginning of a period) by the number that died during a given interval. The *death rate* is determined by dividing the number that died during a period of time by the average number of individuals living during that time. Figure 6.8 shows mortality figures for the population of the United States by age classes and sexes. Note that between age forty and eighty the mortality rate of men is higher than that of women. After age eighty, the rate is similar or may slightly favor the remaining men.

Since the number of survivors is more important to a population than is the number that died, mortality is often better expressed in terms of **survivorship** (the percentage of the maximum lifespan of a species achieved by members of a particular population) or **life expectancy** (the probable number of years of survival for an individual of a given age). If more organisms in a population die than are replaced in a given time, the population will decrease. If mortality is low compared to natality, on the other hand, the population will grow. Between these two broad generalizations, many combinations of mortality and natality rates create very different patterns of population growth.

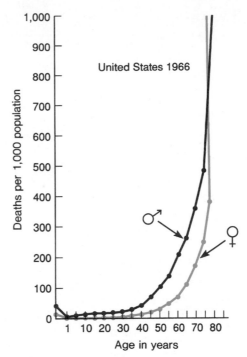

FIGURE 6.8 Mortality curves by age class in the United States, 1966. Note the high death rates for males in the forty- and seventy-year-old age class. It is not yet clear whether this is due to biologic factors or simply life-style. Most mammals have similar mortality curves, although some mammals, such as rabbits and moose, have higher death rates in early life.

Source: U.S. Bureau of the Census.

To understand how these factors interact, we need to define some more population terms.

Life span is the longest period of life reached by a given type of organism. The process of living entails wear and tear that eventually overwhelm every organism, but maximum age is dictated primarily by physiological aspects of the organism itself. There is an enormous difference in life span between different species. Some microorganisms live their whole life cycles in a matter of hours or minutes. Bristlecone pine trees in the mountains of California, on the other hand, have life spans up to 4,600 years.

Most individuals in a population do not live anywhere near the maximum life span for their species. The major factors in early mortality are predation, parasitism, disease, accidents, fighting, and environmental influences, such as climate and nutrition. Important differences in relative longevity among various types of organisms are reflected in the survivorship curves shown in figure 6.9.

Four general patterns of survivorship can be seen in this figure. Curve a is the pattern of organisms that tend to live their full physiological life span if they reach maturity and then have a high mortality rate when they reach old age. This pattern is typical of many large mammals, such as whales, bears, and elephants (when not hunted by humans), as well as humans in

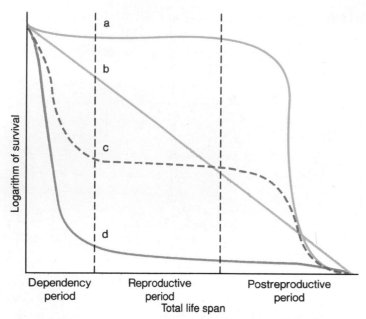

FIGURE 6.9   Four basic types of survivorship curves for organisms with different life histories. Curve (a) represents organisms, such as humans or whales, which tend to live out the full physiological life span if they survive early growth. Curve (b) represents organisms, such as sea gulls, in which the rate of mortality is fairly constant at all age levels. Curve (c) represents such organisms as white-tailed deer, moose, or robins, which have high mortality rates in early and late life. Curve (d) represents such organisms as clams and redwood trees, which have a high mortality rate early in life but live a full life if they reach adulthood.

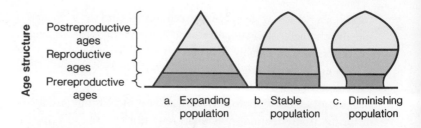

Type of population

FIGURE 6.10   Each of these shapes contains a theoretical population, and the zones divide each population into segments based on age. The different shapes are due to the relative number of individuals in each age class. Note the different age structures in (a) expanding, (b) stable, and (c) diminishing populations.

developed countries. Interestingly, some very small organisms, including predatory protozoa and rotifers (small, multicellular, freshwater animals) have similar survivorship curves even though their maximum life spans may be hundreds or thousands of times shorter than those of large mammals. In general, curve a is the pattern for top consumers in an ecosystem, although many annual plants have a similar survivorship pattern.

Curve b represents the survivorship pattern for organisms for which the probability of death is generally unrelated to age. Sea gulls, for instance, die from accidents, poisoning, and other factors that act more or less randomly. Their mortality rate is generally constant with age, and their survivorship curve is a straight line.

Curve c is characteristic of many songbirds, rabbits, members of the deer family, and humans in less developed countries (see chapter 7). They have a high mortality early in life when they are more susceptible to external factors, such as predation, disease, starvation, or accidents. Adults in the reproductive phase have a high level of survival. Once past reproductive age, they become susceptible again to external factors and the number of survivors falls quite rapidly. The moose on Isle Royale clearly fall into this category.

Curve d is typical of organisms at the base of a food chain, or those especially susceptible to mortality early in life. Many tree species, fish, clams, crabs, and other invertebrate species produce a very large number of highly vulnerable offspring, few

of which survive to maturity. Those individuals that do survive to adulthood, however, have a very high chance of living most of the maximum life span for the species.

Age Structure

An outcome of the interaction between mortality and natality is that growing or declining populations usually will have very different proportions of individuals in various age classes. Figure 6.10 shows three hypothetical population profiles that are distinguished by differences in distribution among prereproductive, reproductive, and postreproductive age classes. Pattern a is characteristic of a rapidly expanding population, like the moose of Isle Royale shortly after their arrival on the island. Young animals make up a large proportion of the population. This large number of prereproductive individuals represent **population momentum** because there is potential for rapid increase in natality once the youngsters reach reproductive age. The age structure of human populations in rapidly growing, developing countries generally resembles pattern a.

When natality comes into balance with mortality, the population enters a stationary phase having an age structure pattern similar to b. This pattern is characteristic of nations with a stable population size, such as many European countries.

The age structure shown by pattern c is characteristic of a diminishing population, where natality has fallen to a level lower than the replacement number. The bulge in the upper age classes represents adults still living but reproducing at a lower rate. When those individuals die, the population will be much smaller.

Emigration

Emigration, the movement of members out of a population, is the second major factor that reduces population size. The dispersal factors that allow organisms to migrate into new areas are important in removing surplus members from the source population. Emigration is also a form of insurance for the species. If the original population is destroyed by some catastrophic change in their environment, their genes still are carried by descendants in other places. Many organisms have very specific mechanisms to facilitate migration of one or more of each generation of offspring.

FACTORS THAT REGULATE POPULATION GROWTH

So far, we have seen that differing patterns of natality, mortality, life span, and longevity can produce quite different rates of population growth. The patterns of survivorship and age structure created by these interacting factors show us not only how a population is growing but also can indicate what general role that species plays in its ecosystem. They also reveal a good deal about how that species is likely to respond to disasters or resource bonanzas in its environment. But what factors *regulate* natality, mortality, and the other components of population growth? In this section, we will look at some of the mechanisms that determine how a population grows.

Various factors regulate population growth, primarily by affecting natality or mortality, and can be classified in different ways. They can be intrinsic (operating within individual organisms or between organisms in the same species) or extrinsic (imposed from outside the population). Factors can also be either **biotic** (caused by living organisms) or **abiotic** (caused by nonliving components of the environment). Finally, the regulatory factors can act in a *density-dependent* manner (effects are stronger or a higher proportion of the population is affected as population density increases) or *density-independent* manner (the effect is the same or a constant proportion of the population is affected regardless of population density).

In general, biotic regulatory factors tend to be density dependent, while abiotic factors tend to be density independent. There has been much discussion about which of these factors is most important in regulating population dynamics. In fact, it probably depends on the particular species involved, its tolerance levels, the stage of growth and development of the organisms involved, the specific ecosystem in which they live, and the way combinations of factors interact. In most cases, density-dependent and density-independent factors probably exert simultaneous influences. Depending on whether regulatory factors are regular and predictable or irregular and unpredictable, species will develop different strategies for coping with them.

Density-Independent Factors

In general, the factors that affect natality or mortality independently of population density tend to be abiotic components of the ecosystem. Often weather (conditions at a particular time) or climate (average weather conditions over a longer period) are among the most important of these factors. Extreme cold or even moderate cold at the wrong time of year, high heat, drought, excess rain, severe storms, and geologic hazards—such as volcanic eruptions, landslides, and floods—can have devastating impacts on particular populations.

Abiotic factors can have beneficial effects as well, as anyone who has seen the desert bloom after a rainfall can attest. Fire is a powerful shaper of many biomes. Grasslands, savannahs, and some montane and boreal forests often are dominated—even created—by periodic fires. Some species, such as jack pine and Kirtland's warblers, are so adapted to periodic disturbances in the environment that they cannot survive without them.

In a sense, these density-independent factors don't really regulate population *per se*, since regulation implies a homeostatic feedback that increases or decreases as density fluctuates. By definition, these factors operate without regard to the number of organisms involved. They may have such a strong impact on a population, however, that they completely overwhelm the influence of any other factor and determine how many individuals make up a particular population at any given time.

Density-Dependent Factors

Density-dependent mechanisms tend to reduce population size by decreasing natality or increasing mortality as the population size increases. Most of them are the results of interactions between populations of a community (especially predation), but some of them are based on interactions within a population.

Interspecific Interactions

As we discussed in chapter 3, a predator feeds on—and usually kills—its prey species. While the relationship is one-sided with respect to a particular pair of organisms, the prey species as a whole may benefit from the predation. For instance, the moose that gets eaten by wolves doesn't benefit personally, but the moose *population* is strengthened because the wolves tend to kill old or sick members of the herd. Their predation helps prevent population overshoot, so the remaining moose are stronger and healthier.

Sometimes predator and prey populations oscillate in a sort of synchrony with each other as is shown in figure 6.11. This is a classic study of the number of furs brought into Hudson's Bay Company trading posts in Canada between 1840 and 1930. As you can see, the numbers of Canada lynx fluctuate on about a ten-year cycle that is similar to, but slightly out of phase with, the population peaks of snowshoe hares. When the hare population is high and food is plentiful, lynx reproduction is very successful and lynx populations grow rapidly. Predation diminishes the number of hares available until eventually lynx do not have enough to eat and their population sizes fall. As the predators die off, the prey begin to make a comeback and the cycle starts over again.

Parasitism is a form of predation in which the parasite draws nutrients or other benefits at the expense of its host but usually does not kill it. The mosquito or tick that helps itself to a blood meal from your epidermis is a good example. Often, a parasite and its host develop a sort of balance with each other, something like the predator/prey relationship we have just discussed. If the parasite is too virulent and kills its host, food for its offspring will be unavailable. Wherever a parasite and host combination have

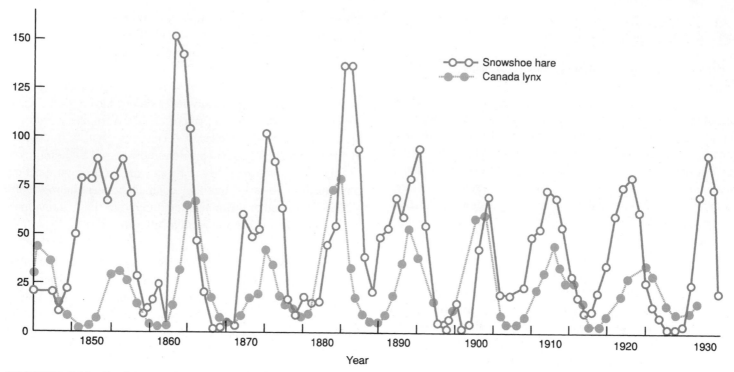

FIGURE 6.11 Oscillations in the populations of snowshoe hare and lynx in Canada suggest the close interdependency of this prey-predator relationship. This data is based on the number of pelts received by the Hudson Bay Company. Both predator and prey show a ten-year cycle in population growth and decline.

Source: Data from D. A. MacLulich, *Fluctuations in the Number of the Varying Hare (Lepus Americus)*. Toronto: University of Toronto Press, 1937, reprinted 1974.

lived together for a long time, the host is likely to have developed some resistance to its parasite so that enough individuals survive to continue the population. If parasites are introduced to new ecosystems, however, or new host species are introduced into the parasite's ecosystem, the effects can be disastrous. On the plains of Africa, for instance, small biting flies called tsetse flies carry the trypanosome that causes sleeping sickness. The native species of hoofed mammals are mostly resistant to this protozoan parasite, but domestic cattle, which have not been part of the community long enough to adapt, may be killed or become seriously debilitated if they are exposed.

In interspecific competition, two species compete for the same environmental resources in an ecosystem, but this competition rarely involves outright struggles or confrontations. It more often involves a contest in which the species tend to eat faster or grow faster or grab a bigger share of the resources in some other way. As a result, competing populations may work out a balance of resource partitioning (chapter 4), or one population may eclipse the other.

Not all interspecific interactions are harmful to one of the species involved. Mutualism and commensalism, for instance, are interspecific interactions that are beneficial or neutral in terms of population growth (chapter 4).

Intraspecific Interactions

Individuals within a population also compete for resources. When population density is low, resources are likely to be plentiful and the population growth rate will approach the maximum possible for the species, assuming that individuals are not so dispersed that they cannot find mates. As population density approaches the carrying capacity of the environment, however, one or more of the vital resources becomes limiting. The stronger, quicker, more aggressive, more clever, or luckier members get a larger share, while others get less and then are unable to reproduce successfully or survive.

Territoriality is a principal way many animal species control access to environmental resources. The individual, pair, or group that holds the territory will drive off rivals if possible, either by threats, displays of superior features (colors, size, dancing ability), or fighting equipment (teeth, claws, horns, antlers). Members of the opposite sex are attracted to individuals that are able to seize and defend the largest share of the resources. From a selective point of view, these successful individuals presumably represent superior members of the population and the ones best able to produce offspring that will survive.

Stress and Crowding

Stress and crowding also are density-dependent population control factors. When population densities get very high, organisms often exhibit symptoms of what is called **stress shock** or **stress-related diseases.** These terms describe a loose set of physical, psychological, and/or behavioral changes that are thought to result from the stress of too much competition and too close proximity to other members of the same species. There is a considerable controversy about what causes such changes and how important it is in regulating natural populations. The strange behavior and high mortality of arctic lemmings or snowshoe hares during periods of high population density may be a manifestation of stress shock. On the other hand, they could simply be the result of malnutrition, infectious disease, or some other more mundane mechanism at work.

Some of the best evidence for the existence of stress-related disease comes from experiments in which laboratory animals, usually rats or mice, are grown in very high densities with plenty of food and water, but very little living space (table 6.4). A variety of symptoms are reported, including reduced fertility, low resistance to infectious diseases, and pathological behavior, such as hypoactivity, hyperactivity, aggression, lack of parental instincts, sexual deviance, and cannibalism. Dominant animals seem to be affected least by crowding, while subordinate animals—the ones presumably subjected to the most stress in intraspecific interactions—seem to be the most severely affected.

The clinical symptoms reported in these studies include enlargement of the adrenal glands, reduction of the thymus and reproductive glands, and deterioration of the heart, blood vessels, kidney, and liver. The most probable explanation for these changes is that the adrenal gland is the source of epinephrine (adrenalin), the "fight or flight" hormone that raises energy levels and stimulates an animal to respond to danger or stress. Ordinarily, the burst of epinephrine released by fear, anger, or excitement is short-lived. When the stimulation is very frequent, however, the adrenal glands become enlarged in response to the continual call for epinephrine, and hormone production is elevated to a constant high level. This puts stress on the heart and circulatory system. In addition to epinephrine, the adrenal glands secrete sex hormones that affect metabolism and behavior. They also secrete corticosteroids that regulate mineral levels, energy metabolism, and the immune systems. Abnormally high levels of these hormones damage the liver and kidneys and can lead to a variety of symptoms including hypertension, arthritis, arteriosclerosis, gastrointestinal ulcers, and reduced resistance to disease. All these factors will tend to increase mortality and decrease fertility.

Are there equivalent stress-related diseases that cause aberrant behavior in *human* societies? Some people suggest that the high concentrations of social problems in dense urban centers might result from crowding stresses similar to those observed in laboratory animals. They point out that inner cities have much higher rates of disease and social problems, such as mental illness, family dysfunction, alcohol and drug abuse, and criminal behavior, than do other areas of the country. The inner city also has much higher levels of poverty, neglect, undereducation, unemployment, and other social conditions, however, that may be more important than the population density problems found there. While it is tempting to extrapolate conclusions from laboratory experiments to natural populations where a complex of factors are interacting, the conclusions can be misleading. In chapter 26, we discuss the socioeconomic disparities in our urban centers and their effects on human populations.

TABLE 6.4 The influence of density on fecundity in the house mouse (*Mus musculus*)

	Sparse	Medium	Dense	Very dense
Average number/m³	34	118	350	1600
Average percentage pregnant	58.3	49.4	51.0	43.4
Average number per litter	6.2	5.7	5.6	5.1

From C. Southwick, "Population Characteristics of House Mice Living in English Corn Ricks: Density Relationships" in *Proc. Zool. Soc. Lond.* 131:163–175, 1958. Copyright © 1958 The Zoological Society of London, London, England. Reprinted by permission.

SUMMARY

Population dynamics play an important role in determining how ecosystems work. Biological organisms generally have the ability to produce enough offspring so populations can grow rapidly when resources are available. Given optimum conditions, populations of many organisms can grow exponentially, that is, they grow at a constant rate of increase so that the population size doubles in some regular interval of time. If conditions appropriate for this kind of growth persist and other factors don't intervene to reduce abundance, exponential growth can produce astronomical numbers of organisms. We describe the rapidly rising curve of an exponentially growing population as a **J** curve.

Some populations will grow exponentially until they overshoot the carrying capacity of the environment. Mortality rates rise as resources become limited and the population may crash, often dying back as rapidly as it rose. Other species have intrinsic mechanisms that regulate their population growth rate so that they reach an equilibrium at or near the carrying capacity of the environment.

The most important components of population dynamics are natality, fertility, fecundity, life span, longevity, mortality, immigration, and emigration. The sum of all additions to and subtractions from the population determines the net rate of growth. Mortality rates and longevity are often expressed as survivorship rates that reveal much about a species' place in its ecosystem and the kinds of hazards that eliminate members from the population.

The factors that regulate population dynamics can be either intrinsic to the population or extrinsic. They can be caused by biotic or abiotic forces, and they can act on the population in either a density-dependent or density-independent fashion. The most important abiotic regulatory factors are usually climate and weather. The most important biological factors are usually competition (both interspecific and intraspecific), predation, and disease.

Often, organisms develop specific behavioral patterns to reduce conflict and facilitate resource partitioning. In some cases of extreme crowding, it is thought that physiological responses to excessive intraspecific interaction can result in stress-related disease that can lead to aberrant behavior, reduced reproductive success, increased susceptibility to disease, and high rates of mortality. Whether these mechanisms operate in humans at high population densities is a matter of controversy.

■ Review Questions

1. Why did moose populations grow so rapidly on Isle Royale in the 1920s?
2. What is the difference between exponential and arithmetic growth?
3. Given a growth rate of 3 percent per year, how long will it take for a population of 100,000 individuals to double? How long will it take to double when the population reaches 10 million?
4. What is environmental resistance? How does it affect populations?
5. What is the difference between fertility and fecundity?
6. Describe four different types of survivorship patterns and explain what they show about the role of the species in an ecosystem.
7. What are the main interspecific population regulatory interactions? How do they work?
8. What are the suspected causes and effects of stress shock or stress-related disease?

■ Questions for Critical Thinking

1. What is the advantage to organisms in having the potential for exponential growth?
2. What are the disadvantages of exponential growth?
3. Are humans a Malthusian or a logistic species? What are the implications for population control suggested by these strategies?
4. Why are traits that encourage immigration or emigration likely to be beneficial to a species?
5. What is the difference between life span and longevity? Why are there such great differences between life span and longevity within and between species?
6. Why are abiotic factors likely to be density independent in regulating populations, while biotic factors are more often density dependent?
7. Compare and contrast the reproductive strategies that would be most advantageous for populations that are regulated primarily by density-dependent versus density-independent factors.
8. Why do we say that infectious diseases are really forms of predation or parasitism?
9. How might behavior help a population to partition resources in the most efficient manner for perpetuation of the species?
10. What factors in human societies might create stress shock or stress-related disease?
11. What factors in human societies might mitigate stress shock in high population densities?
12. What implications for human population control might we draw from our knowledge of basic biological population dynamics?

■ Key Terms

abiotic (p. 93)
arithmetic growth (p. 93)
biotic (p. 99)
biotic potential (p. 94)
carrying capacity (p. 92)
dieback (p. 95)
emigration (p. 98)
environmental resistance (p. 95)
exponential growth (p. 93)
fecundity (p. 97)
fertility (p. 97)
geometric growth (p. 93)
irruptive or Malthusian growth (p. 95)
J curve (p. 94)

life expectancy (p. 97)
life span (p. 97)
logistic growth (p. 95)
mortality (p. 97)
natality (p. 96)
overshoot (p. 95)
population crash (p. 91)
population explosion (p. 91)
population momentum (p. 98)
S curve (p. 95)
stress-related diseases (p. 101)
stress shock (p. 101)
survivorship (p. 97)

SUGGESTED READINGS

■ Begon, M., and M. Mortimer. *Population Ecology: A Unified Study of Animals and Plants*. London: Blackwell Scientific Publications, 1981. A clear, well-illustrated basic text.

■ Berryman, A. *Population Systems*. New York: Plenum Publishing Corp. 1981. A general introduction to dynamics, controls, and analysis of population systems.

■ Christian, J. "Endocrine adaptive mechanisms and the physiologic regulation of population growth." In *Physiological Mammalogy*, edited by W. Mayer and R. Van Gelder, vol. 1. New York: Academic Press, 1963. A good discussion of stress-related population effects.

■ Ehrlich, P., and J. Roughgarden. *The Science of Ecology*. New York: Macmillan Publishing Co., 1987. An excellent general ecology text with good coverage of population biology.

■ Hedrick, P. *Population Biology: The Evolution and Ecology of Populations*. Boston: Jones & Bartlett Publishers, 1984. A good general text.

■ Krebs, C. K. *The Message of Ecology*. New York: Harper & Row, 1988. An introduction to ecology with a strong emphasis on population biology.

■ Pearl, R., and L. Reed. 1945. "On the rate of growth of the population of the United States since 1790, and its mathematical representation." *PNAS*. 6:275. A classic use of statistical demographics.

■ Smith, R. *Elements of Ecology and Field Biology*. New York: Harper & Row, 1990, fourth edition. A classic textbook of ecology updated.

■ Wayne, R. K. et al. 1991. "Conservation Genetics of the Endangered Isle Royale Gray Wolf." *Conservation Biology* 5(1):41. Genetic evidence suggests that all the wolves on Isle Royale are descendents of a single female founder. Implications for management of this endangered population are discussed.

■ Williamson, M. *Island Populations*. New York: Oxford University Press, 1981. Population dynamics in the closed ecological systems of islands.

C H A P T E R 7

Human Populations

CONCEPTS

The world human population is in a **J** curve of exponential growth. In 1987, the population passed 5 billion, with a doubling time of only thirty-five years. Further population growth may be disastrous for humans and the other species with which we share the planet.

Industrialization in the developed countries has been accompanied by a demographic transition from high birth and death rates to low birth and death rates, resulting in a stable population size. Whether a similar transition will occur in the countries now developing remains to be seen.

Some people consider poverty and insecurity to be the underlying causes of excess population growth. They argue that a more equitable distribution of resources is essential to stabilize populations and protect the environment. When infant mortality drops and economic and political development progresses so that people can be sure of a secure future, they will have fewer children. Other people believe inequitable distribution of resources is inevitable and that only by reducing the number of people trying to share limited resources can we increase the share available for each person.

Many more options for birth control are available now than in the past, but successful family planning requires major changes in society and attitudes. There may be other benefits from these changes in addition to stable populations.

INTRODUCTION

Human populations have grown at remarkable, and many people think frightening, rates over the past three hundred years. In 1650, nearly 500 million people lived on Earth. By 1990, the world population reached 5.3 billion (figure 7.1), and it is doubling about every thirty-nine years. We are now adding about 140 million people per year to the population. From our understanding of natural population patterns, we have reason to fear that this population explosion, unless checked immediately, will bring disaster of an unknown scale. Robert McNamara, former president of the World Bank said, "Rampant population growth more certainly threatens humanity than has any catastrophe the world has yet endured." Many people have called for immediate, drastic birth control programs to reduce the rate of population growth and eventually stabilize the total number of humans on Earth so that our population curve will assume a logistic pattern based on the carrying capacity of the environment.

In contrast, others believe that expanding populations can give rise to technological change and economic development that make life better for everyone. They argue that humans will not destroy life-support systems on which we all depend because our ingenuity and effort can expand the carrying capacity of the environment, at least in regard to human populations. According to this view, further population growth is the only way to provide resources that will allow all people in the world to raise their standard of living to a modern level.

Should we be concerned, even alarmed, by our current population curve? Once again, there are differing attitudes. Many people believe overpopulation is the root cause of crowding, poverty, violence, and environmental degradation. In this view, too many people are trying to share limited resources, and these resources are being used in a way that reduces their availability to future generations. Only by reducing the number of people trying to share limited resources can we protect the environment and achieve social goals. Others argue that there are enough global resources for everyone but greed, waste, oppression, and mismanagement deny people access to resources they need. The issue is equitable distribution rather than resource availability. In this view, poverty and insecurity created by unfair social, political, and economic systems are responsible for crowding, scarcity, and violence. Furthermore, they may themselves contribute to excess population growth. Concerned proponents of both views agree, however, that the only permanent solution to resource scarcity and distribution problems is to reform human attitudes and human systems.

Whether human populations will continue to grow at present rates and what that would mean for environmental quality and the quality of human life are among the most central and pressing questions in environmental science. In this chapter, we will look at some causes of human population growth, as well as how populations are measured and described. We will examine

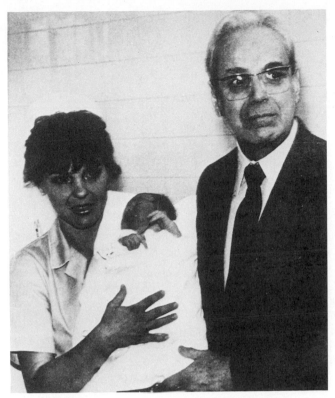

FIGURE 7.1   The world population grows one baby at a time. Young Matej Gaspar, pictured above, was declared the 5 billionth person in the world when he was born in Zagreb, Yugoslavia, on July 11, 1987. It took all of history up to about the year 1800 to build the population to 1 billion. The last billion took about seventeen years.

factors that encourage growth and forces that restrict it. We also will look at some opinions of what the optimum human population is and how we might reach that level. Family planning and birth control are essential for stabilizing populations. The number of children that a couple decides to have and the methods they use to regulate fertility, however, are strongly influenced by culture, religion, politics, and economics, as well as basic biological and medical considerations. We will look at how some of these factors impinge on human demographics.

HUMAN POPULATION HISTORY

Throughout most of history, human populations have been relatively small and slow-growing. Studies of hunting and gathering societies indicate that the total world population was probably only a few million people before the invention of agriculture and the domestication of animals around ten thousand years ago. The larger and more secure food supply made available by the agricultural revolution allowed the human population to grow, reaching about 50 million people by 5000 B.C. For thousands of years, the number of humans continued to increase very slowly. Archaeological evidence and historical descriptions suggest that only about 300 million people were living at the time of Christ.

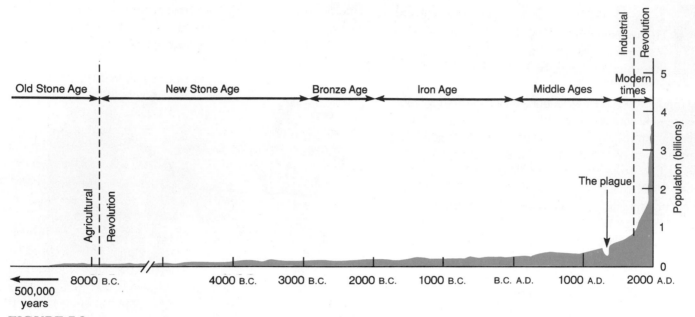

FIGURE 7.2 Human population levels through history. Since about A.D. 1000, our population curve has assumed a J shape. Are we on the upward slope of a population overshoot; or will we be able to adjust our population growth to an S curve; or can we just continue the present trend indefinately?

TABLE 7.1 World population growth and doubling times

Date	Population	Doubling time
5000 B.C.	50 million	?
800 B.C.	100 million	4,200 years
200 B.C.	200 million	600 years
1200 A.D.	400 million	1,400 years
1700 A.D.	800 million	500 years
1900 A.D.	1,600 million	200 years
1965 A.D.	3,200 million	65 years
1990 A.D.	5,300 million	38 years
2020 A.D. (estimate)	8,230 million	55 years

Population Reference Bureau, Inc., Washington, D.C.

Population checks in the form of plagues, famines, and wars kept human numbers from expanding rapidly during the next millennium and a half. Among the most destructive of these calamities were the bubonic plagues that periodically swept across Europe between 1348 and 1650. During the worst plague years, between 1348 and 1350, it is estimated that 25 percent of the European population perished. In 1650, at the end of the last great plague, there were only about 600 million people in the world.

As figure 7.2 shows, the number of humans began to increase rapidly after 1600 A.D. Many factors contributed to this rapid growth. Increased sailing and navigating skills stimulated commerce and communication between nations. Agricultural developments, better sources of power, and better health care and hygiene also played a role. Together, all these advances can be considered part of the scientific and technological revolution brought about by the Renaissance and the Age of Reason.

Clearly our population entered an exponentially increasing pattern of growth—a J curve—at about this time, and the doubling times diminished correspondingly (table 7.1). Recall from chapter 6 that a population that grows by 2 percent per year doubles in 35 years.

LIMITS TO GROWTH: SOME OPPOSING VIEWS

Few people believe that human populations can continue to grow at current rates for very long into the future. Clearly, the physical world has finite limits that must restrict growth sooner or later. The question is whether we already are approaching that limit. And if we are—or when we do—what will happen to the habitability of our environment? Furthermore, what factors will bring our numbers into balance with resources? Will we recognize when we are approaching the limits of our world and control growth ourselves, or will we be controlled by external forces?

Biological Models

As chapter 6 explains, many species are adapted to grow rapidly when resources are available. These species depend on predators or other external regulatory agents to prevent them from overwhelming the carrying capacity of the environment. Unfettered growth can produce astronomical population densities that quickly use up resources and result in catastrophic diebacks. Many scientists see a similarity in the present exponential growth of human populations and the growth phases of irruptive "boom and bust" cycles that result from out-of-control growth of some other organisms.

Not all populations grow until they exhaust their resources, however. Some species have homeostatic mechanisms that are sensitive to environmental resistance. These organisms follow an **S** curve that brings them into a stable equilibrium with the environment. The question is, "Which type of organism are we? Are we headed toward a stable future or toward disaster?"

Malthusian Checks on Population

Most discussions of human population growth begin with *An Essay on the Principle of Population as It Affects the Future Improvement of Society, with Remarks on the Speculations of Mr. Godwin, M. Condorcet, and Other Writers*, by Rev. Thomas Malthus. He wrote this essay in 1798 to refute utopian points of view stimulated by the French Revolution a decade earlier. The essence of his argument was that ". . . the power of population is indefinitely greater than the power in the Earth to produce subsistence for man. Population when unchecked, increases in a geometrical ratio. Subsistence increases only in an arithmetical ratio."

Malthus concluded that achieving population stability under these circumstances implies either a "positive check," which is any factor that contributes to shorten the natural duration of life, or a "preventative check," including all those factors that prevent human birth. Among the preventative checks, he suggested "moral restraint," including later marriage and the practice of celibacy until a couple can afford to support children.

If Malthus's views of the consequences of geometric population growth were dismal, the corollary he drew was even more bleak. He feared that measures to feed and assist the poor would increase their fertility and thereby perpetuate the problem of starvation and misery. Not surprisingly, his essay provoked the great social and economic debate of his generation. Karl Marx was one of his strongest opponents. The debate had a great impact on public policy and on the formulation of ideas by later demographers, philosophers, and scientists, including Charles Darwin, who drew his ideas about the struggle for scarce resources and survival of the fittest from Malthus's essay.

Malthus's theories about checks and balances on human populations remain controversial because they have become associated with social, political, and economic philosophies that many people consider elitist and inhumane. To this day, "Malthusian" sometimes is used as a term to imply an insensitive and exploitative attitude toward human poverty and misery.

On the other hand, his basic premises may be sound. Stripping away the overtones of social and class privilege common in his day, we find that his basic articulation of population regulation by resource-driven checks and balances has validity in natural populations. Most biologists and many social scientists believe that human populations are also subject to the same principles. This current point of view is often called **neo-Malthusian.**

Can Technology Make the World More Habitable?

An alternative to Malthus's pessimistic predictions is the view that technology and human inventiveness can expand the available resources of the natural environment. Figure 7.3 shows his-

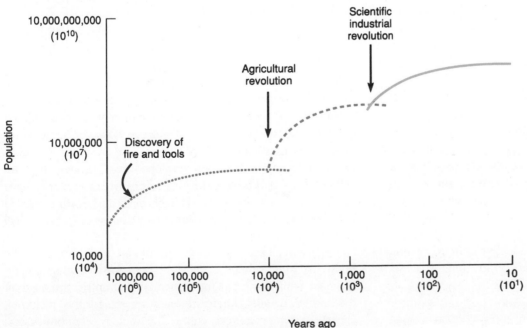

FIGURE 7.3 Human population plotted on a log-log scale. Plotted this way, population growth is seen as occurring in three bursts associated with revolutions in culture, agriculture, science, and technology.

torical human population growth plotted on a logarithmic scale. Seen this way, the data show several episodes in which technological progress has made it possible to support larger populations. The first of these, about a million years ago, was the discovery of fire and the invention of tools that enabled our ancestors to be more effective predators. The second population increase corresponds to the domestication of plants and animals about ten thousand years ago.

The third burst of growth, of which we are a part, was stimulated by the scientific and industrial revolution. Progress in agricultural productivity, engineering, information technology, commerce, medicine, sanitation, and other achievements of modern life have made it possible to support approximately one thousand times as many people per unit area as was possible ten thousand years ago.

Whether we can sustain this pace of technological development remains to be seen. Much of our progress in the past three hundred years has been based on availability of easily acquired natural resources and—especially—cheap, abundant fossil fuels. Whether we can develop substitute, renewable energy sources in time to avert disaster when current fossil fuels run out is a matter of great concern (chapters 18–20). Furthermore, our tradition has been mainly exploitation of resources rather than management of them. Many biologists fear that the basic life-supporting systems of the biosphere are becoming so threatened by overuse and global pollution that they may fundamentally limit the number of humans that can inhabit Earth.

Can More People Be Beneficial?

There can be benefits as well as disadvantages in growing populations. More people mean larger markets, more workers, and the efficiencies of scale in mass production of goods. More people mean more intelligence and enterprise to overcome problems such as underdevelopment, pollution, and resource limitations. In a sense, human ingenuity and intelligence can create new resources through substitution of new materials or new ways of doing things for old materials and old ways. For instance, utility companies are finding it cheaper and more environmentally sound to pay for insulation and energy-efficient appliances for their customers rather than building new power plants. The effect of saving energy that was formerly wasted is comparable to creating a new fuel supply. Similarly, making cars with half as much iron and steel as we now use has the same effect as doubling our iron ore supply. In a growing population there are more scientists, inventors, and workers to bring about these changes, and more customers whose demands for goods and services stimulate the introduction of new technology. We will return to the question of what constitutes a resource and which resources are most likely to limit further growth of human populations in subsequent chapters.

Many economists now consider human resources more important than natural resources in determining a nation's future.

Materials play a less important role in modern economies, while information, communication, and capital play an increasingly important role. An educated, hard-working, experienced labor force is a form of human capital that is an essential ingredient in this equation. Unfortunately, most countries that are experiencing rapid population growth have not yet entered the industrial age, let alone the "information age" of computers, telecommunications, robots, and international trade. It may be, however, that young people at the turn of the century will have more energy, newer ideas, and better ability to adapt to a changing world than their parents. They may be the key to moving their countries into the modern world.

Economist Julian Simon represents the optimistic view of population growth, arguing that people are the "ultimate resource" and that there is no evidence that pollution, crime, unemployment, crowding, the loss of species, or any other resource limitations will worsen with population growth. This view is, of course, controversial. Most biologists argue that human damage to the environment ultimately will limit the world's carrying capacity.

HUMAN DEMOGRAPHY

Demography is derived from the Greek word *demos* (people) + *graphos* (to write or to measure). It encompasses vital statistics about people, such as births, deaths, and marriages. In this section, we will survey ways human populations are measured and described, and discuss demographic factors that contribute to population growth.

More and Less Developed Countries

Demographers usually divide the world into two groups: the more developed countries (MDC) and the less developed countries (LDC). The **more developed countries** (North America, Europe, Japan, Australia, New Zealand, and the Soviet Union) generally have (1) high per capita incomes, (2) low birth and death rates, (3) low population growth rates, and (4) high levels of industrialization and urbanization. Their technological and economic power gives them a strong voice in world affairs. An income of $3,000 (U.S.) per capita per year is considered the threshold for inclusion in this category; however, some countries—such as Poland and Hungary—are, by virtue of geography or history, considered developed even though they are below this level. The **less developed countries** generally have (1) low per capita incomes, (2) high birth and death rates, (3) high population growth rates, and (4) low levels of technological development (table 7.2).

Almost all less developed countries are in the tropics or the Southern Hemisphere in Asia, Africa, Latin America, and Oceania, while almost all more developed countries are north of the thirtieth parallel. There are many reasons for this, including climate, natural resources, culture, technology, economics, and

TABLE 7.2	Demographic conditions in more developed and less developed countries in 1990	
Characteristic	Less developed	More developed
Average per capita GNP	710	15,830
Crude birth rate (per 1,000)	31	15
Crude death rate (per 1,000)	10	9
Natural increase (percent)	2.1	0.5
Infant mortality (per 1,000)	81	16
Total fertility rate (per 1,000)	4.0	2.0
Percent urban	32	73
Current population (millions)	4,107	1,214
Projected population (millions)		
By the year 2000	5,018	1,274
By the year 2020	6,878	1,350

Source: Population Reference Bureau, Inc., Washington, DC, April 1990.

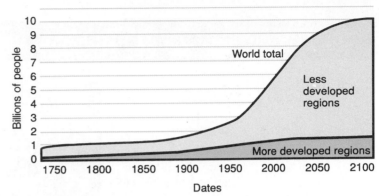

FIGURE 7.4 Human population growth, 1750–2100, in less developed and more developed regions. The human population is now in an exponential growth phase. More than 90 percent of all growth in this century and projected for the next is in the less developed countries. Total population is expected to stabilize at about 10 billion in the next century.

politics. All of these factors complicate any consideration of human population growth. The less developed countries have about 80 percent of the world's population, but constitute only about 20 percent of world trade and commercial economic activity. As figure 7.4 shows, 95 percent of all population growth in the next century—the addition of about three billion more people—will occur in the less developed world.

Another way to describe countries is by their economic system. The **First World** is composed of the capitalist, free-market countries of North America, western Europe, Australia, New Zealand, and Japan. The **Second World** consists of the socialist, centrally-planned countries such as the Soviet Union and its allies in eastern Europe. The **Third World** is made up of all the nonaligned countries whose economies are neither purely capitalist nor socialist. Most of the less developed countries fall in the Third World category.

Some demographers feel that to describe the less-developed countries as Third World sounds derogatory, as if suggesting they are third rate. Others counter that Third World is a description generated by the nonaligned nations themselves to show that they are independent of both the major power centers. Some people feel that to call countries less developed or underdeveloped is belittling. These countries may feel that they are culturally, religiously, socially, or even economically more developed than we are. This is a difficult argument to resolve. We will use both "less developed" and "Third World" in this book and hope that we don't offend anyone or suggest any kind of value judgment.

How Many of Us Are There?

Even in this age of information technology and communication, we do not know exactly how many people there are in the world.

As late as 1970, about 200 million people lived in countries in which a census had never been taken. Some countries still have such confusing conditions that estimates from different sources vary by as much as 50 percent.

The problems of recording the world population are illustrated by the 1982 census of China. It required nearly six million census takers and 100,000 clerks to enumerate the enormous population. It took more than a year to record all the data, during which time at least 30 million babies were born and nearly that many people died. The reliability of the count was probably no better than plus or minus 10 million people, and may have been considerably worse. This was in a country with a very strong central government and a relatively uniform and orderly people who were not at war or suffering exceptional upheavals at the time. Imagine how much more uncertain the data are from Africa, Central America, or the Middle East, where conditions are much less stable and more governments are involved!

Although there is considerable uncertainty about the exact number of people in some places, most demographers agree that about 5.3 billion people live in the world. Table 7.3 shows the population of the ten most populous nations. Together, these countries have about 3.3 billion inhabitants, or nearly two-thirds of the world's population. Only three of the top ten nations (the United States, the Soviet Union, and Japan) are among the more developed countries of the world. China, now the most populous nation with 20 percent of the world's population, is expected to be surpassed by India in the next fifty years because of different success rates of population control programs in the two countries (Box 7.1).

Figure 7.5 shows the density of human populations around the world. Notice the high densities supported by fertile river valleys of the Nile, Ganges, Yellow, Yangtze, and Rhine Rivers and the well-watered coastal plains of India, China, and Europe.

BOX 7.1   China's One-Child Family Program

It's difficult to comprehend the number of people in China. In 1990, the population was estimated to be 1,130 million, nearly one-fifth of all the people in the world. The 22 million Chinese babies born each year are nearly equal to the combined metropolitan populations of New York, Los Angeles, *and* Chicago. More than four-fifths of China's population are rural peasants. The per capita income is about $330 per year, placing it among the less developed countries in the world, and yet it appears to be achieving a demographic transition to a stable population under conditions that experts had predicted were impossible (box figure 7.1).

For many centuries, China followed a repeating cycle of disasters, famines, and political upheavals. It was widely believed that China was so large and unmanageable that this cycle could never be broken. In the 1950s, as a result of the misguided "Great Leap Forward Program," some 20 million people died of starvation. Some experts predicted that China would never be able to feed itself. It now provides an adequate, if spartan, nutritional level for all of its people, however. Medical care, housing, education, and social security have also improved markedly. Between 1950 and 1980, the death rate dropped from twenty to eight per thousand, and average life expectancy increased from forty-seven to seventy years. By comparison, the United States has a death rate of nine per thousand and a life expectancy of seventy-four years.

Profamily traditions and public policies caused explosive population growth following the Socialist Revolution especially in the postfamine recovery in the early 1960s (box figure 7.2). Former Chairman, Mao Zedong believed, as do many Marxists, that labor is the source of all wealth. He proclaimed that "revolution plus production will solve all problems." After Mao's death in 1976, however, new leaders reversed those policies and sought to bring population growth under control. Premier Deng

BOX FIGURE 7.1   China's one child per family policy has resulted in a dramatic decline in birth rates. Many people object, however, to the means employed to enforce this policy. Others claim it has created a generation of "Little Emperors," spoiled, single children.

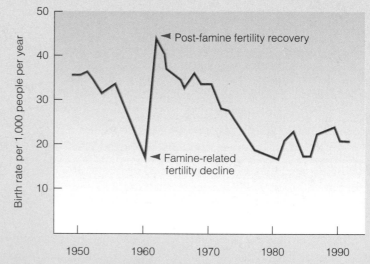

BOX FIGURE 7.2   China's crude birth rate, 1950–1986.

Xiaoping saw rampant population growth as the main obstacle to an improved standard of living. He also worried that modernization of agriculture would make hand labor obsolete, displacing as much as half of the rural population and creating a monumental unemployment problem.

Chinese demographers believe that scarcity of resources (only 11 percent of the land is arable) will limit the population that China can support to a maximum of 1.2 billion people, or only about 100 million more than the present population. Because of the high proportion of young people in the society now, China's population has an internal momentum that will carry it well past the target of 1.2 billion in the year 2020 unless the average number of children per couple immediately falls below the replacement level of 2.2. For the past decade, therefore, China has carried on an extensive campaign to persuade couples to have only one child. Although there have been problems, the program has made dramatic progress. Between 1968 and 1988, the crude birth rate fell from about forty per thousand to twenty per thousand. The Chinese birth rate is now only a little higher than that of the United States, which has more than fifty times the per capita income of China. In Shanghai, Beijing, and other urban centers, 80 to 90 percent of all births are first children.

The success of this campaign undoubtedly lies in the unique social organization in China. Communes and production teams are responsible for jobs, health care, and political organization. Officials in these units pursue the national birth control goals through a combination of social pressure, incentives, and disincentives that many of us would find repressive. Newly married couples are asked to pledge that they will have only one child. Birth control specialists in the production team monitor all married women, and couples must be granted permission to be part of the quota of births allowed in any year. Free contraceptives and medical advice are provided. All methods of birth control are used, including pills, condoms, and abortion, but major reliance is placed on the intrauterine device (IUD), in spite of serious questions about its health effects.

Couples who sign the one-child pledge are guaranteed free delivery of that child when they are granted leave to give birth. The child receives preference in education and job placement as it grows up. The parents get better housing, longer vacations, and an extra month's pay each year. Because of a traditional preference for boys to carry on the family name and a tendency of couples to want to try again if they have a girl, special privileges are accorded to parents of a girl. Penalties for unsanctioned pregnancies include official reprimands, pay cuts, and public censure. Because each cadre is anxious to make a good showing in meeting its birth quotas, zealous local officials may use coercive techniques to demand abortions in cases of unapproved pregnancies and to control the behavior of individuals.

In contrast to most countries where family planning is only effective in urban upper- or middle-class families with a high level of wealth and education, birth control in China is nearly as effective among the masses as in the more affluent classes. There has been resistance to this plan, however, in rural areas and among ethnic minorities. Allowing peasant families to farm private plots has created a demand for large families to work the fields. In many areas couples have been permitted to have a second or even third child.

In spite of these setbacks, however, the relative success of China's one-child program gives us encouraging evidence that a demographic transition can occur without concomitant industrialization and a shift to a high standard of living. If other developing countries can achieve similar results, the human population might stabilize much sooner, and at much lower levels than demographers have previously thought possible.

TABLE 7.3	The world's ten most populous nations in 1990 with prospects for 2020	
Country	Estimated 1990 population (in millions)	Projected 2020 population (in millions)
China	1,130	1,496
India	853	1,375
Soviet Union	291	355
United States	251	294
Indonesia	189	287
Brazil	150	234
Japan	124	124
Nigeria	119	273
Bangladesh	115	201
Pakistan	115	231

Source: Population Reference Bureau, Inc., Washington, DC, April 1990.

Natality and Birth Rates

The most accessible demographic statistic of fertility is usually the **crude birth rate,** the number of births in a year divided by the midyear population. It is usually expressed as crude birth rate per thousand persons, and is statistically "crude" in the sense that it is not adjusted for population characteristics. The **general fertility rate** is a more accurate representation because it reflects age structure and fecundity of the population. It is expressed as the crude birth rate multiplied by the percentage of fecund women (between fifteen and forty-four years of age) multiplied by one thousand. Fertility rate is based on the number of women in a population because the number of potential births is determined by the number of women of reproductive age but is relatively independent of the number of men. The **total fertility rate** is the number of children born to an average woman in a population during her entire reproductive life.

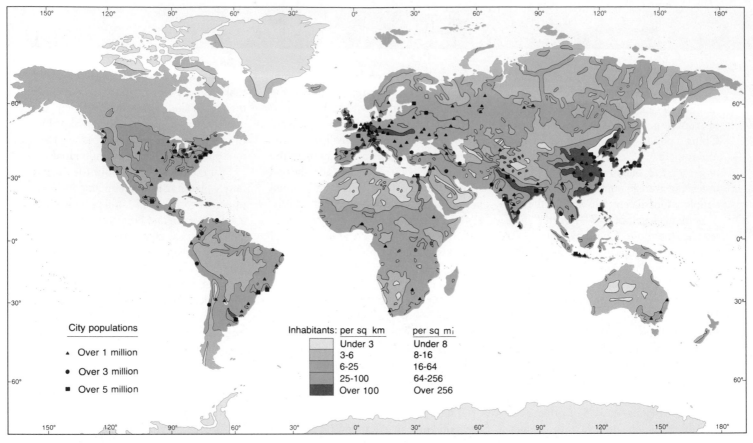

FIGURE 7.5 World population distribution. Highest population densities are found in the fertile river valleys and well-watered coastal plains of temperate zones.

City populations

▲ Over 1 million

● Over 3 million

■ Over 5 million

Inhabitants:	per sq km	per sq mi
	Under 3	Under 8
	3-6	8-16
	6-25	16-64
	25-100	64-256
	Over 100	Over 256

In analyzing population growth, we usually express birth rate in terms of the number of children per couple. The **zero population growth** (ZPG) rate (also called the *replacement* level of fertility) is the number of births at which people are just replacing themselves. In the more highly developed countries, where infant mortality rates are low, this rate is usually about 2.2 children per couple. If this is a "replacement" rate, why is it more than two? It is higher than two because the extra fraction also includes people who are infertile, have children who do not survive, or choose not to have children. In the less developed countries, the replacement birth rate is often six children per couple.

Tables 7.4 and 7.5 show the crude birth rates, infant mortality rates, and total fertility rates for the ten fastest growing and slowest growing countries in the world. Note that the most rapidly growing countries are in the less developed world, whereas all of the slowest growing countries are in the developed world.

Mortality and Death Rates

A traveler to a foreign country once asked a local resident, "What's the death rate around here?" "Oh, the same as anywhere," was the reply, "about one per person." In demo-

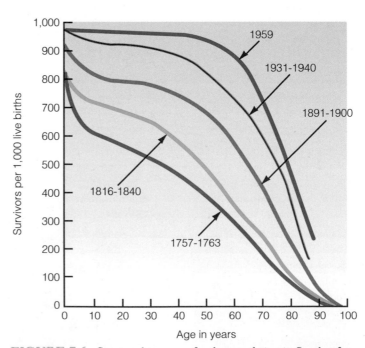

FIGURE 7.6 Survivorship curves for the population in Sweden from 1757 through 1959.

TABLE 7.4 The ten most rapidly growing countries in the world

Country	Crude birth rate (per 1,000)	Crude death rate (per 1,000)	Natural increase (percent)	Total fertility rate (per woman)	Infant mortality rate (per 1,000)	Per capita GNP (U.S. $)
Gaza	50	7	4.3	7.0	55	n.a.
Iraq	46	7	3.9	7.3	67	n.a.
Syria	45	7	3.8	6.8	48	1,670
Kenya	46	7	3.8	6.7	62	360
Ivory Coast	51	14	3.7	7.4	96	740
Maldives	46	9	3.7	6.6	76	410
Rwanda	51	16	3.7	8.5	122	290
Zambia	51	14	3.8	7.2	80	290
Tanzania	51	14	3.7	7.1	106	160
Uganda	52	17	3.6	7.4	107	280

Source: *1990 World Population Data Sheet* (Washington, D.C.: Population Reference Bureau, Inc., April 1990).

n.a. = not available

TABLE 7.5 The ten most slowly growing countries (1990 Figures)

Country	Crude birth rate (per 1,000)	Crude death rate (per 1,000)	Natural increase (percent)	Total fertility rate (per woman)	Infant mortality rate (per 1,000)	Per capita GNP (U.S. $)
Hungary	12	12	−0.2*	1.8	15.8	2,460
West Germany	11	11	−0.0*	1.4	7.5	18,530
Denmark	12	12	−0.0*	1.6	7.8	18,470
East Germany	13	13	0.0	1.7	8.1	n.a.
Austria	12	11	0.0	1.4	8.1	15,560
Bulgaria	13	12	0.1	2.0	13.5	n.a.
Italy	10	9	0.1	1.3	9.5	13,320
Belgium	12	11	0.2	1.6	9.2	14,550
Luxemborg	12	10	0.2	1.4	8.7	22,600
Sweden	14	11	0.2	2.0	5.8	19,150

Source: *1990 World Population Data Sheet* (Washington, D.C.: Population Reference Bureau, Inc., April 1990).

n.a. = not available

graphics, however, **crude death rates** (or crude mortality rates) are expressed in terms of the number of deaths per thousand persons in any given year. If people live longer, the number of deaths in any one thousand people per year will be lower. Crude death rates and longevity (the probability of any given individual staying alive from year to year) are thus inversely proportional. The number of deaths in a population is sensitive to the age structure of the population. Note in tables 7.4 and 7.5 that some rapidly growing, less developed countries have lower crude death rates than do the more developed, slowly growing countries. This is because there are proportionately more youths and fewer elderly people in a rapidly growing country than in a more slowly growing one.

Age-specific death rates are a measure of the mortality for specific age classes. This gives a clearer picture of the sur-

vivorship patterns, or the percentage of maximum life span achieved by members of a population. Figure 7.6 shows the survivorship patterns for Sweden between 1757 and 1959. The pattern in the eighteenth century shows a high early mortality rate and has some resemblance to the pattern for such animals as deer and rabbits (chapter 6). By the twentieth century, as is shown in the top curve in figure 7.6, most people in Sweden were living most of the maximum life span.

Life Span and Life Expectancy

Life span is the oldest age to which humans are known to survive. Although there are many claims in ancient literature of kings living for one thousand years or more, the oldest age that can be certified by written records is that of Pierre Jourbet, a French-Canadian who lived to be 114. The aging process is still

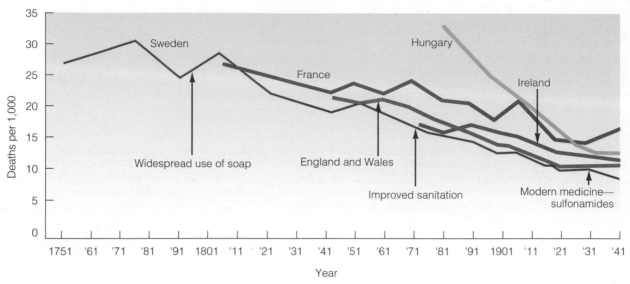

FIGURE 7.7   Mean annual death rates in England and Wales, Sweden, France, Ireland, and Hungary from 1751.

a medical mystery, but it appears that cells in our bodies have a limited ability to repair and reproduce themselves. At some point they simply wear out, and we fall victim to disease, degeneration, accidents, or senility. There are reports that some cells may be preprogrammed to expire after a certain number of reproductive cycles, but the evidence is not conclusive. Although the average life expectancy has nearly doubled in the past two centuries in the more highly developed countries, the maximum life span does not appear to have been changed at all by modern medicine.

Life expectancy, in human terms, is the average age that a newborn infant can expect to attain in any given society. It is another way of expressing the average age at death. For most of human history, we believe that the average life expectancy has been between thirty-five and forty years, and the crude death rate has been about thirty-five per one thousand people. This doesn't mean that no one lived past age forty, but rather that so many deaths at earlier ages (mostly early childhood) balanced out those who managed to live longer. It once was widely believed that differences in life span between ethnic groups were biological and therefore difficult to change. We now know that social and environmental, not biological, factors are responsible for most variations in mortality.

Declining mortality, not rising fertility, is the primary cause of most population growth in the past three hundred years. In Western Europe, crude death rates began falling in most countries in the late 1700s. The major factors in this decline were better food and sanitation. Figure 7.7 shows the declining death rates in Sweden, France, Ireland, England, Wales, and Hungary between 1750 and 1940. Note that records begin at different times, and that the rates of change are also different. Hungary, which started this transition later than other countries, also

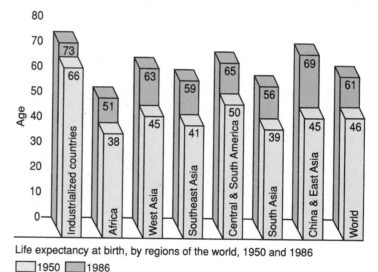

Life expectancy at birth, by regions of the world, 1950 and 1986
□ 1950   ■ 1986

FIGURE 7.8   Life expectancy at birth by regions of the world, 1950 and 1986.

dropped to a low death rate more rapidly than others as it benefited from already established life-lengthening factors.

Note also in figure 7.7 that most of the drop in death rates in these countries came long before the advent of modern medicine in the twentieth century. The advance in longevity has come largely from better food and better sanitation. Figure 7.8 shows the change in life expectancy at birth in major regions of the world between 1950 and 1986. The greatest gains have been in the developing countries, especially in China and East Asia, where life expectancies are approaching those of the developed world.

Figure 7.9 shows how life expectancy has changed during this century in the United States. Life expectancy for white men

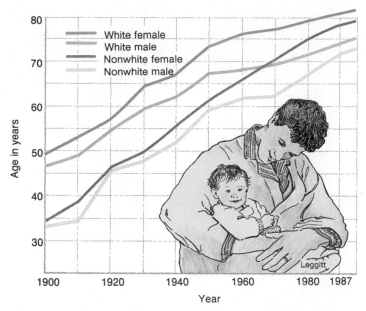

FIGURE 7.9 Life expectancy in the United States during the twentieth century.

and women in the United States has been increasing at about the same rate as in Western Europe, but nonwhites continue to have significantly lower life expectancies. Differences between the sexes are also becoming more pronounced. Some demographers believe that we are approaching a plateau of maximum life span. Others predict that advances in biology and medicine might extend longevity markedly. There are predictions that an infant born in 1990 can expect to live at least one hundred years, on average. If that turns out to be true, it will have profound effects on society. In 1970, the median age in the United States was thirty. By 2100 the median age could be over sixty. If workers continue to retire at sixty-five, half of the population could be retired, and retirees might be facing thirty-five or forty years of retirement. We may need to find new ways to structure our lives.

Living Longer: Demographic Implications

A population that is growing rapidly by natural increase has more young people than does a stationary population. One way to show these differences is to graph age classes in a histogram as is shown in figure 7.10. In Mexico, which is growing at a rate of 2.5 percent per year, 42 percent of the population is in the prereproductive category (below age fifteen) (figure 7.10a). Even if total fertility rates were to fall abruptly, the total number of births and population size would continue to grow for some years as these young people enter reproductive age. This phenomenon is called population momentum.

A population that has recently entered a lower growth rate pattern, such as the United States, will have a bulge in the age classes for the last high-birth-rate generation (figure 7.10b). A country that has had a stable population for many years, such as Sweden, will have approximately the same numbers in all age

classes (figure 7.10c). Notice that there are more females than males in the older age groups in both the United States and Sweden because of differences in longevity between the sexes. These countries also have a high percentage of retired people (17 percent in Sweden, for instance) because of long life expectancy.

Countries with a high percentage of children (such as Mexico) and countries with a high percentage of old people (such as Sweden) have a problem known as **dependency ratio,** or the number of nonworking compared to working individuals in a population. Mexico has a high number of children to be supported by each working person. In Sweden, although there are fewer children, a large number of retired persons must be supported by a small working population. There is considerable worry that not enough workers will be available to support retirement systems in countries where population growth has slowed and life expectancies have increased. If life expectancies approach one hundred years, as some gerontologists suggest, and people continue to retire at the age of sixty-five, the United States could reach a point where retired people outnumber employed people. Considering the number who are too young to work, or are unemployed for other reasons, we could have two or three nonworkers for every worker. The Social Security system couldn't continue to operate. Retirement might have to be postponed or eliminated altogether.

Population Growth Rates

Crude death rate subtracted from crude birth rate gives the **natural increase** of a population. This is distinguished from the total increase or total growth rate, which includes immigration and emigration, as well as births and deaths. The natural increase is usually expressed as a percent (number per one hundred people) rather than per thousand. As is discussed in chapter 6, a growth rate of 1 percent per year means a doubling rate of about seventy years. Kenya, which is growing at a rate of 3.8 percent per year, is doubling in about eighteen years (table 7.4). The United States, which has a crude birth rate of sixteen per one thousand and a crude death rate of nine per one thousand, has a natural increase of 0.7 percent. Based solely on natural increase, the United States population will double in one hundred years. Actually, because of immigration, it is growing twice as fast. Italy, with a growth rate of 0.1 percent, is doubling in about seven hundred years.

Effects of Mobility

Humans tend to relocate, both within their own countries and between countries. Such movements affect the social, economic, and political conditions of both the point of origin and the destination.

Emigration and Immigration

Humans are highly mobile, so emigration and immigration play a larger role in human population dynamics than they do in those

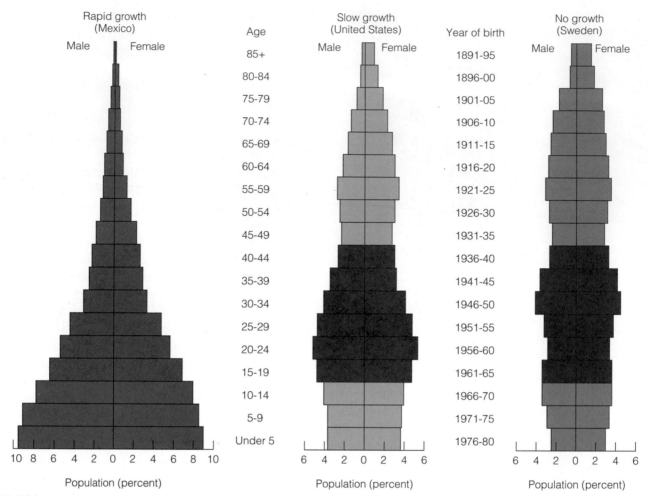

FIGURE 7.10 Population by age and sex in rapidly growing, slowly growing, and no-growth countries. Colored region represents individuals in reproductive ages. Note the high proportion of children in the rapidly growing population and the high proportion of elderly in the stable, no-growth population.

of many species. Figure 7.11 shows the top ten sources of new arrivals to the United States between 1820 to 1985. The peak of legal immigration into the United States was the decade between 1900 to 1909 when 800,000 new residents arrived, mainly from Europe. In interpreting this graph, remember that Canada, like the United States, is largely populated by the descendants of immigrants from many nations, especially European countries. Canada's bar on the graph, therefore, may be regarded, to some extent, as representing secondary immigration levels from Europe—it is more heterogenous than the other bars on the graph. The departure of this "surplus" population from Europe helped to keep population densities from growing during a period of high natural increase rates. It also played a major role in the population growth of the United States. The total immigration into the United States over the past 165 years has been 52.5 million people, or 25 percent of total population growth. Currently, about 570,000 people immigrate legally to the United States each year, but at least twice that number enter illegally. The largest number of both legal and illegal entrants are from

Mexico, followed by the Philippines, Korea, China, and the West Indies.

Immigration is a controversial issue in the United States, as well as in other wealthy countries. "Guest workers" and immigrants perform much heavy, dangerous, or disagreeable work that citizens are unwilling to do. In many countries, including the United States, these people are often of a different racial or ethnic background than the majority in the country to which they go in search of work. The treatment of migrants and guest workers often is appalling. They are paid low wages and given substandard housing, poor working conditions, and few rights. Local people complain that immigrants take away jobs, overload social services, and ignore established rules of behavior or social values. There are often undertones of racism and xenophobia in relations with newly arrived immigrants. Ironically, the next most recently arrived immigrants are usually the most adamant about closing doors to future migration.

Immigrants also benefit the countries to which they move. They contribute energy, ingenuity, cultural diversity, and vigor.

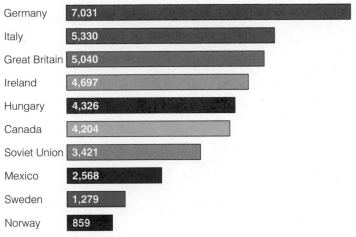

Germany	7,031
Italy	5,330
Great Britain	5,040
Ireland	4,697
Hungary	4,326
Canada	4,204
Soviet Union	3,421
Mexico	2,568
Sweden	1,279
Norway	859

FIGURE 7.11 Source of immigration to the United States by country of last residence, 1820 to 1985. Immigrants from Canada may have immigrated from another country.

The newest immigrants often are the hardest working members of the population. The top students in many schools in the United States are immigrants or first generation Americans.

Internal Relocation

Shifts of the population from one place to another within a nation can cause marked demographic dislocations. In the United States, for instance, there has been internal migration during this century from (1) rural areas to cities, (2) North to South, and (3) East and Midwest to West and Southwest. In the 1970s, more than 3 million people moved from the "Frost Belt" of the Great Lakes and New England States and the "Rust Belt" of the Ohio River Valley and Mid-Atlantic States to the "Sun Belt" of the southern and southwestern tier of states. Another million or so moved to the Intermountain West and West Coast. The fastest growing states during this decade were Arizona, Florida, Wyoming, Utah, Hawaii, and Alaska.

POPULATION GROWTH: OPPOSING FACTORS

Various social and economic pressures affect decisions about family size, which in turn affects the population at large.

Pronatalist Pressures

Many pressures and interests make people want to have babies. These influences are called **pronatalist pressures**. Raising a family can be enjoyable and rewarding. Children can be a source of pleasure, pride, and comfort. They can help support parents in old age, especially in countries without a social security system. Where infant mortality rates are high, couples may need to have six children or more to be sure that at least one or two will survive to take care of them when they are old. Where there is little opportunity for upward mobility to which people can aspire,

children give status in society, express parental creativity, and provide a sense of continuity and accomplishment otherwise missing from life. Our response to babies (especially our own) and our urge to reproduce must have at least some basis in an instinctive need to insure survival of the species.

Society also has an impetus to survive, and the need to replace members who die or become incapacitated has centuries-old historic roots. This need often is codified in cultural or religious values that encourage bearing and raising children. Families with few or no children are looked upon in some societies with pity and horror. The idea of deliberately controlling fertility may be shocking, even taboo. Women who are pregnant or have small children are given special status and protection. Boys frequently are more valued than girls because they carry on the family name and are expected to support their parents in old age. Couples may have more children than they really want in an attempt to produce a son.

Male pride sometimes is linked to having as many children as possible. In some cultures, this is accomplished by having multiple wives. In societies where a woman has little status or control over her own life, she may be subjected to demands of her husband, family, religion, or social group in determining how many children she will have. Even though a woman might desire fewer children, she may be powerless in making decisions for herself. In many societies, a woman has no status outside of her role as wife and mother. She has no way to support herself if she does not have children.

Figure 7.12 shows a model for the variables determining fertility. Three primary factors interact in this model: the biological supply of children determined by fecundity and infant mortality, the demand for children determined by economics and social values, and the regulation of fertility determined by knowledge, attitudes, and access to birth control. The combined interaction of these factors and their intervening variables determine fertility.

Birth Reduction Pressures

In more highly developed countries, many pressures tend to reduce fertility. Higher education and personal freedom for women often result in decisions to limit childbearing. The desire to have children is offset by a desire for other goods and activities that compete with childbearing and childrearing for time and money. When women have opportunities to earn a salary, they are less likely to stay home and have more children. Not only are the challenge and variety of a career attractive to many women, but the money that they can earn outside the home becomes an important part of the family budget. Thus, education and socioeconomic status are usually inversely related to fertility in richer countries. In less developed countries, however, fertility is likely to be positively related to educational levels and socioeconomic status. As income increases, families are better able to afford the children they want; higher income means that

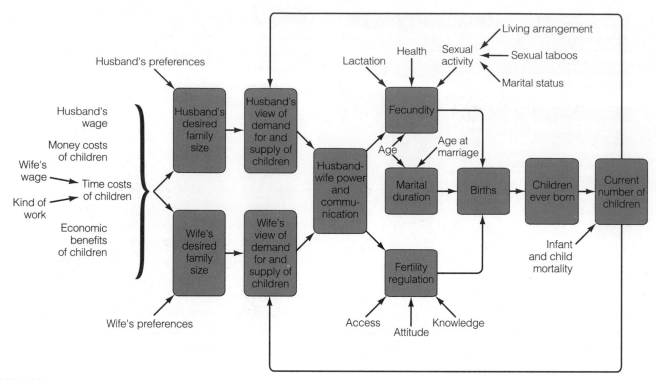

FIGURE 7.12 A model for the variables determining fertility. Three major factors interact in this model: the biological supply of children, the demand for children, and the regulation of fertility. Other intervening factors have either positive or negative effects on the interaction of these factors.

women are likely to be healthier and therefore better able to conceive and carry a child to term.

In less developed countries, where feeding and clothing children can be a minimal expense, adding one more child to a family usually doesn't cost much. An additional child can be taken care of by older siblings and can begin to do simple household chores or help in the fields at an early age. By contrast, raising a child in an upper-class family in the United States might easily cost hundreds of thousands of dollars by the time he or she is through school and is independent. Parents under these circumstances are more likely to choose to have one or two children on whom they can concentrate their time, energy, and financial resources.

Figure 7.13 shows United States birth rates between 1910 and 1988. As you can see, birth rates have fallen and risen in a complex pattern. The period between 1910 and 1930 was a time of industrialization and urbanization. Women were getting more education than ever before and entering the work force. The Great Depression in the 1930s made it economically difficult for families to have children, and birth rates were low. The birth rate increased during World War II (as it often does in wartime). For reasons that are unclear, a higher percentage of boys are usually born during war years.

At the end of the war, there was a "baby boom" as couples were reunited and new families started. This high birth rate persisted through the times of prosperity and optimism of the 1950s,

but began to fall in the 1960s. Part of this decline was caused by the small number of babies born in the 1930s. This meant fewer young adults to give birth in the 1960s. Part was due to changed perceptions of the ideal family size. Where women typically reported that they wanted four children or more in the 1950s, the norm dropped to one or two (or no children) in the 1970s. A small "echo boom" occurred in the 1980s as people born in the 1960s began to have babies, but changing economics and attitudes seem to have permanently altered desired ideal family sizes in the United States.

Birth Dearth?

In some countries, birth rates have fallen below replacement rates. Most of these countries are found in Europe (table 7.5), but Japan and Singapore are also facing a "child shock" as fertility rates have fallen below the replacement level of 2.1 children per couple. There are concerns about falling military strength (lack of soldiers), economic power (lack of workers), and declining social systems (not enough workers and taxpayers) if these low birth rates persist or are not balanced by immigration. Economist Ben Wattenberg warns that this "birth dearth" might seriously erode the powers of western democracies in world affairs. He points out that Europe and North America accounted for 22 percent of the world's population in 1950. By the 1980s, this number had fallen to 15 percent, and by the year 2030, Europe and North America will make up only 9 percent of the world's

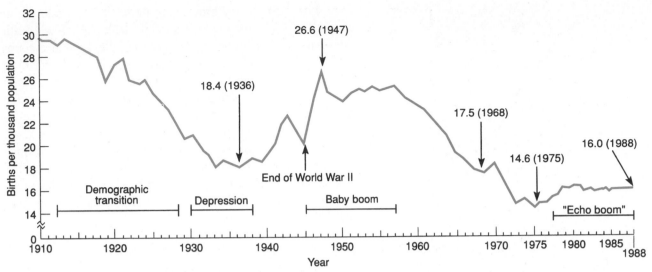

FIGURE 7.13 Birth rates in the United States, 1910 to 1988. The falling birth rate from 1910 to 1929 represents a demographic transition from an agricultural to an industrial society. Note that this decline occurred before the start of the Great Depression. The "baby boom" following the Second World War lasted from 1945 to 1957. A much smaller "echo boom" occurred around 1980 when the "baby boomers" started to reproduce, but it produced far fewer births than anticipated.

Sources: Adapted from *Population profile*, Population Reference Bureau, March 1967. Courtesy of the Population Reference Bureau, Inc., Washington, D.C. Recent data from U.S. Bureau of the Census.

population, Germany, Hungary, Denmark, and the Soviet Union now offer incentives to encourage women to bear children. Some Asian countries are also worried about a birth dearth. Japan offers financial support to new parents and Singapore provides a dating service to encourage marriages among the upper classes as a way of increasing population.

Some authors respond to fears of a "birth dearth" by pointing out that if the population of European people falls to about 10 percent of the world population, it will be nearly the same as it was before the population explosion of the eighteenth and nineteenth centuries. Furthermore, since Europeans and North Americans use so much more resources per capita than most other people in the world, a reduction in the population of these countries will do more to spare the environment than would a reduction in population almost anywhere else.

DEMOGRAPHIC TRANSITION

In 1945, demographer Frank Notestein pointed out that there is a typical pattern of falling death rates and birth rates due to improved living conditions that usually accompanies economic development. He called this pattern the **demographic transition** between high birth and death rates to lower birth and death rates. An idealized version of this transition is shown in figure 7.14.

Industrialization and Development

Phase I in this figure represents the conditions in a premodern society. Food shortages, malnutrition, lack of sanitation and medicine, accidents, and other hazards keep death rates high.

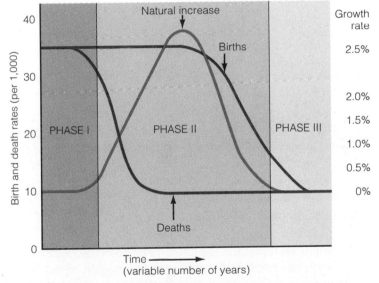

FIGURE 7.14 Theoretical birth, death, and growth rate curves during a demographic transition. Phase I—pre-modern society with high birth and death rates and stable population size. Phase II—developmental stage in which death rate has fallen, but birth rate remains high. Population grows rapidly. Phase III—modern society in which birth and death rates come into equilibrium again but at a much lower level than before. Population size stabilizes in this phase but at a much higher level than before.

Birth rates are correspondingly high to keep population densities relatively constant.

Phase II represents the conditions in a developing country. Better jobs and more income improve the standard of living. More food is available, and sanitation and medicine improve longevity. As a result, death rates fall—often very rapidly. Birth rates

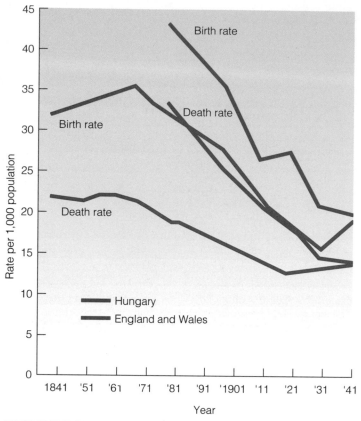

FIGURE 7.15 Mean annual birth and death rates for Hungary, and for England and Wales combined.

also fall, but only after a lag of several decades because it takes a generation or so before people believe that the improvements in society are permanent. It also takes some years to change attitudes and values. One generation clings to the values it grew up with, and it takes a new generation to adapt to new realities. During phase II, the population grows rapidly as the rate of natural increase (birth rate minus death rate) rises. Depending on how long it takes to complete this phase, the population will grow exponentially and may go through one or more rounds of doubling before coming into balance again.

Phase III represents the conditions in a developed country. Both birth rates and death rates are low, often one-third or less than those in the predevelopment era. The population comes into a new equilibrium in this phase, but at a much higher density than before. Most of the countries of northern and western Europe went through a demographic transition in the nineteenth or early twentieth century, although the shifts were not as uniform or pronounced as the theoretical model. Figure 7.15 shows actual data for England and Wales, and Hungary. Notice that the transition began in Britain about the middle of the nineteenth century and took fifty or sixty years to be completed. Hungary began the transition later but modernized faster as technology was transferred from already developed countries.

The most rapidly growing countries in the world are in Phase II of this demographic transition (table 7.4). Their death rates have fallen close to the rates of the fully developed countries, but birth rates have not fallen correspondingly. In fact, birth rates are higher than those in most European countries when they started industrializing three hundred years ago. The large disparity between birth and death rates means that these countries are growing at 3 to 4 percent per year. It is this high rate of growth in the less developed countries that will boost total world population to 10 billion or more before the end of the next century. Perhaps the most important questions in this whole chapter are "Why are birth rates not falling in these countries?" and "What can be done about it?"

A Demographically Divided World

Table 7.6 shows the birth and death rates in broad geographical regions of the world. As many authors have pointed out, the world is becoming increasingly divided demographically. The developed nations of Europe, North America, the Soviet Union, and Japan have completed, or nearly completed, the demographic transition, and have stable or slowly growing populations. The less developed countries of Africa, Asia, and Latin America have started the demographic transition. Improved health care, sanitation, and food supplies have lowered death rates, but the improved income, security, and changed values that are necessary to alter birth rates have not yet affected the majority of the population.

An Optimistic View

Some authors claim that demographic transition is in progress in most developing nations, and we just have to be patient until it runs its course. Evidence suggests birth rates are beginning to fall worldwide, and most demographers believe the world population will stabilize sometime in the next century. What factors might facilitate transition to a stable world population?

1. Growing prosperity and social reforms that accompany development should reduce the need and desire for large families in most countries.
2. Technology is available to bring advances to the developing world much more rapidly than was the case a century ago, and the rate of technology transfer is much faster than it was when Europe and North America were developing.
3. Less developed countries have historic patterns to follow: they can benefit from our mistakes and chart a course to stability more quickly than they might otherwise do.
4. Modern communications (especially television) have caused a revolution of rising expectations that act as a stimulus to spur change and development.

TABLE 7.6 Estimates of birth rates, death rates, and rates of natural increase of population in different regions of the world (1990)

	Birth rate (per 1,000)	Death rate (per 1,000)	Rate of natural increase (percentage)	Years to double population
Europe	13	10	0.3	266
More developed regions				
Northern Europe	14	11	0.2	286
North America	16	9	0.7	98
Soviet Union	19	10	0.9	80
All more developed regions	15	9	0.5	128
Less developed regions				
Africa	44	15	2.9	24
Asia	27	9	1.9	37
Latin America	28	7	2.1	33
All less developed regions	31	10	2.1	33
World	27	10	1.8	39

Source: *1990 World Population Data Sheet* (Washington, D.C.: Population Reference Bureau, Inc., April 1990).

A Pessimistic View

Economist Lester Brown of the Worldwatch Institute presents a more pessimistic view of the demographic situation. He warns that many of the poorer countries of the world appear to be caught in a "demographic trap" that leads to ever higher population densities and limits their ability to escape from phase II of the demographic transition. According to this view, population growth is overwhelming local environmental support systems and causing environmental deterioration, economic decline, and political instability. As these factors feed on one another, the whole country plunges ever deeper in a downward spiral that is harder to break. Once the population expands to the point where human demands exceed the sustainable yield of local forests, grasslands, croplands, or aquifers, it begins to consume the resource base itself.

Many people see this downward cycle as evidence that the only way to break out of the demographic trap is to immediately and drastically reduce population growth by whatever means are necessary. They argue strongly for birth control education and bold national policies to encourage lower birth rates.

Some agree with Malthus that helping the poor will simply increase their reproductive success and further threaten the resources on which we all depend. What can we do? Will our approach be what author Garret Hardin calls lifeboat ethics? "Each rich nation," he says, "amounts to a lifeboat full of comparatively rich people. The poor of the world are in other much more crowded lifeboats. Continuously, so to speak, the poor fall out of their lifeboats and swim for a while, hoping to be admitted to a rich lifeboat, or in some other way to benefit from the goodies on board. . . . We cannot risk the safety of all the passengers by helping others in need. What happens if you share space in a lifeboat? The boat is swamped and everyone drowns. Complete justice, complete catastrophe."

A Social Justice View

A third view is that **social justice** is the real key to successful demographic transitions. The world has enough resources for everyone, according to this view, but inequitable social and economic systems cause maldistributions of those resources. Hunger, poverty, crowding, unemployment, violence, and overpopulation are symptoms of a lack of social justice rather than a strain on resources. The way to solve these problems is to establish fair systems, not to blame the victims. Figure 7.16 expresses the opinion of many in the less developed countries about the relation between resources and population.

It is significant, in this view of the world, that nearly all developed countries with a high standard of living and slow population growth were (or still are) colonial powers. Their colonies supplied raw materials for industry, absorbed surplus population, manufactured goods, and provided capital for transformation from an agrarian to an industrial society.

By contrast, those countries suffering from the most extreme social and economic problems were (or still are) colonies of more powerful nations. Many countries that are now among the poorest in the world, such as India, Ethiopia, and Mali, were once rich in resources and had an adequate food supply for their populations. The biological and mineral wealth that made them attractive targets for colonization was stripped away and sent to the richer countries to pay for progress there. Yet we tend to blame victims of this system for the exhausted resource base and chaotic political systems that resulted from their colonial past.

Exploitation of the poor by the rich did not end with the demise of the nineteenth century colonial empires. "Superpowers" of the world still dominate their less powerful neighbors, but control now is mainly held through "economic imperialism" rather than "gunboat diplomacy." Bank loans, political pressures, foreign aid, market systems, and monetary policies have

FIGURE 7.16 Controlling our population and resources—there may be more than one side to the issue.

Asian Cultural Forum on Development. Used with permission.

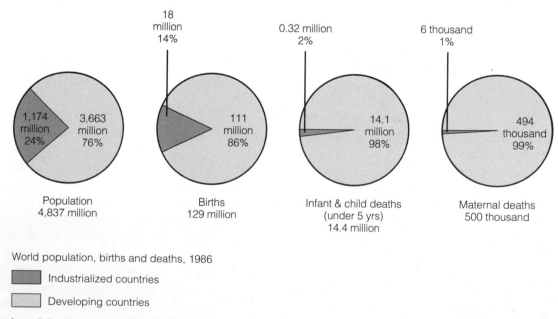

FIGURE 7.17 Births and deaths in industrialized and developing countries. Better hygiene and nutrition in the more developed countries result in low infant mortality and maternal deaths. Child survival has a direct, inverse relationship to birth rates.

Sources: United Nations Population Division, United Nations Statistical Office, and World Health Organization.

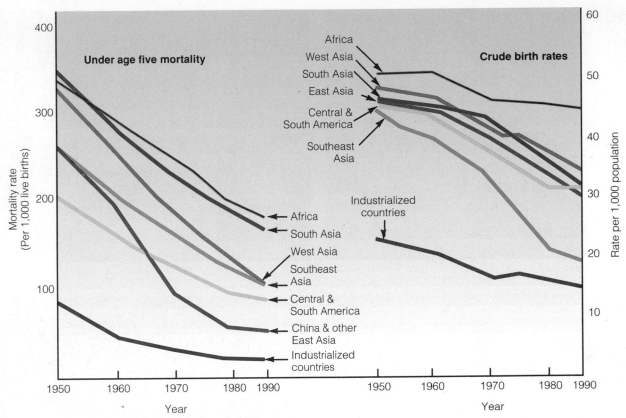

FIGURE 7.18 Under age five mortality rates and crude birth rates by region, 1950–1986. Falling birth rates have paralleled falling infant mortality rates in most regions of the world except Africa, where other cultural, political, and economic factors complicate demographics.

been established that favor the richer nations and keep the poorer countries in a state of instability and dependency. If we really want the less developed countries to finish the demographic transition and stabilize their populations, we may need to establish the "new economic order" between the north and the south called for by less developed countries (chapter 8).

Child mortality is one of the most critical factors in stabilizing population. When infant mortality rates are high, as they are in much of the developing world, parents tend to have high numbers of children to insure that some will survive to adulthood. There has never been a sustained drop in birth rates that was not first preceded by a sustained drop in infant and child mortality. One of the most important distinctions in our demographically divided world is the high proportion of deaths under five years of age in the developing countries (figure 7.17). Improved health care, simple oral rehydration therapy, and immunization against infectious diseases (chapter 21) have brought about dramatic reductions in child mortality rates that have been accompanied in most regions by falling birth rates (figure 7.18).

FAMILY PLANNING AND FERTILITY CONTROL

Whatever the causes of social and economic injustice and its relationship to overpopulation, it is clear that in many cases people have more children than is good for their health, their environment, or the future of the world. In this section, we will survey some methods that can be used to limit or space births, and some government programs that have been designed to encourage birth control and family planning.

Reasons to Control Fertility

Family planning allows couples to determine the number and spacing of their children. It doesn't necessarily mean fewer children—people may use family planning to have the maximum number of children possible—but it does imply that they will take control of their reproductive lives and make rational, conscious decisions about how many children they will have and when those children will be born, rather than leaving it to chance.

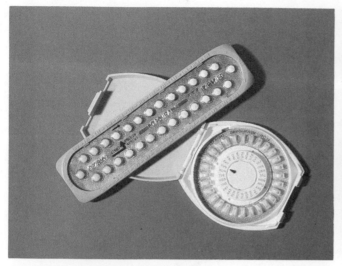

a.

d.

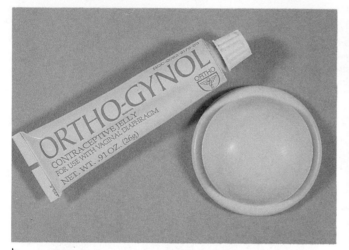

b.

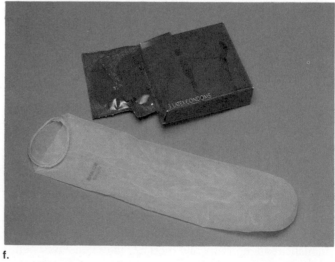

e.

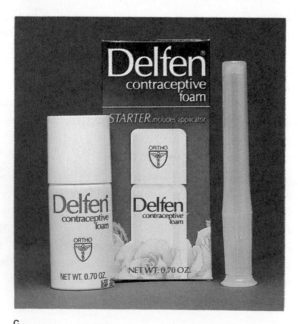

c.

f.

FIGURE 7.19 (*a*) Daily birth control pills; (*b*) spermicidal jelly and diaphram; (*c*) spermicidal foam; (*d*) subdermal slow-release progestin; (*e*) intrauterine device; (*f*) condom.

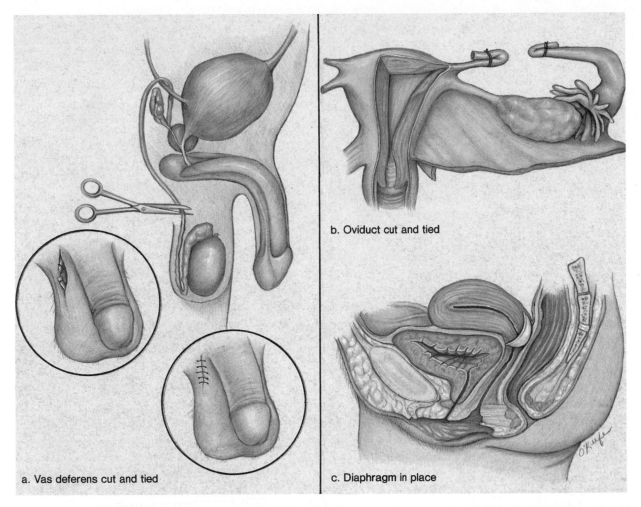

b. Oviduct cut and tied

a. Vas deferens cut and tied

c. Diaphragm in place

FIGURE 7.20 Surgical birth control procedures include vasectomy for men and tubal ligation for women. (*a*) Vasectomy requires only a small incision under local anesthesia to cut and tie the sperm-conducting *vas deferens*. (*b*) Tubal ligation in women is a more difficult operation and generally requires general anesthesia. Neither procedure affects sex life or sexual characteristics, but neither is easily reversible. (*c*) The cervical cap or diaphragm is an example of a mechanical barrier. It covers the mouth of the cervix and prevents sperm from entering the uterus. While not as dependable as surgical methods, mechanical barriers are easily reversible.

Current Birth Control Methods

Modern medicine gives us many more options for controlling fertility than were available to our ancestors. Some of these techniques are safer, easier, and more pleasant to use. The major categories of birth control techniques include: (1) avoidance of sex during fertile periods (celibacy, using changes in body temperature or cervical mucus color and viscosity to judge when ovulation will occur); (2) mechanical barriers that prevent contact between sperm and egg (condoms, spermacides, diaphragm, cervical cap, and vaginal sponge) (figure 7.19 and 7.20*c*); (3) surgical methods that prevent release of sperm or egg (sterilization: tubal ligation or use of the Filshie clip in females; vasectomy in males) (figure 7.20*a* and *b*); (4) chemicals that prevent maturation or release of sperm or eggs or implantation of the embryo in the uterus (the pill: estrogen + progesterone, progesterone alone, or RU486 for females; gossypol for males); (5) physical barriers to implantation (IUD); and (6) abortion.

None of these methods is perfect and none suits every contraceptive need. About 10 million American women now use the pill because it is easy to use and readily reversible. The pill is not recommended for women over age thirty-five who smoke because of the risk of strokes and other side effects involving blood vessels. The use of pills with low estrogen levels or progesterone alone can reduce these risks. In fact, there is evidence that at least brief use of the pill protects against some forms of cancer.

As economic development and medical advances make it less necessary to have a large family to ensure that enough children survive to keep the family going, most parents find it desirable to have fewer children and to invest more time, energy, and money on each child. As the desire for smaller families becomes more common, birth control becomes an essential part of family planning in most cases. In this context, **birth control** usually means any method used to reduce births, including cel-

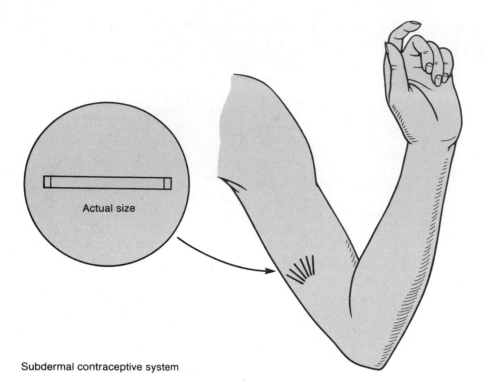

Actual size

Subdermal contraceptive system

FIGURE 7.21 In a subdermal contraceptive system, six flexible, match-sized capsules are implanted under the skin in the woman's arm. They release a slow but steady supply of birth control hormones over about five years, or they can be removed before then if the woman wants to become pregnant.

ibacy, delayed marriage, contraception, methods that prevent implantation of embryos, and induced abortions.

Sometimes people in the more developed countries believe that residents of the poorer, less technological countries are too ignorant or primitive to understand fertility and birth control. This may or may not be true. People who do not practice birth control may be making a very deliberate rational decision, given their life circumstances. Evidence suggests that people in every culture and every historic period have used a variety of techniques to control fertility. Studies of hunting and gathering people, such as the Kungi of the Kalahari Desert in southwest Africa, indicate that our early ancestors had stable population densities, not because they starved to death regularly, but because they used a variety of methods to control fertility. For instance, women in many primitive cultures breast-feed children for three or four years. The high caloric requirement of lactation depletes body fat stores and suppresses ovulation. When coupled with taboos against intercourse while breast-feeding, it is an effective way of spacing children. Other ancient techniques to control population size include celibacy, polygamy, folk medicines, abortion, and infanticide. We may find some or all of these techniques backward, unpleasant, or morally unacceptable, but we shouldn't assume that other people are too ignorant or too primitive to make decisions about fertility.

Although people have used a variety of methods to control birth in many cultures, high infant mortality rates made it necessary to institute practices that encouraged fertility and child care. Over the years, those attitudes and practices became

embedded in cultural and religious values that cannot or will not adapt quickly enough to keep up with changes brought about by modern technology. Finding ways to bring current societal needs into balance with traditional customs and values is an important aspect of family planning programs in most countries.

New Developments in Birth Control

Some new methods of birth control presently being studied may have great promise. Flexible, matchstick-sized, silicon-rubber implants containing slow-release progestin (the analog of progesterone) were approved for use in the United States in 1991. They are inserted under the skin where they will release hormones for up to five years (figure 7.21). These implants avoid having to keep track of and take daily pills. The "morning-after" drug, RU486 (mifepristone or mifegyne), can be taken any time up to ten days after a missed period. It blocks the effects of progesterone in maintaining the lining of the uterine wall and prevents the embryo from implanting. It is reported to cause fewer physical and psychological side effects than abortion after implantation has occurred. Interestingly, RU486 also appears to be effective treatment for breast cancer, brain cancer, diabetes, and hypertension. Because some groups consider its action to be a form of abortion, however, its use—even in research—has been blocked in the United States.

Development of simple, inexpensive, do-it-yourself tests for levels of estrogen and progesterone in urine may make the rhythm method easier to follow and more reliable for women who cannot or would rather not use other methods of birth control. Some antipregnancy vaccines are being tested that would use the immune system to prevent pregnancy, but they are several years away from the market. The sperm suppressant gossypol is being tested as a male contraceptive in some countries, but there are side effects, and males often are reluctant to take responsibility for reproduction. Finally, potent new spermicides are being tested, as are new methods of using them.

GOVERNMENT PROGRAMS TO ENCOURAGE FAMILY PLANNING

Coordinated, effective family planning requires governmental action, ranging from policies to assistance. The methods used to promote family planning and the ethics of social intervention and population control are often controversial.

Conditions for Successful Family Planning Programs

What conditions must be in place for a family planning program to be successful? Sometimes it takes deep changes in a culture to make these conditions acceptable and widespread. Among the most important of these are (1) improved social, educational and economic status for women—successful birth control and women's rights are often interdependent; (2) improved status for children—strange as it may seem, placing a higher value on children may mean that fewer are born as parents come to regard them as valued individuals rather than possessions; (3) acceptance of calculated choice as a valid element in life in general and in fertility in particular—when people feel that fate or the gods make it impossible or unnecessary to plan for the future, they are unlikely to use family planning; (4) social security and political stability that give people the means and the confidence to plan for the future; (5) knowledge, availability, and use of effective and acceptable means of birth control.

Methods of Promoting Family Planning

Governments can encourage family planning and fertility control in a variety of ways. Education is important. Classes on contraceptive techniques give women and men the information to make informed choices and to carry them out. Public opinion campaigns on radio, television, posters, signboards, and other public media can change public attitudes towards contraception. This is useful not only in persuading people to practice birth control who might not otherwise do so, but it also allows people who want to control fertility to do so by making it publicly acceptable. Many governments distribute free birth control materials at such public places as railroad stations and shopping areas, or through family planning workers who make home visits. Payments may be given to individuals who agree to be sterilized. One to two extra weeks of wages (only $10 to $12 for agricul-

tural workers in India) to compensate for lost time can make these programs more palatable and feasible for poor people. Cash payments, preferred jobs, better housing, extra food rations, guaranteed schooling for their children, savings accounts, or low-interest credit are among the incentives offered by some governments to encourage citizens to limit the number of children they have.

Some approaches are public, rather than private. Offering a reward, such as a new school, a village well, or a television set to a whole community that meets family planning quotas may bring effective peer pressure to bear on those of reproductive age. It may also inspire people to work together for the common good. Extra taxes, loss of the incentives listed above, public criticism, shame, and punishment are sometimes used to discourage people from having more than the approved number of children. These measures are considered oppressive and counterproductive by some people. The most controversial methods of imposing birth control include forced sterilization, forced abortions, and reproductive police who pry into intimate details of private life and intimidate and terrorize those who fail to cooperate fully with government programs.

Ethics of Population Control

Probably no other area of environmental science involves such difficult ethical questions as population control. For many people, the right to reproduce is a sacred and fundamental human right. To interfere with this most sensitive and private part of a person's life is considered a terrible thing. On the other hand, overpopulation may threaten the well-being—and even the survival—of us all. Given that ominous prospect, do the needs and rights of the individual supersede the rights and needs of society as a whole? If society's needs supersede those of the individual, what methods and tactics are justified in striving to reach its goals?

THE FUTURE OF HUMAN POPULATIONS

How many people will be in the world a century from now? Most demographers believe that world population will stabilize sometime during the next century. The total number of humans, when we reach that equilibrium, probably will be between 6 and 16 billion people, depending on the success of family planning programs and the multitude of other factors affecting human populations (figure 7.22). Figure 7.23 presents some of the factors affecting population levels that have been discussed in this and previous chapters, and that will be expanded in chapters that follow.

As we have discussed in this chapter, some demographers believe that 16 billion humans are far too many for Earth to support. They argue that unless we have immediate drastic re-

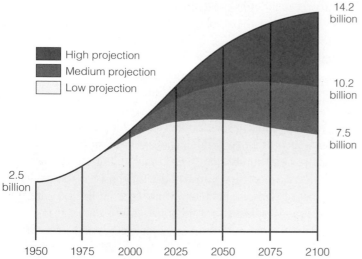

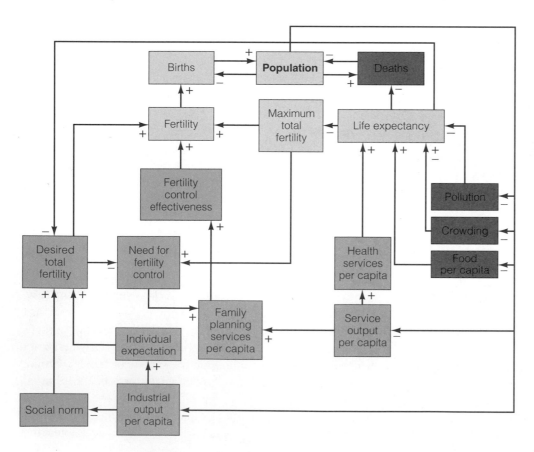

FIGURE 7.22 Long-term projections of world population result in different scenarios, depending on assumptions about the pace of development, birth control, and subsequent demographic transitions. The United Nations projection, shown in the middle scenario above, assumes that world fertility will decline to two children per woman in 2035, resulting in a population of 10.2 billion or exactly twice that of today.

ductions in birth rates, we will exhaust or degrade Earth's life-support systems and cause a catastrophic decline, not only in our own species, but also in biological diversity and stability of the biosphere. Others who are more optimistic believe that the world can safely sustain more people (some say far more) than are alive today. These optimists expect further technological advances to extend the carrying capacity of Earth, and rising prosperity will automatically limit population growth without requiring drastic intervention other than to increase equity, jobs, opportunity, and education for all.

In subsequent chapters of this book, we will discuss many resource issues in greater detail. You can learn about the resources available, the problems associated with their use, and some suggestions for conserving or protecting these resources and the values they offer. We hope that this information will help you form your own opinions about these important questions concerning human population and the environment.

FIGURE 7.23 The large computer model represents the interactions of human population within the environment. Plus signs (+) indicate factors that cause an increase in the next factor; minus signs (−) indicate an effect that tends to diminish the next factor. Each of us creates a demand upon the environment for natural resources. The more of us there are, the more is the demand. Obtaining raw resources and converting them to useful products nearly always has adverse side effects on the natural environment and, potentially, on us. Thus, there has been an historical tendency for increased population to cause increased pollution, via resource acquisition and use.

Human populations have grown at an unprecedented rate over the past three centuries. By 1990, the world population stood at 5.3 billion people. If the current growth rate of 1.8 percent per year persists, the population will double in thirty-nine years. Most of that growth will occur in the less developed countries of Asia, Africa, and Latin America. There is a serious concern that the number of humans in the world and our impact on the environment will overload the life-support systems of Earth.

The crude birth rate is the number of births in a year divided by the average population. A more accurate measure of growth is the general fertility rate, which takes into account the age structure and fecundity of the population. The crude birth rate minus the crude death rate gives the rate of natural increase. When this rate reaches a level at which people are just replacing themselves, zero population growth is achieved.

In the more highly developed countries of the world, growth has slowed or even reversed in recent years so that without immigration from other areas, populations would be declining. The change from high birth and death rates that accompanies industrialization is called a demographic transition. Many less developed countries have begun this transition. Death rates have fallen, but birth rates remain high. Some demographers believe that as infant mortality drops and economic development progresses so that people in these countries can be sure of a secure future, they will complete the transition to a stable population. Others fear that excessive population growth and limited resources will catch many of the poorer countries in a demographic trap that could prevent them from ever achieving a stable population or a high standard of living.

Some people argue that rather than being a burden, more people represent a valuable resource of energy, intelligence, and enterprise that will make it possible to overcome resource limitation problems. They believe that a more equitable distribution of wealth might reduce both excess population growth and environmental degradation.

There are many more options now for controlling fertility than were available to our ancestors. Some techniques are safer than those available earlier; many are easier and more pleasant to use. Sometimes it takes deep changes in a culture to make family planning programs successful. Among these changes are improved social, educational, and economic status for women; higher values on individual children; acceptance of calculated choice as a valid element in life; social security and political stability that give people the means and confidence to plan for the future; and knowledge, availability, and use of effective and acceptable means of birth control.

■ Review Questions

1. At what point in history did the world population pass its first billion? What factors restricted population before that time, and what factors contributed to growth after that point?

2. How might growing populations be beneficial in solving development problems?
3. Why do some economists consider human resources more important than natural resources in determining the future of a country?
4. What are the major demographic differences between the more developed countries and the less developed countries?
5. Define crude birth rate, general fertility rate, total fertility rate, crude death rate, and zero population growth.
6. What is the difference between life expectancy and longevity?
7. What is dependency ratio, and how might it affect the United States in the future?
8. What pressures or interests make people want to have babies?
9. What pressures or interests discourage people from having babies?
10. Describe the conditions that lead to a demographic transition.
11. Describe the mechanisms by which the major categories of birth control work.
12. What societal changes are necessary for family planning to work?

■ Questions for Critical Thinking

1. What do you think is the maximum population of Earth?
2. What do you think would be the optimum population?
3. Do you believe that technology can make the world more habitable or that it will make the world less habitable?
4. What similarities or differences do you see in the ten most rapidly growing countries and in the ten most slowly growing countries?
5. Do you think that continued immigration would be a benefit or a burden to the United States?
6. What factors do you think will lead to or prevent a demographic transition in the developing countries of the world?
7. What countries of the world do you think might be headed for a demographic trap? Is there anything that we can do to help them get out of this situation?
8. Are you optimistic or pessimistic about the demographic situation of the world?
9. What role do you think social justice plays in population growth?
10. What role do you think the United States should play in family planning in the less developed countries of the world?

■ Key Terms

age-specific death rates (p. 113)
birth control (p. 125)
crude birth rate (p. 111)
crude death rates (p. 113)

demographic transition (p. 119)
dependency ratio (p. 115)
family planning (p. 123)
First World (p. 109)
general fertility rate (p. 111)
less developed countries (p. 108)
more developed countries (p. 108)
natural increase (p. 115)
neo-Malthusian (p. 107)
pronatalist pressures (p. 117)
Second World (p. 109)
social justice (p. 121)
Third World (p. 109)
total fertility rate (p. 111)
zero population growth (p. 112)

SUGGESTED READINGS

Berreby, D. "The numbers game." *Discover* 11(4) (April 1990):42. An interesting comparison of the views of Paul Ehrlich and Julian Simon.

- Blaxter, K. L. *People, Food and Resources*. Cambridge: Cambridge University Press, 1986. An economist looks at demographics and resources. Well written.
- Brown, L. R., and J. L. Jacobson. December 1986. "Our demographically divided world." *Worldwatch Paper #74*. Washington, D.C.: Worldwatch Institute. A sobering analysis of our demographic future.
- Caldwell, J. C., and P. Caldwell "High fertility in sub-Saharan Africa." *Scientific American* 262(5) (May 1990):118. Birth rates and population growth have begun to decline everywhere else in the world. What makes this region different?
- Cole, H., et al. *Models of Doom: A Critique of the Limits to Growth*. New York: Universe Books, 1973. An answer to the computer forecasts of Meadows, et al.
- Eberstadt, N. *Fertility Decline in the Less Developed Countries*. New York: Praeger Scientific, 1981. Excellent survey of demographic transitions.
- Ehrlich, A. H., and P. R. Ehrlich. Spring 1987. "Why do people starve?" *The Amicus Journal* 9(2):42. What would be the carrying capacity of a world full of saints?
- Ehrlich, P. R. 1968. *The Population Bomb*. New York: Ballantine Books, 1968. A landmark in public concern about population problems.
- Greenhalgh, S., and J. Bongaarts. March 6, 1987. "Fertility policy in China: future options." *Science* 235(4793):1167. An excellent overview of current conditions in China.
- Gustlee, C. "The coming world labor shortage." *Fortune* (April 1990):71. Describes the baby bust facing industrialized economies. Warns that retirement may be impossible in the future.
- Hardin, G. *Exploring Ethics for Survival: The Voyage of the Spaceship Beagle*. New York: The Viking Press, 1972. Controversial ideas from one of our most challenging thinkers.

- Jacobsen, J. June 1983. "Promoting population stabilization: incentives for small families." *Worldwatch Paper #54*. Washington, D.C.: Worldwatch Institute.
- Keyfitz, Nathan. September 1989. "The growing human population." *Scientific American* 261(3):119. This distinguished population biologist predicts that human population will stabilize but that the planet's life-support capacity may not support all of us.
- Kolata, G. February 13, 1987. "Wet-nursing boom in England explored." *Science* 235(4790):745. A fascinating study of why English upper-class women had as many as thirty children.
- McKeown, T. R., G. Brown, and R. G. Record. 1972. "An interpretation of the modern rise of population in Europe." *Population Studies* 26:3. An interesting historical analysis of possible causes of accelerating population growth in Europe in the eighteenth and nineteenth centuries.
- Mamdani, M. *The Myth of Population Control: Family, Caste and Class in an Indian Village*. New York: Monthly Review Press, 1972. An intriguing study of why externally imposed population policies don't work.
- Mass, B. *Population Target: The Political Economy of Population Control in Latin America*. Toronto: Canadian Women's Educational Press, Latin American Working Group, 1976. Demographics in a social and political perspective.
- Meadows, D. H., et al. *Limits to Growth*. 2d ed. New York: Universe Books, 1974. A computer model of population-resource interactions sponsored by the Club of Rome.
- Population Reference Bureau. *1987 World Population Data Sheet of the Population Reference Bureau, Inc.* Washington, D.C.: PRB. An excellent, annually updated source of demographic data.
- Preston, S. H. March 1986. "Population growth and economic development." *Environment* 28(2):6. Links these two important factors.
- Repetto, R. *The Global Possible*. New Haven: Yale University Press, 1985. An excellent overview of the current situation with an emphasis on ways we can improve.
- Simon, J. L. *The Ultimate Resource*. Princeton: Princeton University Press, 1981. A cornucopian view that more people means more geniuses, bigger markets, and a better life for all.
- Tierney, J. January/February 1986. "State of the species: Fanisi's choice." *Science 86* 7(1):26. A personal view of why women in less developed countries might choose to have eight children.
- Ulmann, A., et al. "RU486." *Scientific American* 262(6):42 (June 1990). An excellent description of both the science and the politics of this controversial method of birth control.

PART THREE

Resources and Resource Economics

C H A P T E R *8*

Environmental and Resource Economics

C O N C E P T S

Economics is a science that deals with resource allocations; how we should use (or not use) resources, and for whom and by whom they should be used. Economic growth is an increase in the total wealth of a nation, while economic development is an increase in the real income per person. Many economists believe that economic development is essential to solve societal and environmental problems.

Ecologists often warn that environmental damage caused by industrial activity and human populations is threatening life-support systems of the biosphere. Many call for a steady-state economy characterized by a stable population and use of renewable resources.

Scarcity of a limited resource can stimulate the search for new supplies or substitute materials and technological developments that extend the resource or the goods and services it supplies. Paradoxically, tangible physical resources, such as metals, that are present in finite, limited supply in the earth often can be extended by recycling so that they become practically limitless. Renewable resources, such as biological species, can be misused or overused to the point of being destroyed.

A cost/benefit analysis that evaluates direct, internal costs and benefits, as well as external, intangible costs and benefits, can help determine who pays and who benefits from resource use.

INTRODUCTION

Is there an impending collapse of ecological, economic, and social systems in the world unless we take drastic steps to curtail growth of all sorts? Are we using up resources that will be irreplaceable in the future? Are we approaching an age of scarcity in which population growth and lack of resources are leading to inescapable shortages of basic necessities, unemployment, stagnant or declining economic conditions, and social disruption? Questions such as these about economic development, resource scarcity, and population growth underlie much of our global debate about environmental policy.

In this chapter, we will survey some major economic theories and show how they apply to environmental science. Accurately reflecting the real cost of external environmental benefits, widespread but low-level health effects, and the value of future use of resources often is considered the greatest problem in bringing about pollution abatement and resource conservation. We will look at some contrasting views about the relationship between resource values, scarcity, economic development, and limits to growth. Deciding which of these theories is correct (all or in part) is a difficult, controversial question that each of us must examine in planning for the future.

Ecology and economy are derived from the same root words and concerns. *Oikos* or *Ecos* is the old Greek word for household. Economy is the *nomos* or counting of the household goods and services. Ecology is the *logos* or logic of how the household works. In both disciplines, household is expanded to include the whole world—the household of humans. Although ecologists and economists study the same world, their approaches and interpretations are so different and the language they use to describe what they see is so specialized and mutually incomprehensible that a gulf of mistrust and misunderstanding often separates the two disciplines.

True understanding of how our environment works and how we interact with it requires some understanding of the world views of both economics and ecology. Much of this book focuses on biological, chemical, and physical aspects of Earth. This chapter, however, surveys some principles of environmental economics. We will look at resources, scarcity, growth, and development from an economic viewpoint to see how we have used the environment in the past, and what future scenarios can be predicted for our world household.

ECONOMIC CONTEXT

Money and politics are the languages of most policy planners and decision makers. They ask, "How much will it cost?" and "What are the benefits?" Economists try to answer those questions. Basically, economics deals with resource allocation, either on the "micro" scale of personal purchases and decisions or on

FIGURE 8.1 Are natural resources like canned goods on a shelf awaiting our use until they are consumed, or can they be extended and created by technology and enterprise? Economics deals with resource values. What should we produce, for whom, and at what cost?

the "macro" scale of national policy and world economic systems. It is a description of how valuable goods and services are to us, and how we make decisions about the relative abundance and desirability of competing wants. Humans have unlimited wants. If resources were equally unlimited, there would be no conflicts between these wants and the available goods and services. There would be no need to choose between alternatives, and there would be no need for a system of economics.

In a finite world, however, we face constant economic decisions about how to use time and resources (figure 8.1). Consider these questions (1) What should we produce: "guns" or "butter?" (2) For whom should we produce goods? How shall we allot our productive capacity among competing wants? Is ability to pay for goods the best mechanism to determine distribution? (3) How will we make those things that we desire? What level of pollution and environmental disruption is acceptable? What trade-off between labor and technology or resource extraction and recycling represents the best use of resources, human potential, and capital?

In figures 8.2 and 8.3, "guns" represents production of defense-related goods and services, and "butter" represents food and other consumer goods and services. Figure 8.2 shows a simplified trade-off between production of competing goods and services. The maximum output at different levels of production of two commodities is called the **production frontier.** Any point inside this curve, such as point U in this figure, represents resources and production capacity that are not being fully employed in the most efficient manner. However, there may be

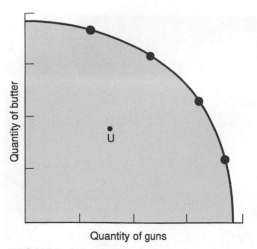

FIGURE 8.2 Production possibilities: "guns" or "butter." An economy can produce all "guns" or all "butter," or some combination of the two at a given level of technology. The curve represents balanced production possibilities. Any point inside the curve, such as U, indicates that resources are not being fully employed in the most efficient possible manner.

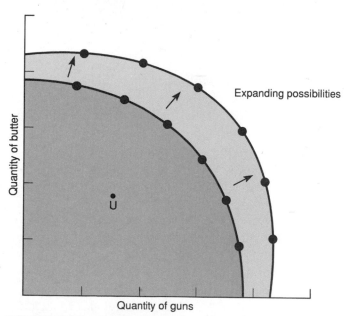

Expanding possibilities

FIGURE 8.3 Compare this figure to figure 8.2. Economic growth brings expanding possibilities for both "guns" and "butter" or any other pair of competing goods and services.

other societal or ecological reasons to not fully use resources at a given time.

Economic growth is an increase in the total wealth of a nation regardless of the population. If the population grows faster than the economy, there may be real growth but the share per person will decline. **Economic development** is a rise in the real income *per person*. It does not guarantee that the average citizen is better off, however. Development often involves the introduction of labor-saving machines that create greater profits but displace workers. As a result, the per capita income may rise, but

the actual standard of living for much of the population may decline.

Figure 8.3 shows how growth in an economic system expands the production frontier. With better access to resources, improved technological efficiency, larger populations, or more investment in production facilities (all considered to be forms of capital), we can produce more of both guns and butter (or whatever else people want). In most economic systems, continued growth is recognized as essential to maintain full employment and prevent class conflict that arises from inequitable distribution. In recent years, however, many people have become concerned that incessant growth of both human populations and productive systems soon will exhaust natural resources and surpass the capacity of natural systems to withstand disruption, either by effluents and wastes or by harvesting more stock than can be replaced.

Many ecologists and a few economists have called for a transition to a **steady-state economy** characterized by low birth and death rates, use of renewable energy sources, recycling of materials, and emphasis on durability, efficiency, and stability rather than high throughput of materials. Perhaps the first economist to advocate this new way of looking at the world was John Stuart Mill, whose (1857) *Principles of Political Economy* preceded Karl Marx. More recently, Kenneth Boulding's (1966) essay, "The Economics of the Coming Spaceship Earth" and Herman E. Daly's (1991) *Steady-State Economics* have argued that continual growth cannot be sustained. We will discuss the implications of steady-state economic theory further after looking at some of the questions of resource scarcity that make it seem necessary.

RESOURCES AND RESERVES

What are the resources that economic systems are set up to manage and allocate? How can we determine the available amount of a specific resource, given a particular economic system and technology? These are vital questions in the field of environmental studies because much of our concern about population growth hinges on a continual supply of resources.

Defining Resources

Simply defined, a resource is any useful information, material, or service. Within this broad generalization, we can differentiate between **natural resources** (goods and services supplied by our environment) and **human resources** (human wisdom, experience, skill, and enterprise). It is also useful to distinguish between exhaustible, renewable, and intangible resources.

In general, **exhaustible resources** are Earth's geologic endowment: the minerals, nonmineral resources (chapter 9), fossil fuels, and other materials present in fixed amounts in the

FIGURE 8.4 The geological resources of Earth's crust, such as oil from this well are nonrenewable. Often the limit to our use of these resources is not so much the absolute amount available as the energy required to extract the resources and the environmental consequences of doing so.

FIGURE 8.5 Biological resources are renewable in that they replace themselves by reproduction, but if overused or misused, populations die. When a whole species is lost, it is unlikely that it will ever be recreated. These northern fur seals were almost totally exterminated by overhunting in the nineteenth century.

environment. In theory, these exhaustible resources place a strict upper limit on the number of humans the environment can support and the amount of industrial activity we can carry on. Predictions that we are in imminent danger of running out of one or another of these exhaustible resources are abundant. In practice, however, the available amount of many physical resources, such as metals, can be effectively expanded by more efficient use, recycling, substitution of one material for another, and extraction of materials from dilute or dispersed supplies (figure 8.4).

Renewable resources include the sun—our ultimate source of energy—and the biological organisms and biogeochemical cycles that it sets in motion. In theory, these resources are infinite. Biological organisms, especially, are self-renewing. Theoretically, we can harvest surplus plant and animal populations indefinitely without reducing the available supply. As human activities expand, however, we are finding that when we deplete or disturb environmental cycles or populations of organisms past a certain point, resources can be pushed into a catastrophic downward spiral (figure 8.5). The assimilative and rejuvenative capacities of natural systems can be overwhelmed and effectively destroyed, threatening the organisms in the systems, eventually including humans.

Abstract or **intangible resources,** including open space, beauty, serenity, genius, information, diversity, and satisfaction, are also important to us (figure 8.6). Strangely, these resources are both infinite and exhaustible. There is no upper limit to the amount of beauty, knowledge, or satisfaction that can exist in the world, yet they can be easily destroyed. Unlike tangible re-

FIGURE 8.6 Many of the intangible resources of nature enrich the quality of our lives. Our enjoyment of these resources may be enhanced by sharing them with others. On the other hand, too many other people, or people engaged in conflicting activities, can destroy the values we seek.

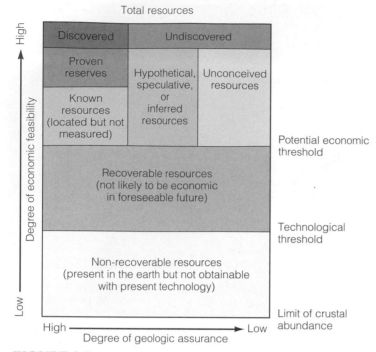

FIGURE 8.7 Categories of natural resources. These categories, based on degree of geological assurances and economic feasibility, were intended to describe mineral resources. With some modification, these categories can describe many types of nonrenewable resources.

sources that usually are reduced by use or sharing, intangible resources often are increased by use and multiplied by sharing. Sometimes these resources are identified as ones that "enrich the quality of life." When they attract or deter settlement of people and their businesses, a tangible economic impact on a human community may result.

Economic Categories

Although we have defined a resource as anything that is useful, we should distinguish between economic usefulness and the total abundance of a material or service that is available. Vast supplies of potentially important materials are present in Earth's crust, for instance, but they are useful to us only if they can be recovered in reasonable amounts with available technology and with acceptable environmental and economic costs. Within the aggregate total of any natural resource, we can distinguish categories on the basis of economic and technological feasibility, as well as the resource location and quantity (figure 8.7).

How do you interpret the terminology of "resource" and "reserve" discussions in newspapers and magazines? *Proven reserves* of a resource are those that have been thoroughly mapped and are economical to recover at current prices with current technology. You've probably heard or read about proven reserves of oil and natural gas, for instance. *Known resources* are those that have been located but are not completely mapped. Their recovery or development may not be economical now, but they are likely to become economical in the foreseeable

future. *Undiscovered resources* are only speculative or inferred from similarities between known deposits or conditions and unexplored ones. There are also unconceived resources that may be economic but have not even been thought of yet. *Recoverable resources* are accessible with current technology, but are not likely to be economic in the foreseeable future, whereas *nonrecoverable* resources are so diffuse or remote that they are not ever likely to be technologically accessible.

There is a vast difference between the first and last of these categories, at least in terms of Earth's crustal resources. The U.S. Geological Survey estimates the ultimate recoverable reserves of minerals to be approximately 0.01 percent (one hundredth of one percent) of the total material in the upper one km of Earth's crust. Chapter 9 discusses these crustal resources in greater detail, and chapter 18 gives some practical examples of proven, known, and ultimately recoverable fuel resources.

ECONOMIC DEVELOPMENT AND RESOURCE USE

The supplies of economically accessible natural resources and the use of these resources by a society depend on economic and technological development. To describe the nature and extent of resources, therefore, we should define the kind of economy in which those resources will be used. We generally see the categories of *frontier*, *industrial*, and *postindustrial* economy as stages in a developmental sequence. Sometimes, however, a country or a group of people will stop in one of the stages for one reason or another.

Frontier Economy

Reflect for a moment on the history of U.S. resource use. Vast eastern forests, enormous flocks of passenger pigeons, and thundering herds of bison seemed to be in unlimited supply and were easily exploited during our Frontier Era. When supplies of natural resources are plentiful relative to capital or labor, those resources typically are regarded as expendable and of little value. Because resource stocks are relatively large, their market values are nearly constant over time, unaffected by either contemporary rates of use or cumulative use. Both capital and labor supplies are limited on a frontier, however, and wasteful and profligate use of resources may be substituted to facilitate development.

Pollution levels tend to be relatively low in a frontier economy because population density and industrial activities tend to be low compared to the absorptive capacity of the environment. Resulting wastes can be processed by natural systems without overloading them.

Industrial Economy

The Industrial Revolution was one of the most significant developments in human history. Substitution of water power and

FIGURE 8.8 In a postindustrial society, information, education, research, communication, and technology replace material goods or natural resources as the main sources of wealth and power. Service also becomes a major segment of the economy.

fossil fuels for muscle power and the introduction of machines and mass production techniques have brought great changes to both human society and the environment.

In an industrial economy, greater production capacity increases the rate of resource use and depletes stocks of readily available resources. This, in turn, causes rising extraction and shipping costs because poorer and more remote sources are used. Factory owners are willing to pay more for raw materials because they can produce more valuable goods from those materials than they could have done in a frontier society. Workers receive higher hourly wages because technology and economy of scale make them more productive. This means that they can afford to pay more for goods and services. Pollution levels tend to be high in an industrial society due to the high rate of materials that flow through the system.

Postindustrial Economy

Most of the developed world has undergone a dramatic change in the source of wealth and power in the past century. More than 50 percent of the real economic growth in the United States in that period has been in organizational, educational, and technological areas. We have gone beyond an industrial-based economy to an information-based economy. A century ago, 90 percent of the American work force was engaged in farming, harvesting resources (logging, mining, etc.), or manufacturing. Now only 2 percent farm and 20 percent harvest resources or produce material goods. The rest are involved in "service" work, at least two-thirds of which involves transferring, storing, or processing information of some kind.

Information has become our most important resource and is, therefore, the main source of wealth and power. Unlike most resources, it is not depleted through use. It is expandable, shareable, transportable, substitutable, and storable (figure 8.8). The

more widely information is disseminated and used, the more valuable it becomes, for the most part, and the more we have. Although some people may try to restrict information flow to gain power or financial benefits, there is no scarcity rent (charge for using a depletable resource) on information that reflects diminishing supplies. Since few materials are used in an information and service economy, pollution levels are lower than those in an industrial economy. Table 8.1 compares some characteristics and strategies of an industrial or material economy to those of an information economy.

POPULATION AND RESOURCE SUPPLIES

As the preceding discussion shows, the supply of a particular natural resource available for human use is not determined so much by the absolute amount present on Earth as by economic, social, and technological factors.

Supply, Price, and Demand Relationships

Supply depends on (1) which raw materials can supply a service using present technology, (2) the availability of those materials in various quantities, (3) the costs of extracting, shipping, and processing them, (4) competition for those materials by other uses and processes, (5) feasibility and cost of recycling already used material, and (6) social and institutional arrangements in force.

In a market system, most of the considerations previously mentioned are expressed in terms of market price for a good or service—the amount it sells for. The available quantity of a resource or opportunity usually increases as the price rises. For example, the U.S. Bureau of Mines estimated in 1968 that the supply of uranium oxide (U_3O_8) available at the then current price of $8/ton would be 191,000 tons. At $30/ton, the supply would increase 50 percent to be 285,000 tons. If the price were to rise to $70/ton, about 2.6 million tons would be economically recoverable. The reason for an increasing supply was not that uranium was being created, but that it would become worthwhile to extract lower quality ore bodies as prices rose. If the price were to get too high, however, substitute material might become more attractive than uranium, and the effect would be as if a whole new resource had been created.

In economic terms, the relationship between available supply of a commodity or service and its price is known as a **supply/demand curve.** Demand is the amount of a product that consumers are willing and able to buy at various possible prices during some time period, other things being equal. Supply is the quantity of that product being offered for sale. If sellers increase the quantity of a product as the price increases, or if buyers want to purchase more of a product at a lower price and less at a higher price, we say that the product has **price elasticity.** The inverse relationship between these factors is shown

TABLE 8.1 Comparison of material and information economies

Expansive: Cheaper resources favor centralized manufacturing and large-scale distribution. An idea becomes a shop, a shop becomes a chain, a chain becomes an industry; General Motors or McDonalds, for example.

Replicative: Large-scale production requires product uniformity and long runs to achieve maximum efficiency. Regional differences and advantages are washed out.

Accretive: Wealth and power are derived from amassing resources or by dominating markets.

Consumptive: Consumption is encouraged to keep industrial systems operating. Higher productivity raises wages, inflating demands for goods and services.

Entropic: High consumption of resources and energy produces waste, pollution, and toxicity.

High profits and wages: High rate of resource extraction and energy consumption brings high profits for owners and high wages for workers.

Specialized: Expansion and competition favor a narrow set of skills.

Contractive: When resources are no longer limitless or inexpensive, contraction becomes the key to survival. Contraction makes consumers smarter and business leaner.

Differentative: Contraction causes large markets to break up. Production must become more flexible to meet specific needs. Production efficiency is not as important as utility. "Does the product really work for me?"

Cooperative: Interrelated society requires cooperation to maintain living standards. Competition is disruptive; power and wealth come from bringing groups together.

Conservative: Overconsumption and pollution threaten well-being by reducing environmental quality and resource availability for all. A conservative ethic is imperative.

Steady-state: Stable population, conservation, and use of renewable resources produces a sustainable system. Knowledge, information, and communication are the basis of the economy.

Lower profits and wages: Conservation and lower consumption means lower profits and stabilizing wages. People make or grow what they need. Barter and trade become a larger part of the economy.

Generalized: People need to have a broad range of skills to be more self-sufficient and less reliant on specialists to produce the goods and services they need.

From *Sustainable Communities* by Sim Van Der Ryn and Peter Calthorpe. Copyright © 1986 by Van Der Ryn and Calthorpe. Reprinted with permission of Sierra Club Books.

in a supply/demand curve (figure 8.9). As the price rises, the supply increases and the demand falls. The reverse holds as the price decreases. The intersection between these two curves is called the **market equilibrium.** The price will tend to be maintained at this point by natural market forces in a competitive free market.

Market Efficiencies and Technological Development

In a frontier economy, procedures for gaining access to resources and turning them into useful goods and services tend to be primitive and inefficient. As markets develop, however, experience accumulates in obtaining and working with a particular resource. Specialization and experimentation lead to discovery of new, more efficient technology, making it possible to produce larger quantities of goods at lower prices. The supply/demand curve shifts to a new equilibrium point, as is shown in figure 8.10. At each successive stage in this development process, a larger quantity of product is available at a lower price. The effect is that the standard of living increases—at least in economic terms.

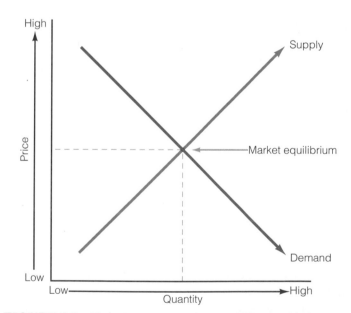

FIGURE 8.9 Classic supply-demand curves. When price is low, supply is low and demand is high. As prices rise, supply increases, but demand falls. The market equilibrium is the price at which supply and demand are equal.

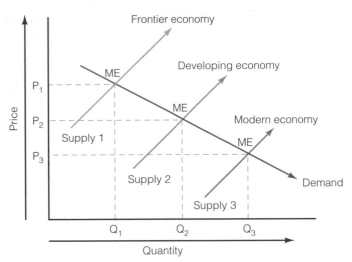

FIGURE 8.10 Supply and demand curves at three different stages of economic development. At each stage there is an equilibrium point at which supply and demand are in balance. As the economy becomes more efficient, the equilibrium shifts so there is a larger quantity available at a lower price than before. (P = price, Q = quantity, ME = market equilibrium.)

The Law of Diminishing Returns

In the early stages of economic development, adding more laborers to any particular business tends to *increase* the output per person up to a maximum level of production, as shown in figure 8.10. Adding more workers or more labor beyond that optimum point, however, tends to decrease efficiency. This is known as the **law of diminishing returns** because any change from the optimum level of input tends to reduce the efficiency of the operation (figure 8.11).

A similar effect applies to extraction of a limited resource. As the easily accessible reserves are depleted, we must extract raw materials from ever more remote or dilute deposits. This means that a larger share of society's investment capital must be diverted to this process and less is available for consumption and real growth. Eighty years ago, getting oil out of the ground in the United States required little more than pounding in a pipe. Now we spend billions of dollars to drill offshore or in the Arctic to get the same product.

Population Effects

There can be a double effect of diminishing returns on a national level with increasing population. As the number of workers increases, a point may be reached at which there are not enough jobs to employ everyone efficiently. As a result, the productivity per person will decline and wages will fall. This is intensified by the pressure on resources created by a growing population. More people use more resources, and we must look to less accessible or desirable supplies. Raw materials, therefore, become more ex-

pensive, as do the prices of goods and services provided by those resources; thus, workers have less money to buy more expensive items, and the standard of living declines. This "iron law of diminishing returns" led Thomas Malthus (chapter 7) to predict that unrestrained population growth would inevitably cause the standard of living to decrease to a subsistence level where poverty, misery, vice, and starvation would make life permanently drab and miserable.

Growing populations also place a strain on economic development by diverting the capital necessary for growth. In a rapidly growing country, a large proportion of the population is made up of children who require social overhead expenditures, such as new housing, schools, and roads, that contribute little to development. The creation of new jobs to employ a growing population also traps capital in conventional industries and lessens investments in new technology that might provide a real improvement in the standard of living. This diversion of investment capital is called the **population hurdle.**

However, growing populations can create markets that encourage specialization, innovation, and capital investment that result in efficiency. They also can bring young, energetic, and better trained workers into the work force and make changes possible in traditional patterns of doing things. Some demographers argue that growing populations cause problems, but they also generate more human ingenuity, energy, and cooperation to solve those problems. Where are we now in the process of economic development and population growth? Are we on a curve of diminishing returns, or are we benefiting from economy of scale in terms of human populations and environmental problems?

Factors That Mitigate Scarcity

Human social systems can adapt to resource scarcity in a number of ways. Some economists point out that scarcity provides the stimulus that catalyzes innovation and change (figure 8.12). As materials become more expensive and difficult to obtain, it becomes cost-efficient to try to discover new supplies or to use the ones we have more carefully; thus, we may be better off in the long run because of these developments.

Several factors can alleviate the effects of scarcity:

1. Technological inventions can increase efficiency of extraction, processing, use, and recovery of materials.
2. Substitution of new materials or commodities for scarce ones can extend existing supplies or create new ones. For instance, substitution of aluminum for copper, concrete for structural steel, grain for meat, and synthetic fibers for natural ones all remove certain limits to growth.

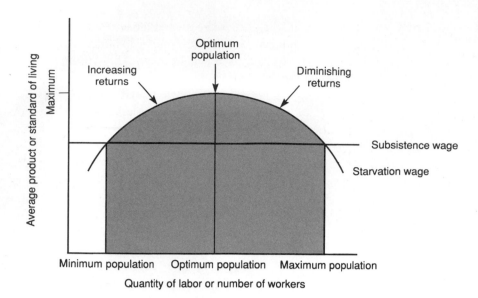

FIGURE 8.11 Graph of diminishing returns at a given level of resources and technology. Increasing labor input raises productivity until optimum population is reached, then diseconomies set in and returns per unit of labor diminish. When production falls below subsistence wage level, workers will starve or find something else to do, thus reducing the work force.

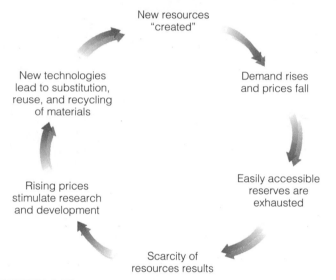

FIGURE 8.12 Scarcity/development cycle. Paradoxically, resource use and depletion of reserves can stimulate research and development, substitution of new materials, and the effective creation of new resources.

3. Trade makes remote supplies of resources available and may also bring unintended benefits in information exchange and cultural awakening.

4. Discovery of new reserves through better exploration techniques, more investment, and looking in new areas becomes rewarding as supplies become limited and prices rise.

5. Recycling becomes feasible and accepted as resources become more valuable. Recycling now provides about 37 percent of the iron and lead, 20 percent of the copper, 10 percent of the aluminum, and 60 percent of the antimony that we consume each year in the United States.

Increasing Environmental Carrying Capacity

Economist Julian Simon says that in spite of recurring fears of scarcity of natural resources, technological advances, market development, and the economic efficiencies of mass production have made every commodity cheaper in real terms over time. In fact, responding to the growing scarcity of resources actually enables us to increase the carrying capacity of the environment for humans. Figure 8.13 shows the change in real Gross National Product (GNP) (as a percent of that in 1890) during the period of industrial growth and transformation to a postindustrial society. There has been about a 500 percent increase in per capita GNP during this century even though population has tripled and the easily accessible resources largely have been used up. Most countries have not shared in the rapid increase of wealth enjoyed by the United States. Figure 8.14 shows GNP and population by region. As you can see, North America and Europe have 65 percent of the world GNP but only 18 percent of the world population. Asia, Latin America, and Africa have 74 percent of the population, but only 16 percent of the GNP. This may be a very real example of the population hurdle discussed in chapter 7.

LIMITS TO GROWTH

We return again and again in environmental studies to the underlying question of whether continued population growth and/or economic growth would be good or bad. At what point will continued growth in the number of people or of economic activities that impinge on our environment bring disaster, not only to humans, but to the whole life-supporting system of the biosphere? On the other hand, how will we improve the standard of living in less developed countries and clean up damage done to the environment by earlier stages of economic and technical

RESOURCES AND RESOURCE ECONOMICS

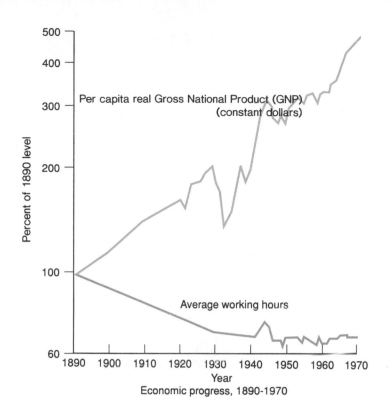

Economic progress, 1890-1970

FIGURE 8.13 Technological improvements, capital investments, and more highly trained labor have raised production faster than population growth in the United States. Between 1890 and 1970, real per capita GNP rose fourfold even though average working hours fell nearly 40 percent. In other words, our increased productivity has given us both more output and more leisure.

Source: U.S. Department of Commerce.

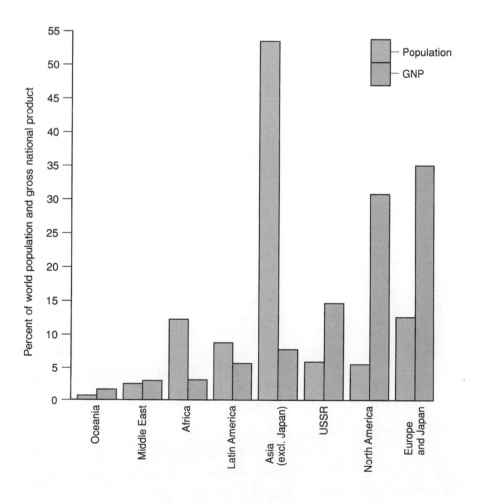

FIGURE 8.14 Percent of world population and Gross National Product (GNP) by region. The more developed countries (MDC) have 65 percent of the world GNP, but only 18 percent of the world's population. The less developed countries (LDC) have 74 percent of the world's population but only 16 percent of the total GNP.

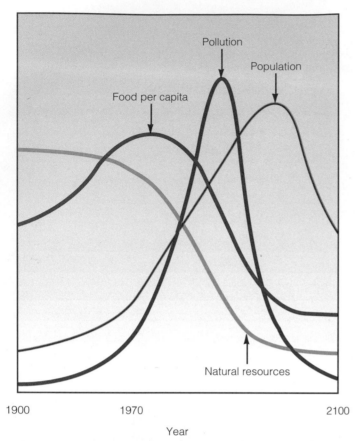

FIGURE 8.15 The standard run of World Computer Model 3 developed by the Club of Rome. This model assumes no major changes in technology, resource availability, or food production. Natural resources diminish as population rises until increasing death rates cause a massive dieback or population crash.

development if growth stops? The crux of this argument is whether resources are finite or can be effectively expanded through human ingenuity and enterprise. Let's look at what some economic models predict might be the result of further economic and population growth.

Computer Models of Resource Use

In the early 1970s, an influential study of resource limitations funded by the Club of Rome was undertaken by a team of scientists from the Massachusetts Institute of Technology headed by Donnela Meadows. The results of this study were published in 1972 in *Limits to Growth*. Several computer models of world economy were used to examine various scenarios of resource depletion, growing population, pollution, and economic response. In every model tested, a catastrophic economic collapse is predicted to occur within the next one hundred years unless population growth is immediately reduced and we adopt a steady-state economic system based on renewable resources and nonpolluting production and consumption techniques.

Figure 8.15 is a typical example of a run of one of the world models. Note that population continues to rise for some time after resources and food supplies begin to diminish. Even-

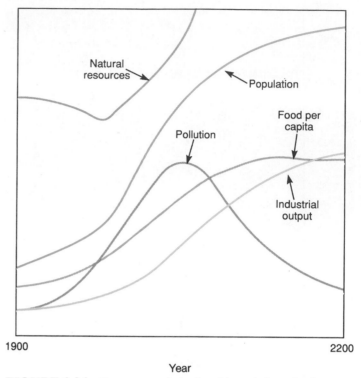

FIGURE 8.16 Computer analysis of world trends from the Sussex group. This model uses the same consumption and output patterns as figure 2.15 but assumes technological progress. Note that resources increase and pollution declines, while other factors stabilize.

tually, the population crashes. There is a strong similarity between this growth curve and the irruptive (Malthusian) curves observed in natural populations (chapter 6).

Some authors have criticized *Limits to Growth* because the computer models used by the Meadows group discounted the factors that might mitigate the effects of scarcity. An alternative set of computer models was presented in the book *Models of Doom* by the Science Policy Research Unit of the University of Sussex, England. They point out that by "building into the Meadows model technical progress in natural resource discovery, pollution abatement, and agricultural developments at rates that have been achieved historically for large parts of the world, all modes of collapse are avoided indefinitely into the future."

An example of the alternative computer models run by the Sussex group is shown in figure 8.16. The assumptions of technological progress built into this model make natural resources, food supplies, and industrial output rise rather than fall. Population doubles or quadruples (depending on how much progress is assumed). Neither of these computer models prove that technological progress will, or will not, occur. They only show what the outcomes might be if it does, or does not, happen.

Why Not Conserve Resources?

Even if large supplies of resources remain, or the possibility of technological advances to mitigate scarcity exists, wouldn't it be better to reduce our use of natural resources so they will last as long as possible? Author Paul Ehrlich stated this proposition in *The Population Bomb* in 1968; "If I'm right [and we stop population growth], we will save the world. If I'm wrong, people will be better fed, better housed, and happier, thanks to our efforts. Will anything be lost if it turns out later that we can support a much larger population than seems possible today?"

This makes sense if the assumption is correct that resources are limited, cannot be replenished, or that their extraction inevitably degrades the environment. Using them more slowly and sharing them with fewer people will give each of us a larger share, will be kinder to the environment, and will make our supply last longer. Most economists, however, look at resources as means to an end, rather than having inherent value in themselves. They think of resources as being like money. If you bury your savings in a jar in the backyard, it will last a long time but may not be worth much when you dig it up. If you invest it productively, you will have much more in the future than you do now. Furthermore, there may be a window of opportunity for investment that is open now but may not be later.

RESOURCE ECONOMICS

Some of the most crucial commodities that may limit our future growth are not represented by monetary values in the marketplace. Groundwater, sunlight, clean air, biological diversity, and other common resources are treated as public goods that anyone can use freely. Our economic system does not charge for using the absorptive capacity of the environment to dispose of wastes. In theory, these resources are self-renewing, but as we will show in other chapters of this book, many of these vital environmental goods and services are threatened by human activities. If we damage basic life-support systems of the biosphere, we cannot simply substitute another material or service for the ones that have become limited. The crux of this question is the way we manage resources in a market system. Let's look now at how our economic system handles internal and external costs and intergenerational justice.

Internal and External Costs

Internal costs are the expenses (monetary or otherwise) that are borne by those who use a resource. Often, internal costs are limited to the direct out-of-pocket expenses involved with gaining access to the resource and turning it into a useful product or service.

External costs are the expenses (monetary or otherwise) that are borne by someone other than the individuals or groups who use a resource. External costs often are related to public goods that, in turn, are related to a resource. Some examples of external costs are the general environmental or health effects of using air or water to dispose of wastes. Since these effects usually are diffuse and difficult to quantify, they do not show on the ledgers of the responsible parties. They are likely to be ignored

in private decisions about the costs and benefits of a project. One way to use the market system to optimize resource use is to make sure that those who reap the benefits of resource use also bear all the external costs. This is referred to as **internalizing costs.**

A controversial provision of the 1990 revision of the Clean Air Act allows companies to market emission quotas as a way of reducing pollution in the most efficient and least costly way possible (Box 8.1). This would have the effect of internalizing external costs but is regarded by its opponents as merely a license to pollute.

Intergenerational Justice

In setting values for discount rates and future worth of natural resources, we are making decisions not only for ourselves, but also for future generations. Although having access to clean groundwater or biological diversity one hundred years from now isn't worth much to me—assuming that I will be long gone by that time—those resources might be quite valuable to those alive at the time. Future citizens are very much involved in the choices we make today, but they don't have a vote. Our decisions about how to use resources raise difficult questions about justice between generations. How shall I weigh their interests in the future against mine right now?

Even if we discount our own future, is it reasonable to also discount that of following generations? How do we know what resources will be worth to them? What obligations do we have to leave resource supplies for our progeny? You might say "What has posterity ever done for me?" Perhaps economist Kenneth Boulding has the best answer for this question. He says, "The welfare of the individual depends on the extent to which he can identify himself with others, and the most satisfactory identity is not only with a community in space, but also extending over time from the past to the future."

Cost/Benefit Ratios

One way to evaluate the outcomes of large-scale public projects is to analyze the costs and benefits that accrue from them. The assumption that marginal cost/benefit analysis can be applied to present and future values of a resource, given proper criteria and procedures, is one of the main conceptual frameworks of resource economics. This process is usually controversial, however, because it deals with vague and uncertain values and compares costs and benefits that are as different as apples and oranges. Yet we continue doing these analyses because we don't have better ways to allocate resources.

Figure 8.17 presents a flowchart for preparing a cost/benefit analysis. As you can see, several different tributary paths come together to determine the final outcome of this process. The easiest parts of the equation to quantify are the direct costs and benefits to the developer or agent who has proposed the project: the out-of-pocket expenses and the immediate profits that will

BOX 8.1   Marketing Pollution Allowances

What is the most efficient and economical way to eliminate pollution? Some people argue that we should simply say to polluters, "Stop it! You can't dump effluents into the air or water anymore." Many economists, however, view absolute injunctions and rigid limits as counterproductive and illogical. They would prefer to rely on market mechanisms to balance costs and effects and to reduce pollution.

The 1990 revisions of the Clean Air Act (chapter 23) contain provisions for marketing pollution reductions. This is how they might work. One of our goals is to reduce sulfur dioxide pollution by 50 percent. We could simply require the major sources (power plants and smelters) to reduce their emissions by half. Some companies might be able to reach this goal easily and cheaply. Some, in fact, might be able to achieve more than 50 percent reduction, but it is unlikely that they will spend the money to do so if it isn't required. Others might face exorbitant costs and real hardships in trying to reach the required goal.

Suppose, for instance that we have two neighboring power companies. Each burns high-sulfur coal and emits 75,000 tons of sulfur dioxide per year. Company A has a new power plant that could be easily retrofitted with sulfur scrubbers. For $50 million, it could eliminate half its sulfur emissions (37,500 tons/year). For another $50 million, it could reduce emissions by 98 percent (another 36,000 tons/year). Company B has an old, decrepit plant. It would cost $75 million to remove just half of its sulfur.

One alternative would be to require both utilities to reach 50 percent sulfur reduction. This would cost a total of $125 million and remove a total of 73,000 tons/year. It would also make electricity much more expensive for customers of Company B. Another alternative would be to allow Company B to pay Company A $50 million to install 98-percent efficient scrubbers. This plan would accomplish nearly the same pollution reduction but

BOX FIGURE 8.1 Allowing firms to buy and sell pollution allowances or "rights to pollute" may be the most efficient way to reduce pollution on a regional level. It could have unfortunate local impacts, however.

would cost only $100 million and spread the costs more equitably.

Yet another alternative might be for Company B to abandon its old power plant and install some alternative energy source (such as solar) that produces no pollution. Suppose a solar facility costs $100 million. Company B can't afford that investment on its own. Instead of installing scrubbers in its own plant, Company A might contribute $50 million to help its neighbor convert to solar energy. Company B would save $25 million and sulfur emissions would be reduced by an extra 1,000 tons/year over having scrubbers on both existing power plants.

This plan would mean that Company A could continue to pollute even though it could install scrubbers. Purists find this idea reprehensible. They believe that every factory should be required to reduce pollution as much as possible. To allow pollution because it is economically

efficient is repugnant in this view. Others argue that only the total amount of pollution matters, not where it comes from. If we can meet our goals in a less costly way, doesn't it make sense to do so?

This economic analysis assumes that everyone in a given neighborhood suffers equally from pollution and benefits equally from economic efficiency. Obviously, this is not always the case. In our hypothetical example, residents near Company A may not benefit much from cleaner air around their neighboring utility. Nor is it an advantage to them that Company B saves $25 million. On the other hand, the air is cleaner, on average, and the cost is lower, on average. This approach also assumes that there is only a certain amount of pollution that it is economical or necessary to remove. It rejects the idea of zero emissions as unworkable and unreasonable. What do you think?

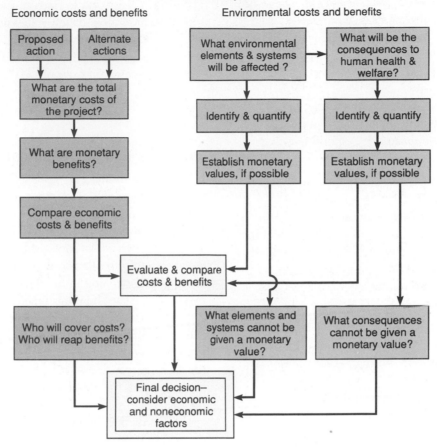

FIGURE 8.17 A flowchart for a cost/benefit analysis.

result from this investment. These direct monetary costs and benefits are usually the most concrete and accurate components in the analysis. It is important that they not outweigh other factors more difficult to ascertain but of equal importance.

The other branch of the flowchart involves analysis of more diffuse, nonmonetary factors, such as environmental quality, ecosystem stability, human health impacts, historic importance of the area to be affected, scenic and recreational values, and potential future uses. It is difficult to quantify these values. It is even more difficult to express them in monetary terms. How much are beauty or tranquility worth? What are the benefits of ethical behavior? How much would you pay for good health?

Some project costs and benefits simply cannot be expressed in monetary terms. These invaluable (in a positive sense) factors bypass the mathematical stages of comparison and are considered—we hope—in the final decision-making process, which is more political than mechanical. Also factored in at this stage are distributional considerations (i.e., who will bear the costs of the project and who will reap the benefits). If these are different people, as they usually are, questions of justice arise that need to be resolved, perhaps in some other venue.

The criticisms and complications of this process include the following:

1. *Absence of standards.* Each person assesses costs and benefits by his or her own criteria, often leading to conflicting conclusions about the comparative values of a project. It has been suggested that an agency or influential group set specifications for how factors should be evaluated.

2. *Inadequate attention to alternatives.* To really understand the true costs of a project, including possible loss of benefits from other uses, it is essential to evaluate alternative uses for a resource and alternative ways to provide the same services. This is often slighted.

3. *Assigning monetary values to intangibles and diffuse or future costs and benefits.* Some critics of this process claim that we should not even try. They believe that attempting to express all values in monetary terms leads people to believe that only monetary gains and losses are important. This is why, in cases of ethics and morals, we more often say; "Thou shalt not . . ." rather than "You may do that if you want, but it will cost you. . . ."

4. *Acknowledging the degree of effectiveness and certainty of alternatives.* Sometimes speculative or even hypothetical results are given specific numerical values and treated as if they were hard facts. We should use risk-assessment techniques to evaluate and compare uncertainties in the process.

5. *Justification of the* status quo. Agencies may make decisions to go ahead with a project for political reasons and then manipulate the data to correspond to preconceived conclusions.

Tangible and Intangible Assets

Can the market work in a socially responsible way in setting priorities for use and protection of intangible assets? Sometimes it does, but there are many instances where it does not. It is important to recognize when we can and cannot depend on market forces to manage natural resource economics. Consider the following points:

1. In theory environmental services, such as pollution absorption, are available to everyone and are not bought and sold. There is, therefore, no market for them and no price-setting mechanism.

2. In making decisions about whether it would be more profitable to use a resource now or save it for the future, private interests almost always choose to use it quickly because they benefit directly from immediate use, whereas the benefits from future use may be spread among many users.

3. Private discount rates for future benefits do not consider social objectives and, therefore, do not reflect complete values of *in situ* resources.

4. The costs for destruction of a resource are often external, that is, they accrue to individuals other than those who benefit from its use.

Since cost/benefit ratios are often the deciding factors in public works projects and setting priorities regarding natural resources, students of environmental science should understand how these values are determined. First, we have to decide who benefits and who pays. Then, we have to set prices on the goods and services provided by the environment. This may be fairly straightforward for tangible assets, but it is very difficult when considering opportunity costs or existence values. What is the market value of good health, or seeing a grizzly bear in the wild, or just knowing that a herd of caribou exists in Alaska?

The Office of Management and the Budget regards overmature trees in the forest as a loss of harvestable board feet of lumber and suggests that the Forest Service should declare more timber sales. This viewpoint ignores such concepts as ecosystem and biological community. What about their value as habitat for wild orchids, flowering shrubs, insects, birds, and fungi?

How much is it worth to have an undisturbed forest for wilderness recreation? Recreation values are usually set by estimating how much people are willing to pay to get to an area or to experience its pleasures. If it costs $100 per day to sail a luxury cruiser on a reservoir but only $10 per day to paddle your own canoe, does that mean that motorboating is ten times more worthwhile than canoeing? Is one motorboater balanced by ten canoeists?

International Development

No single institution has more influence on the financing and policies of developing countries than the World Bank. Of the $24 billion approved for Third World projects by multinational development banks in 1987, $17 billion came from the World Bank. For every dollar invested by the World Bank, two dollars are attracted from other sources. If you want to have an impact on what is happening in the developing countries, it is imperative to understand how this huge enterprise works.

The World Bank was founded in 1945 to provide aid to war-torn Europe and Japan. In the 1950s, its emphasis shifted to development aid for Third World countries. This aid was justified on humanitarian grounds, but providing markets and political support for Western capitalism was an important by-product.

The Bank is jointly owned by 150 countries, but one-third of its support comes from the United States. The Bank president has always been an American. The bulk of its $66.8 billion capital comes from private investors who buy instruments (bonds and debentures) on the open market. Loan applications are first screened by the professional staff and then voted on by a council of all member countries. No loan has ever been turned down once it makes it to the full council. So far, one hundred countries have borrowed a total of $118 billion from the World Bank for a wide variety of development projects.

Many World Bank projects have been environmentally destructive and highly controversial. In Botswana, for example, $18 million was provided to increase beef production for export by 20 percent, despite already severe overgrazing on fragile grasslands. The project failed, as did two previous beef production projects in the same area. In Ethiopia, rich floodplains in the Awash River Valley were flooded to provide electric power and irrigation water for cash export crops. More than 150,000 subsistence farmers were displaced and food production was seriously reduced. In India, the sacred Narmada River is being transformed by thirty large dams and 135 medium-sized ones financed by the World Bank. About 1.5 million hill people and farmers will be displaced as a result.

These are only a few of the projects that have aroused concern and protest. They typify the environmental and social costs of many development programs. Loans from the World Bank tend to favor large-scale agricultural, transportation, and

BOX 8.2   Microlending at the Grammeen Bank

To foster economic development in the poorer nations of the world, the richer nations have generally relied on mega-projects funded through bilateral development banks or the World Bank. The billion-dollar projects financed by these loans are highly political and unwieldy. They often do more social, environmental, and economic harm than good. An exciting alternative called microlending has been pioneered by Muhammad Yunus of the Grammeen Bank in Bangladesh. This unconventional bank is one of the most heartening success stories in the field of developmental economics.

The purpose of microlending is to provide credit directly to the poorest people, those who have no collateral and no steady source of income. The 400,000 customers of the Grammeen Bank are mostly women who would never have had a chance to borrow money from an ordinary bank. Their loans are small, averaging only $67. They are enough, however, to buy a used sewing machine, a bicycle, a loom, a cow, some garden tools—something to start a small business and provide needed family income.

These are people to whom most bankers wouldn't even consider giving

BOX FIGURE 8.2 Community development banks make "micro-loans" to individuals and small companies to buy tools, such as this spinning wheel used by this woman in India. Collective management of these projects provides mutual support and education in financial management. Grass roots development is often both more democratic and more effective than huge multinational projects.

credit, but the recovery rate on Grammeen loans is an astonishing 98 percent. Borrowers are organized into five-member peer groups that act both as mutual aid societies and collection agencies. Payments must be made in regular weekly installments. If one member of the group defaults, the others must repay the loan. Having some dignity, respect, and independence is highly empowering and encourages responsibility and self-reliance.

Microlending and self-help programs are now being used around the world, including the United States. They may be doing more to help people improve their lives than all the billions invested in development programs by conventional banks and international agencies.

energy projects, which make up more than half of all funds it distributes. In part, this is because both the World Bank and debtor nations find it easier to manage one big project than many small ones. Furthermore, both the World Bank and the governments borrowing money prefer large, impressive, modern projects to show how the money is being spent. Small cooperatives or cottage industries might do more for the people than a big dam, but they don't look as impressive. Recently, however, a near and viable model has been demonstrated (Box 8.2).

In 1984, the U.S. Congress insisted that all loans be reviewed for environmental and social effects before being approved. It asked for assurance that each project (a) use renewable resources and not exceed the regenerative capacity of the environment, (b) not cause severe or irreversible environmental deterioration, and (c) not displace indigenous people.

In 1986, only six of the nearly seven thousand staff members were assigned to do ecological assessments on $17 billion in development projects. Pressure from environmental groups has forced the World Bank to add about 100 ecologists and environmental specialists to the staff. Critics charge that many of these people are merely recycled economists and that business is going on as usual. Loans have been canceled, however, for road building that would lead to tropical forest destruction and loans have been made for environmental restoration projects in Brazil. The World Bank continues to be criticized, at least partly because it's easy to aim at a large target. A fundamental question remains whether technological and economic growth can bring a better standard of living for everyone without causing unacceptable environmental disruption.

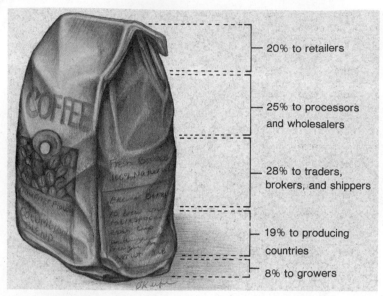

- 20% to retailers
- 25% to processors and wholesalers
- 28% to traders, brokers, and shippers
- 19% to producing countries
- 8% to growers

FIGURE 8.18 What do we really pay for when we purchase a pound of coffee?

International Trade

International issues further complicate questions of resource management. Much of the vast discrepancy between richer and poorer nations is related to past economic and political history, as well as to current international trade relations. The banking and trading systems that control credit, currency exchange, shipping rates, and commodity prices were set up by the richer and more powerful nations in their own self-interest. These systems tend to keep the less developed countries in a perpetual role of resource suppliers to the more developed countries. The producers of raw materials, such as mineral ores or agricultural products, get very little of the income generated from international trade (figure 8.18). Furthermore, they suffer both from low-commodity prices relative to manufactured goods and from wild "yo-yo" swings in prices that destabilize their economies and make it impossible to either plan for the future or to accumulate the capital for further development.

The result of this market system is that the Third World nations of Asia, Africa, and Latin America have 74 percent of the world's population, but only 16 percent of the total GNP. The more developed countries of Europe, North America, and the Soviet Union have only 24 percent of the world's population, but have 80 percent of the GNP.

In 1974, the United Nations General Assembly adopted Resolution Number 3281 calling for a "new international economic order" in which "every State has and shall freely exercise full permanent sovereignty, including possession, use and disposal over all its wealth, natural resources and economic activities . . . to regulate foreign investment and transnational corporations . . . and to nationalize, expropriate or transfer ownership of foreign property." This idea of a new relationship

between the nations of the North and South is developed further in the "Brandt Commission Report" of 1980, which was chaired by Willy Brandt, former mayor of West Berlin.

SUSTAINABLE DEVELOPMENT: THE CHALLENGE

One of the most difficult, and as yet unanswered, questions in this chapter is whether further growth in either population or economic activity is tolerable. On one hand, many environmental scientists warn that human activities are overwhelming the basic life-support systems of the biosphere. They call for a steady-state economic system that will minimize our impact on the environment. On the other hand, many people see economic growth as a way to bring the underdeveloped nations of the world up to a higher standard of living without calling on those of us in the richer nations to give up the wealth we now enjoy. Economic growth is also advocated as necessary to provide the funds necessary to clean up the environmental damage caused by earlier, more primitive technologies and misguided resource uses. It is estimated to cost $150 billion per year to control population growth, develop renewable energy sources, stop soil erosion, protect ecosystems, and provide a decent standard of living for the world's poor. This is a great deal of money but is small compared to the $1 trillion per year spent on wars and military equipment.

An intermediate position between the extremes of no growth versus unlimited growth is **sustainable development** based on use of renewable resources in harmony with ecological systems. Economist Robert Repetto of the World Resources Institute outlined some of the changes necessary for a transition to a more stable world in *Global Possible: A Report of the Global Possible Conference*. Some of his principles and strategies for achieving these goals are listed in tables 8.2 and 8.3.

TABLE 8.2 Principles for sustainable development

1. A demographic transition to a stable world population of low birth and death rates.
2. An energy transition to high efficiency in production and use, coupled with increasing reliance on renewable resources.
3. A resource transition to reliance on nature's "income" without depleting its "capital."
4. An economic transition to sustainable development and a broader sharing of its benefits.
5. A political transition to global negotiation grounded in complementary interests between North and South, East and West.

Source: Robert Repetto, *Global Possible: A Report of the Global Possible Conference*. World Resources Institute, Washington, DC

RESOURCES AND RESOURCE ECONOMICS

TABLE 8.3 Strategies for sustainable development

1. *Greater attention to the problems of the poor people of the world*. In many cases, environmental damage is caused by people who have no other alternatives in their struggle for existence. Modest improvements in their income, health status, political freedom, and access to education, capital, and technology could have major impacts on society and the environment.

2. *Local input in planning and managing resource development*. Often local people have valuable ecological knowledge that is overlooked by planners. Giving them a better share of proceeds from resource development provides an incentive to protect and conserve resources.

3. *Proper resource pricing*. Internalizing external costs shows the real trade-offs in resource development and gives resource users the incentive to minimize *all* costs, not just those that directly affect themselves.

4. *Demand management*. Dramatic and far-reaching improvements in resource availability can be obtained by reducing or eliminating wasteful or unnecessary resource uses. This can be accomplished either by pricing mechanisms or by changes in regulations and institutional mechanisms. Sometimes it is faster and more effective to pay for new efficient equipment (insulation, new furnaces for the poor) than to depend on market forces or laws.

5. *Better management capability*. We need technical personnel, information, and legal and administrative systems to plan and guide resource use so that market forces or legal mechanisms will work as they should to protect and sustain vital resources.

6. *Minimize throughput*. By emphasizing durable goods, recycling and reuse of materials, efficiency, and lower consumption, we can lower our consumption of resource "capital" and reduce our impact on the environment.

7. *Redistribute wealth and power*. We can no longer justify inequality as necessary for savings, investment, and growth. As economist Herman Daly points out, "sustainable development will make fewer demands on our environmental resources, but greater demands on our moral resources."

8. *Develop nondestructive resource uses*. By focusing on activities that use intangible resources, such as information, creativity, communications, leisure, and art, we can have a life rich in values but with minimal impact on environmental resources.

Source: Robert Repetto, *Global Possible: A Report of the Global Possible Conference*. World Resources Institute, Washington, DC

Will we accomplish the transition to an economy based on sustainable development, intergenerational justice, an equitable distribution of resources, and a harmony with nature? We will succeed only if we bring the wisdom and knowledge of both ecology *and* economics to bear on our problems. The future will be what we make it.

> *It is scarcely necessary to remark that a stationary condition of capital and population implies no stationary state of human improvement. There would be just as much scope as ever for all kinds of mental culture and moral and social progress; as much room for improving the art of living and much more likelihood of its being improved when minds cease to be engrossed by the art of getting on.*
> John Stuart Mill
> *Principles of Political Economy*, 1857

S U M M A R Y

In this chapter, we have reviewed some economic theories of the effects of natural resource scarcity. A resource is defined as any useful material or service. This includes tangible, physical assets and intangible services of the environment. Resources are defined by their economic and technological feasibility, as well as their location and physical size. We distinguished between known, proven, inferred, and unconceived resources, and between economically important and technically accessible resources.

Two main mechanisms determine who shall benefit from natural resources. In the case of private goods, we depend on the marketplace to set prices that determine supply and demand. For public goods, where costs and benefits are widely spread and difficult to evaluate in a market price, we use the political process to reflect social values and to make a fair distribution of resources. For a number of reasons, the market may fail to act in an optimum way to set priorities in conservation or utilization of natural resources. Among the most important of these reasons are inadequate reflection of external costs, the public good, and future values in the price system.

Questions about the scarcity of resources and their effects on economic development are important in determining what kind of society we have. Three theories about economics and the role of resources form the basis for market or centrally planned economic systems. These systems are developmentally progressive, based on attitudes toward—and use of—natural resource stocks. These systems are frontier, industrial, and postindustrial economies. Some people have called for a transition to a steady-state economic system, one in which there is sustainable development.

It is important for us to decide, individually as well as collectively, how we should use our resources and how we can best reach the goal of a just, sustainable society.

■ Review Questions

1. Define a resource and distinguish between tangible and intangible resources.
2. What is the difference between economic growth and economic development?
3. List four economic categories of resources and describe the differences among them.
4. Describe the relationship between supply and demand.
5. What causes diminishing returns in natural resource use? How does population growth affect this phenomenon?
6. Describe how cost/benefit ratios are determined and how they are used in natural resource management.

7. Distinguish between a material-based and an information-based economy.
8. Describe how frontier, industrial, and postindustrial economies use resources, capital, and labor.
9. Why does the marketplace sometimes fail to optimally allocate natural resource values?
10. What would be some of the characteristics of a sustainable economic system?

■ Questions for Critical Thinking

1. What do you think are the most serious resource limitations related to economic activities that may limit further economic growth?
2. Do you believe that human ingenuity and enterprise can increase the carrying capacity of Earth over the long-term? If so, how? If not, why?
3. If you could retroactively stabilize economic growth or population growth at some point in the past, when would you choose for that to be?
4. Do you think that technological progress is likely to continue in the future as it has in the past? Why or why not?
5. How do you think we should determine the values and uses of environmental resources? How should we distribute the benefits of their use?
6. How might we achieve a sustainable economic system?
7. What would be the effect on the developing countries of the world if we were to change to a steady-state economic system? How could we achieve a just distribution of resource benefits while still protecting environmental quality and future resource use?
8. What costs of our industrial and societal processes are now treated as external costs? How might we internalize those costs?
9. What do you think would be a fair way to bring about intergenerational justice? How much use of resources would you be willing to forego now in order to leave more for future generations?
10. Do you think that economic development is necessary for environmental protection, or is development a threat to the environment?

■ Key Terms

economic development (p. 134)
economic growth (p. 134)
exhaustible resources (p. 134)
external costs (p. 143)
human resources (p. 134)
intangible resources (p. 135)
internal costs (p. 143)
internalizing costs (p. 143)
law of diminishing returns (p. 139)

market equilibrium (p. 138)
natural resources (p. 134)
population hurdle (p. 139)
price elasticity (p. 137)
production frontier (p. 133)
steady-state economy (p. 134)
supply/demand curve (p. 137)
sustainable development (p. 148)

■ SUGGESTED READINGS

■ Anderson, F. R., et al. *Environmental Improvement through Economic Incentives*. Washington, D.C.: Resources for the Future, 1978. Examines possibilities for a mixture of regulatory measures and economic incentives to control pollution problems.

■ Boulding, K. "The economics of the coming spaceship earth." In *Environmental Quality in a Growing Economy*. Baltimore: Johns Hopkins University Press, 1966. A comparison of "cowboy" economics and the "spaceship earth model" of steady-state economics.

■ Brant Commission. *Common Crisis: North-South Cooperation for World Recovery*. Cambridge, Massachusetts: MIT Press, 1983. A plan for reducing Third World poverty and disparities between rich and poor nations.

■ Carpenter, R. A., and J. A. Dixon. 1985. "Ecology meets economics: A guide to sustainable development." *Environment* 27(5):6. How the classic confrontation between economic development and environmental protection might be solved by combining ecology and economics.

■ Chandler, W. U. 1986. "The changing role of the market in national economies." *Worldwatch Paper #72*. Washington, D.C.: Worldwatch Institute. Comparison of market and centrally planned economies in a changing world.

■ Daly, H. E., 1991. *Steady-State Economics*. Washington D.C.: Island Press. A collection of essays on steady-state systems by economists, ecologists, and others.

■ Dorfman, R. "An economist's view of natural resource and environmental problems." In *Global Possible*, edited by R. Repetto. New Haven: World Resources Institute, Yale University Press, 1985. p. 67.

■ Georgescu-Roengen, N. 1977. "The steady-state and ecological salvation: A thermodynamic analysis." *BioScience* 27(4): 266. An insightful analysis on the ultimate limits to growth and the necessity for steady-state systems.

■ Harden, G. 1968. "The tragedy of the commons." *Science* 162: 1243. A classic article and one of the most influential metaphors in environmental science in the past twenty years.

■ Howe, C. W. *Natural Resource Economics: Issues, Analysis and Policy*. New York: Wiley, 1979. A good introductory text in basic resource economics.

■ Krutilla, J. V., and A. C. Fisher. *The Economics of Natural Environments: Studies in the Valuation of Commodity and Amenity Resources*. Washington, D.C.: Resources for the Future, 1985. A good discussion of natural resources and applied economics.

■ Larson, E. D., M. H. Ross, and R. H. Williams. June 1986. "Beyond the era of materials." *Scientific American* 254(6):34. Economic growth in industrial nations is no longer accompanied by increased consumption of basic materials.

Mill, J. S. *Principles of Political Economy*. Vol. II. London: J.W. Parker & Son, 1857. Contrary to most economists of his time, Mill questioned the need for and wisdom of continual growth.

Pearson, C. S. *Multinational Corporations, Environment, and the Third World*. Washington, D.C.: World Resources Institute, Duke University Press, 1987. Discusses pollution havens, hazardous exports, the workplace environment, and Bhopal.

Portney, P. R., ed. *Current Issues in Natural Resource Policy*. Washington, D.C.: Resources for the Future, 1982. Do environmental controls inhibit or facilitate economic efficiency?

Repetto, R., and T. Holmes. 1983. "The role of population in resource depletion in developing countries." *Population and Development Review* 9:609. What is the relation between economics, population, and environment?

Riddel, R. *Ecodevelopment: An Alternative to Growth Imperative Models*. Hampshire, England: Gower Publishers, 1981. Discusses ways that less developed countries can use ecological principles to direct sustainable development.

Schumacher, E. F. *Small is Beautiful: Economics as If People Mattered*. New York: Harper & Row, 1973. A highly readable and insightful discussion on Buddhist economics and good work from the father of appropriate technology.

Shrybman, S. 1990. "International Trade and the Environment" *Ecologist* 20(1):30. A useful and timely environmental assessment of the General Agreement on Tariffs and Trade (GATT).

Swartzman, D., et al., eds. *Cost-Benefit Analysis and Environmental Regulations: Politics, Ethics and Methods*. Washington, D.C.: The Conservation Foundation, 1982. Theory and application of cost-benefit analysis with case studies.

Walter, E. *The Immorality of Limiting Growth*. New York: State University of New York Press, 1981. Arguments for continuing economic growth.

Yudelman, M. *The World Bank and Agricultural Development*. Washington, D.C.: World Resources Institute, 1985. Insights into the World Bank's Rural Development Program by a former director.

CHAPTER 9

Earth and Its Crustal Resources

And this our life, exempt from public haunt,

finds tongues in trees, books in running brooks,

sermons in stones, and good in everything.

William Shakespeare: *As You Like It*

CONCEPTS

Earth is a layered sphere with a dense, extremely hot core composed mainly of molten iron and nickel. Surrounding the core is a pliable layer called the mantle. The outermost solid layer, the crust, is composed of lighter elements that crystallize to form rocks. It is broken into a mosaic of slowly moving tectonic plates whose wanderings, collisions, and recombinations play a large role in shaping the landscape and determining our climate.

The three main rock types are igneous, sedimentary, and metamorphic. Minerals are formed into rocks and then broken down and transformed into new types by weathering, sedimentation, leaching, and other geologic processes. Collectively, these processes are known as the rock cycle.

We all use minerals. The global network of mineral trade connects all nations and makes each nation economically valuable to and dependent upon the others. Consumption of traditionally important minerals, such as copper, iron, and aluminum, has decreased in recent years with the introduction or invention of new materials and processes for conservation and recycling.

Recycling metals relieves pressure on our remaining mineral supplies. It is also more efficient than processing raw materials, allowing large savings of both money and energy.

Earthquakes, landslides, and volcanoes have always threatened human populations. New information is being used to predict these events and decrease human losses from their hazards.

INTRODUCTION

Just what is Earth? What materials does it give us, and how do we use them? These are some of the questions an earth scientist or geologist asks and that we will address in this chapter. First, we will look at Earth's composition, structure, and dynamic physical processes. Then, we will consider what happens when we disrupt these features of Earth by extracting and using mineral resources (figure 9.1). Finally, we will look at some of the geologic hazards created by the forces that shape Earth's surface.

By now you recognize that Earth and its biosphere make up a complex, interactive system. The solid, abiotic part of this system (sometimes called the lithosphere) provides the raw materials and the supporting surface on which many of the processes of life depend. Learning how this part of our ecosystem works and how the resources on which we depend came to be in their present locations and forms is important in understanding our environment.

EARTH: A DYNAMIC PLANET

The solid ground on which you stand probably seems to be the very essence of permanence and stability. If you could see our planet from a cosmic perspective, however, you might be surprised to find that most of it is fluid or semisolid, and its surface is flexible and dynamic, bulging, wrinkling, and cracking as it slides around on the underlying plastic layers. What causes this restless, global motion?

A Layered Sphere

The **core,** or interior of Earth (figure 9.2), is composed of a dense, intensely hot mass of molten metal—mostly iron and nickel—thousands of kilometers in diameter. Semisolid in the center but more fluid in the outer core, this immense mass is stirred by convection currents that are thought to generate the magnetic field that shields us from cosmic radiation. These currents, and the heat that drives them, also provide much of the force that shapes and modifies the surface of our world. These dynamic processes create a significant amount of the geological resources on which we depend.

What heats the core and keeps it molten? Some heat comes simply from the dense packing of material in the core and some is left over from Earth's formation, that is, it hasn't finished cooling yet from the Big Bang that scientists theorize started our universe. Most of the heat, however, is generated by radioactive decay of unstable elements, such as uranium, thorium, and potassium (see chapter 19). Eventually, in a few billion years, this heat will all be dissipated, and Earth probably will be as cold and lifeless as the moon.

Surrounding the molten core is a hot, pliable layer of rock called the **mantle.** The mantle is much less dense than the core because it contains a high concentration of lighter elements, such as oxygen, silicon, and aluminum. Cooling as it nears the surface, the mantle becomes stiffer and more solid, eventually crystallizing into the lower margins of the lithosphere, or rock layer.

The outermost layer of the lithosphere is the cool, lightweight, brittle **crust** of rock that floats on the mantle something like the "skin" on a bowl of warm chocolate pudding. The oceanic crust, which forms the seafloor, has a composition somewhat like that of the mantle, but it is richer in silicon. Continents are thicker, lighter regions of crust rich in calcium, sodium, potassium, and aluminum. The continents rise above both the seafloor and the ocean surface. You might imagine them as marshmallows embedded in the surface of your chocolate pudding, surrounded by puddles of milk. Although the interior of the pudding is still soft, warm, and semisolid, the surface has cooled to a crust that hardens, cracks, and traps material floating on its surface. Table 9.1 compares the composition of the whole Earth (dominated by the dense core) and the crust.

TABLE 9.1 Eight most common minerals (percent)

Whole World		Crust	
Iron	33.3	Oxygen	45.2
Oxygen	29.8	Silicon	27.2
Silicon	15.6	Aluminum	8.2
Magnesium	13.9	Iron	5.8
Nickel	2.0	Calcium	5.1
Calcium	1.8	Magnesium	2.8
Aluminum	1.5	Sodium	2.3
Sodium	0.2	Potassium	1.7

Tectonic Processes and Shifting Continents

Convection currents and uneven heat flows passing through the core and mantle break the overlying crust into a mosaic of huge blocks called **tectonic plates** (figure 9.3). These plates slide slowly across Earth's surface like immense icebergs, in some places breaking up into smaller pieces, in other places crashing ponderously into each other to create new, larger landmasses. Ocean basins form where continents crack and pull apart. Earthquakes are caused by the grinding and jerking as plates slide past each other. Mountain ranges are pushed up at the margins of colliding continental plates. The Atlantic Ocean is growing slowly as Europe and Africa drift away from the Americas. The Himalayas are still rising as the Indian subcontinent smashes into Asia. Southern California is slowly sailing north toward Alaska. In a few million years, Los Angeles will pass San Francisco, if either still exists.

When an oceanic plate collides with a continental landmass, the continental plate usually rides up over the seafloor, and the oceanic plate is subducted, or pushed down into the mantle,

FIGURE 9.1 The world's largest open-pit mine, Bingham Canyon, Utah. Over $6 billion worth of copper ore has been extracted from this giant hole, which covers 400 ha (1,000 a) and is 670 m (220 ft) deep.

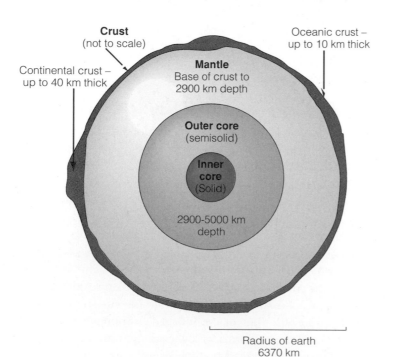

FIGURE 9.2 A layered Earth. The intensely hot, liquid or semisolid core is made mostly of molten metal. Around it is a mantle of lighter elements that cool and crystallize near the surface. Floating on top of the mantle is a thin crust of rock that breaks up into large, slowly-moving tectonic plates. The crust appears ten times thicker in this drawing than it is in reality.

where it melts and rises back to the surface as **magma,** or molten rock (figure 9.4). Deep ocean trenches mark these subduction zones, and volcanoes form where the magma erupts through vents and fissures in the overlying crust. All around the Pacific Ocean rim from Indonesia to Japan to Alaska and down the West Coast of the Americas is the so-called "ring of fire" where the Pacific plate is being subducted under the continental plates. This ring is the source of more earthquakes and volcanic activity than any other place on Earth.

Over millions of years, the drifting plates can move long distances. Antarctica and Australia once were connected to Africa, for instance, somewhere near the equator and supported luxuriant forests (figure 9.5). Geologists suggest that several times in Earth's history most or all of the continents have gathered to form a single supercontinent surrounded by a single global ocean. Every few hundred million years, this supercontinent breaks up into many smaller pieces through a process called seafloor spreading. Eventually, the forces driving the continents apart dissipate—perhaps due to cooling of the thin crust in spreading regions—and they slide back together again. These massive rearrangements undoubtedly have profound effects on Earth's climate and may help explain the periodic mass extinctions of organisms marking the divisions between many major geologic periods.

RESOURCES AND RESOURCE ECONOMICS

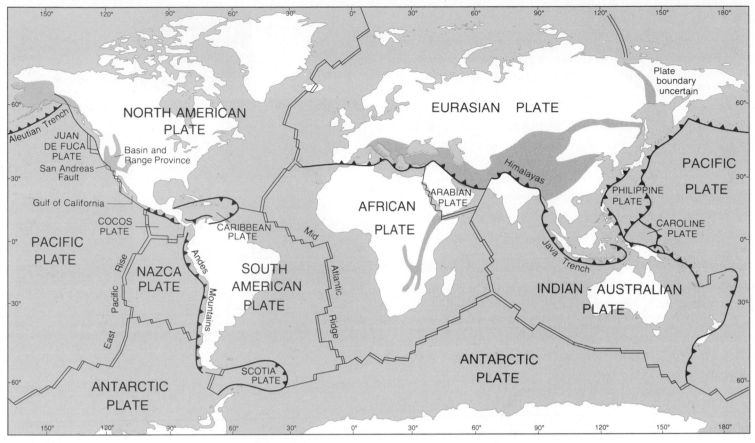

FIGURE 9.3   Map of tectonic plates. Plate boundaries are dynamic zones, characterized by earthquakes and volcanism, formation of great rifts and of mountain ranges.

Sources: U.S. Department of the Interior, U.S. Geological Survey.

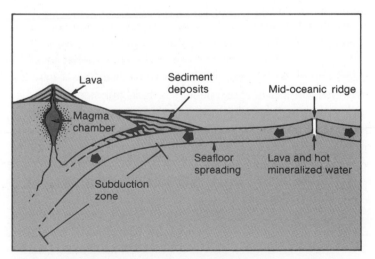

FIGURE 9.4   The rock cycle interpreted in plate-tectonic terms. Old seafloor and sediment deposits are melted in subduction zones. The magma rises to erupt through volcanoes or to recrystallize at depth into igneous rocks. Weathering breaks down surface rocks and erosion deposits residue in sedimentary formations. The pressure and heat caused by tectonic movements causes metamorphism of both sedimentary and igneous rocks.

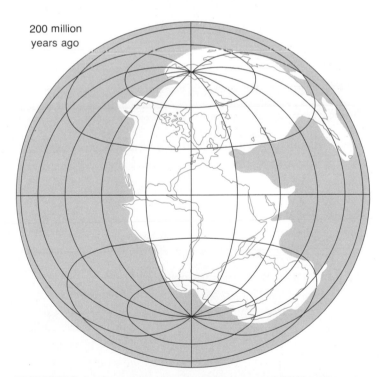

200 million years ago

FIGURE 9.5   Pangea, the ancient supercontinent of 200 million years ago, combined all the world's continents in a single landmass.

From "The Supercontinent Cycle" by R. Damian Nance, Thomas R. Worsley, and Judith B. Moody. Copyright © 1988 by Scientific American, Inc. All rights reserved.

THE ROCK CYCLE

What could be harder and more permanent than rocks? Like the continents they create, rocks are also part of a relentless cycle of formation and destruction. They are made and then torn apart, cemented together by chemical and physical forces, crushed, folded, melted, and recrystallized by dynamic processes related to those that shape the large-scale features of the crust. We call this cycle of creation, destruction, and metamorphosis the **rock cycle** (figure 9.6). Understanding something of how this cycle works helps explain the origin and characteristics of different types of rocks, as well as how they are shaped, worn away, transported, deposited, and altered by geologic forces.

Rocks and Minerals

The many different types of rocks are classified according to their internal structure, chemical composition, physical properties, and mode of formation. Minerals are the basic materials of all rocks. A **mineral** is a naturally occurring, inorganic, crystalline solid that has a definite chemical composition and characteristic physical properties. There are thousands of known and described minerals in the world, but most of the rocks we see every day are composed mainly of just a few dozen different minerals, such as quartz, feldspar, biotite, and calcite. These minerals, in turn, are composed mainly of just a few elements, such as silicon, oxygen, iron, magnesium, aluminum, and calcium.

Rock Types and How They Are Formed

There are three major rock classifications: igneous, sedimentary, and metamorphic. In this section, we will look at how they are made and some of their properties.

Igneous Rocks

Most rocks on Earth are crystalline minerals solidified from magma, molten liquid from Earth's interior. These rocks are classed as **igneous rocks** (from *igni*, the Latin word for fire). Magma extruded to the surface from volcanic vents cools to make basalt, rhyolite, andesite, and other fine-grained rocks. Magma that cools slowly in subsurface chambers or is intruded between overlying strata makes granite, gabro, or other coarse-grained crystalline rocks depending on its specific chemical composition.

Weathering and Sedimentation

Most of these crystalline rocks are extremely hard and durable, but exposure to air, water, changing temperatures, and reactive chemical agents slowly breaks them down in a process called **weathering.** *Mechanical weathering* is the physical breakup of rocks into smaller particles without a change in chemical composition. You have probably seen mountain valleys scraped by glaciers or river and shoreline pebbles that are rounded from being rubbed against one another as they are tumbled by waves and currents. *Chemical weathering* is the selective removal or alteration of specific components that leads to weakening and

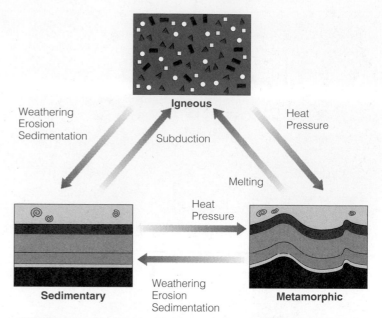

FIGURE 9.6 The rock cycle consists of processes of creation, destruction, and metamorphosis. Each of the three major rock types can be converted to either of the other types.

disintegration of rock. Among the more important chemical weathering processes are oxidation (removal of electrons from atoms) and hydrolysis (addition of water to other molecules). The products of these reactions are more susceptible to both mechanical weathering and to dissolving in water. For instance, when carbonic acid (formed when CO_2 and H_2O combine) percolates through porous limestone layers in the ground, it dissolves the calcium and creates caves and sinkholes.

Particles of rock loosened by wind, water, ice, and other weathering forces are carried downhill, downwind, or downstream until they come to rest again in a new location. The deposition of these materials is called **sedimentation.** Waterborne particles from sediments cover ocean continental shelves and fill valleys and plains. Most of the American Midwest, for instance, is covered with a layer of sedimentary material hundreds of meters thick in the form of glacier-borne till (broken rock rubble), wind-borne loess (fine dust deposits), river-borne sand and gravel, and ocean deposits of sand, silt, and clay. Deposited material that remains in place long enough, or is covered with enough material to compact it, may once again become stone. Some examples of **sedimentary rock** are shale (compacted mud), sandstone (cemented sand), breccia (volcanic ash), and conglomerates (aggregates of sand and gravel).

Much of this settling is due to gravity, but chemical, evaporative, and biogenic sedimentation are also important mechanisms of rock formation. *Chemical sedimentation* occurs when soluble chemicals react to make insoluble products that precipitate from water, creating mineral deposits. Hydrothermal vents in ocean spreading zones are important regions of chemical sedimentation that create economically important deposits of min-

FIGURE 9.7 The biological origin of this limestone is easily visible.

Metal	Use
TABLE 9.2 Primary uses of some major metals consumed in the United States	
Aluminum	Packaging foods and beverages (38%), transportation, electronics
Chromium	High-strength steel alloys
Copper	Building construction, electric and electronic industries
Iron	Heavy machinery, steel production
Lead	Leaded gasoline, car batteries, paints, ammunition
Manganese	High-strength, heat-resistant steel alloys
Nickel	Chemical industry, steel alloys
Platinum-group	Automobile catalytic converters, electronics, medical uses
Gold	Medical, aerospace, electronic uses; accumulation as monetary standard
Silver	Photography, electronics, jewelry

erals. *Evaporative sedimentation* occurs in warm, shallow bodies of water where evaporation is great and outflow is small. The evaporating water leaves behind any dissolved minerals it may have been carrying. Some economically important deposits include halite (rock salt) and gypsum ($CaSO_4$), a common building material. Evaporative sedimentation also can result in accumulation of harmful minerals, such as arsenic and selenium. Many soils in the western United States have high levels of these toxic sediments. Irrigation run-off from these soils has contaminated rivers, lakes, and wetlands including a number of wildlife refuges. The Kesterson marsh in California is one of the most dramatic and tragic examples, resulting in deaths and developmental deformities in waterfowl that feed and breed there. *Biogenic sedimentation* is caused by living organisms. The most important biogenic sedimentary rock is limestone, calcium carbonate ($CaCO_3$), formed from the skeletons and shells of marine organisms (figure 9.7). Iron deposits also can be created by biogenic sedimentation.

Metamorphic Rocks

Both igneous and sedimentary rocks can be modified by heat, pressure, and chemical reagents to create new forms called **metamorphic rock.** Deeply buried rock strata are subjected to great heat and pressure by deposition of overlying sediments or while they are being squeezed and folded by tectonic processes. Chemical reactions can alter both the composition and structure of metamorphic rocks. Some common metamorphic rocks are marble (from limestone), quartzite (from sandstone), and slate (from mudstone and shale). Metamorphic rocks are often the source of metal ores, such as gold, silver, and copper.

ECONOMIC MINERALOGY

Economic mineralogy is the study of minerals that are heavily used in manufacturing (mainly metal ores) and are, therefore, an important part of domestic and international commerce. Nonmetallic economic minerals are mostly graphite, some feld-

spars, quartz crystals, diamond, and many other crystals that are valued for their beauty and/or rarity. Minerals have been so important in human affairs that major epochs of human history are commonly known by the dominant materials and the technology to use them (e.g., Stone Age, Bronze Age, and Iron Age). The mining, processing, and distribution of these minerals have broad and varied implications both for culture and our environment. Most economically valuable minerals exist everywhere in small amounts; the important thing is to find them concentrated in economically recoverable levels.

Metals

How has the quest for mineral supplies affected global development? We will focus first on world use of metals, earth resources that always have received a great deal of human attention. The availability of metals and the methods to extract and use them have determined technological developments, as well as economic and political power for individuals and nations. We still are strongly dependent on the unique lightness, strength, and malleability of metals.

The metals consumed in greatest quantity by world industry include iron and steel (740 million metric tons annually), manganese (22.4 million metric tons), copper and chromium (8 million metric tons each), aluminum (4.8 million metric tons) and nickel (0.7 million metric tons). Most of these metals are consumed in the United States, Japan, and Europe, in that order. They are produced primarily in South America, South Africa, and the Soviet Union. It is easy to see how these facts contribute to a worldwide mineral trade network that has become crucially important to the economic and social stability of all nations involved (figure 9.8). Figure 9.9 shows where the world's largest known reserves of metals are located, and table 9.2 gives their primary uses.

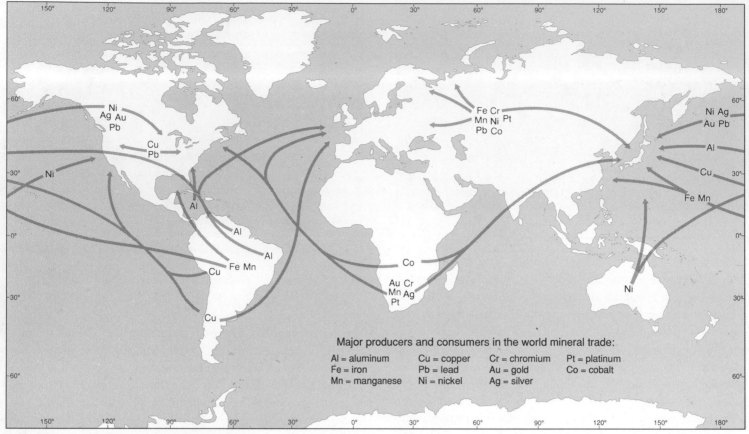

FIGURE 9.8 Global mineral trade. Metals produced in South Africa, South America, and the Soviet Union are shipped to markets in the United States, Europe, and Japan, creating a global economic network on which both consumers and producers depend.

Nonmetal Mineral Resources

Nonmetal minerals is a broad class that covers resources from silicate minerals (gemstones, mica, talc, and asbestos) to sand, gravel, salts, limestone, and soils. Sand and gravel production comprise by far the greatest volume and dollar value of all nonmetal mineral resources. Sand and gravel are used mainly in brick and concrete construction, paving, as loose road filler, and for sandblasting. High-purity silica sand is our source of glass. These materials usually are retrieved from surface pit mines and quarries, where they have been deposited by glaciers, winds, or ancient oceans.

Limestone, like sand and gravel, is mined and quarried for concrete and crushed for road rock. It also is cut for building stone, pulverized for use as an agricultural soil additive that neutralizes acidic soil, and roasted in lime kilns and cement plants to make plaster (hydrated lime) and cement.

Evaporites are mined for halite, gypsum, and potash. These are often found at or above 97 percent purity. Halite, or rock salt, is used for water softening and melting ice on winter roads in some northern states. Refined, it is a source of table salt. Gypsum (calcium sulfate) now makes our plaster wallboard, but it has been used for plaster ever since the Egyptians plastered

the walls of their frescoed tombs along the Nile River some five thousand years ago. Potash is an evaporite composed of a variety of potassium chlorides and potassium sulfates. These highly soluble potassium salts have long been used as a soil fertilizer.

Sulfur, in the form of pyrite (FeS_2), is mined mainly for sulfuric acid production. In the United States, sulfuric acid use amounts to more than 200 lbs per person per year, mostly because of its use in industry, car batteries, and some medicinal products.

Strategic Minerals

World industry depends on about eighty minerals, some of which exist in plentiful supplies. Three-fourths of the eighty minerals are abundant enough to meet all of our anticipated needs or have readily available substitutes. At least eighteen minerals, including tin, platinum, gold, silver, and lead, are in short supply.

Of these eighty minerals, between one-half and one-third are considered "strategic" minerals. **Strategic minerals** are minerals that a country uses but cannot produce itself. The United States is dependent on chromium, manganese, cobalt, aluminum, tantalum, tin, and platinum-group metals. Since we

RESOURCES AND RESOURCE ECONOMICS

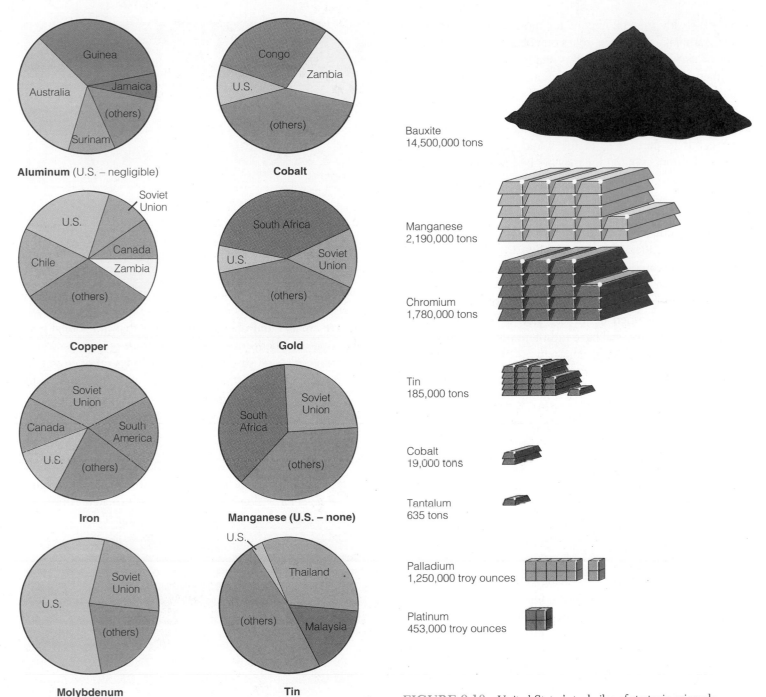

FIGURE 9.9 Proportions of world reserves of some nonfuel minerals controlled by various producers.

Sources: Data from J. E. Fergusson, *Inorganic Chemistry and the Earth.* Copyright © 1982 Pergamon Press, New York, NY and *Mineral Commodity Summaries 1985*, U.S. Bureau of Mines.

FIGURE 9.10 United States' stockpiles of strategic minerals.

have no significant domestic supplies of these metals and must depend on foreign sources for them, they are considered strategic. We also depend on molybdenum, which is scarce worldwide. Since 60 percent of world molybdenum reserves are found in the United States, however, it is not considered a strategic mineral.

As the term *strategic* suggests, these are minerals that a government considers capable of crippling its economy or military strength if unstable global economics or politics were to cut off supplies. For this reason, wealthy industrial nations stockpile strategic minerals in times when prices are low and a supply is available. Figure 9.10 shows some of the major stockpiles of strategic minerals in the United States.

For less wealthy mineral-producing nations, there is another side to strategic minerals. Many less developed countries depend on steady mineral exports for most of their foreign exchange. Zambia, for instance, relies on cobalt production for 50 percent of its national income. If a steady international market is not maintained, such producer nations could be devastated. From the small-nation producer's point of view, mineral exports, like concentration on any single product or industry, are an unstable economic foundation. Often no option exists, however, if a producer is to participate in the world economy. Environmental consequences of mining may not be a high priority under these circumstances.

Mineral resources were among the prizes the European colonial system was set up to acquire. Access to mineral resources was a motivating factor in exploration and colonialism as far back as the fifteenth century, when the Portuguese and Spanish went in search of El Dorado, the mythical land of gold in the New World. Gold from South and Central America made European countries rich and financed much of the rise of industrialism. Resource-rich countries, such as Namibia, Zaire, South Africa, and Chile, still have turbulent politics complicated by the desire of various groups and nations to control access to their mineral supplies.

CONSERVING EARTH'S MINERAL RESOURCES

There is great potential for extending our supplies of economic minerals and reducing the effects of mining and processing through recycling. The advantages of recycling are significant: less waste to dispose of, less land lost to mining, and less consumption of money, energy, and water resources.

Recycling

Some waste products already are being exploited, especially for scarce or valuable metals. Aluminum, for instance, must be extracted from bauxite by electrolysis, an expensive and energy-intensive process. Recycling waste aluminum, such as beverage cans, on the other hand, consumes one-twentieth of the energy of extracting new aluminum. The energy cost of extracting other metals is shown in table 9.3. Platinum, the catalyst in automobile catalytic exhaust converters, is valuable enough to be regularly retrieved and recycled from used cars (figure 9.11). Other metals commonly recycled are gold, silver, copper, lead, iron, and steel. The latter three of these are readily available in a pure and massive form, including copper pipes, lead batteries, and steel and iron auto parts. Gold and silver are valuable enough to warrant recovery, even through more difficult means.

Steel and Iron Recycling: Minimills

While total U.S. steel production has fallen in recent decades—largely because of inexpensive supplies from new and efficient Japanese steel mills—a new type of mill subsisting entirely on a

TABLE 9.3 Energy requirements in producing various materials from ore and raw source materials

| Product | Energy requirement (MJ/kg) | |
	New	From scrap
Glass	25	25
Steel	50	26
Plastics	162	n.a.
Aluminum	250	8
Titanium	400	n.a.
Copper	60	7
Paper	24	15

Source: E. T. Hayes, *Energy Implications of Materials Processing.*

readily available supply of scrap/waste steel and iron is a growing industry. Minimills, which remelt and reshape scrap iron and steel, are smaller and cheaper to operate than traditional integrated mills that perform every process from preparing raw ore to finishing iron and steel products. Minimills produce steel at between $225 and $480 per metric ton, while steel from integrated mills costs from $1,425 to $2,250 per metric ton on average (figure 9.12). The energy cost is likewise lower in minimills: 5.3 million BTU/ton of steel compared to 16.08 million BTU/ton in integrated mill furnaces. Minimills are expected to produce 26 million metric tons in 1990, about 25 percent of U.S. steel production, and to peak at 30 to 40 percent sometime before 1995. Recycling is slowly increasing as raw materials become more scarce and wastes become more plentiful. Table 9.4 shows the change in recycling rates of some materials in the United States between 1970 and 1985.

TABLE 9.4 Recycled scrap as a percentage of total consumption in the United States

	Ni	Co	Mg	Zn	Al	Cr	Cu	Sn	Hg	Au	Fe	Ag	Pb	Sb	Pt
1970	–	–	3	4	4	15	17	18	25	25	30	33	35	60	–
1985	11	5	–	8	16	18	21	21	–	53	25	19	48	–	33

See appendix A for names of elements.
Source: *Mineral Commodity Summaries 1985*, U.S. Bureau of Mines.

Substituting New Materials for Old

Mineral consumption can be reduced by new materials or new technologies developed to replace traditional minerals and mineral uses. This is a long-standing tradition, as when bronze replaced stone technology and iron replaced bronze. More recently, the introduction of polyvinylchloride (PVC) plastic pipe has decreased our consumption of copper, lead, and steel pipes. In the same way, the development of fiber optic technology and satellite communication reduce the need for copper telephone wires.

FIGURE 9.11 The richest ore we have—our mountains of scrapped cars—offers a rich, inexpensive, and ecologically beneficial resource that can be "mined" for a number of metals.

Iron and steel have been the backbone of heavy industry, but we are now moving toward other materials. One of our primary uses for iron and steel has been machinery and vehicle parts. In automobile production, steel is being replaced by polymers (long-chain organic molecules similar to plastics), aluminum, ceramics, and new, high-technology alloys. All of these reduce vehicle weight and cost, while increasing fuel efficiency. Some of the newer alloys that combine steel with titanium, vanadium, or other metals wear much better than traditional steel. Ceramic engine parts provide heat insulation around pistons, bearings, and cylinders, keeping the rest of the engine cool and operating efficiently. Plastics and glass fiber-reinforced polymers are used in body parts and some engine components.

Electronics and communications (telephone) technology, once major consumers of copper and aluminum, now use ultrahigh-purity glass cables to transmit pulses of light, instead of metal wires carrying electron pulses. Once again, this technology has been developed for its greater efficiency and lower cost, but it also affects consumption of our most basic metals.

ENVIRONMENTAL EFFECTS OF RESOURCE EXTRACTION

Resource extraction involves physical processes of mining and physical and chemical processes of separating minerals from ores.

Mining

Mineral extraction is done by several different techniques depending on the accessibility of the ore. All of these methods have environmental hazards. Native metals deposited in the gravel of streambeds can be washed out hydraulically in a process called placer mining. This not only destroys stream beds but fills the water with suspended solids that smother aquatic life. Larger or

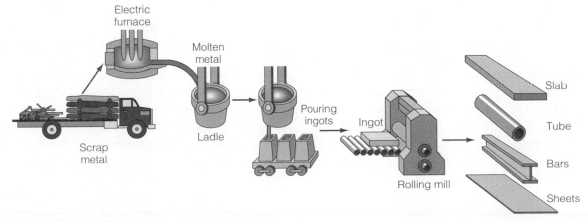

FIGURE 9.12 "Minimills" remelt and reshape scrap iron and steel. They not only extend our mineral resources by recycling discarded materials, they also conserve energy and are cheaper to operate than traditional integrated mills that depend on virgin ore.

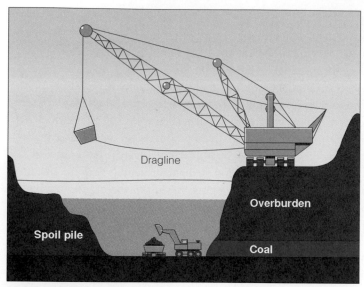

FIGURE 9.13 Giant draglines strip off as much as 75 m (250 ft) of overburden to expose a coal seam. Topsoil is supposed to be kept separate from subsoil and rock so that spoil piles can be leveled and revegetated. Reclamation efforts are often unenthusiastic and ineffectual.

deeper ore beds are extracted by strip mining or open pit mining where overlying material is removed by large earth-moving equipment. The resulting pits can be many kilometers across and hundreds of meters deep (figure 9.13). Even deeper deposits are reached by underground tunneling, an extremely dangerous process for mine workers.

Old tunnels occasionally collapse, or subside. In coal mines, natural gas poses dangers of explosion. Uncontrollable fires, producing noxious smoke and gases, sometimes ignite coal-bearing scrap heaps stored inside or outside the mine. Surface waste deposits called tailings can cause acidic or otherwise toxic runoff when rainwater percolates through piles of stored material. Tailings from uranium mines give rise to wind scattering of radioactive dust.

Water leaking into mine shafts also dissolves metals and other toxic material. When this water is pumped out or allowed to seep into groundwater aquifers (chapter 16), pollution occurs. Some critics claim that nearly every river and stream in western Colorado has been poisoned to some degree by mine drainage.

The process of strip-mining involves stripping off the vegetation, soil, and rock layers, removing the minerals, and replacing the fill. The fill is usually replaced in long ridges, called spoil banks, because this is the easiest way to dump it cheaply and quickly. Spoil banks are very susceptible to erosion and chemical weathering. Rainfall leaches numerous chemicals in toxic concentrations from the freshly exposed earth, and the water quickly picks up a heavy sediment load. Chemical- and sediment-runoff pollution becomes a major problem in local watersheds. Acid runoff had contaminated 6,700 miles of streams in the United States by 1980. Problems are made worse by the

fact that the steep spoil banks are very slow to revegetate. Since the spoil banks do not have natural topsoil, succession, soil development, and establishment of a natural biological community occur very slowly.

The 1977 federal Strip-Mining Reclamation and Control Act (SMRCA) requires better restoration of strip-mined lands, especially of land classed as prime farmlands, but restoration is difficult and expensive. Even if soil is carefully replaced, it will take centuries, if not millenia, to regain its former fertility by natural causes. Topsoil is dispersed and often buried by the activity of heavy machinery working to resculpture the land. Compaction disrupts air and water flow through soil, restricts root growth, and causes poor drainage, resulting in wet, stagnant soil conditions. The difficulties of reestablishing vegetation in dry climates means that reclamation must be considered essentially impossible where rainfall is less than 25 cm per year.

The monetary expense of reclamation is also high. Minimum reclamation costs about $1,000 per acre, while "complete" restoration (where it is possible) costs $5,000 an acre. Nevertheless, 50 percent of United States coal (225–270 million metric tons per year) is strip-mined. Nearly a million acres of land in the United States have been devastated by strip-mining.

Processing

Processing minerals, which usually means smelting and purifying with chemical solvents, can be even more environmentally hazardous than extraction. Solvents used to separate and float desired minerals are usually highly toxic, and safe disposal sites are rare. Often solvents and other chemicals leak into surface and groundwater supplies, either directly or through the leaching of ore tailings. In Brazil, *garimperos* (independent gold miners) were estimated to have dumped over 100 metric tons of mercury into the Amazon River and its tributaries in 1990. This threatens to poison the whole ecosystem, perhaps the biologically richest on earth.

Smelting causes most of its damage by releasing huge amounts of toxic gases into the air (figure 9.14). One copper smelter, the Phelps Dodge smelter in Douglas, Arizona, once belched out an average of 905 metric tons of sulfur per day through the 1960s and 1970s. Because the prevailing winds blew from the smelter toward the Colorado Rockies, pollution problems were especially damaging and noticeable. When the smelter closed in the early 1980s, acid rain levels in the downwind region dropped dramatically. Altogether, ore smelters release 1.7 million tons of SO_2 annually. For further discussion of atmospheric pollutants see chapter 23.

Biological mineral recovery is a technique that may be an attractive alternative to either smelting or chemical extraction. To do biological processing, ore is crushed and spread in shallow, football field-sized pans. A culture of special bacterial or algal cells is mixed in the rock, and water is sprayed on the surface.

FIGURE 9.14   A luxuriant forest once grew on this now barren hillside near Ducktown, Tennessee. Smelter fumes killed all the vegetation nearly a century ago, and erosion has washed away all the topsoil.

The organisms used are especially selected (or created via genetic engineering) to absorb and concentrate specific metals. After a suitable time (months or years), the cells are collected and minerals are extracted from them. This is much easier, less polluting, and less energy-intensive than traditional methods.

GEOLOGICAL HAZARDS

Earthquakes and volcanoes are normal Earth processes, events that have made our Earth what it is today. However, when they occur in proximity to human populations, their consequences can be among the worst and most feared disasters that befall us. For thousands of years people have been watching and recording these hazards, trying to understand them and learn how to avoid them.

Earthquakes

Earthquakes, especially, have received much attention. They have always seemed mysterious, sudden, and violent, coming without warning and leaving little sign of their passing except

ruined cities and dislocations in the landscape (table 9.5). Volcanoes smoke and steam before they erupt, and they usually occur in predictable places where one can easily see they have occurred before. Earthquakes leave no long-standing reminder, such as a mountain, and any pretremor activity is usually too subtle to be detected by humans (Box 9.1).

Earthquakes generally consist of a principal tremor followed by aftershocks from reverberations within Earth's crust. Sometimes the aftershocks are even more severe than the principal tremor because they can be amplified as they travel through the ground. On October 17, 1989, an earthquake measuring 7.1 on the Richter scale rumbled through the San Francisco Bay area. The shock was amplified by soft fill along the coast in San Francisco and Oakland. Apartment buildings tipped over and a double-decked segment of Interstate 880 collapsed, crushing cars on the lower level like a trash compactor. Sixty-seven people died and 24,000 were left homeless. This was not the most calamitous earthquake in history, however. A tremor of equal magnitude killed 10,000 people in Mexico City in 1985 (figure 9.15). Another is thought to have killed 650,000 people in China in 1977.

BOX 9.1   When Will "The Big One" Hit?

In 1989, a strong earthquake (magnitude 7.1 on the Richter scale) rocked the San Francisco Bay area and reminded us of the dangers of geological forces. Many people began to wonder when and where the next big tremor would hit. Tragic as the San Francisco earthquake was, it was not by any means the largest or most catastrophic that might occur. Geologists warn that an earthquake of equal or greater power is likely along the south end of the San Andreas Fault near Los Angeles. An earthquake in the Los Angeles basin would probably cause more death and destruction than in San Francisco because more people live in the area and because geologic formations there would transmit and magnify the earth's motion to a greater degree.

Surprisingly, the West Coast is not the only place in the United States where a geological cataclysm might occur. People who think that earthquakes are strictly a Californian phenomenon might be amazed to learn that the most powerful earthquake in recorded American history occurred in the middle of the country near New Madrid, (pronounced MAD-rid) Missouri. Between December 16, 1811, and February 7, 1812, about 2,000 tremors shook Southeastern Missouri and adjacent parts of Arkansas, Illinois, and Tennessee. The largest of these earthquakes is thought to have had a magnitude of 8.8 on the Richter scale, making it one of the most massive ever recorded.

Witnesses reported shocks so violent that trees two meters thick were snapped like matchsticks. More than 60,000 ha (150,000 acres) of forest were flattened. Fissures several meters wide and many kilometers long split the earth. Geysers of dry sand or muddy water spouted into the air. A trough 240 km (150 mi) long, 64 km (40 mi) wide, and up to 10 meters (30 ft) deep formed along the fault line. The town of New Madrid sank about four

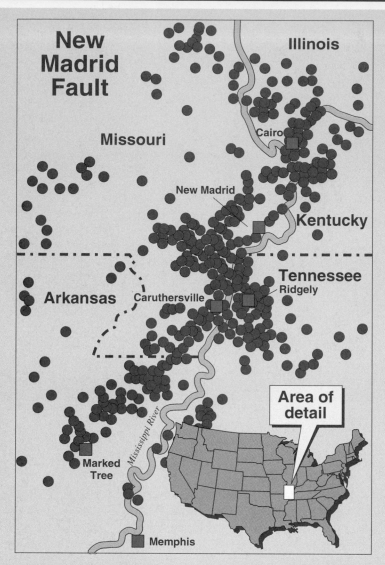

BOX FIGURE 9.1 Earthquakes along the New Madrid fault, 1974–1987. Although few people were aware of it before the 1990 earthquake scare, this is one of the most seismically active areas of the country.

meters (12 ft). The Mississippi River reversed its course and flowed north rather than south past New Madrid for several hours. Many people feared that it was the end of the world.

One of the most bizarre effects of the tremors was soil liquefaction. Soil with high water content was converted instantly to liquid mud. Buildings tipped over, hills slid into the valleys, and animals sank as if caught in quicksand. Land

surrounding a hamlet called Little Prairie suddenly became a soupy swamp. Residents had to wade for miles through hip-deep mud to reach solid ground. The swamp wasn't drained for nearly a century.

Some villages were flattened by the earthquake, others were flooded when the river filled in subsided areas. The tremors rang bells in Washington, D.C., and shook residents out of bed in Cincinnati.

Since the country was sparsely populated in 1812, however, few people were killed.

The situation is much different now, of course. The damage from an earthquake of that magnitude would be calamitous. Much of Memphis, Tennessee, only about 100 mi from New Madrid, is built on landfill similar to that in the Mission District of San Francisco where so much damaged occurred. St. Louis had only 2,000 residents in 1812. Nearly a half million live there now. Scores of smaller cities and towns lie along the fault line and transcontinental highways and pipelines cross the area. Few residents have been aware of earthquake dangers or how to protect themselves. Midwestern buildings generally are not designed to survive tremors.

Anxiety about earthquakes in the Midwest was aroused in 1990 when climatologist Iben Browning predicted a 50–50 chance of an earthquake 7.0 or higher on or around December 3, in or near New Madrid. Browning based his prediction on calculations of planetary motion and gravitational forces. Many geologists were quick to dismiss these techniques, pointing out that seismic and geochemical analyses predict earthquakes much more accurately than the methods he used. Although there were no large earthquakes along the New Madrid fault in 1990, the probability of a major tremor there remains high.

While the general time and place of some earthquakes have been predicted with remarkable success, mystery and uncertainty still abound concerning when and where "the next big one" will occur. Will it be in California? Will it be in the Midwest? Or will it be somewhere entirely unsuspected? Meanwhile, residents of New Madrid are planning emergency exit routes and stocking up on camping gear and survival supplies. How about you? What would you do if the ground around you began to shake?

TABLE 9.5 Worldwide frequency and effects of earthquakes of various magnitudes

Richter scale magnitude*	Description	Average number per year	Observable effects
2–2.9	Unnoticeable	300,000	Detected by instruments, but not usually felt by people.
3–3.9	Smallest felt	49,000	Hanging objects swing, vibrations like passing of light truck felt.
4–4.9	Minor earthquake	6,200	Dishes rattle, doors swing, pictures move, walls creak.
5–5.9	Damaging earthquake	800	Difficult to stand up. Windows, dishes break. Plaster cracks, loose bricks and tile fall, small slides along sand or gravel banks.
6–6.9	Destructive earthquake	120	Chimneys and towers fall, masonry walls damaged, frame houses move on foundations, small cracks in ground. Broken gas pipes start fires.
7–7.9	Major earthquake	18	General panic. Frame houses split and fall off foundations, some masonry buildings collapse, underground pipes break, large cracks in the ground.
8–8.9	Great earthquake	1 or 2	Catastrophic damage. Most masonry and frame structures destroyed. Roadways, dams, dikes collapse. Large landslides. Rails twist and bend. Underground pipes rupture.

From B. Gutenberg in *Earth* by F. Press and R. Seiver. Copyright © 1978 W. H. Freeman & Company Publishers, New York, NY. Reprinted by permission.

*For every unit increase in the Richter Scale, ground displacement increases by a factor of 10, while energy release increases by a factor of 30. There is no upper limit to the scale, but the largest earthquakes recorded have been 8.9.

Modern contractors in earthquake zones are attempting to prevent damage and casualties by constructing buildings that can withstand tremors. The primary methods used are heavily reinforced structures, strategically placed weak spots in the building that can absorb vibration from the rest of the building, and pads or floats beneath the building on which it can shift harmlessly with ground motion.

One of the most notorious effects of earthquakes is the **tsunami.** These giant seismic sea swells (sometimes improperly called tidal waves) can move at 1,000 km per hour (600 mph) or faster, away from the center of an earthquake. When these swells approach the shore, they can easily reach 15 m or more. They may reach 65 m (nearly 200 ft). A 1960 tsunami coming from a Chilean earthquake still caused 7-meter breakers when

FIGURE 9.15 Collapsing like a house of cards, the Juarez hospital in Mexico City became a tomb for patients and hospital workers alike on September 19, 1985, when an earthquake of magnitude 8.1 hit the city.

it reached Hawaii fifteen hours later. Tsunamis also can be caused by underwater volcanic explosions. The eruption of the Indonesian volcano, Krakatoa, in 1883 created a tsunami 40 m (130 ft) high that killed 30,000 people on nearby islands.

While most earthquakes occur in known earthquake-prone areas, sometimes they strike in unexpected places. In the United States, major quakes have occurred in South Carolina, Missouri, Massachusetts, Alaska, Nevada, Texas, Utah, Arizona, and Washington, as well as in California.

Volcanoes

Volcanoes and undersea magma vents are the sources of most of Earth's crust. Over hundreds of millions of years, gaseous emissions from these sources formed the Earth's earliest oceans and atmosphere. Many of the world's fertile soils are weathered volcanic materials. Volcanoes have also been an ever-present threat to human populations. One of the most famous historic volcanic eruptions was that of Mount Vesuvius in southern Italy, which buried the cities of Herculaneum and Pompeii in A.D. 79. The mountain had been giving signs of activity before it erupted, but many citizens chose to stay and take a chance on survival.

On August 24, the mountain buried the two towns in ash. Thousands were killed by the dense, hot toxic gases that accompanied the ash flowing down from the volcano's mouth. It is still erupting.

Nuees ardentes (French for "glowing clouds") are deadly, denser-than-air mixtures of hot gases and ash like those that inundated Pompeii and Herculaneum. Temperatures in these clouds may exceed 1,000°C, and they move at more than 100 km/hour (60 mph). Nuees ardentes destroyed the town of St. Pierre on the Caribbean island of Martinique on May 8, 1902. Mount Pelee released a cloud of nuees ardentes that rolled down through the town, killing somewhere between 25,000 and 40,000 people within a few minutes. All the town's residents died except a single prisoner being held in the town dungeon.

Mud slides are disasters sometimes associated with volcanoes. The 1985 eruption of Nevada del Ruiz, 130 km (85 mi) northwest of Bogata, Colombia, caused mud slides that buried most of the town of Amero and devastated the town of Chinchina. An estimated 25,000 people were killed. Heavy mud slides also accompanied the eruption of Mount St. Helens in Washington in 1980. Sediments mixed with melted snow and the

expelled 175 km³ of dust and ash, more than fifty-eight times that of Mount St. Helens. These dust clouds circled the globe and reduced sunlight and air temperatures enough so that 1815 was known as the year without a summer. It is not only a volcano's dust that blocks sunlight. Sulfur emissions from volcanic eruptions combine with rain and atmospheric moisture to produce sulfuric acid (H_2SO_4). The resulting droplets of H_2SO_4 interfere with solar radiation and can significantly cool the world climate. One of several theories about the extinction of the dinosaurs 65 million years ago is that they died off in response to the effects of acid rain and climate changes caused by massive volcanic venting in the Deccan Plateau of India.

SUMMARY

The earth is a complex, dynamic system. Although it seems stable and permanent to us, the crust is in constant motion. Tectonic plates slide over the surface of the ductile mantle. They crash into each other in ponderous slow motion, crumpling their margins into mountain ranges and causing earthquakes. Sometimes one plate will slide under another, carrying rock layers down into the mantle where they melt and flow back toward the surface to be formed into new rocks.

Rocks are classified according to composition, structure, and origin. The three basic types of rock are igneous, metamorphic, and sedimentary. These rock types can be transformed from one to another by way of the rock cycle, a continuous process of weathering, transport, burying in sediments, metamorphosis, melting, and recrystallization.

During the cooling and crystallization process that forms rock from magma, minerals often distill into concentrated ores that become economically important reserves if they are close enough to the surface to be reached by mining. Hot, mineral-laden water flowing up through deep sea thermal vents also makes rich hydrothermal mineral deposits from dissolved minerals transported up from the mantle. Biogenic and chemical sedimentation, placer action, evaporation, and weathering of surface deposits also create valuable mineral deposits.

For reasons that are not entirely clear, a few places in the world are especially rich in mineral deposits. South Africa and the Soviet Union contain most of the world's supply of several strategic minerals. Less developed countries, most of which are in the tropics or the Southern Hemisphere, are often the largest producers of ores and raw mineral resources for the strategic materials on which the industrialized world depends. The major consumers of these resources are the industrialized countries.

Worldwide, only a small percentage of minerals are recycled, although it is not a difficult process technically. Recycling saves energy and reduces environmental damage caused by mining and smelting. It reduces waste production and makes our mineral supplies last much longer. Substitution of materials usually occurs when mineral supplies become so scarce

FIGURE 9.16 Mount St. Helens in Washington State continued to pour out ash and steam months after its big erruption in 1980. More than 3 km³ of ash were spewed into the air. Hot gasses and mud slides destroyed thousands of hectors of forest and killed dozens of people.

waters of Spirit Lake at the mountain's base and flowed many kilometers from their source. Extensive damage was done to roads, bridges, and property; but because of sufficient advance warning, there were few casualties.

Volcanic eruptions often release large volumes of ash and dust into the air. Mount St. Helens expelled 3 km³ of dust and ash, causing ash fall across much of North America (Figure 9.16). This was only a minor eruption. An eruption in a bigger class of volcanoes was that of Tambora, Indonesia, in 1815, which

that prices are driven up. Many of the strategic metals that we now stockpile may become obsolete when newer, more useful substitutes are found.

Both mining and extraction of minerals have environmental effects. Mine drainage has polluted thousands of kilometers of streams and rivers. Fumes from smelters kill forests and spread pollution over large areas. Surface mining results in removal of natural ecosystems, soil disruption, creation of trenches and open pits, and accumulation of tailings. It is now required that strip-mined areas be recontoured, but revegetation is often difficult and limited in species composition. Smelting and chemical extraction processes also create pollution problems.

Earthquakes and volcanic events are natural geological hazards that are a result of movements of Earth's restless core and mantle. Big earthquakes are among the most calamitous natural disasters that befall people, sometimes killing hundreds of thousands in a single cataclysm.

■ Review Questions

1. Describe the layered structure of Earth.
2. What heats Earth and keeps the core molten?
3. What are tectonic plates and why are they important to us?
4. Why are there so many volcanoes and earthquakes along the "ring of fire" that rims the Pacific Ocean?
5. Describe the rock cycle and name the three main rock types that it produces.
6. Distinguish between gravitational, chemical, and biogenic sedimentation. Give an example of each.
7. Give some examples of strategic minerals. Where are the largest supplies of these minerals located?
8. Give some examples of nonmetal mineral resources and describe how they are used.
9. What are some of the advantages of recycling minerals?
10. Describe some ways we recycle metals and other mineral resources.
11. What are some environmental hazards associated with mineral extraction?
12. Describe some of the leading geologic hazards and their effects.

■ Questions for Critical Thinking

1. Look at the walls, floors, appliances, interior, and exterior of the building around you. How many earth materials were used in their construction?
2. What is the geologic history of your town or county?
3. Is your local bedrock igneous, metamorphic, or sedimentary? If you don't know, who might be able to tell you?
4. What would life be like without the global mineral trade network? Can you think of advantages as well as disadvantages?
5. Suppose a large mining company is developing ore reserves in the small, underdeveloped country where you live. How

will revenues be divided fairly between the foreign company and local residents?
6. How could we minimize the destruction caused by geologic hazards? Should people be discouraged from building in volcanic or earthquake-prone areas?
7. What might be the climatic effects of having all the continents clustered together in one giant supercontinent?
8. How would your life be affected if your country were to run out of strategic minerals?
9. What effect do you think our need for strategic minerals has had on our foreign policy toward South Africa and the Soviet Union?
10. How could our government encourage more recycling and more efficient use of minerals?
11. What is the potential for geologic hazards where you live?
12. What can you do to protect yourself from geologic hazards?

■ Key Terms

core (p. 153)
crust (p. 153)
igneous rocks (p. 156)
magma (p. 154)
mantle (p. 153)
metamorphic rock (p. 157)
mineral (p. 156)

rock cycle (p. 156)
sedimentary rock (p. 156)
sedimentation (p. 156)
strategic minerals (p. 158)
tectonic plates (p. 153)
tsunami (p. 165)
weathering (p. 156)

SUGGESTED READINGS

■ Chiles, J. Fall 1986. "Standing up to earthquakes." *Invention and Technology.* 2(2):56. Discusses current efforts to build earthquake-proof buildings and cities.

■ Johnson, A. C., and L. R. Kanter. March 1990. "Earthquakes in stable continental crust." *Scientific American.* 262(3):68. A theory of how earthquakes can occur well away from the edges of tectonic plates.

■ Liedl, G. October 1986. "The science of materials." *Scientific American.* 255(4):126. The entire issue of the journal is devoted to new materials technology.

■ McPhee, J. *Basin and Range.* New York: Farrar, Straus & Giroux, Inc., 1980. An eloquent essay on geology.

■ ———. *Rising from the Plains.* New York: Farrar, Straus & Giroux, Inc., 1987. A highly readable description of geology and geologists.

■ Montgomery, C. *Physical Geology.* Dubuque, Iowa: Wm. C. Brown Publishers, 1990. A good basic geology text to add to this chapter's discussion.

■ Plotkin, S. January/February 1986. "From surface mine to cropland." *Environment.* 28(1):16. The current state of mine reclamation efforts: Hopeless or full of promise?

■ Postel S., and C. Flavin "Reshaping the Global Economy." In *State of the World 1991.* New York: W. W Norton Co. Page 170. Calls for aid for sustainable development, redirecting government incentives, and a change from growth to sustainable progress.

Rona, P. January 1986. "Mineral deposits from sea-floor hot springs." *Scientific American*. 254(1):84. An excellent introduction to ocean thermal vents.

Simon, J., and H. K. Herman, eds. *The Resourceful Earth*. New York: Basil Blackwell, 1984. The minerals section is a rebuttal to the resource shortage philosophy of Global 2000.

Tanji, K., A. Lauchli, and J. Meyer. July/August 1986. "Selenium in the San Joaquin Valley: A challenge to Western irrigation." *Environment*. 28(6):6. Good overview of the selenium problems at Kesterson Reservoir.

U.S. Bureau of Mines. 1987. *Mineral Commodity Summaries, 1987*. Washington, D.C.: Government Printing Office. A good annual reference to current supply and consumption information.

Vogely, W. "Non-fuel minerals and the world economy." In *The Global Possible*, edited by Robert Repetto. New Haven: Yale University Press, 1985. pgs. 457–73. Opposing views of impending world mineral shortages versus plentiful mineral supplies discussed.

C H A P T E R *10*

Human Nutrition, Food, and Hunger

The battle to feed all of humanity is over. In the 1970s the world will undergo famines— hundreds of millions of people are going to starve to death in spite of any crash programs embarked upon now. At this late date nothing can prevent a substantial increase in the world death rate, although many lives could be saved through dramatic programs to stretch the carrying capacity of the earth by increasing food production. But these programs will only provide a stay of execution unless they are accompanied by determined and successful efforts at population control.

Paul Ehrlich

Properly managed, the Earth's more fertile lands and its forests could meet everyone's food and wood needs abundantly and indefinitely. The persistent undernourishment of some half a billion people today does not stem from a global scarcity of resources; even as tens of thousands of babies die each day from diseases exacerbated by malnutrition, over one third of the world's grain is fed to livestock to supply the meat-rich diet of the affluent.

Erick Eckholm

Food supplies have been growing more rapidly than human populations in every major region of the world except sub-Saharan Africa. There is enough food to more than meet minimum dietary requirements of everyone now living. In spite of these increasing food supplies, 750 million people in the world suffer from chronic undernutrition or malnutrition. Some 18 to 20 million people (mostly children) die each year from diseases related to undernutrition and malnutrition. Chronic undernutrition and malnutrition can cause mental retardation, developmental abnormalities, permanently stunted growth, and increased susceptibility to infectious diseases in children.

Scientific and technical advances in crop genetics, farm management, and food distribution have significantly increased production and availability of food. There is considerable potential for further increases, but the realization of that promise remains to be seen.

Only a few dozen species and varieties of agricultural crops, domestic livestock, and seafood make up the bulk of the human diet, worldwide. Other crops offer a potential for diversifying food sources and improving our dietary intake.

The immediate causes of famine are usually bad weather, insects, or some other natural disaster; but the underlying causes are economics, politics, environmental degradation, or other human actions. While natural disasters and large-scale famine attract much media attention, far more people die of chronic hunger and malnutrition caused by poverty and unjust distribution of resources.

INTRODUCTION

We live in perplexing times. World food supplies have increased at spectacular rates in the last forty years. Between 1950 and 1990, world agricultural output rose more than 2.5 fold. Food supplies have risen faster than population in every continent except Africa; yet, there are more chronically undernourished people now—both in total numbers and as a percentage of the world population—than in 1950. Many countries that have had the biggest gains in food production also have the greatest numbers and largest percentages of malnourished people. Some places that have the most heart-rending scenes of starvation and wasted human potential also grow billions of dollars worth of luxury foods to be exported to the industrialized countries of Europe and North America.

The Food and Agricultural Organization (FAO) of the United Nations estimates that at least 750 million people (15 percent of the total world population) have less than the minimum caloric intake necessary to sustain a productive working life. Nearly half of those people do not have enough food for normal growth and development in children or to forestall serious health risks. Some 18 to 20 million people (three-quarters of them children) starve to death or die of diseases aggravated by malnourishment each year.

Even the richest countries have high numbers of hungry people (figure 10.1). The United States, which pays farmers around $25 billion a year not to grow crops and to store mountains of surplus food, has at least 25 million undernourished people, mostly children. How can this occur when we have an embarrassing surplus of food? If we cannot feed everyone now, in this time of relative plenty, what will happen in the future if populations continue to grow and resources become more limited?

In this chapter, we will review human nutritional needs. Then, we will look at world food supplies, how production has increased in recent years, and what the prospects are for future crop production. We will survey the major food crops of the world, where they are grown and consumed, and how they contribute to the global economy and land-use patterns. Finally, we will look at some of the reasons why people are still hungry in a world with a surplus of food.

HUMAN NUTRITION

Our bodies require a constant supply of energy and raw materials to maintain vital functions and to rebuild cellular structures and tissues worn out in the day-to-day processes of living. In this section, we will study the major nutrients needed by humans to remain strong and healthy, and we will look at some of the most serious nutritional deficiencies from which people suffer.

FIGURE 10.1 Even in the United States, a land of plenty, some people are unable to feed themselves. Soup kitchens and relief shelters, such as this one, are often inadequate to take care of the need for food.

Meeting Energy Needs

The amount of energy each of us needs to carry on an active, healthy life depends on body weight, climate, state of health, stress level, and basic metabolism. A small, sedentary adult living in a warm climate might require less than 2,000 calories per day to remain healthy, whereas a large, muscular person living a vigorous outdoor life in a cold climate might need 6,000 to 7,000 calories per day to stay warm and active. The Food and Agriculture Organization (FAO) of the United Nations estimates that the average minimum daily caloric intake over the whole world is about 2,500 calories per day.

People who receive less than 90 percent of their minimum dietary intake on a long-term basis are considered **undernourished.** While not starving to death, they tend not to have enough energy for an active, productive life. Lack of energy and nutrients also tends to make them more susceptible to infectious diseases.

People who receive less than 80 percent of their minimum daily caloric requirements are considered **seriously undernourished.** Children who are seriously undernourished are likely to suffer from permanently stunted growth, mental retardation, and other social and developmental disorders. Infectious diseases that are only an inconvenience for well-fed individuals become lethal threats to those who are poorly nourished. Diarrhea rarely kills a well-fed person, but children weakened by nutritional deficiencies are highly susceptible to this and a host of other diseases. One child in four in the developing world—10 to 12 million children per year—dies of diseases that could be prevented with a better diet, clean water, and simple medicines. One child dies every two seconds from problems caused by lack of food.

Table 10.1 shows the per capita daily caloric intake for the major geographical regions of the world. The *average* world caloric intake is now above 2,600 calories per day, but at least one billion people—one-fifth of the world's population—have less than this amount. We will discuss the causes of hunger and where it occurs in the world later in this chapter after looking at more specific nutritional requirements.

TABLE 10.1 Food available for human consumption

| | Calories per person per day | |
	1961–1963	1984–1986
World Total	2,300	2,690
Africa (sub-Sahara)	2,040	2,060
North Africa/Near East	2,240	3,050
Asia	1,830	2,430
Latin America	2,380	2,700
North America	3,180	3,620
Western Europe	3,090	3,380
Eastern Europe	3,140	3,410

Source: Data from *World Resources 1990–1991.*

In the richer countries of the world, the most common dietary problem is **overnutrition** from getting too many calories. The average daily caloric intake in North America and Europe is above 3,500 calories, nearly one-third more than is needed for adequate nutrition. At least 20 percent of Americans are seriously overweight. Overnutrition contributes to high blood pressure, heart attacks, strokes, and other cardiovascular diseases that have become the leading causes of death in most developed countries since infectious diseases have been reduced or controlled by better sanitation and health care.

Meeting Nutritional Needs

In addition to calories, we need specific nutrients, such as proteins, vitamins, and minerals in our diet. It is possible to have excess food and still suffer from **malnourishment,** a nutritional imbalance caused by a lack of specific dietary components or an inability to absorb or utilize essential nutrients. In richer countries, people often eat too much meat, salt, and fat and too little fiber, vitamins, trace minerals, and other components lost from highly processed foods. In poorer countries, people often lack specific nutrients because they cannot afford more expensive foods (usually high-protein foods) that would provide a balanced diet. Let's look now at some of the essential dietary requirements and some of the important types of malnutrition.

Carbohydrates

For most people in the world, carbohydrates make up the bulk (up to 80 percent) of daily calorie intake, mainly as starches. Figure 10.2 shows the relative contribution of fats, oils, carbohydrates, and protein in diets as a function of income. As income goes up, the percentage of dietary protein, fat, and sugar increases, and the percentage of calories from complex carbohydrates decreases.

There are no indispensable dietary carbohydrates, since we can synthesize all we need from other foods we eat; however, our digestive system seems to require a diet rich in complex carbohydrates to absorb toxins and flush out wastes. Many people in more developed countries suffer from too little plant fiber in their diet.

Proteins

Proteins are essential to life. Enzymes keep the metabolic machinery of cells operating, and enzymes are proteins. Cellular structure also requires protein building blocks. The average adult human needs about 40 g (1.5 oz) of protein per day. Pregnant women, growing children, and adolescents need up to twice as much to create new cells and build growing tissues. Not only is the total *amount* of protein crucial, but the *quality* of that protein also is of vital importance. We depend on our diet to supply us with the ten essential amino acids we cannot make for ourselves (table 10.2). Furthermore, these are not interchangeable. They have to be present in a balanced ratio in order for us to synthesize functional proteins. Corn, for instance, has a reasonably high protein content but is generally low in lysine and arginine. Adding beans, peas, or milk to a corn-based diet can help balance the amino acid content.

TABLE 10.2 Human amino acid requirements met through diet

Essential amino acids*

Arginine
Histidine
Isoleucine
Leucine
Lysine
Methionine
Phenylalanine
Threonine
Tryptophane
Valine

*Amino acids that cannot be synthesized in the body are an *essential* part of the diet.

The two most widespread human protein deficiency diseases are kwashiorkor and marasmus. **Marasmus** (from the Greek "to waste away") is caused by a diet low in both calories and protein. A child suffering from severe marasmus is generally thin and shriveled like a tiny, very old starving person. **Kwashiorkor** is a West African word meaning "displaced child." It occurs when people eat a starchy diet that is low in protein or has poor-quality protein, even though it may have plenty of calories. Children with kwashiorkor often have reddish-orange hair

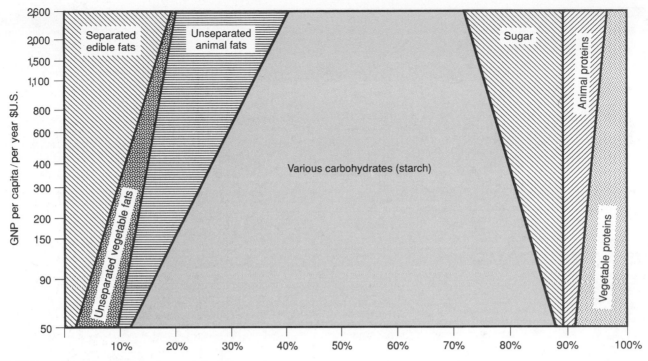

FIGURE 10.2 Calories derived from fats, carbohydrates, and proteins as percent of total calories in terms of income per capita in eighty-four countries. As GNP increases, so does the proportion of the diet made up of refined products and meat.

Percent of total calories

and puffy, discolored skin and a bloated belly (figure 10.3). They become anemic and listless, and have low resistance to even the mildest diseases and infections. Even if they survive childhood diseases, they are likely to suffer from stunted growth, mental retardation, and other developmental problems. Providing quality dietary protein is one of the world's greatest problems.

Lipids and Oils

Lipids (fats, oils, and related compounds) are an important source of energy, containing about twice as many calories per gram as proteins or carbohydrates. Such foods as peanut butter, margarine, butter, and vegetable oils are rich in calories. The main function of fat in our body—other than as a form of stored energy—is to form cellular membranes. Most of us get adequate levels of lipids in our diet, but very poor people in some areas of the world exist on a diet of starchy food, such as manioc, that does not give them enough oils and lipids. There are certain lipids we can't synthesize for ourselves and must get from our diet.

Minerals

Humans also require inorganic nutrients, both for building cellular structures and for regulating many cellular reactions. Table 10.3 describes some important major and trace minerals along with their dietary sources. We generally need only small amounts

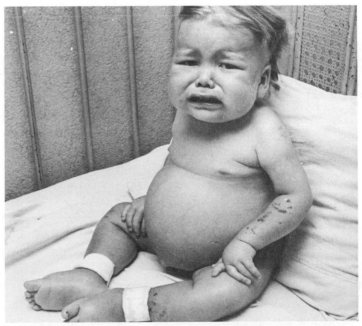

FIGURE 10.3 Kwashiorkor is a protein-deficiency disease resulting from a diet low in protein or a diet lacking the high quality protein necessary for normal growth and development. A bloated abdomen, puffy skin, and reddish-orange bleached hair are characteristic of this disease. Children are often listless, apathetic, and very susceptible to infectious diseases. Persistent kwashiorkor can result in mental and developmental retardation or death.

TABLE 10.3 Some important minerals and their sources

Major minerals	Food sources	Roles in body
Sodium (Na)	Table salt (NaCl), meat, processed foods	Transmission of electrochemical signals in muscles and nerves
Potassium (K)	Whole grains, fruits (esp. bananas), meat, legumes	Regulation of nerve and muscle action, protein synthesis
Calcium (Ca)	Milk, cheese, leafy green vegetables, egg yolk, nuts, whole grains, legumes	Matrix of bones and teeth, blood clotting, muscle relaxation, cell membrane function, intracellular signaling
Magnesium (Mg)	Whole grains, nuts, meat, legumes, milk	Constituent of bones and teeth, enzyme cofactor
Phosphorus (P)	Cheese, milk, meat, egg yolk, whole grains, nuts	Bone formation, energy metabolism, synthesis and use of carbohydrates, component of genetic material
Chlorine (Cl)	Table salt	Hydrochloric acid in stomach, water balance, membrane function
Sulfur (S)	Meat, eggs, cheese, milk, nuts, legumes	Component of some amino acids and proteins, regulates protein structure
Trace minerals		
Iron (Fe)	Liver, egg yolk, whole grains, dark green vegetables, legumes	Component of hemoglobin in blood, energy metabolism in cells
Copper (Cu)	Meat, seafood, whole grains, legumes, nuts	Present in some enzymes, bone, maintenance of nervous tissue
Iodine (I)	Iodized salt, seafood	Component of thyroid hormone
Manganese (Mn)	Legumes, cereals, nuts, tea, coffee	Enzyme function, protein metabolism
Cobalt (Co)	Vitamin B_{12} in meat	Essential for red blood cell formation
Zinc (Zn)	Widely distributed in foods	Constituent of some enzymes
Molybdenum (Mo)	Organ meats, milk, leafy vegetables, grains, legumes	Constituent of some enzymes

TABLE 10.4 Some important vitamins

Vitamin	Sources	Deficiency symptoms
Water-soluble vitamins		
Thiamine (B_1)	Fruits, cereal grains, milk, green vegetables	Beriberi, neuritis, fatigue, heart failure, edema
Riboflavin (B_2)	Milk, cheese, eggs	Sores on lips, bloodshot eyes
Niacin (B_3)	Whole grains, kidney, liver, fish, yeast	Pellagra, skin eruptions, fatigue, digestive disturbances
Pyridoxine (B_6)	Eggs, liver, yeast, milk, fish, grains	Anemia, convulsions, dermatitis, impaired immune response
Pantothenic acid	Most foods, meat	Retarded growth, mental instability
Folic acid	Egg yolk, liver, yeast	Anemia, impairment of immune system
Cobalamin (B_{12})	Meat, fish, eggs, milk, cheese, tempeh	Pernicious anemia, defective DNA synthesis
Ascorbic acid	Citrus fruits, tomatoes, green leafy vegetables	Scurvy (loose teeth, hemorrhage, sterility)
Fat-soluble vitamins		
A (retinol)	Green and yellow vegetables, dairy products, fish oils	Night blindness, dry, flaky skin, and dry mucous membranes
D (calciferol)	Fish oils, liver, egg yolk, milk and dairy products, action of sunlight on lipids in skin	Rickets (deformed bones)
E (tocopherol)	Widely distributed: oils, grains, lettuce, eggs, beef	Red blood cell fragility, sterility in male rats, aging, susceptibility to environmental oxidants
K (menadione)	Green leafy vegetables, synthesis by intestinal bacteria	Slow blood clotting, hemorrhage

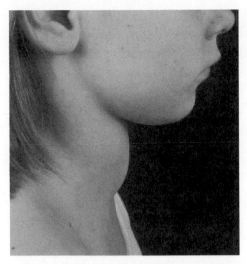

FIGURE 10.4 Goiter, a swelling of the thyroid gland at the base of the neck, is often caused by an iodine deficiency. It is a common problem in many parts of the world.

of most of these minerals, but deficiencies can have serious health effects. The most common mineral deficiencies worldwide are for calcium, iodine, and iron. Calcium deficiency causes irritability, muscle cramps, and bone defects. Iron deficiency leads to **anemia** (low levels of hemoglobin in the blood), which more often is caused by an inability to absorb iron from food than from a lack of iron in the diet.

The main symptoms of iodine deficiency are goiter (swollen thyroid glands) (figure 10.4) and **hypothyroidism** (listlessness and other metabolic symptoms due to low thyroid hormone levels). Hypothyroidism in early childhood can cause developmental abnormalities, such as mental retardation and deaf-mutism. Goiter was quite common in northern Europe and the United States before the introduction of iodized salt. It is estimated that 180 million people worldwide have symptoms of iodine deficiency and that 3 million suffer from cretinism (severe mental deficiency), mostly in South and Southeast Asia. In some villages, 15 to 20 percent of the children are brain damaged due to iodine deficiency. Since the human body needs only a teaspoonful of iodine in a whole lifetime, this problem is both technically simple and inexpensive to solve. Adding potassium iodate to salt costs only a few cents per year per person and is highly successful.

The best cure for most mineral deficiencies is usually a well-balanced diet with a good variety of foods. Whole grains, legumes, milk, eggs, leafy vegetables, and fruits are all good sources of essential minerals and vitamins. The people most likely to have mineral deficiencies are those who eat a diet of highly processed foods or who subsist on a single starchy food, such as manioc or white rice.

Vitamins

Vitamins are organic molecules essential for life (*vita* = life), but that we cannot make for ourselves and must get from our diet. They generally act as cofactors of enzymes that metabolize energy or build essential cellular components. Table 10.4 shows the vitamins essential for humans. We usually require only minute amounts (milligrams per day) of vitamins, and get all we need from a varied diet of fruits, vegetables, whole grains, and dairy products. Highly processed foods (white bread, cane sugar, and snack foods) often have lost their nutrients and need to be supplemented with additional vitamins.

Vitamin deficiencies that used to be very common in the United States are now much reduced. In less developed countries, however, vitamin deficiencies still are prevalent. Maize (called corn in North America) not only is deficient in tryptophan and lysine, but also is low in usable niacin. A deficiency of tryptophan and niacin results in **pellagra,** the symptoms of which include lassitude, torpor, dermatitis, diarrhea, dementia, and death. Pellagra used to be very common in the southern United States, where poor people subsisted mainly on corn. It still is tragically common in parts of India and Africa where jowah (a sorghum species) is the only food available to very poor people. Vitamin B$_{12}$ also is usually lacking in a strict vegetarian diet. Animal tissues and products are the ordinary sources of this essential dietary ingredient, but it is also present in tempeh, a cultured soybean food from Indonesia.

Vitamin A deficiency causes xerophthalmia (literally, dry eyes) and retinal degeneration, especially in children. Over five hundred thousand children in less developed countries lose their sight every year because of these diseases. Within a few weeks of becoming blind, 60 to 70 percent of these children die. An additional 6 to 7 million children show signs of moderate vitamin A deficiency and, therefore, are more vulnerable to infectious diseases.

Other vitamin deficiency diseases, such as scurvy, beriberi, anemia, and rickets (table 10.4), are still a problem for people in countries where diets are limited to starchy food, such as manioc, polished rice, maize, or wheat noodles. Enrichment of flour and milk with vitamins A and D has largely eliminated deficiency problems in developed countries. In fact, now there are concerns about excess vitamins in our diets in richer countries.

WORLD FOOD RESOURCES

Of the thousands of edible plants and animals in the world, only about a dozen types of seeds and grains, three root crops, twenty or so common fruits and vegetables, six mammals, two domestic fowl, and a few fish and other forms of marine life make up almost all the food humans eat. Table 10.5 shows some of the most important food crops in human diets.

TABLE 10.5 Some important food crops

Crop	1986 Yield (million metric tons)	Remarks
Wheat	522	Temperate climates. Staple food of 1/3 of world population. 8–15% protein.
Rice	469	Mostly Asian, both dry and wet cultivation. Staple food of 1/2 of world population. 8–9% protein; excellent food.
Maize (corn)	449	U.S. is largest producer; mainly livestock feed. Staple in South America, Central America, and Africa. 10% protein, but deficient in lysine and tryptophane.
Potatoes	290	Cool, moist-climate crop. Staple in north Europe and Andean countries. Very productive.
Barley and oats	178	Used for animal feed and malting for beer and whiskey in developed countries.
Cassava and sweet potato	140	Grown in warm, wet, tropical climates in South America, Africa, Southeast Asia, and Oceania. Mostly starch.
Sugar (cane and beet)	100	Provides up to 20% of total calories for many in MDCs*. Little nutritional value.
Pulses (legumes)	96	Soybeans, peas, and edible beans. 50–60% protein. Used mainly as animal food in MDCs, but important protein source in LDCs+.
Sorghum and millet	72	Drought resistant. Used as animal feed in developed countries. Staple in Africa.
Vegetable oils	47	Soybeans, sunflower, and rape seeds; olives, coconut, and corn oil.
Vegetables and fruits	347	About 15 vegetables and 20 fruit crops provide a valuable source of vitamins, minerals, and roughage.
Meat and milk	140	MDCs are 20% of world population, but consume 80% of all meat and milk. LDCs have 60% of all livestock and 80% of all population, but only 20% of meat and milk.
Fish and seafood	70	One-third consumed in Asia. Major source of protein in many countries.

+LDC = less developed countries
*MDC = more developed countries

Major Crops

The three crops on which humanity depends for the majority of its nutrients and calories are wheat, rice, and maize. Together, about 1,440 million metric tons of these three crops are grown each year, roughly half of all agricultural crops. Wheat and rice are especially important since they are the staple foods for most of the 4 billion people in the developing countries of the world. These two grass species supply around 60 percent of the calories consumed directly by humans.

Potatoes, barley, oats, and rye are staples in mountainous regions and high latitudes (northern Europe, north Asia) because they grow well in cool, moist climates. Cassava, sweet potatoes, and other roots and tubers grow well in warm, wet areas and are staples in Amazonia, Africa, Melanesia, and the South Pacific. Sorghum and millet are drought resistant and are staples in the dry regions of Africa.

Meat and milk are highly prized by people nearly everywhere, but their distribution is highly inequitable. Although the industrialized, more highly developed countries of North America, Europe, and Japan make up only 20 percent of the total population, they consume 80 percent of all meat and milk in the world. The 80 percent of the world's people in less developed countries raise 60 percent of the 3 billion domestic ruminants and 6 billion poultry in the world, but consume only 20 percent of all animal products. The grazing lands that support many of these domestic animals are discussed further in chapters 12 and 13.

Fish and seafood contribute about 100 million metric tons of high-quality protein to the world's diet, about two-thirds as much as that from land animals. This is an important source of protein in many countries, contributing up to one-half of the animal protein and one-fourth of the total dietary protein in Japan, for instance.

There are indications that we have already surpassed the sustainable harvest of fish from most of the world's oceans. Our current fishing methods are wasteful, destroying habitat and young fish by indiscriminate trawling. Several tons of unwanted species are discarded for every ton of seafood brought to market. Much of what is discarded could be eaten, but it is the wrong size, too young, out of season, or a variety that we do not like to eat. Changing a name from dogfish or hagfish to ocean perch or sea trout may make an unwanted species acceptable. Tragically, much of the catch that is tossed back into the sea is too badly injured by decompression or being crushed in the net to survive.

What are our alternatives? Obviously, we could use more selective harvest methods and less selective eating preferences. There also is great potential for raising fish and crustaceans in ponds or in estuaries. Under controlled conditions, even environmentally sensitive fish like trout can be raised in high-density

FIGURE 10.5 Aquaculture—growing fish, and other edible species in confined ponds—can be highly productive and add a valuable source of high-quality protein to our diets.

TABLE 10.6 Major crops in leading countries of production, 1984 (million metric tons)				
	Wheat	Rice	Maize	% World production
World total	522	469	449	100
China	88	181	73	24
United States	71	6	195	19
India	45	91	8	10
Soviet Union	76	3	13	6

From *Food and Agriculture Production Yearbook*, Vol. 39, 1985. Copyright © 1985 Food and Agriculture Organization of the United Nations, Rome, Italy. Reprinted by permission.

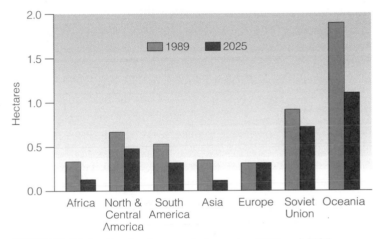

FIGURE 10.6 Cropland per capita by region 1989 and 2025 (projected). Growing populations will mean less land per person everywhere except Europe in the next century.

ponds (figure 10.5). Genetic engineering techniques are being used to breed "superfish" that will grow faster and increase yields in the way that the "green revolution" has done for plants.

Major Crop-Producing Countries

Where is most of the world's food grown? Look at table 10.6. As might be expected, the four countries that grow most of the three major crops are the biggest countries in the world, both in terms of land area and population. They are China, India, the Soviet Union, and the United States. Together, these four giants have one-half of the world's population and grow 60 percent of all crops on one-half of the world's croplands. The Soviet Union has by far the largest amount of arable land—2.3 billion hectares, or 16 percent of the total world cropland. The United States is second with 1.9 billion hectares to feed 250 million people. By contrast, China feeds 1.1 billion people from about 1 billion hectares of land. India has almost as much land as the United States, but has three times as many people and gets only one-third the yield from its land. Growing populations will mean that most regions of the world will have less arable land in the future than they do now (figure 10.6).

In spite of its relative land shortage, China is the world's largest producer of both wheat and rice. Wheat (in the form of noodles) is the staple of the colder, drier regions of north China; rice is the staple of the wetter, warmer southern regions. The United States is the largest producer of maize, growing about 8 billion bushels (nearly 200 million metric tons) or slightly less than half of the world total. Most of the grain we produce is not consumed directly by humans. About 90 percent is used to feed dairy and beef cattle, hogs, poultry, and other animals. It is used especially to fatten beef cattle during the last three months before they are slaughtered.

As chapter 2 shows, there is a great loss of energy with each step up the food chain. This means we could feed far more people if we ate more grain directly rather than feeding it to livestock. Every 16 kg of grain and soybeans fed to beef cattle in feedlots produces 1 kg of edible meat. The other 15 kg are used by the animal for energy or body parts we do not eat or it is eliminated. If we were to eat the grain directly, we would get 21 times more calories and 8 times more protein than we get by eating the meat it produces. Hogs and poultry are about 2 and 4 times as efficient, respectively, as cattle in converting feed to edible meat.

Although India is the third largest producer of staple crops, an estimated 300 million Indians are undernourished. This represents one-third of the country's population or about 40 percent of all hungry people in the world. In spite of serious shortages, India *exported* about 24 million metric tons of grain in 1986. This would have been enough to raise the caloric intake of most of India's hungry people to the minimum requirements for normal life. Instead, this grain went primarily to feed livestock in Europe and the Middle East.

The Soviet Union is so far north (it lies at the same latitude as Canada) and so large that much of its land is too cold and dry to produce good crop yields. It is the second largest producer of wheat, but it is also the largest importer of both wheat and maize. Much of its imported grain is used as livestock feed. The decision in the early 1970s to introduce more meat into the Russian diet set off an agricultural boom in North America that has had severe environmental repercussions as marginal land was put into production and land was farmed more intensively than is sustainable.

Increases in World Food Production

Contrary to what Thomas Malthus predicted in 1789 (chapter 7), crop yields have increased faster over the past two centuries than have human populations, even though we have experienced unprecedented population growth. Some people see this as a second agricultural revolution, a radical change that could change the course of human history.

Over the past 40 years, food supplies increased dramatically both in absolute terms and in food available per person worldwide. This success was due to a variety of factors including increased use of irrigation, fertilizers, and pesticides, expanded croplands, and new high-yielding crop varieties. Table 10.7 shows gains in major food categories over the past 23 years. World grain production grew nearly 70 percent during that time and the developing countries of the world generally doubled their output. Pulses (legumes), fruits, melons, vegetables, and oil production also increased by about 70 percent and meat, milk, and fish supplies grew by nearly 60 percent.

Only root crops failed to match these impressive growth rates, and there is great potential for introducing new varieties of potatoes, cassava, yams, and other highly productive crops in tropical countries where food shortages are worst. Whether continued improvements will be possible remains one of the most important questions in environmental science. Many resources, such as water, fertilizer, and land necessary for agricultural production, are becoming limited. Furthermore, environmental and cultural impacts of intensive agriculture may be unsustainable in the future (chapter 11).

The recent gains in world food production mean that everyone could have 2,600 calories per day if the food were equally divided. Unfortunately, food is not equitably distributed. Even such countries as the United States with huge food surpluses have large numbers of citizens who are undernourished or malnourished. Some areas have especially severe food shortages. The conditions in sub-Saharan Africa are of great concern. Although the total amount of food produced there has doubled over the past 30 years, population has grown even faster, so food available per person remains below acceptable levels in 31 out of 41 countries in the area (figure 10.7).

Year	Developing countries	Developed countries	World total
TABLE 10.7 Crop production 1965–1988 (millions of metric tons per year)			
Cereal grains			
1965	470	536	1,006
1988	969	774	1,743
Root crops			
1965	246	243	489
1988	376	196	572
Meat, milk, fish			
1965	115	387	502
1988	250	530	780
Pulses, fruits, melons, vegetables, oils			
1965	284	229	513
1988	555	322	877

Source: Data from *World Resources 1990–1991*.

FIGURE 10.7 Thirty-one of forty-one countries in Sub-Saharan Africa have population growth rates that exceed their growth rate of food production. Food shortages have been common in a broad, horseshoe-shaped band, as can be seen on the map.

Source: The World Bank.

If present trends continue, the African population will triple by 2025, and food supplies will fall even further below necessary levels. Helping Africans stabilize populations, protect the environment, and grow enough food for their needs surely should be a world priority.

TABLE 10.8 Net cereal importers and exporters (1987)			
Importers		**Exporters**	
Country	Million metric tons	Country	Million metric tons
Soviet Union	29	United States	83
Japan	27	Canada	28
China	16	France	26
Egypt	9	Australia	18
Korea	9	Argentina	9
Saudi Arabia	8	Thailand	6
Iran	6	United Kingdom	4
Italy	5	South Africa	2
Mexico	5	Denmark	1
Iraq	4	New Zealand	0.2

Source: Food Agricultural Organization of the United Nations.

The Green Revolution

Most of the recent growth in world food supply has come from higher yields per hectare of land. Figure 10.8 shows the maize yields per hectare over the past century in the United States. As you can see, there has been nearly a six-fold increase in yield during this time. This represents a growth from an average of less than 40 bushels per acre fifty years ago to more than 200 bushels per acre today. The two main contributors to this remarkable change have been better-yielding crop varieties and the use of fertilizer. Much of the progress in new hybrid maize varieties, fertilizers, farming procedures, and machines to carry out high-intensity agriculture came from research and training at U.S. land-grant universities.

Starting shortly after the Second World War, the Ford and Rockefeller Foundations (along with a number of governments and other agencies) set up agricultural research stations to breed tropical wheat and rice varieties that could provide yield gains similar to those obtained in the United States (table 10.8). Collectively, these are known as the Consultative Group on International Agricultural Research (CGIAR) (table 10.9). The results have been spectacular for many crops. "Miracle" strains of rice and wheat have made it possible to triple or quadruple yields per hectare (figure 10.9). This dramatic yield increase from new crop varieties has been called the **green revolution.**

These new strains are really "high responders" rather than high yielders. That is, they respond more efficiently to increases in fertilizer and water and have a higher yield under optimum conditions than do traditional, native species. The average, worldwide grain yields using these new varieties have increased from 1.1 to 2.6 metric tons per hectare. The benefits of these higher yields, however, are not accessible to everyone. Poor farmers cannot afford the seed, fertilizer, water, pesticides, fuel,

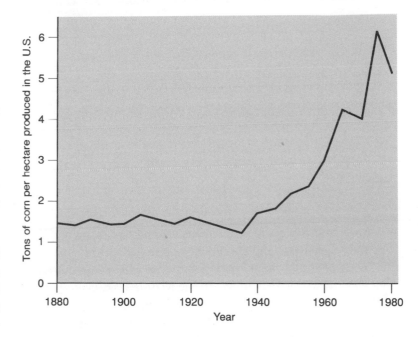

FIGURE 10.8 United States maize (corn) yields in metric tons per hectare, 1880 to 1980. Declining yields in the late 1970s were due to land taken out of production for environmental protection and to reduce surpluses.

Source: USDA.

and farm equipment necessary to cultivate the new strains. Often only the most prosperous farms are able to participate in the green revolution. The crop surpluses they produce are likely to drive prices down, so the marginal farmers are even worse off than before.

There is a worry about whether crop breeders can continue to produce new varieties that will maintain these high yields.

Center	Location	Year established	Major crop research
International Center of Tropical Agriculture	Cali, Colombia	1966	Cassava, rice, beef
International Center of the Potato	Lima, Peru	1971	Potato
International Center of Improvement of Maize and Triticale	Mexico City, Mexico	1943	Maize, wheat, barley, triticale
International Board for Plant Genetic Resources	Rome, Italy	1974	Germplasm preservation
International Center for Agricultural Research in the Dry Areas	Aleppo, Syria	1977	Dryland agriculture
International Crops Research Institute for Semiarid Tropics	Andhra Pradesh, India	1972	Semiarid tropics crops
International Food Policy Research Institute	Washington, D.C., U.S.A.	1974	Government food policy
International Institute of Tropical Agriculture	Ibadan, Nigeria	1967	Yam, cowpea, cassava, rice, maize, beans
International Laboratory for Research on Animal Disease	Nairobi, Kenya	1974	Animal diseases
International Livestock Center for Africa	Addis Ababa, Ethiopia	1974	Livestock production
International Rice Research Institute	Los Banos, Philippines	1960	Rice (paddy)
International Service for Agricultural Research	The Hague, The Netherlands	1980	National research systems
West Africa Development Association	Monrovia, Liberia	1971	Rice (upland)

Source: Food and Agricultural Organization of the United Nations, 1990.

Throughout the world, native crop varieties are being replaced by these new types, and the genetic resource available to breeders is being seriously diminished. Fifty years ago several hundred varieties of wheat were grown in the Middle East, each adapted through centuries of selection to a particular set of environmental conditions. Now, a few miracle varieties have displaced most of those indigenous species. There is a great danger that an epidemic might sweep through the fields when they are all planted with a single variety, resulting in a major famine. A hint of what this could mean was experienced in the United States in 1970, when nearly all hybrid corn in the country shared a set of genes called Texas male sterile that made them all susceptible to a fungal disease called southern corn leaf blight. An epidemic swept across the country and threatened the entire crop.

Plant conservators are busy hunting for native crop varieties that might be disease resistant or able to grow in a unique set of environmental conditions. For instance, many irrigated soils in dry climates tend to accumulate surface salt and mineral deposits. Some wild strains are salt tolerant, and there is a possibility that they could be foundation plants for salt-resistant domestic varieties. **Gene banks** have been set up to store many seed varieties for future breeding experiments.

Most of our commercial crops originated in the tropics or semiarid lands of the Middle East, Africa, or Central America. Some of the less developed countries that still have wild ancestors of cultivated crops have raised objections to having scientists from the developed countries collecting samples of their flora and fauna. They feel entitled to a share of the profits that may come from the use of their native species for future crop development. There are threats of "gene wars" over control of this resource. Who is right? Should genetic resources be considered as a salable commodity, like mineral and crop resources, to benefit their native country, or are they a globally shared resource?

New or Unconventional Food Supplies

Agricultural economist Edward C. Wolf of the Worldwatch Institute points out that although at least three thousand species of plants have been used for food at one time or another, most of the world's food now comes from only sixteen widely grown crops. There are many new or unconventional crops that might

FIGURE 10.9 Short-stemmed "semidwarf" wheat (*right*) developed by Nobel prize winner, Norman Borlang, is a high responder; that is, it can utilize high levels of fertilizer and water to produce high yields without growing so tall that the stalks fall over and are impossible to harvest. Notice how much denser the seed heads are on the semidwarf than its normal size relative (*left*).

make valuable contributions to human food supplies, however, especially in areas where conventional crops are limited by climate, soil, pests, or other problems. Some of the plants now being studied as potential additions to our crop roster include the following:

- *Winged beans.* A perennial plant that bears well in hot climates where other beans will not grow, is totally edible (pods, mature seeds, shoots, flowers, leaves, and tuberous roots), and enriches the soil.

- *Amaranth.* A staple seed crop of the Aztecs and the Incas with edible leaves and seeds; one of the most nutritious grains known that produces high yields in a variety of conditions.

- *Triticale.* A hybrid between wheat (*Triticum*) and rye (*Secale*) that grows in light, sandy, infertile soil, is drought resistant, has nutritious seeds, and is being tested for salt tolerance for growth in saline soil or irrigation with seawater.

- *Wax gourd vegetable.* A creeping vine from Africa that looks like a pumpkin but is easier to grow and more

nutritious and can be eaten at any stage of growth as a cooked vegetable, soup base, or food extender; its waxy coat resists microorganisms, and it can be stored up to one year without refrigeration.

Single-cell protein (SCP) has been proposed by some scientists as a potential new food source for feeding the world's population. SCP is obtained from unicellular organisms (bacteria, algae, or yeast) grown in culture under controlled conditions. Many microbes are able to grow on waste products, such as crop or food processing residue or wood by-products, and transform them into useful protein for human consumption. Whether factory-grown single-cell protein will prove economical or acceptable as a dietary supplement remains to be seen.

ECONOMICS OF FOOD SUPPLY

As a result of the yield increases over the past three or four decades, the world food supply at present is best described as a state of abundance, rather than shortage. In fact, many countries now have embarrassing food surpluses.

Food Surpluses and Agricultural Subsidies

The United States food reserves reached a historic high of more than 200 million metric tons in 1987. This was nearly 60 percent of all the food stored in the world.

Finding places to store all this excess has become a serious problem. Country grain elevators are full to the brim. The federal government is paying farmers to put up rows of gleaming metal storage bins on their land. Some farmers make more money by withholding land from production and storing surpluses than by growing new crops. Grain is being held in caves, in barges on the Mississippi River, and in abandoned school houses. During harvest, mountains of corn and wheat sit out in the open where they are exposed to rain, insects, birds, and rodents because there is no other storage place (figure 10.10). The total cost of price supports, subsidies, and storage of surpluses in the United States was around $25 billion in 1987. The droughts of 1987–88 reduced U.S. crop reserves to 84 million metric tons in 1989, causing some people to fear impending shortages in the future.

One effect of increasing crop production in recent years is that crop prices have declined steadily (in constant dollars) about 1.5 percent per year since 1950. The more developed countries have had to struggle to protect their farm sector from competition from cheap foreign imports, setting off trade wars, costly subsidy programs, destruction of surpluses, unmarketable food supplies, and a substantial reduction in use of marginal farmland.

In 1972, then Secretary of Agriculture Earl Butz called on American farmers to increase their production to "feed the world." Farmers were urged to plow up pastures, drain wetlands, and plant fence-row to fence-row. Butz predicted an unlimited world market for U.S. agricultural exports. The price of farmland doubled—even quadrupled in some areas—in the 1970s. Farmers borrowed money to buy new machinery and expand their operations. The total farm debt in the United States exceeded $230 billion—more than the foreign debt of Brazil and Mexico, the two biggest Third World debtor nations, together.

Exports also rose dramatically: from $2.5 billion in 1970 to $43.8 billion in 1981. The oil price shocks of 1973, 1979, and 1989, however, coupled with increased agricultural productivity in other countries, have sent U.S. agriculture into a tailspin. Many developing countries have decreased imports, both because they can't pay for them and because they don't need them anymore. The cost of raising crops by energy-, water-, and fertilizer-intensive techniques has risen dramatically. Other countries can raise their productivity with less costly investment than we can. American farmers are suffering. About half of the farms that expanded during the 1970s have either gone out of business or are teetering on the brink of financial disaster.

FIGURE 10.10 After a good harvest in the Midwest there is often no place to store grain because so much storage space is already occupied by previous year's surplus. When grain is piled on the ground in the open, it is subject to birds, insects, and spoilage.

World Agricultural Trade and Politics

We hope you're beginning to appreciate how humans everywhere are part not only of a global biosphere system, but also of a global economic and political network. As a system, world agricultural trade is full of inefficiencies, intrigue, and uncertainties that are difficult to unravel.

The number of countries dependent on foreign food supplies has increased dramatically since the Second World War. In 1939, only ten countries imported more than 1 million tons of food. In 1980, this number had risen to forty countries. More

RESOURCES AND RESOURCE ECONOMICS

FIGURE 10.11 Cash crops, such as these coffee beans being picked in Colombia, often replace production of staples necessary to feed the local population of developing countries. Government policies encourage export commodities to raise foreign exchange.

importantly, world food trade now accounts for more than 20 percent of all food consumed. The world is becoming more and more interdependent. What are the reasons for this increase? One is, of course, that there are more hungry people who need food. Another is that the massive shipments of personnel and supplies during World War II established the merchant fleet, trade routes, and loading and unloading facilities necessary for food shipments. Furthermore, recent monetary agreements and trade policies have helped make international trade possible.

World trade in agricultural commodities has brought both benefits and problems for farmers and planners in many countries besides the United States. On one hand, international trade makes up the food deficit in many countries, staving off famine and social disruption. On the other hand, nations that become dependent on imported food supplies may divert so much of their foreign earnings to buying expensive imports that they can never develop indigenous resources. They also may become politically indentured to the countries that sell them food. As former

U.S. Secretary of Agriculture John Block said, "food is a weapon."

Food-producing countries also can become captive to international trade. Often a country finds it expedient to specialize in a few crops that fit its climate and work force. It then can trade some of its surplus for other basic foods that are grown more efficiently elsewhere. The foreign exchange generated by these **cash crops** can provide funds necessary to buy tools, information, and technology needed for development. Too often, however, land for growing these export crops is taken out of staple food production for local consumption. This necessitates importing foreign food and drives up local food prices.

Often, the export crops grown by developing countries are not food supplies for needy neighbor nations, but rather luxury foods, drugs, and other nonessential items to be shipped to the rich countries of the world (figure 10.11). Little of the money generated by these crops trickles down to the underclass of the producing countries. Most of it goes to landowners, money

Human Nutrition, Food, and Hunger

lenders, export brokers, or political leaders. In many countries, all these functions are carried out by the same small group of people. In eighty-three developing countries of the world, 85 percent of the land is owned or controlled by 5 percent or less of the population. This land-owning oligarchy usually finds that it can make more profits from the land by growing cash crops for export rather than growing low-priced cereals and staples for local populations. Much of the money generated by this international trading now is used to buy weapons, support repressive governments, and resist land reform or other changes in the social, political, or economic system.

These export-oriented land-use patterns, which are often the legacy of colonial power structures and market relationships, contribute significantly to malnutrition problems in many countries. In Guatemala, for example, 97 percent of the citrus crop is exported, while 60 percent of the local population suffers from vitamin C deficiency. In Central America, beef production and export increased nearly six-fold in the 1960s and 1970s, while local per capita meat consumption fell by 50 percent. Land use is discussed further in chapter 12.

Cash-crop agriculture also contributes to the problems of rapidly growing Third World cities. Subsistence farmers and tenant farmers are forced off the land to make way for more profitable, large-scale, and industrial agriculture. These displaced workers move into the cities in search of food, jobs, and shelter, adding to the burdens of already overcrowded urban areas (chapter 25). Governments of underdeveloped countries often keep basic food prices artificially low to reduce unrest in politically powerful urban populations. This reduces the incentives for maximum production by farmers, lowers rural income, and increases poverty and hunger.

Are there any constructive ways to deal with these crushing problems? One way to reduce urban growth and simultaneously reduce malnutrition would be to provide more rural jobs and raise rural wages so people can stay on the land and afford to buy the food they need. If people have money to buy food, local farmers will have an incentive to grow more, and food supplies will increase. If there are alternatives to farming as a way to earn a living, there will be more land per farmer and better profits, allowing farmers to buy better seed, fertilizer, water, and equipment to further increase local food supplies.

Another problem faced by Third World countries is that international monetary exchange rates, credit payments, shipping costs, and commodity values are determined by a market and banking system established by the richer countries of the world (chapter 8). This system usually follows policies and practices that benefit its most powerful members. Commodity prices fluctuate wildly, making it difficult for producing countries to plan their budgets or accumulate the capital necessary for development that might make them more independent.

International **food aid** by the more developed countries is often selective and highly political. We give aid to those governments that support our policies rather than to the countries with the most hungry people. The United States, for example, gives Egypt $5.4 billion and Israel $5.2 billion in food and agricultural aid each year. The total aid to India, Bangladesh, the Philippines, and Indonesia, which together have more than 1 billion people (about half of whom are undernourished), is less than half as much as the food aid to Israel, which has a total population of only 4.4 million. Much of the food aid we do give hurts rather than helps. It doesn't reach the poor. It drives down prices for local farmers, creates conditions that trap some countries in permanent dependence on additional welfare, and helps support unpopular and repressive governments.

WORLD HUNGER

The nutritional deficiencies we discussed earlier can affect individuals anywhere in the world who don't get an adequate diet either because wholesome food isn't available, they can't afford it, or they don't know how to choose the right things to eat. What about larger-scale food shortages and their causes? Food and water are our barest necessities to remain alive. Can you imagine what it would be like to be without them?

Famines

The images that probably first come to your mind when you think about food shortages are the tragic scenes of mass starvation that come from places like Ethiopia and Somalia in the Horn of Africa, Bangladesh, Angola, or Biafra. These acute shortages, or **famines,** are characterized by large-scale loss of life, social disruption, and economic chaos. Starving people eat their seed grain and slaughter their breeding stock in a desperate attempt to keep themselves and their families alive. Even if better conditions return, they have sacrificed their productive capacity and will take a long time to recover. Famines are characterized by mass migrations as starving people travel to refugee camps in search of food and medical care (figure 10.12). Many die on the way or fall prey to robbers.

In the preceding section, we suggested some economic and political factors that affect food supplies, but what are the causes of famine? Crop failure and interruption of food supplies from producing regions to the places where people need food are the immediate causes. Adverse weather, insect infestations, and other natural disasters often work with human-caused calamities to cause crop failure and create food shortages. During droughts, floods, and wars, people can neither grow their own food nor find jobs to earn money to buy the food they need. The African nations that suffered the most widespread and long-lasting famines during the early 1980s were Angola, Mozambique, Sudan, Ethiopia, and Chad, all of which suffered not only severe drought, but also continuing wars. Neighboring countries that experienced comparable weather did not have as severe problems as these unfortunate countries.

FIGURE 10.12 Floods, drought, insects, and other natural disasters are usually the immediate cause of famine; however, politics, economics, and environmental mismanagement are often contributing or underlying causes. Some of the countries with the largest numbers of starving people export cash crops.

BOX 10.1   Feeding a Fifth of the World

Twenty years ago, many people were predicting that China would never be able to feed all its people. They said that conditions were hopeless, that "bombs would be kinder than food aid" because the people were going to starve to death anyway. China was one of several countries, including India, Egypt, and Libya, categorized as "can't be helped" by those who believed that population growth had already outstripped the carrying capacity of the land.

In fact, the situation in China did seem impossible. In his 1959 "Great Leap Forward," Chairman Mao Tse Tung ordered collectivization of farms, clearing of forests, planting of all land—regardless of its potential for farming—as well as standardization of crops, planting schedules, and agricultural planning. Bureaucratic bungling disrupted planting and harvesting. Peasants who no longer controlled their land, nor benefited from working on it, resisted the changes. Farm production fell disastrously.

Between 1959 and 1961, China suffered the greatest famine in the world's history. It was kept secret from the western world, and even today we don't know the full extent of the disaster. Bad weather played a part. Droughts in some parts of the country and floods elsewhere ruined crops. Some areas had food but couldn't send it to market because of government policies. Farmers were ordered to plant the wrong crops at the wrong time. Crops were planted on steep hillsides that quickly eroded, smothering good bottomland in mud and silt. Estimates of the number who died range from 10 to 60 million, but most experts agree that a realistic number is 30 million deaths. There surely have never been so many untimely deaths in a single country in so short a time.

Amazingly, just twenty years later, the whole country reversed its course just as suddenly. After Mao's death, Chairman Deng broke up the communes and returned farms to family units. Chinese researchers succeeded in adapting the miracle rice strains for growth in the cooler climate of the North China Plain. A modified market incentive system was reestablished and the results were amazing. Between 1975 and 1985, the total rice yield more than doubled—from 80 to 180 million metric tons per year. Almost half of that increase occurred in just four years, from 1980 to 1984. China is now number one in both rice *and* wheat production.

Nearly all of China's 800 million farmers now cultivate small family plots, getting 50 percent more crops per hectare than did the communes (box figure 10.1). Each crop requires an average of sixty worker days, compared to 250–320 commune worker days a year to produce comparable yields. In addition to a diversified mix of crops, farmers are producing commodities, such as fish, chickens, pigs, ducks, and rabbits. An ample and varied diet has replaced thin rice porridge as standard fare, and meat consumption has more than quadrupled.

China now has enough food for everyone in the country to have an adequate, if limited, diet. Fewer than 3 percent of the Chinese people are underfed—a lower percentage than in the United States, where about 10 percent of the population goes hungry. China has even begun to export grain to other countries. The country that was thought to be a living example of Malthus's dismal predictions that population would inevitably outstrip food production now seems to have enough food to feed everyone.

BOX FIGURE 10.1   Recent decollectivization and return to traditional farming practices have more than doubled China's farm yields and improved nutrition.

Government policies also can add to food shortages and dislocations. People are allowed or even encouraged to cut down forests, plow up fragile prairie or savannah soils, overgraze rangelands and steep hillsides, and conduct other environmentally damaging practices. Could the tragedy of famine be avoided if governments would plan ahead for bad weather and poor harvests? The worst famine in the twentieth century and probably the worst famine in human history occurred in China in the late 1960s as a result of adverse weather and political decisions on land use and farming (Box 10.1).

Chronic Undernutrition and Malnutrition

Although we most often think of food shortages in terms of catastrophic famine and starvation, far more people in the world are affected by **chronic food shortages** than by acute famines. About 1 million people are thought to have died in sub-Saharan Africa in the early 1980s as a result of the drought-related famine. By contrast, approximately 18 to 20 million people—mostly children—die *each year* from the effects of chronic undernutrition and malnutrition. We will now turn our attention to the causes and effects of chronic food shortages.

Although world food production has increased dramatically over the last forty years and the average per capita daily calorie intake is slightly above the minimum recommended for good health, there are still many hungry people in the world. As we saw in table 10.1, while North America and Europe have nearly 1,000 calories above the minimum level of 2,500 calories per person per day, sub-Saharan Africa has only about 80 percent, on average, of the minimum daily caloric requirement.

Table 10.10 shows an estimate by the World Bank of the number of hungry people in the world's less developed countries. The 730 million hungry people in these countries make up about 18 percent of the total population: *about one person in six in the world.* The largest number of hungry people live in South and Southeast Asia. India has the largest number of hungry residents—at least 300 million people, or about one-third of its population. Bangladesh probably has the highest proportion of hungry people; 80 percent of its population has less than the minimum daily caloric intake. Other nations with high levels of hunger are Indonesia, the Philippines, and Thailand, all of which are major agricultural exporters *and* major beneficiaries of new, high-yield crop varieties.

CAUSES AND SOLUTIONS FOR WORLD HUNGER

As we discussed in chapter 7, food supplies often are considered among the main limiting factors that determine how many people the world can support. Many of those who study environmental resources debate the root causes of chronic world hunger and what can be done about it. Most authors agree that population

TABLE 10.10 Prevalence of energy-deficient diets in developing countries (1980)

Insufficient calories for an active working life: 90% of FAO/WHO minimum requirements

Region	Number of countries	Population (millions)	% of total population	% change since 1970
Sub-Saharan Africa	37	150	44	+30
East Asia and Pacific	8	40	14	−57
South Asia	7	470	50	+38
Latin America and Caribbean	24	50	13	−15
Middle East and North Africa	11	20	10	−62
Total	87	730	18	

Insufficient calories to prevent stunted growth and serious health risks: Below 80% of FAO/WHO requirements

Region	Number of countries	Population (millions)	% of total population	% change since 1970
Sub-Saharan Africa	37	90	25	+49
East Asia and Pacific	8	20	7	−57
South Asia	7	200	21	+47
Latin America and Caribbean	24	20	6	−21
Middle East and North Africa	11	10	4	−68
Total	80	340	9	

FAO = Food and Agricultural Organization of the United Nations
WHO = World Health Organization

Source: World Bank (in *World Resources* 1986).

growth, environmental degradation, and inequitable distribution of resources are linked in a vicious cycle that results in food shortages and human suffering. There is much disagreement, however, about where the most effective point is to break this cycle.

Most of the countries with the highest numbers of hungry people also suffer rapid population growth that outstrips food supplies and leads to environmental degradation that lowers the carrying capacity of environmental and agricultural systems. Many people warn that human populations are approaching—or may already have surpassed—the carrying capacity of the planet. They point out the need for strong, effective birth control programs to bring populations into balance with resources. Some argue that drastic actions are needed, such as limiting food aid to only those countries that are meeting population-control quotas.

Other people argue that technology and human ingenuity have increased the world's capacity to provide food so that existing resources could support a much larger world population than now exists. It is true that during the past one hundred years, food supplies have risen faster than population growth, not so much because more agricultural land has become available, but because yields per unit of land have increased dramatically. Estimates vary as to how many people the world could support if all available croplands were farmed intensively and produced yields comparable to those obtained in the developed countries. These estimates range from twice to more than ten times the current population (10 billion to 50 billion people). Those who are optimistic about the power of technology stress the need to encourage development in order to eliminate hunger.

Not everyone agrees that it would be that simple. Some people see technology as the cause of many problems rather than the solution. They argue that technology (or the misuse of it) allows humans to conduct massively destructive wars and to degrade the environment so that it becomes uninhabitable. Furthermore, there are serious doubts about whether high crop yields based on intensive use of fertilizers, water, pesticides, new crop varieties, and fossil fuel energy can be maintained. While some people believe that Earth's carrying capacity (for humans) can be increased by technology, others believe that a "technical fix" may cause more problems than it solves.

Some people believe that both population growth and misapplied technology are merely symptoms of deeper underlying issues. They argue that hunger and many other problems are really caused by poverty, powerlessness, and maldistribution of resources. In this view, neither reducing populations nor growing more food will necessarily help those people at the bottom of the social scale, because we already have a more than adequate supply of both land and food. What people need is access to resources to grow their own food, or money to buy what they need and freedom to do so. Proponents of this view argue that if people have security, education, control of their own lives, and equitable access to the world's resources, they can and will resolve both hunger and population growth themselves. Whether humans and human systems can be perfected to this degree remains to be seen.

Perhaps these theories on the causes of hunger are not mutually exclusive. In reality, hunger, poverty, environmental degradation, social problems, and population growth are parts of a complex, interconnected web. Each is a cause, as well as a consequence of the others.

In chapter 11, we will continue this discussion with a survey of the agricultural systems that produce our food. As part of that discussion, we will look at the resource "inputs" of soil, water, fertilizer, and energy that are necessary to sustain crop production and support future human populations. In subsequent chapters, we will discuss other resource limitations that bear on this all-important question: How many people—and at what level of civilization and environmental quality—can the world support?

SUMMARY

World food supplies have been rising at an unprecedented rate, and have grown faster than populations in every continent except Africa. There is enough food to supply everyone in the world with more than the minimum daily food requirements, but food is inequitably distributed. The FAO estimates that 750 million people are chronically undernourished or malnourished, and 18 to 20 million (mostly children) die each year from diseases related to malnutrition. Additional millions survive on a deficient diet, but suffer from stunted growth, mental retardation, and developmental disorders.

Among the essential dietary ingredients for good health are proteins, lipids (especially unsaturated ones), vitamins, and minerals. Marasmus and kwashiorkor are protein-deficiency diseases; anemia and goiter are often caused by mineral deficiencies; and pellagra, scurvy, beriberi, and rickets are vitamin-deficiency diseases that affect millions of people worldwide.

The three major crops that are the main source of calories and nutrients for most of the people are rice, wheat, and maize. About a dozen other types of seeds and grains, a few root crops, twenty or so fruits and vegetables, six mammals (and their milk), a few domestic fowl, and a variety of seafoods comprise nearly all the food that humans eat. Some new crops or unrecognized traditional crops hold promise for increasing nutritional status of the poorer people of the world. Scientific improvement of existing crops and modernization of agriculture (irrigation, fertilizer, and better management) are potential sources of greater agricultural production.

Over the past thirty-five years, the total amount of food in the world has grown faster than the average rate of population growth, so there is now about 25 percent more food per person than there was in 1950, even though the total number of people has doubled. The biggest gains have been in Asia, North America, and Latin America, all of which have nearly tripled their food production. The only major region in which food production has failed to keep pace with population growth has been sub-Saharan Africa, where adverse weather, insect infestations, wars, inept governments, social and religious factors, economics, and international politics have intervened.

World food trade and international food aid help transfer food from areas of abundance to areas of shortage, but they also undercut local food supplies by encouraging the conversion of land from production of food for local consumption to production of cash crops for export. They also widen economic and social disparities that increase poverty and powerlessness and make it even more difficult for the poorest people to feed themselves.

Hunger, poverty, population growth, environmental degradation, and social problems form a complex interconnected web. Each is a cause, as well as a consequence, of the others. One of the most important questions in environmental science is, "How many people, and at what level of civilization and environmental quality, can the world feed?"

■ Review Questions

1. How many calories does the average human adult need to meet minimum daily requirements to carry out a healthy, active life?
2. What are the consequences of chronic undernutrition or malnutrition?
3. What is the difference between undernutrition and malnutrition?
4. Why is protein so essential in our diet?
5. What are the three major grain crops of the world? What are some of the other major crops?
6. What are some alternative or new food crops being considered? Describe some advantages or disadvantages of these crops.
7. Why is it important to preserve wild ancestors and traditional varieties of domestic crops?
8. Why are less developed countries eager to grow cash crops? What are some of the effects of export agriculture on local food supplies and internal economy of the producing country?
9. What countries have the largest numbers of hungry people in the world?
10. How can human actions change the carrying capacity of the world? What are some of the environmental consequences of those actions?

■ Questions for Critical Thinking

1. Why do you suppose so few food crops and species of domestic animals make up such a large proportion of the human diet? Why do we choose to grow these particular crops?
2. What might be the causes and consequences of a worldwide epidemic in rice, wheat, or maize?
3. Describe the benefits and dangers of the green revolution in agriculture. Do you think that "miracle" crops will help or hurt the marginal farmers and farm laborers of the less developed countries? Why?
4. What effects will the green revolution have on world demand for fertilizer, irrigation, farm machinery, and pesticides?
5. Should plant collectors from more developed countries pay a royalty when they collect wild plant or animal species or traditional crop varieties?
6. Why do we have such large food surpluses? What could or should we do about them?
7. In what ways do international food aid and agricultural trade hurt rather than help the hungry people of the world?

8. How do government policies and politics contribute to problems of food shortages? Why does India have so many hungry people while China has relatively few?
9. How are hunger, poverty, population growth, and environmental degradation interconnected?
10. What can we do individually about problems of world hunger?

■ Key Terms

anemia (p. 175)
cash crops (p. 183)
chronic food shortages (p. 187)
famines (p. 184)
food aid (p. 184)
gene banks (p. 180)
green revolution (p. 179)
hypothyroidism (p. 175)

kwashiorkor (p. 172)
malnourishment (p. 172)
marasmus (p. 172)
overnutrition (p. 172)
pellagra (p. 175)
seriously undernourished (p. 171)
undernourished (p. 171)
vitamins (p. 175)

SUGGESTED READINGS

■ Batie, S., and R. Healy. 1986. *Beyond Oil: The Threat to Food and Fuel in the Coming Decades*. Washington, D.C.: Carrying Capacity, Inc. A Malthusian view of the future.

■ Blaxter, K. *People, Food and Resources*. Cambridge, England: University of Cambridge Press, 1986. An economist's view of agriculture and resources.

■ Brown, J. L. February 1987. "Hunger in America." *Scientific American* 256(2):36. Report of a study by Physician's Task Force on Hunger in America.

■ Brown, L. R. 1987. "The New World Order." In *State of the World 1991*. New York: Norton and Co. A good survey of the constraints and potential of world agriculture.

■ Consultative Group on International Agricultural Research (CGIAR). 1990. *1989 Annual Report*. Washington, D.C. Reports current progress in the green revolution.

■ Crosson, P. R., and N. J. Rosenberg. "Strategies for agriculture." *Scientific American*. 261(3):128. Agricultural research will probably yield many new technologies for expanding food production. The challenge will be getting farmers to use them.

■ Ehrlich, A. H., and P. R. Ehrlich. Spring 1987. "Why do people starve?" *The Amicus Journal* 9(2):42. A discussion of limits in the world's carrying capacity for humans.

■ Glantz, M. June 1987. "Drought in Africa." *Scientific American*. 256(6):34. Drought is a recurrent and often devastating feature of the sub-Saharan climate. If leaders were to treat drought as recurrent, they could deal with it in ways that would stabilize the region's farm production.

■ Hildyard, N. et al. April 1991. "Declaration of the International Movement for Ecological Agriculture." *The Ecologist* 21(2)1107. The conclusion of a special issue devoted to a critique of the FAO and modern, high technology agriculture with strategies for encouraging traditional indigenous forms of food production.

■ Hrabovszky, J. "Agriculture: the land base." In *Global Possible*, edited by R. Repetto. Pages 211–54. New Haven: Yale University Press, 1985. An optimistic appraisal of the potential for feeding the world.

■ Kent, G. 1985. "Food trade: the poor feed the rich." *The Ecologist* 15(5/6):232. A discussion of the imbalances in world food trade.

■ Lappe, F., and J. Collins. *World Hunger, 12 Myths*. San Francisco: Food First Institute for Food and Development Policy, 1986. An argument that maldistribution of resources is responsible for famine.

■ Nabhan, G. *Gathering the Desert*. Tucson: University of Arizona Press, 1985. An interesting account of wild foods and traditional native crops in the desert Southwest.

■ Nearing, H., and S. Nearing. *Living the Good Life*. New York: Schocken Books, 1948. A practical description of how to live sanely and simply in a troubled world.

■ O'Brien, P. 1985. "Agricultural productivity and the world market." *Environment*. 27(9):15. An argument that there would be plenty of food if it were shared more equitably.

■ Omvedt, G. 1975. *The Political Economy of Starvation: Imperialism and the World Food Crisis*. Bombay: Scientific Socialist Education Trust. This author argues that colonialism—past and present—is the cause of hunger.

■ Pluckett, D., and N. Smith. January 1986. "Sustaining agricultural yields." *Bioscience*. 36(40):5. How can we feed the world's people on a sustainable basis?

■ Shiva, V. April 1991. "The Failure of the Green Revolution." *The Ecologist* 21(2):57. Criticizes the green revolution as ecologically destructive and socially inequitable. Argues that the evidence produced in its favor is little more than myth.

■ United Nations Food and Agriculture Organization (FAO). *Agriculture: Toward 2000*. Rome: FAO, 1981. A detailed review of world food supplies.

■ Wittwer, S., et al. *Feeding a Billion*. East Lansing: Michigan State University Press, 1987. A comprehensive look at how China accomplished its remarkable improvement in food production by an American farm expert and three high-ranking Chinese colleagues.

■ Wolf, E. 1986. "Beyond the green revolution: new approaches for Third World agriculture." *Worldwatch Paper #73*. Washington, D.C.: Worldwatch Institute. Alternative agriculture for the developing world.

■ *World Agriculture Statistics: FAO Statistical Pocketbook 1985*. 1986. Lanham, Maryland: Bernan-Unipub. Good worldwide statistics on agriculture and food production.

CHAPTER *11*

Soil Resources and Sustainable Agriculture

Barring nuclear disaster, soil erosion is the most certain threat to the long-term future of civilization. . . . The new paradigm for farming in the future is stewardship—an approach to land use that recognizes we are not absolute owners, but caretakers of a portion of creation that should not be diminished during our tenure.

Joe and Nancy Paddock

CONCEPTS

Soil is an essential resource. Our existence on Earth depends on maintaining fertile, tillable soil for crops. There are many different soil types, each with its own composition of inorganic minerals, air, water, dead organic matter, and living organisms.

In a healthy ecosystem, soil is a renewable resource. Under the best conditions, it can accumulate 1 mm deep per year. Under adverse conditions, it may take thousands of years for the same accumulation.

Human-caused erosion is generating soil losses that exceed replacement rates in many areas. An estimated 25 billion tons of soil are lost from croplands each year, resulting in seriously diminished crop production.

Ecologically inappropriate farming and grazing practices are largely responsible for the destruction of our precious soil resources. These techniques may be used because of tradition, lack of information, or social and economic conditions that encourage destructive, short-term gain instead of long-term planning for a sustainable agriculture.

Many alternative methods could be used in farming to reduce soil erosion, avoid dangerous chemicals, and improve yields. These techniques for sustainable agriculture stem from traditionally successful farming practices, as well as recent scientific discoveries.

INTRODUCTION

Of all Earth's crustal resources, the one we take most for granted is soil. We are terrestrial animals and depend on soil for life, yet most of us think of it only in negative terms. English is unique in using "soil" as an interchangeable word for earth and excrement. "Dirty" has a moral connotation of corruption and impurity. Perhaps these uses of the word enhance our tendency to abuse soil without scruples; after all, it's only dirt.

The truth is that **soil** is a marvelous substance, a resource of astonishing beauty, complexity, and frailty. It is a complex mixture of weathered mineral materials from rocks, partially decomposed organic molecules, and a host of living organisms. It can be considered an ecosystem by itself. Soil is an essential component of the biosphere (figure 11.1).

There are at least 20,000 different soil types in the United States alone; perhaps hundreds of thousands worldwide. They vary by origin, parent materials, age, and climate. There are **young soils** that, because they have not weathered much, are rich in silicon, iron, and aluminum. There are **old soils,** like the red soils of the tropics, from which rainwater has borne away most of the soluble aluminum and iron and left behind clay and rust-colored oxides. Some soils have exotic origins, such as the midwestern loess deposits that contain silts blown all the way from Asia.

In chapter 10, we examined the dramatic gains in agricultural productivity over the past century. Sustaining that productivity depends on healthy soil, water, fertilizer, energy to run equipment, pest control, and stable ecological systems that maintain biological and genetic diversity. Present agricultural systems are depleting these resources at unsustainable rates. One of the most important questions in environmental science and resource management is "How can we establish a sustainable agricultural system that provides an adequate diet for everyone and also maintains a healthy environment?" In this chapter, we will look at the resource base that has made productivity increases possible. We will examine the potential for expanding agriculture, and we will look at some ways our agricultural systems can become sustainable and self-renewing.

WHAT IS SOIL?

To understand the potential for feeding the world on a sustainable basis we need to know how soil is formed, how it is being lost, and what can be done to protect and rebuild good agricultural soil.

A Renewable Resource

With careful husbandry, soil can be replenished and renewed indefinitely. Many modern farming techniques deplete soil nutrients, however, and expose the soil to the erosive forces of wind

TABLE 11.1 Soil particle sizes

Classification	Size
Gravel	Greater than 1 mm
Sand	0.05 to 1 mm
Silt	0.002 to 0.05 mm
Clay	Less than 0.002 mm

and moving water. As a result, we are essentially mining this resource and using it much faster than it is being replaced.

Building good soil is a slow process. Under the best circumstances, good topsoil accumulates at a rate of about 10 tons per hectare (2.5 acres) per year—enough soil to make a layer about 1 mm deep when spread over a hectare. Under poor conditions, it can take thousands of years to build that much soil. Perhaps one-third to one-half of the world's current croplands are losing topsoil faster than it is being replaced. In some of the worst spots, erosion carries away about 2.5 cm (1 in.) of topsoil per year. With losses like that, agricultural production has already begun to fall in many areas.

Soil Composition

Most soil is about half mineral. The rest is air and water mixed with a little organic matter from plant and animal residue. The mineral particles are derived either from the underlying bedrock or from materials transported and deposited by glaciers, rivers, ocean currents, windstorms, or landslides. The weathering processes that break rocks down into soil particles are described in chapter 9.

Particle sizes affect the characteristics of the soil (table 11.1). The spaces between sand particles give sandy soil good drainage and usually allow it to be well aerated (figure 11.2). Silt particles pack together more tightly, making silty soil less permeable to water and air than sandy soil. Tiny clay particles have a large surface area and a high ionic charge that makes them stick together tenaciously, giving clay its slippery plasticity, cohesiveness, and impermeability. As a result, clay is rock hard when dry and difficult to work with when wet. Soils with a high clay content are called "heavy soils," in contrast to easily worked "light soils" that are composed mostly of sand or silt. Varying proportions of these mineral particles occur in each soil type (figure 11.3). Farmers usually consider sandy loam the best soil type for cultivating crops.

The organic content of soil can range from nearly zero for pure sand, silt, or clay, to nearly 100 percent for peat or muck. Much of the organic material in soil is humus, a sticky, brown, insoluble residue from the bodies of dead plants and animals. Humus is much more important to soil quality than its proportion indicates. It gives soil its structure, a description of how the

FIGURE 11.1 Good soil is both a gift and a responsibility. With work, care, and understanding, we can use it productively and pass it on to the next generation more fertile than when we received it.

soil particles clump together. Humus coats mineral particles and holds them together. By binding particles in loose crumbs, humus gives the soil a spongy texture that holds water and nutrients needed by plant roots, and maintains the spaces through which delicate root hairs grow in search of sustenance.

Soil Organisms

Without soil organisms, Earth would be covered with sterile mineral particles far different from the rich, living soil ecosystems on which we depend for most of our food. The activity of the myriad organisms living in the soil creates its structure, fertility, and tilth (structure suitable for tilling or cultivation) (figure 11.4).

Soil organisms usually stay close to the surface, but that thin living layer can contain thousands of species and billions of individual organisms per hectare. Algae, bacteria, and fungi flourish in the top few centimeters of soil. A single gram of soil (about a teaspoonful) can contain hundreds of millions of these microscopic cells. Algae and blue-green bacteria capture sunlight and make new organic compounds. Bacteria and fungi decompose organic detritus and recycle nutrients that plants can use for additional growth. The sweet aroma of freshly turned

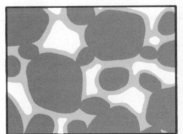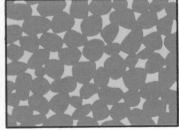

FIGURE 11.2 Soil characteristics depend on pore spaces and particle sizes. The soil on the left, composed of particles of various sizes, has spaces for both air and water. The soil on the right, composed of uniformly small particles, is more compacted and has less space for either air or water.

soil is caused by actinomycetes, bacteria that grow in funguslike strands and give us the antibiotics streptomycin and tetracyclines.

Roundworms, segmented worms, mites, and tiny insects swarm by the thousands in that same gram of soil from the surface. Some of them are herbivorous, but many of them prey upon one another. Soil roundworms (nematodes), for instance, attack plant rootlets and can cause serious crop damage. A carnivorous fungus snares nematodes with tiny loops of living cells

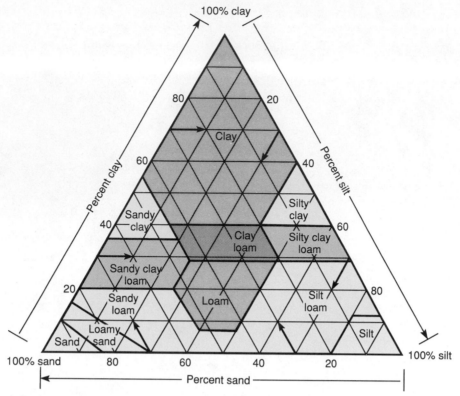

FIGURE 11.3 Soil texture depends on the percentage of clay, silt, and sand particles in the soil. Soils with the best texture for most crops are loams.
Source: Soil Conservation Service.

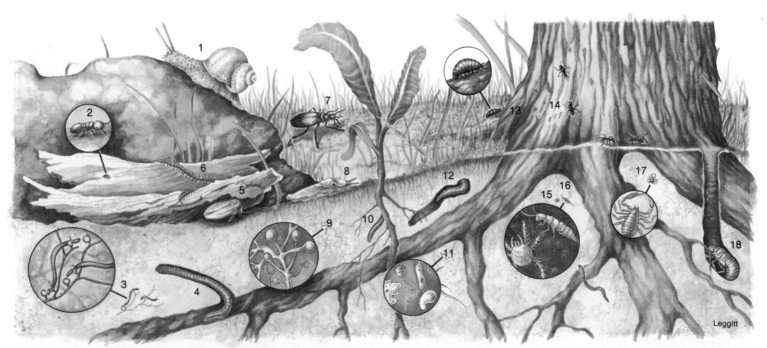

FIGURE 11.4 Soil ecosystems include numerous consumer organisms, as depicted here: (1) snail (2) termite (3) constricting fungus killing nematodes (4) earthworm (5) wood roach (6) centipede (7) carabid beetle (8) slug (9) soil fungus (10) wireworm (11) soil protozoans (12) earthworm (13) sow bug (14) ants (15) mite (16) springtail (17) pseudoscorpion (18) cicada nymph.

that constrict like a noose when a worm blunders into it. Burrowing animals, such as gophers, moles, insect larvae, and worms, tunnel deeper in the soil, mixing and aerating it. Plant roots also penetrate lower soil levels, drawing up soluble minerals and secreting acids that decompose mineral particles. Fallen plant litter adds new organic material to the soil, returning nutrients to be recycled.

Soil Profiles

Most soils are stratified into horizontal layers called **soil horizons** that reveal much about the history and usefulness of the soil. The thickness, color, texture, and composition of each horizon are used to classify the soil. A cross-sectional view of the horizons in a soil is called a soil profile. Figure 11.5 shows the series of horizons generally seen in a soil profile. Soil scientists give each horizon a letter name. Not all soils have all of these horizons. One or more may be missing, depending on the soil type and history of a specific area.

The soil surface is often covered with a layer of leaf litter, crop residues, or other fresh or partially decomposed organic material. Under this organic layer is the first true soil layer, called **topsoil** (or A horizon), where organic material is mixed with mineral particles. The topsoil horizon ranges from a thickness of 1 meter or more under virgin prairie to zero in some deserts. Topsoil contains most of the living organisms and organic material in the soil, and it is in this layer that most plants spread their roots to absorb water and nutrients. The topsoil horizon blends into a layer that is subject to leaching (removal of soluble nutrients) by water that percolates through it. This **zone of leaching** may have a very different appearance and composition from the layers above or below it.

Beneath the topsoil is the **subsoil,** which usually has a lower organic content and higher concentrations of fine mineral particles. Soluble compounds and clay particles carried by water percolating down from the layers above often accumulate in the subsoil. Subsoil particles can become cemented together to form hardpan, a dense, impermeable layer that blocks plant root growth and prevents water from draining properly.

Beneath the subsoil is the **parent material,** made of relatively undecomposed mineral particles and unweathered rock fragments with very little organic material. Weathering of this layer produces new soil particles for the layers above. About 97 percent of all the parent horizon material in the United States was transported to its present site by geologic forces (glaciers, wind, and water) and is not directly related to the **bedrock** below it.

Soil Types

Soils are classified according to their structure and composition into orders, suborders, great groups, subgroups, families, and series. There are hundreds of thousands of specific types within this taxonomic system! Figure 11.6 shows the major soil types for North and South America. Most other soil types are limited for one reason or another in their usefulness to agriculture. The richest farming soils are the mollisols (formed under grasslands) and alfisols (formed under moist, deciduous forests). North America is fortunate to have extensive areas of these fertile soils.

WAYS WE USE (AND ABUSE) SOIL

One of the main limitations to maintaining current, high levels of agricultural production is degradation of arable lands by erosion, toxification, desertification, and conversion to nonagricultural uses (chapter 12).

Measuring land degradation is a difficult and subjective task. Evidence is often anecdotal and subjective. Experts disagree on the relative importance of human actions and natural forces in changing land conditions. Perhaps the best definition of land degradation is a reduction in biological productivity. On farmland, this results in reduced crop yields. On grazing lands, fewer livestock can be maintained. In nature preserves, wildlife abundance and diversity are reduced.

Agricultural Land Degradation

The Food and Agriculture Organization (FAO) of the United Nations estimates that total world cropland losses amount to an area equal in size to the United States and Mexico combined (11 million hectares or 27 million acres) *every year*. Conversion to nonagricultural uses—urbanization, highways, industrial sites, strip-mining, abandonment of marginal farmland—is responsible for one-fourth of that loss. Toxification by hazardous wastes, chemical spills, salinization, alkalinization, and waterlogging of irrigated lands, misapplication of pesticides, and deposition of atmospheric pollutants degrade about 2 million hectares (5 million acres) each year. Finally, around 6 million hectares of cropland are rendered unproductive each year by severe erosion problems.

Erosion: The Nature of the Problem

Erosion is an improtant natural process, resulting in the redistribution of the products of geologic weathering, and is part of both soil formation and soil loss. The world's landscapes have been sculpted by erosion. When the results are spectacular enough, we enshrine them in National Parks like the Grand Canyon. Where erosion has worn down mountains and spread soil over the plains, or deposited rich alluvial silt in riverbottoms, we gladly farm it. Erosion is a disaster only when it occurs in the wrong place at the wrong time.

In some places, erosion occurs so rapidly that anyone can see it happen. Deep gullies are created where water scours away the soil, leaving fenceposts and trees sitting on tall pedestals as the land erodes away around them. In most places, however,

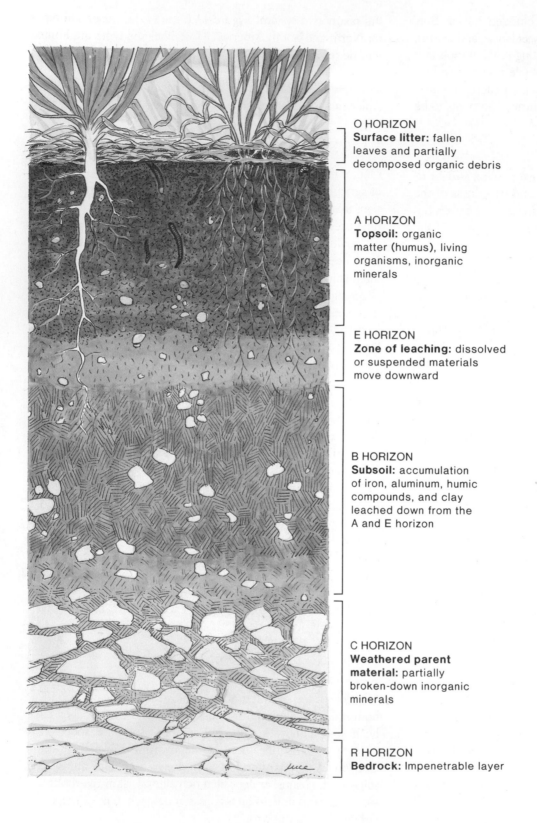

O HORIZON
Surface litter: fallen leaves and partially decomposed organic debris

A HORIZON
Topsoil: organic matter (humus), living organisms, inorganic minerals

E HORIZON
Zone of leaching: dissolved or suspended materials move downward

B HORIZON
Subsoil: accumulation of iron, aluminum, humic compounds, and clay leached down from the A and E horizon

C HORIZON
Weathered parent material: partially broken-down inorganic minerals

R HORIZON
Bedrock: Impenetrable layer

FIGURE 11.5   Soil profile showing possible soil horizons. The actual number, composition, and thickness of these layers varies in different soil types.

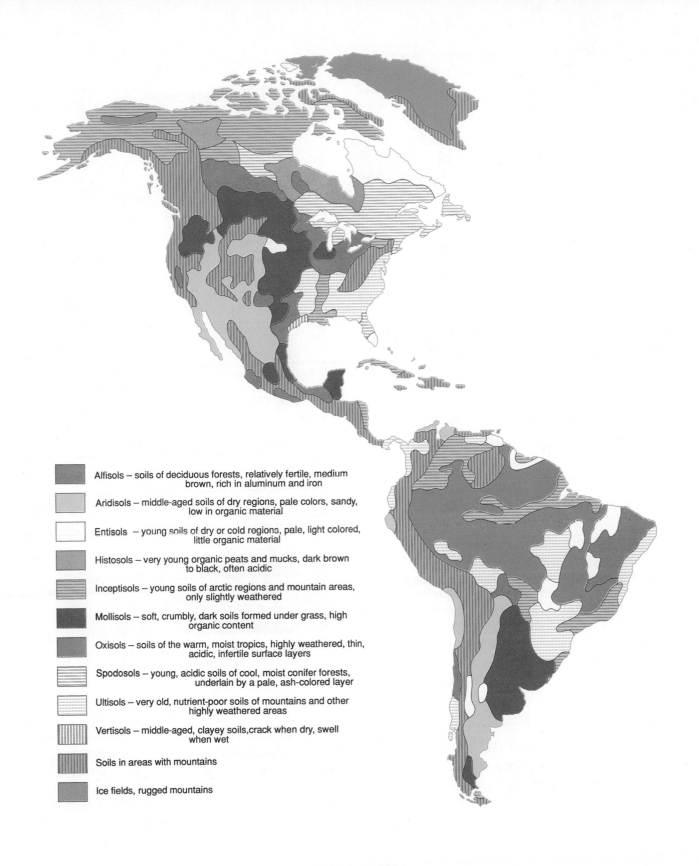

Alfisols – soils of deciduous forests, relatively fertile, medium brown, rich in aluminum and iron

Aridisols – middle-aged soils of dry regions, pale colors, sandy, low in organic material

Entisols – young soils of dry or cold regions, pale, light colored, little organic material

Histosols – very young organic peats and mucks, dark brown to black, often acidic

Inceptisols – young soils of arctic regions and mountain areas, only slightly weathered

Mollisols – soft, crumbly, dark soils formed under grass, high organic content

Oxisols – soils of the warm, moist tropics, highly weathered, thin, acidic, infertile surface layers

Spodosols – young, acidic soils of cool, moist conifer forests, underlain by a pale, ash-colored layer

Ultisols – very old, nutrient-poor soils of mountains and other highly weathered areas

Vertisols – middle-aged, clayey soils, crack when dry, swell when wet

Soils in areas with mountains

Ice fields, rugged mountains

FIGURE 11.6 Soil types of North and South America. Alfisols and mollisols make the best farmland.

erosion is more subtle. It is a creeping disaster that occurs in small increments. A thin layer of topsoil is washed off fields year after year until eventually nothing is left but poor-quality subsoil that requires more and more fertilizer and water to produce any crop at all.

The net effect, worldwide, of this general, widespread topsoil erosion is a reduction in crop production equivalent to removing about 1 percent of world cropland each year. Many farmers are able to compensate for this loss by applying more fertilizer and by bringing new land into cultivation. Continuation of current erosion rates, however, could reduce agricultural production by 25 percent in Central America and Africa and 20 percent in South America by the year 2000. The total annual soil loss from croplands is thought to be 25 billion metric tons. About twice that much soil is lost from rangelands, forests, and urban construction sites each year.

In addition to reduced land fertility, this erosion results in sediment-loading of rivers and lakes, siltation of reservoirs, smothering of wetlands and coral reefs, and clogging of water intakes and waterpower turbines. It makes rivers unnavigable, increases the destructiveness and frequency of floods, and causes gullying that turns good lands into useless wastelands.

Mechanisms of Erosion

Wind and water are the main agents that move soil around. Thin, uniform layers of soil are peeled off the land surface in a process called **sheet erosion.** When little rivulets of running water gather together and cut small channels in the soil, the process is called **rill erosion** (figure 11.7). When rills enlarge to form bigger channels or ravines that are too large to be removed by normal tillage operations, we call the process **gully erosion** (figure 11.8). **Streambank erosion** refers to the washing away of soil from the banks of established streams, creeks, or rivers, often as a result of removing trees and brush along streambanks and by cattle damage to the banks.

Most soil erosion on agricultural land, particularly cropland, is sheet and rill erosion. Large amounts of soil can be transported by these mechanisms without being very noticeable. A farm field can lose 20 metric tons of soil per hectare in winter and spring runoff in rills so small that they are erased by the first spring cultivation. That represents a loss of only a few millimeters of soil over the whole surface of the field, hardly apparent to any but the most discerning eye. It doesn't take much mathematical skill to see that if you lose soil twice as fast as it is being replaced, eventually it will run out.

Wind can equal or exceed water in erosive force, especially in a dry climate and on relatively flat land. When plant cover and surface litter are removed from the land by agriculture or grazing, the wind will lift loose soil particles and sweep them away (figure 11.9). Wind-borne dust is sometimes transported from one continent to another. Scientists in Hawaii can tell when

FIGURE 11.7 Sheet and rill erosion caused by water flowing across an unprotected field. Cover crops, crop residue, and terracing are all effective means of reducing this problem.

spring plowing begins in China because dust from Chinese farmland is carried by winds all the way across the Pacific Ocean. Similarly, summer dust storms in the Sahara Desert of North Africa create a hazy atmosphere over islands in the Caribbean Sea, 5,000 km (3,000 mi) away. It has been estimated that winds blowing over the Mississippi River basin have one thousand times the soil-carrying capacity of the river itself.

Figure 11.10 shows wind and water erosion rates by state. Note that these are averages over a whole state and mask local variations. Wind erosion per acre is highest in the western states where soil is dry and winds are strong. Water (sheet and rill) erosion is greatest in the hilly areas of Missouri, Tennessee, Ken-

FIGURE 11.8 Severe gullying has cut deep trenches in this land in western Tennessee. It is ruined for farmland or any other use.

FIGURE 11.9 Father and sons walking in the face of a dust storm, Cimarron County, Oklahoma.
Source: Arthur Rothstein, 1936.

tucky, and Hawaii where soils are soft, rainfall is abundant, and monoculture farming leaves the soil depleted and exposed to the elements. Figure 11.11 shows total sheet and rill erosion in different regions of the United States. Notice that the Corn Belt, because it has so much land in cultivation, has by far the largest total soil loss of any region.

Erosion in the United States

The total amount of soil lost to erosion in the United States appears to be the highest in the world. The United States Department of Agriculture reported in 1989 that 69 million hectares (170 million acres) of U.S. farmland are eroding at rates that are reducing long-term productivity. About 2.5 billion metric tons of soil wash away or are blown from cultivated croplands in this country each year, and 3.3 billion metric tons are lost from forests, pastures, stream banks, and construction sites. Altogether, that's enough soil to fill about 50 million box cars. Imagine 500,000 trains, each 100 cars long, carrying topsoil away from fields and forests and dumping it into rivers, lakes, and oceans. Soil erosion exceeds tolerable losses on nearly 40 percent of U.S. cropland. A frequently quoted estimate is that two bushels of soil

are lost from Iowa farmland for every bushel of corn produced. We are essentially mining the soil to produce crops.

When European settlers moved onto the American prairies in the nineteenth century, the topsoil depth averaged 25 cm (10 in.). A thick mat of grass roots held the topsoil in place, and each year a new crop of organic matter was added to the surface horizon to be shredded, tilled, and converted to new soil by the multitude of organisms living there. With the help of the steel moldboard plow invented by John Deere, the homesteaders broke the tough prairie sod and planted their crops. The first few years of farming brought a bonanza. The corn grew "as high as an elephant's eye," and stories of the rich harvests brought a flood of land-hungry immigrants from Europe.

With the sod broken and the surface litter removed, the soil was open to erosion. The fertile earth, accumulated over thousands of years, began to wash down the Mississippi River and its tributaries to the Gulf of Mexico. In a century of farming, we have lost half of the topsoil, on average, and perhaps two-thirds of the soil carbon that had been built up under forests and grasslands in the United States. In some places, there is no longer *any* topsoil left. Farmers plant their crops in rocky subsoil. Even

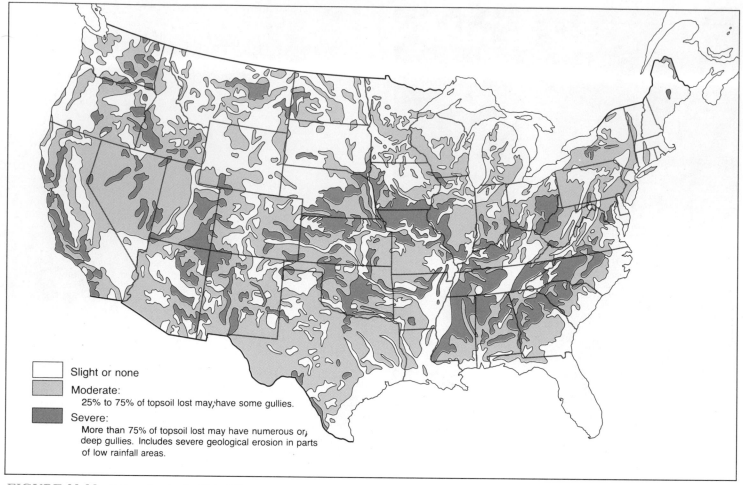

FIGURE 11.10 Soil erosion in the United States.

Source: Based on data from Reconnaissance Erosion Survey of the
United States and other soil conservation surveys by the Soil
Conservation Service—U.S. Soil Conservation Service.

with increasing amounts of expensive fertilizer and water, yields
are getting smaller and smaller.

The rate of soil loss appears to be getting worse. When
Hugh Bennett founded the Soil Conservation Service in 1935
during the dust bowl era, the total soil loss each year was about
3 billion tons. It is now nearly twice that much. The United
States General Accounting Office says that 84 percent of U.S.
farms have soil losses greater than 5 tons per acre (11 metric
tons per hectare). Five tons per acre is generally considered the
soil tolerance level, or the rate at which the ecosystem can replace soil.

Intensive farming practices are largely responsible for this
situation. Row crops, such as corn and soybeans, leave soil exposed for much of the growing season. Deep plowing and heavy
herbicide applications create weed-free fields that look neat but
are subject to erosion. Many farmers now plow in the fall to save
a few days at spring planting time, leaving fields lie bare and
exposed to high winds all winter. Because big machines cannot
easily follow contours, they often go straight up and down the

hills, creating ready-made gullies for water to follow. Farmers
sometimes plow through grass-lined watercourses and have pulled
out windbreaks and fencerows to accommodate the large machines and to get every last meter into production. Consequently, wind and water carry away the topsoil.

Pressed by economic conditions, most farmers have abandoned traditional crop rotation patterns and the custom of resting
land as pasture or fallow every few years. Continuous monoculture cropping can increase soil loss tenfold over other farming
patterns. A soil study in Iowa showed that a three-year rotation
of corn, wheat, and clover lost an average of only 2.7 short tons
per acre (6 metric tons per hectare). By comparison, continuous
wheat production on the same land caused nearly four times as
much erosion and continuous corn cropping resulted in seven
times as much soil loss as the rotation with wheat and clover.

Erosion in Other Countries

Data on soil condition and soil erosion in other countries are less
complete and less easily accessible than U.S. data, but it is evi-

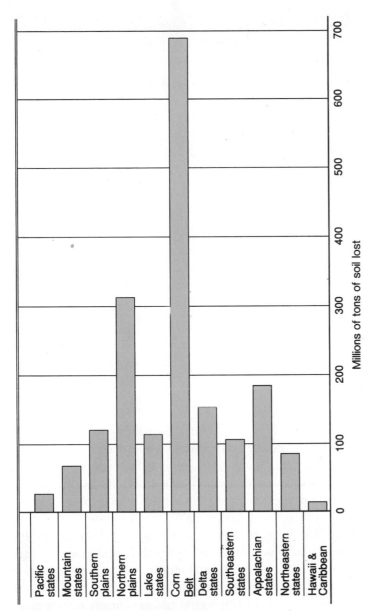

FIGURE 11.11 Estimated total annual sheet and rill erosion on cropland, by crop production region. Although the rate of erosion on farmland is not as high per acre in the Corn Belt as in the border states, there is so much more land under cultivation (95 percent of all land in Iowa is cropland by our estimate) that the total impact is higher there.

Source: U.S. Department of Agriculture.

dent that many places have problems as severe as, or perhaps worse than, the United States. China, for example, has a large area of loess (wind-blown silt) deposits on the North China Plain that once was covered by forest and grassland. The forests were cut down and the grasslands were converted to cropland. This plateau is now scarred by gullies 30 or 40 m (100 to 130 ft) deep, and the soil loss is thought to be at least 480 metric tons per hectare per year. This would be equivalent to 3 cm (1.2 in.) of topsoil per year.

One way to estimate soil loss is to measure the sediment load carried by rivers draining an area. Table 11.2 shows the

TABLE 11.2 Estimated annual soil erosion in selected river basins

River name	Principal country	Drainage basin (thousand km²)	Annual average suspended load (10^6 metric tons)	Estimated annual soil erosion (tons/ hectare)
Congo	Zaire	4,014	65	3
Nile	Egypt, Sudan	2,978	111	8
Mississippi	United States	3,220	365	10
Amazon	Brazil	5,776	363	13
Mekong	Laos, Kampuchea, Vietnam	795	170	43
Irrawaddy	Burma	430	290	139
Ganges	India, Nepal	1,076	1,455	270
Huang (Yellow)	China	668	1,600	479

Source: World Resources Institute, 1987.

average annual suspended solids in some of the major rivers of the world. The highest concentration of sediment in any river is in the Huang (Yellow) River that originates in the loess plateau of China. Although its drainage basin is only one-fifth as big as that of the Mississippi River, the Huang carries more than four times as much soil each year. This suggests that the average soil loss per hectare in China may be twenty times that in the United States.

In its middle reaches, the Huang carries about 700 kg of silt per cubic m of water, or 50 percent by weight—just under the level classified as liquid mud. As the river winds through northern China, much of this sediment settles out, raising the river bottom above the level of the surrounding countryside. Only by building dikes to contain the river have the Chinese been able to keep it in its course. In some places, the river bed is now 10 m (30 ft) above the farmland through which it flows. If the dikes give way during the summer rice season, millions of people would starve to death as a result of lost crops.

Next after the Huang in annual sediment load is the Ganges River, which carries 1,455 million metric tons of mud to the Bay of Bengal every year. Much of this sediment comes from the hill country of northern India, Nepal, and Bangladesh. Population pressures and preemption of good bottomlands for cash crop production have forced farmers to try to grow crops on steep, unstable slopes. Fuelwood shortages also cause local people to cut down the forests that stabilize mountain soils. When the monsoon rains come, they wash whole hillsides away, destroying villages and farmland below.

Soil Resources and Sustainable Agriculture

Perhaps the worst erosion problem in the world, per hectare of farmland, is in Ethiopia. Although Ethiopia has only 1/100 as much cropland in cultivation as the United States, it is thought to lose 2 billion metric tons of soil each year to erosion. This high rate of erosion is both a cause and consequence of famine, poverty, and continued social unrest in that country.

Haiti is another country with severly degraded soil. Once covered with lush tropical forest, the land has been denuded for fire wood and cropland. Erosion has been so bad that some experts now say the country has absolutely *no* topsoil left, and poor peasant farmers have difficulty raising any crops at all. Economist, Lester Brown, of Worldwatch Institute warns that the country may never recover from this ecodisaster.

OTHER AGRICULTURAL RESOURCES

Soil is only part of the agricultural resource picture. Agriculture is also dependent upon water nutrients and favorable climates to grow crops, and upon mechanical energy to tend and harvest them.

Water

All plants need water to grow. Agriculture accounts for the largest single share of global water use. Some 73 percent of all fresh water withdrawn from rivers, lakes, and groundwater supplies is used for irrigation (chapter 16). This is about six times the total annual flow of the Mississippi River. Although estimates vary widely (as do definitions of irrigated land), about 15 percent of all cropland, worldwide, is irrigated.

Table 11.3 shows the percentage of irrigated farmland by region. As you can see, the amount varies from 31 percent of all cropland in Asia to 4 percent in Oceania. Some countries have a low level of irrigation because their agricultural areas are adequately watered naturally. Other countries, like Ethiopia, have a dry climate and need irrigation but can't afford it.

Some countries are water rich (chapter 16) and can readily afford to irrigate farmland, while other countries are water poor and must use water very carefully. The efficiency of irrigation water use is rather low in most countries. High evaporative and seepage losses from unlined and uncovered canals often mean that as much as 80 percent of water withdrawn for irrigation never reaches its intended destination. Farmers tend to over-irrigate because water prices are relatively low and because they lack the technology to meter water and distribute just the amount needed.

Most farmers assume it is better to overwater than to underwater. This may not be true. Profligate use not only wastes water; it often results in **waterlogging.** Waterlogged soil is saturated with water, and plant roots die from lack of oxygen. **Salinization,** in which mineral salts accumulate in the soil and kill plants, occurs particularly when soils in dry climates are irrigated

TABLE 11.3 Grain yield per hectare, irrigation, and fertilizer use by region

Region	Cropland (millions of ha)	Percent irrigated	Fertilizer use (kg/ha)	Grain yield (tons/ha)
World	1,474	15	91	2.5
Africa	185	6	19	1.1
North and Central America	274	9	83	3.8
South America	142	6	39	2.0
Asia	451	31	93	2.5
Europe	140	12	228	4.2
Soviet Union	233	9	114	1.6
Oceania	49	4	34	1.6

Source: *World Resources*, 1990–1991.

profusely. Excessive irrigation accelerates movement of silt and chemicals into surface waters. The largest source of nonpoint chemical water pollution in the United States is runoff from farm fields.

Irrigation problems are a major source of land degradation and crop losses. The Worldwatch Institute reports tht 60 million ha (150 million acres) of cropland worldwide have been damaged by salinization and waterlogging. India has the world's largest total area of irrigated land (55 million ha). About one-third of that land is degraded and 7 million ha have been abandoned. China, the second largest irrigator, also has about 7 million ha (17 million acres) of saline and alkaline land. Pakistan and the Soviet Union have about half as much land degraded by irrigation. The environmental and human health costs of irrigation abuse can be severe (see Box 16.2).

Water conservation techniques can greatly reduce problems arising from excess water use. Conservation also makes more water available for other uses or for expanded crop production where water is in short supply. The most efficient way to water crops is **drip irrigation** (figure 11.12). In drip irrigation, a series of small perforated tubes are laid across the field, at or just under the surface of the soil. The tubes deliver the amount of water that each plant needs, directly onto its roots where the water will do the most good, and with a minimum of evaporative loss or oversoaking of the soil. (For more discussion of water conservation techniques, see chapter 16.)

Plant Nutrients

In addition to water, sunshine, and carbon dioxide, plants need small amounts of inorganic nutrients for growth. The major elements required by most plants are nitrogen, potassium, phosphorus, calcium, magnesium, and sulfur. Calcium is usually plentiful in soil, but nitrogen, potassium, and phosphorus avail-

FIGURE 11.12 Drip irrigation delivers water directly to plant roots. Water consumption is reduced and salinization of the soil is reduced. This system can also deliver fertilizer very efficiently.

TABLE 11.4 Grain yield in selected countries

Country	Average grain yield (kilograms/hectare) 1974–1976	1981–1983	Percent increase
Netherlands	4,771	6,357	33
Japan	5,620	5,278	−6
Egypt	3,921	4,254	8
United States	3,339	4,075	22
China	2,479	3,399	37
Argentina	1,971	2,353	19
Soviet Union	1,466	1,488	1
India	1,179	1,435	22
Ethiopia	966	1,280	33
Sudan	645	602	−7
Niger	395	408	3

From *Food and Agriculture Production Yearbook*, Vol. 39. Copyright © 1985 Food and Agriculture Organization of the United Nations Rome, Italy. Reprinted by permission.

ability often limits plant growth. Addition of these elements in fertilizer usually stimulates growth and greatly increases crop yields. A good deal of the doubling in worldwide crop production since 1950 has come from increased inorganic fertilizer use, mostly nitrogen, phosphorus, and potassium. In 1950, the average amount of fertilizer used per acre was 20 kg per hectare. In 1990, this had increased to an average of 91 kg per hectare. This change represents an increase in total fertilizer use from 30 million metric tons in 1950 to 134 million metric tons in 1990.

Table 11.3 also shows the range in fertilizer use by region. As you can see, there is an even wider variation in fertilizer use than in irrigation. European countries are relatively wealthy but have little cropland. They can afford to use twelve times as much

fertilizer per hectare as Africa, producing four times higher crop yields in Europe than in Africa. Notice that the relationship between fertilizer and yield is not linear. Individual differences in fertilizer use are even greater than regional averages. New Zealand, for instance, can afford to apply one thousand times as much fertilizer per hectare as Niger. The variations in average grain yields shown in table 11.4 are largely due to these differences in fertilizer use.

Fertilizer could be used more efficiently in many cases. Farmers often overfertilize because they are unaware of the specific content of their soils or the needs of their crops. In Illinois, for instance, it has been reported that farmers use 40 percent more fertilizer than is needed. Applying fertilizer in several small doses rather than one big dose, using the precise mixture needed by a specific crop on a specific field, and timing application to the exact time that plants need it saves fertilizer and reduces runoff.

Overfertilization wastes money and has negative environmental impacts. Phosphates and nitrates from farm fields and cattle feedlots are a major cause of aquatic ecosystem pollution. Fertilizer runoff percolates through the soil to contaminate groundwater supplies. Nitrate levels in groundwater have risen to dangerous levels in many areas where intense farming is practiced. England, France, Denmark, Germany, the Netherlands, and the United States have reported nitrate concentrations above the safe level of 11.3 mg per liter in drinking water in farming areas. Young children are especially sensitive to the presence of nitrates. Using nitrate-contaminated water to mix infant formula can be fatal for newborns.

Much of the fertilizer used in the developed world comes from petroleum fuels or mineral deposits. Petroleum fuels are used to reduce gaseous nitrogen to ammonia, so synthetic nitrate fertilizers have a substantial hidden energy cost. Mineral deposits mined for fertilizer include potash for potassium and phosphate rocks for phosphates. There does not appear to be a shortage of phosphate and potassium resources, but as fossil fuel energy becomes increasingly scarce, we may be unable to produce as much inorganic nitrogen fertilizer as we have in the past.

What are some alternate ways to fertilize crops? Manure, crop residues, ashes, composted refuse, and green manure (crops grown specifically to add nutrients to the soil) are important natural sources of soil nutrients. Nitrogen-fixing bacteria living symbiotically in root nodules of legumes (a broad group of plants including peas, beans, vetch, alfalfa, and leucaena trees) are valuable for making nitrogen available as a plant nutrient (chapter 3). Interplanting or rotating beans or some other leguminous crop with such crops as corn and wheat are traditional ways of increasing nitrogen availability. In some cases, growing and then plowing under a leguminous green manure crop, such as alfalfa, clover, or vetch, every third or fourth year is necessary to maintain soil fertility.

In many Third World countries, there is a conflict between using crop residues and animal manure for fertilizer or as an energy source. These fuels are the only energy sources many poor people have for cooking and space heating; however, their use for heat energy means a reduction in food production. One metric ton of dung used as fertilizer will increase cereal crop yields by 50 kg in most agricultural systems. This means that the 400 million metric tons of dung used in Africa and Asia as fuel could produce 20 million metric tons of food if used as fertilizer. This is enough food to sustain 40 to 50 million people. Energy issues are discussed further in chapters 18, 19, and 20.

What is the potential for increasing world food supply by increasing fertilizer use in low-production countries? Africa, for instance, uses an average of only 19 kg of fertilizer per hectare (17 lb per acre), or about one-fourth of the world average. India also has a relatively low average fertilizer level (30 kg per hectare). It has been estimated that both these areas could at least triple their crop production by raising fertilizer use to the world average. Other increases could be achieved by using currently idle land, by introducing new high-yield crop varieties, and by investing in irrigation where water is available. All these steps, however, will provide only illusory relief unless careful thought is given to how agriculture is to be made stable and renewable, how food is to be distributed to those who need it, and how population growth can be brought into line with realistic future planning.

A number of studies have estimated the maximum number of people the world could support given predictions of fertilizers, water, arable cropland, and other factors of agricultural production. The results have ranged from pessimistic warnings that there are already more people than we can feed to optimistic claims that the world could support thirty to fifty times the present pop-

TABLE 11.5	Population potential in developing countries with various levels of agricultural inputs			
Region	1990 population (millions)	Potential population with varying agricultural inputs		
		Low	Medium	High
Africa	661	840	3,248	9,492
Southwest Asia	1,324	490	797	1,226
Southeast Asia	445	265	723	1,229
South America	296	867	3,513	8,408
Central America	152	107	281	770
Total	2,888	2,569	8,562	21,125

Source: Food and Agriculture Organization of the United Nations.

ulation. One set of estimates by the FAO is shown in table 11.5. This study predicts that at a low level of agricultural inputs, developing countries could only feed about 2.5 billion people and the maximum world population would be as much as four times the present 5.3 billion people. As inputs are increased in this model, the maximum population also increases to around 30 or 40 billion people. This says nothing about the quality of life at various levels, however, or the effects on other species of conversion of all possible land to crop production. Notice that even with high levels of input, southwest Asia could not support many more people than it does now. Africa and South America, on the other hand, potentially could support far more people if agriculture were modernized and more land were converted to agricultural uses.

Climate

Climate is a critically important ingredient in agriculture. Global climate change caused by the greenhouse effect (chapter 17) could have severe impacts on world food production. Crop yields could increase in some areas. For instance, Canada and the Soviet Union would benefit from warmer temperatures and longer growing seasons. Some crops grow better with higher atmospheric carbon dioxide levels. Soybeans, for instance, may have up to 30 percent higher yields if carbon dioxide doubles. The most serious impact of climate warming would likely be decreasing rainfall at high latitudes and in midcontinent regions. More frequent and more severe droughts in the American Midwest, central Soviet Union and China would surely have devastating effects on world food supplies. Some people fear terrible famines might ensue. The costs in social, ecological, and economic terms could be catastrophic.

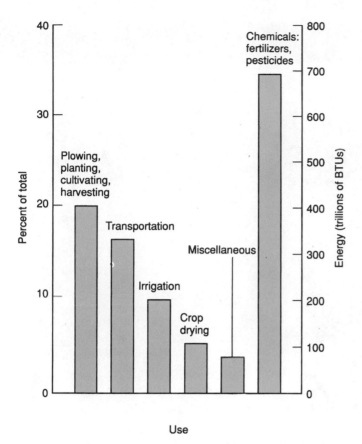

FIGURE 11.13 Energy in agriculture. Energy is used in running farm machinery, irrigation pumps, and grain driers, but the largest energy input is in producing and applying synthetic fertilizers, pesticides, and other agricultural chemicals.

Source: Data from L. Brown, *State of the World*, 1987. Worldwatch Institute, DC.

Energy

Farming as it is generally practiced in the industrialized countries is highly energy-intensive. Fossil fuels supply almost all of this energy. The largest energy consumption is for liquid fuels for farm machinery used in planting, cultivating, harvesting, and transporting crops to market (figure 11.13). The second largest energy cost is the energy contained in chemical stocks used to synthesize fertilizers, pesticides, and other agricultural chemicals. In the western United States, water pumping for irrigation is a major energy consumer, representing a *triple* mining of resources. We use oil, gas, and coal to pump groundwater that is used to grow crops that deplete the soil!

Total agricultural energy use in the United States, with a population of about 250 million, is now about 11.5 Quads $(12 \times 10^9$ GJ). By contrast, the 2.9 billion people in the less developed countries of the world are supported by agricultural systems that use only about 16 Quads $(17 \times 10^9$ GJ). This means that U.S. agriculture uses about twelve times as much energy per person as do these poorer countries.

After crops leave the farm, additional energy is used in food processing, distribution, storage, and cooking. It has been

estimated that the average food item in the American diet travels 2,000 km (1,250 mi) between the farm that grew it and the person who consumes it. The energy required for this complex processing and distribution system may be five times as much as is used directly in farming. Altogether the food system in the United States consumes about 16 percent of the total energy we use. Most of our foods require more energy to produce, process, and get to market than they yield when we eat them (table 11.6).

Clearly, unless we find some new sources of energy, our present system is unsustainable. As fossil fuels become more scarce, we may need to adopt farming methods that are self-supporting. Is it possible that we may need to go back to using draft animals that can eat crops grown on the farm? Could we reintroduce natural methods of pest control, fertilization, crop drying, and irrigation? Or can we develop alternative energy sources, like solar energy, to run our equipment and machinery?

TOWARD A JUST AND SUSTAINABLE AGRICULTURE

How, then, shall we feed the world? Can we make agriculture compatible with sustainable ecological and social systems? Having discussed some of the problems that beset modern agriculture, we will now consider some suggested ways to overcome problems and make farming and food production just and lasting enterprises. This goal is usually termed **sustainable agriculture** or **regenerative farming** (Box 11.1).

TABLE 11.6 Estimated energy intensiveness of food crops

Crop	Calories invested per calorie of food produced
Distant fishing	12
Feedlot beef	10
Fish protein concentrate	6
Grass-fed beef	4
Coastal fishing	2
Intensive poultry production	2
Milk from grass-fed cows	1
Range-fed beef	0.5
Intensive corn or wheat	0.5
Intensive rice	0.25
Hunting and gathering	0.1
Traditional wet-rice culture	0.05
Shifting agriculture	0.02

Note: Any crop with a ratio above 1 consumes more energy, in the form of tractor fuel, fertilizer, drying, processing, and shipping, than it yields. Crops with a ratio below 1 yield more energy than is put in from human-controlled sources.

Source: U.N. Food and Agriculture Organization, 1987.

BOX 11.1   Land Stewardship; and Sustainable Agriculture in America

Land stewardship is a concept that has been missing for many years in large-scale American agriculture. Recently, however, a few farmers who were disillusioned with modern chemical farming, the cycle of overproduction and falling prices, and the economic uncertainties of depending on a single large crop every year, have been trying to find new and sometimes old alternatives. They are looking for farming methods that protect the land as well as their own health. They are rediscovering the pride of treating the land well, and they are learning that the land rewards good stewardship.

Dick and Sharon Thompson of Boone, Iowa, (box figure 11.1) are two of those farmers seeking alternatives to chemical-dependent farming. After earning a master of science degree in animal husbandry from Iowa State University, Dick Thompson inherited his family's farm in 1957 and began practicing the modern version of farming that he had learned at the University—removing fences, expanding fields, and specializing on a single crop. Soils are unable to support the same crops year after year without rest, because nutrients

BOX FIGURE 11.1   Dick and Sharon Thompson of Boone, Iowa, practice sustainable farming to produce healthy crops that are economically successful..

drawn from the soil by plants are lost by harvesting. Farmers were advised to replace those nutrients using chemical fertilizers. Pesticides were to be used to keep all harmful insects away from crops. Liberal amounts of herbicides were to be applied to prevent competition from weeds and keep the fields "clean" between the rows.

These modern farming techniques are practiced by the vast majority of American farmers, and agricultural

Bringing New Land under Cultivation

How much land is available for agriculture? Are we running out of land, or are we about to? Over the whole world, some 3.2 billion hectares (about 25 percent of all ice-free land) is potentially suitable for agriculture. We currently use half of that land for crops. This represents 0.28 hectares (0.69 acres) per person, worldwide. Table 11.7 shows the general distribution of present world cropland. Asia has the most cultivated land in total, but the least land per capita. With 451 million hectares to support 3.1 billion people, each person has only 0.15 hectares (about one-third acre).

Japan is the most "land-poor" country in the world (except for such city-states as Singapore and Monaco). Only 20 percent of this mountainous country is arable. With its high population density, Japan has 0.06 hectares (0.15 acres) per person to grow food. This is a plot about 20 by 30 m (66 by 100 ft, the size of an average American city lot) per person. Australia, by contrast, has the largest amount of land per capita (2.1 hectares). North

America and the Soviet Union are relatively "land-rich," with about 1 hectare (2.47 acres) per person.

In the developed countries, 95 percent of recent agricultural growth in this century has come from improved crop varieties or land-use intensity, rather than from bringing new land into production. In fact, less land is being cultivated now than one hundred years ago in North America, or six hundred years ago in Europe. As more effective use of labor, fertilizer, water, and improved seed varieties have increased in the more developed countries, productivity per unit of land has increased, and marginal land has been retired, mostly to forests and grazing lands. Land continues to be a cheaper resource and new land is still being brought under cultivation, mostly at the expense of forests and grazing lands. Still, at least two-thirds of recent production gains have come from new crop varieties and more intense cropping rather than expansion into new lands.

The largest increases in cropland over the last 30 years occurred in South America and Oceania where forests and

schools continue to teach them. Unfortunately, some modern farming techniques have brought many farmers to bankruptcy, and they are causing alarming rates of erosion, sending millions of tons of precious, irreplaceable topsoil downriver to the sea every year.

Dick and Sharon Thompson practiced high-intensity, single-crop farming with pesticides and fertilizers for ten years, but they felt that something was not right. Their hogs and cattle were sick. Fertilizer, pesticide, and petroleum prices were rising faster than crop prices. They began looking for a better way. In 1967 a friend took them to a program on natural farming. For the Thompsons, it was almost a religious revelation to hear that natural, organic farming methods could be the answer to their problems. They stopped using chemicals and started developing what Dick Thompson calls "regenerative agriculture."

The Thompsons have become experts at ridge-tilling without herbicides, using special machinery that turns the ground into rows of tiny ridges and valleys. Before fall harvest, they plant oats, hay, or hairy vetch (a soil-enriching legume) in the valleys. During the cold, windy winter, these plants protect the soil from erosion, and in spring they inhibit weeds until the new crop is planted on the ridges. Any spring weeds in the valleys are left as ground cover until they can be removed easily with a cultivator. Unlike his neighbors, Dick Thompson doesn't leave his fields bare all winter by plowing in the fall, a practice he found encouraged weed growth thus requiring extra herbicides, and which left the soil exposed to winter wind erosion.

The Thompsons grow a variety of crops and practice crop rotation, allowing the soil to rest and regenerate itself. They pasture their cattle, an unusual idea in Iowa, and feed them hay and grain crops from their own fields. They carefully collect and compost manure to re-enrich the fields from which the cattle feed has been taken. Their annual yields have remained high, and their costs have fallen considerably.

Dick and Sharon Thompson have done a lot of experimentation over the years. They have used new technology, such as ridge-tilling, and ancient ideas, such as regular crop rotation. They have tried many different methods and combinations, but they have never retreated from regenerative agriculture. They actively work to recruit other farmers to their methods. Every September they hold a field day, inviting farmers from across the nation to see their work in action and begin thinking about working out their own methods of regenerative farming and land stewardship.

The Thompsons have found many rewards in their practice of sustainable agriculture. Soil fertility is improving, and the earthworms that can live again in their pesticide-free soil are bringing back birds. They no longer are contaminating their groundwater and surface water with toxic chemicals. They are being energy-efficient and cost-effective, and they are getting recognition and praise from environmental and agricultural organizations around the country. The Thompsons are especially proud that they are fostering an attitude of caring for the land so that it will remain rich and fertile.

grazing lands are rapidly being converted to farms (table 11.7). Many developing countries are reaching the limit of lands that can be exploited for agriculture without unacceptable social and environmental costs, whereas other areas still have considerable potential for opening new agricultural lands. East Asia, for instance, already uses 72 percent of its potentially arable land. Most of its remaining land has severe restrictions for agricultural use. Further increases in crop production will probably have to come from higher yields per hectare. Latin America, by contrast, uses only about one-fifth of its potential land, and Africa uses only about one-fourth of the land that theoretically could grow crops. However, there would be serious ecological tradeoffs in putting much of this land into agricultural production.

While land surveys tell us that much more land in the world *could* be cultivated, not all of that land necessarily *should* be farmed. Much of it is more valuable in its natural state. The soils over much of tropical Asia, Africa, and South America are old, weathered, and generally infertile. Most of the nutrients are in the standing plants, not in the soil. In many cases, clearing land for agriculture in the tropics has resulted in tragic loses of biodiversity and the valuable ecological services that it provides. Ultimately, much of this land is turned into useless scrub or semidesert.

On the other hand, there are also large areas of rich, subtropical grassland and forest that are well watered, have good soil, and could become productive farmland without unduly reducing the world's biological diversity. Argentina, for instance, has pampas grasslands about twice the size of Texas that closely resemble the American midwest a century ago in climate and potential for agricultural growth. Some of this land could probably be farmed with relatively little ecological damage if it were done carefully.

The FAO estimates that a growth rate of 1 percent per year in total agricultural land is not unreasonable for the im-

TABLE 11.7 Cropland per capita, percent increase (1990–1991), and percent of arable land by region

Region	Population (millions)	Cropland (10^6 ha)	Hectare/ person	Percent change (1968–1990)	Percent of potentially arable land
World	5,300	1,474	0.28	+9.1	40
Africa	661	185	0.28	+14.0	25
North and Central America	396	274	0.69	+7.7	58
South America	330	142	0.43	+35.0	19
Asia	3,116	451	0.15	+4.2	72
Europe	501	140	0.28	−5.0	81
Soviet Union	291	233	0.80	+1.3	65
Oceania	27	49	1.82	+26.6	30

Source: *World Resources*, 1990–1991.

mediate future. The world could add 250 million hectares (625 million acres) of new cropland by the end of this century. The major potential expansion areas, according to the FAO, are tropical forests, savannahs, seasonally flooded alluvial zones (flood plains), and hill land. Each of these environmental types has special problems that must be carefully considered if social and biological disasters are to be avoided.

Soil Conservation

With careful husbandry, soil is a renewable resource that can be replenished and renewed indefinitely. Since agriculture is the area in which soil is most essential and also most often lost through erosion, it is also in agriculture where there is the greatest potential for soil conservation and rebuilding. Some rice paddies in Southeast Asia, for instance, have been farmed continuously for a thousand years without any apparent loss of fertility. The rice-growing cultures that depend on these fields have developed management practices that return organic material to the paddy and carefully nurture the soil's ability to sustain life.

What can be done to put other agricultural systems on an equally permanent basis? Among the most important considerations in a soil conservation program are topography, ground cover, climate, soil type, and tillage system.

Managing Topography
Water runs downhill. The faster it runs, the more soil it carries off the fields. Comparisons of erosion rates in Africa have shown that a 5 percent slope in a plowed field has three times the water runoff volume and eight times the soil erosion rate of a comparable field with a 1 percent slope. Water runoff can be reduced by **contour plowing,** plowing across the hill rather than up and down. Contour plowing is often combined with **strip-farming,** that is, planting different kinds of crops in alternating strips along the land contours (figure 11.14). When one crop is

harvested, the other is still present to protect the soil and keep water from running straight downhill. The ridges created by cultivation make little dams that trap water and allow it to seep into the soil rather than running off. In areas where rainfall is very heavy, **tied ridges** are often useful. This involves a series of ridges running at right angles to each other, so that water runoff is blocked in all directions and is encouraged to soak into the soil.

Terracing involves shaping the land to create level shelves of earth to hold water and soil (figure 11.15). The edges of the terrace are planted with soil-anchoring plant species. This is an expensive procedure, requiring either much hand labor or expensive machinery, but makes it possible to farm very steep hillsides. The rice terraces in the Chico River Valley in the Philippines rise as much as 300 m (1,000 ft) above the valley floor. They are considered one of the wonders of the world.

Planting **perennial species** (plants that grow for more than two years) is the only suitable use for some lands and some soil types. Establishing forest, grassland, or crops such as tea, coffee, or other crops that do not have to be cultivated every year may be necessary to protect certain unstable soils on sloping sites or watercourses (low areas where water runs off after a rain).

Providing Ground Cover
Often, the easiest way to provide cover that protects soil from erosion is to leave crop residues on the land after harvest. They not only cover the surface to break the erosive effects of wind and water, but they also reduce evaporation and soil temperature in hot climates and protect ground organisms that help aerate and rebuild soil. In some experiments, 1 ton of crop residue per acre (0.4 hectare) has increased water infiltration 99 percent, reduced runoff 99 percent, and reduced erosion 98 percent. Leaving crop residues on the field also can increase the potential for disease and pest problems, however, and may require increased use of pesticides and herbicides.

FIGURE 11.14 Contour plowing and strip cropping on these Wisconsin farms protect the soil from erosion and help maintain fertility as well as forming a beautiful landscape. With care and good stewardship, we can increase the carrying capacity of the land and create a sustainable environment.

FIGURE 11.15 Rice terraces on Java, Indonesia. Some rice paddies have been cultivated for hundreds or even thousands of years without any apparent loss of productivity.

Where crop residues are not adequate to protect the soil or are inappropriate for subsequent crops or farming methods, such **cover crops** as rye, alfalfa, or clover can be planted immediately after harvest to hold and protect the soil. These cover crops can be plowed under at planting time to provide green manure. Another method is to flatten cover crops with a roller and drill seeds through the residue to provide a continuous protective cover during early stages of crop growth.

In some cases, interplanting of two different crops in the same field not only protects the soil, but also makes more efficient use of the land, providing double harvests. Native Americans and pioneer farmers, for instance, planted beans or pumpkins between the corn rows. The beans provided nitrogen needed by the corn, pumpkins crowded out weeds, and both crops provided foods that nutritionally balance corn. Traditional swidden (slash and burn) cultivators in Africa and South America often plant as many as twenty different crops together in small plots. The crops mature at different times so that there is always something to eat, and the soil is never exposed to erosion for very long.

Mulch is a general term for a protective ground cover that can include manure, wood chips, straw, seaweed, leaves, and other natural products. For some high-value crops, such as tomatoes, pineapples, and cucumbers, it is cost-effective to cover the ground with heavy paper or plastic sheets to protect the soil, save water, and prevent weed growth. Israel uses millions of square meters of plastic mulch to grow crops in the Negev Desert.

Using Reduced Tillage Systems

Farmers have traditionally used a moldboard plow to till the soil, digging a deep trench and turning the topsoil upside down. In the 1800s, it was shown that tilling a field fully—until it was "clean"—increased crop production. It helped control weeds and pests, reducing competition; it brought fresh nutrients to the surface, providing a good seedbed; and it improved surface drainage and aerated the soil. This is still true for many crops and many soil types, but it is not always the best way to grow crops. We are finding that less plowing and cultivation often makes for better water management, preserves soil, saves energy, and increases crop yields.

There are three major **reduced tillage systems** (figure 11.16). *Minimum till* uses a disc or chisel plow rather than a traditional moldboard plow. A chisel plow is a curved chisel-like blade that gouges a trench in the soil in which seeds can be planted. It leaves up to 75 percent of plant debris on the surface between the rows, preventing erosion. *Conserv-till* farming uses a coulter, a sharp disc like a pizza cutter. It slices through the soil, opening up a furrow or slot just wide enough to insert seeds. This disturbs the soil very little and leaves almost all plant debris on the surface. *No-till* planting is accomplished by drilling seeds into the ground directly through mulch and ground cover. This allows a cover crop to be interseeded with a subsequent crop.

Farmers who use these conservation tillage techniques often must depend on pesticides (insecticides, fungicides, and herbicides) to control insects and weeds. Increased use of toxic agricultural chemicals is a matter of great concern (Box 11.2).

BOX 11.2 Focus on Pesticides

What do you do when you discover cockroaches in your cupboard or dandelions in your yard? Do you look for a chemical pesticide to take care of the problem? If so, you are reacting as most Americans do. You are also contributing to a serious environmental problem. The EPA ranks pesticide pollution as the most urgent concern facing the United States. Some medical experts think pesticide contamination of air, water, and food is a major cause of chronic health problems in the United States and around the world. Pesticide residues contaminate groundwater drinking supplies in about half of the United States, and pesticide production is probably the largest source of toxic wastes in this country.

What is a pest, and what are pesticides? A **pest** is a troublesome animal, plant, microbe, or person. Put another way, a pest is any organism that reduces the availability, quality, or value of resources useful to humans. Both definitions are highly subjective. The same organism that may be a pest in one case may be considered a friend in another. Only about one hundred species of plants and animals cause 90 percent of all damage to foods and crops. A **pesticide** is any chemical that kills, controls, drives away, or modifies the behavior of a pest so that it is no longer troublesome. Pesticides are identified by their target organism, such as insecticides, rodenticides, fungicides, or herbicides (box table 11.1).

Synthetic chemical pesticides have brought us many benefits. Most agronomists believe that the rapid increases in food productivity in this century would have been impossible without effective pest control. Even with our extensive use of chemicals, pests and diseases reduce or destroy crops amounting to about half the world's food supply. In 1945, crop losses in the United States caused by insects, diseases, and weeds amounted to 32 percent of the harvest. In 1980, these losses had risen to 37 percent of the harvest, despite the use of 450,000 metric tons (1 billion pounds) of pesticides. The USDA estimates that food in the United States would cost 30 to 50 percent more than it does now if we had no chemical pesticides. The control of disease-carrying

insects has spared billions of people from diseases, such as malaria, yellow fever, typhus, sleeping sickness, river blindness, and filariasis. Millions of lives have been saved. The reductions in medical expenses, social disruption, and days lost from work probably outweigh the direct costs of pesticide application by a thousand to one.

There have been problems associated with widespread use of pesticides, however. It soon was discovered that pests tend to reproduce rapidly and develop pesticide resistance very quickly. Natural predators are killed and ecological balances that controlled pests are upset by broad-spectrum chemicals. Pests then rebound to higher levels than ever before and secondary pests are created when organisms that were not previously a problem begin to grow out of control. Opponents of pesticide use claim that we are on a pesticide treadmill where ever-increasing amounts of toxins are required just to keep even with the pests. Persistent chemicals (such as DDT) are bioaccumulated and magnified in food chains, eventually reaching toxic levels in such top carnivores as bald eagles, peregrine falcons, brown pelicans, salmon, seals, and humans. One of the main threats faced by many endangered species is pesticide poisoning.

New types of pesticides, such as organophosphates and carbamates, have been synthesized to combat pest resistance and reduce environmental accumulation (box table 11.1). These chemicals break down much more quickly than most chlorinated hydrocarbons and, therefore, do not bioaccumulate as much. They are fast acting and extremely toxic to target species. Unfortunately, they also are very toxic to nontarget species, including humans. Factory workers who manufacture these compounds often have severe or acute reactions including neurological damage, muscle paralysis, skin burns and severe acne, hallucinations, tremors, memory loss, and sterility. Farm workers who apply pesticides or work in fields where they have been used have similar problems. The World Health Organization estimates there are 2 million pesticide poisonings in the world each

year, and at least 10,000 people die of immediate pesticide effects.

Many pesticides are known or suspected to be mutagens, teratogens, carcinogens, and cumulative metabolic poisons in low chronic doses. No one knows how much these chemicals contribute to cancer, birth defects, genetic diseases, reproductive failure, and other long-term human health problems. Of the six hundred active pesticide ingredients on the market, the EPA has completed a preliminary assessment of only 20 percent. These active ingredients are combined with another nine hundred chemical solvents, thickeners, propellants, stabilizers, adsorbents, and other "inert" ingredients to make more than 50,000 commercial products. The EPA estimates that it will take twenty years to test all the products now on the market. Meanwhile, these chemicals are accumulating in the environment and in our bodies, and new ones are being introduced. In a 1985 survey by the National Institute of Environmental Health Sciences, 100 percent of the Americans tested had detectable DDT residues in their body, and 90 percent also had traces of chlordane, heptachlor, aldrin, dieldrin, or hexachlorobenzene.

Herbicides make up nearly two-thirds of all the pesticides used in the United States. They reduce cultivation costs for farmers, increase crop yields, and allow intensive, monoculture farming; but, the clean, bare fields produced by these techniques result in high soil erosion rates. Corn and soybeans account for about one-third of all herbicide use. Kansas farmers who work regularly with herbicides have a sixfold increase in the rate of non-Hodgkin lymphoma (a cancer of the lymphoid system) compared to those who are not exposed to these chemicals. The 20 to 30 percent of herbicides not applied to crops are spread on golf courses, parks, roadsides, and lawns. Much of our herbicide use could be eliminated by other control methods or by changing our attitudes about what is aesthetically pleasing.

Insecticides account for about one-quarter of our pesticide use. Cotton is one of our most intensely treated crops, con-

BOX TABLE 11.1 Pesticides

Type	Examples	Characteristics	Type	Examples	Characteristics
Insecticides					
Inorganic chemicals	Mercury, lead, arsenic, copper sulfate	Highly toxic to many organisms, persistent, bioaccumulates	Plant products and synthetic analogs	Nicotine, rotenone, pyrethrum, allethrin, decamethrin, resmethrin, fenvalerate, permethrin, tetramethrin	Natural botanical products and synthetic analogs, fast acting, broad insecticide action, low toxicity to mammals, expensive
Organo-chlorines	DDT, methoxychlor, heptachlor, HCH, pentachloraphenol, chlordane, toxaphene, aldrin, endrin, dieldrin, lindane	Mostly neurotoxins, cheap, persistent, fast acting, easy to apply, broad spectrum, bioaccumulates, biomagnifies	Microbes	*Bacillus thuringensis*	Kills caterpillars
				Bacillus popilliae	Kills beetles
				Viral diseases	Attack a variety of moths and caterpillars
Organo-phosphates	Parathion, malathion, diazinon, dichlorvos, phosdrin, disulfoton, TEPP, DDVP	More soluble, extremely toxic nerve poisons, fast acting, quickly degraded, toxic to many organisms, very dangerous to farm workers	*Fungicides*	Captan, maneb, zeneb, dinocap, folpet, pentachlorphenol, methyl bromide, carbon bisulfide, chlorothalonil (Bravo)	Most prevent fungal spore germination and stop plant diseases; among most widely used pesticides in United States.
Carbamates and urethanes	Carbaryl (Sevin), aldicarb, carbofuran, methomyl, Temik, mancozeb	Quickly degraded, do not bioaccumulate, toxic to broad spectrum of organisms, fast acting, very toxic to honey bees	*Fumigants*	Ethylene dibromide, dibromochloro-propane, carbon tetrachloride, carbon disulfide, methyl bromide	Used to kill nematodes, fungi, insects, and other pests in soil, grain, fruits; highly toxic, cause nerve damage, sterility, cancer, birth defects
Formamidines	Amitraz, chlordimeform (Fundal and Galecron)	Neurotoxins specific for certain stages of insect development, act synergistically with other insecticides	*Herbicides*	2,4 D; 2,4,5 T; paraquat, dinoseb, diaquat, atrazine, Silvex, linuron	Block photosynthesis, act as hormones to disrupt plant growth and development, or kill soil micro-organisms essential for plant growth
Microbes	*Bacillus thuringensis*	Kills caterpillars			
	Bacillus popillae	Kills beetles			
	Viral diseases	Attack a variety of moths and caterpillars			

suming about one-quarter of all insecticides and half of all chlorinated hydrocarbons used in the United States. Mosquito and termite control are major civil and domestic uses. Because insecticides are generally more acutely toxic than herbicides, they account for a far larger proportion of immediate illnesses in spite of being used in far smaller quantities.

Fungicides make up approximately one-sixth of the pesticides we use. They protect crops from fungal diseases, such as wheat rust, mildew, smut, and various blights, and are used in paints and wood preservatives. A major use is to prevent spoilage of fruits and vegetables. Much,

perhaps most, of the produce you buy in a grocery store has been treated with a fungicide either before or after harvest. Seven of the ten pesticides listed by the EPA as the greatest health concern in the United States are fungicides (Captan, Maneb, Zeneb, Mancozeb, Captifol, Folpet, and Chlorothalonil). The other three are a herbicide (Linuron) and two insecticides (Permethrin and Chlordimeform).

The U.S. National Academy of Sciences estimates that tighter controls on these compounds would reduce the risk of cancer from pesticides in the United States by 80 percent. The fifteen foods considered the greatest risk to the public

in terms of pesticide-related cancer are (in order of importance): tomatoes, beef, potatoes, oranges, lettuce, apples, peaches, pork, wheat, soybeans, beans, carrots, chicken, corn, and grapes.

What can you do to reduce your exposure to these dangerous chemicals? Try growing your own pesticide-free food. By joining a food-buying co-op or shopping at a local farmer's market you can buy "organic" food and support growers who practice pesticide-free farming. Become politically active. Write to your congressional representatives, and ask them to support legislation that regulates pesticide residues in food and supports alternative, sustainable farming practices.

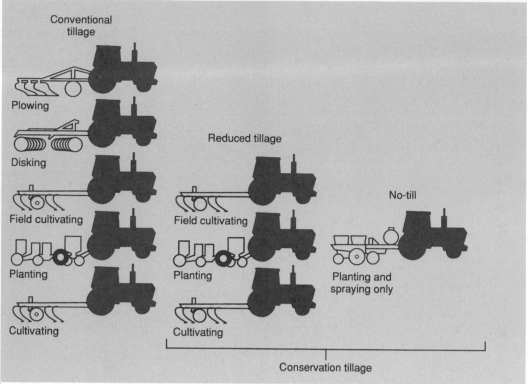

FIGURE 11.16 Reduced tillage and no-till farming protect soil from erosion, save energy, and reduce soil compaction. These methods often depend on increased use of pesticides, however.

Massive use of pesticides is not, however, a necessary corollary of soil conservation. It is possible to combat pests and diseases with crop rotation, trap crops, natural repellents, and biological controls.

Integrated Pest Management

In many cases, improved pest management techniques can cut pesticide use between 50 and 90 percent without reducing crop production or suffering increasing diseases. This approach is called **integrated pest management (IPM)**, a flexible, ecologically based pest-control strategy that uses a combination of techniques applied at specific times, aimed at specific crops and pests. IPM often involves such biological controls as predators (wasps, ladybugs, praying mantises) or pathogens (viruses, bacteria, fungi) to control pests safely. Releasing sterile males to interfere with insect pest reproduction, hormones that upset development, or using sex attractants to bait traps also can be effective.

Cultural practices also are a part of IPM. Crop rotation (growing a different crop a field each year in a four to six year cycle) keeps pest populations from building up. Flooding fields before planting or burning crop residues and replanting with a cover crop can suppress both weeds and insect pests. Home gardeners often plant a border of insect-repelling plants, such as garlic or marigolds. Where there is no alternative to using a chemical toxin for pest control, a single heavy dose of a non-

persistent pesticide might be applied just at the time insects or weeds are most vulnerable. Trap crops, a small area planted a week or two before the main crop, are also useful. This plot matures before the rest of the field and attracts pests away from other plants. The trap crop then is sprayed heavily with pesticides—enough so that no pests are likely to escape. The trap crop is destroyed so that workers will not be exposed to the pesticide and consumers will not be at risk. The rest of the field should be mostly free of both pests and pesticides.

IPM programs are already in use all over the United States on a variety of crops. Massachusetts apple growers, for instance, who use IPM have cut pesticide use by 43 percent in the past 10 years, while maintaining per-acre yields of marketable fruit equal to that of farmers who use conventional techniques. Some of the most dramatic IPM success stories have come in the Third World. In Brazil, pesticide use has been reduced up to 90 percent on soybeans. In Costa Rica, use of IPM on banana trees has eliminated pesticides altogether in one region. In Africa, mealybugs were destroying up to 60 percent of the cassava crop (the staple food for 200 million people) before IPM was introduced in 1982. A tiny wasp that destroys mealybug eggs was discovered and now controls this pest in over 65 million hectares (160 million acres) in thirteen countries.

One of the most important IPM programs now underway is in Indonesia, where an insect called the brown planthopper has developed resistance to virtually every insecticide and is threatening the country's hard-won self-sufficiency in rice. In 1986, President Suharto banned 56 of 57 pesticides previously used in Indonesia and declared a crash program to educate farmers about IPM and the dangers of pesticide use. Researchers found that farmers were spraying their fields habitually—sometimes up to three times a week—regardless of whether or not fields were infested. By allowing natural predators to combat pests and only spraying when absolutely necessary with chemicals specific for planthoppers, Indonesian farmers using IPM have had higher yields than their neighbors using normal practices and they have cut pesticide costs by 75 percent. In 1988, only two years after its initiation, the program was declared a success and is being extended to the whole country. Since nearly half the people in the world depend on rice as their staple crop, this experiment could have important implications for everyone.

Low Input Sustainable Agriculture

Many farmers are choosing to farm in a less intensive style by using less inorganic fertilizer, fewer or no pesticides, less water, less heavy machinery, and less fossil fuel energy. Although the yields are sometimes lower with this style of farming, so are the input costs in terms of commercial purchases of supplies and materials and in health and well-being. This type of agriculture is often called organic farming but there are so many different definitions of that term that many farmers prefer the terms low-input sustainable agriculture (LISA) or regenerative farming because they emphasize reliance on natural agricultural ecosystems and building long-term soil fertility.

There have been many success stories published recently that show both the practicality and profitability of sustainable agriculture. A 1989 National Research Council report highlights several case studies in regenerative farming of fruit, livestock, vegetables, and cereal crops. These alternative approaches are healthy both for the soil and for the farmers and their families.

Fair and Ecologically Sound Farming Systems

There is a great need for agricultural research focused on sustainable systems, rather than exclusively on higher production. Some authors suggest that we need to be concerned more about ecology and less about chemistry if we want to establish farming systems that will last for many generations. We need to explore new crops that may have useful nutritional attributes or pest resistance. We especially need to investigate tropical species that are suitable for low-input farming in Africa, southern Asia, and South America. Already, some progress has been reported in selecting and breeding such crops as yams, sorghum, millet, and rye that grow in particularly stressful environments or fit customary diets.

Some people believe the best way to stabilize both food production and population is to establish social, political, and economic systems based on a just distribution of resources. They encourage land reform that would enable those who work the land to reap the benefits from their labor, providing incentives for farmers to increase productivity while protecting the land. Land reform and more personal freedom for rural people have boosted farm output in free-market and centrally-planned economies (chapter 12).

Paying a fair price for farm products and farm labor is another way to stabilize farming and stimulate food production. This not only is a more humane treatment of rural people, but it also gives them the capital necessary to improve their land and invest in better machinery, new crop varieties, conservation practices, and other ways to increase productivity and protect and sustain resources.

In much of the Third World, agricultural issues are inextricably associated with women's rights. Women grow most of the family food in many cultures and also do much of the field labor in cash crop production; yet, they tend to lack economic and political equality with men. The health effects of pesticide use, fuelwood and water shortages, diversion of land and resources towards export crops, wages, and the costs of agricultural inputs are primarily women's issues in many places. Where women are denied access to credit, tools, education, respect, and power to participate in land-use decisions, families often go hungry.

Nutritional, economic, sociological, political, and population issues are interwoven. One way to increase nutritional levels for rural poor is to provide more rural jobs so people can have money to buy a variety of nutritional foods. Reducing the surplus of rural labor allows wages to rise so that those who remain in agriculture are better paid and, therefore, better fed. The availability of jobs also seems to decrease birth rates. Women who are valued more at work than at home tending a family are inclined to have fewer children. Parents with higher incomes and more financial security don't need to depend as much on having many children to take care of them in old age. Furthermore, a better educated population is more likely to be able to plan reproduction and avoid unwanted pregnancies.

According to the FAO, one of the most cost-effective ways to improve the nutritional status of the world is to provide clean water and better medical care for everyone. People who are sick from one of the many waterborne infectious diseases that inflict residents of the poorest countries cannot absorb the nutrients in the food they do have. Furthermore, they don't have the energy to grow more food or to work to buy what they need.

WHAT CAN WE DO PERSONALLY?

Probably the first thing that each of us should do with regard to agricultural and other environmental problems is to become educated about the issues involved. We can use that knowledge to develop good nutritional habits so that we eat only those foods

FIGURE 11.17 Growing some of your own food can be healthful and fun. Even a small balcony or patio can be turned into a minigarden.

and amounts we really need. For instance, most of us could benefit ourselves and the world at large by eating lower on the food chain. We don't really need as much meat and animal fat as we consume. It is not only wasteful, it is not very good for us.

Many people could grow more of their own food. Even a small yard can be used to grow delicious vegetables. Think of how much time, money, and resources are spent growing grass and shrubs that are used primarily for ground cover and decoration. Suburban lawns are often the most heavily fertilized, herbicided, watered, and cared-for lands in the country, using dozens or even hundreds of times as much energy, water, and chemicals per unit area as the most intensely farmed cropland. A well-tended garden also can be attractive. Working in your garden is healthful, rewarding, and can produce valuable and wholesome produce. Even if you live in a city, a few vegetables can be grown in pots on a windowsill, a balcony, or rooftop (figure 11.17). Cities have many acres of level rooftops that can serve admirably for gardens, as is done in some European countries. A plastic bag full of dirt with a couple of holes punched in it can grow a surprising harvest of tomatoes, lettuce, or squash.

For the foods that you need to buy, consider shopping at a farmer's market. The produce is fresh, and profits go directly to the person who grows the crop. A local food co-op or owner-operated grocery store also is likely to buy from local farmers and to feature pesticide-free foods.

Accept fruits and vegetables that may not be absolutely perfect. You can trim away a few bad spots or wash off an insect or two. If you overlook a worm or a bug, it may be less harmful to you than the toxic chemicals that are used to make supermarket foods sterile and cosmetically perfect!

S U M M A R Y

Fertile, tillable soil for growing crops is an indispensable resource for our continued existence on Earth. Soil is a complex system of inorganic minerals, air, water, dead organic matter, and a myriad of different kinds of living organisms. There are hundreds of thousands of different kinds of soils, each produced by a unique history, climate, topography, bedrock, transported material, and community of living organisms.

It is estimated that 25 billion tons of soil are lost from croplands each year because of wind and water erosion. Perhaps twice as much is lost from rangelands and permanent pastures. This erosion causes pollution and siltation of rivers, reservoirs, estuaries, wetlands, and offshore reefs and banks. The net effect of this loss is worldwide crop reduction equivalent to losing 15 million hectares (37 million acres), or 1 percent of the world's cropland each year.

The United States has one of the highest total rates of soil erosion in the world. Soil erosion exceeds soil formation on at least 40 percent of U.S. cropland. About one-half of the topsoil that existed in North America before European settlement has been lost.

Other areas with high erosion rates are China, India, the Soviet Union, and Ethiopia. Worldwide, about 25 percent of the land (about twice as much as we now use) has the potential for agricultural use. Putting much of that land into agricultural production would mean loss of valuable forests and grasslands and would cause loss of biodiversity and result in major ecological destruction. Large expanses of land, however, could be converted to agricultural use or could be used more intensively (but carefully) without causing great damage. It is possible that food production could be expanded considerably, even on existing farmland, given the proper inputs of fertilizer, water, high-yield crops, and technology. This will be essential if human populations continue to grow as they have during this century. Whether it will be possible to supply agricultural inputs and expand crop production remains to be seen. Global climate change could have devastating effects on world food supplies and might necessitate conversion of forests and grasslands to feed the world's population.

There are great differences in fertilizer use between the more developed and less developed countries. Many farmers in the developed countries use more fertilizer than is needed. Excess chemicals are a major source of water pollution and threaten health in some areas. Pesticides used in agriculture are also a serious problem. They disrupt ecosystems, cause about 500,000 cases of pesticide poisoning per year (especially in farm workers), and leave residues on many of the foods we eat.

Many new and alternative methods could be used in farming to reduce soil erosion, avoid dangerous chemicals, improve yields, and make agriculture just and sustainable. Some alternative methods are developed through scientific research; others are discovered in traditional cultures and practices nearly forgotten in our mechanization and industrialization of farming. Some authors advocate returning to low-input, regenerative, "organic" farming that may be more sustainable and more healthful than our current practices.

■ Review Questions

1. What is the composition of soil? What is humus? Why are soil organisms so important?

2. What are the four major problems we have maintaining high levels of agricultural production?

3. What are four kinds of erosion? Is erosion ever beneficial? Why is it a problem?

4. What is drip irrigation? Why is it beneficial in some cases? Why don't we always use drip irrigation?

5. What are some possible effects of overirrigation?

6. What can farmers do to increase agricultural production without increasing land use?

7. What is the estimated potential for increasing world food supply by increasing fertilizer use in low-production countries?

8. What is sustainable agriculture?

9. What is a perennial species? Why is coffee a good crop for rugged and hilly country?

10. What are some elements of integrated pest management?

■ Questions for Critical Thinking

1. Should farmers be forced to use ecologically sound techniques that serve farmers' best interests in the long run, regardless of short-term consequences? How could we mitigate hardships brought about by such policies?

2. In a crisis, when small farms are in danger of being lost, should preserving the soil be the farmer's first priority?

3. Should we encourage (and subsidize) the family farm? What are the advantages and disadvantages (economic and ecological) of the small farm and the corporate farm?

4. How many people do you think the world could support? What do you think would be the ideal number?

5. Should we try to increase food production on existing farmland, or should we sacrifice other lands to increase farming areas?

6. Some rice paddies in Southeast Asia have been cultivated continuously for thousands of years without losing fertility. Could we, and should we, adapt these techniques to our own country?

7. Should we continue our current rate of pesticide use? What are the advantages and disadvantages of continuing such use?

■ Key Terms

bedrock (p. 195)
contour plowing (p. 208)
cover crops (p. 209)
drip irrigation (p. 202)
gully erosion (p. 198)
integrated pest management (IPM) (p. 212)
mulch (p. 209)
old soils (p. 192)
parent material (p. 195)
perennial species (p. 208)
pest (p. 210)
pesticide (p. 210)
reduced tillage systems (p. 209)
regenerative farming (p. 205)
rill erosion (p. 198)
salinization (p. 202)
sheet erosion (p. 198)
soil (p. 192)
soil horizons (p. 195)
streambank erosion (p. 198)
strip-farming (p. 208)
subsoil (p. 195)
sustainable agriculture (p. 205)
terracing (p. 208)
tied ridges (p. 208)
topsoil (p. 195)
waterlogging (p. 202)
young soils (p. 192)
zone of leaching (p. 195)

SUGGESTED READINGS

■ Altieri, M. *Agroecology: The Scientific Basis of Alternative Agriculture*, 2d ed. Boulder: Westview Press, 1986. Describes ecological basis of agriculture, soil conservation, nutrient restoration, and biological pest control.

■ Barbier, E. B. (1989). "Sustaining Agriculture on Marginal Land." *Environment* 31(9):12. Describes national policies that emphasize low-input technology, land reform, and economic incentives to encourage sustainable systems on marginal land.

■ Berry, W. *The Gift of Good Land*. San Francisco: North Point Press, 1981. A series of essays on a variety of topics ranging from Amish farms to sheep and scythes by one of America's best environmental writers.

■ Berry, W. *The Unsettling of America: Culture and Agriculture*. San Francisco: Sierra Club Books, 1977. A thoughtful and passionate argument against the misuse of the land and destruction of rural culture.

■ Brown, Lester R., and John E. Young. "Feeding the World in the Nineties." *State of the World 1990*. Washington, D.C.: Worldwatch Institute. Warns of increasing land degradation, declining harvests, and impending food shortages in coming decades.

■ Fukuoka, M. *One Straw Revolution: An Introduction to Natural Farming*. Emmaus: Rodale Press, 1978. An account of one man's successful experiments in organic gardening.

■ Gipps, T. *Breaking the Pesticide Habit*. Minneapolis: International Alliance for Sustainable Agriculture, 1990. Presents alternatives to the "dirty dozen" most dangerous pesticides in pest control.

■ Hightower, J. *Hard Tomatoes, Hard Times: The Hightower Report*. Cambridge: Schenkman Press, 1972. A report of the Agribusiness Accountability Project inquiry into the land-grant college research program showing how agricultural research has been focused almost exclusively on helping industrial agribusiness rather than helping family farms or increasing nutritional food values.

■ Jackson, W. *New Roots for Agriculture*. San Francisco: Friends of the Earth, 1980. A prescription for new attitudes and new approaches to agriculture in America.

■ Jackson, W., W. Berry, and B. Colman. *Meeting the Expectations of the Land: Essays in Sustainable Agriculture and Stewardship*. San Francisco: North Point Press, 1984. An eloquent and insightful discussion of sustainable farming.

■ National Research Council. *Alternative Agriculture*. National Academy Press, Washington, D.C., 1989. Success stories in a variety of applications of low-input regenerative agriculture.

■ Paddock, J., N. Paddock, and C. Bly. *Soil and Survival: Land Stewardship and the Future of American Agriculture*. San Francisco: Sierra Club Books, 1987. This book argues that barring nuclear disaster, erosion is the greatest threat to civilization.

■ Parry, M. L., T. R. Carter, and N. T. Konijin (eds.) *The Impact of Climatic Variations on Agriculture*. Dordrecht, the Netherlands: Kluwer Academic Publishers, 1988. An important analysis of the effects of global climate change on agriculture.

■ Postel, S. "Saving Water for Agriculture." *State of the World 1990*. Washington, D.C.: Worldwatch Institute: A good discussion of water use in agriculture and its associated problems.

■ Power, J., and R. Follet. March 1987. "Monoculture." *Scientific American*. 256(3):78. The practice of growing the same crop on the same land repeatedly has some advantages for the farmer but may not always be good agronomy.

■ Reagnold, J. P., R. I. Papendick, and J. F. Parr. "Sustainable Agriculture." *Scientific American* 262(6):112. Describes environmental and economic rewards of alternative agriculture.

■ Rodale, R. *Regenerative Farming Systems*. Emmaus: Rodale Press, 1985. Organic agriculture described by one of the leaders in the field.

■ Ronfeldt, D. *Atencingo: The Politics of Agrarian Struggle in a Mexican Ejido*. Stanford: Stanford University Press, 1973. A description of the struggles of a traditional communal village in Mexico.

■ Simon, J. "Some False Notions about Farmland Preservation." In *The Vanishing Farmland Crisis*, edited by John Baden. Bozeman: The Political Economy Research Center, 1984. An argument that cropland losses are overestimated and that there is no crisis of productive capacity.

■ World Bank. *Sub-Saharan Africa: From Crisis to Sustainable Growth*. The World Bank, Washington, D.C., 1989. Policies and practical examples of sustainable agriculture under difficult conditions.

CHAPTER *12*

World Land Use

We abuse land because we regard it as a commodity belonging to us. When we see land as a community to which we belong, we may begin to use it with love and respect.

Aldo Leopold

CONCEPTS

About 60 percent of Earth's land is used by humans in one way or another. Three major uses of this land are croplands, forests, and rangelands. The remaining 40 percent of the land is too steep, rocky, inhospitable in climate, or otherwise unsuitable for human use. Much of this land is valuable, however, as wilderness, refuge, wildlife habitat, or simply open space.

Forests cover a little more than one quarter of Earth's land area, providing a variety of useful products, such as lumber, pulpwood, and firewood. Tropical rain forests are in danger of destruction by overuse and exploitation, threatening extinction of millions of species, exposing land to erosion, and affecting global atmospheric and climatic patterns.

Range and pasture lands occupy a little more than one-fifth of Earth's land area. When not managed properly, these lands can be degraded and turned permanently into desert. Croplands occupy about 10 percent of Earth's land area, much of it also suffering the effects of poor land management.

We have made considerable progress in recent years creating parks, wildlife refuges, and wilderness areas. Debate continues over management and expansion of these protected areas. Land reform is an essential part of sound land-use management. Fair distribution of land and its benefits encourages good stewardship, increased food production, sustainable agriculture, and social justice.

INTRODUCTION

The ways we use land clearly indicate our priorities and values. A series of land-use decisions—often seemingly small and insignificant, but generally irreversible—have shaped our history and our future. In a real way, we create our environment through care or abuse of the land (figure 12.1). This chapter surveys major global categories of land resources by human land use: forests, grazing lands, croplands, parks, and nature preserves. We will look at some of the problems in land use and the ways this irreplaceable resource can be and is being protected.

Throughout history, land ownership has been the traditional source of human wealth and power. Although land no longer is the most important resource in most industrialized countries, its use and control still shape societies and economies in many parts of the world. Inequitable patterns of land ownership lie at the root of many social and environmental problems in those countries. We will study some of the reform movements around the world that seek to improve the distribution of resources and the use of land.

WORLD LAND AREA CHARACTERISTICS

Earth's total land area is about 144.8 million sq km (55.9 million sq mi), or about 29 percent of the surface of the globe. Table 12.1 and figure 12.2 show the area devoted to four major land-use categories by continents.

Growing populations and expanding forestry and agricultural operations have brought about massive changes in land use in recent years. Table 12.2 shows broad categories of land conversion between 1970 and 1980 in the more developed and less developed countries. Notice the large reduction in forests, es-

TABLE 12.1 Present world land use (millions of hectares)

Region	Total land	Crop-land	Pasture and range	Forest and woods	Other
Africa	2,968	185	782	684	1,317
North and Central America	2,139	274	362	802	701
South America	1,753	142	457	858	296
Asia	2,679	451	644	492	1,092
Europe	473	140	85	159	89
Soviet Union	2,227	233	374	929	691
Oceania	843	49	455	158	181
Antarctica	1,400	0	0	0	1,400
Total	14,482	1,474	3,159	4,082	5,767

Source: Data from World Resources Institute 1990–1991.

TABLE 12.2 World land use changes between 1970 and 1980 (areas in million hectares)

	Developed world		Developing world	
	Area in 1970	Change in 10 years	Area in 1970	Change in 10 years
Land area	5,484		7,591	
Arable area	650	− 3	678	+30
Permanent crops	23	0	64	+ 6
Pasture	1,279	− 8	1,842	− 3
Forest	1,861	−26	2,348	−76
Other land	1,671	+37*	2,647	+44*

*Mainly land that has been changed to wasteland as a direct result of human activities.

From K. Blaxter, *People, Food and Resources.* Copyright © 1986 Cambridge University Press, New York, NY. Reprinted by permission.

pecially in less developed countries. About one-third of this forest area has been converted to cropland, but nearly two-thirds has been degraded to desert or useless scrub by wasteful practices.

Much of the world land that falls into the residual "other" category is naturally tundra, marsh, desert, scrub forest, bare rock, and ice or snow. About one-third of this land is so barren that it lacks plant cover altogether. Land in this category is not generally used intensively by humans but plays an important role, nonetheless, in biogeochemical cycles and as a refuge for biological diversity. Presently, only about 3 percent of the world's land surface is formally protected in parks, wildlife refuges, and nature preserves. Some land-use planners suggest that at least 10 percent of the land should be set aside to protect natural ecosystems and endangered species.

The distribution of human populations on Earth's land masses varies greatly. Oceania and Australia have the lowest population densities of the inhabited continents, averaging three persons per sq km (multiply by 2.59 to determine the number of people per sq mi). North America and the Soviet Union each have about twelve persons per sq km. South America and Africa have twenty-four and thirty-six persons per sq km, respectively. Europe and Asia are the most crowded continents, with 101 and 106 people per sq km, respectively. Other than city-states—such as Singapore, Monaco, and Hong Kong—Bangladesh and Bahrain are the most densely populated countries with about 5,200 people per sq km in 1988. Interestingly, Singapore, with about 26,000 people per sq km, and Hong Kong, with about 36,000 people per sq km, still produce much of their own food.

CROPLANDS

Most of the food on which humans depend comes from just a handful of species (chapters 10 and 11) grown on cultivated **croplands,** making up only one-tenth of the world's land. Notice how much more of the total area of Europe is devoted to crop-

FIGURE 12.1 The large clear cut patches visible in this aerial photo provide valuable wood products but preempt other uses and land values.

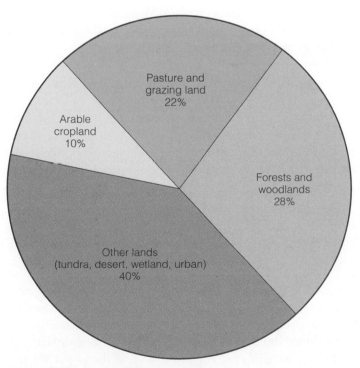

FIGURE 12.2 World land use, 1990.
Source: Data from *World Resources 1990–1991*.

land (30 percent) than in other regions. This is mainly because Europe is fortunate to have much land suitable for agriculture. Some places, such as South America, could convert more land to crops if they had capital and human resources to do so, but the effects on biodiversity would be severe. Other regions have little potential for new farm lands.

Some agricultural experts estimate that as much as 3.2 billion ha (7 billion acres) of the 7.2 billion ha (18 billion acres) of present forests and grazing lands could be converted to crop production, given the proper inputs of water, fertilizer, erosion control, and mechanical preparation. This land could feed a vastly larger human population (perhaps ten to thirty times the present population), but sustained intensive agriculture could result in serious energy and environmental limitations, such as depletion of groundwater, soil salinization and mineralization, soil erosion, loss of biological diversity, and even long-term climatic changes. Since agriculture and biological resources are discussed in detail in chapters 11 and 14, we will focus here primarily on the forest and grazing lands being converted to croplands.

FORESTS

Forests play a vital role in regulating climate, controlling water runoff, providing shelter and food for wildlife, and purifying the air. They produce valuable materials, such as wood and paper

TABLE 12.3 World forest resources (millions of hectares)

Region	Forest and woodland			Average annual deforestation	Average annual reforestation
	Closed	Open	Total		
Africa	220	465	684	3.8	0.4
North and Central America	541	261	802	1.3	2.6
South America	654	204	858	11.2	0.7
Asia	409	83	493	4.4	5.7
Europe	137	22	159	n.a.	1.0
Soviet Union	792	137	929	n.a.	4.5
Oceania	86	72	158	0.03	0.1
Total	2,839	1,243	4,082	n.a.	n.a.

Source: Data from World Resources Institute 1990–1991.

pulp, on which we all depend. Furthermore, forests have scenic, cultural, and historic values that deserve to be protected. In this section, we will look at forest distribution, use, and management.

Forest Distribution

Before large-scale human disturbances of the world began many thousands of years ago, forests probably covered 6 billion ha (15 billion acres). Since then, nearly *one-third* of that area has been converted to cropland, pasture, settlements, or unproductive wastelands. The 4 billion ha still forested covers about 28 percent of Earth's surface, nearly three times as much as all croplands. About two-thirds of the forest is classified as **closed canopy** (where tree crowns spread over 20 percent or more of the ground) and has potential for commercial timber harvests. The rest is **open canopy forest** or **woodland,** in which tree crowns cover less than 20 percent of the ground.

The major forest resources of the world are described in table 12.3. The main vegetation zones of the world's forest are shown in figure 12.3. As you can see, the largest areas of closed forests are in South America and the Soviet Union. Most of the Soviet Union's forests are temperate deciduous or boreal coniferous forests, whereas the forests of South America are mainly broad-leaved—evergreen or seasonal—deciduous, moist tropical forests. Africa has the largest amount and the largest percentage of open woodlands in the world, mainly in the dry savannas and thorn brush of the sub-Saharan region.

By far the largest tropical forests in the world are in the Amazon basin of South America. North America has large areas of temperate deciduous forests (mainly in the "lower 48" states) and boreal forests (mainly in Canada and Alaska). The tropical forests of Asia (primarily Southeast Asia and the large islands of Indonesia and the Philippines) are moist, broad-leaved evergreen or monsoon forests (chapter 5). The rapid clearing of these rich and important forests leads experts to predict that at least

one-half of them will be destroyed or seriously degraded by the end of this century.

South America also is losing forests at a rapid rate (figure 12.4). The loss of species that results from this forest clearing rivals the rate of extinction at the end of the Cretaceous period 65 million years ago when the dinosaurs disappeared.

Forest Products

Wood plays a part in more activities of the modern economy than any other commodity. There is hardly any industry that does not use wood or some wood products somewhere in its manufacturing and marketing processes. Furthermore, consider the impact our information explosion has had on use of paper, a wood product. Total wood consumption is about 3.2 billion metric tons or three billion cubic meters annually.

Industrial timber (or roundwood) used for lumber, plywood, veneer, particleboard, and chipboard, accounts for slightly less than one-half of worldwide wood consumption (about 1.5 billion tons per year). This exceeds the use of steel and plastics combined. International trade in wood and wood products amounts to more than $100 billion each year. Paper pulp accounts for only about 6 percent of the annual wood harvest. Developed countries produce approximately 60 percent of all industrial wood and account for about 80 percent of its consumption. Less developed countries, mainly in the tropics, produce the other 40 percent of industrial wood but use only about 20 percent.

Table 12.4 lists the world's leading producers of forest products. Note that the United States, the Soviet Union, and Canada are the largest producers of both industrial wood (lumber and panels) and paper pulp. Most of the wood harvested in China, India, Brazil, and Indonesia is used for fuel or charcoal for cooking or metal smelting. Japan is by far the largest net importer of wood in the world, purchasing about 43 million cubic

RESOURCES AND RESOURCE ECONOMICS

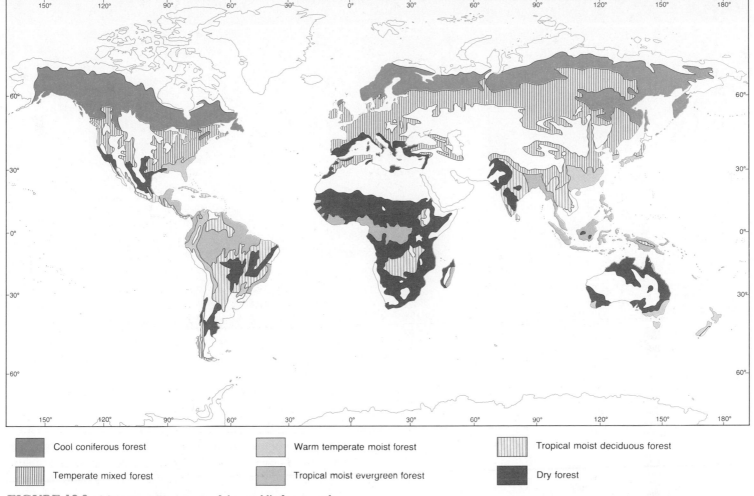

▓	Cool coniferous forest	▓	Warm temperate moist forest	▓	Tropical moist deciduous forest
▓	Temperate mixed forest	▓	Tropical moist evergreen forest	▓	Dry forest

FIGURE 12.3 Main vegetation zones of the world's forests under natural conditions.

meters per year. China is the second largest net importer, buying some 7 million cubic meters. Notice that the largest increase in wood harvest between 1975 and 1985 was not Brazil or Nigeria, but the United States.

More than *half* of the people in the world depend on firewood or charcoal as their principal source of heating and cooking fuel (figure 12.5). Consequently, **fuelwood** accounts for slightly more than one-half of all wood harvested worldwide. Unfortunately, burgeoning populations and dwindling forests are causing wood shortages in many less developed countries. About 1.5 billion people who depend on fuelwood as their primary energy source have less than they need. At present rates of population growth and wood consumption, the deficit is expected to increase from 400 million cubic meters (m^3) in 1990, to 2,600 m^3 in 2025. At that point, the demand will be twice the available fuelwood supply. The average amount of wood used for cooking and heating in sixty-three less developed countries is about 1 m^3 per person per year, roughly equal to the amount of wood that each American consumes as paper products alone. Whereas de-

veloped countries utilize only 13 percent of the fuelwood harvested each year, less developed countries consume 87 percent.

Too often, reforestation has been "top down," with urban bureaucrats and large corporations trying to impose plans on rural communities. The result is that villagers have little incentive to protect or care for new plantations. They poach firewood or allow animals to browse on young trees. When the woodlots are communally owned, however, and planted with a mixture of species that will produce a sustained yield of products useful to the villagers, they have reasons to be involved. Community woodlots can be planted on wasteland or along roads or slopes too steep to plow so they do not interfere with agriculture. They protect watersheds, create windbreaks, and provide such useful products as building materials, forage for animals, firewood, fruits, nuts, mushrooms, and materials for handicrafts on a sustained-yield basis.

Do you personally have anything to do with worldwide depletion of forest resources? Consider your consumption patterns. On average, each American uses 275 kg (600 lb) of paper

TABLE 12.4 World's leading producers of forest products

Country	Roundwood production (millions of cubic meters)			Paper production (millions of metric tons)	Annual average net trade (millions of cubic meters)	Percent change 1975–1985
	Total	Fuel and charcoal	Industrial			
United States	486	106	380	64.0	18.6	45
Soviet Union	374	87	287	10.0	17.7	−3
China	269	174	95	9.6	−7.0	28
India	250	226	24	1.8	−0.4	23
Brazil	238	172	66	4.0	−1.8	39
Canada	180	7	173	15.0	0.4	34
Indonesia	158	130	28	0.6	0.8	22
Nigeria	99	91	8	0.1	0.6	44
Sweden	53	5	48	7.4	5.6	−1
Finland	41	3	38	7.7	4.3	27

Source: Data from World Resources Institute 1990–1991.

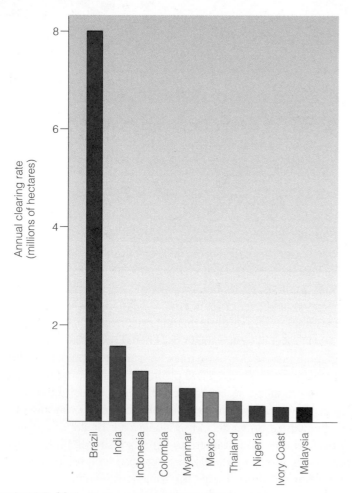

FIGURE 12.4 Deforestation rates in tropical countries.

each year, while a typical person in a less developed country uses about 5 kg per year. We probably could use far less without discomfort. During World War II, we recycled one-half of the paper we used, whereas the amount recycled now is closer to 25 percent. One of the impediments to greater recycling is that paper mills are not set up to handle recycled pulp. Furthermore, commodity prices, taxes, and shipping rates are skewed so that virgin (new) pulp is usually cheaper than recycled material. We also could reduce our consumption of imported beef, tropical fruits, and other products linked to forest destruction. In addition, we can contribute funds for forest research, rational management, and reforestation projects. At present, less than 1 percent of international development assistance funds are for forestry.

Forest Management

Roughly 25 percent of the world's forests are managed scientifically for wood production. **Forest management** involves planning for sustainable harvests, with particular attention to forest regeneration.

Aside from human use, what are some factors that contribute to forest loss? Fires, insects, and diseases damage up to one-quarter of the annual growth in temperate forests. Recently, reduced forest growth and sudden die-off of certain tree species in industrialized countries have caused great concern. It is thought that long-range transport of air pollutants (chapter 23) is contributing to this sudden forest death, but not all the causes and solutions are yet understood.

Most countries replant far less forest than is harvested or converted to other uses each year, but there are some outstanding examples of successful reforestation. China, for instance, cut down most of its forests one thousand years ago and

FIGURE 12.5 More than half of the world's people rely on firewood or charcoal as their main source of fuel. If present trends continue, the demand for wood will be twice the available supply, and a severe energy shortage for the poorest people will result.

has suffered centuries of erosion and terrible floods as a consequence (chapter 11). Recently, however, a massive reforestation campaign has been started. An average of 4.5 million hectares per year were replanted during the last decade. South Korea also has had very successful forest restoration programs. After losing nearly all its trees during the civil war thirty years ago, the country is now about 70 percent forested again.

The Soviet Union also has a comprehensive program of forest management, replanting roughly the same area each year as China and three times as much as the United States. Nearly every area in the Soviet Union has a management plan, and tree replanting equals timber harvesting in most places. In spite of being the world's largest net importer of wood, Japan has increased forests to approximately 68 percent of its land area. Strict environmental laws and constraints on the harvesting of local forests encourage imports so that Japan's forests are being preserved while it uses those of its trading partners. It is estimated that two-thirds of all tropical hardwoods cut in Asia are shipped to Japan.

Most reforestation projects involve large plantations of single-species, single-use, intensive cropping called **monoculture agroforestry.** Although this is an efficient approach, high density of a single species encourages pest and disease infestations. High levels of pesticides and herbicides often are required. This type of management lends itself to clear-cut harvesting, which not only saves money and labor, but also tends to leave soil exposed to erosion. Monoculture also often requires higher fertilizer inputs than does a mixed-species forest. Obviously, the biological community is different than it would be in the natural state. Where profits from these agroforest projects go to absentee landlords or government agencies, local people have little incentive to prevent fires or keep grazing animals out of newly

planted areas. In some countries, such as the Philippines and El Salvador, government reforestation projects have been targets for destruction by antigovernment military forces, with devastating environmental impacts.

Very promising alternative reforestation plans are being promoted by conservation and public service organizations, such as The New Forest Fund and Oxfam. These groups encourage people to plant community woodlots of fast-growing, multipurpose trees, such as *Leucaena*. Millions of seedlings have been planted in hundreds of self-help projects in Asia, Africa, and Latin America. *Leucaena* is a legume, so it fixes nitrogen and improves the soil. Its nutritious leaves are good livestock fodder. It can grow up to 3 m per year and quickly provides shade, forage for livestock, firewood, and good lumber for building. A well-managed *Leucaena* woodlot can yield up to 50 tons per hectare on a sustained basis (figure 12.6).

Tropical Forest Use

The richest and most diverse terrestrial ecosystems on Earth are the tropical forests (chapter 5). Although they now occupy less than 10 percent of Earth's land surface, these forests are thought to contain more than two-thirds of all plant biomass and at least one-half of all plant, animal, and microbial species in the world.

The Diminishing Forests

While temperate forests are expanding slightly due to reforestation and abandonment of marginal farmlands, tropical forests are shrinking rapidly (table 12.5 and figure 12.7). Chapter 14 discusses the causes and consequences of species loss in more detail. Briefly, the worldwide rate of extinction has accelerated from perhaps one species lost per decade in prehistoric times to one or two species lost *per day* now, largely due to tropical forest destruction. This is a human-caused environmental calamity of unprecedented proportions.

At the beginning of this century, an estimated 20 million square km of tropical lands were covered with closed-canopy forest. This was an area about twice as big as the United States. Nearly 25 percent of that forest has been seriously degraded and another 25 percent has been completely destroyed by human activities, mostly in the last thirty to forty years. Noted tropical forest ecologist Norman Meyers estimates that current rates of forest clearing are about 200,000 sq km (an area the size of Kansas) each year. Half of this destruction is the work of farmers who move into the jungles in search of agricultural land. Most of the rest is caused by commercial logging or fuelwood gathering (figure 12.7).

Brazil has by far the highest rate of deforestation in the world. This forest destruction is often accompanied by burning that is a major contributor of carbon dioxide to the atmosphere (chapter 17). Environmentalists argue that the forests should be preserved as extractive reserves (Box 12.1). Both the biological

FIGURE 12.6 Tree planting or reforestation can never replace the primal forest, but it can provide fuel, fodder, and other useful products for villages in the Third World. It can also help reduce global climate warming.

TABLE 12.5 Deforestation rates for closed tropical forests in selected countries (thousands of hectares)

Country	Total closed forest area	Annual clearing rate	Annual reforestation	Percent lost annually
Brazil	357,480	8,000	561	2.2
India	36,540	1,500	173	4.1
Indonesia	113,895	900	164	0.8
Colombia	46,400	820	11	1.8
Myanmar	31,941	677	n.a.	2.1
Mexico	46,250	595	28	1.3
Thailand	9,235	397	31	2.5
Nigeria	5,950	300	32	5.0
Ivory Coast	4,458	290	8	6.5
Malaysia	20,996	255	25	1.2

Source: Data from World Resources Institute 1990–1991.

and social costs of forest destruction are disastrous. The coastal forests of Ecuador, Sierra Leone, Ghana, Madagascar, Cameroon, Liberia, and Brazil, for instance, already have been mostly destroyed. Haiti was once 80 percent forested; today, essentially all that forest has been destroyed and the land lies barren and eroded. India, Burma, Kampuchea, Thailand, and Vietnam all have little virgin lowland forest left. In Central America, nearly two-thirds of the original moist tropical forest has been destroyed, mostly within the last thirty years and primarily due to conversion of forest to cattle range (figure 12.8).

Swidden Agriculture

Indigenous forest people often are blamed for tropical forest destruction because they carry out shifting agriculture that requires forest clearing. Actually, this is an ancient farming technique that can be an ecologically sound way of obtaining a sustained yield

FIGURE 12.7 Cutting and burning of tropical rainforest results in wildlife destruction, habitat loss, soil erosion, rapid water runoff, and waste of forest resources. It also contributes to global climate change.

FIGURE 12.8 Cattle graze on recently cleared tropical rain forest land in Costa Rica. About two-thirds of the forest in Central America has been destroyed, mostly in the past few decades as land is converted to pasture or cropland. Unfortunately, the soil is poorly suited to grazing or farming, and these ventures usually fail in a few years.

from fragile tropical soils if it is done carefully and in moderation. This practice used to be called "slash and burn" by people who didn't realize how complex and carefully balanced this method of farming actually can be. It is now called **milpa** or **swidden agriculture** from local names for fallow fields.

In this system, the farmer clears a new plot (about a hectare, or 2.5 acres) each year. Small trees are felled and large trees are killed by girdling (cutting away a ring of bark) so that sunlight can penetrate through to the ground. After a few weeks of drying, the branches, leaf litter, and fallen trunks are burned to prepare a rich seedbed of ashes. Fast-growing crops, such as bananas and papayas, are planted immediately to control erosion and to shade root crops, such as cassava and sweet potato, which anchor soil. Maize, rice, and up to eighty other crops are planted in a riotous profusion. Although they would not recognize the terms, these indigenous people are practicing **mixed perennial polyculture**.

The diversity of the milpa plot mimics that of the jungle itself, even though the species representation is more restricted. When managed well, the soil is covered with vegetation. This variety means that crops mature in a staggered sequence and there is almost always something to eat from the plot. It also helps prevent irruptive insect infestations that would plague a monoculture crop. Yields from a single hectare can be as high as 6 tons of grain (maize or rice) and another 5 tons of roots, vegetable crops, nuts, and berries each year. This yield is very comparable to the best results with intensive row cropping and about one thousand times as much food as is produced from the same land when it is converted to cattle pasture.

After a year or two, the forest begins to take over the garden plot again. The farmer will continue to harvest perennial crops for a while and will hunt for small animals that are attracted to the lush vegetation. Ideally, the land then will be allowed to remain covered with rain forest vegetation for ten to fifteen years

while nutrients accumulate before it is cleared and replanted again. Recently, growing populations and displacement of farmers from other areas have forced shifting agriculturalists to reuse their traditional plots on shorter and shorter rotation. When plots are farmed every year or two, nutrients are lost faster than they can be replaced, the forest doesn't regrow as vigorously as it once did, and additional species are lost. Eventually, erosion and overuse reduce productivity so much that the land is practically useless.

Logging and Population Resettlement

The other major source of forest destruction is usually the result of logging and subsequent invasion by land-hungry people from other areas. The loggers often are interested only in "creaming" the most valuable hardwoods, such as teak, mahogany, sandalwood, or ebony. Although only one or two trees per hectare might be taken, widespread devastation usually results. Because the canopy of tropical forests is usually strongly linked by vines and interlocking branches, felling one tree can easily bring down a dozen others. Tractors dragging out logs damage more trees, and construction of roads takes large land areas. Insects and infections invade wounded trees. Tropical trees, which usually have shallow root systems, are easily toppled by wind and erosion when they are no longer supported by their neighbors. Up to three-fourths of the canopy may be destroyed for the sake of a few logs (figure 12.9). Obviously, the complex biological community of the layered canopy is severely disrupted by this practice.

What happens next? Bulldozed roads make it possible for large numbers of immigrants to move into the forest in search of land to farm. Some governments are encouraging such resettlement (Box 12.1). People with little experience or understanding of the complex rain forest ecosystem try to turn it into farms. Instead of farming small temporary clearings as successful mixed perennial polyculture, they plant large fields of annual

BOX 12.1   Murder in the Rainforest

Francisco "Chico" Mendes was born in 1944 on the far upper reaches of the Amazon River in Acre Province of northwestern Brazil. Like his father before him, Chico was a *seringueiro*, one of about one hundred thousand independent rubber tappers who live with their families in remote areas of the tropical rain forest. Like the other *seringueiros*, Chico grew a few crops in the small clearing around his house. His main income, however, came from several hundred wild rubber trees in the forest that he visited every week. By carefully making two long V-shaped cuts in the bark of the rubber tree, Chico was able to collect a cup or two of milky latex sap every week fjrom each tree. The latex was dried to make natural rubber (box figure 12.1).

With only two cuts per week, the *seringueiro* doesn't hurt the tree, and production can continue for many years. From his father, Chico inherited trees that had been tapped for fifty years or more. In addition to rubber, the *seringueiros* collect Brazil nuts, wild fruits, and other natural products from the forest. This sustainable harvest allows them to live there without destroying the forest or reducing its natural diversity.

Other people have plans for the forest, however, that don't include *seringueiros* or wild rubber trees. Land speculators and big ranching companies can make quick profits by cutting down the valuable tropical hardwoods and converting the forest to cattle pasture. The

BOX FIGURE 12.1 Chico Mendes was a leader of the *seringueiros*, or rubber tappers, in Acre province of the Brazilian Amazon. His murder in 1988 became an international scandal and led directly to establishment of extractive forest reserves that will preserve both the forest and the way of life of his people.

first year after forest clearing and burning, the grass grows luxuriantly. Soon, however, pounding tropical rains wash away nutrients, so that after a few years it takes 2 to 5 hectares (5 to 10 acres) to support a single cow.

The ranchers need to clear enormous areas each year to maintain their herds. Satellite photos showed that 12 million hectares (30 million acres) of forest—an area the size of South Carolina—were cleared and burned in Brazil in 1988. This destruction was encouraged by low land prices, favorable tax rates, and direct subsidies to ranchers to convert virgin forest to pasture. In addi-

rowcrops. The result is ecological disaster. Rains wash away the topsoil and the tropical sun bakes the exposed subsoil into an impervious hardpan that is nearly useless for farming.

Degradation of rivers is another disastrous result of forest clearing. Tropical rivers carry two-thirds of all freshwater runoff in the world. In an undisturbed forest, rivers are usually clear, clean, and flow year-round because of the "sponge effect" of the thick root mat created by the trees. When the forest is disrupted and the thin forest soil is exposed, erosion quickly carries away the soil, silting river bottoms, filling reservoirs, ruining hydroelectric and irrigation projects, filling estuaries, and smothering coral reefs offshore. In Malaysia, sediment yield from an undis-

tion, roads were built and maintained by the government to open up the forest and make it possible to ship cattle to market. In the early 1970s, Brazil built highway BR 364 into Rondonia Province, immediately to the south of Acre, with the help of a $200 million loan from the World Bank. A flood of loggers, land speculators, ranchers, and landless peasants poured into the area. Within a decade, more than a third of the Rondonian forest had been destroyed. Native tribal people and *seringueiros* were driven out.

By the 1980s, the developers turned their attention to Acre Province. Brazil asked the Inter-American Development Bank (IDB) for another $200 million to extend BR 364 into the Xapuri region of western Acre where Chico lived. Chico and other *seringueiros* organized to resist this development. They proposed that instead of cutting down the forest it should be declared an "extractive reserve" where traditional patterns of harvest could continue unhindered.

Ecologists and plant biologists supported the idea of sustainable forest yield. They found that logging brought a one-time profit of $3,184 per hectare, on average, followed by only about $150 per year as cattle pasture. By contrast, the native wild fruits, rubber, and other natural products gathered by the *seringueiros* from that same hectare had a potential value of $6,820 every year.

In 1986, Chico Mendes ran for a seat in the legislature as a Worker's Party can-didate on a forest-protection platform. His opponent was heavily financed by loggers, ranchers, and land speculators. Chico won the popular vote but lost in the electoral count. Charges of fraud and vote buying were widespread. Shortly after the election, paving of the Acre road began.

With the help of American conservation organizations, Chico made a trip to Washington, D.C., in 1987 to lobby the United States Congress on behalf of forest protection. His arguments were persuasive. The Senate Appropriations Committee denied $200 million (out of a total of $258 million) requested by the IDB because of the environmental destruction caused by building road BR 364. The IDB immediately suspended payments to Brazil.

When Chico returned home, he was invited to meet with the Interior Ministry. In a major victory for the *seringueiros*, several extractive reserves were established in 1988, ending almost all forest clearing in Acre. The ranchers were furious, especially Darli Alves da Silva, who claimed a large area in the Seringal Cachoeira extractive reserve where Chico had grown up. He vowed that Chico would not live out the year.

On the evening of December 22, 1988, Chico was playing cards at home with his wife, Ilzmar, and the two policemen who were assigned to guard him. He took a towel and went out the back door to the shower in the outdoor bathroom. As he opened the door, Chico was hit in the chest by a shotgun blast from the backyard. He staggered back into the house as the policemen ran out the front door. Although the police station was right across the street, no one responded to Ilzmar's cries for help. Chico died immediately.

The local police said they had no clues or suspects in the case. A storm of protest, both local and international, forced the Brazilian government to intervene. Investigators soon followed a trail of evidence to the ranch of Darli da Silva. In 1990, Darli, his son Darci Pereia da Silva (one of his thirty children), and a ranchhand, Jerdeir Pereia, were convicted for murder. Evidence suggests that Darli ordered the murder, Darci went along to supervise, and Jerdeir actually pulled the trigger.

Responsibility for the killing of Chico Mendes may lead beyond Darli, Darci, and Jerdeir, however. Other prominent ranchers and land brokers have been accused of paying Darli to kill Chico. Even further, it can be argued that politics, economics, and world trade played a role. Certainly international development loans and cash crops that encourage forest destruction are factors. Perhaps each of us bears some responsibility as well. Do you know where the wood veneer on products you buy comes from, or where the beef in your fast-food burger or hot dog was grown? Is is possible that each of us may have played some small part in the death of a *seringueiro* deep in the Amazonian forest of Brazil?

turbed primary forest was about 100 cubic m per sq km per year. After forest clearing, the same river carried 2,500 cubic m per sq km per year.

Increased runoff after forest clearing also leads to disastrous floods. In 1978, for instance, India and Bangladesh suffered some of the worst floods in their history after forest clearing in the Himalayan foothills allowed monsoon rains to run off unchecked. The Ganges River was swollen forty times larger than normal. Sixty thousand villages were flooded, 10,000 people drowned, and 40,000 cattle were lost. The total monetary damage was more than $2 billion.

FIGURE 12.9 Even selective logging causes severe damage to the tropical rain forest.

Forest Protection

What can be done to stop this destruction and encourage careful management? While there is much discouraging news, there are also some hopeful signs for tropical forest conservation. Many tropical countries have realized that forests represent a valuable resource, and they are taking steps to protect them. Indonesia has announced plans to preserve 100,000 sq km, one-tenth of its original forest. Zaire and Brazil each plan to protect 350,000 sq km (about the size of Norway) in parks and forest preserves. Costa Rica has one of the best plans for forest protection in the world. Attempts are being made there to not only **rehabilitate land** (utilitarian program to make an area useful to humans) but also to **restore ecosystems** (reinstate an entire community of organisms to naturally occurring association). One of the best known of these projects is Dan Janzen's work on restoration of the dry tropical forest of Guanacaste National Park (Box 12.2).

Financing nature protection is often a problem in developing countries where the need is greatest. One promising approach is called *debt-for-nature swaps*. Banks, governments, and lending institutions now hold nearly $1 trillion in loans to developing countries. There is little prospect of ever collecting much of this debt, and banks are often willing to sell bonds at a steep discount—perhaps as little as 10 cents on the dollar. Conservation organizations buy debt obligations on the secondary market at a discount and then offer to cancel the debt if the debtor country will agree to protect or restore an area of biological importance.

The first such swap was made in 1987. Conservation International bought $650,000 of Bolivia's debt for $100,000—an 85 percent discount. In exchange for canceling this debt, Bolivia agreed to protect nearly 1 million hectares (2.47 million acres) around the Beni Biosphere Reserve in the Andean foothills. Ecuador and Costa Rica have had a different kind of debt-for-nature swap. They exchanged debt for local currency bonds that are used to fund activities of local private conservation organizations in the country. This has the dual advantage of building and supporting indigenous environmental groups while also protecting the land.

Agreements have been reached with Madagascar and Zambia to swap debts for nature, and negotiations are under way with Peru, Mexico, and Tanzania. Critics charge that these swaps compromise national sovereignty and that they will do little to reduce Third World debt or change the situations that lead to environmental destruction in the first place.

RESOURCES AND RESOURCE ECONOMICS

BOX 12.2 Restoring a Dry Tropical Forest

When the Spanish *conquistadores* arrived in Central America in the sixteenth century, about 5.5 million hectares (21,000 square miles) of dry tropical forest stretched along the Pacific coast from Columbia to Mexico. In contrast to the evergreen rain forests and cloud forests on the Atlantic side of the isthmus, dry forests have distinct seasons. During the wet summer months, the vegetation is dense and lush. In the winter, however, when rains are sparse, trees and bushes lose their leaves, and the whole forest becomes open and desertlike.

This dry forest was much easier to convert to farms and ranches than the moist forests. Its climate is healthier, and its soil was more fertile and conducive to agriculture. Today, only about 1 percent of Central America's dry forest remains in anything like its original condition, making it one of the most threatened ecosystems in the world. As the forest has disappeared, many of its unique plant and animal species have become rare and endangered. If much more forest is lost, hundreds or even thousands of species will become extinct.

An exciting project is currently underway in Costa Rica where scientists and local residents have joined together to restore about 700 square kilometers (28,000 acres) of dry tropical forest to approximately its original condition. A new national park called Guanacaste (named after the Costa Rican national tree that once grew in this forest) is being created from private lands, an existing park, and other public land holdings. Under the leadership of entomologist, Dan Janzen, attempts are being made to understand the ecosystem and to reintroduce native plants and animals in an effort to restore—rather than just rehabilitate—the forest.

How is this possible after the land has been abused and degraded for centuries? Isn't it long past the point at which it can be rescued? Fortunately, according to Janzen, most of the original flora and fauna have not been completely eliminated, only reduced. Small areas con-

BOX FIGURE 12.2 Former cowboys guard recovering pasture lands in the new Guanacaste National Park in Costa Rica where they once herded cattle. By involving local people in park management, the park service benefits both from their knowledge of the area and their support of ecosystem restoration.

taining most of the indigenous (native) species remain scattered across the countryside. The challenge is to find these species and create habitats where they can thrive and recreate the forest.

Fire is one of the greatest threats to the forest. Every year during the dry season local people accidentally or deliberately start fires that sweep across the land destroying native species and converting forest to grassland full of non-native invaders. Creating breaks to control the spread of fire and persuading residents to fight fires rather than set them is the first step toward restoring the forest.

Contrary to what you might expect, grazing animals are not excluded from Guanacaste National Park. In fact, they are encouraged because they are efficient seed dispersers. The forest probably coevolved with a fauna that included large, hooved grazing animals before humans arrived, so many plant species actually depend on animals for regeneration. Horses, monkeys, goats, birds, and even turtles eat fruits and pass their seeds through their digestive system days or weeks later. This not only distributes seeds to new locations, it also provides fertilizer for their initial growth. Furthermore,

some seeds have tough outer coverings that are weakened by digestive acids and enzymes, facilitating germination. Being able to use the new national park for grazing during restoration makes the whole process much more attractive to its neighbors.

Involving local people in the project and making the park economically beneficial to them is another of the essential keys to successful restoration (box figure 21.1). When they see how a park will help them, residents will be enthusiastic participants. Native people, with their knowledge of the forest and their skills as land stewards, can be an invaluable resource in the restoration process.

Once Guanacaste National Park is reconstituted, locals can work as guides and rangers or provide services to tourists who come to visit and view wildlife. Providing jobs in the area will help stem the tide of urbanization and also preserve local culture. Biodiversity and cultural heritage can be saved simultaneously. This exciting project may serve as an inspiration and guide to similar efforts in many areas of the world where bad land use practices threaten both wildlife and indigenous people (box figure 12.2).

FIGURE 12.10 Women of the *Chipko Andolan* movement in India organize to protect local forests from logging.

People on the grass-root level also are working to protect and restore forests. India, for instance, has a long history of non-violent, passive resistance to protest unfair government policies. These *stayagrahas* go back to the beginning of Indian culture and often have been associated with forest preservation. Gandhi drew on this tradition in his protests of British Colonial rule in the 1930s and 1940s. During the 1970s, commercial loggers began large-scale tree felling in the Garhwal region in the state of Uttar Pradesh in northern India. Landslides and floods resulted from stripping the forest cover from the hills. The firewood on which local people depended was destroyed, and the way of life of the traditional forest culture was threatened. In a remarkable display of courage and determination, the village women wrapped their arms around the trees to protect them, sparking the *Chipko Andolan* movement (literally, movement to hug trees). They prevented logging on 12,000 km² of sensitive watersheds in the Alakanada basin. Today, the *Chipko Andolan* movement has grown to more than four thousand groups working to save India's forests (figure 12.10).

RANGELANDS

Pasture (generally enclosed domestic meadows or managed grazing lands) and **open range** (unfenced, natural grazing lands) occupy about 24 percent of the world's land surface. Tables 12.1

TABLE 12.6 World pasture and rangelands (millions of hectares)

Region	Permanent pasture	Open woodland	Other lands	Total	Percent of land surface
North America	265	275	746	913	50
Latin America	590	248	239	964	47
Europe	86	22	91	153	33
Soviet Union	373	137	702	861	39
Africa	778	508	1,317	1,945	65
Asia	1,137	569	1,919	2,666	53
Oceania	460	76	182	672	75
Total	3,157	1,372	4,384	6,721	51

Source: Data from World Resources Institute 1987.

TABLE 12.7 World grazing animals (millions)

Region	Cattle	Sheep and goats	Buffaloes and camels	Horses
Africa	180	326	16.6	17
North and Central America	167	33	0.0	27
South America	256	130	0.1	21
Asia	385	613	138.0	44
Europe	128	152	0.3	5
Soviet Union	121	148	0.6	6
Oceania	32	224	0.0	1
Total	1,269	1,626	156.6	122

Source: Data from U.N. Food and Agriculture Organization, 1990.

and 12.6 show the area and distribution of grazing lands around the world. The main commercial use of this land is raising livestock. More than three billion domestic grazing animals turn grass and forage into protein-rich meat and milk that make a valuable contribution to human nutrition. Table 12.7 shows how the major domestic grazing animals are distributed. Oceania has a surprisingly large number of sheep because of Australia and New Zealand. The 3 billion ha (12 million sq mi) of permanent grazing land (both pasture and open range) in the world is about twice the area of all agricultural crops. When you add to this more than 1 billion ha of open woodlands and 4 billion ha of other lands (e.g., desert, tundra, marsh, and scrub) that are used seasonally or in favorable years for raising livestock, nearly one-half of the total landmass of Earth is used at least occasionally as grazing land for domestic animals.

Range Management

By carefully monitoring the numbers of animals and the condition of the range, ranchers and pastoralists (people who live by herding animals) can adjust to variations in rainfall, seasonal plant conditions, and nutritional quality of forage to keep livestock healthy and avoid overusing any particular area. Conscientious management can actually improve the quality of the range.

Some nomadic pastoralists who follow traditional migration routes and animal management practices produce admirable yields from harsh and inhospitable regions. They can be ten times more productive than dryland farmers in the same area and come very close to maintaining the ecological balance, diversity, and productivity of wild ecosystems on their native range. Nomadic herding requires large open areas, however, and wars, political problems, travel restrictions, incursions by agriculturalists, growing populations, and changing climatic conditions on many traditional ranges have combined to disrupt an ancient and effective way of life. The social and environmental consequences often are tragic.

Overgrazing and Desertification

About one-third of the world's range is severely degraded by overgrazing (figure 12.11). Among the countries with the most damage and the greatest area at risk are Pakistan, Sudan, Zambia, Somalia, Iraq, and Bolivia. Usually, the first symptom of improper range management is elimination of the most palatable herbs and grasses. Grazing animals tend to select species they prefer and leave the tougher, less tasty plants. When native plant species are removed from the range, weedy invaders move in. Gradually, the nutritional value of the available forage declines. As overgrazing progresses, hungry animals strip the ground bare, and their hooves pulverize the soil, hastening erosion.

The process of denuding and degrading a once-fertile land initiates a desert-producing cycle that feeds on itself and is called **desertification.** With nothing to hold back surface runoff, rain drains off quickly before it can soak into the soil to nourish plants or replenish groundwater. Springs and wells dry up. Trees and bushes not killed by browsing animals or humans scavenging for firewood or fodder for their animals die from drought. When the earth is denuded, the microclimate near the ground becomes inhospitable to seed germination. The dry barren surface reflects more of the sun's heat, changing wind patterns, driving away moisture-laden clouds, and leading to further desiccation. Deserts have been called the footprints of civilization.

This process is ancient, but in recent years it has been accelerated by expanding populations and political conditions that force people to overuse fragile lands. Those places that are most severely affected by drought are the desert margins, where rainfall is the single most important determinant in success or failure of both natural and human systems (figure 12.12). In good years, herds and farms prosper and the human population grows. When

FIGURE 12.11 Sheep grazing on land in Guatemala that was once tropical forest. Soils in the tropics are often thin and nutrient-poor. When forests are felled, heavy rains carry away the soil and fields quickly degrade to barren scrubland. In some areas, land clearing has resulted in climate changes that has turned once lush forests into desert.

drought comes, there is no reserve of food or water and starvation and suffering are widespread. Can we reverse this process? In some places, people are reclaiming deserts and repairing the effects of neglect and misuse.

Forage Conversion by Domestic Animals

Ruminant animals, such as cows, sheep, goats, buffaloes, camels, and llamas, are especially efficient at turning plant material into protein because bacterial digestion in their multiple stomachs allows them to utilize cellulose and other complex carbohydrates that many mammals (including humans) cannot digest. As a result, they can forage on plant material from which we could otherwise extract little food value. Many grazers have very dif-

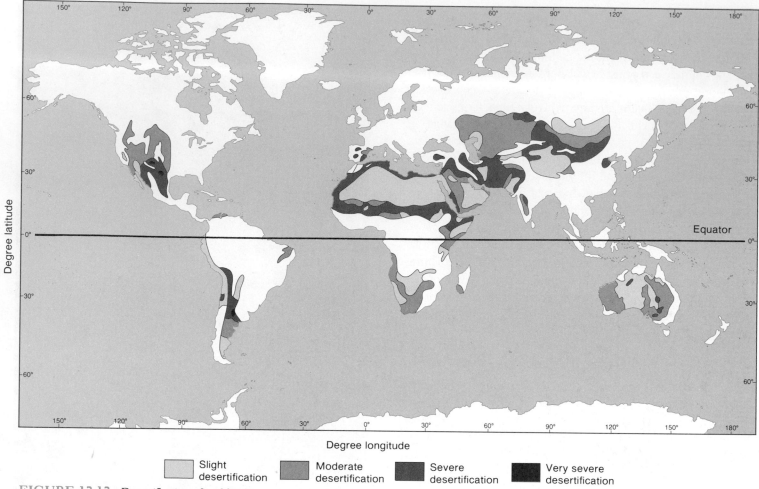

FIGURE 12.12 Desertification of arid lands.

ferent feeding preferences and habits. Often the most effective use of rangelands is to maintain small mixed-species herds so that all vegetation types are utilized equally and none is overgrazed. Cattle and sheep, for instance, prefer grass and herbaceous plants, goats will browse on low woody shrubs, and camels can thrive on tree leaves and larger woody plants.

Worldwide, 85 percent of the forage for ruminants comes from native rangelands and pasture. In the United States, however, only 15 percent of livestock feed comes from native grasslands. The rest is made up of crops grown specifically for feed, particularly alfalfa, corn, and oats. Grain surpluses and our taste for well-marbled (fatty) meat have shifted livestock growing in the developed countries to feedlot confinement and high-quality diets. In the United States, roughly 90 percent of our total grain crop is used for livestock feed.

Harvesting Wild Animals

A few people in the world still depend on wild animals for a substantial part of their food. About one-half of the meat eaten in Botswana, for instance, is harvested from the wild. There are good reasons to turn even more to native species for a meat

source. On the African savannah, researchers are finding that springbok, eland, impala, kudu, gnu, oryx, and other native animals forage more efficiently, resist harsh climates, are more pest- and disease-resistant, and fend off predators better than domestic livestock. Native species also are members of natural biological communities and demonstrate niche diversification, spreading feeding pressure over numerous plant populations in an area. A study by the U.S. National Academy of Sciences concluded that the semiarid lands of the African Sahel can support only about 20 to 28 kg (44 to 62 lb) of cattle per ha, but can produce nearly three times as much meat from wild ungulates (hooved mammals) in the same area (figure 12.13). In the United States, native bison and elk are raised as "novelty" meat sources. Might there be a greater future role for ranching of wild animals on our own range lands?

WORLD PARKS AND PRESERVES

The idea of setting aside nature preserves has spread rapidly since World War II as the world has become aware of the rapid disappearance of wildlife and wild places.

FIGURE 12.13 Range grazers are a valuable part of grassland ecosystems; they may be the best way to harvest biomass for human consumption. Native species often forage more efficiently, resist harsh climates, and are more pest- and disease-resistant than domestic livestock.

A Positive Trend

So far, about 427 million ha (3 percent of Earth's land) have been preserved in parks and wildlife refuges (table 12.8 and figure 12.14). Arctic biomes comprise nearly 30 percent of all protected areas and are especially well represented in Alaska, Canada, Greenland, and Scandinavia. Deserts and tropical dry forests also are well represented, making up another 30 percent of all protected areas. Other important biogeographical regions are badly underrepresented, however. The International Union for the Conservation of Nature and Natural Resources (IUCN) has identified three thousand areas totaling about 3 billion ha as worthy of national park or wildlife refuge status. The most significant of these areas are designated World Biosphere Reserves or World Heritage Sites. So far, only 144 of these special refuges have been selected. Table 12.9 shows the parks and wildlife areas identified by the IUCN as "most threatened."

The tropics are suffering the greatest destruction and loss of species in the world, especially in tropical rain forests. People in some of the affected countries are beginning to realize that the biological richness of their environment may be their most valuable resource and that its preservation is vital for sustainable development. Tourism alone can bring more into a country over the long term than extractive industries, such as logging and mining.

Among the countries with the most admirable plans to protect natural values are Costa Rica, Tanzania, Rwanda, Botswana, Benin, Senegal, Central Africa Republic, Zimbabwe, Butan, and Switzerland, each of which has designated 10 percent or more of its land as ecological protectorates. This is a larger percentage than the United States has set aside in parks, wilderness, and wildlife refuges combined. Brazil has even more ambitious plans, calling for some 231,600 km² or 18 percent of the country to be protected in nature preserves.

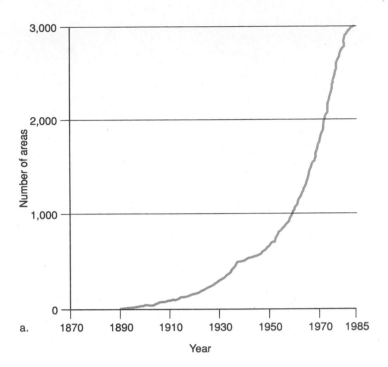

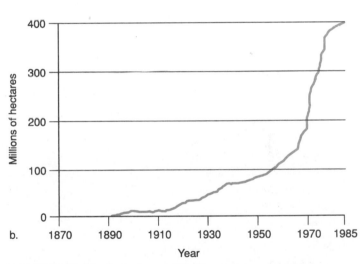

FIGURE 12.14 Both the number of areas (*a*) and the total area protected as parks, wildlife refuges, biological preserves, etc., (*b*), have increased dramatically over the past half century. Still, many species, and even whole ecosystems, are threatened by human encroachment.

Does park or preserve status effectively protect a wild area? Present levels of protection and management cannot always be adequately administered and are subject to changes in political climate. Many problems threaten natural resources and environmental quality in the parks. Dam building for hydroelectric and irrigation projects, deforestation, oil drilling and pipelines, gold prospecting, timber and wildlife poachers, air pollution, and off-road vehicles all menace natural values in these areas. Many of the countries with the most important biomes lack funds, trained personnel, and experience to manage some of the areas

TABLE 12.8 Protected areas by biogeographical classification

Biome type	Number of areas	Protected area (millions of hectares)
Tropical moist forest	280	39.1
Temperate rain forests and woodlands	275	22.4
Temperate coniferous forests	175	38.8
Tropical dry forests	581	65.5
Temperate deciduous forests	483	11.5
Evergreen broad-leaved forests	475	12.0
Warm deserts	161	41.6
Cold deserts	76	13.5
Tundra	39	115.4
Tropical grasslands and savannas	30	9.1
Temperate grasslands	90	1.9
Mountain biomes	436	32.5
Mixed island systems	267	15.8
Lake systems	10	0.5
Total	3,510	427.6

From International Union for Conservation of Nature and Natural Resources (IUCN), 1985. Used with permission.

TABLE 12.9 IUCN "Most Threatened" national parks

Araguaia National Park, Brazil
Juan Fernandez National Park, Brazil
Krkonose National Park, Czechoslovakia
Kutai Game Reserve, Indonesia
Tai National Park, Ivory Coast
Manu National Park, Peru
Mount Apo National Park, Philippines
Ngorongoro Conservation Area, Tanzania
John Pennekamp Coral Reef State Park, United States
Key Largo National Marine Sanctuary, United States
Durmitor National Park, Yugoslavia
Garamba National Park, Zaire

From International Union for Conservation of Nature and Natural Resources (IUCN), 1985. Used with permission.

under their control. Management of biological resources and endangered species is discussed further in chapter 15.

The IUCN has developed a **world conservation strategy** for natural resources that includes the following three objectives: "(1) To maintain essential ecological processes and life-support systems (such as soil regeneration and protection, recycling of nutrients, and cleansing of waters) on which human survival and development depend. (2) To preserve genetic diversity, on which

TABLE 12.10 IUCN Ecological plan of action

1. Launch a consciousness-raising exercise to bring the issue of biological resources to the attention of policy makers and the public at large.
2. Design national conservation strategies that take explicit account of the values at stake.
3. Expand our network of parks and preserves to establish a comprehensive system of protected areas.
4. Undertake a program of training in the fields relevant to biological diversity to improve scientific skills and technological grasp of those charged with its management.
5. Work through conventions and treaties to express the interest of the community of nations in the collective heritage of biological diversity.
6. Establish a set of economic incentives to make species conservation a competitive form of land use.

From United Nations Environment Program and World Wildlife Fund, 1986. Reprinted by permission.

depend the breeding programs necessary for protection and improvement of cultivated plants and domesticated animals. (3) To ensure that any utilization of species and ecosystems is sustainable." These goals are further elaborated in the ecological plan of action adopted by the IUCN and shown in table 12.10.

Conservation and Economic Development

Many of the most seriously threatened species and ecosystems of the world are in the developing countries, especially in the tropics. This situation concerns all of us because these countries are the guardians of biological resources that may be vital to all our futures. Unfortunately, where political and economic systems fail to provide people with land, jobs, and food, disenfranchised citizens turn to legally protected lands, plants, and animals for their needs. Immediate human survival needs always will take precedence over long-term environmental goals. Clearly the struggle to save species and unique ecosystems cannot be divorced from the broader struggle to achieve a social order or global homeostasis in which the basic needs of all are met.

At the 1982 World Congress on National Parks on the island of Bali, Indonesia, five hundred scientists, managers, and politicians discussed the design and location of biological reserves and the ecological, economic, and social factors that impinge on wildlife preservation. They concluded that conservation and rural development are not necessarily incompatible. In many cases, sustainable production of food, fiber, medicines, and water in rural areas *depends on ecosystem services derived from adjacent conservation reserves.* Tourism associated with wildlife watching and outdoor recreation can be a welcome source of income for underdeveloped countries. If local people share in the benefits of saving wildlife, they probably will cooperate and programs will be successful. To paraphrase Thoreau's famous dictum: "In broadly shared economic progress is preservation of the world."

TABLE 12.11	Landless and near-landless* rural households in selected Asian and Latin American countries	
Country	Number of landless households (millions)	Percent of all rural households
Asia		
Bangladesh	11.85	75
India	86.00	53
Java, Indonesia	9.39	85
Philippines	4.43	78
Sri Lanka	1.89	77
Latin America		
Bolivia	0.61	85
Brazil	9.72	70
Colombia	2.40	66
Ecuador	0.86	75
El Salvador	0.53	80
Guatemala	0.66	85
Mexico	4.50	60
Peru	1.48	75

*Those who do not have enough land to support a family, usually less than 0.5 hectares.

Source: Data from World Resource Institute 1987.

FIGURE 12.15 Landless rural peasants still work most of the land in Asia, Latin America, and Africa while they live in poverty.

LAND OWNERSHIP AND LAND REFORM

Patterns of land ownership and control have shaped human relationships in nearly every culture. Some land tenure patterns have supported or even created inequality, ignorance, and social stagnation, while other systems of land use have promoted equality, liberty, and progress.

Who Owns How Much?

Whatever political and economic system prevails in a given area, the largest landowners usually wield the most power and reap the most benefits, while the landless and near landless suffer at the bottom of the scale. This is a hard fact of life. The World Bank estimates that nearly 800 million people live in "absolute poverty . . . at the very margin of existence." Three-fourths of these wretched people are rural poor who have too little land to support themselves (table 12.11). Most of the other 200 million absolute poor are urban slum dwellers, many of whom migrated to the city after being forced off the land in their rural homes (chapter 25).

In many countries inequitable land ownership is a legacy of colonial estate systems on which landless peasants still work in virtual serfdom (figure 12.15). For instance, the United Nations reports that only 7 percent of the landowners in Latin America own or control 93 percent of the productive agricul-

tural land. Table 12.11 lists some of the countries with the highest percentages of rural poor. By far the largest share of landless poor in the world are in South Asia. The 86 million landless households in India comprise about 500 million people; they are two-thirds of both the total population of India and of all the landless rural people in the world. Notice that Brazil has by far the largest number of landless rural people in South America. Not surprisingly, the countries with the widest disparity in wealth are often the most troubled by social unrest and political instability. As Adam Smith said in *The Wealth of Nations* in 1776, "No society can surely be flourishing and happy, of which the far greater part of the members are poor and miserable."

Land tenure is not just a question of social justice and human dignity. Political, economic, and ecological side effects of inequitable land distribution affect those of us far from the immediate location of the problem. For example, people forced from their homes by increasing farm mechanization and cash crop production in developing countries often are forced to try to make a living on marginal land that should not be cultivated. The local ecological damage caused by their struggle for land can, collectively, have a global impact. The contribution of concentrated land ownership to these environmental problems receives less attention than do the threatening ecological trends themselves.

Throughout history, some land reform movements have been successful and others have not. In many countries, significant land reforms have occurred only after violent revolutionary movements or civil wars. An important factor in revolutions in Cuba, Mexico, China, Peru, the Soviet Union, and Nicaragua was the breaking of feudal or colonial land-ownership patterns. This struggle still continues in many countries. Hundreds of large and small peasant movements demand land redistribution, economic security, and political autonomy in Latin America, Africa, and Asia. Entrenched powers respond with arrests, torture, open warfare, and "disappearances." Millions of people have been killed in such struggles in recent centuries alone.

Consider Brazil: 340 rich landowners possess 46.8 million ha (117 million acres) of cropland, while 9.7 million families (70 percent of all rural people) are landless or nearly landless. Studies have shown that 13 percent of the land on the biggest estates is completely idle, while another 76 percent is unimproved pasture. When President José Sarney took over the government in 1985 after twenty-one years of right-wing military rule, he ordered sweeping land reform to redistribute land to the rural poor. Although the government has offered to pay for expropriated land, the large landowners are resisting and have spent millions of dollars to raise and arm private armies to protect the *status quo*.

In some countries, land reform has been relatively peaceful and also successful. In Taiwan, only 33 percent of the farm families owned the land they worked in 1949, whereas 59 percent of rural families farm their own land today. South Korea also has carried out sweeping land redistribution. In the 1950s, more than half of all farmers were landless, but now 90 percent of all South Korean farmers own at least part of the land they till. In general, those countries that have had successful land reform also have stabilized population growth (chapter 7) and have begun industrialization and development (chapter 8).

What are the ecological implications of land ownership? Absentee landlords have little personal contact with the land, may not know or care about what happens to it, and often won't let sharecroppers cultivate the same land from year to year for fear that they may lay claim to it. This gives tenant farmers little incentive to protect or improve the land because they can't count on reaping any long-term benefits. In fact, if tenants do improve the land or increase their yields, their rents may be raised. Also, where land is aggregated into collective farms or large estates worked by landless peasants, productivity and health of the soil tend to suffer.

Far from being a costly concession to the idea of equality, land reform often can provide a key to agricultural modernization. In many countries, the economic case for land reform rivals the social case for redistributive policies. Many studies have shown that the productivity of owner-operated farms is significantly higher than that of corporate or absentee land owner farms. In-

FIGURE 12.16 Native tribal people, such as these Yanomano from Brazil, are threatened by tropical forest destruction. These people have lived in harmony with nature for thousands of years. If their culture is lost, valuable knowledge about the forest will be lost as well.

dependent farmers, especially those who have only a few hectares to grow crops, tend to lavish a great deal of effort and attention on their small plots, and their yields per hectare often are twice those of larger farms.

Resettlement Projects

Mass movements of people into new settlement areas often have serious adverse environmental consequences. In the 1960s, the Soviet Union tried to open new lands in Siberia for grain farming. Much of this land was too dry and too cold for agriculture, and crop failures, environmental damage, and human hardship were widespread.

Several current transmigration projects are causing large-scale tropical forest destruction and disruption of indigenous cultures. In Brazil, for instance, the Polonoroeste project, financed in part by development funds from the World Bank (chapter 8), built a road 1,500 km long into the western Amazonian provinces of Matto Grosso and Rondonia. The immigration and jungle clearing stimulated by this road could result in destruction of an area of virgin tropical rain forest about the size of California before the end of this century. About thirty-four Indian tribes, some that have never had contact with the Western world, are being eliminated along with the forests. As usual, social disruption and ecological damage go hand in hand (figure 12.16).

S U M M A R Y

Land has traditionally been the source of wealth and power. The ways we use this limited resource shape our lives and futures. About one-third of Earth's surface is too inhospitable for agriculture, livestock, or forestry, but is vitally important for such purposes as wilderness preservation and recreation. We grow crops on about 10 percent of the total land area. With proper preparation, we could expand cropland in some areas. Most of Earth's land, however, is inappropriate for agriculture and is ruined by attempts to cultivate it.

Forests and woodlands occupy about 28 percent of Earth's total land area. Forest resources are in great demand. Lumber and other wood products are a foundation of our modern economy. More than half the people in the world depend on firewood or charcoal for heating and cooking. Our valuable forests are being converted to cropland at an increasing rate.

Northern forests are growing faster in most areas of the world than they are being cut and seem in little danger of being exhausted. Tropical forests, on the other hand, are in critical danger. Irreplaceable ecosystems that are home to as many as half of all biological species are being destroyed. Quick profits encourage this exploitation, but hidden costs, such as lost wildlife habitat, erosion and devaluation of exposed land, and other disastrous environmental damage will follow from small short-term gains.

Twenty-two percent of Earth's land is used as range and pasture. Three billion grazing animals convert roughage to protein on poor land that could not otherwise be used to produce food for humans. When herds are managed properly, they actually can improve the quality of their pasture. Unfortunately, about one-third of the world's rangelands are degraded by overgrazing, with disastrous environmental consequences similar to forest destruction.

Some land should be preserved in its natural state, safe from human exploitation. In recent years, we have made heartening progress establishing parks, wildlife refuges, wilderness areas, and other types of nature preserves. Biological diversity can be preserved and studied in these areas, while allowing people to experience a part of the unspoiled natural world. Heated debate continues over which areas most need protection, how much land is needed in each preserve, and how to manage such land. The world conservation strategy of the ICUN lists three objectives in preserving natural areas: (1) to maintain ecological life-support processes, (2) to preserve genetic diversity, and (3) to ensure sustainable utilization of living resources.

Patterns of land ownership dictate our patterns of land use. Land reform is essential in developing a sustainable land-use plan for a healthy future. Inequitable land ownership so common today in much of the world forces the poor to use land unsuited to agriculture, while good land is monopolized by the rich. Dividing land more fairly could increase agricultural productivity and sustainability of the land because farmers who own their own plot generally use their land more efficiently and more carefully than absentee landlords.

Review Questions

1. Which type of land use occupies the greatest land area?
2. What continent has the highest density of people?
3. What are the advantages and disadvantages of monoculture agroforestry?
4. Describe milpa (or swidden) agriculture. Why are these techniques better for fragile rain forest ecosystems than other types of agriculture?
5. What are some results of deforestation?
6. How does overgrazing encourage undesirable forage species to flourish?
7. Why can grazing animals generate food on land that would otherwise be unusable?
8. Why do Americans prefer to raise corn-fed cows in feedlots instead of grazing them in fields?
9. What are the three objectives of the conservation strategy proposed by the International Union for the Conservation of Nature?
10. What is the relationship between fair land distribution and appropriate land use?

Questions for Critical Thinking

1. Some forest and rangeland would be suitable as cropland. Do you think it should be converted? Why or why not?
2. What could we do to reduce or redirect the demand for wood products?
3. Brazil needs cash to pay increasing foreign debts and to fund needed economic growth. Why shouldn't it harvest its forests and mineral resources to gain the foreign currency it wants and needs? If we want Brazil to save its forests, what can or should we do to encourage conservation?
4. Why do you suppose that biomes such as the desert and the arctic are so well protected in parks, while biomes in more temperate climates remain less well preserved?
5. Five thousand landless peasants build a squatter settlement and begin mining for gold in officially protected Brazilian rain forest. Should we force them back to the slums of Rio de Janeiro?
6. Do you think milpa (or swidden) agriculture could be adopted on a large scale for profit instead of small scale subsistence farming? Should such a plan be adopted?
7. Suppose American or Canadian oil companies have discovered a massive deposit of oil near an island in the North Sea that happens to be the only breeding grounds for a rare seal species. Drilling for this precious oil may endanger, or even destroy, this species. Suppose further that we are experiencing an energy crisis and that gasoline prices are rising rapidly. At what price would you favor drilling in the North Sea?

Key Terms

closed canopy (p. 220)	open canopy forest (p. 220)
croplands (p. 218)	open range (p. 230)
desertification (p. 231)	pasture (p. 230)
forest management (p. 222)	land rehabilitation (p. 228)
fuelwood (p. 221)	ecosystem restoration
industrial timber (p. 220)	(p. 228)
milpa agriculture (p. 225)	swidden agriculture (p. 225)
mixed perennial polyculture	woodland (p. 220)
(p. 225)	world conservation strategy
monoculture agroforestry	(p. 234)
(p. 223)	

SUGGESTED READINGS

Bandyopadhyay, J., and V. Shiva. 1987. "Chipko: Rekindling India's Forest Culture." *The Ecologist.* 17(1):26. A description of the peasant forest protection movement in India.

Cartwright, J. "Conserving Nature, Decreasing Debt." *Third World Quarterly.* May 1989.

Caufield, C. *In the Rainforest.* New York: Alfred A. Knopf, 1985. A highly readable description of world rain forests and their destruction.

Colchester, M., et al. 1986. "Banking on Disaster: International Support for Transmigration." *The Ecologist.* 16(2/3):71. A special double issue on the ecological, social, and economic effects of the transmigration in Indonesia. See update by Colchester in *The Ecologist.* 17(1)35.

Douglas, J., and R. A. Hart. *Forest Farming: Towards a Solution to Problems of World Hunger and Conservation.* Boulder: Westview Press, 1985. A description of agrosilvaculture, or tree farming, based on ecological principles, appropriate technology, and local community involvement.

Franke, R. *Seeds of Famine: Ecological Destruction and the Development Dilemma in the West African Sahel.* Totowa: Rowman & Allanheld, 1980. A description of how transformation of agricultural land to cash crops has pushed subsistence farmers and pastoralists onto marginal land in the Sahel that cannot support them in bad years.

George, Susan. *Debt: The Profit of Doom.* San Francisco: The Institute for Food and Development Policy, 1988. An insightful analysis of the relation between international debt, ecological destruction, and social injustice.

Hecht, S., and A. Cockburn. *Fate of the Forest.* New York: Harper Collins, 1991. Traces European exploitation of the Amazon starting with the rubber boom of the last century and continuing to the present. Argues that only "socialist ecology" can save the forest.

Marshall, G. 1990. "The Political Economy of Logging: A Case Study in Corruption." *The Ecologist* 20(5):174. Examines logging practices and corruption in Papau, New Guinea.

Medea, B., and R. Buell. *The Coalition of Ejidos of the Valleys of Yaqui and Mayo, Sonora State, Mexico.* San Francisco: Institute for Food and Development Policy, 1985. A description of traditional nonauthoritarian collective agriculture in Mexico.

Nations, J. D., and D. I. Komer. 1983. "Rainforests and the Hamburger Society." *Environment.* 25(3):12. An outspoken indictment of the destructive farming and land management practices in Central America.

Repetto, R. 1990. "Deforestation in the Tropics." Scientific American 262(4):36. Discusses policies to encourage forest preservation and to stop destruction of an irreplaceable resource.

Revkin, A. *The Burning Season: The Murder of Chico Mendez and the Fight for the Amazon Rain Forest.* New York: Houghton Mifflin, 1991. An excellent account of the complex fight to save the rain forest and of the rise to prominence and martyrdom of Chico Mendez.

Shane, D. R. *Hoofprints on the Forest; Cattle Ranching and the Destruction of Latin America's Tropical Forest.* Philadelphia: Institute for the Study of Human Issues, 1986. A vivid account of the role that cattle ranching has played in the destruction of tropical forests in Latin America.

Shoumatoff, A. *The World is Burning: Murder in the Rain Forest.* New York: Little Brown, 1991. A frenzied and rather self-centered account of the reporter's personal travels in Brazil. Captures some of the flavor of the rain forest and its inhabitants.

Spears, J., and E. S. Ayensu. 1985. "Resources, Development and the New Century: Forestry." In *Global Possible*, edited by R. Repetto. New Haven: Yale University Press.

Stone, R. D. *Dreams of Amazonia.* New York: Viking-Penguin, Inc., 1985. A travelogue of local color, sights, sounds, and smells of the world's largest rain forest written by a former reporter and present officer of the World Wildlife Fund.

Sutton, S. L., T. C. Whitmore, and A. C. Chadwick, eds. *Tropical Rain Forest: Ecology and Management.* London: Blackwell Scientific Publications, 1983. Research papers and reviews from a symposium on the biology and politics of tropical rain forests.

Tropical Forests: A Call for Action. Part I. 1985. Washington, D.C.: World Resources Institute. Report of an International Task Force of the World Resources Institute, the World Bank, and the United Nations Development Program.

United Nations Food and Agriculture Organization (FAO). 1984. *Land, Food, and People.* Rome: FAO. A useful analysis of the relation between land use, land tenure, food production, and poverty around the world.

American Forestry Association. (November/December 1988) "Tropical Deforestation." *American Forests* 94(11–12). A special double issue of *American Forests* devoted entirely to analyses of causes, effects, and solutions for tropical deforestation.

CHAPTER *13*

U.S. Land Resources: Forests, Rangelands, and Wilderness

C O N C E P T S

The abundance of resources within the boundaries of the United States has played a major role in this country's growth. Early homesteading practices transferred land from public to private ownership, reinforcing democratic ideals and opening the West to development. The resulting destruction of resources, however, brought about a conservation movement that was divided between those advocating the preservation of nature for altruistic reasons and those arguing for a utilitarian use of resources. Conservation efforts moved from setting aside land for reserves and wildlife refuges to attempting to undo the damage of frontier practices.

Current environment protection focuses on long-term management designed to protect and maintain the environment for its aesthetic value and practical use. In the actual practice of managing forests, rangelands, national parks, and wilderness areas, results have been criticized by both sides of the environmental debate. Conservationists argue that not enough land is being preserved in its natural state, while developers claim that too much land is being protected from development.

The National Forest Service manages roughly one-third of all forest lands. Controversy surrounds many of its management policies concerning forest-harvesting techniques, road building, reforestation, herbicide and pesticide use, preservation of virgin and old-growth timber, and budget priorities. Public rangelands are controlled by the Bureau of Land Management (BLM). Much of this land is in poor condition because of overgrazing, erosion, and the loss of native forage species.

The idea of preserving national parks in a wilderness state as a refuge for wildlife and a pleasuring ground for ordinary citizens is a unique American contribution to land-use management. Unfortunately, in most parks, entertainment value has priority over keeping natural features and wildlife intact. The practice of catering to tourists has resulted in an excess of lodges, roads, and concession stands, as well as pollution, the destruction of quiet and solitude, and unsustainable management policies.

The American wilderness, which symbolizes strength, beauty, and challenge, has strongly influenced our culture and character. By the beginning of the twentieth century, as conservationists recognized the eventuality of a destroyed wilderness, the challenge to conquer the wilderness became a challenge to save it. This movement culminated in the 1964 Wilderness Act, which defined the nature of areas to be set aside for government protection. The U.S. Forest Service and the BLM were designated to evaluate which areas fit the wilderness definition and submit proposals to Congress. Designation and management of our remaining wilderness, however, are still vulnerable to political and economic challenges and require constant debate and affirmation.

Be it resolved that none of us know, or care to know, anything about grasses outside of the fact that for the present there are lots of them and we are after getting the most of them while they last.

Colorado Cattlemen's Association, 1900

INTRODUCTION

The first Europeans to arrive in America found a continent rich in resources. A magnificent, dense forest of conifers and hardwoods covered nearly half the continent. A vast empire of grass rolled across the Great Plains. Rich soil, a temperate climate, and ample rainfall favored agriculture in many areas (figure 13.1). Mineral and fossil fuel deposits were abundantly available for future industrial development. These resources have played a large role in making the United States rich and powerful. To a large extent, access to and development of natural resources has been determined by land-use and land management policies. An introduction to the policies and practices that are part of our heritage will help you understand some of the intricacies of how lands are managed and by whom. This information is an important key to effective citizen participation in environmental issues and their resolutions.

In this chapter, we will study land use in the United States, particularly forests, grazing lands, parks, and wilderness areas. Agricultural lands are discussed in chapter 11, wilderness refuges are discussed in chapter 15, and urban land use is discussed in chapters 25 and 26. Here we will focus on the management policies governing public land, and we will look at the agencies managing federal lands.

A BRIEF HISTORY OF U.S. LAND OWNERSHIP AND USE

Land-use patterns in the United States are an integral part of our history and show an evolution of attitudes about ownership and stewardship of resources.

The Tribal Era

Approximately 5 million native tribal people lived in North America before European settlers arrived. They shared the land communally, and their hunting, gathering, and agriculture had relatively low impact—for the most part—on their surroundings. As settlers moved west, the native people lost more and more land; they now own about 4 percent of their original territory (figure 13.2). Modern technology has made it possible to support a population at least fifty times larger than the original indigenous peoples, but many human and environmental values have been lost in the process.

The Frontier Era

For most of its first century of existence, the United States followed a policy of transferring land and natural resources from the public domain to private ownership as quickly as possible. Settling the land, exploiting resources, taming the frontier, and building towns and roads were regarded not only as means of personal enrichment, but also a public service. Although there were many abuses in this system, policies that encouraged land

FIGURE 13.1 The ways we use and abuse the land shape our culture, ourselves, and our environment.

ownership by ordinary citizens probably did more than anything else in our history to create democracy. The decision by our founding fathers to distribute public lands to independent family farms was a radical departure from the European pattern of large estates worked by tenant farmers and landless laborers.

Numerous important legislative acts have shaped our national land use policy over the past two centuries (table 13.1). One of the most important of these was the **Homestead Act** of 1862, which allowed any citizen (or applicant for citizenship) who was over twenty-one and head of a family to acquire 160 acres of public land by living on it and cultivating it for five years (figure 13.3).

Although these land disposal policies had many positive social and economic benefits, they also set off a frenzy of land speculation and fraud. Some of the laws that opened the door to outrageous plundering of the public domain included the Swamp Lands Act of 1850, the Minerals and Mining Act of 1866, the Desert Lands Act of 1877, and the Timber and Stone Act of 1878. Corrupt politicians and compliant land office agents sold land to cronies for as little as ten cents per acre. Good farmland was claimed as worthless swamp by men who swore they had crossed it in a boat—neglecting to mention that the boat was in the back of a wagon. Logging companies hired "entrymen" by the trainload to file for good forestlands as homesteads or mining claims. A glass of whiskey and a few dollars bought the timber rights on these claims. After the trees were stripped off, the land was abandoned.

The Conservation Era

By 1900, almost half the country had been given away, sold, stolen, or otherwise transferred to private ownership. Throughout the United States, ruthless and wasteful practices of the pioneer era left much of the land worn out and desolate. Forests had been stripped and burned, leaving vast graveyards of blackened stumps and worthless brush. Poor farming practices (chapter 11)

RESOURCES AND RESOURCE ECONOMICS

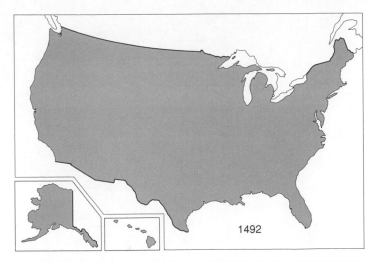

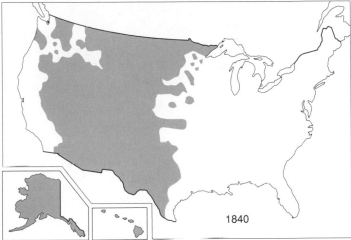

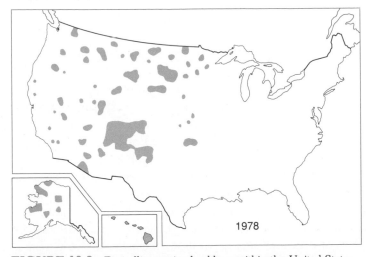

FIGURE 13.2 Dwindling native land base within the United States. Tribal people once owned all the land but now have only 4 percent.

had caused erosion and depleted the soil. Rivers and harbors were clogged with debris. Floods devastated towns, such as Pittsburgh and Johnstown, Pennsylvania.

A few voices were raised in protest. One of the first and most important of these early conservationists was George Per-

TABLE 13.1 Some major U.S. land-use decisions

Date	Name of act	Effects
1785	Land Ordinance	Sold land for $1/acre; opened up West to development
1841	State Land Grants	Provided funds for public education and improvements
1862	Homestead Act	Gave free land to settlers
1872	Yellowstone National Park	World's first national park; still the largest in the United States
1891	General Revision	Established forest preserves
1902	Reclamation	Bureau of Reclamation provides cheap water and power
1916	National Park "Organic" act.	Established National Park Service
1934	Taylor Grazing	Closed to homesteading most of remaining public domain. Established Grazing Service.
1956	Public Highways	Started Interstate Highway System
1960	Forest Multiple Use	Requires sustained yield and uses other than timber
1964	Wilderness	Established National Wilderness System
1968	Wild and Scenic Rivers	Named first Wild and Scenic Rivers
1971	Alaska Native Claims	Grants 18 million ha (44 million acres) to native corporations
1972	Coastal Zone Management	Requires regulation of coastal development
1976	Federal Lands Policy and Management (FLPMA)	Closed public domain; many management provisions require wilderness review of BLM lands (often called RARE II)
1977	Surface Mining	Regulates strip mining, requires reclamation
1980	Alaska National Interest Conservation Lands (ANICLA)	Preserved 44 million ha (109 million acres) in parks, refuges, and wilderness areas

kins Marsh, whose 1864 book, *Man and Nature*, has been called the fountainhead of the American conservation movement. Marsh pointed out the effects of upsetting the balance of nature and called for an end to wanton destruction and profligate waste of resources. Largely as a result of his book, national forest reserves were established in 1873 to protect dwindling timber supplies and endangered watersheds.

Among those influenced by Marsh's observations and predictions were President Theodore Roosevelt and his chief conservation advisor, Gifford Pinchot. Teddy Roosevelt did more than any other president to conserve resources (figure 13.4). During his eight years in office (1901 to 1909), he withdrew approximately 93 million ha (230 million acres) of timberland, range, and mineral and energy reserves from the public domain to save them from exploitation and waste. Using his power of presidential proclamation, he quadrupled the existing forest pre-

FIGURE 13.3 Homesteaders in front of their log cabin in Utah in 1870. Low-cost land made it possible for working class people to acquire their own farms.

FIGURE 13.4 Teddy Roosevelt (*center*) in Yosemite National Park in 1903. John Muir is on Roosevelt's left; Gifford Pinchot is behind Roosevelt.

serves and established the first wildlife refuges (more than fifty) and five new parks, including Grand Canyon and Olympic National Parks.

A champion of the populist, progressive movement, Roosevelt established the Bureau of Reclamation and put management of the public domain on an honest, professional, and scientific basis for the first time in our history. In 1905, when he made the Forest Service a separate agency in the Agriculture Department, Roosevelt appointed Gifford Pinchot to be its first chief. In 1908, Pinchot organized and chaired the White House Conference on Natural Resources, perhaps the most prestigious and influential American environmental meeting ever held.

Conservation or Preservation?

The basis of Roosevelt's and Pinchot's policies was pragmatic **utilitarian conservation.** They argued that the purpose of saving forest was "not because they are beautiful or because they shelter wild creatures of the wilderness, but only to provide homes and jobs for people." They fought against waste and corruption, and for scientific, modern, professional management of resources "for the greatest good for the greatest number for the longest time." These principles continue to guide the U.S. Forest Service today.

John Muir (see figure 13.4), author and first president of the Sierra Club, strenuously opposed Pinchot's influence and policies. Muir's philosophy has been called **altruistic preservation** because he argued that nature deserves to exist for its own sake, whether or not it is of use to us. He stressed aesthetic and spiritual values in pleading for protection of nature. The National Park Service, an agency of the U.S. Department of the Interior, was organized in 1916 by John Muir's disciple, Stephen Mather, and has always been oriented toward nature preservation. It has often been at odds with Pinchot's utilitarian Forest Service.

The Land Management Era

In the last half of this century, public policy has turned from simply trying to protect and preserve land and resources to long-term management for the public good. One of the most important laws of this era is the National Forest Multiple Use and Sustained Yield Act (1960), which mandates that forests be managed for **multiple use** (many different uses that occur simultaneously) and **sustained yield** (a continuing production of wood and other goods and services on a long-term basis). The implementation of this policy has been controversial. Critics point out that some uses, such as wilderness recreation and strip mining, are simply incompatible. The lowest, least sensitive use drives out the higher, more sensitive uses. Sustainable yield, which is a good concept, can place too much emphasis on economic benefits while neglecting biological diversity and intangible values, such as beauty, historic significance, and ethical considerations.

Another important step in environmental protection was the **Wilderness Act** (1964), which recognized that leaving land in its natural state may be the highest and best use in some areas. A number of amendments and additions to this act have ex-

FIGURE 13.5 Winter storms have eroded the beach and undermined the foundations of homes on this barrier island. Breaking through protective dunes to build such houses damages sensitive plant communities and exposes the whole island to storm and erosion. Coastal zone management attempts to limit development on fragile sites.

panded and improved the national wilderness system and have protected especially valuable wild and scenic rivers.

The **Coastal Zone Management Act** (1972) was another landmark in land-use planning. It gave federal money to thirty seacoast and Great Lakes states for development and restoration projects, contingent on satisfactory state performance in regulation of coastal development. This act has protected fragile coastlines, scenic areas, wetlands, estuaries, and barrier dunes. It also has raised a great deal of controversy over issues of state's rights, private property, and the power of the federal government to regulate land use (figure 13.5).

A National Land-Use Policy

Many environmentalists believe that a comprehensive national land-use policy would be the best way to ensure rational, fair, and uniform management of land resources. Legislation to establish such a law was introduced in Congress every year between 1968 and 1976 and was close to enactment in 1974 when it was passed by the Senate but defeated by four votes in the House. During the following years, however, when the country was involved with Vietnam and Watergate, interest in environmental protection began to wane. Many of the provisions sought by supporters of a comprehensive bill have been incorporated into other laws, and the idea of a national land-use policy seems to have been abandoned for the present.

CURRENT LAND USE

How do we use the 9 million sq km (3.5 million sq mi) of land in the United States? Forests still occupy 31 percent of the country, and another 30 percent is grazing and pasture. Farm fields occupy nearly 20 percent of U.S. land, approximately twice the proportion in the world overall. The abundance and fertility of this farmland has played an important role in the economic strength of the United States and in its ability to feed its own

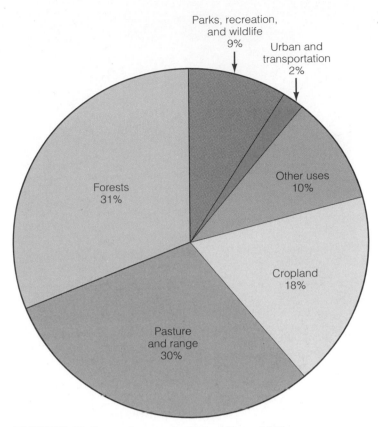

FIGURE 13.6 Land use in the United States, 1990.

people abundantly and cheaply. Nine percent of our land has been reserved as parks, wildlife refuges, and wilderness areas. Although only 2 percent of the land is used for urban housing, manufacturing, and shopping centers, more than 80 percent of Americans live and work in urban areas (figure 13.6).

Land Ownership

The current value of all land in the United States is about $1.4 trillion (1988 resale value), approximately 15 percent of the nation's wealth. Of the 890 million ha (2.2 billion acres) of land in this country, almost all has been owned—at one time or another—by the federal government. Land sales and land grants have distributed most of the public domain to private or corporate ownership. Currently 59 percent of the United States is privately owned, 7 percent belongs to state or local government, and 4 percent has been granted to native peoples as tribal lands or reservations. The federal government still owns some 268 million ha (662 million acres), or about 30 percent of the land (figure 13.7).

Administration of Public Lands

The agencies responsible for managing public domain are identified in table 13.2. By far the largest area is assigned to the Department of the Interior, which includes the Bureau of Land Management (BLM), the Fish and Wildlife Service (FWS), the

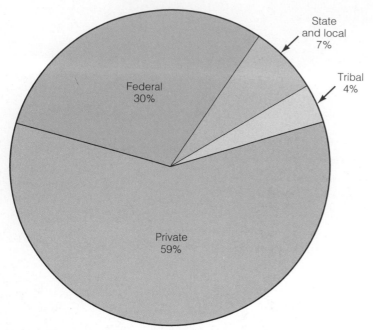

FIGURE 13.7 Land ownership in the United States. Most land is now in private hands. The federal government owns one-third of all land, mostly in the West.

TABLE 13.2 Administration of U.S. federal lands

Agency	Area (millions of hectares)	Administrative responsibilities
Department of Agriculture Forest Service	74.6	National Forests
Department of Defense Army, Navy, Air Force	10	Forts, bases, shipyards, bombing ranges
Army Corps of Engineers	5	Flood control and navigation projects
Department of Energy and Tennessee Valley Authority (TVA)	1.3	Flood control, navigation, and hydroelectric power projects
Department of the Interior Bureau of Land Management	111.5	National Interest Lands
Fish and Wildlife Service	34	Wildlife Refuge System
National Park Service	28	Parks, Monuments, Wild Rivers, Historic Sites
Bureau of Reclamation	1.7	Irrigation, grazing, energy, and flood-control projects
Other Agencies, including Department of Transportation, National Aeronautical and Space Administration, Housing and Human Services	1.4	Miscellaneous
Total	**267.5**	

Source: U.S. Statistical Abstracts 1990.

National Park Service (NPS), and the Bureau of Reclamation (BuRec). Altogether, these agencies manage 175 million ha (432 million acres).

Where are these federal lands located? Ask a westerner! More than 85 percent of all federal land is in the twelve states lying across or west of the Rocky Mountains (figure 13.8). Public lands make up 60 percent of the total area, on average, in these western states. By contrast, no state east of Colorado has more than twelve percent. New York, Iowa, Maine, Connecticut, and Rhode Island each have less than one percent. Many westerners feel that federal control of lands unfairly restricts their right to economic development, private ownership of property, and recreational opportunities. These concerns were expressed in the "**Sagebrush Rebellion**," a coalition of cattlemen, miners, loggers, developers, farmers, politicians, and others who wanted to see more local control over land management and natural resources. Members of this group played an important role in the Reagan administration in the 1980s and reversed many of the conservation practices established in the previous eighty years.

FORESTS

Much of the original forest that once existed in North America was cut down so that land could be converted to farm fields. Millions of hectares of croplands in New England, Appalachia, and the Great Lakes states that were cleared with great effort in the nineteenth century have since been abandoned and are returning to forest. Loggers cut enormous quantities of lumber during the frontier era, often using incredibly wasteful and de-

structive techniques (figure 13.9). The land they left behind often regrew to "inferior scrub and brush," actually early stages in secondary succession. Large tree species have gradually replaced the volunteer growth, and some of those forests are once again becoming commercially useful. In many areas, however, the seed supply for the most valuable tree species was eliminated, and the grand primeval forest will probably never be restored.

Approximately half of America's native forestlands still have productive forests, however, and they make the United States the world's largest producer of wood and wood products. The annual harvest is some 485 million cubic meters, more than all produced in Africa or South America put together. Most of this harvest is industrial roundwood (lumber, paneling, etc.), but about 100 million cubic meters is used for charcoal or firewood. Another 64 million cubic meters is used to make paper pulp—nearly one-third the total world supply. The United States is also the world's largest net exporter of wood and wood products (chapter 12). While these exports help balance our trade deficit, much of our trade is in raw logs or wood chips. We could save both jobs and exchange currency by manufacturing finished products from the wood before selling it.

RESOURCES AND RESOURCE ECONOMICS

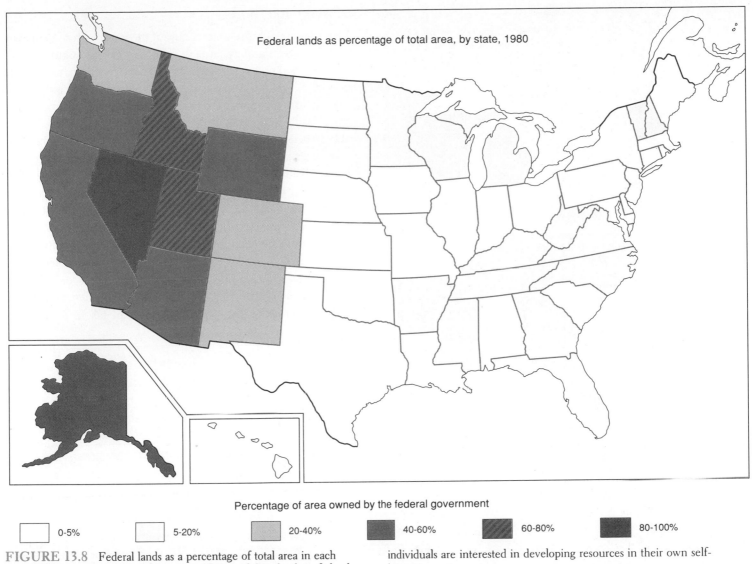

Federal lands as percentage of total area, by state, 1980

Percentage of area owned by the federal government

| | 0-5% | | 5-20% | | 20-40% | | 40-60% | | 60-80% | | 80-100% |

FIGURE 13.8 Federal lands as a percentage of total area in each state. Western states have a far larger proportion of their lands in federal ownership than do eastern states. This brings benefits in terms of federal management programs and support funds, but it also causes anger when individuals are interested in developing resources in their own self-interest.

Source: Bureau of Land Management—Department of the Interior.

Forest Distribution and Administration

The U.S. Forest Service (USFS) now manages 171 national forests (figure 13.10) and 18 national grasslands, totaling some 75 million ha (185 million acres). It has 39,000 employees and an annual budget of about $2 billion. Although the Forest Service owns nearly one-third of all forestlands in the United States, only 23 percent of the commercial timber harvest comes from its forests (figure 13.11). This is partly because most of the best timber land was transferred to private ownership during the frontier era, and partly because private forests are more intensively managed than public lands.

The Forest Service is at the center of debates over how public forests should be used. About 30 million ha (80 million acres) of the national forests are designated as wilderness, and 41 percent of all recreation on federal land occurs in the forests, twice as much as in national parks. The management policies of the USFS are criticized by both developers and preservationists. Loggers feel that too much timber is "locked up" in wilderness, and regulations on their activities are too restrictive. Some environmentalists believe that not enough land is preserved as wilderness—particularly areas that would sustain long-term damage or are unique in some ways—and that the USFS is little more than a welfare agency for logging companies.

Some of the more controversial issues in current forest management include harvest methods, road building, reforestation rates, herbicide and pesticide spraying, cutting of very old tree stands, and budget priorities of the federal government. Let's look more closely at each of these issues.

FIGURE 13.9 Early logging practices were highly wasteful and destructive. Billions of board feet of lumber were stripped off the land to build the burgeoning cities of America. Within a few decades, the once vast coniferous forests of pine in New England and the Great Lakes states had fallen to the logger's ax. The logging companies moved Westward to the lush forests of the Pacific Northwest.

Harvest Methods

Loggers generally prefer to **clear-cut** the forest—that is, cut every tree in a given area regardless of species or size (figure 13.12). This method makes it possible to use large machines to fell, trim, and haul logs, but it often wastes many small trees and can disrupt natural regeneration. Because it disrupts protective ground cover and gouges ruts in the soil, clear-cutting exposes vulnerable slopes to erosion. It also destroys the biological community and drives out wildlife and noncommercial plant species. Forests of successional species, such as aspen, jack pine, and lodgepole pine, often respond well to clear-cutting if the blocks harvested are less than 5 ha (12 acres), but loggers usually find such small cuts uneconomical. The lush Douglas fir forests of the rainy Pacific Coast Range also regenerate quickly after clear-cutting. The problem is, however, that many of these forests are on steep slopes where erosion is a serious problem.

An alternative harvest method is **selective cutting,** in which only mature trees of certain species are taken. This method

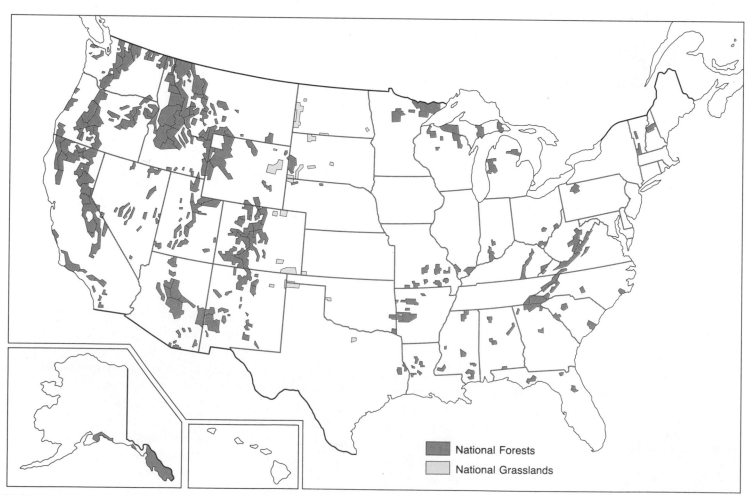

FIGURE 13.10 Distribution of lands administered by the U.S. Department of Agriculture, especially the U.S. Forest Service

Source: U.S. Department of Agriculture.

RESOURCES AND RESOURCE ECONOMICS

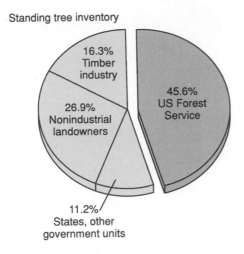

Standing tree inventory

16.3% Timber industry

26.9% Nonindustrial landowners

45.6% US Forest Service

11.2% States, other government units

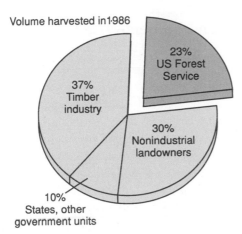

Volume harvested in 1986

23% US Forest Service

37% Timber industry

30% Nonindustrial landowners

10% States, other government units

FIGURE 13.11 The U.S. Forest Service controls more timber land (*top*) but produces less timber (*bottom*) than private and corporate landowners.

Source: U.S. Forest Service.

causes much less disruption to the forest, preserves species and habitat diversity, and can produce sustained and even yields for many years. Productivity in a mixed forest may be lower, however, than in a monoculture, and this form of harvesting is more expensive to administer and operate. Selective cutting can cause environmental problems because it requires more roads and more frequent intrusion into the forest by loggers. An alternative to both clear cutting and selective harvest has been advocated by some foresters and ecologists. They call it partial cutting or "sloppy clear cutting."

This new forestry mimics how forests recover from natural disasters. Logs and debris are left on site to protect soil and provide nutrients. Twenty to seventy percent of the living trees are left standing including some huge specimens that provide shade and shelter for wild life. Natural regeneration is allowed to occur rather than controlled burning, rock-raking, herbicide applications, and replanting as is the case on clear cuts. These techniques have been successfully tested on the Siskiyou and

FIGURE 13.12 Clear-cutting and road cutting on steep, unstable slopes expose soil to erosion, damage watersheds, displace wildlife, and make forest regeneration difficult, if not impossible.

Willamette National Forests in Oregon and will be extended to other forests in the future.

Roads and Trails

Since World War II, the USFS has engaged in a massive program of road building. Logging roads have been expanded nearly ten-fold in the last forty years to a current system of 547,000 km (340,000 mi). This is ten times the size of the Interstate Highway System. Current plans call for building another 100,000 km (60,000 mi) of roads, mostly into *de facto* wilderness areas. In many cases, building these roads costs more than is obtained from the timber they access. The Forest Service claims it needs roads to "open up" the country for fire control, to treat insect infestations, to distribute hunters, and to use lands that would otherwise "go to waste."

Critics argue that road building is a subterfuge to preempt wilderness designation. They claim that much of the land being exploited is too steep and erodible to support road building and logging. These forests serve better as wildlife habitat, watershed protection, and primitive recreation than for tree production. Wilderness advocates would rather see the funds used for hiking trail restoration. Total trail mileage has decreased by nearly 30 percent, while trail use has more than doubled in the last two decades. Many trails are overused, overcrowded, and badly in need of repair.

Reforestation Rates

Each year about 4 million ha (10 million acres) of U.S. forests are harvested and an equal area is damaged by insects, diseases, and fire (figure 13.13). Less than one-fifth of this area is re-

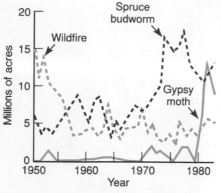

Forest acres damaged, by selected causes

Forest acres planted and seeded

FIGURE 13.13 Forest acres damaged, planted, and seeded, 1950–1982.

Source: U.S. Forest Service.

FIGURE 13.14 Aerial spraying of herbicides and insecticides is a controversial forest management practice. Toxic chemicals often contaminate water supplies, and spray can drift over inhabited areas.

planted or managed for effective natural revegetation. Much of what is replanted resembles other types of monoculture agriculture in both methods and effects. Certain harvesting techniques, such as cutting strips of forest rather than large blocks, facilitate natural reseeding, but they are not often used.

Most of the public forests in the eastern United States are poorly managed, overcrowded, and underproductive by modern forestry standards. The original climax forest that was removed by the logging industry is being replaced by less desirable successional species. According to some estimates, approximately one-third of present forestland is "unstocked" (either bare or growing noncommercial trees). Better management and reforestation could increase our timber production while also sparing wilderness areas and virgin forests from further incursions.

Herbicide and Pesticide Use

Aerial spraying of herbicides and pesticides is widely used to combat forest insect infestations (e.g., gypsy moths and spruce budworms) and to protect young conifers from competition with broad-leaved species (figure 13.14). This spraying has been highly criticized because pesticides often disrupt natural checks on pest populations by killing natural predators and competitors. Even where pesticides succeed in reducing pest numbers, the effect may be to increase the food available for survivors, thus resulting in more or less continuous infestations. Furthermore, there is widespread fear (and some evidence) that toxic chemicals used in forest spraying may have harmful effects on wildlife and humans living in or near the treated areas.

Cutting Old-Growth Timber

Many people believe that the few remaining groves of virgin (never harvested) and old-growth timber in the national forests should be preserved for wildlife habitat, scientific study, and wilderness recreation. Such forests are characterized by complex community structures, niche diversity, and homeostasis. Some endangered species, such as spotted owls and tree voles, and marbled murrelets, live only in mature forests. Logging companies, on the other hand, want to harvest the large trees of the last virgin forests. They resent interference in what they regard as their right to do business. From the foresters' point of view, a young, vigorously growing forest stand is more productive than an "overly mature" stand, which they may consider "a waste." The irony of this reasoning is that the one-time harvesting of these magnificent forest remnants is a substantial public expense, subsidized in the form of management, development, and administrative costs (Box 13.1).

Below-Cost Timber Sale

The Forest Service spends a majority of its efforts on resource exploitation, as can be seen in its budget allocations. The 1990

USFS budget, for example, allocated $600 million for timber harvest and grazing, but only $170 million for all other programs, including wildlife and fish, recreation, soil and water conservation, wilderness management, and research. According to The Wilderness Society, the Forest Service loses an enormous amount of money on below-cost timber sales. The costs of administration, road building, and reforestation on these sales are sometimes one hundred times as much as is paid in "stumpage fees." This means returns as low as one cent per dollar on the federal investment. Losses over the past decade have been about $2.3 billion, an expensive subsidy for the forest industry.

RANGELANDS

Much of the western half of the United States is too dry and/or cold to support forests. A vast sea of grass once covered most of the plains between the Mississippi River and the Rocky Mountains. This region supported huge herds of grazing animals that rivaled the plains of Africa in both wildlife abundance and diversity. Although most native prairie has been degraded or converted to agriculture, the lands west of the one hundredth meridian still represent a major source of rangeland for the United States.

Rangeland Distribution, Ownership, and Management

The United States has approximately 319 million ha (788 million acres) of **rangeland** (grasslands and open woodlands suitable for livestock). Most of this rangeland is in the West, and about 60 percent is privately owned. Of the 120 million cattle and 20 million sheep in the United States, only about 2 percent of the cattle and 10 percent of the sheep graze on public rangelands. Federal lands, thus, are not very important, overall, in livestock production, but they do have important local economic and environmental ramifications. The BLM controls 84 million ha (200 million acres) of grazing lands and the U.S. Forest Service manages about one-fourth as much.

The BLM manages more land than any other agency in the United States, but it is little known outside of the western states where most of its lands are located. It was created in 1946 by a merger of the Grazing Service and the General Land Office. Its formation signaled an end to public land disposal and a commitment to permanent management by the government. The BLM has such a strong inclination toward resource utilization that critics claim the initials really stand for the "bureau of livestock and mining." While only 25 percent of BLM land is considered suitable for grazing, its policies have an important effect on the economy and environment of western states.

State of the Range

The health of most public grazing lands in the United States is not good. Political and economic pressures encourage managers

FIGURE 13.15 A lightly grazed pasture consisting of a dense growth of grasses and small, broad-leaved plants is on the left side of this fence. On the right side is overgrazed rangeland, where the vegetation has become sparse.

to increase grazing allotments beyond the carrying capacity of the range. Lack of enforcement of existing regulations and limited funds for range improvement have resulted in overgrazing, loss of native forage species, and erosion. A 1986 National Resources Defense Council survey concluded that only 30 percent of public rangelands are in "fair" condition, and 55 percent are "poor" or "very poor" (figures 13.15 and 13.16).

Overgrazing has allowed populations of unpalatable or inedible species, such as sage, mesquite, cheatgrass, and cactus, to build up on both public and private rangelands. Furthermore, competing herbivores, such as grasshoppers, jackrabbits, prairie dogs, and feral burros and horses, further damage the range and reduce the available forage. (A **feral** animal is a domestic animal that has taken up a wild existence.) Control programs using poison baits, traps, and hunting to reduce the numbers of these competing native and feral herbivores have been highly controversial. They are seen as both inhumane and dangerous. The case of wild burros and mustangs is especially difficult. They reduce critical winter forage needed by both native wildlife and domestic stock. Most ecologists and land managers believe that the herds of feral animals must be controlled, but attempts to reduce their numbers or remove them from the range meet with vigorous opposition from those for whom they are a symbol of the freedom and romance of the West.

Grazing Fees

Another controversial aspect of federal range management is that fees charged for grazing on public lands are far below market value and represent an enormous hidden subsidy to western ranchers. The 1990 charge for grazing permits on BLM or USFS land was only $2.00 per animal unit month (AUM). (One AUM is enough to feed an average cow or five sheep for one month.) Comparable private land rented for an average of $7.12 that year. The 31,000 permits on federal range bring in only $15 million in grazing fees but cost $47 million per year for

BOX 13.1   Ancient Forests of the Pacific Northwest

Just a century ago, most of the coastal ranges of Washington, Oregon, British Columbia, and southeastern Alaska were clothed in a lush forest of huge, ancient trees. The moist, mild climate and rich soil of the lowland valleys nurtured magnificent stands of Douglas fir, western red cedar, hemlock, and Sitka spruce. Trees 600 to 1,000 years old, 3 to 4 m (9 to 12 ft) in diameter and 90 m tall (as high as a 20 story building) were commonplace.

Imagine walking through one of these magical old-growth forests. Your first impression probably would be quiet serenity and immense antiquity. The widely spaced tree trunks rise like cathedral columns to the dense canopy high overhead. Although you can see and hear the wind stirring the branches far above, where you are the air is calm, humid, and peaceful. Occasional shafts of sunlight pierce the canopy to dapple the forest floor. Enormous moss-covered logs lie scattered around, slowly decaying into a rich nursery bed in which new seedlings can germinate and grow. The ground on which you walk is soft and moist, carpeted with a thick layer of moss and conifer needles and decorated with occasional clusters of sword fern, Oregon grape, huckleberry, and spindly rhododendrons. Many people find visiting these forests an epiphanic experience.

Before loggers and settlers came, there were probably 12.5 million ha (31 million acres) of old-growth, temperate

BOX FIGURE 13.1 The Quinauet Rain Forest on the Olympic Peninsula in Washington is an excellent example of the old-growth forests of the Pacific northwest. Huge Sitka spruce, Douglas fir, hemlock, and western cedar form a dense, multilayered, closed canopy that shelters a number of rare and endangered plant and animal species. The forest floor is dark, moist, and quiet, cushioned with a layer of fallen needles, moss, and ferns. Downed tree trunks serve as a nursery bed for new growth. Unfortunately, these temperate rain forests are disappearing at an alarming rate and only scattered remnants remain.

rain forest in the Pacific Northwest. Less than 10 percent remains now. Nearly all big trees on private lands have been cut. Most remaining virgin forest is on federal land, and more than 80 percent of that land is scheduled to be harvested in the near future. At the current harvest rate of 24,000 ha (100 sq mi) per year, all the remaining unprotected old growth forest in Washington and Oregon will be gone in about 50 years.

The forest products industry is a powerful force in the Pacific Northwest. The 5.4 billion board foot annual harvest creates approximately 150,000 jobs and adds $7 billion per year to the economy. This is about one-fifth of the gross state product in Oregon. Many small towns in

administration and maintenance. The $32 million difference amounts to a massive "cow welfare" system of which few people are aware.

The Office of Management and Budget concluded that increasing grazing fees to a fair market value would reduce the livestock herd less than 10 percent and would result in removing animals only from the most marginal areas, which probably should not be grazed anyway. The Reagan administration declined to raise fees, however, because of the "political sensitivity" of the issue.

Buffalo Commons

In 1990, Frank and Debora Popper of Rutgers University stirred up a heated controversy on the Great Plains by suggesting that some 3.6 million sq km (139,000 sq mi) of farms and ranches be turned into a wildlife preserve named the Buffalo Commons. Citing the dwindling water supplies, soil erosion, harsh climate, diminishing human population, and economic decline of the high plains, they argued that attempts to settle and domesticate the area have been the "largest, longest-running social and environmental miscalculation in our history."

the area depend almost entirely on wood products for their economic life. A reduction in logging would mean loss of homes, businesses, and a culture of skills, knowledge, and history.

To the loggers, it's unthinkable that their way of life might come to an end. They view leaving trees to decay in the forest as unreasonable, even immoral. They describe the old-growth forest as decadent, over mature, wasteful, decrepit, and terminally ill. As one logger put it, "Bugs or fire or humans are going to harvest the trees; they don't last forever. Why shouldn't we make something useful out of them instead of letting them fall and rot in the woods?"

To wildlife, however, the old-growth forest is not decadent or wasteful. The dead snags are home for ospreys, bald eagles, pileated woodpeckers, and flying squirrels. The complex root system and deep duff store water to keep salmon and trout streams flowing clear and cool throughout the year. Fallen logs recycle nutrients and shelter a myriad of worms, insects, and fungi. For highly specialized species, such as the Northern Spotted Owl, Vaux's Swift, and Marbled Murlets, the ancient forest is the only place they can live.

Biologists are just beginning to understand how this complex ecosystem works. Some species of birds, amphibia, insects, and even mammals live their entire lives high in the canopy and, consequently, have rarely been studied or even observed before. Furthermore, detailed studies of the invertebrate life of the forest floor are revealing a richness and diversity of species akin to that of tropical rain forests. A single site in an Oregon temperate rain forest may be home for more than 8,000 species. Removing the big old trees and replacing them with young, fast-growing saplings might increase the primary productivity of the forest, but it would eliminate a whole assemblage of plants and animals that require old growth for existence and are part of the rich tapestry of this biome.

In 1989, environmentalists sued the U.S. Forest Service over plans to clearcut a high proportion of the remaining old growth forest. They argued that spotted owls are endangered and that destroying their habitat was pushing them into extinction. They claimed, further, that spotted owls are an indicator species. That means when owls disappear, it indicates that the whole community of which they are members is in jeopardy. A federal judge agreed and ordered logging halted on about half the timber sales planned for the year in Washington and Oregon. Outrage in the logging communities was loud and clear. Convoys of logging trucks converged on protest sites while angry crowds burned environmentalists in effigy. Bumper stickers urged "Save a logger; eat an owl."

After much angry debate and political maneuvering, a compromise was reached in 1990. The U.S. Forest Service agreed to preserve about half the existing owl habitat and to allow logging on the rest. The Fish and Wildlife Service listed spotted owls as an endangered species but agreed to protect only about half the 2,400 existing pairs.

Setting aside 1 million ha (2.5 million acres) for owls seems outrageous to the loggers. They claim it will cost 40,000 jobs per year in Washington and Oregon. Environmentalists agree that jobs are disappearing, but claim it is mostly due to mechanization, a naturally dwindling resource base, and the shipping of raw logs and wood chips to Japan for processing rather than doing it here. They argue that only a few thousand jobs will be lost by protecting spotted owls and ancient forests. Besides, they argue, the big old trees will be gone in a few years if cutting continues. Loggers and mill workers will have to learn new jobs and new ways of making products anyway. Why not change now and save the last few big old trees?

Ultimately, it comes down to a question of values. How much forest should we save to protect beauty, inspiration, or the home of wildlife? Is an old growth forest valuable for the scientific information it may yield on how life on our planet operates, or should it serve a short-term economic goal? How much do we owe to other species or future generations of our own species? How many owls (and other species) should we try to preserve? How many can we afford to save—or lose?

The Poppers propose turning about one-quarter of the area between the Rocky Mountains and the one hundredth meridian back into open range stocked with native species, such as bison, antelope, elk, and deer. These native grazers have varied feeding patterns that spread the demands for food among grasses and perennial broadleaved plants and are thus more in balance with the native vegetation than those of cattle. This commons would be owned and managed by a consortium of public and private agencies. The few remaining residents might earn more money by harvesting wild game and guiding tourists than they do now in conventional farming and ranching.

Restoring something approaching the natural ecosystem could help adapt to potential climate change and avoid dust bowl conditions when drought comes again. It might also help preserve Native American culture by offering an alternative to reservation life. Interestingly, this proposal is very similar to the recommendations of artist George Catlin in 1832 and of explorer and geologist John Wesley Powell in his 1878 *Report on the Arid Lands*. It is no surprise, however, that many ranchers, farmers, and developers strongly oppose the idea of a Buffalo Commons.

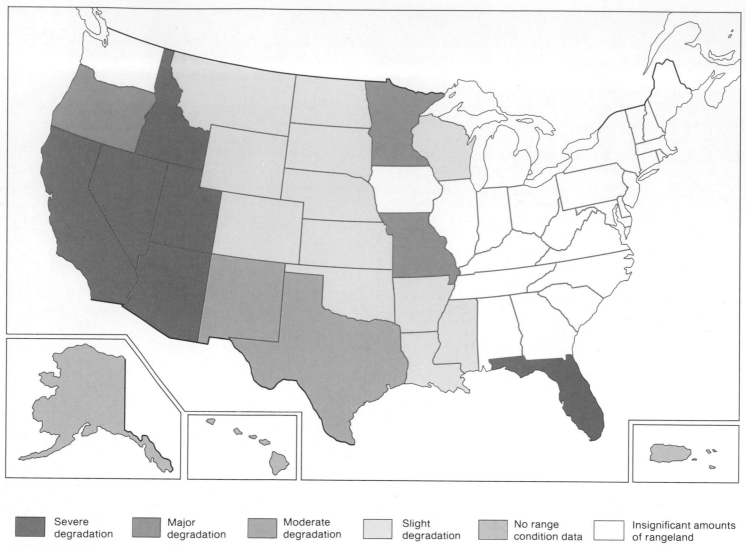

FIGURE 13.16   Range degradation on nonfederal land.

Legend:
Severe degradation
Major degradation
Moderate degradation
Slight degradation
No range condition data
Insignificant amounts of rangeland

NATIONAL PARKS

The idea of establishing parks where nature is preserved for its own sake and where common people can enjoy wildlife, scenery, and outdoor recreation in a natural setting is a unique American contribution to the world. Parks have existed in other countries, of course, but they were usually menageries of exotic animals (and sometimes people), bizarre natural features, or private playgrounds for the rich and powerful.

The Origin of American Parks

Yosemite Park (established in 1864) and Yellowstone Park (1872) are generally regarded as the first national parks in the world set aside to protect the beauties of nature. Yosemite was authorized by President Abraham Lincoln in the midst of the American Civil War, surely a remarkable gesture of optimism and vision (figure 13.17).

From the earliest days, the parks needed protection from the people for whom they were established. Army patrols were assigned to guard the park's resources, but hunters slaughtered game and timber thieves cut down the forests. Vandals wantonly destroyed natural wonders. One group, hoping to witness a spectacular eruption, dumped a thousand pounds of rubbish into Old Faithful geyser in 1890. By the 1920s, auto tourism had become the most popular way to visit the parks, and a flood of people overloaded the limited facilities. Concessionaires, sensing lucrative business possibilities, sold the entertainment value of the parks. Park Service priorities stressed the importance of visitors over natural features or wildlife. The evening entertainment at Camp Curry in Yosemite, for instance, drew two thousand tourists and featured jazz bands, vaudeville acts, a bear-feeding show, and dancing until midnight. One critic said that the honky-tonk atmosphere reminded him of Coney Island.

RESOURCES AND RESOURCE ECONOMICS

FIGURE 13.17 Yosemite National Park, established in 1864, was the world's first public park set aside to protect natural beauty—a unique American contribution to the world.

The National Park System Today

Our national park system has grown to more than 280,000 sq km (108,000 sq mi) in 341 parks, monuments, historic sites, and recreation areas. Each year we spend about 300 million visitor days in this system. The most heavily visited units are the urban recreation areas, parkways, and historic sites (figure 13.18). The jewels of the park system, however, and what most people imagine when they think of a national park, are the great wilderness parks of the West: Yellowstone, Yosemite, Glacier, Rocky Mountain, Grand Canyon, Olympic, and Canyonlands. Passage of the Alaska Lands Act of 1980 nearly doubled the national park system.

Current Problems

Although the national parks are a wonderful resource, offering natural beauty and the potential for a peaceful and meaningful experience, they have their problems. Too many people concentrated in an area too small and coming to the parks for the wrong reasons remain problems in most parks. Roads, trails, buildings, and natural communities have suffered from years of neglect. In 1988, the General Accounting Office reported that the parks needed $1.9 billion merely for repair and restoration. Little money was allocated for land acquisition in the 1980s. In 1978, the budget was $681 million; for 1989, the Reagan administration requested only $17 million.

Yosemite National Park is a good example of park problems: 95 percent of the overnight visitors never leave the 18 sq km (7 sq mi) valley floor, which is less than 1 percent of the total park area. They fill the valley with noise and smoke, trample fragile meadows to dust, and spend hours in traffic jams. Park rangers have become traffic cops and crowd-control specialists rather than naturalists. On Cape Cod National Seashore and in the new California Desert Park, dune buggies, dirt bikes, and off-road vehicles (ORV) run over fragile sand dunes, disturbing vegetation and wildlife, and destroying the aesthetic experience of those who come to enjoy nature (figure 13.19).

Pollution also has come to the parks. The haze over the Blue Ridge Parkway is no longer blue, but gray-brown because of air pollution carried in by long-range transport. Visitors to the Grand Canyon could once see mountains 160 km (100 mi) away; now the air is so smoggy that you can't see from one rim to the other during one-third of the year. The main culprits are power plants in Utah and Arizona that supply electricity to Los Angeles, Phoenix, and other urban areas. Acid rain threatens sensitive lakes in the high mountains of the West, as well as in the Great Lakes and eastern states. Ozone is damaging the giant redwoods in California's Sequoia National Park, and unknown agents (probably air pollutants) are killing trees in the Great Smoky Mountains.

Mining and oil interests continue to push for permission to dig and drill in the parks, especially on the 3 million acres of

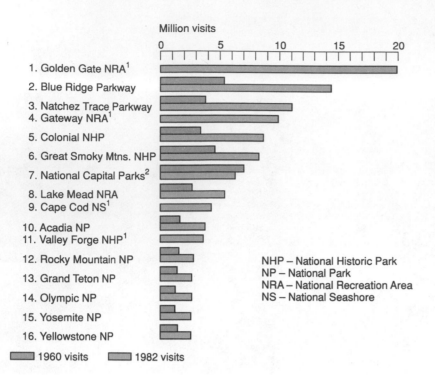

Million visits

1. Golden Gate NRA[1]
2. Blue Ridge Parkway
3. Natchez Trace Parkway
4. Gateway NRA[1]
5. Colonial NHP
6. Great Smoky Mtns. NHP
7. National Capital Parks[2]
8. Lake Mead NRA
9. Cape Cod NS[1]
10. Acadia NP
11. Valley Forge NHP[1]
12. Rocky Mountain NP
13. Grand Teton NP
14. Olympic NP
15. Yosemite NP
16. Yellowstone NP

NHP – National Historic Park
NP – National Park
NRA – National Recreation Area
NS – National Seashore

◼ 1960 visits   ◻ 1982 visits

FIGURE 13.18 Visits to some of the most popular units of the National Park System in 1960 and 1982.

Source: National Park Service.

In 1971, the Park Service began counting recreation and nonrecreation visits separately. Figures for 1982 include only recreation visits; thus, 1960 visitation is overstated relative to 1982.

[1] Established after 1960

[2] National Capital Parks had higher visitation in 1960 than 1982, due to redesignation of several units formerly included in National Capital Parks as separate units of the park system.

FIGURE 13.19 Off-road machines, such as motorcycles, dune buggies, and four-wheel drive vehicles can cause extensive and long-lasting damage to sensitive ecosystems. Tracks can persist for decades in deserts and wetlands where recovery is slow. Wildlife vegetation, solitude and the natural beauty for which parks and recreation areas were set aside are affected negatively.

private inholdings in the parks. These forces were successful in excluding mineral lands in the Misty Fjords and Cape Kruzenstern National Monuments in Alaska. Placer mines, which wash sediments out of hillsides with high-pressure hoses in Denali National Park, dump thousands of tons of sediment each day into valuable salmon streams. Uranium mines at the edge of the Grand Canyon threaten to contaminate the Colorado River and the park's water supply with radioactive contaminants. The wreck of the *Exxon Valdez* in 1989 contaminated hundreds of miles of Alaskan coastline including Katmai and Kenai National Parks (see Box 22.1).

In Florida's Everglades National Park, water flow through the "river of grass" has been disrupted and polluted by encroaching farms and urban areas. Wading-bird populations have declined by 90 percent, down from 2.5 million in the 1930s to 250,000 now.

Park Wildlife: A Controversy

Wildlife is at the center of many arguments regarding whether the purpose of the parks is to preserve nature or to provide entertainment for visitors. In the early days of the parks, "bad" animals (e.g., wolves and mountain lions) were killed so that

RESOURCES AND RESOURCE ECONOMICS

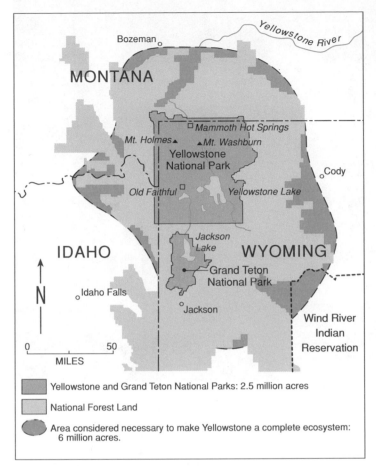

FIGURE 13.20 This map shows the Yellowstone ecosystem complex or biogeographical region, which extends far beyond the park boundaries. Park managers ecologists believe that it is necessary to manage the entire region if the park itself is to remain viable.

Map legend:
- Yellowstone and Grand Teton National Parks: 2.5 million acres
- National Forest Land
- Area considered necessary to make Yellowstone a complete ecosystem: 6 million acres.

populations of "good" animals (e.g., deer and elk) would be high. Rangers cut trees to improve views, put out salt blocks to lure animals to good viewing points close to roads, and otherwise manipulated nature to make a more enjoyable experience for the guests.

Critics of this policy claim that favoring some species over others has unbalanced ecosystems and created only a sad illusion of a natural system. They claim that excessively large elk populations in Yellowstone and Grand Teton National Parks, for instance, have degraded the range so badly that other species are being crowded out. Park rangers tried hiring professional hunters to reduce the elk herd, but a chorus of protest was raised. Sportsmen want to be able to hunt the elk themselves, animal lovers don't want them to be killed at all, and wilderness advocates don't like the precedent of hunting in national parks. The Park Service has retreated to a policy of "natural regulation," meaning that they intend to let nature take its course. When elk starve to death, as hundreds did after the Yellowstone fires in 1988 and as they will inevitably do elsewhere, new protests and appeals will result. An ideal solution might be to reintroduce

predators, but ranchers who raise livestock on lands adjacent to the parks are adamantly opposed to the idea. Wildlife issues are discussed further in chapters 14 and 15.

One of the most difficult problems for many parks is that they are not self-contained ecosystems. Their boundaries were established by political compromise between local landowners, loggers, miners, and others with a variety of interests in the area. Lands around the parks that once provided a buffer against what happened outside are being developed. Airsheds, watersheds, viewsheds, and animal migration routes extend far beyond park boundaries and yet profoundly affect conditions within the parks. We need to manage the whole **biogeographical area** (an entire ecosystem and its associated land, water, air, and wildlife resources) to preserve natural conditions within a park (figure 13.20).

New Directions

What is being done to enhance the visitor's experience to a national park while maintaining the natural ecosystem as much as possible? Several parks have removed facilities that conflict with natural values. In Yellowstone, a big laundry and a cabin ghetto that stood next to Old Faithful were torn down in 1986. Shabby hotels and filling stations that once stood at Norris Junction and Yancy's Hole are gone. The golf course, slaughterhouse, and tent camps at Mammoth Hot Springs also have been removed. In Yosemite, Grand Canyon, and Denali, tourists park their cars and take shuttle buses into the park to reduce congestion and pollution. A poorly performing concessionaire was denied renewal of a contract for Yosemite in 1979, something that had never before happened.

There are proposals that a number of parks be closed to cars, and some areas might be closed to tourists altogether to protect wildlife and fragile ecosystems. Most parks limit the number of overnight visitors. Backpackers must apply for permits and may camp only in designated places. People are turned away from some parks when visitor limits are reached. The time may come when park permits will need to be reserved years in advance and visits to certain parks will be limited to once in a lifetime!

Proposals for New Parks

One solution to congestion and overuse of our parks is to create new ones to distribute the load. This would also allow us to protect and enjoy other magnificent areas that deserve preservation in their own right. The Great Basin Park in Nevada was established in 1988 to protect a portion of the intermountain desert and two groves of ancient bristlecone pines. The new Mojave National Park in California will protect sand dunes, Joshua tree "forests," Indian petroglyphs, fossils, and critical habitat for rare and endangered species, such as the desert pupfish and desert tortoise.

FIGURE 13.21 The Absaroka-Beartooth Wilderness, north of Yellowstone National Park, preserves a piece of relatively untouched natural ecosystem. Areas such as this serve as a refuge for endangered wildlife, a place fo routdoor recreation, a laboratory for scienitific research, and a source of wonder, awe, and inspiration.

Some other areas that have been proposed for park status are the Tallgrass Prairie in Oklahoma and Kansas, the Columbia Gorge in Washington and Oregon, Big Sur and Lake Tahoe in California, some of the Florida Keys, and the Hudson Valley and Thousand Island in New York. In 1991, the U.S. National Park Service was considering $343 million worth of new park lands.

WILDERNESS AREAS

American culture and mythology are strongly influenced by our recent history as a wilderness country. As historian Frederick Jackson Turner pointed out in a series of articles and speeches around the turn of the century, Americans have traditionally believed that the wilderness was not only a source of wealth, but also a symbol of strength, self-reliance, wisdom, and character. The frontier was seen as a source of continuous generation of democracy, social progress, economic growth, and national energy.

By 1900, the last frontier was closed, and people began to realize that the wilderness was endangered. The first official wilderness area in the United States was established in 1924 on the Gila National Forest in New Mexico, where noted author Aldo Leopold started his career as a young forest ranger in 1909. By 1939, under the leadership of wilderness advocate Bob Marshall, the USFS had designated seventy-five large Wilderness Areas and forty-two smaller Wild Areas totaling nearly 8.4 million ha

(20.7 million acres). These were only administrative classifications, however, that could be changed at any time by the chief of the Forest Service. In 1956, Howard Zahnister, executive director of The Wilderness Society (founded by Leopold and Marshall), began to work for legislative protection of these and other primitive areas. This campaign culminated in the passage in 1964 of the Wilderness Act, which legally defined **wilderness** (figure 13.21) as "an area of undeveloped land which is affected primarily by the forces of nature, where man is a visitor who does not remain; it contains ecological, geological, or other features of scientific or historic value; it possesses outstanding opportunities for solitude or a primitive and unconfined type of recreation; and it is an area large enough so that continued use will not change its unspoiled, natural conditions."

Today, 264 units of the National Wilderness System encompass nearly 36 million ha (88 million acres). Almost two-thirds of that total was added to the system by passage of the Alaska National Interest Conservation Land Act in 1980. Additional wilderness areas are being evaluated for protected status. The USFS was instructed by the 1964 act to carry out a roadless area review and evaluation (RARE) on all *de facto* wilderness areas under its jurisdiction. Using a deliberately "pure" interpretation that excluded all lands with any history of roads or development (even if all traces of human impact were long gone), it finally decided in 1979 that only about one-fourth of its 23 million ha (56 million acres) of roadless areas qualified for protection.

These proposed wilderness areas now are being considered by Congress on a state-by-state basis. Each bill usually pits environmental groups who want more wilderness against loggers, miners, ranchers, and others who want less wilderness. The arguments for saving wilderness are that it provides (1) a refuge for endangered wildlife, (2) an opportunity for solitude and primitive recreation, (3) a baseline for ecological research, and (4) an area where we have chosen simply to leave things in their natural state. The arguments against more wilderness are that timber, energy resources, and critical minerals contained on these lands are essential for economic development. To people who live in remote areas, jobs, personal freedom, and local control of resources seem more important than abstract values of wilderness. Wilderness proponents point out that 96 percent of the country already is open for resource exploitation; the 4 percent that has been set aside is mostly land that developers didn't want anyway.

The last large areas still being studied for wilderness preservation are on BLM land. The 1976 Federal Land Policy and Management Act (FLPMA) ordered the BLM to review all its roadless areas for wilderness potential. Applying the same "pure" standards used by the Forest Service, the BLM found only one-fourth of its 112 million ha (276.6 million acres) suitable for wilderness. Preservationists argue that twice that much land should be considered.

LAND-USE PLANNING

Differing opinions on how land should be used often come down to philosophical questions of values and beliefs. Among the most important points of controversy are the following questions:

(1) Should the government continue to transfer resources wherever possible to private hands or should it retain land and manage resources in public ownership? Is private enterprise more efficient and effective than public management?

(2) To what extent should governments interfere with the use of private lands? At what point does public welfare take precedence over individual freedom?

(3) How should competing and incompatible uses, such as wildlife preservation, recreation, logging, mining, energy extraction, agriculture, and aesthetic values, be balanced? How much land should be devoted to each use?

(4) Is local control or central planning more effective in determining how land is to be used? How should competing interests be reconciled?

Central versus Local Planning

Many environmentalists, like reformers in other social movements, tend to be political progressives who favor public planning and centralized authority in dealing with societal problems. They suspect that local units of government are unlikely to be aware of or concerned with regional ecosystems, ecological principles, or long-range environmental values. Universal, impersonal application of laws eliminates the special privileges of the elite and powerful, whereas, in the view of some land reformers, local regulation is the developer's best friend.

Critics of central planning, on the other hand, believe that local units of government understand local conditions and the needs of residents best, and perceive the federal government as distant, impersonal, and cumbersome. They fear it will grow larger and larger, constantly adding needless regulations that crush private enterprise, reduce profits, and interfere with traditional ways. Many rural people highly value self-sufficiency and freedom from outside interference, favoring local control to take care of their problems.

Local Planning Boards and Zoning Commissions

How do citizens determine land use in our democratic system? Most land-use planning in the United States is done by city or county planning and zoning commissions. Although some decisions may seem routine and trivial, they gradually shape how and by whom land resources are used. The location of roads, sewers, water supplies, shopping centers, and freeway exchanges usually amount to *de facto* land-use decisions, many of which are essentially irreversible. A farm or forest that becomes a shopping center, power plant, or housing development is unlikely to revert to its former use. Furthermore, a decision concerning one parcel of land can greatly affect others nearby. If a neighbor puts in a hog farm or a junkyard, for instance, the options for use of your land may be limited.

While even the earliest settlements attempted to prevent the misuse of common resources and the development of public nuisances, building and zoning ordinances didn't come into general use until the 1920s. Most local units of government, such as cities and counties, have some kind of land-use plan, but they are sometimes little more than a statement of good intentions. Boards that administer these plans often tend to be dominated by the economic elite of the community who likely derive much of their income from the land. As businesspersons, they are inclined to support the entrepreneur's right to speculate and profit, and they place a very high priority on the immediate economic health of the community. Controlled growth and preservation of such intangible values as solitude, biological diversity, or scenic beauty are often neglected or ignored altogether.

Regional Land-Use Planning

Regional planning has been successfully accomplished in a few areas. One of the most stringent, complex, and comprehensive regional plans in the United States is the responsibility of the New York State Adirondack Park Agency. The Adirondack Park, established in 1892, is the largest wilderness area in the East, and the State Constitution requires that it be held "forever wild."

Covering an area as large as Vermont, the planning region includes state forests and parks, mining and timber company lands, 2,300 lakes, thousands of private cabins and cottages, several towns, and 107 local units of government.

The land-use plan adopted in 1972 for this complex area designates land for wilderness, canoe recreation, travel corridors, timber harvest, farming, housing, and commercial use. Agency approval is required for any large developments, and funds and assistance are provided for local land-use controls. However, the plan is highly controversial. Some residents regard the bureaucracy involved as unwarranted interference; others welcome the regulation of lands within the park but deplore the garish commercial developments around its perimeter.

Siting Industrial Facilities

Among the most controversial and difficult land-use decisions in most states are the siting of power plants and power lines, strip mines, landfills, and disposal areas for hazardous, toxic, or radioactive waste. If located in or near urban areas, these facilities may create a health hazard for large numbers of people. If located in rural areas, they can disrupt farmland, scenic beauty, recreational opportunities, and wildlife. Although most people agree that these installations are necessary somewhere, the general reaction is "not in my backyard . . . anyplace but here." No matter how rational and orderly the siting process is, conflicting interests and emotional reactions are bound to arise whenever a particular site is proposed. The challenge is to design regulatory processes and pollution-control equipment that we can be reasonably confident will work, and to fashion land-use planning procedures that will fairly represent all of the many competing interests and values of different people and different areas. This seems a daunting prospect, but perhaps no traits are so human as the abilities to compare, discuss, remember, plan, argue, negotiate, and compromise.

> *"A land ethic, then, reflects the existence of an ecological conscience, and this, in turn, reflects a conviction of individual responsibility for the health of the land. Health is the capacity of the land for self-renewal. Conservation is our effort to understand and preserve this capacity. We are remodeling the Alhambra with a steamshovel, and we are proud of our yardage. We shall hardly relinquish the shovel, which after all has many good points, but we are in need of gentler and more objective criteria of its successful use."*
> Aldo Leopold

SUMMARY

Forests cover nearly 31 percent of the land in the United States, grazing lands and pastures make up 30 percent, and croplands occupy nearly 20 percent (about twice the proportion of arable land in the world overall). The current U.S. land value is about $1.2 trillion, only 15 percent of the nation's wealth. While nearly all of the 900 million ha (2.2 billion acres) in the United States was once owned by the federal government, about two-thirds have been given away, sold, or transferred to states, communities, and private citizens. Fifty-nine percent of the land is privately owned, 30 percent is federal, 7 percent is state or local, and 4 percent remains in the hands of native people.

The decision to distribute public lands to small farmers was a radical step that has had an immense impact on American society and the environment. There were many abuses of the land-distribution system, but there were many advantages as well. By 1900, it was recognized that land resources, including forests, grazing lands, and wilderness, were rapidly disappearing. Much of the early history of the conservation movement was concerned with land use and preservation. There was, and continues to be, a great difference between the utilitarian conservationists, who argue that resources should be saved so they can be used for the greatest good for the greatest number for the longest time, and the altruistic preservationists, who argue that natural resources deserve to be saved for their own sake.

Although we have no comprehensive land-use policy in the United States, many federal land-management acts and regulations have been established. The U.S. Forest Service manages 171 national forests and 19 national grasslands, encompassing some 75 million ha of land. Timber harvest methods, road building, reforestation, herbicide and pesticide spraying, cutting old-growth forests, and wilderness preservation are all highly controversial aspects of forest management policies.

The Bureau of Land Management (BLM) administers some 112 million ha of grassland, desert, and range, about one-quarter of all grazing land in the United States. Little known outside of the western states where most of its lands are located, the BLM tends to work closely with local ranchers, miners, and timber harvesters. Consequently, 85 percent of all federal grazing lands are in only fair or poor condition.

The establishment of national parks where nature is saved for its own sake, and where common people can enjoy wildlife, scenery, and outdoor recreation in a natural setting is a unique American idea. Started more than one hundred years ago, our national parks preserve some of the most beautiful and inspiring places in America. In 1990 nearly 300 million visitor days were spent in these national treasures. Pollution, crowding, aging facilities, conflicts between humans and wildlife, and incompatible uses are among the problems that plague our national parks.

Land-use planning is a subject of much controversy, based on differing philosophical values and beliefs. Questions of public control over private lands, central versus local planning, and choices between competing uses are central issues. Most communities now have some form of land-use planning, usually in the form of zoning ordinances drawn up by city or county planning commissions. A few areas have successful regional plans that integrate multiple local authorities, environmental quality, economic activities, housing, transportation, and the many other aspects of modern society. Siting industrial facilities, such as power plants, chemical factories, strip mines, and toxic waste dumps, is one of the most difficult and devisive land-use decisions we face.

RESOURCES AND RESOURCE ECONOMICS

Review Questions

1. What is the land area of the United States, and how is it divided among major uses?
2. How much land is publicly owned and how much is privately owned in the United States?
3. Who were some of the early conservationists in the United States and what philosophy did they espouse?
4. What are multiple uses and sustainable yield in forest resources? Why has this been controversial?
5. What is the official definition of wilderness in the United States? What are the current issues regarding wilderness preservation and wilderness management?
6. What is the difference between clear-cutting and selective cutting in forests? Describe some advantages and disadvantages of each.
7. How does the condition of public grazing lands in the United States compare to the condition of private lands? What factors or forces contributed to the degradation of public rangelands?

Questions for Critical Thinking

1. How might our history have been different if we had a different pattern of land ownership? What has been the experience of countries with different land-use policies?
2. What are the differences between utilitarian conservation and altruistic preservation? Describe how these two philosophies might differ in the use, protection, and management of some specific natural resources. Where do you stand?
3. Why do you suppose it has been so difficult to pass a comprehensive land-use policy act in the United States? What might be some of the benefits and disadvantages of having a national land-use policy?
4. Describe some of the conflicts that might arise out of the Coastal Zone Management Act. Why do you suppose this is so controversial? Would you support coastal zone management?
5. Do you think the United States has too much, not enough, or about the right amount of officially designated wilderness? Why?
6. Why do some people oppose road-building in national forests? What is your position on this issue?
7. Do you think feral animal populations should be controlled? If so, how?
8. Many national park visitors would like to have shopping centers, golf courses, swimming pools, high-speed highways, night clubs or bars, and amusement parks within the parks. What activities and facilities do you think are acceptable?
9. How would you feel about banning all private automobiles from national parks?
10. Grand Teton National Park supports a wintering herd of some sixteen thousand elk. They have to be fed hay and alfalfa pellets to prevent mass starvation. Hunters would like an open hunting season; park officials would like to hire professional hunters to thin the herds; some biologists want to let nature (starvation and disease) take its course; some people want to continue to feed them. What would you do?

Key Terms

altruistic preservation (p. 242)
biogeographical area (p. 255)
clear-cut (p. 246)
Coastal Zone Management Act (p. 243)
feral (p. 249)
Homestead Act (p. 240)
multiple use (p. 242)
rangeland (p. 249)
"Sagebrush Rebellion" (p. 244)
selective cutting (p. 246)
sustained yield (p. 242)
utilitarian conservation (p. 242)
wilderness (p. 256)
Wilderness Act (p. 242)

SUGGESTED READINGS

- Arrandale, T. *The Battle for Natural Resources.* Washington, D.C.: Congressional Quarterly Books, 1983. Excellent overview of the political history of public land-use policy.
- Berger, J. J. *Restoring the Earth: How Americans are Working to Renew our Damaged Environment.* New York: Alfred A. Knopf, 1985. A heartening collection of case studies in resource protection and restoration.
- Chase, A. *Playing God in Yellowstone: Destruction of America's First National Park.* New York: Atlantic Monthly Press, 1986. A strong criticism of National Park management. Controversial, but well worth reading.
- Connelly, J. May/June 1991. "The Big Cut" *Sierra* 76 (3):42. A scathing denunciation of clear cutting and exploitation of British Columbia's old-growth forest.
- Ferguson, D., and N. Ferguson. *Sacred Cows at the Public Trough.* Bend: Maverick Publications, 1983. An excellent report on the state of the range and misuse of the public domain by the cattle industry.
- Fishman, David J. 1989. "America's Ancient Forests." *E Magazine* vol. 1 no. 1. A critique of current forest management practices and a call for a new Siskiyou National Park to protect old-growth forests.
- Fox, S. *John Muir and His Legacy: The American Conservation Movement.* Boston: Little, Brown & Company, 1981. Excellent biography of one of America's foremost environmental philosophers and the history of environmental protection. One of the best analyses available of conservation organizations.
- Lamm, R. D., and M. McCarthy. 1982. *The Angry West: A Vulnerable Land and Its Future.* Boston: Houghton Mifflin Company, 1982. A vivid and controversial, but highly readable, statement of a western view of land resources.

Leopold, A. *A Sand County Almanac*. New York: Oxford University Press, 1949. A classic in American nature writing. Clear, lucid, compassionate. Usually coupled with Leopold's essays on land ethic.

Libecap, G. D. *Locking Up the Range: Federal Land Control and Grazing*. San Francisco: Pacific Institute for Public Policy Research, 1986. The Westerner's view of land-use policy.

Marsh, G. P. *Man and Nature: or, Physical Geography as Modified by Human Action*. Cambridge: Harvard University Press, 1864 (reprinted 1965). Called "the fountainhead of the conservation movement" by historian Lewis Mumford.

McHarg, I. L. *Design with Nature*. Garden City: Doubleday/Natural History Press, 1971. A classic in land-use planning. Eloquently committed to environmental values.

Mitchell, J. T. "War in the Woods II: West Side Story." *Audubon* January 1990. An exploration of the various sides of old-growth forest protection in Washington and Oregon.

Nash, R. *Wilderness and the American Mind*. New Haven: Yale University Press, 1987. An outstanding survey of the history and philosophy of American conservation. An environmental classic.

National Audubon Society. *Audubon Wildlife Report 1986*. New York: National Audubon Society, 1986. Comprehensive survey of the state of wildlife refuges, park forests, and other public lands. Good summary of public land management agencies and their operation.

Nelson, R. Winter 1990. "Alaska: A Glint in the Raven's Eye." *Wilderness* 54(191). The introduction to a special issue devoted to the history of the Alaska National Interest Lands Act.

Norse, E. A. 1990. *Ancient Forests of the Pacific Northwest*. Washington D.C.: The Wilderness Society. An excellent overview of the biology of the ancient temperate rain forest with recommendations for new forestry techniques and management principles.

Repetto, R. 1990. "Deforestation in the tropics." *Scientific American* 262 (4):36. Proposes policies to encourage forest protection and stop the destruction of irreplaceable resources.

Sax, J. *Mountains Without Handrails: Reflections on the National Parks*. Ann Arbor: University of Michigan Press, 1980. Eloquent analysis of the state of our parks and how they should be managed to preserve their highest values.

Scott, Douglas. January/February 1991. "The Fight to Save Alaska" *Sierra* 76 (1):38. An excellent history of the national campaign to save Alaskan lands and an update on continuing problems.

Stegner, W. September/Ocotober 1989 "Our Common Domain." *Sierra* 74(5). An overview of Bureau of Land Management lands and management policies by the dean of western writers.

Turner, F. J. *The Frontier in American History*. New York: H. Holt Company, 1920. Describes the frontier as the source of American democracy and values, a theory he first described in the Atlantic Monthly, September 1896.

World Resources Institute, World Bank, and United Nations Development Program. *Tropical Forests: A Call for Action*. Washington, D.C.: World Resources Institute, 1985. A plan for protecting and restoring the world's endangered tropical forests.

Wuerther, G. Spring 1991. "How the West was Eaten." *Wilderness* 54 (192):28. The state of the range on the public domain is a disgrace. Grazing leases controlled by a small number of cattle and sheep barons pay only a fraction of their true costs while wasting much of our common inheritance.

C H A P T E R *14*

Biological Resources

It is not the decimation of large animals or known endangered species that threatens the systems that support our lives. The loss of myriad unsung populations of unspectacular organisms is the major threat to our well-being.

Paul Ehrlich

Any species of bug is an irreplaceable marvel, equal to the works of art which we religiously preserve in museums.

Claude Levi-Strauss

C O N C E P T S

Scientists estimate that between 3 and 30 million different species inhabit Earth. So far, 1.7 million species have been identified. We derive benefits, such as food, medicine, clothing, and building materials, from wildlife. Additional wild species not yet studied by science may have potential as future sources of valuable products, if we can prevent their destruction.

Even obscure organisms contribute to essential ecological processes of the biosphere and play a role in Earth's life-support systems. Through species extinction, unique and complex characteristics may be permanently lost. Natural causes of wildlife destruction include evolutionary replacement and mass extinction. Periods of mass extinction have opened niches, allowing for evolution and new growth. Humans are now causing a species extinction that could have disastrous consequences.

Humans threaten wildlife directly by overharvesting animals and plants for food and commerce, and by using pesticides. We threaten wildlife indirectly by altering or destroying habitats with exotic species, disease, and pollution.

Zoos and animal parks play an important role in public education and serve as refuges for some species whose habitat in the wild is threatened. Biological diversity can be saved through nature preserves, national parks, and other areas where whole ecosystems can be protected. Considerable progress has been made in establishing world biological reserves.

INTRODUCTION

As far as we know, our planet is the only place in the universe that supports life, yet there are few places on Earth that are not home to some kind of organism. From the most arid desert to the dripping rain forest, from the highest mountain peak to the deepest ocean trench, life occurs in a marvelous spectrum of sizes, colors, shapes, life cycles, and interrelationships. Think, for a moment, how remarkable, varied, abundant, and important the other living creatures are with whom we share Earth (figure 14.1).

In previous chapters, we have described some of the diversity and complexity of Earth's populations, communities, and ecosystems. You have learned ways in which the vast number and amazing variety of living organisms move nutrients and energy through ecosystems and modify the environment to make Earth habitable. Although our understanding of Earth's organisms—its **biological resources**—is still imperfect, there is no doubt that the many different kinds of organisms provide abundant benefits and make our world a beautiful and interesting place to live.

How do we fit into this diverse world of life? For most of the approximately 3 million years humans have lived on Earth, we have lacked the strength and numbers to have much effect on neighboring organisms or our environment. In recent times, however, we have become a serious threat both to ourselves and other life-forms. Rapidly expanding human populations and activities, amplified by the power of technology, threaten to eliminate much of the diversity of the biosphere. Furthermore, we have become a geological force, leveling mountains, diverting rivers into new channels, and causing soil erosion on the order of 25 billion metric tons worldwide per year (chapter 12).

Extinction, the irrevocable elimination of species, is a normal process of the natural world. Species die out and are replaced by others, often their own descendents, as part of evolutionary change. In undisturbed ecosystems, the rate of extinction appears to be about one species lost every ten years. In this century, however, human impacts on populations and ecosystems have accelerated that rate, causing dozens to hundreds of species, subspecies, and varieties to become extinct every year. If present trends continue, we may destroy *millions* of kinds of plants and animals in the next few decades.

In this chapter, we will look at some ways humans are causing extinction, both directly and indirectly. First, we will consider some benefits we derive from other organisms to illustrate why we should care about the fate of the species threatened by our activities. We will focus on wild (nondomesticated) plant and animal species because they are most threatened by extinction.

FIGURE 14.1   Every year, some 1.5 million wildebeests migrate across the Serengetic Plain of East Africa in one of the world's greatest wildlife spectacles. Can we coexist with these vast herds? How shall we weigh the rights and values of other species?

BIOLOGICAL RESOURCES

We benefit from other organisms in many ways, some of which we may not appreciate until the species that provides the benefit is gone. Even seemingly obscure and insignificant organisms can play an irreplaceable role in the complex web of life that makes up the life-support system on which we all depend. Although it seems that Earth teems with an inexhaustable diversity of life, that may not be the case.

How Many Species Are There?

At the end of the great exploration era of the nineteeth century, some scientists confidently declared that we had found and named most of the kinds of living things on Earth. Since then, however, additional, more detailed investigation has revealed that millions of new species and varieties remain to be studied scientifically.

We now believe that the approximately 1.7 million species presently known (table 14.1) are only a small fraction of the total number that exist. Taxonomists (scientists who classify species) estimate that there are somewhere between 3 million and 30 million different species. In fact, there may be 30 million species of tropical insects alone. Most of the organisms yet to be discovered and classified will probably be plants, invertebrates, and microbes.

Of all the world's species, roughly 10 to 15 percent live in North America and Europe. We're pretty certain that most of these plants, vertebrates, and larger invertebrates have been studied and classified. It is a rare occurrence to find a new species

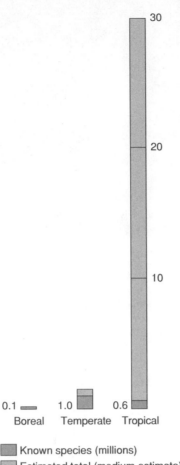

| 30 |
| 20 |
| 10 |
| 0.1 | 1.0 | 0.6 |
| Boreal | Temperate | Tropical |

■ Known species (millions)
▢ Estimated total (medium estimate)

FIGURE 14.2 Number of known and estimated species by climatic zones. Boreal regions have large populations of a few species (all known). Temperate regions have more species (mostly known), and tropical regions are thought to have 80 to 90 percent of all species on Earth, less than 10 percent of which have been identified.

Source: Data from *World Resources 1987*.

TABLE 14.1	Number of living species by taxonomic group	
Groups	**Identified species**	**Estimated species**
Mammals	4,170	4,300
Birds	8,715	9,000
Reptiles	5,115	6,000
Amphibians	3,125	3,500
Fishes	21,000	23,000
Invertebrates	1,300,000	4,400,000*
Vascular Plants	250,000	280,000
Nonvascular Plants	150,000	200,000
Total†	1,742,000	4,926,000

*This figure is a minimum. Recent research suggests there could be as many as 30 million insect species in tropical forests alone.

†Totals are rounded.

Source: Data from the World Resources Institute 1986.

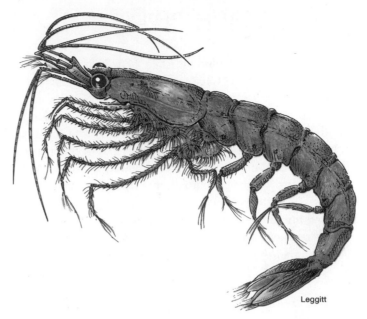

FIGURE 14.3 Krill are thought to be the most numerous organisms on Earth. These small (1–5 cm), shrimplike crustaceans (perhaps one quadrillion individuals) inhabit the Antarctic oceans where they are eaten by whales, penguins, fish, and birds in teeming multitudes.

of higher plant or animal in the developed countries of the world. By contrast, tropical countries have an incredible diversity of organisms, most of which have not even been seen by scientists. Ecuador, for instance, which is nearly the size of Colorado, has twice as many bird species as the United States and Canada combined. In figure 14.2, we see that only 600,000 of an estimated 8.6 million species have been identified in tropical climates. The Malay Peninsula has at least 8,000 species of flowering plants, while Britain, with twice as large an area, has only 1,400 species. There are probably more botanists in Britain than there are species of higher plants! By comparison, South America has fewer than one hundred botanists to study perhaps 200,000 species of plants.

Biological Abundance

If it is difficult to imagine the number of species alive today, it is staggering to try to comprehend the total number of living organisms in the world. Many of the most abundant species are invisible to the unaided eye or are in remote places that most of us never visit. The most numerous multicellular organism is thought to be krill: small, shrimplike crustaceans (figure 14.3) that live in Antarctic oceans and form the base of a rich food web that includes whales, seals, penguins, and fish. Summer krill swarms are estimated to contain 650 million metric tons of biomass and close to 10^{15} (one million billion) individual organisms. Humans probably make up the next largest biomass for a single species (about 250 million metric tons), even though we are not nearly as numerous as some others.

BENEFITS FROM BIOLOGICAL RESOURCES

Why should we care about the state of wild populations, including their extinction? Chapter 3 describes the ecological services that keep environments habitable. Chapters 15 and 27 consider some ethical arguments that govern our relationships with other organisms. In this chapter, we will look at some practical benefits (direct and indirect) we derive from other species, such as food, clothing, medicine, and building supplies. We will focus on **wildlife** (plants, animals, and microbes that live independently of humans) since these are the species most endangered by human activities. *Keep in mind that wild species represent a genetic library of information that has developed over millenia, something that we can never recreate once they are gone.*

Food

All of our food comes from other organisms. Although most of what we in developed countries eat comes from domesticated plants and animals, we still get at least part of our diet from wild species. Most seafood, for instance, is harvested from free-roaming wild organisms. Aquaculture (the cultivation of aquatic species) is growing in importance as a food source in many areas but contributes only 10 percent of the roughly 70 million metric tons of seafood harvested worldwide each year. Seafood isn't a very large percentage of the total caloric intake for most people; however, it constitutes a valuable, easily digested protein source that makes the difference between adequate nutrition and malnourishment in many developing countries—especially for women and children, who need more protein. Wild amphibians, reptiles, birds, and mammals also are used as food, to varying degrees, around the world.

Wild plants also make up a welcome dietary supplement for people in many countries. Even those of us in developed countries may eat wild products without being aware of it. Brazil nuts, for instance, are harvested only from wild forest trees. The coconut oil in which your french fries are cooked comes, to a large extent, from noncultivated palm trees. In certain areas of the United States, the harvesting of wild mushrooms is so intense that some of the most highly prized species, such as morels and chantrelles, are becoming locally threatened.

Our domestic food supply also benefits from wild species. Agricultural plant and animal breeding programs often incorporate genetic traits from wild stocks to improve yields, increase disease resistance, and incorporate other desirable traits (chapter 10).

Many plant species have great potential as human food sources because they are culturally acceptable and are much better adapted and more appropriate to local conditions than the dominant western crops. Noted tropical ecologist Norman Meyers estimates that as many as eighty thousand edible wild

FIGURE 14.4 Mangosteens from Indonesia have been called the world's best-tasting fruit, but they have never been cultivated on a large scale and are practically unknown beyond the islands where they grow naturally. There may be thousands of other traditional crops and wild food resources that could be equally valuable but are threatened by extinction.

plant species could be utilized by humans. Villagers in Indonesia, for instance, are thought to use some four thousand native plant and animal species for food. Few of these species have been explored for their potential for domestication or more widespread cultivation. One of these, the mangosteen, has been called the "world's best-tasting fruit" (figure 14.4). A 1975 study by the National Academy of Science (U.S.) pointed out that New Guinea has two hundred and fifty-one edible fruits, only forty-three of which have been cultivated.

Unfortunately, potentially valuable food species and the wild ancestors of our domestic crops are being destroyed by forest clearing, grazing, and conversion of wild lands to domestic crops before they can be identified and their genes can be preserved. Chapter 15 looks at some of the programs underway to find useful wild species and preserve them in gene banks, botanical gardens, zoos, and nature preserves.

Industrial and Commercial Products

Wood is one of the most valuable commercial products we obtain from wild species. Although trees can be cultivated in plantations, the growth rate of most tree species is so low that we depend on harvest of natural forests for much of our 3 billion cubic meter/year wood consumption. This is especially true for hardwoods (nonconifer species) and plywood veneer logs, almost all of which come from old-growth forest trees. Commercial growers cannot afford to wait for a tree to grow big enough to make a veneer log. More than 85 percent of all hardwoods harvested each year come from virgin tropical moist forests, mostly in Southeast Asia. About one-half of the wood harvested each year is simply burned as fuel for cooking fires. The growing shortage of firewood in developing countries is becoming acute (chapter 12).

TABLE 14.2 Some natural medicinal products

Product	Source	Use
Penicillin	Fungus	Antibiotic
Bacitracin	Bacterium	Antibiotic
Tetracycline	Bacterium	Antibiotic
Erythromycin	Bacterium	Antibiotic
Digitalis	Foxglove	Heart stimulant
Quinine	Chincona bark	Malaria treatment
Diosgenin	Mexican yam	Birth-control drug
Cortisone	Mexican yam	Anti-inflammation treatment
Cytarabine	Sponge	Leukemia cure
Vinblastine, vincristine	Periwinkle plant	Anticancer drugs
Reserpine	Rauwolfia	Hypertension drug
Bee venom	Bee	Arthritis relief
Allantoin	Blowfly larva	Wound healer
Morphine	Poppy	Analgesic

FIGURE 14.5 The rosy periwinkle from Madagascar provides recently discovered anticancer drugs that now make childhood leukemias and Hodgkin's disease highly remissible.

We get additional valuable products from forests and other natural communities. Rattan, cane, sisal, rubber, pectins, resins, gums, tannins, vegetable oils, waxes, and essential oils are among the products gathered in the wild. Many wild species have potential for production of a variety of useful products. Guayule, for instance, is a shrub native to the deserts of Texas and New Mexico. It produces latex that is essentially identical to that harvested from rubber (*Hevea*) trees. In 1910, guayule accounted for 10 percent of the world supply of rubber and 50 percent of the United States supply. It grows well in poor soil and lends itself to genetic improvement through breeding programs. The plants live up to fifty years and represent a stockpile of rubber. Species in the Euphorbia family (milkweeds, etc.) are also being investigated as a source of rubber, alkaloids, and other valuable organic chemicals.

These are only a few of the many valuable commercial products we derive from wild species of plants and animals. There are probably thousands of potential products that have not yet been recognized. These products alone are a good reason to protect biological diversity.

Medicine

Wild plants and animals are sources of drugs, analgesics (pain killers), pharmaceuticals, laxatives, antibiotics, heart regulators, anticancer and antiparasite drugs, blood pressure regulators, anticoagulants, enzymes, and hormones (table 14.2). More than half of all prescriptions in the United States contain some natural products. The total value of drugs from natural sources amounts to nearly $3 billion per year in the United States alone. Patent medicines, such as laxatives and cough and cold remedies from natural sources, add an equal amount to the economy.

What about future uses of wild plant and animal resources? A sea squirt (tunicate) has recently been discovered that produces a transplant antirejection drug more effective and one hundred times less toxic than the one now used. A sponge contains compounds that could revolutionize arthritis treatment. Since 1960, the National Cancer Institute (U.S.) has tested approximately thirty thousand plants for anticancer drugs. About three thousand species make potentially useful compounds, fifteen have had clinical trials, and five are expected to be released soon for general use. The species tested represent less than 10 percent of the total number of flowering plants. More than 70 percent of the untested species live in the tropics where co-evolution of plants and animals has produced an immensely complex chemical mosaic. Unfortunately, thousands, perhaps even millions, of potentially useful plants are being eliminated by tropical forest destruction before they have even been identified by science, much less investigated for useful products.

Consider the relatively recent success story of vinblastine and vincristine. These anticancer alkaloids are derived from the Madagascar periwinkle (*Catharanthus roseus*) (figure 14.5). They bind to microtubules in the cell and prevent cell division, thus, they inhibit the growth of cancer cells. Twenty years ago, before these drugs were introduced, childhood leukemias were invariably fatal. Now the remission rate for some childhood leukemias is 99 percent. Hodgkin's disease has gone from 98 percent fatal to about 40 percent fatal, thanks to these compounds. The total value of the periwinkle crop is roughly $15 million per year, although Madagascar gets little of those profits (chapter 15).

FIGURE 14.6 African violet (*Saintpaulia ionantha*). These familiar house plants have almost totally disappeared from their native habitat in South Tanzania, Africa. While the species is in no danger of extinction, valuable genetic traits and ecological relationships in the wild population may have been lost.

Ecological Benefits

Natural biological communities play important ecological roles in producing and sustaining habitable environments. Soil formation, waste disposal, air and water purification, nutrient cycling, solar energy absorption, and management of biogeochemical and hydrological cycles all depend, to a significant extent, on nondomestic plants, animals, and microbes (chapter 3). Earth's ecosystems represent the culmination of historic evolutionary processes of immense antiquity and majesty. They have resulted from billions of years of evolution under conditions that may never have occurred anywhere else in the universe. Wild species maintain ecological processes at no cost to us, and they represent a genetic library of information we could never reproduce.

Wild species also provide a valuable but often unrecognized service in suppressing pests and disease-carrying organisms. It is estimated that 95 percent of the potential pests and disease-carrying organisms in the world are kept under control by other species that prey upon them or compete with them in some way. In most cases, we are not even aware of those interactions. We find out how valuable natural predators are when we try to control systems with synthetic chemicals because broad-spectrum biocides kill both pests and natural predators. As a result, pest populations often surge to higher levels than before (chapter 11). By preserving natural areas and conserving wild species, we utilize the stabilizing diversity of nature that keeps pest organisms in balance.

Aesthetic and Cultural Benefits

Wild species add to our appreciation and enjoyment of the environment in many ways. All of our familiar domestic plants are derived from wild ancestors, many of which are now endangered in their native habitat (figure 14.6)

Gathering wild food is no longer a necessity for most people in developed countries, but many millions enjoy hunting, fishing, mushroom picking, nut gathering, and other outdoor activities

FIGURE 14.7 One maned lion in an African National Park can bring in a half million dollars in tourism over its lifetime, but only a few thousand dollars for hunting. For many people, just knowing that wild species exist has value.

that involve wild species. These activities provide healthful physical exercise, a chance to renew pioneer skills, and a reason to enjoy a wild environment. Other people find enjoyment in just observing and photographing wild species. There are some 8 million birdwatchers and 42 million hunters and anglers in the United States. The total amount spent each year in America on these activities is probably several billion dollars.

In many cases, land is more valuable as a wildlife preserve than it would be if converted to cropland. Wildlife biologist Michael Soulé estimates that one maned lion in Kenya's Amboseli National Park is worth $515,000 for tourist viewing but is worth only $8,500 for hunting (figure 14.7). The economic yield from tourists who come to see a lion in Africa is equal to the income from a herd of 30,000 cows. Unfortunately, many marvelous wildlife resources are being lost to land development and illegal hunting. We will discuss how commercial trade in furs, ivory, and pets contributes to destruction of wildlife resources later in this chapter.

For many people, the value of wildlife goes beyond the opportunity to shoot or photograph or even see a particular species. They argue that **existence value,** simply knowing that a species exists, is reason enough to protect and preserve it. We contribute to programs to save bald eagles, redwood trees, whales, whooping cranes, desert pupfish, and a host of other rare and endangered organisms because we like to know they still exist somewhere, even if we may never have an opportunity to see them. A particular species or community of organisms may have emotional value for a group of people because they feel that their identity is inextricably linked to the natural components of the environment that shaped their culture. This may be expressed as a religious value, or it may be a psychological need for access to wildlife. In either case, we often place a high value on the preservation of certain wild species.

RESOURCES AND RESOURCE ECONOMICS

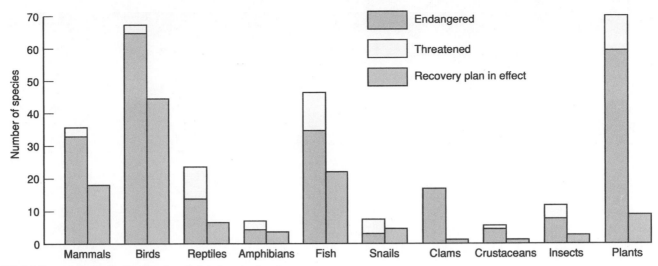

FIGURE 14.8 Numbers of officially classified endangered and threatened species in the United States. The fact that so many more birds and mammals are on this list than insects represents our preference for some life-forms and our relative indifference to insects, rather than actual numbers of endangered species.

Source: U.S. Fish and Wildlife Service.

DESTRUCTION OF BIOLOGICAL RESOURCES

Biological resources are diminished or destroyed in a number of ways. Natural changes in the environment eliminate once successful species or reduce their numbers to mere remnant populations. Humans disrupt ecosystems and extirpate species, both deliberately and accidentally.

Kinds of Losses

Three kinds of losses are of major concern: (1) *loss of abundance* of a once plentiful species, (2) *species extinction*, and (3) *ecosystem disruption*. There is an increasing gradient of seriousness in these losses. If a once abundant species is depleted, it probably can be restored if its proper habitat still exists and if its ecological niche hasn't been usurped. The herds of deer, elk, antelope, and buffalo that once roamed the American plains have been replaced by even larger numbers of domestic cattle and sheep. Since the wild species are still preserved in wildlife refuges, however, they could be reintroduced if we chose to do so. Figure 14.8 shows the numbers of species listed as endangered or threatened in the United States. To some extent, the numbers represent how interested we are in various groups rather than the actual number in trouble.

Extinction represents a permanent loss. Each species represents a unique set of characteristics resulting from long interaction between the genetic system and its environment. Because past evolutionary conditions can never be reproduced exactly, each species represents an irreplaceable resource. When a species is destroyed, not only is that particular set of characteristics lost, but also all of the potential adaptations and developments that might have appeared in its offspring.

Ecosystem disruption is even more serious. In a healthy ecosystem there is usually enough diversity so that a single species, if lost, can be replaced by other organisms that use the same resources. Beyond a certain point, however, losses can leave an ecosystem so impoverished that major ecological processes are disrupted and some niches cannot be filled. A catastrophic **decline spiral** can be set in motion that not only destroys a particular community but may even affect the whole biosphere. There is a worry that human disruptions of natural systems, such as by deforestation and marine pollution, could have worldwide effects in some cases.

Natural Causes of Extinction

Extinction is neither a new phenomenon nor a process caused only by humans. Studies of the fossil record suggest that more than 99 percent of all species that ever existed are now extinct. Most of those species disappeared long before humans came on the scene. Species arise through processes of mutation, isolation, and natural selection. Evolution can proceed gradually over millions of years or may occur in large jumps when new organisms migrate into an area or when environmental conditions change rapidly. New forms replace their own parents or drive out less adapted competitors. In a sense, species that are replaced by their descendants are not completely lost. The tiny *Hypohippus*, for instance, has been replaced by the much larger modern horse, but most of its genes probably still survive in its distant offspring.

TABLE 14.3 Characteristics that affect chances of extinction

Endangered	Example	Safe	Example
Individuals of large size	Florida cougar	Individuals of small size	Bobcat
Predator	Hawk	Grazer, scavenger, insectivore, etc.	Vulture
Narrow habitat tolerance	Orangutan	Wide habitat tolerance	Chimpanzee
Valuable fur, hide, oil, etc.	Chinchilla	Limited source of natural products and not exploited for research purposes	Gray squirrel
Hunted for market or for sport where there is no effective game management	Green Jungle Fowl	Commonly hunted for sport in game management areas	Mourning dove
Has a restricted distribution: island, desert watercourse, bog, etc.	Bahamas parrot	Has broad distribution	Yellow-headed parrot
Lives largely in (or migrates across) international boundaries	Green sea turtles	Has populations that remain largely within the territory(ies)	Loggerhead sea turtle
Intolerant of the presence of humans	Grizzly bear	Tolerant of humans	Black bear
Species reproduction in 1 or 2 vast aggregates	West Indian flamingo	Reproduction by pairs or in small groups	Bittern
Long gestation period, 1 or 2 young per litter, and/or prolonged maternal care	Giant panda	Short gestation period, more than 2 young per litter, and/or young become independent early	Raccoon
Has behavioral idiosyncracies that are nonadaptive today	Red-headed woodpecker: flies in front of cars	Has behavior patterns that are particularly adaptive today	Burrowing owl: highly tolerant of noise and low-flying aircraft; lives near runways of airports

Reprinted from *Biological Conservation*, by David Ehrenfeld (Holt, Rinehart & Winston, 1970). Reprinted by permission of the author.

Mass Extinctions

The geological record shows that a number of widespread biological catastrophes have caused mass extinctions of plants and animals over the course of geological history. The best known of these cycles occurred at the end of the Cretaceous period when dinosaurs disappeared, along with at least 50 percent of existing genera and 15 percent of marine animal families (chapter 9). An even greater disaster occurred at the end of the Permian period about 250 million years ago when two-thirds of all marine species and nearly half of all plant and animal families died out over a period of about 10,000 years—a short time by geological standards. The extent of these catastrophes makes us consider what global climate change might do (chapter 17).

Although extinction is obviously disastrous for the organisms to which it occurs, it provides an opening for the next stages of evolution. We can speculate that the end of the dinosaurs made possible the age of mammals, of which we are the beneficiaries. An important lesson learned from fossil studies is that evolutionary change is based on the species and characteristics that happen to be available in a particular environment at a particular time. Neither the individual organisms present in a given ecosystem nor the ecosystem itself are necessarily perfectly designed or ideally adapted to existing conditions. Table 14.3 shows some characteristics that affect the probability of extinction for a species.

Evolutionary history suggests that this may not be the best of all possible worlds. We shouldn't assume that every ecosystem is a perfect and fragile balance of irreplaceable species. On the other hand, we can't assume that if we exterminate existing life-forms, they will be replaced with equal or better ones. As Aldo Leopold said, "The first rule of intelligent tinkering is to save all the pieces."

Human-Caused Extinction

Humans have a long history of biological resource depletion (figure 14.9). Stone age human hunters may have been responsible for the extermination of the "megafauna" of both America and Eurasia during the Pleistocene era, some 20,000 years ago. Mastodons, mammoths, giant bison, ground sloths, early horses, and cameloids all disappeared from North America about the time humans first appeared here. Climatic change may have been partially or primarily responsible, but vast "boneyards" in Europe and villages constructed entirely of mammoth bones in Siberia suggest the hunting prowess of our ancestors.

Ancient civilizations were probably responsible for the extinction of many species. Misuse of the land and destruction of biological resources have played a role in the demise of every major civilization from the Babylonian, Sumerian, Indus, and Mayan cultures to ancient Greece, Rome, and China.

RESOURCES AND RESOURCE ECONOMICS

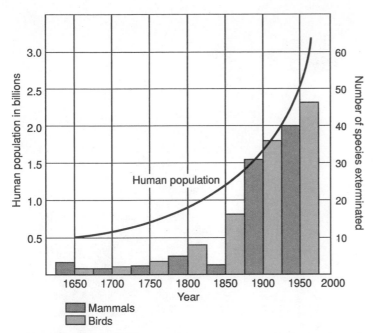

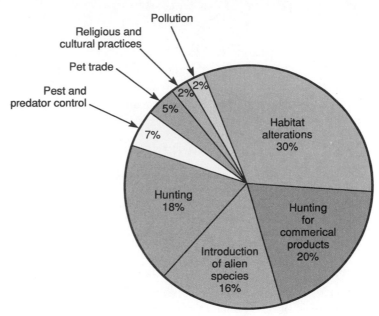

FIGURE 14.9 As human population increases, so do the numbers of extinct species of mammals (light bars) and birds (dark bars). Species are lost both through direct actions, such as hunting, and indirect actions, such as habitat destruction.

FIGURE 14.11 Estimated contribution of various human activities in wildlife extinction. More than one activity may be involved in extermination of a particular species.

Source: Data from World Wildlife Fund.

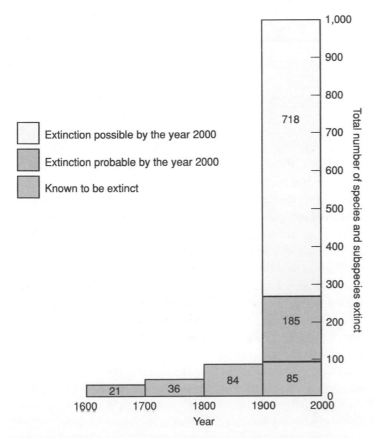

FIGURE 14.10 Extinction forecast. Rates of extinction appear to have increased at least threefold and may reach one hundred times natural rates by the end of this century.

Current Extinction Rates

The rate at which species have been lost appears to have increased dramatically over the last one hundred years (figure 14.10). Before humans became a major factor, extinction rates from natural causes appear to have been one species lost every five to ten years. Between A.D. 1600 and 1900, human activities seem to have been responsible for the extermination of about one species per year. Harvard entomologist E. O. Wilson estimates that we now are pushing 20,000 species a year into extinction. We cannot be absolutely sure of these rates because many parts of the world haven't been thoroughly explored, and many species may have disappeared before they were studied and classified by biologists. In North America, sixty-one species of flowering plants and six bird species are known to have been eliminated since Europeans arrived.

The main reason for the current increase in extinctions is habitat loss (figure 14.11). Destruction of tropical forests, coral reefs, estuaries, marshes, and other biologically rich ecosystems threatens to eliminate thousands or even millions of species in a human-caused mass extinction that could rival those of geologic history. By destroying habitat, we eliminate not only prominent species, but also many obscure of which we may not even be aware. It has been suggested that half of all existing species could be endangered or extinct by the middle of the next century if this destruction continues. Many of the species now being lost have never been studied by scientists and may have valuable attributes that we will never be able to utilize.

WAYS HUMANS CAUSE EXTINCTION

How do we cause extinction? People rarely intend to completely destroy other species; but through overhunting, overfishing, habitat destruction, and introduction of diseases and exotic competitors, we have caused extinction of many species. In this section, we will look at some of the ways we extirpate species.

Direct Destruction of Wildlife

First, we will examine ways that humans destroy species through direct actions.

Hunting and Food Gathering

Overharvesting of food species is probably the most obvious type of human destruction of biological resources. The great auk was a large, flightless ocean bird that formed huge colonies on the islands and shores of the North Atlantic from Newfoundland to Scandinavia (figure 14.12). The French called the auks "sitting ducks," *pingouin* (from which the name for antarctic penguins was derived). They were so easy to catch that hunters could kill hundreds in a few minutes. Millions of birds and eggs were collected for food and oil. In 1844, Icelandic fishermen killed the last two known birds to provide skins for a collector.

Some other well-known overhunting cases include the near extermination of the great whales (figure 14.13) and the American bison (or buffalo) and extinction of the passenger pigeons (Box 14.1). In 1850, some 60 million bison roamed the western plains. Forty years later, there were only about one hundred and fifty wild bison and two hundred and fifty in captivity. Many were killed only for their hides or tongues, leaving as much as a ton of meat in each carcass to rot on the plains. Much of the destruction was carried out by the U.S. Army to vanquish native peoples who depended on bison for food, clothing, and shelter and thereby force them onto reservations. Native people also killed many bison wastefully. In 1832, George Catlin reported that a group of Indians brought more than one thousand bison tongues to a trading post along the Missouri River to trade for a barrel of whisky and a few trinkets.

Fishing

Fish stocks have been seriously depleted by overharvesting in many parts of the world. A classic case of the collapse of a fishery is the California sardine. Commercial fishing for this once abundant species began in the 1920s. In 1936, 750,000 tons were harvested. Twenty years later, in 1956, the total catch was 1 ton.

Modern fishing techniques are especially destructive. Huge trawlers churn up spawning beds and feeding banks, sucking up everything on the bottom, including baby fish, invertebrates, and noneconomic species. Only large fish are kept, but after the trauma of sudden decompression and entrainment in the pumps, most of what goes back into the ocean is dead or dying. Another ominous development are the huge, super-efficient drift nets now

FIGURE 14.12 The great auk once formed huge colonies on North Atlantic islands. Overhunting made them extinct by 1844. How many other species have also vanished?

in wide-spread use around the world. Each net can be up to 48 km (30 mi) long and 10 m (32.5 ft) deep. As they sweep across the oceans, these "walls of death" trap and drown millions of sea birds and marine mammals, and kill uncounted tons of unwanted fish species. The Asian squid-fishing fleets lay 3 million km (2 million mi) of drift nets annually in the Pacific. Thousands of miles of drift nets are lost or abandoned every year. These ghost nets will continue to catch and kill fish and diving birds for years as they drift across the oceans. In 1990, the U.N. General Assembly called for a worldwide moratorium on drift net fishing. Whether this ban will be effective remains to be seen.

Shrimp trawlers catch several tons of "trash fish" for every ton of shrimp they keep. They also catch and drown an estimated 10,000 endangered sea turtles in their nets each year. The National Marine Fisheries Service invented a metal and net box called a turtle excluder device (TED) to save turtles from drowning in shrimp nets. The TED has slats that let shrimp into the net, but push turtles up and out a trap door. In 1989, shrimpers were ordered to install and use TEDs but many refused to comply because they claimed it reduces the shrimp catch.

BOX 14.1 Passenger Pigeons

Two hundred years ago the passenger pigeon (*Ectopistes migratorius*) was the world's most abundant bird. Although the species was found only in eastern North America, the population of 3 to 5 billion birds amounted to about one-fourth of all North American land birds. Early in the nineteenth century, John James Audubon witnessed a flock of birds that took three days to pass overhead. He estimated it to be several miles wide and hundreds of miles long, containing perhaps a billion birds. It seemed impossible to early settlers that this resource could ever be exhausted; yet the vast flocks disappeared in a twenty-year period after the Civil War. How did this happen? The extinction of this abundant and once-successful species is a case study in biological resource mismanagement.

Colonial behavior was the key to the passenger pigeon's success in the wild and to its extinction by human hunting pressures. The birds fed on rich crops of acorns, beechnuts, and chestnuts in the eastern deciduous forest. The trees didn't yield nuts uniformly, however. Their reproductive pattern was to produce small crops for several years and then produce a bumper crop that satiated local nut predators, ensuring that some seedlings could survive and germinate. The pigeons were highly mobile and could forage widely and concentrate quickly in areas with large nut crops.

The pigeons also found safety in numbers, nesting and traveling in huge flocks. In 1871, a nesting colony was reported in Wisconsin that covered around 1,500 sq km (850 sq mi) and contained an estimated 135 million adult birds. Every suitable tree in the area had nests, sometimes as many as one hundred per tree. The flocks rarely nested in the same area two years in a row, so local predator populations could not build up enough to reduce the size of the bird population. Because of passenger pigeons' high reproductive success, each pair of birds produced only one egg per year.

In Audubon's day, hunters had little impact on the pigeons. Nesting in a particular area was a bonanza for local residents, but the number of pigeons that

BOX FIGURE 14.1 Shooting passenger pigeons in Iowa. In the nineteenth century huge flocks of pigeons darkened the skies for days as they passed overhead. Market hunters traveling by train destroyed so many of these birds that nesting failed and the species became extinct.

individuals could eat or store was limited. The development of the railroads and the telegraph changed the situation, however. The telegraph kept market hunters informed of the location of the flocks, and railroads made it possible to travel hundreds of miles overnight to reach colonies and to ship birds to market. Although there probably never were more than one thousand professional pigeoners, they took a frightful toll. Birds were killed with guns, nets, dynamite, and traps. Young birds were knocked from nests with poles, and nest trees were burned or cut down.

The impact of hunting was devastating. In 1878, approximately 15 million birds were shipped to market from a single colony near Petosky, Michigan, and perhaps twice that many died. It is thought that not a single nestling survived this slaughter. In spite of the immense number of birds, there probably were never more than a dozen nesting areas active in any one year, and heavy hunting pressures could disrupt every one of them.

The population of passenger pigeons plummeted abruptly as a result of

reproductive failure that reduced the numbers below the threshold for viability. Because the birds were adapted to colonial life, they probably were dependent on high densities to stimulate mating and nesting. Lacking the protection of their former numbers, the birds were more vulnerable to natural predators and less able to find suitable foraging areas. In the 1890s, several states passed laws to protect the pigeons from overhunting, but it was already too late. The last known wild bird was shot in 1900, and the last passenger pigeon, a female named Martha, died in 1914 in the Cincinnati Zoo.

It is not clear whether passenger pigeons could have survived to modern times. The forests that sheltered and fed them are mostly gone, but perhaps if wildlife resources had been managed more carefully in the nineteenth century, at least a remnant of the vast flocks of passenger pigeons might have remained for us to see.

FIGURE 14.13 A blue whale (*Balenoptera musculus*). Reaching a length of 33 m (100 ft) and a weight of 200 metric tons, these leviathans are bigger than dinosaurs and are the largest animals ever known. Their population before commercial whaling was estimated to be about 200,000; now, only about 10,000 remain. When spread through all the world's oceans, the whales may be too dispersed to find each other to mate.

Cashing in on Animal Products

Undoubtedly, the most valuable nonfood products from wild animals are furs and skins. Fur coat sales amounted to some $1.5 billion worldwide in 1985. The fur industry says that only 10 percent of the furs sold in the United States each year come from wild animals, but trapping opponents say that the total is closer to 50 percent. The U.S. Department of the Interior reports that 100 million steel-jaw, leghold traps are used in the United States each year and that approximately 25 million animals are trapped. Two-thirds of those animals are discarded, however, because they are the wrong species or are too badly mangled or decomposed to be useful. Publicity and protests by environmental and animal rights groups have had a dramatic impact on the killing of baby harp and hooded seals. The annual kill dropped from some 200,000 in 1983 to about 5,000 in 1985. Nevertheless, there is still a very damaging traffic in skins and furs of endangered species that we will discuss later in this chapter.

In spite of progress in wildlife protection and international conventions against trade in endangered species, overharvesting of many species continues today because the economic pressures are great. Tiger or leopard fur coats can bring $100,000 in Japan or Europe. The population of African black rhinos dropped from approximately 65,000 in 1973 to about 3,000 in 1990 because of a demand for rhino horn handles for jambiyya (traditional curved daggers) of the Middle East or for folk medicine in Africa and Asia.

Perhaps no animal symbolizes wasteful exploitation as much as the African elephant. In 1979, there were an estimated 1.3 million elephants in Africa. Ten years later, only 625,000 remained (figure 14.14). It is estimated that 100,000 elephants are killed by poachers each year for the ivory trade (figure 14.15). It is little wonder that natives are tempted by this opportunity for wealth. When wholesale prices for ivory jumped from $10 to over $100 per kg (2.2 lbs) in the 1970s, a medium-sized set of tusks was equal to ten years' income for a subsistence farmer. The Convention on International Trade in Endangered Species (CITES) voted in 1989 to ban all ivory trade. The ban seems to be working, and the world ivory market has collapsed.

Harvesting Wild Plants

Plants also are threatened by overharvesting. Wild ginseng has been nearly eliminated in many areas because of the Asian demand for roots that are used as an aphrodisiac and folk medicine. Cactus "rustlers" steal cacti by the ton from the American southwest and Mexico. In 1977, some 10 million cactus plants were shipped from Texas to both domestic and foreign markets. With prices ranging as high as $1,000 for rare specimens, it's not surprising that many are now endangered. TRAFFIC (a program of the World Wildlife Fund) reported that some twenty thousand illegal orchids were exported by smugglers in 1980.

Pet and Scientific Trade

The trade in wild species for pets is an enormous business. Worldwide, some 5 million live birds are sold each year for pets, mostly in Europe and North America. In 1980, pet traders imported (often illegally) into the United States some 2 million reptiles, one million amphibians and mammals, 500,000 birds, and 128 million tropical fish. Keeping an aquarium is one of the most popular hobbies in America. About 75 percent of all saltwater tropical aquarium fish sold come from the marvelously rich coral reefs of the Philippines. These fish are caught by divers who use plastic squeeze bottles of cyanide to stun their prey. Far more fish die with this technique than are caught. Worst of all, it kills the coral animals that create the reef. A single irresponsible diver can destroy all of the life on 200 square meters of reef in a day. Altogether, thousands of divers now working destroy about 50 km² of reefs each year. Net fishing would prevent almost all this destruction. It could be enforced if pet owners would insist on net-caught fish.

Smuggling of rare and endangered species is particularly lucrative. Bird collectors will pay $10,000 for a hyacinth macaw from Brazil or $12,000 for a pair of golden-shouldered parakeets from Australia. A rare albino python might bring $20,000 in Germany. The destruction in this live animal trade is enor-

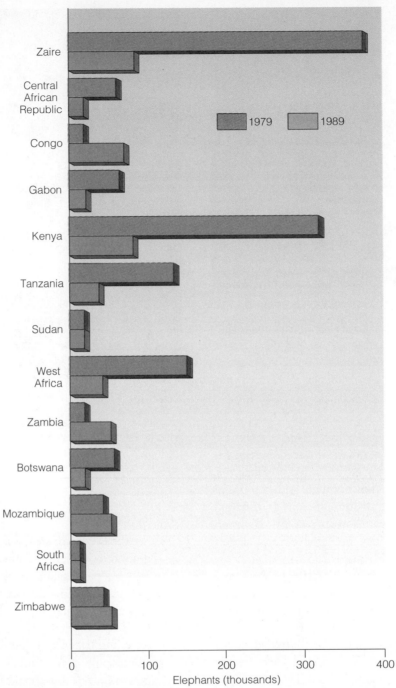

FIGURE 14.14 Changes in African elephant populations, 1979–1989. More than half the elephants in Africa were killed during this 10-year period, mainly by poachers who wanted to sell the ivory tusks.

Source: *TRAFFIC USA*, World Wildlife Fund, Washington, DC, 1989 (personal communication).

mous. It is generally estimated that fifty animals are caught or killed for every live animal that gets to market. It is the buyers, many of whom say they love animals, who keep this trade going.

Every year, approximately 100 million animals are used in scientific research. Most of those animals are lab-reared rats and mice, but some experiments require chimpanzees, monkeys, and other species obtained from the wild, where they are becoming rare. Unfortunately, one of the main techniques for capture is to kill a mother and take her baby. Many young animals don't survive capture and shipping. We might be able to spare many of these deaths if we develop alternate techniques for medical research, such as human cell cultures and organ cultures.

FIGURE 14.15 Soldiers load illegal elephant tusks seized from poachers. In 1989, Kenya burned 2,400 confiscated tusks worth more than $3 million in an attempt to end the global ivory trade that threatens elephants with extinction.

Source: Data from *TRAFFIC USA*, World Wildlife Fund, Washington, DC, 1989 (personal communication).

FIGURE 14.16 Sarah, a Hungarian Komondor guarding dog, protects a flock of goats from predators. Guarding dogs are not trained to herd sheep or goats; they simply stay with them to keep predators away. This not only saves livestock but avoids trapping, shooting, or poisoning the predators.

Predator and Pest Control

Some animal populations have been greatly reduced or even deliberately exterminated because they are regarded as dangerous to humans or livestock or because they compete with our use of resources. The first recorded case of human-caused extinction was extermination of the European lion in A.D. 80. Between 1937 and 1970, U.S. government predator control agents trapped, poisoned, or shot 23,800 bears, 7,255 mountain lions, 1,574 gray wolves, 50,283 red wolves, 477,104 bobcats, and 2,823,056 coyotes. Uncounted millions of animals were killed unintentionally by poisoned bait or misplaced traps, or intentionally by private individuals for bounty or sport.

Predator control programs demonstrate the paradox that can exist when human-wildlife conflicts are not examined in a holistic manner. In 1990, predator-related livestock losses (both confirmed and unconfirmed) in the U.S. amounted to $27.4 million. The cost of predator control that year by the Office of Animal Damage Control (as it is now called) amounted to $38 million. It would have been much cheaper to leave the wild animals alone and pay farmers and ranchers for their losses.

Three main techniques are used in predator and pest control:

1. Poison baits, including poisoned grain, suet pellets, and carcasses laced with such compounds as arsenic, strychnine, and 1080 (sodium flouroacetate) are distributed widely for rodent and predator control. In some areas, baits are distributed by airplane.

2. Leghold traps or steel-jaw traps are buried in the ground or hidden in the water to catch larger mammals. When a passing animal steps on the trigger pan, the jaws snap closed. Abandoned traps may remain active for years, catching pets and other nontarget species.

3. Getters are spring-activated or explosive-charged poison cartridges baited with animal scent or carrion. When an animal pulls at the bait, the getter fires a charge of poison into its mouth and face.

Many poisons used in pest and predator control are wide-spectrum, long-lasting biocides. Once they contaminate soil or water or get into the food chain, they persist, moving from organism to organism and affecting many nontarget species. One of the most controversial of these agents is 1080. First introduced in 1945, this powerful, persistent poison is capable of killing not only the primary target animal, but also a whole series of scavengers and decomposers that eat the poisoned carrion as it spreads through the food chain. Over 150,000 baits laced with 1080 were distributed before its use was banned in 1972. Recently, ranchers have been pushing for permission to use 1080 collars on livestock. The premise is that wolves and coyotes will

TABLE 14.4 Some destructive introduced exotic species

Common name	Origin	Introduced to	Effects
Nile perch	Egypt	Lake Victoria, Africa	Eating hundreds of rare fish species
Grass carp	Taiwan	Southern United States	Competes with local species; eats native plants
Domestic pigs	Europe	Hawaii	Eats ground-nesting bird eggs; roots up soil and vegetation
African killer bees	Africa	South, Central, and North America	Aggressive behavior; dangerous to wildlife, livestock, and humans
Zebra mussel	Europe	Great Lakes	Clogs drinking water intakes; depletes fish food; covers spawning beds
English "sparrows"	England	Entire United States	Competes with native birds for food and nesting space
Purple loosestrife	Europe	Northern wetlands	Fills wetlands and crowds out native species and wildlife
Eurasian milfoil	Europe	Northern lakes	Chokes lakes, makes boating and swimming difficult, and smothers fish and other plants
Tamarisk trees	West Africa	Desert southwest	Crowds out native plants; depletes soil moisture
Multiflora rose	Asia	Eastern United States	Creates impenetrable briar patches; takes over woods and pastures

grab their prey by the neck, puncturing the collar and getting a lethal dose of poison. Each collar, however, contains enough poison to kill 185 coyotes or six men. A much better solution would be to employ more herders and guard dogs to watch the sheep and cattle (figure 14.16).

Indirect Damage

Although hunting and some other human pursuits that kill wildlife directly often have disastrous effects on specific species, processes that destroy habitat and disrupt ecosystems can endanger far more species than any other human actions. We now turn to some indirect ways humans affect plants and animals.

Habitat destruction

There are many historic examples of human disturbances of natural systems. In much of the Middle East, Asia, and the Americas, once-fertile areas have become deserts because of unsound forestry, grazing, and agricultural practices (chapter 10). The whole Mediterranean Basin, in fact, has been described as a "goatscape" because of the destruction caused by domestic animals. Technology now makes it possible for us to destroy vast areas even faster than in the past. Native forests are cut down, marginal land is plowed for crops, and plant cover that protects the land from erosion is destroyed with herbicides or tractors dragging huge chains.

Undoubtedly the greatest current losses in terms of biological diversity and unique species occur when tropical moist forests are disrupted. Of some 10 million sq km of original tropical forests, more than a third has been disturbed to some extent by humans. Approximately 11 million ha of tropical forests—an area larger than Austria—are disrupted every year by agriculture, firewood collecting, rural development, and logging (chapter 12). It is thought that more than half of all the species on Earth live in the tropics. Upsetting these complex ecosystems could endanger millions of species.

If the global climate changes as rapidly and to the extent predicted by the worst-case scenarios for the "greenhouse effect" (see Chapter 17) there could be massive losses of species unable to adapt to new conditions in the next century. This could rival the catastrophic die-offs at the ends of geologic epocs in the past.

Exotic Species Introductions

Human tinkering with natural environments often has disastrous consequences. When a species in introduced into a new environment, it may be released from restraints imposed by predators and competitors. A mild-mannered, useful organism can become a monster when introduced deliberately or accidentally (Table 14.4). One of the best examples is the introduction of exotic or alien species. In 1859, for instance, a shipment of twenty-four wild English rabbits arrived in Victoria, Australia, and they were set free to provide food and sport hunting. Finding an ideal habitat and no natural predators, the rabbit population exploded. Within six years, the landowner who first released them killed twenty thousand on his own land and still had an equal number left. The rabbits stripped vegetation from pastures and depleted the range, starving both livestock and the native kangaroos (figure 14.17). A 2,000 mile fence was erected across the country to keep the rabbits out of western Australia, but before the fence was finished, the rabbits were on the other side. Mixomatosis, a viral disease, was introduced to kill off the rabbits. Many rabbits were killed initially, but a virus-resistant population quickly developed. It looks as if they are going to be a permanent problem.

FIGURE 14.17 Rabbits gather to drink in the Australian outback. Introduced for sport and hunting in 1849, these wild English rabbits had no predators in Australia and quickly overran the entire continent. Now a destructive and ubiquitous pest, they are there to stay.

Many nuisance birds of our cities, such as starlings, English sparrows, and common pigeons, are aliens that were introduced by "acclimatizers" who spent considerable sums of money to import these species from Europe. These birds have prospered because they are opportunists that do well in urban areas, but they also have driven out more desirable native species in many places.

A classic example of meddling with nature was the importation of the mongoose to Caribbean and Hawaiian Islands. Because sugar cane fields in Cuba, Jamaica, and Hawaii were heavily infested with roof rats in the nineteenth century, plantation owners imported mongooses from India to kill the rats. However, the rats were nocturnal and arboreal, while the mongooses were mostly diurnal and terrestrial; thus, instead of attacking the rats, the mongooses quickly exterminated many defenseless native birds.

Plants also can multiply explosively when introduced into a new environment. Kudzu, a mild-mannered Asiatic legume, has been called "the vine that swallowed Georgia" because it blankets trees, houses, and hillsides. Water hyacinths are floating water plants with waxy green leaves and beautiful blue flower spikes. They were brought to the United States from Argentina in 1880 for an exhibition in New Orleans. People took plants home and either deliberately put them in streams and lakes or just tossed them out. Within ten years, this prolific weed had spread across the South from Florida to Texas, completely covering the surface of many waterways, blocking boat traffic, and smothering fish and native plants. Mechanical harvesters and poisons were tried, but the plant population doubled every two weeks. Ironically, there now appear to be some benefits to this pest that we have fought for nearly a century. The water hyacinth has been proposed as a source of biomass for energy production since it grows so fast. It is also useful in water purification, absorbing large amounts of contaminants, such as heavy metals and toxic organic compounds.

Diseases

Disease organisms, or pathogens, may be considered predators. To be successful over the long term, a pathogen must establish a balance in which it is vigorous enough to reproduce, but not so lethal that it completely destroys its host. When a disease is introduced into a new environment, however, this balance may be lacking and an epidemic may sweep through the area.

The American chestnut was once one of the major trees in the Eastern deciduous forest, comprising some 25 percent of the forest canopy. Often over 45 m (150 ft) tall, 3 m (10 ft) in diameter, fast growing, and able to sprout quickly from a cut stump, it was a forester's dream. Its nutritious nuts were important for birds (like the passenger pigeon), forest mammals, and humans. The wood was straight grained, light, and as rot-resistant as redwood. It was used for everything from fence posts to fine furniture. In 1904, a shipment of nursery stock from China brought a fungal blight to the United States, and within about forty years the American chestnut had all but disappeared from its native range. Efforts are now underway to transfer blight-resistant genes into the few remaining American chestnuts that weren't reached by the fungus or to find biological controls for the fungus that causes the disease.

A similar disaster has all but exterminated the American elm (figure 14.18). A shipment of European elm logs introduced a fungal disease fatal to the American elm. The loss is most noticeable in prairie towns of the Midwest where elms formed arching colonnades over the streets and provided an oasis of shade from the summer sun. As Dutch elm disease swept across the country, some towns lost all of their trees in just a few years. Des Moines, Iowa, had to remove some 250,000 trees in five years at a cost of about five million dollars.

Pollution

We have known for a long time that toxic pollutants can have disastrous effects on local populations of organisms (chapters 23 and 24). The publication of *Silent Spring* by Rachel Carson in 1962 alerted the public to a much more insidious and widespread effect of pollution. Carson pointed out that pesticides and other pollutants were causing declines in many nontarget or-

FIGURE 14.18 Dutch elm disease (American elm wilt) destroyed virtually all the shade trees in many Midwestern towns. Top: a street in Waukegan, Illinois, as it appeared in the summer of 1962. Bottom: the same view after this fungal disease swept through the town.

ganisms and warned that they might even drive certain species into extinction. The book and its warnings were attacked by many as being unfounded, but subsequent studies have shown that many of Carson's concerns are valid.

Persistent pesticides like DDT (dichloro-diphenyl-trichloro-ethane) accumulate in food chains and are especially dangerous to top carnivores, such as hawks, eagles, and game fish. Largely because of DDT accumulated from their prey, peregrine falcons and brown pelicans disappeared from large parts of their former ranges in the United States. DDT disrupts hormone regulation of calcium levels and causes thin eggshells that break when nesting parents try to incubate them, resulting in reproductive failure. Since DDT was phased out in the early 1970s, its levels in the environment have been falling and both falcons and pelicans have been making a comeback.

Lead poisoning is another major cause of mortality for many species of wildlife. Bottom-feeding waterfowl, such as

ducks, swans, and cranes, ingest spent shotgun pellets that fall into lakes and marshes. They store the pellets, instead of stones, in their gizzards and the lead slowly accumulates in their blood and other tissues. The U.S. Fish and Wildlife Service (USFWS) estimates that 3,000 metric tons of lead shot are deposited annually in wetlands and that between 2 and 3 million waterfowl die each year from lead poisoning. Scavengers, such as condors and bald eagles, eat birds and mammals that have lead shot or fragments of bullets in their bodies. In 1987, the last five wild California condors were trapped and put into zoos. Lead poisoning was thought to be a major cause of their high mortality in the wild. In 1976, the USFWS banned the use of lead shot in certain areas with high accumulations of spent shot. Hunters are required to use nontoxic steel shot, but they complain that it causes excessive wear in gun barrels and has poor ballistic characteristics.

Genetic Assimilation

Some rare and endangered species are threatened by **genetic assimilation** because they crossbreed with closely related species that are more numerous or more vigorous. Opportunistic plants or animals that are introduced into a habitat or displaced from their normal ranges by human actions may genetically overwhelm local populations. For example, the rare southern red wolf crossbreeds with the more prolific coyote, whose range has been expanding because of its successful adaptation to human presence. The red wolf was nearly extinct by 1970 but is now being reintroduced through captive breeding programs (chapter 15). Black ducks have declined severely in the eastern United States in recent years. Hunting pressures and habitat loss are factors in this decline, but so is interbreeding with mallards forced into black duck habitat by destruction of prairie potholes in the Midwest.

SUMMARY

In this chapter, we have briefly surveyed world biological resources and the ways humans benefit from wildlife. Natural causes of wildlife destruction include evolutionary replacement and mass extinction. We also have studied the ways human actions threaten our wildlife resources. Among the direct threats are overharvesting of animals and plants for food and various industrial and commercial products, such as rubber, wood, and oil. Millions of live wild plants and animals are collected for pets, houseplants, and medical research. We also exterminate many plants and animals deliberately or inadvertently when we use pesticides to destroy "pest" species. Examples of indirect threats to biological resources are habitat destruction, the introduction of exotic species and diseases, pollution of the environment, and genetic assimilation.

The potential value of the species that may be lost if environmental destruction continues could be enormous. It is also possible that the changes we are causing could disrupt vital ecological services on which we all depend for life.

■ Review Questions

1. Why don't we grow hardwood trees on commercial plantations?
2. Is a single lion more valuable to the hunting industry or the tourism industry in Kenya?
3. Define extinction. What is the natural rate of extinction in an undisturbed ecosystem?
4. What is aquaculture?
5. Describe some useful products we get from wild plant species.
6. What are the three categories of ecological damage to species that are of major concern?
7. Describe the decline spiral.
8. Why did passenger pigeons become extinct?
9. Describe the three main techniques used in predator control.
10. What is genetic assimilation?

■ Questions for Critical Thinking

1. Why are there so many species on Earth? (Review chapters 4 and 5 to answer this.) Why only 30 million rather than 30 billion or 30 trillion?
2. What would our world be like if there were only a few species of organisms?
3. Why are krill so abundant? Why in the Antarctic?
4. According to our understanding of evolution, the stronger species dominate and flourish at the expense of weaker species. Since we, as humans, are now strong enough to exploit our environment, do we have any practical or ethical reasons to restrain ourselves?
5. Why do villagers in Indonesia use so many different species as food and medicine? Do you think that we might benefit from exploring that traditional knowledge?
6. Describe some of the ecological benefits derived from wild species.
7. Should the well-being of a species be important to us only if it affects us in some way, or should we consider species also for their existence value?
8. In the past, mass extinction has allowed for new growth, including the evolution of our own species. Should we assume that another mass extinction would be a bad thing? Could it possibly be beneficial to us? To the world?
9. Do you believe that humans might have contributed to the mass extinction of the New World megafauna at the end of the last ice age? Could you kill a woolly mammoth or a saber-toothed tiger?
10. Philippine divers need the cash they get from cyanide-poisoning of tropical reef fish for the pet trade. Do we have a right to interfere with their livelihood? How could we help preserve the reefs while still providing an income for the divers? How would you enforce such a program?

■ Key Terms

biological resources (p. 262)
decline spiral (p. 267)
existence value (p. 266)
extinction (p. 262)
genetic assimilation (p. 277)
wildlife (p. 264)

SUGGESTED READINGS

■ Abramovitz, J. N. 1991. *Investing in Biological Diversity*. Washington D.C.: World Resources Institute. Analyzes projects in 100 developing countries designed to protect and restore biodiversity.

■ Baker, R. *The American Hunting Myth*. New York: Vantage Press, 1985. Presents arguments for and against hunting.

■ Campbell, T. 1991. "Net Losses." *Sierra* 76 (2):48. A chilling investigative report on the destructive practice of driftnet fishing in the North Pacific.

■ Durant, M., and M. Saito. July 1985. "The Hazardous Life of our Rarest Plants." *Audubon*. 87 (4):50. Far more endemic races, varieties, and subspecies of higher plants are threatened with extinction than are higher animals. This is a good description of some examples of plant losses with lovely paintings of endangered species.

■ Durell, L. *State of the Ark: An Atlas of Conservation in Action*. Garden City: Doubleday, 1986. A popular and colorful overview of wildlife conservation.

■ Ehrlich, P. R., and A. H. Ehrlich. *Extinction: The Causes and Consequences of the Disappearance of Species*. New York: Random House, 1981. A highly readable discussion of the worldwide problem of loss of biological diversity.

■ Kirch, P. 1982. "The Impact of the Prehistoric Polynesians on the Hawaiian Ecosystem." *Pacific Science*. 36 (1):1. The first Hawaiians appear to have exterminated dozens of indigenous species.

■ Laycock, G. April 1978. "The Importation of Animals." *Sierra*. 63 (3):20. A description of deliberate introduction of exotic species by "acclimatizers" and their effects.

■ Leopold, A. *Game Management*. New York: Scribner's, 1933. The first textbook of wildlife management published in America. A classic by one of America's greatest naturalists.

■ McNeely, J. A. et al. 1990. *Conserving the World's Biodiversity*. New York: International Union for the Conservation of Nature. A valuable overview of the effects of human actions on biodiversity.

■ Miller, K. R., et al. "Issues on the Preservation of Biological Diversity." In *The Global Possible*, edited by R. Repetto. World Resources Institute. New Haven: Yale University Press, 1985. An overview of the problems of preservation of biological diversity with suggested actions.

- Myers, N. *The Sinking Ark*. Oxford: Pergamon Press, 1979. A discussion of habitat loss and species extinction by one of the world's leading experts on tropical rain forest destruction.
- Peters, R. L. ed. 1991. *Consequences of the Greenhouse Effect for Biological Diversity*. New Haven: Yale University Press. A comprehensive series of articles linking two of the most important environmental problems of our day.
- Soulé, M. 1986. *Conservation Biology: the Science of Scarcity and Diversity*. Sunderland: Sinauer Associates, 1986. A conservation biology textbook with contributions by many authors on theory as well as applications in specific case studies.
- U.S. Office of Technology Assessment. March 31, 1987. *Technologies to Maintain Biological Diversity*. Washington, D.C.: Government Printing Office. A unique compilation of information on the threat to biological diversity and the approaches by which species can be protected.
- Wilson, E. O. 1988. *Biodiversity*. Washington D.C.: National Academy Press. A compilation of articles by leading experts on biodiversity. An important landmark in this field.
- World Wildlife Fund. 1990. *The Official World Wildlife Guide to Endangered Species of North America*. New York: Beacham Publishers. A two-volume set describing each of the officially listed endangered species in the United States. Volume I treats plants and mammals. Volume II describes other species.
- Yates, S. July 1984. "On the Cutting Edge of Extinction." *Audubon*. 86 (4):62. The story of birds of Hawaii, how they got there, and how humans have threatened their existence.

CHAPTER *15*

Managing Biological Resources and Endangered Species

The nobler animals have been exterminated here—the cougar, panther, lynx, wolverine, wolf, bear, moose. . . . I seek acquaintance with Nature—to know her moods and manners. Primitive Nature is most interesting to me. I take infinite pains to know all the phenomena of the spring, for instance, thinking that I have here the entire poem, and then, to my chagrin, I hear that it is but an imperfect copy that I possess and have to read, that my ancestors have torn out many of the first leaves and grandest passages, and mutilated it in many places. I should not like to think that some demigod had come before me and picked out some of the best stars. I wish to know an entire heaven and entire earth.

Henry David Thoreau, 1856

Wildlife preservation has utilitarian benefits that provide us with food, medicine, and other commercial products. It also brings aesthetic benefits, such as increased knowledge and emotional satisfaction. Other species also have a right to exist because of their inherent or intrinsic value.

Popular attitudes toward wildlife and its importance in our lives have changed through history. During the nineteenth century, it became apparent that seemingly inexhaustible supplies of wildlife were rapidly disappearing. An attitude of human superiority changed to a sense of compassion for the rights of all creatures. Urbanization and industrialization stimulated the romantic movement in art and literature, a nostalgia for rural, natural life, and an appreciation of natural beauty.

One of the most important and enduring priorities of conservation organizations has been to preserve wildlife. Marine mammal protection, laboratory animal treatment, and farm animal abuse have recently emerged as important issues.

Passage of the Endangered Species Act in 1973 classified disappearing species into three categories: endangered, threatened, and extinct. An important test of this Act was the case of the snail darter versus Tellico Dam.

There is much controversy about the best ways to manage endangered species. Many wildlife professionals favor "hands-on" management that brings all tools of technology to bear on captive breeding programs and other intervention measures. Others believe we should adopt a hands-off policy that emphasizes reducing human impacts and preserving habitat. Zoos and animal preserves work positively to educate people about wildlife protection; however, they also contribute to wildlife destruction when they acquire specimens from the wild.

Debate extends to the questions of how many species we can save and which species are most important. Are large ecosystems with many diverse, though perhaps more obscure, species more valuable than individual species? Should we prefer to save the "good" over "less desirable" species?

Responsibility for saving endangered species often falls to the developing countries where species diversity is large. Should wealthier developed nations support preservation efforts in other countries? How valuable is this conservation?

The Man and Biosphere Program demonstrates an integration of human and conservation needs by dedicating the core of a reserve for strict preservation, while allowing peripheral areas to be managed for sustainable resources.

INTRODUCTION

Chapter 14 describes the diversity and abundance of wildlife, the benefits we derive from biological resources, and ways human actions threaten to deplete or destroy them. In this chapter, we will look at how endangered species are protected and managed to sustain those benefits (figure 15.1). We will trace the changing attitudes toward other living creatures that are the basis of wildlife protection laws, zoos, parks, refuges, breeding programs, and restoration projects. Most of this chapter will focus on wild species because domestic species rarely are threatened with extinction.

Protecting and managing endangered species require asking some difficult questions. It probably isn't necessary, possible, or maybe even desirable to try to preserve every subspecies and local variant of plant, animal, and microorganism in the world. Some will disappear by natural causes without human intervention. How many species do we need or want to save? More significantly, *which* species should be saved by our active efforts? How many can we afford to save? How shall we choose those that are or may become the most valuable? Should some organisms and ecosystems just be left alone?

Finally, what moral and ethical considerations intermingle with pragmatic and scientific concerns? A reverence towards life lies at the heart of most philosophical and religious systems of the world, but the application of this principle has been far from universal. For many people, it applies only to those organisms (or other humans) most like themselves. Most people agree that "good" plants and animals—the cute, useful, well-behaved ones—should be preserved. What about the slimy or ugly creatures or those that bite or sting? Do we need to save everything? Is it biologically, economically, or politically possible to do so?

WILDLIFE PROTECTION

Over the years, we have gradually become aware of the harm done by overharvesting, destroying habitat, and otherwise disturbing sensitive wildlife populations. Still, it has taken a long time to incorporate that knowledge into our society.

Hunting and Fishing Laws

In 1874, federal legislation was introduced to protect the American bison, whose numbers were already falling dangerously. This legislation failed, however, because most legislators believed that bison were too numerous to ever be depleted by human activity. They also felt that hunting restrictions were an infringement on personal freedom and the frontier way of life.

By the 1890s, however, most states had enacted hunting and fishing restrictions. Regulations specified the season, time of day, bag limit (number taken), and ways in which wildlife could be harvested. The general idea behind these laws was to conserve the resource for future human use rather than to preserve

FIGURE 15.1 Elephants have disappeared from much of their former range in Africa and Asia. They are endangered or threatened nearly everywhere. Will we be able to preserve this magnificent and interesting species?

wildlife for its own sake. When wildlife was plentiful, restrictions were minimal, but the regulations became more strict as wildlife became scarce. Often the laws weren't fully accepted until the species in question were nearly or completely extinct (figure 15.2).

The wildlife refuges and hunting and fishing laws established in the first half of this century have enabled many wildlife species to increase in abundance. At the turn of the century, there were an estimated 500,000 white-tailed deer in the United States; now there are some 14 million. Wild turkeys and wood ducks were near extinction fifty years ago. By restoring habitats, planting food crops in appropriate areas, transplanting breeding stock, protecting these birds during breeding season, and other conservation measures, the numbers of these beautiful and interesting birds have been restored to several million each. Snowy egrets, which were feared to be near extinction eighty years ago, are now common.

Protective laws also embrace invertebrates that are gathered for food, shells, or other purposes. At various times lobsters, crabs, oysters, clams, and shrimp have been depleted by overharvesting in some areas; they are now more closely regulated.

The Endangered Species Act

Passage of the Endangered Species Act of 1973 represents a sweeping advance over previous wildlife protection efforts in the United States. Earlier legislation had focused primarily on "game" mammals, birds, and fish. Laws were mostly restrictive, telling hunters and harvesters what they could not do in terms of killing specific animals and regulating trade in such items as pets, skins, feathers, and ivory.

"Had to bag one, Harry—in case this damned Conservation thing doesn't work and they become extinct."

FIGURE 15.2 This cartoon alludes to a vicious circle in which rare game animals may be caught.

© Punch/Rothco.

Basic Provisions

The Act of 1973 offers protection to any member of the plant or animal kingdom, including subspecies, races, and local populations, that is threatened by extinction. It created two categories of protected species: endangered and threatened. It prohibits hunting, killing, capturing, selling, importing, or exporting flowers, hides, pelts, feathers, or other products from any endangered species. It also protects habitat critical for the survival of threatened species and requires plans to restore endangered species. It forbids units of government from undertaking any project that would damage endangered species. Finally, it requires the government to prepare a list of all species that are endangered or threatened.

When the act was passed, only 109 species were included on the endangered list. By 1990, some 550 species had been formally listed as endangered, and 560 were categorized as threatened. Another 3,000 have been nominated for inclusion in the "little red book" of endangered species in the United States.

TABLE 15.1 Number of endangered or threatened species in the United States (1990)

Group	Number
Mammals	395
Birds	301
Fish	85
Reptiles	137
Amphibians	23
Insects	18
Snails	13
Clams	35
Crustaceans	6
Plants	103
Total	1,116

The group that makes classification decisions is known as the "God Committee" because of the life and death consequences of its rulings. The framers of this legislation probably didn't fully realize how far reaching the effects of their law would be. Table 15.1 gives the numbers of species listed by group as endangered or threatened in the United States.

Defining Endangered, Threatened, and Extinct

Endangered species are those considered in imminent danger of extinction (figure 15.3). In 1989, the International Union for Conservation of Nature and Natural Resources listed 2,214 bird species, 1,666 kinds of mammals, 858 reptiles, and 129 amphibians as endangered worldwide. These numbers probably represent only a fraction of the species that need protection because they are only the well-known and well-documented cases. It is estimated that at least 30,000 plant species fall into this category. No one knows how many insects, worms, mollusks, and other invertebrates may be endangered, but the number may be in the millions. Table 15.2 lists 34 species known to have become extinct in the United States while waiting to be added to the endangered list.

A **threatened species** is one that, while still abundant in parts of its territorial range, has declined significantly in total numbers and may be on the verge of extinction in certain regions or localities. In the United States, this classification includes the gray wolf, grizzly bear, sea otter, bald eagle, and a number of native orchids and other rare plants.

An **extinct species** is, of course, one that has been completely eliminated. It's not always possible to be sure that a species is unequivocally gone. There are several cases in which a plant or animal was rediscovered long after it was thought to be extinct. Perhaps the most spectacular example is the coelacanth, a primitive fish of a group thought to have died out 90 million years ago. In 1939, however, fishermen caught several specimens in the deep ocean off South Africa. Now there is so

FIGURE 15.3 Endangered species. (a) Bornean orangutan (b) cotton top tamarin (c) ocelot (d) Reddell's Tooth Cave harvestman (e) Northwestern salamander (f) California condor (g) spotted owl.

283

TABLE 15.2 Animal and plant species that have been declared extinct in the United States since 1980

Animals

Common name	Historic habitat
Insular long-tongued bat	Puerto Rico
Woodland caribou	Montana
Penasco least chipmunk	New Mexico
Guam rufous fronted fantail	Guam
Independence Valley tui chub (fish)	Nevada
Relic leopard frog	Arizona, Nevada, Utah
Anastasia Island cotton mouse	Florida
Chadwick Beach cotton mouse	Florida
Carolina elktoe mussel	North Carolina
Sherman's southeastern pocket gopher	Georgia
Goff's southeastern pocket gopher	Florida
Phantom shiner (fish)	Texas, New Mexico
Rio Grande bluntnose shiner (fish)	New Mexico
Longstreet spring snail	Nevada
Fish Springs pond snail	Utah
Texas Henslow's sparrow	Texas
Louisiana prairie vole	Louisiana, Texas

Plants

Plant	
Clarkia, Mosquin's	California
Cyanea linearifolia	Hawaii
Cyanea pycnocarpa	Hawaii
Delissea rivularis	Hawaii
Erigeron perglaber	Arizona
Greensword	Hawaii
Lycopodium haleakalae	Hawaii
Myrsine mezii	Hawaii
Phyllostegia hillebrandii	Hawaii
Phyllostegia knudsenii	Hawaii
Rollandia purpurellifolia	Hawaii
Sphaeralcea procera	New Mexico
Spiderflower, wild	Hawaii
Stenogyne viridis	Hawaii
Styrax portoricensis, Palo de jazmin	Puerto Rico
Tetramolopium consanguineum	Hawaii
Wikstroemia hanalei	Hawaii

Source: U.S. Fish and Wildlife Service.

much competition for aquaria and museums to capture coelacanths that there are fears they really will be hunted into extinction.

Occasionally, although a few individuals remain, a species may be extinct for all practical purposes. The rarest known species in the United States is a type of Indian paintbrush (*Castilleja uliginosa*). A single plant of this species lives in remote Pitkin Marsh in California. With no way to reproduce, it is only a matter of time until it will be extinct.

Application of the Act
One of the most controversial features of the Endangered Species Act has been its power to stop government projects, such as roads and dams, to protect threatened habitat. Opponents of the Act claim that obscure organisms are used as an excuse to stall development that is really opposed for many other reasons.

A celebrated test of the Endangered Species Act was the case of *The snail darter versus Tellico Dam*. In the 1960s, the Tennessee Valley Authority (TVA) announced intentions to

FIGURE 15.4 The snail darter, a paperclip-sized fish whose protection held up the Tellico Dam in Tennessee in one of the most famous tests of the Endangered Species Act.

FIGURE 15.5 The American alligator was once endangered throughout the United States. Protection has restored its numbers to several million and hunting is now allowed again.

dam the Little Tennessee River for power, flood control, and recreation. Opponents fought against this project for years, arguing that the dam was not needed, the cost was too high, the payback was too low, and it destroyed valuable forests, wildlife habitat, Native American archeological sites, and recreational, historic, and scenic values. They were unsuccessful until the Endangered Species Act of 1973 became law. It was revealed that the only known habitat of a small paperclip-sized fish called the snail darter (*Percina tanasi*) (figure 15.4) would be destroyed if the river valley was flooded. The snail darter thus became the most widely publicized of the many reasons that the dam was opposed.

The district court agreed that the snail darter might be harmed if the dam was completed, but declined to stop construction, which was already 80 percent completed. The circuit court of appeals overturned the district court, however, and granted an injunction on further work on the dam. In June 1978, the Supreme Court upheld the injunction, but practically invited Congress to amend the Endangered Species Act. The amendment, sponsored by Tennessee Senator Howard Baker, exempted projects where no "reasonable and prudent" alternatives exist and when benefits "clearly outweigh" those of alternatives.

The Tellico Dam has been completed at a cost of more than $100 million, and its reservoir is being filled. However, the snail darter will not be exterminated after all because populations have been found elsewhere in the river system.

The case of the snail darter raises some interesting questions. How would you have voted if you had been the judge hearing this case? Suppose the snail darter had not been found elsewhere. Would preserving one species justify abandoning this costly project? As it happens, there are several other closely related species of darters in the Little Tennessee and other rivers. They are so similar as to be indistinguishable to the untrained eye. Many of these darters are abundant, and the group as a whole is in no danger of extinction. Does that change your feeling about saving the snail darter? Try using the cost/benefit analysis

process in figure 8.18 to imagine the considerations that would go into a decision such as this. How would the competing values add up in your estimation?

The original Endangered Species Act of 1973 was authorized for only ten years. Reauthorization was delayed for more than four years because of objections by a few senators who feared that listing endangered species would stand in the way of development projects or predator control in their states. The biggest problem was a 1984 decision by U.S. District Judge Miles Lord who agreed with environmentalists that a trapping or hunting season on Minnesota's endangered timber wolves would violate the Act. This decision was used to block killing of grizzly bears in Montana, especially adjacent to Yellowstone National Park. Angry Western ranchers persuaded their senators to block the reauthorization bill from consideration. Finally, in 1988, the Senate passed the bill on a 93 to 2 vote. Funding for endangered species protection was increased from $39 million to $66 million per year. Civil and criminal penalties for violation of the law were increased, and a loophole was closed that had allowed the destruction of endangered plants on private property.

Recovery Programs

Once a species has been listed as threatened or endangered, the USFWS is required to prepare a **species recovery plan** that spells out how the species can be restored to numbers that permit its delisting. The format and timing are left to the discretion of the Secretary of the Interior. As of 1990, only 203 recovery plans had been filed for the 1,116 species listed as endangered or threatened. Some recovery plans cover more than one species so that 40 percent of all listed species are covered to some extent. Only 10 percent of the species managed by these plans are stable or are increasing in numbers. The other 90 percent are either declining or are not well enough known to make a judgment.

Some recovery programs have been very successful. The American alligator (figure 15.5) was listed as endangered in 1967 because of hunting (for skins, meat, and sport) and because of

BOX 15.1   California Sea Otter Recovery

Who could fail to love an otter? Few animals have as many endearing features as these cute, cuddly creatures. Their round fuzzy faces with dark button eyes and a huge walrus mustache seem to expresss perpetual curiosity, amusement, and good nature. They are beautiful swimmers: graceful, swift, and supple in the water. Hours every day are spent playing in small family groups or lolling about in kelp beds, basking in the sun, and rocking gently on the waves. Sea urchins and shellfish are a large part of their diet. Floating on its back in the water, an otter holds a clam or urchin on its belly with one stubby forepaw and hammers on the shell with a rock held in its other paw (box figure 15.1).

Sea otters (*Enhvdra lutris*) once ranged from Japan along the coast of Siberia, across the Aleutian chain and down the coast of North America as far as Baja California. By 1850, however, the population was so low that it was nearly extinct over most of its range. Because they don't have a layer of blubber under their skin like seals and whales, otters depend entirely on their thick fur coat to keep them warm in the cold North Pacific

BOX FIGURE 15.1 California sea otters were thought to have been completely destroyed by fur hunters early in this century. A few otters escaped, however, and the population has now regrown to several thousand individuals.

waters. That wonderful fur coat was almost their downfall. Humans prized it so much that sea otters were hunted nearly to extinction in the eighteenth and nineteenth centuries. The southern population along the California coast was thought to be entirely gone by 1911,

when Russia, Japan, Canada, and the United States finally agreed to stop hunting them.

In 1938, however, a few otters were spotted along a rugged and isolated section of the California coast called the Big Sur. Somehow they had escaped detec-

habitat destruction. Protection has been so effective that the species is now plentiful. Florida officials estimate that their state alone now has 1 million alligators. In 1987, the alligator was removed from the endangered species list throughout its entire southeastern range. The crocodile, a close relative of the alligator, is still endangered throughout the world. Another success story is that of the Arabian oryx. This pure white, long-horned desert antelope (perhaps the origin of unicorn legends) once was listed among the world's twelve most critically endangered species and was extinct in the wild. Individuals from zoos have been reintroduced in their native habitat, and the wild population is growing satisfactorily. The California sea otter was also thought to be extinct early in this century but the population is now about 100,000 (Box 15.1).

Protecting Habitat

Often the best way to preserve a species is to protect its habitat. This can require careful consideration of many factors. For instance, consider the case of the Kirtland's warbler. This small blue, gray, and yellow songbird nests in jackpine stands in northern Michigan. The trees must be young and not too closely packed to be acceptable for nesting. Any given stand of trees is used for only a few years. As the forest matures, the birds move on to new territory. Formerly, the forest was regenerated at regular intervals by forest fires. In the early 1900s when fires were controlled, however, the Kirtland's warblers began to disappear.

Wildlife biologists studied the nesting and feeding patterns of the species and determined that prescribed burning is necessary to maintain their preferred habitat. A forest management

tion and survived two centuries of persecution and neglect. Now protected, the otter population has rebounded to about 100,000 individuals in the Aleutian Islands and southeast Alaska. Another 2,000 animals occupy about 320 km (200 mi) of coast between Santa Cruz and San Luis Obispo, California. Management of this charismatic but controversial species has become the subject of intense debate in recent years.

Friends of the otters advocate expanding the southern population both in numbers and territory. They point out that more than 100 million barrels of oil are shipped along the California coast every year. If an oil spill like that of the Exxon *Valdez* in 1989 occurred near Big Sur, the entire California otter population might be wiped out. If there were another group of otters further down the coast, some might survive. To accomplish this, biologists propose catching about 250 otters and transplanting them to islands south and west of Los Angeles.

Fishermen and shellfish collectors oppose this plan angrily. They claim that there are too many otters already and that the otters' appetite for valuable fish, abalone, lobsters, oysters, and crabs cuts into their catch. They threaten to kill any otters moving into "their" territory. Of the few otters transplanted so far, at least half have disappeared mysteriously. The guns, traps, and nets of fishermen are thought to be a major source of mortality.

Even some people who want otter populations expanded don't like seeing them caught and transplanted. How would you like to be chased by a motor boat, caught in a big fish net, trussed up like pig bound for market, and then exiled to a strange new place far from your familiar home? Many die from stress or are lost trying to find their way back home again. Of 63 relocated to San Nicolas Island in 1988, only 15 were still there and alive one year later.

Another controversy of "hands-on" management concerns researchers who want to know where the transplanted otters go and what happens to them. Tracking the animals involves putting radio transmitters on them. Because otters are too sleek and sinuous to wear radio collars around their necks, transmitters are surgically implanted in their abdomens. Mortality from the operation is high, however, and critics of this research argue that it isn't justifiable. They would prefer to just leave the otters alone and appreciate them from a distance.

An intriguing benefit of expanding otter populations is recovery of the sea kelp forests in which they live. When otters were removed, sea urchin populations exploded. Urchins eat kelp; too many of them can completely destroy the kelp and the whole biological community of which it is the dominant species. Without otters to control the urchins, much of the coast had become denuded "urchin barren" with no kelp and few fish or other species. By reintroducing otters, the kelp has been protected and the ecosystem is recovering. This is called "trickle-down" conservation where protecting one key species helps many other obscure organisms.

If you were designing an otter recovery program, how much active intervention would you allow? Otter research and recovery now costs about $300,000 per year. Is that too much? How would you weigh the otters right to exist against the interests of fishermen? How many otters are enough?

program was begun, but warbler numbers continued to drop. It was discovered that the coastal scrub in the wintering area in the Bahamas, on which the warblers depended, was disappearing because of seaside development. This winter habitat is just as critical as the summer nesting area. Land is now being purchased for nature reserves to provide a winter haven for this small but colorful summer resident. Forest destruction in Central and South America is similarly affecting many of our favorite songbirds and summer residents.

CITES

The 1975 Convention on International Trade in Endangered Species (CITES) agreement was a significant step toward worldwide protection of endangered flora and fauna and regulating trade in living specimens and products derived from listed species, but it is not foolproof. Species are smuggled out of countries where they are threatened or endangered (figure 15.6), and documents are falsified to make it appear they have come from areas in which the species is still common. Investigations and enforcement are especially difficult in developing countries where wildlife is disappearing most rapidly. Still, eliminating markets for endangered wildlife is an effective way of stopping poaching. In 1988, the World Conservation Centre reported that 430 metric tons of ivory were sold on world markets. The 1990 world ban on ivory trade dried up almost all of that traffic.

ZOOS, PRESERVES, REFUGES, AND GENETIC BANKS

We have many different ways to preserve species that we value. Each has advantages and disadvantages. In this section, we will look at both the problems and the successes in conservation and restoration programs.

FIGURE 15.6 Endangered wildlife products. These leopard and cheetah skins were seized by United States Fish and Wildlife agents. Smugglers of products from endangered species are subject to heavy fines and/or jail terms. Unfortunately, however, as long as there are consumers for these products, some hunters and dealers will be encouraged to take the risks.

FIGURE 15.7 Newer zoos are making special efforts to provide comfortable, naturalistic habitats. It can be difficult, however, to find a balance between accessibility to visitors and privacy for animals.

Zoos

Records of wildlife collections date back to the beginning of history. The ancient Chinese, Egyptians, and Romans had menageries of bizarre and exotic wild animals. Even today, many zoos appear to be sideshows or amusement parks more than refuges for endangered species or places to learn about nature. Most modern zoos and gardens, however, have a strong commitment to public education and can be powerful forces for wildlife understanding and conservation. There is a movement toward establishing breeding programs that preserve endangered species and display them in conditions as close as possible to their natural habitats (figure 15.7). These facilities aren't always popular, however, because some visitors want to see trained animal acts and spectacular, exotic species, such as elephants, lions, pandas, and boa constrictors. When enclosures are large enough for animals to hide away as they would normally choose to do, zoo visitors are disappointed because there is nothing to see. Who cares about prairie star grass, five-lined skinks, or Przewalski's horses, especially if you can't see them clearly?

In the past, most plants and animals in zoos and botanical gardens were obtained from the wild, often illegally. Institutions now raise their own specimens in **captive breeding** programs, and buy or trade for captive-bred organisms with other zoos and gardens to avoid depleting wild populations. This is a praiseworthy new direction. In some cases, zoos have been able to replenish or even reestablish wild populations with surplus breeding stock.

Recent successful captive breeding programs give hope for reintroduction of species that are endangered or even extinct in the wild. Some of the animals whose numbers have been built up by captive breeding include whooping cranes, peregrine falcons, salmon, green sea turtles, and Puerto Rican parrots. Many species are difficult or impossible to breed artificially, however.

Bats, whales, and many reptiles rarely reproduce in captivity. You may have heard about the birth of baby killer whales in captivity, but you rarely hear what happens to them later. Of six whales born in 1985, all died within a few weeks, either from infection or simply because their mothers did not know how to care for them.

Zoos, however, never will be able to protect the complete spectrum of biological variety. Worldwide, zoos house some 500,000 individual mammals, birds, reptiles, and amphibians. This represents only about 900 species in all. Few of these species are being bred and raised in captivity in numbers sufficient to preserve the species if the wild populations are lost. If all the space in U.S. zoos were used for captive breeding, only about one hundred species of large mammals could be maintained indefinitely. The cost of such a program, if it were independent of wild populations, would be around one billion dollars a year. Plants take less space than big mammals, but we will never recreate the total diversity of a tropical rain forest. Clearly, preserving native habitat is the only way to sustain the full diversity of the biosphere.

Wildlife Refuges in the United States

The fifty-one **wildlife refuges** established by President Teddy Roosevelt in 1901 were the first in what has become an important but troubled system for wildlife preservation. We now have 410 wildlife refuges in the United States, encompassing nearly 40 million ha of land and water and representing every major biome in North America. The units in this system range in size from less than 1 ha (2.5 acres) for the tiny Mille Lacs Refuge in Minnesota to 7 million ha (18 million acres) in the Arctic Wildlife Refuge of Alaska. Altogether, about 1 percent of the United States surface area is designated as wildlife refuge (figure 15.8).

RESOURCES AND RESOURCE ECONOMICS

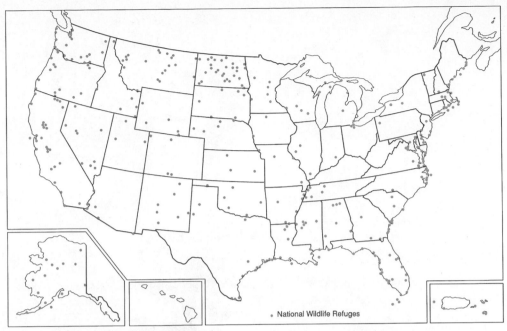

FIGURE 15.8 Locations of the 410 wildlife refuges in the United States. Nearly every state has at least one refuge.

Source: U.S. Department of the Interior.

Between 1930 and 1934, President Franklin D. Roosevelt and Harold Ickes, his Secretary of the Interior, took advantage of low land prices during the depression to build up the refuge system. The largest additions of land protected for wildlife came in 1980 when President Jimmy Carter signed the Alaska National Interest Land Act and added 22 million ha (54 million acres) of new refuges to the 12.5 million ha (31 million acres) already existing, about two-thirds of which was in Alaska.

Over the years, a number of improbable and incompatible uses have become accepted in wildlife refuges. In order to raise money for wetland protection, the idea of selling duck stamps was proposed; but in a compromise to get the necessary legislation passed, it was agreed that refuges would be open to hunters (figure 15.9). Although hunting in a refuge seems like a contradiction in terms, it has become firmly established. In recent years, wildlife in the refuges have had to compete with oil drilling, cattle grazing, snowmobiling, motorboating, and off-road vehicle use, as well as timber harvesting, hay cutting, trapping, and camping. Nevada has bombing and gunnery ranges and even a brothel on wildlife refuges. Drainage of polluted irrigation water into the Kesterson Refuge in California turned it into a death trap for wildlife rather than a sanctuary. Subsequent research has shown that at least nine wildlife refuges in Western states have dangerously high selenium levels, mostly resulting from agricultural drainage.

The biggest battle over wildlife refuges in the late 1980s and early 1990s has been over proposals for oil and gas drilling in the 7 million ha (18 million acre) Arctic National Wildlife Refuge on the north slope of Alaska's Brooks Range. In 1987, Secretary of the Interior Donald Hodel recommended that oil companies be allowed to do exploratory drilling and seismic mapping of the 600,000 ha (1.5 million acre) coastal plain bordering the Beauford Sea. A harsh and inhospitable wilderness in winter, this arctic tundra biome is a lush summer wetland of innumerable potholes and marshes that teem with wildlife. It is the central calving ground of the 180,000 caribou of the Porcupine herd, one of the world's largest caribou herds (figure 15.10). It is also an important habitat for thousands of snow geese and other migratory waterfowl, muskoxen, Alaskan brown bears, polar bears, and arctic wolves, along with unimaginable clouds of insects and many other species of rare plants and animals.

The Persian Gulf War in 1991 raised fears of depending on foreign oil and stimulated demands for exploratory drilling on the arctic plain. Energy companies are extremely interested in this area because any oil found could be pumped out through the existing oil pipeline. The Department of the Interior environmental impact statement (EIS) estimated a 19 percent chance of finding at best a 200-day supply for the United States. The same EIS predicted the loss of 40 percent of the caribou, 25–50 percent of the muskox, and 50 percent of the snow geese. In spite of this pessimistic outlook, the Bush administration recommended that drilling commence.

FIGURE 15.9 Hunters on their way to duck blinds in a wildlife refuge. Although it seems a contradiction in terms, half of all refuges allow hunting.

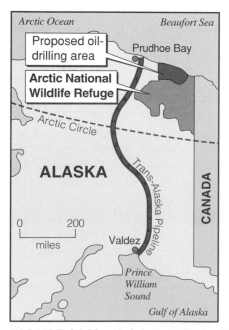

FIGURE 15.10 Alaska's Arctic National Wildlife Refuge, home to one of the world's largest caribou herds, has been called the crown jewel of the entire refuge system. The 600,000 ha coastal plain is the nesting ground for countless migratory birds and the calving ground of one of the world's largest caribou herds. Oil companies have proposed exploratory drilling here, but environmentalists fear irrepairable damage to the fragile ecosystem.

TABLE 15.3 Protected natural areas and known endangered mammal species

Region	Number of areas	Millions of hectares	Percent of land area	Number of endangered mammal species
Africa	521	102	3	559
North and Central America	890	194	9	182
South America	453	80	5	252
Asia	1,126	61	2	225
Europe	1,347	31	7	299
Soviet Union	171	21	1	78
Oceania	774	40	5	70
World total	5,282	529	4	1,645

Source: Data from World Conservation Monitoring Center 1990.

Opponents to this project point out that scars left by mechanized equipment, blasting for seismic mapping, and drilling will persist in the tundra for centuries. They feel that wildlife and wilderness values that the refuge was set up to protect should be paramount over short-term economic interests. Others feel that the potential for a valuable energy source overrides biological importance, and that, furthermore, wildlife will not be adversely affected if drilling is done carefully. In 1991, competing bills were being considered in the U.S. Congress that would either strictly protect this area as a wilderness or open it up to industrial development.

Refuges and Nature Preserves in Other Countries

In recent years, other countries have recognized the value of their biological resources and have set up preserves, refuges, and national parks to protect wildlife and habitat (table 15.3). Notice that North America and Europe have relatively high percentages of land set aside in protected areas, while Asia, Africa, and the Soviet Union have lower percentages than the world average. Note also that the area preserved does not correlate with the number of known endangered species. The discrepancy may be worse than indicated since most endangered species are known in Europe, but probably only a small share of those in Asia, Africa, and South America have been identified.

Some countries, including Costa Rica, Brazil, Indonesia, and Tanzania, have plans to set aside a larger percentage of their total area than has the United States. Many of these plans are mostly wishful thinking, however. Many of the countries with the greatest wealth of biological diversity, abundance, and importance are tropical, Third World countries that are among the world's poorest, least developed, and most politically unstable. These countries also have some of the most rapidly growing human populations in the world. Maintaining parks in these countries is tremendously difficult.

An outstanding example of both the promise and the problems in managing parks in the less developed countries is seen in the Serengeti ecosystem in Kenya and Tanzania (figure 15.1). This area of savannah, thorn woodland, and volcanic highland lying between Lake Victoria and the Great Rift Valley in East Africa is home to the highest density of ungulates (hoofed grazing animals) in the world. Over 1.5 million wildebeests (blocky, bearded members of the antelope family; gnus) graze on the savannah in the wet season, when grass is available, and then migrate through the woodlands into the northern highlands during the dry season. The ecosystem also supports hundreds of thousands of zebras, gazelles, impalas, giraffes, and other beautiful and interesting animals. The herbivores, in turn, support lions and a variety of predators and scavengers, such as leopards, hyenas, cheetahs, wild dogs, and vultures. This astounding diversity and abundance is surely one of the greatest wonders of the world.

Tanzania's Serengeti National Park was established in 1940 to protect 15,000 sq km (5,700 sq mi), an area about the size of Connecticut or twice as big as Yellowstone Park. It is bordered on the east by the much smaller Ngorongoro Conservation Area and Lake Manyara National Park. Kenya's Masa Mara National Reserve borders the Serengeti on the north. Rapidly growing human populations push against the boundaries of the park on all sides. Herds of domestic cattle compete with wild animals for grass and water. Agriculturalists clamor for farmland, especially in the temperate highlands along the Kenya-Tanzania border.

FIGURE 15.11 A guard protects a black rhinoceros from poachers. His ancient rifle is no match, however, for the powerful automatic weapons with which the poachers are now armed.

The worst problem in the reserve is **poachers,** illegal hunters, who massacre wildlife for valuable meat, horns, and tusks. As recently as 1970, a healthy population of 65,000 black rhinoceroses roamed in Africa, 1,000 of which were in the Serengeti. In 1990, there were about 3,000 black rhinos in Africa and only 20 in the Serengeti (figure 15.11). The rest were killed for their horns.

Elephants are under a similar assault. Thirty years ago there were no elephants in the Serengeti, but perhaps 3 million in all of Africa. Since then, more than 80 percent of the African elephants have been hunted and killed, mainly for their ivory, at the rate of 100,000 each year. The 2,000 elephants presently in Serengeti National Park have been driven there by hunting pressures elsewhere. Fortunately, the elephants find refuge in the park and add to the pleasure of tourists who come to see the wildlife; however, they are changing the ecosystem. They smash down the acacia trees, turning the woodland and mixed sa-

vannah into continuous grassland. This is beneficial for some animals, but not for others.

The poachers continue to pursue the elephants and rhinos, even in the park. Armed with high-powered rifles and even machine guns and bazookas from the many African wars in the last decade, the poachers take a terrible toll on the wildlife. Park rangers try to stop the carnage, but they usually are outgunned and outmanned by the poachers. The parks themselves are beginning to resemble war zones, with fierce and lethal fire fights.

Size and Location of Nature Preserves

If you were responsible for setting up biological reserves, especially in tropical countries where ecosystems are not well understood, where would you begin? There are many issues to consider when determining the size and location of parks and refuges. The challenge begins with obtaining basic information about the organisms and ecosystems to be protected. Large carnivores that are normally rare or organisms that have unusual and limited habitat requirements may require a large area to preserve a viable population. Tigers, for instance, may have a home range of 40,000 km², requiring a very large wilderness to support a viable population. The same principle is true for condors or grizzly bears. It would be ideal to set aside whole ecosystems with natural boundaries, such as entire watersheds, but that is rarely possible (Chapter 13).

Some closely related species have a checkerboard pattern of distribution. In New Guinea, four species of cuckoo-doves occur in similar habitat, but only certain combinations can occur together because the presence of one species "locks out" certain others. This means that it is necessary to have several similar but separate areas to save all four species. Some animal species range widely during seasons when food is plentiful but depend on a single species or even a single individual for food or shelter during hard times (figure 15.12). In a tropical rain forest, where individuals of the larger tree species occur infrequently, it may take a large area to sustain dependent species during migration or during the dry season when food is scarce.

How much land is "enough"? Diseases or unexpected natural or human-caused disasters can wipe out a species if all representatives are concentrated in a single reserve. This argues for multiple, dispersed, and large reserves. Some organisms are adapted to specific successional stages and depend on a continual renewal; thus, only a large, diverse reserve will have a balance of successional stages necessary to preserve such organisms.

Finally, how many individuals make up a sustainable population? This is a problem in captive breeding programs as well as wild populations. Species that are reduced below a minimum size (usually between fifty and one hundred individuals for most species) will suffer from what is called **inbreeding depression.** This is an accumulation of harmful genetic traits through random

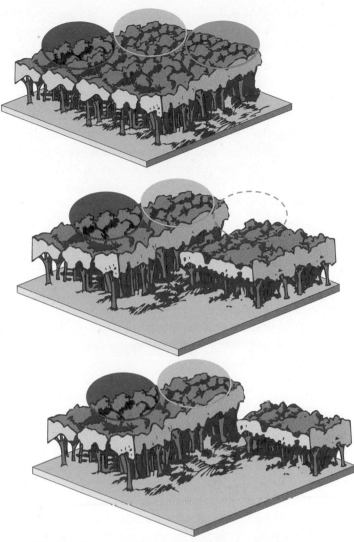

FIGURE 15.12 Biogeographical effects. The colored circles represent ranges of mutually exclusive species. A nature preserve must be above a certain minimum size to protect certain species. In general, a 90 percent reduction in habitat size results in a 50 percent loss in the number of species present.

mutations and natural selection that lowers viability and reproductive success of enough individuals to affect the whole population. Ordinarily, in a large population of freely inter- breeding plants or animals, the chances are small that two or more harmful genes will occur in the same individual in a way that affects fitness. When a population is small and interbreeding of closely related individuals is frequent, there is a much greater chance that deleterious genes will occur together and will result in less fit individuals.

Populations that have originated from a small number of founders suffer from what is called a "founder bottleneck" caused by a lack of genetic diversity. The flightless nene goose, for instance, numbered around 25,000 when Captain Cook landed on Hawaii. By 1940, only 43 were left. A captive breeding program has built the population back up to 3,000 birds (mostly in

England), but male infertility has become a serious problem because of lack of genetic diversity. Cheetahs also suffer from founder bottleneck. Genetic tests suggest that all existing cheetahs originated from a single female ancestor some time in the past. Infertility is high (60 to 70 percent) and the species may be genetically doomed.

Island effects are reductions in species diversity that accompany reductions in ecosystem areas. This concept is derived from the biogeographical work of R. H. MacArthur and E. O Wilson. When a new island forms, organisms migrate to it from nearby areas. The first species to arrive may be highly successful at colonization because they lack competition or predators. As colonization proceeds, however, and niches fill, subsequent arrivals are less successful (figure 15.13a). On a more densely populated island, increasing competition and decreasing resources also cause higher losses of existing species.

Eventually, the number of successful new colonizations and the number of extinctions reach an equilibrium that depends on the size of the island (figure 15.13b) and the distance from the sources of new species. A large island can support more species than a small one, and a nearby island receives more species than a distant one. Island effects have been demonstrated in many places. For instance, Cuba in the West Indies is one hundred times as large and has approximately ten times as many species of amphibians as its neighbor, Monserrat. These effects also apply to "islands" of undisturbed rain forest in the midst of large clear cuts in the Amazon jungle. In general, a 90 percent reduction in ecosystem size results in a 50 percent loss of species.

These observations have important implications for wildlife management. It has been estimated that the national parks of Africa could lose one-fourth of their species in 50 years and 90 percent in 5,000 years because of genetic drift and island biogeographical effects.

Social and Economic Factors

Preservation of wilderness parks is also closely tied to appropriate rural development in the area surrounding the park. Traditional seasonal migration patterns often include corridors through areas outside established parks. Human use of these areas will not unduly affect wild species as long as wild populations are not too large and competition for food, water, and living space is not too severe. Predation on domestic livestock and gardens may be the price of sharing an ecosystem with wild inhabitants. Governments can selectively remove the most troublesome individual predators and compensate farmers for their losses as a way to make coexistence with wildlife acceptable.

If local people benefit from tourism and harvesting of surplus wildlife, they will be more supportive of programs to preserve wilderness and protect wildlife. Some promising programs attempt to integrate human needs and wildlife preservation. In 1986, UNESCO started its Man and Biosphere (MAB) pro-

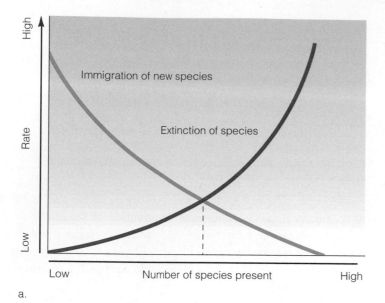

a.

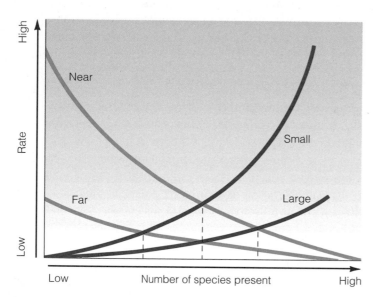

b.

FIGURE 15.13 Theoretical models of island colonization and species extinction. (a) On a single island the rate of successful colonization of new species decreases while the rate of extinction of existing species increases with increasing number of species present. Given stable conditions and sufficient time an equilibrium will be established. (b) Compare the rate of successful colonization (green curves) for islands far from or near to the mainland and rates of extinction (red curves) for large and small islands. Note that each of these combinations gives a different equilibrium.

gram. In this creative approach to serving both human and conservation needs, the core of a reserve or park is dedicated to strict preservation. Peripheral areas are managed for sustainable resource use, so human needs are met in ways compatible with resource conservation. Traditional native cultures are encour-

aged and protected in these peripheral areas. Mexico's 545,000 ha (2,100 sq mi) Sian Ka'an Reserve on the Caribbean coast of the Yucatan Peninsula, for example, is an exemplary MAB reserve. The core area includes 112 km of coral reef and adjacent bays and marshes. Sustainable resource use in the rest of the reserve includes lobster fishing, farming, and a small coconut industry. The *Amigos de Sian Ka'an*, a local community organization, played a central role in establishing the reserve and is working to establish sustainable resource use while it improves living standards for local people.

Germ Plasm Banks

Because so many species are rapidly disappearing, there is great interest in preserving genetic material (**germ plasm**) for future agricultural, commercial, and perhaps ecological values. Techniques for storing frozen plant or animal eggs, sperm, and embryos are widely used in agriculture and are highly successful for many species. Whether they will be suitable for wild species and for long-term storage remains to be seen.

Plants already have an embryo storage system worked out—it's called a seed! Many seeds are easy to collect, remain viable for years, and don't require elaborate storage facilities. Several plant germ plasm banks have been established around the world (figure 15.14). The main center in the United States is in Fort Collins, Colorado. By 1985, this facility had more than 200,000 different seed samples and was running out of room.

Seeds don't live forever, unfortunately. In general, larger seeds, such as beans, last up to four or five years, while small seeds, such as mustard and thistle, last only one or two years. Plants that grow in harsh and uncertain climates have seeds that persist many years, while seeds from a tropical rain forest must germinate in a few weeks. Seeds from some desert plants can survive in the dry soil for decades or even centuries until suitable rains come. Since most seeds lose their viability in storage, they must be taken out periodically and grown under appropriate conditions to produce new fresh seeds. Each time this is done, the sample changes slightly. Initially, natural variations arise by mutation and genetic recombination. Also, some variations survive better than others under cultivated conditions and are overrepresented in the next generation. How does this inadvertent selection occur? The growth conditions in the gardens of the research center cannot be exactly like those where the seeds are native. As a result, the species changes slowly and potentially valuable characteristics may be lost.

NOAH PRINCIPLE: HOW MANY SHOULD BE SAVED?

Stories something like Noah's Ark of Genesis are common in many cultures. There are many legends of disasters in which the world's wildlife is saved by climbing on the sacred turtle's back,

FIGURE 15.14 Seed banks are part of an effort to preserve diversity of biological resources. Even the best seed bank is no match for a natural ecosystem in preserving species, however.

floating in a hollow reed, or going, two by two, into an ark. We now find ourselves in a position much like Noah's. Humans are both the cause of much wildlife extinction and the only ones who can save this resource. Which organisms should we save? Which ones can we save? Clearly, we can't preserve every race, variety, or subspecies on Earth in our germ plasm banks. Should we only save "clean animals" as Noah did? It's not difficult to organize support and collect funds to help big, beautiful, friendly, colorful organisms. We are especially partial to organisms that resemble us in some way.

What about inconspicuous life-forms and the slimy, squishy, crawly things? It's hard to drum up support for things that bite, smell badly, or are just plain ugly. Should we preserve mosquitoes, cockroaches, and fleas? Do we need leprosy, gonorrhea, or amoebic dysentery? Would anyone mind if they disappeared? Are the roles that they play important and worthwhile? Should we protect and preserve species that are dangerous, that compete with us, or that interfere with our access to resources (Box 15.1)? How do we make decisions about species that presently have no direct economic use? Are we willing to set aside some part of the world for a particular species or biological community even if it means we can never go there? These are some of the questions that are the foundation of our future policy-making, in which you will be a participant, either actively or passively.

WHO OWNS BIOLOGICAL RESOURCES?

One of the central problems in wildlife preservation is that we treat wild species as common property. A cow can be owned, and ownership encourages care and protection. A farm that can be passed on to one's children is conserved for the sake of the family. Common property, such as wildlife, air, and water, can suffer because it is owned and preserved by no one. Immediate profits for those with direct access to and political control of the resource outweigh diffuse, long-term benefits for the many. Our economic systems, social institutions, and political processes are failing to ensure sustained management of biological resources. This situation should alarm you.

Who should pay for preservation? Who should benefit from future uses of biological diversity and biological resources? Some people feel that biological diversity is a resource that belongs to the country where it occurs and should be paid for just as oil or minerals are. Plant hunters from the developed countries are searching the remote areas of less developed nations for

FIGURE 15.15 Plant collectors explore the mysteries of a tropical rain forest in search of new species. Who knows what useful properties might be found in the thousands of undiscovered species that live here?

wild relatives of domesticated species or new varieties that may someday be useful (figure 15.15). Some countries with great biological diversity complain they are being exploited in this process. If a new crop is developed from the discoveries of these explorers, the less developed countries will be forced to pay high prices to buy seeds. The plant hunters, on the other hand, pay no royalties for the specimens they gather or the valuable products, such as medicines, derived from them. If those species are not collected and preserved in seed banks or gardens in the developed world, however, many of them soon will be lost. It is, therefore, beneficial to both the native country and the rest of the world to have them preserved.

One reason for saving wildlife is the utilitarian benefit of food, medicine, commercial products, and ecological services we derive from wild species and ecosystems. Wildlife preservation is also based on ethics, compassion, and aesthetics. Some people feel that all other species have a right to exist, regardless of their benefit to us. Others believe that our superiority (technological, at least) over other creatures obliges us to treat them with care and compassion. At the very least, we have come to realize that it makes practical sense to protect other living beings on which we depend.

During the nineteenth century, it suddenly became apparent that the seemingly inexhaustible supplies of wildlife that once filled this continent were rapidly disappearing. The fledgling conservation movement began to work for wildlife protection, and, over the last century or so, has made remarkable progress in preserving and restoring wildlife in many areas. Many problems remain, however. Populations of great whales were hunted to dangerously low levels. Although a hunting ban has been promised, it is not yet clear whether some species of whales will be able to survive.

A major step in United States wildlife protection was the passage of the 1973 Endangered Species Act. An important test of this act was the case of *The snail darter versus Tellico Dam.* The 1975 CITES agreement was an equally important step on the international level. Zoos have been effective in educating people about wildlife problems and have carried out successful breeding programs for a number of threatened and endangered species. Zoos also contribute to wildlife destruction when they buy specimens from unscrupulous hunters who kill large numbers of animals to obtain a few to sell.

Some people argue that it is better to leave wildlife in their native habitats rather than manipulate them with studies and breeding programs. An instructive example of this controversy is the California condor. Whether this rare and magnificent species should be managed actively or passively is a question that deeply divides the conservation community.

Debates have raged about the value of preserving particular species. Are some species worth more than others? Should we use available funds to save large numbers of obscure species by preserving whole ecosystems, or is it better to save a few species that are of particular interest? By focusing our efforts on certain key species and their habitats, perhaps other less popular species will survive as well.

We are left with questions about how many species and races of plants and animals is it necessary or even desirable to save. Some people believe we should save only the "good" organisms. Others argue that we have an equal obligation to protect the ugly, creepy, and disagreeable things. What about dangerous organisms or ones that require huge areas to survive? Who should pay for this preservation?

Some people believe we should pay Third World countries to preserve their heritage because it concerns all of us. They argue that when we use biological resources, we should pay royalties and reparations, as mining and lumber companies do for their use of resources.

Review Questions

1. What are the basic provisions of the Endangered Species Act. How does it protect endangered species?

2. What is the difference between endangered and threatened?

3. Give some examples of species that were thought to be extinct but then were rediscovered. Is this likely to be common?

4. How has habitat loss affected Kirtland's warblers? Why is this a classic example of an endangered species?

5. What are the issues at stake in protection or exploitation of the Arctic National Wildlife Refuge?

6. What effect does refuge size have on species number? Why is it better to have a single large reserve for some species than several smaller ones? When might we prefer several small refuges over one large one?

7. What is a founder bottleneck? How has it affected nene geese and cheetahs?

8. What is germ plasm? How does a germ plasm bank work?

9. Why was the Tellico Dam in Tennessee temporarily stopped by the Endangered Species Act? What were the issues in this case and how were they resolved? Do you agree with the court decision?

10. Protection and reintroduction are two types of recovery plans. Describe the difference in approach between these two concepts.

11. Describe the problems of the "common good" aspect of wildlife preservation. Are countries where biological diversity is abundant sometimes at a disadvantage? Why?

Questions for Critical Thinking

1. Why are there so many more birds and mammals on the endangered species list than insects or snails? Does the list represent actual numbers of endangered species?

2. Do you believe it is ethical to hold animals in zoos? Why?

3. Should we protect species that are dangerous and who compete with our own access to resources? Should we be willing to set aside habitats for particular species, even if it means we can't go to these areas?

4. Who should pay for the preservation of species and who will benefit from the biological diversity? Should richer nations contribute to Third World countries so they can preserve their biological heritage?

5. Some wildlife experts argue that if we were to set up a legal trading system for rhinoceros horns and ivory, we might disrupt the channels for smuggled products and drive poachers out of business. Others argue that this would just hasten extinction of elephants and rhinos. What do you think? Would you favor cutting off the rhinos' horns to save them from poachers?

6. If you were a poor African farmer making $100 per year and you were offered $30,000 to kill a rhino and deliver its horn to a black-market dealer, what would you do?

7. How can we ensure that local residents will enjoy the benefits of new national parks in developing countries? What role should the United States play in these parks?

8. How do we decide which species are most important to save? Are some species more important than others? Is it more important to save large numbers of obscure species or fewer numbers of more select species?

Key Terms

captive breeding (p. 288)
endangered species (p. 282)
germ plasm (p. 294)
inbreeding depression (p. 292)
island effects (p. 293)
poachers (p. 292)
species recovery plan (p. 285)
threatened species (p. 282)
wildlife refuges (p. 288)

SUGGESTED READINGS

■ Craighead, F., Jr. *Track of the Grizzly*. San Francisco: Sierra Club Books, 1979. A gripping account of grizzly bear biology by one of the pioneers in the field.

■ Davis, S., et al. *Plants in Danger: What Do We Know?* Cambridge, England: Conservation Monitoring Center, International Union for Conservation of Nature and Natural Resources, 1986. A survey of the status of endangered plant species.

■ Doherty, J. July 1983. "Refuges on the Rocks." *Audubon*. 85 (4):74. Describes some of the contradictions and problems in managing our National Wildlife Refuges.

■ Eckholm, E. 1978. "Disappearing Species: The Social Challenge." *Worldwatch Paper #22*. Washington, D.C.: The Worldwatch Institute. A valuable contribution that links questions of social justice and human needs with conservation of resources.

■ Ehrenfeld, D. *Biological Conservation*. New York: Holt, Rinehart & Winston, 1970. An overview of world biological resources and how they are threatened.

■ International Union for the Conservation of Nature (IUCN). 1980. *World Conservation Strategy*. New York: Unipub. A strategy for conserving biological diversity from an important international organization. A landmark statement.

■ Luoma, J. *A Crowded Ark: The Role of Zoos in Wildlife Conservation*. Boston: Houghton Mifflin, 1987. A good discussion of zoos in history and today.

■ McNeely, J. A. et al. 1990. *Conserving the World's Biological Diversity*. Washington D.C.: International Union for Conservation of Nature. A good discussion of ways to preserve biological resources.

■ Packer, C. et al. 1991. "Case Study of a Population Bottleneck: Lions of the Ngorongoro Crater." *Conservation Biology* 5 (2):219. All of the lions in the crater are descended from 15 founder animals. The lack of genetic diversity is causing reproductive failure.

Reid, W. V. C., and K. R. Miller. 1989. *Keeping Options Alive*. Washington D.C.: World Resources Institute. Examines questions of how much biodiversity should be preserved and how this might be accomplished.

Servheen, C. 1985. "The Grizzly Bear." In *Audubon Wildlife Report*, edited by R. DiSilvestro, New York: The National Audubon Society. A description of the status of grizzly bears in the United States with recommendations for management and recovery plans.

Soule, M. E., and B. A. Wilcox. 1980. *Conservation Biology: An Evolutionary-ecological Perspective*. A classic textbook in the field. Sunderland, MA.: Sinauer and Associates.

Tennesen, M. 1991. "Poaching, Ancient Traditions, and the Law." *Audubon* July–August 1991: 90. A good description of the problem of wildlife poaching for traditional medicine and international trade.

Tuan, Yi-fu. *Dominance and Affection: The Making of Pets*. New Haven: Yale University Press, 1984. An interesting discussion about why we have pets and what that reveals about us by a brilliant scholar and writer.

Western, D., and M. C. Pearl. 1989. "Conservation for the Twenty-First Century" New York: Oxford University Press. An excellent collection of articles by leading conservation biologists on wildlife problems and an agenda for future protection of endangered species.

Wilcove, D. June/July 1987. "Back from the Brink." *Nature Conservancy News*. 37 (3):4. Case studies of restoration of Kirtland's warblers, red wolf, osprey, and other endangered species.

Wilson, E. O. 1989. "Threats to Biodiversity." *Scientific American* 261 (3):108. An excellent discussion of the ethical and ecological implications of destruction of biological diversity.

Yonzon, P. B., and M. I. Hunter. 1991. "Cheese, Tourists, and Red Pandas in the Nepal Himalayas." *Conservation Biology* 5 (2):196. An interesting case study showing how cheese consumption by tourists encourages deforestation by local yak herders and threatens the endangered red panda.

CHAPTER *16*

Water Resources, Use, and Management

C O N C E P T S

Under heaven nothing is more soft and yielding than water,

Yet for attacking the solid and strong, nothing is better;

It has no equal.

The weak can overcome the strong.

The supple can overcome the stiff.

Everyone knows this,

Yet no one puts it into practice.

Lao Tsu, 500 B.C.

Water is fundamental for life and essential for nearly every human endeavor. Although there is a great deal of water in the world, only about 2 percent is fresh, and only a tiny fraction of that is surface water easily available for human use.

Fresh water is a renewable resource, constantly purified and redistributed by the hydrologic cycle, but the distribution is uneven. Much global precipitation falls when or where it is not useful to humans. Uneven distribution, inequitable access, and increasing pollution of water supplies may become the next major environmental crisis. Conflict between regions for limited water supplies could cause social, political, and economic disruptions.

Only about 10 percent of annual precipitation is economically accessible to humans. Worldwide, 69 percent of the water withdrawn is used in agriculture. Eight percent is used for direct human consumption, but in many places, clean, fresh drinking water is in desperately short supply. The United Nations estimates that at least 2 billion people, or 40 percent of the total world population, do not have access to adequate water supplies.

In developed countries, industry appropriates up to half of all water withdrawn. The largest single industrial water use is for cooling, especially in power plants. Little of this water is actually consumed; most of it could be recycled and reused for other purposes.

More than thirty times as much fresh water is stored in underground reservoirs as is in all surface freshwater sources put together. Groundwater is a valuable resource in arid lands where precipitation is undependable or insufficient for human uses, but where replenishment rates are slow, groundwater reservoirs are often depleted. Excessive groundwater pumping can result in destructive surface subsidence.

Water transfer projects involving dams, artificial reservoirs, pumping stations, and canals can stabilize water supplies and greatly increase the carrying capacity of dry lands, but environmental, social, and economic costs often outweigh the benefits of these projects.

We can conserve and protect water supplies in many ways. Some simple techniques can save significant amounts of water. More efficient water use in agriculture and industry could result in even greater savings. One way to encourage conservation might be to allow marketing of water in arid areas and to charge users for the true costs of water supply projects, rather than to subsidize them with public funds.

INTRODUCTION

Water is a marvelous substance—flowing, rippling, swirling around obstacles in its path, seeping, dripping, trickling—constantly moving from sea to land and back again (figure 16.1). Water can be clear, crystalline, icy green in a mountain stream or black and opaque in a cypress swamp. Water bugs skitter across the surface of a quiet lake; a stream cascades down a stairstep ledge of rock; waves roll endlessly up a sand beach, crash in a welter of foam, and recede. Rain falls in a gentle mist, refreshing plants and animals. A violent thunderstorm floods a meadow, washing away stream banks. Why do we find water so fascinating? Why are its diverse forms so central to natural and human systems?

Water is essential for life (chapter 2). It is the medium in which all living processes occur. Water dissolves nutrients and distributes them to cells, regulates body temperature, supports structures, and removes waste products. About 60 percent of your body is water. You could survive for weeks without food, but only a few days without water.

We are used to thinking about water as an infinitely available, renewable resource because it is constantly purified and redistributed by the action of the sun, wind, and gravity. But in many parts of the world, water supply is increasingly limited. More people making demands on the resource, natural variations in rainfall, and wasteful or extravagant uses create shortages in many areas. To make matters worse, pollution makes whatever water is available unfit for many uses, further exacerbating supply problems. Eminent hydrologist Luna B. Leopold of the U.S. Geological Survey warns that water shortages might be the environmental crisis of the 1990s and that water conservation might be as much a national priority in a few years as energy conservation was in the 1970s.

Water is increasingly being regarded as an economic commodity. Schemes to transfer massive amounts of water from one area to another have stirred up a spirited debate. Who owns water? The concept of marketing fresh water is already causing conflict between competing regions and among users of water within a region. If all users were required to pay the full market value for the water they use, plus hidden costs and reasonable discount rates, there might be massive dislocations in society. Food prices might rise tenfold for some water-intensive crops. The populations of arid, Sun Belt states might drop precipitously. Many people are concerned that environmental values, such as wildlife habitat, recreation, and scenic beauty, are likely to be neglected in this competition.

In this chapter, we will look at the processes that supply fresh water to the land and how humans access and use it. We will survey major water compartments of the environment and see how they are depleted by human uses and replenished by natural processes. Finally, we will examine some schemes for transferring water from one area to another, along with some techniques for conserving water.

FIGURE 16.1 Water is a magical substance. It is the most unique aspect of our world and is essential for life. But will it always be available in pure form and abundant quantities as it has been in the past? Some authors predict that water shortages will be the environmental crisis of the next decade.

WATER RESOURCES AND RECYCLING PATTERNS

Earth is the only place in the Universe, as far as we know, where liquid water exists in substantial quantities. Oceans, lakes, rivers, glaciers, and other bodies of liquid or solid water cover more than 70 percent of our world's surface.

Earth's Water

The total amount of water on our planet is more than 1,404 million cu km (370 billion gal) (table 16.1). If Earth had a perfectly smooth surface, an ocean about 3 km (1.9 mi) deep would cover everything. Fortunately for us, continents rise above the general surface level, creating dry land over about 30 percent of the planet. It is generally assumed that most of Earth's water has been formed from oxygen and hydrogen released from rocks by volcanic activity.

Other planets have rocks and volcanic activity similar to Earth's. Why don't they have oceans? Earth is unique in having an atmosphere to trap water vapor, and a temperature range that keeps most of it liquid.

The Hydrologic Cycle

"All rivers run into the sea, yet the sea is not full: Unto the place from which rivers come, thither they return again."

Ecclesiastes 1:7

RESOURCES AND RESOURCE ECONOMICS

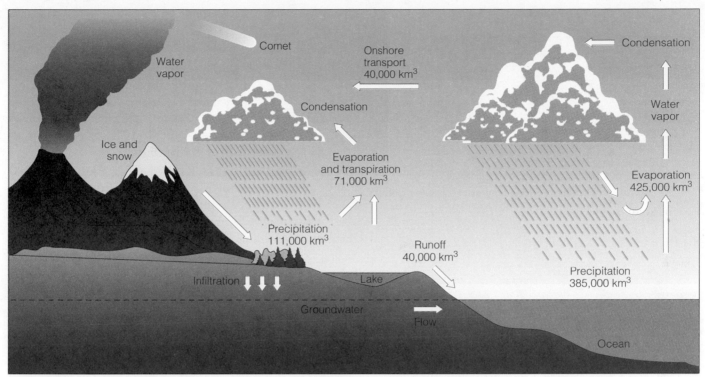

FIGURE 16.2 The hydrologic cycle moves water constantly between aquatic, atmospheric, and terrestrial compartments driven by solar energy and gravity. Total annual flows shown here are in thousands of cubic kilometers. The total annual runoff from land to the oceans is about 10.3×10^{15} gallons.

TABLE 16.1 Some units of water measurement
One cubic kilometer (km³) equals one billion cubic meters (m³), one trillion liters, or 264 billion gallons.
One acre-foot is the amount of water required to cover an acre of ground one foot deep. This is equivalent to 325,851 gallons, or 1.2 million liters, or 1,234 m³, about the amount consumed annually by a family of four in the United States.
One cubic foot per second of river flow equals 28.3 liters per second or 449 gallons per minute.

The hydrologic cycle (water cycle) describes the circulation of water as it evaporates from land, water, and organisms, enters the atmosphere; condenses and is precipitated to Earth's surfaces; and moves underground by infiltration or overland by runoff into rivers, lakes, and seas (chapter 3). The total amount of water on Earth remains about the same from year to year, and the hydrologic cycle simply moves it from one place to another (figure 16.2). This process supplies fresh water to the land masses while also playing a vital role in creating a habitable climate and moderating world temperatures. Movement of water back to the sea in rivers and glaciers is a major geological force that shapes the land and redistributes material. Plants play an important role in the hydrologic cycle, absorbing groundwater and pumping it into the atmosphere by transpiration (transport

plus evaporation). In tropical forests, as much as 75 percent of annual precipitation is returned to the atmosphere by plants.

Solar energy drives the hydrologic cycle by evaporating surface water. **Evaporation** is the process in which a liquid is changed to vapor (gas phase) at temperatures well below its boiling point. Water also can move between solid and gaseous states without ever becoming liquid in a process called **sublimation**. On bright, cold, windy winter days, when the air is very dry, snowbanks disappear by sublimation, even though the temperature never gets above freezing. This is the same process that causes "freezer burn" of frozen foods.

In both evaporation and sublimation, molecules of water vapor enter the atmosphere, leaving behind salts and other contaminants and thus creating purified fresh water. This is essentially distillation on a grand scale. We used to think of rainwater as a symbol of purity, a standard against which pollution could be measured. Unfortunately, increasing amounts of atmospheric pollutants are picked up by water vapor as it condenses into rain. We will have more to say about long-range transport and deposition of pollutants from air in chapter 23.

The amount of water vapor in the air is called humidity. Warm air can hold more water than cold air. When a volume of air contains as much water vapor as it can at a given temperature, we say that it has reached its **saturation point**. Figure 16.3 shows saturation points over a range of temperatures. **Relative humidity** is the amount of water vapor in the air expressed

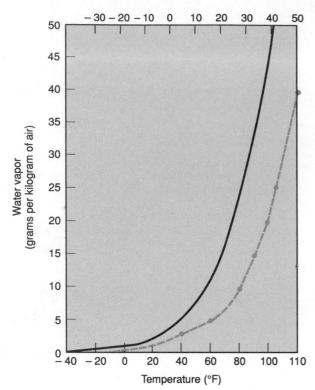

FIGURE 16.3 Water content of air at various temperatures. The darkline represents saturation points (maximum water content). The dotted line represents 50 percent relative humidity.

as a percentage of the maximum amount (saturation point) that could be held at that particular temperature.

When the saturation concentration is exceeded, water molecules begin to aggregate in the process of **condensation.** If the temperature at which this occurs is above 0° C, tiny liquid droplets result. If the temperature is below freezing, ice forms. For a given amount of water vapor, the temperature at which condensation occurs is the **dew point.** Tiny particles called **condensation nuclei** floating in the air facilitate this process. Smoke, dust, sea salts, spores, and volcanic ash all provide such particles. Even apparently clear air can contain large numbers of these particles, which are generally too small to be seen by the naked eye. Sea salt is an excellent source of such nuclei, and heavy, low clouds frequently form in the humid air over the ocean. Some nucleating agents are so efficient at accumulating water that they can cause precipitation even when the air is far below its saturation point.

A cloud, then, is an accumulation of condensed water vapor in droplets or ice crystals. Normally, cloud particles are small enough to remain suspended in the air; but when cloud droplets and ice crystals become large enough, gravity overcomes uplifting air currents and precipitation occurs. Some precipitation never reaches the ground. Temperatures and humidities in the clouds where snow and ice form are ideal for their preservation, but as they fall through lower, warmer, and drier air layers, reevapor-

ation occurs. Rising air currents lift this water vapor back into the clouds, where it condenses again; thus, liquid water and ice crystals may exist for only a few minutes in this short cycle between clouds and air (figure 16.2).

Rainfall and Topography

Rain falls unevenly over the planet. Figure 16.4 shows broad patterns of precipitation around the world. Very heavy rainfall is typical of tropical areas, especially where monsoon winds carry moisture-laden sea air onshore (chapter 17). Mountains act as both cloud formers and rain catchers. As air sweeps up the windward side of a mountain, pressure decreases and temperature falls, causing relative humidity to increase. Eventually, the air is supersaturated with moisture and condensation occurs. Further cooling of the air causes droplets of moisture to coalesce and become too heavy to remain suspended, so they fall as rain. The now cooler and drier air continues over the mountain to the leeward side. It descends and warms once again, reducing its relative humidity even further, not only preventing rainfall there, but absorbing moisture from other sources. As a result, a mountain range generally has two distinct climatic personalities. The windward side is usually cool, wet, and cloudy, while the leeward side is warm, dry, and sunny. We call the dry area on the downwind side of a mountain its **rain shadow.**

A striking example of this dichotomy is found in the Hawaiian Islands (figure 16.5). The windward side of Mount Waialeale on the island of Kauai is one of the wettest places on Earth, with an annual rainfall near 12 m (460 in.). The leeward side, only a few kilometers away, is in the rain shadow of the mountain and has an average yearly rainfall of only 46 cm (18 in.). On a broader scale, some mountain ranges cast rain shadows over vast areas. The Himalaya and Karakorum ranges of south Asia block moisture-laden monsoon winds from reaching central Asia. The Sierra Nevada of California and the coastal ranges of Oregon and Washington intercept moisture-laden Pacific winds, resulting in the arid intermountain Great Basin of the western United States (chapter 4).

Desert Belts

Rising and falling air masses that result from global circulation patterns also help create deserts in two broad belts on either side of the equator around the world. Evaporation is highest near the equator where direct rays of the sun produce the greatest heat budgets. Hot air over the equator rises, cools, and drops its moisture as rain; thus, equatorial regions are areas of high precipitation. As this cooler, drier air moves towards the poles, it condenses and sinks earthward again along the Tropics of Cancer and Capricorn (23° north and south latitude, respectively), warming as it descends.

This hot, dry air causes high evaporative losses in these subtropical regions and creates great deserts on nearly every continent: the Sahara of North Africa, the Takla Makan and Gobi

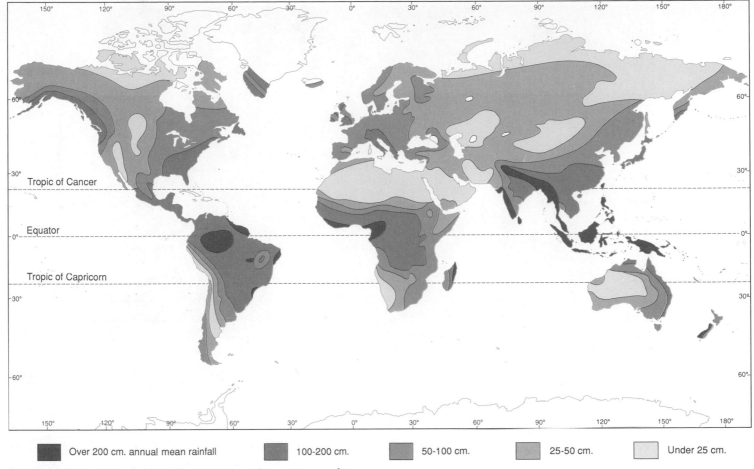

Over 200 cm. annual mean rainfall	100-200 cm.	50-100 cm.	25-50 cm.	Under 25 cm.

FIGURE 16.4 Patterns of world precipitation. Notice wet areas that support tropical rain forests along the equator and dry areas where subsidence occurs along the Tropics of Cancer and Capricorn.

of China, the Sonoran and Chihuahuan of Mexico and the United States, the Kalahari and Namib of Southwest Africa, and Australia's Great Sandy Desert. Dry air cascading down the west side of the Andes creates some of the driest deserts in the world along the coasts of Chile and Peru. Some towns in Chile's Atacama Desert have had no rain in recorded history.

Humans and domestic animals have expanded many of these deserts by destroying forests and stripping away protective vegetation from once fertile lands, exposing the bare soil to erosion. Local weather patterns and water supplies also are adversely affected by this process of desertification (chapter 12). Weather and climate are discussed further in chapter 17.

Balancing the Water Budget

Everything about global hydrological processes is awesome in scale. Each year the sun evaporates approximately 496,000 cu km of water from Earth's surface. More water evaporates in the tropics than at higher altitudes, and more water evaporates over the oceans than over land (figure 16.2). Although the oceans cover about 70 percent of Earth's surface, they account for 86

percent of total evaporation. Ninety percent of the water evaporated from the ocean falls back on the ocean as rain. The remaining 10 percent is carried by prevailing winds over the continents where it combines with water evaporated from soil, plant surfaces, lakes, streams, and wetlands to provide a total continental precipitation of about 111,000 km³.

What happens to the surplus water on land—the difference between what falls as precipitation and what evaporates? Some of it is incorporated by plants and animals into biological tissues. A large share of what falls on land seeps into the ground to be stored for a while (from a few days to many thousands of years) as soil moisture or groundwater. Eventually, all the water makes its way back downhill to the oceans. The 41,000 km³ carried back to the ocean each year by surface runoff or underground flow represents the renewable supply available for human uses and sustaining freshwater-dependent ecosystems.

The global water budget thus is balanced by circulation systems on land, in the atmosphere, and in the oceans that move water from areas of excess to areas of deficit. Rivers that carry water from the land to the sea are balanced by wind currents

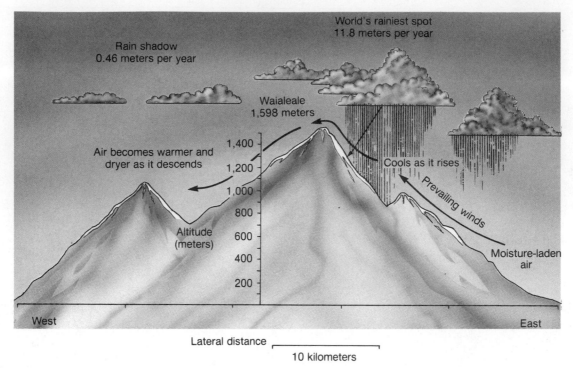

FIGURE 16.5 Rainfall on the east side of Mount Waialeale in Hawaii is more than twenty times as much as on the west side. Prevailing trade winds bring moisture-laden sea air onshore. The air cools as it rises up the flanks of the mountain and the water it carries precipitates as rain—11.8 m (38 ft) per year!

flowing in great swirls above the Earth, moving moisture-laden air from one region to another. Ocean currents are equivalent to vast rivers, carrying warm water from the equator to higher latitudes on the surface and returning cold, nutrient-rich waters in deep currents. The Gulf Stream, which flows along the east coast of North America at a steady rate of 10 to 12 km per hour (6 to 7.5 mph), carries more than one hundred times more water than all the rivers on land put together.

The redistribution of heat that results from the massive evaporation, precipitation, and transport of water is a major factor in keeping world temperatures relatively constant and making the world habitable. Without oceans to absorb and store heat, and wind currents to redistribute that heat in the latent energy of water vapor, Earth would probably undergo extreme temperature fluctuations like those of the moon, where it is 100°C (212°F) during the day, and −130°C (−200°F) at night. Water is able to perform this vital function because of its unique properties in heat absorption and energy of vaporization (chapter 2).

MAJOR WATER COMPARTMENTS OF THE WORLD

The distribution of water often is described in terms of interacting compartments in which water resides for short or long times. Table 16.2 shows the major water compartments in the world.

Oceans

Together, the oceans contain roughly 97 percent of all the *liquid* water in the world. (The water of crystallization in rocks is far larger than the amount of liquid water.) While the ocean basins really form a continuous reservoir, shallows and narrows between them reduce water exchange, so they have different compositions, climatic effects, and even different surface elevations. Oceans play a crucial role in moderating Earth's temperature, but they are generally too salty for most human uses.

In tropical seas, surface waters are warmed by the sun, diluted by rainwater and runoff from the land, and aerated by wave action. In higher latitudes, surface waters are cold and much more dense. This dense water subsides or sinks to the bottom of deep ocean basins and flows towards the equator. Warm surface water of the tropics stratifies or floats on top of this cold, dense water like cream on an unstirred cup of coffee. Sharp boundaries form between different water densities, different salinities, and different temperatures, retarding mixing between these layers.

We mentioned that the hydrologic cycle is an extremely long-term process. The average **residence time** of water in the ocean (the length of time that an individual molecule spends circulating in the ocean before it evaporates and starts through the hydrologic cycle again) is about three thousand years. In the deepest ocean trenches, movement is almost nonexistent and water may remain undisturbed for tens of thousands of years.

RESOURCES AND RESOURCE ECONOMICS

TABLE 16.2 Earth's water compartments—estimated volume of water in storage, percent of total, and average residence time

	Volume (thousands of km³)	% total water	Average residence time
Total	1,403,377	100	2,800 years
Ocean	1,370,000	97.6	3,000 years to 30,000 years*
Ice and snow	29,000	2.07	1 to 16,000 years*
Groundwater down to 1 km	4,000	0.28	From days to thousands of years*
Lakes and Reservoirs	125	0.009	1 to 100 years*
Saline lakes	104	0.007	10 to 1,000 years*
Soil moisture	65	0.005	2 weeks to a year
Biological moisture in plants and animals	65	0.005	1 week
Atmosphere	13	0.001	8 to 10 days
Swamps and marshes	3.6	0.003	From months to years
Rivers and streams	1.7	0.0001	10 to 30 days

*Depends on depth and other factors

Source: Data from U.S. Geological Survey.

Glaciers, Ice, and Snow

Of the 3 percent of all water that is fresh, about three-fourths is tied up in glaciers, ice caps, and snowfields. Glaciers are really rivers of ice flowing downhill very slowly (figure 16.6). They now occur only at high altitudes or high latitudes, but as recently as eighteen thousand years ago about one-third of the continental landmass was covered by glacial ice sheets. Most of this ice has now melted and the largest remnant is in Antarctica. As much as 2 km (1.25 mi) thick, the Antarctic glaciers cover all but the highest mountain peaks and contain nearly 85 percent of all ice in the world.

An ice sheet that is similar in thickness but much smaller in volume covers most of Greenland. There is no landmass at the North Pole. A permanent ice pack made of floating sea ice covers much of the Arctic Ocean. Although sea ice comes from ocean water, salt is excluded in freezing so the ice is mostly fresh water. Together with the Greenland ice sheet, arctic ice makes up about 10 percent of the total ice volume. The remaining 5 percent of the world's permanent supply of ice and snow occurs mainly on high mountain peaks.

Groundwater Aquifers

After glaciers, the next largest reservoir of fresh water is held in the ground as **groundwater.** Precipitation that does not evaporate back into the air or run off over the surface percolates through the soil and into pores and hollows of permeable rocks in a process called **infiltration** (figure 16.7). Upper soil layers that hold both air and water make up the **zone of aeration.** Moisture for plant growth comes primarily from these layers.

Depending on rainfall amount, soil type, and surface topography, the zone of aeration may be very shallow or quite deep. Lower soil layers where all spaces are filled with water make up the **zone of saturation.** The top of this zone is the **water table.** The water table is not flat, but undulates according to the surface topography and subsurface structure. Nor is it stationary through the seasons, rising and falling according to precipitation and infiltration rates.

Porous, water-bearing layers of sand, gravel, and rock are called aquifers. Aquifers are always underlain by impermeable layers of rock or clay that keep water from seeping out at the bottom.

Folding and tilting of Earth's crust by geologic processes can create shapes that generate water pressure in confined aquifers (those trapped between two impervious rock layers). When a pressurized aquifer intersects the surface, or if it is penetrated by a pipe or conduit, an artesian well or spring results from which water gushes without being pumped.

Areas in which infiltration of water into an aquifer occurs are called **recharge zones** (figure 16.8). The rate at which most aquifers are refilled is very slow, however, and groundwater presently is being removed faster than it can be replenished in many areas. Although water use and resource depletion are discussed later in this chapter, it is significant to refer to some present problems with aquifer recharging. Urbanization, road building, and other development often block recharge zones and prevent replenishment of important aquifers. Contamination of surface water in recharge zones and seepage of pollutants through wells has polluted aquifers in many places, making them unfit for most

FIGURE 16.6   An alpine glacier in Alaska. Ice and snow make up only 2 percent of total water, but 99 percent of all surface fresh water in the world.

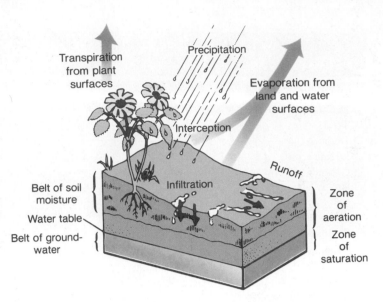

FIGURE 16.7   Precipitation that does not evaporate or run off over the surface percolates through the soil in a process called infiltration. The upper layers of soil hold droplets of moisture between air-filled spaces. Lower layers, where all spaces are filled with water, make up the zone of saturation or groundwater.

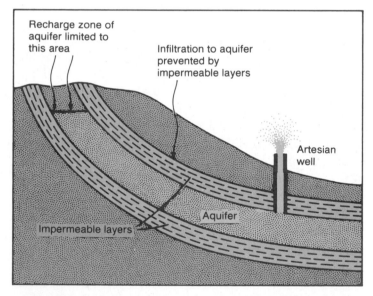

FIGURE 16.8   An aquifer is a porous, water-bearing layer of sand, gravel, or rock. This aquifer is confined between impermeable layers of rock or clay and bent by geological forces, creating hydrostatic pressure. A break in the overlying layer creates an artesian well or spring.

uses (chapter 22). Many cities protect aquifer recharge zones from pollution or development, both as a way to drain off rainwater and as a way to replenish the aquifer with pure water.

Some aquifers contain very large volumes of water (Box 16.1). The groundwater within 1 km of the surface in the United States is more than thirty times the volume of all the lakes, rivers, and reservoirs on the surface (table 16.2). Water can flow through aquifers in massive underground rivers, and large springs sometimes produce billions of liters of water per day.

Rivers and Streams

Precipitation that does not evaporate or infiltrate into the ground runs off over the surface, drawn by the force of gravity back towards the sea. Rivulets accumulate to form streams, and streams join to form rivers. Although the total amount of water contained at any one time in rivers and streams is small compared to the other water reservoirs of the world (table 16.2), these sur-

face waters are vitally important to humans and most other organisms. Most rivers, if they were not constantly replenished by precipitation, meltwater from snow and ice, or seepage from groundwater, would began to diminish in a few weeks.

The speed at which a river flows is not a very good measure of how much water it carries. Headwater streams are usually small and fast, often tumbling downhill in a continuous cascade. As the stream reaches more level terrain, it slows and generally becomes deeper and more quiet. The best measure of the volume carried by a river is its **discharge,** the amount of water that passes a fixed point in a given amount of time. This is usually expressed as liters or cubic feet of water per second. The sixteen largest rivers in the world carry nearly half of all surface runoff on Earth. The Mississippi River, which is the fourth longest river in the world and the largest in North America, carries an average of 14 million liters (450,000 cu ft) per second. Peak flow in the spring can be as high as 45 million liters.

Lakes and Ponds

There isn't a clear distinction between ponds and lakes, but ponds are generally considered to be small temporary or permanent bodies of water shallow enough for rooted plants to grow over most of the bottom. Lakes are inland depressions that hold standing fresh water year-round. Maximum lake depths range from a few meters to over 1,600 m (1 mi) in Lake Baikal (USSR). Surface areas vary in size from less than one-half ha (one acre) to large inland seas, such as Lake Superior (U.S.) and the Caspian Sea (USSR), covering hundreds of thousands of square kilometers. Both ponds and lakes are relatively temporary features on the landscape because they eventually fill with silt or are emptied by cutting of the outlet stream through the barrier that creates them.

While lakes contain nearly one hundred times as much water as all rivers and streams combined, they are still a minor component of total world water supply. Their water is much more accessible than groundwater or glaciers, however, and they are important in many ways for humans and other organisms. The Great Lakes (Superior, Michigan, Ontario, Huron, and Erie) are the largest single reservoir of surface fresh water in the world, containing some two quadrillion liters (526 trillion gal). This is about 20 percent of the world's fresh liquid surface water.

Wetlands

Bogs, swamps, wet meadows, and marshes play a vital and often unappreciated role in the hydrological cycle. Their lush plant growth stabilizes soil and holds back surface runoff, allowing time for infiltration into aquifers and producing even, year-long stream flow. When wetlands are drained, filled, or otherwise disturbed, their natural water-absorbing capacity is lost and surface waters run off quickly, resulting in floods and erosion during the rainy season and dry, or nearly dry, stream beds the rest of the year. This has a disastrous effect on biological diversity and productivity, as well as on human affairs.

The Atmosphere

The atmosphere is among the smallest of the major water reservoirs of Earth in terms of water volume, containing less than 0.001 percent of the total water supply. It also has the most rapid turnover rate. An individual water molecule resides in the atmosphere for about ten days, on average. While water vapor makes up only a small amount (4 percent maximum at normal temperatures) of the total volume of the air, movement of water through the atmosphere provides the mechanism for distributing fresh water over the landmasses and replenishing terrestrial reservoirs.

TABLE 16.3 Distribution of renewable freshwater supplies by continent

	Average annual runoff	Share of global runoff	Share of global population	Share of runoff that is stable
	(cubic km)	(percent)	(percent)	(percent)
Africa	4,225	11	11	45
Asia	9,865	26	58	30
Europe	2,129	5	10	43
North America*	5,960	15	8	40
South America	10,380	27	6	38
Oceania	1,965	5	1	25
Soviet Union	4,350	11	6	30
World	38,874	100	100	36†

*Includes Central America, with runoff of 545 cubic kilometers
†Average

Source: Worldwatch Paper 62: *Water: Rethinking Management in an Age of Scarcity.*

WATER AVAILABILITY AND USE

Clean, fresh water is essential for nearly every human endeavor. Perhaps more than any other environmental factor, the availability of water determines the location and activities of humans on Earth. Table 16.3 shows the average annual **runoff** by continent. Runoff is the excess of precipitation over evaporation and infiltration and represents, in broad terms, the water *available* for human use.

Water Supplies

The richest continents in terms of total water supply are South America and Asia. Each has about 12 percent of the total land area of the world, but receives about one-fourth of the total global runoff. In terms of water available per person, South America has the most abundant supply. Its 27 percent of total runoff is shared by only 6 percent of the world population. However, most of the rainfall and runoff in South America occurs in the jungles of the Amazon basin, where infertile soil and inhospitable conditions limit human habitation. Much of the runoff in Asia does occur in areas suitable for agriculture, which is one of the reasons that Asia has nearly 60 percent of all humans on Earth.

Stable runoff, the fraction that is available year-round, is usually more important than total runoff in determining human uses. In most parts of the world, a majority of the precipitation falls during a limited "wet" season. Water is abundant—perhaps

BOX 16.1 The Ogallala Aquifer: An Endangered Resource

The Ogallala Aquifer System is the largest known underground freshwater reservoir in the world. Lying under parts of eight states in the arid, high plains region (box figure 16.1), this reservoir has supplied water to one of the most important agricultural areas in the United States and has been a significant factor in the high rates of productivity we have enjoyed in recent years. However, the enormous amounts of water now being pumped out of the aquifer are depleting it much faster than it can be recharged from surface infiltration. In some areas, water tables are falling as much as 1 m (3 ft) per year. Many farmers are having to abandon irrigation, either because the aquifer has been exhausted under their land or because the costs of pumping from greater and greater depths are no longer justified by the crops produced.

It is estimated that this vast aquifer once contained about 2,000 cu km (2 billion acre-feet) of water in porous rock layers, ranging in thickness from a few meters around the periphery of the formation to more than 400 m (1,200 ft) in the center of the pool under the sand hills of Nebraska. This is about sixteen times as much water as all lakes, streams, rivers, marshes, and other surface freshwater bodies on Earth. Most of the water in the aquifer is thought to have been left by the glaciers that melted 15,000 years ago. In 1930, the average depth of the water was nearly 20 m (58 ft). In 1987, it was less than 3 m (8 ft), and falling. Some places are essentially out of water.

Exploitation of the high plains groundwater began about one hundred years ago when pioneer farmers and ranchers set up windmills for domestic supplies and to water crops and livestock. Although this early technology played a vital role in the settling of the American West, it had relatively little effect on the enormous amount of water contained in the aquifer. The real change came in the 1960s with the invention of center-pivot

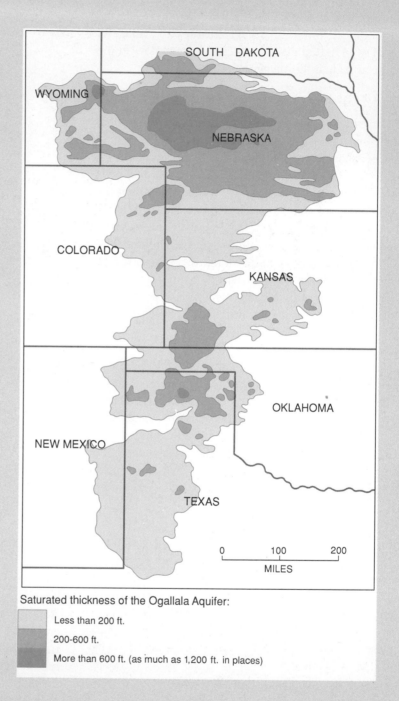

Saturated thickness of the Ogallala Aquifer:

Less than 200 ft.

200–600 ft.

More than 600 ft. (as much as 1,200 ft. in places)

BOX FIGURE 16.1 The Ogallala aquifer, underlying parts of eight plains states, is the largest known body of fresh water in the world. Its water bearing rock and gravel formations range from 10 m (30 ft) to 400 m (1,200 ft) in thickness. Yearly withdrawal from the aquifer (mostly for agriculture) is equal to the total annual flow of the Colorado River. The water table has dropped as much as 50 m (150 ft) in a few decades in some places.

Source: U.S. Geological Survey.

irrigation systems (see figure 16.12). In these systems, a high-capacity well drilled in the center of a field pumps water to large sprinklers that ride on a pipe up to 1 km (0.6 mi) long. The pipes are carried on motor-driven wheels in a huge circle around the well. Because the pipes have flexible joints and the wheels are driven independently, the whole apparatus can traverse hilly land that could not be irrigated by conventional gravity-fed methods.

The amount of irrigated land on the high plains jumped from about 1 million ha (2.5 million acres) in 1950, to 6.5 million ha (16 million acres) in 1980. More than 150,000 wells supplied the water for about 33 percent of all cotton, 50 percent of all grain, and 40 percent of all beef produced in the United States in the early 1970s. More water was being drawn out of the aquifer each year at the peak of irrigation than the entire annual flow of the Colorado River.

In contrast to surface waters, which are tightly controlled by water laws and appropriation rights, there are no limits on how much water a person can pump from the ground, even though it is a shared resource that underlies neighboring land as well. There also are no regulations about how and for what purposes the water can be used. The rule was, and still is, "Those who pump fastest, get most."

What will happen when the aquifer runs dry? Farmers may have to return to dry land farming that typically yields only one-third as much per unit of land as irrigated fields. Alternative sources for municipal and industrial water supplies often do not exist. Cities may become ghost towns unless water is brought in from elsewhere.

Pressure undoubtedly will rise for water transfer projects that can ship water from the Great Lakes states or the Mississippi River valley. The costs of those

projects would be billions of dollars and the price of water delivered might be ten times what farmers comfortably can pay. Should the government bear some or all of those costs? Another interesting question is whether the areas that have plentiful water supplies should share their water. Many of the states that have lost industry and agricultural production in recent years to the western states might prefer that people and businesses move back to where the water is, rather than moving the water to where the people and businesses are. Where millions of dollars are invested in farms and communities, however, there are also human considerations. Should society have restricted population growth in areas where water has always been limited? Could we have foreseen the implications of water overdraft? What can we learn from this situation about the future use of water?

unpleasantly so—during this season, but much of that water quickly drains away to the ocean and isn't available during the succeeding dry season. In India, for instance, 90 percent of annual precipitation falls between June and September.

Another important consideration is the interannual variability of rainfall shown in figure 16.9. In some areas, such as the African Sahel region, abundant rainfall occurs some years but not others. Unless steps are taken to even out water flows, the lowest levels encountered usually limit both ecosystem functions and human activities. Some of the world's earliest civilizations, such as the Sumerians and Babylonians of Mesopotamia, the Harappans of the Indus Valley, and the early Chinese cultures, were based on communal efforts to control water, to divert floods and drain marshes during wet seasons or wet years, and to store water in reservoirs or divert it from streams so that it would be available during the dry seasons or dry years.

Drought Cycles

Rainfall is never uniform in either geographical distribution or yearly amount. Every continent has regions where rainfall is scarce because of topographic effects or wind currents. In ad-

dition, cycles of wet and dry years create temporary droughts. Water shortages have their most severe effect in semiarid zones where moisture availability is the critical factor in determining plant and animal distribution. Undisturbed ecosystems often survive extended droughts with little damage, but introduction of domestic animals and agriculture disrupts native vegetation and undermines natural adaptations to low moisture levels.

In the United States, the cycle of drought seems to be about thirty years. There were severe dry years in the 1870s, 1900s, 1930s, 1950s, and 1970s. The worst of these in economic and social terms were the 1930s. Wasteful farming practices and a series of dry years in the Great Plains combined to create the dust bowl. Wind stripped topsoil from millions of hectares of land and billowing dust clouds turned day into night. Thousands of families were forced to leave farms and migrate to cities. There now is a great worry that the greenhouse effect (Box 17.1) will bring about major climatic changes and make droughts both more frequent and more severe than in the past.

Tables 16.4 and 16.5 compare total and per capita water availability in some water-rich and water-poor countries. Brazil has by far the largest renewable water supply in the world, but Iceland is the wealthiest country in the world in terms of per

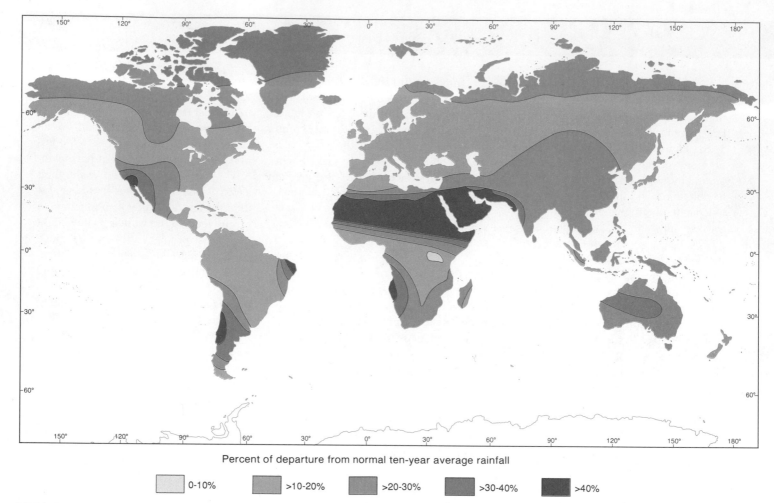

Percent of departure from normal ten-year average rainfall

| | 0-10% | | >10-20% | | >20-30% | | >30-40% | | >40% |

FIGURE 16.9 Variability of annual rainfall as likely percentage departure in any given year from the 10-year average.

TABLE 16.4	Water availability in selected water-poor countries		
Country	Annual renewable reserves (m)	Annual per capita supply (1,000 m³)	Gallons/ capita/yr (×1,000)
Kuwait	0.00	0.00	0
Bahrain	0.00	0.00	0
Egypt	1.80	0.03	7.9
Qatar	0.02	0.06	15.8
Malta	0.03	0.07	18.5
Libya	0.70	0.15	39.6
Barbados	0.05	0.20	52.8
Saudi Arabia	0.60	0.22	58.8
Hungary	6.00	0.57	150.4
Dijibouti	0.30	0.74	195.3
Germany	96.00	1.22	322.0
India	1,850.00	2.17	572.8
China	2,800.00	2.47	650.0

Source: Data from *World Resources 1990–1991*.

TABLE 16.5	Water availability in selected water-rich countries		
Country	Annual renewable resources (km³)	Annual per capita (1,000m³)	Gallons/ capita/yr (×1,000)
Iceland	170	672	177,408
Suriname	200	496	130,944
Papua New Guinea	801	200	52,800
Canada	2,901	109	28,776
Norway	450	96	25,344
Liberia	232	91	24,024
Congo	181	91	24,024
Laos	270	66	17,424
Brazil	5,190	35	9,240
Zaire	1,019	28	7,392
Soviet Union	4,384	15	3,960
Indonesia	2,530	14	3,696
United States	2,478	10	2,640

Source: Data from *World Resources, 1990–1991*.

RESOURCES AND RESOURCE ECONOMICS

capita water supply with some 672,000 cu m of water available per person each year. The most water-poor countries are Kuwait and Bahrain, which have essentially no renewable water supply. China is fourth in the world in terms of total water supply, but its large population makes it below average in terms of per capita availability. Consider how water supplies affect the history and the future of each of these countries.

Types of Water Use

In contrast to energy resources, which are consumed when used, water has the potential for being reused many times. In discussing water appropriations, we need to distinguish between different kinds of uses and how they will affect the water being appropriated.

Withdrawal is the total amount of water taken from a lake, river, or aquifer for any purpose. Much of this water is employed in nondestructive ways and is returned to circulation in a form that can be used again. **Consumption** is the fraction of withdrawn water that is lost in transmission, evaporation, absorption, chemical transformation, or otherwise made unavailable for other purposes as a result of human use. **Degradation** is a change in water quality due to contamination or pollution so that it is unsuitable for other desirable service. The total quantity available may remain constant after some uses, but the quality is degraded so the water is no longer as valuable as it was.

Worldwide, humans withdraw about 10 percent of the total annual runoff and about 25 percent of the stable runoff. The remaining three-quarters of the stable supply is generally either uneconomical to tap (it would cost too much to store, ship, purify, or distribute), or there are ecological constraints on its use. Consumption and degradation together account for about half the water withdrawn in most industrial societies. The other half of the water we withdraw would still be valuable for further uses if we could protect it from contamination and make it available to potential consumers.

We have always treated water as if there is an inexhaustible supply. It has been cheaper and more convenient for most people to dump all used water and get a new supply than to determine what is contaminated and what is not. The natural cleansing and renewing functions of the hydrologic cycle do replace the water we need if natural systems are not overloaded or damaged. Water is a renewable resource, but renewal takes time. The rate at which we are using water now may make it necessary to conscientiously protect, conserve, and replenish our water supply, however.

Quantities of Water Used

The average amount of water withdrawn worldwide is about 646 cu m (170,616 gal) per person per year. This overall average hides great discrepancies in the proportion of annual runoff withdrawn in different areas. As you might expect, those countries with a plentiful water supply and a small population withdraw a

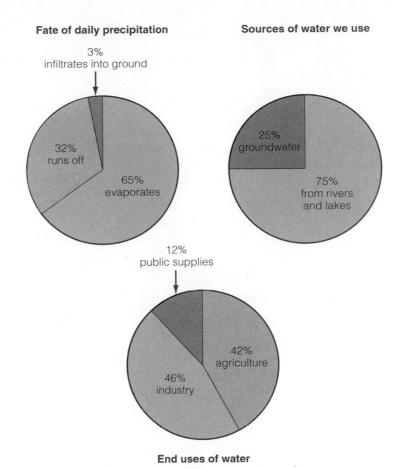

Fate of daily precipitation

Sources of water we use

3% infiltrates into ground

32% runs off

65% evaporates

25% groundwater

75% from rivers and lakes

12% public supplies

42% agriculture

46% industry

End uses of water

FIGURE 16.10 Water sources and uses in the United States. Source: U.S. Water Resources Council.

very small percentage of the water available to them. Canada, Brazil, and the Congo, for instance, withdraw less than one percent of their annual runoff. By contrast, in countries, such as Kuwait, Libya, and Qatar, where water is one of the most crucial environmental resources, groundwater and surface water withdrawal together amount to more than 100 percent of their renewable supply.

The total runoff from precipitation in the United States amounts to an average of 10,430 cu m (2.7 million gal) per person per year. We now withdraw about one-fifth of that amount, or some 5,400 liters (1,400 gal) per person per day. By comparison, the average water use in in less developed countries is only about 45 liters per person per day.

Use by Sector

Water use can be analyzed by identifying three major kinds of use, or sectors: public, industry, and agriculture. Table 16.6 shows the average annual water use in some selected countries by sector. Figure 16.10 shows the origin and distribution of the water we use in the United States. Although the United States does not have the most abundant water supply in the world, we use more water per person than any other country because we can afford

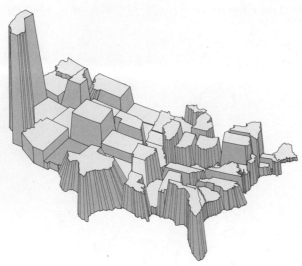

FIGURE 16.11 Regional variations in annual water use in the United States. Note high water use in states with much irrigated agriculture, such as California, Colorado, Texas, and Florida. The heavily industrialized Great Lakes states also use large quantities of water.
Source: U.S. Geological Survey Water Supply Circulation 1001.

FIGURE 16.12 Center-pivot sprinklers allow farmers to irrigate crops on rolling terrain. Wheels driven by electric or gasoline engines circle around a central deep well and can water an area up to 2.6 km² (640 acres). Agriculture now uses 42 percent of all water withdrawn. In some areas, irrigation consumes 90 percent of all available water.

the facilities to extract and use large amounts of water. As this figure indicates, only 12 percent of total water withdrawal in the United States is for municipal supplies that provide water for direct personal consumption. The other 88 percent is divided between industry and agriculture.

Figure 16.11 shows the variation in water use by state. States with higher than average use levels are either heavily industrialized, as in the Ohio River Valley and the Great Lakes States, or have high levels of irrigated agriculture, as in Florida, Texas, Arkansas, Colorado, and Idaho. California, with its large population and intense agriculture, uses almost twice as much water as any other state and about 10 percent of all water used in the United States.

It is interesting to note the variations in water use by different sectors of the economy in other countries in table 16.6. Worldwide, agriculture claims about 69 percent of total water withdrawal, ranging from 93 percent of all water used in India to only 4 percent in Kuwait, which cannot afford to spend its limited water on crops. Canada, where the fields are well watered by natural precipitation, uses only 12 percent of its water for agriculture.

Agricultural water use is notoriously inefficient and highly consumptive. Typically, from 70 to 90 percent of the water withdrawn for agriculture never reaches the crops for which it is intended. The most common type of irrigation is to simply flood the whole field or run water in rows between crops. As much as half is lost through evaporation or seepage from unlined irrigation canals bringing water to fields. Most of the rest runs off, evaporates, or infiltrates into the field before it can be used. The water that evaporates or seeps into the ground is generally lost for other purposes, and the runoff from fields is often con-

taminated with soil, fertilizer, pesticides, and crop residues, making it low quality. Sprinklers (figure 16.12) are more efficient in distributing water evenly over the field and can be used on uneven terrain, but they lose a great deal of water to evaporation.

Worldwide, industry accounts for about one-fourth of all water use, ranging from 70 percent of withdrawal in some European countries, such as Germany, to 5 percent in less industrialized countries, such as Egypt and India. Cooling water for power plants is by far the largest single industrial use of water, typically accounting for 50 to 100 percent of industrial withdrawal. Unlike agriculture, however, only a small fraction of this water is consumed or degraded. Most power plants have a "once-through" cooling system that returns water to its source after it passes through the plant. Typically, only 2 to 5 percent of the cooling water is lost through leaks or evaporation. If care is taken to avoid contamination, this water is not degraded and can be used for other purposes.

A few other industries account for the majority of the remaining industrial water use. In the United States, primary metal smelting and fabrication, petroleum refining, pulp and paper manufacturing, and food processing use about two-thirds of the industrial water not used by power plants. You would probably be surprised to learn how much water is used to manufacture some of the ordinary products that we consume. Table 16.7 shows a sample of products and the water used in their production. Much of the water used by these industries could be recycled and used over again in the factory. This would have benefits both in extending water supplies and in protecting water quality. Although Third World countries typically allocate only about 10 percent of their water withdrawal to industry, this could

RESOURCES AND RESOURCE ECONOMICS

TABLE 16.6 Average annual water use in selected countries

Country	Withdrawal (km³)	Percent of total available	Use by sector (%) Domestic	Industry	Agriculture
Brazil	35.04	1	43	17	40
Canada	36.15	1	18	70	12
Congo	0.04	<1	62	27	11
Egypt	56.40	97	7	5	88
Germany	50.53	26	10	70	20
India	380.00	18	3	4	93
Indonesia	16.59	1	13	11	76
Kuwait	0.01	x	64	32	4
Libya	2.62	374	15	10	75
Norway	2.00	<1	20	72	8
Qatar	0.04	174	36	26	38
United States	467.00	19	12	46	42
Soviet Union	353.00	8	6	29	65
World average	3296.00	8	8	23	69

x = no natural freshwater available.

Source: Data from *World Resources, 1990–1991.*

change rapidly as they industrialize. Water may be as important as energy in determining which countries develop and which remain underdeveloped.

FRESHWATER SHORTAGES

Water is a major limiting factor of the environment, both for biological systems (chapter 2) and human societies. Our growing world population is placing great demands upon natural freshwater sources. The world is faced with increasing pressure on water resources and widespread, long-lasting water shortages in many areas for three reasons: (1) rising demand, (2) unequal distribution of usable fresh water, and (3) increasing pollution of existing water supplies. The Russian hydrologist G. P. Kalinin predicts that by the year 2000 about half of all Earth's renewable water will be in use by humans. Arnon Sofer of Israel's Haifa University warns that Middle Eastern water wars might erupt over competition for this scarce resource. Every continent has areas of water shortage. In this section, we will look at some causes and effects of those water shortages.

A Scarce Resource

About 2 billion people, nearly half the world's population, already do not have an adequate supply of safe drinking water. In many countries, women and children spend a large part of each day walking to and from the nearest water supply or standing in line at the village well (figure 16.13). It has been estimated that as much as 80 percent of the diseases that affect people in the poorest countries of the world are caused by contaminated water

TABLE 16.7 Examples of water use

	Liters	Gallons
Home use:		
Bath	100–150	30–40
Shower	20 per min	5 per min
Washing clothes	75–100	20–30
Cooking	30	8
Flushing toilet (once)	10–15	3–4
Watering lawn	40 per min	10 per min
Agriculture and food processing:		
1 egg	150	40
1 ear corn	300	80
1 loaf bread	600	160
1 pound beef	9,500	2,500
Industrial and commercial products:		
1 Sunday paper	1,000	280
1 pound steel	110	32
1 pound synthetic rubber	1,100	300
1 pound aluminum	3,800	1,000
1 automobile	380,000	100,000

supplies and lack of sanitation. In some cases, these shortages are caused by natural forces: the rains fail, rivers change their courses and take water elsewhere, or dry winds evaporate the moisture that would normally have carried people through the dry season.

In other cases, shortages are human in origin: too many people compete for the resource; overgrazing and inappropriate

FIGURE 16.13 Many people in developing countries have inadequate water supplies. In India, 8,000 villages have no water at all. Women and children, who do almost all the family domestic chores, must walk long distances to the nearest well or river. Often, the water is contaminated and unsafe to drink.

TABLE 16.8 Subsiding cities

	Maximum subsidence (m)	Area affected (km²)
Coastal		
Venice	0.22	150
Shanghai	2.63	121
Bangkok	1.00	800
Tokyo	4.50	3,000
Niigata	2.50	8,300
San Jose	3.90	800
Houston	2.70	12,100
New Orleans	2.00	175
Long Beach/Los Angeles	9.00	50
Inland		
Mexico City	8.50	225
Denver	0.30	320
San Joaquin Valley	8.80	13,500
Baton Rouge	0.30	650

Many of the world's great cities are sinking because of the removal of groundwater or oil and gas beneath them. Unstable city locations include unconsolidated sediments on river floodplains (New Orleans, London, Bangkok), in deltaic coastal marshes (Venice, Houston, Tokyo, Shanghai), or lake beds (Mexico City).

From R. Dolan and H. Goodell, "Sinking Cities" in *American Scientist*, 74:38, 1986. Copyright © 1986 Sigma XI Scientific Research Society, New Haven, CT. Reprinted by permission.

agricultural practices allow water to run off before it can be captured; and lack of sewage systems contaminate existing water supplies. A shortage of money for wells, storage reservoirs, delivery pipes, and other facilities means that people can't use the resources available to them. The United Nations declared the 1980s as the decade for solving problems of clean water supplies and adequate sanitation for all. It is estimated that providing clean water and adequate sanitation for everyone in the world would cost about $300 billion, approximately what the United States spends on military expenses each year.

Unbalancing the Groundwater Cycle

Groundwater is the source of nearly 40 percent of the fresh water for agricultural and domestic use in the United States. Nearly half of all Americans and about 95 percent of the rural population depend on groundwater for drinking and other domestic purposes. Overuse of these supplies causes several kinds of problems, including depletion, subsidence, and saltwater infiltration. The map in figure 16.14 shows where these problems are most serious.

Aquifer Depletion

In many areas of the United States, groundwater is being withdrawn from aquifers faster than natural recharge can replace it. On a local level, this causes a cone of depression in the water table, as is shown in figure 16.15. A heavily pumped well can lower the local water table so that shallower wells go dry. On a broader scale, heavy pumping can deplete a whole aquifer (Box

16.1). Many aquifers have slow recharge rates, so it will take thousands of years to refill them once they are emptied. Much of the groundwater we now are using probably was left there by the glaciers thousands of years ago. It is fossil water, in a sense. When we pump water out of a reservoir that cannot be refilled in our lifetime, we essentially are mining a nonrenewable resource. Covering aquifer recharge zones with urban development or diverting runoff that once replenished reservoirs insures that they will not refill.

Subsidence and Saltwater Infiltration

Withdrawal of large amounts of groundwater causes porous formations to collapse, resulting in **subsidence** or settling of the surface above (see figure 16.4). The United States Geological Survey estimates that the San Joaquin Valley in California has sunk more than 10 m in the last fifty years because of excessive groundwater pumping. Table 16.8 lists some cities that are experiencing subsidence problems. Most of these are coastal cities, built on river deltas or other unconsolidated sediments. Flooding is frequently a problem as these coastal areas sink below sea level. Some inland areas also are affected by severe subsidence. Mexico City is one of the worst examples. Built on an old lake bed, it has probably been sinking ever since Aztec times. In recent years,

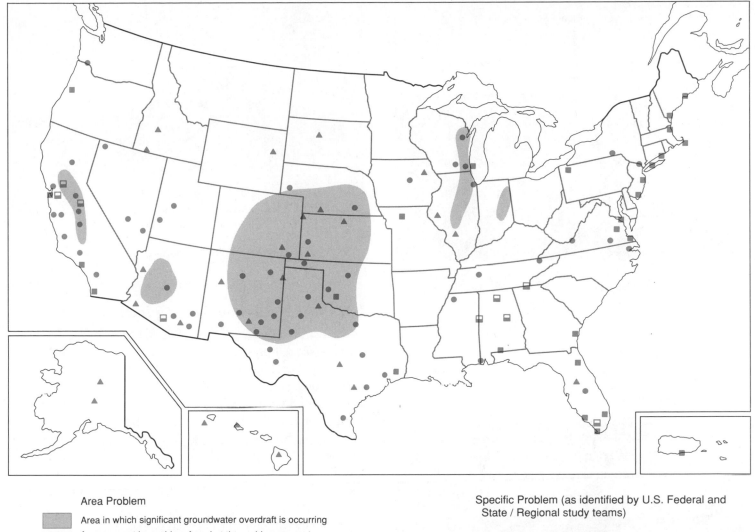

FIGURE 16.14 Declining groundwater levels and related problems in the United States.

Sources: Adapted from V. J. Pye, et al., *Groundwater Contamination in the United States*, © 1983 University of Pennsylvania Press, Philadelphia, PA; and U.S. Water Resources Council, 1990.

Area Problem

Area in which significant groundwater overdraft is occurring

Area may not be problem-free, but the problem was not considered major

Specific Problem (as identified by U.S. Federal and State / Regional study teams)

● Declining groundwater levels

▲ Diminished springflow and streamflow

▭ Formation of fissures and subsidence

■ Saline water intrusion into freshwater aquifers

rapid population growth and urbanization (chapter 26) have caused groundwater overdrafts. Some areas of the city have sunk as much as 8.5 m (25.5 ft). The Shrine of Guadalupe, the Cathedral, and many other historic monuments are sinking at odd and perilous angles.

Sinkholes form when the roof of an underground channel or cavern collapses, creating a large surface crater (figure 16.16). Drawing water from caverns and aquifers accelerates the process of collapse. Sinkholes can form suddenly, dropping cars, houses, and trees without warning into a gaping crater hundreds of meters across. Subsidence and sinkhole formation generally represent permanent loss of an aquifer. When caverns collapse or the pores between rock particles are crushed as water is removed, it is usually impossible to reinflate these formations and refill them with water.

Another consequence of aquifer depletion is saltwater infiltration. Along coastlines and in areas where saltwater deposits are left from ancient oceans, overuse of freshwater reservoirs often allows saltwater to intrude into aquifers used for domestic and agricultural purposes (figure 16.17).

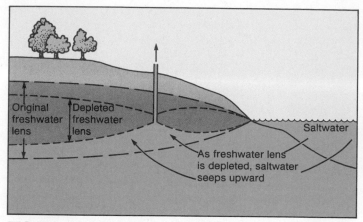

FIGURE 16.17 Saltwater intrusion into a coastal aquifer as the result of groundwater depletion. Many coastal regions of the United States are losing water sources due to saltwater intrusion.

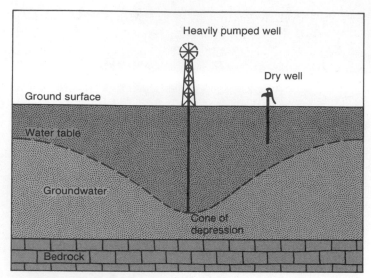

FIGURE 16.15 A cone of depression forms in the water table under a heavily pumped well. This may dry up nearby shallow wells or make pumping so expensive that it becomes impractical.

FIGURE 16.16 Home lost to a sinkhole in Bartow, Florida. This sinkhole was over 150 m (nearly 500 ft) long, 40 m wide, and 20 m deep. It formed without warning in a matter of minutes.

INCREASING SUPPLIES: PROPOSALS AND PROJECTS

Where do present and impending freshwater shortages leave us now? On a human time scale, the amount of water on Earth is fixed, for all practical purposes, and there is little we can do to make more water. There are, however, several ways to increase local supplies.

Seeding Clouds and Towing Icebergs

In the dry prairie states of the 1800s and early 1900s, desperate farmers paid self-proclaimed "rainmakers" in efforts to save their withering crops. Centuries earlier, Native Americans danced and prayed to rain gods. We still pursue ways to make rain. Seeding clouds with dry ice or potassium iodide particles sometimes can initiate rain if water-laden clouds and conditions that favor precipitation are present. In a sense, cloud seeding, if it works, means "robbing Peter to pay Paul" because the rain that falls in one area decreases the precipitation somewhere else. Furthermore, there are worries about possible contamination from the salts used to seed clouds.

Another scheme that has been proposed for supplying fresh water to arid countries is to tow icebergs from the Arctic or Antarctic. Icebergs contain tremendous quantities of fresh water, but whether they can be towed without excessive melting to places where water is needed is not clear. Nor is it clear whether the energy costs to move an iceberg thousands of miles (if that were, in fact, possible) would be worthwhile.

Desalinization

A technology that might have great potential for increasing freshwater supplies is desalinization of ocean water. The most common methods of desalinization are distillation (evaporation and recondensation) or reverse osmosis (forcing water under pressure through a semipermeable membrane whose tiny pores allow water to pass but exclude most salts and minerals). Already over one hundred desalinization plants are in operation in the United States, and several hundred more are at work in other countries. Total annual freshwater production from these plants is more than 10 billion l (2.6 billion gal). The cost of this water is prohibitively high for most purposes. Only in such places as Oman and Bahrain where there is no other access to fresh water does it pay to desalt ocean water with present technology. If a

FIGURE 16.18 Mono Lake in the Owens Valley of eastern California. Diversion of tributary rivers to provide water for Los Angeles has shrunk the lake's surface area by one-third, threatening migratory bird flocks that feed here. These tuft formations were formed underwater where calcium-rich springs entered the brine-laden lake. They show how far the lake surface has been drawn down.

cheap, inexhaustible source of energy were available, however, the oceans could supply all the water we would ever need.

Dams, Reservoirs, Canals, and Aqueducts

People have been moving water around for thousands of years. Some of the great civilizations (Sumeria, Egypt, China, and the Inca culture of South America) were based on large-scale irrigation systems that brought river water to farm fields. In fact, some historians argue that organizing people to carry out large-scale water projects was the catalyst for the emergence of civilization. Roman aqueducts built two thousand years ago are still in use. Those early water engineers probably never even dreamed of moving water on a scale that is being proposed and, in some cases, being accomplished now.

It is possible to trap runoff with dams and storage reservoirs and transfer water from areas of excess to areas of deficit using canals, tunnels, and underground pipes. Some water transfer projects are truly titanic in scale. Los Angeles began importing water in 1913 through an aqueduct from the Owens Valley, 400 km (250 mi) to the north. This project led to a statewide program known as the California Water Plan in which a system of dams, reservoirs, aqueducts, and canals transfer water from the Colorado River on the eastern border and the Sacramento and Feather rivers in the north to Los Angeles and the San Joaquin Valley in the south.

There has been much controversy about this massive project. Some of the water now being delivered to southern California is claimed by other parts of the state, neighboring states, and even Mexico. Environmentalists claim that this water transfer upsets natural balances of streams, lakes, estuaries, and terrestrial ecosystems. Fishing enthuasiasts, whitewater boaters, and others who enjoy the scenic beauty of free-running rivers mourn the loss of rivers drowned in reservoirs or dried up by diversion projects. These projects also have been criticized for using public funds to increase the value of privately held farmland and for encouraging agricultural development and urban growth in arid lands where other uses might be more appropriate.

Mono Lake, a salty desert lake in the Owens Valley east of Yosemite National Park, is an example of environmental consequences of the California Project and is an important legal symbol as well (figure 16.18). Diversion of tributary rivers has shrunk the surface area of this lake by one-third, threatening the populations of resident and migratory birds that seek shelter and food there. In 1985, the California Supreme Court held that the state should reserve enough water to protect wildlife and other environmental values. It ruled that government holds a **public trust,** an obligation to maintain public lands in a natural state. Streams must flow sufficiently to nourish native growth and keep fish alive. This doctrine could have powerful implications in water use and environmental planning.

In China, a huge water transfer project is underway that will carry water from the Chang Jiang (Yangtze) River in the center of the country north to the dry plains around Beijing. The 1,000 km (625 mi) main trunk canal in this system would move about 30 km³ of water per year (three times the annual flow of the Colorado River in the United States). Among the engineering problems to be surmounted are the crossing of several mountain ranges and nearly 150 rivers in its path. The costs of this project include moving earth equivalent to that of about a dozen Panama Canals and an investment equal to 10 million working years. A big question about this project is whether the enormous distribution system required to deliver water to hundreds of thousands of farms and communes can be made to work. An even more disastrous situation has developed in the Soviet Union where river diversions have dried up the Aral Sea to an alarming extent (Box 16.2).

Dam Difficulties

Dams have been useful over the centuries for ensuring a year-round water supply, but they are far from perfect. In many cases, they reduce water availability and destroy both natural and human values. In this section, we will look at some of the disadvantages of dams.

Evaporation, Leakage, and Siltation

The main problem with dams is inefficiency. Some dams built in the western United States lose so much water through evaporation and seepage into porous rock beds that they waste more water than they make available. The evaporative loss from Lake Mead and Lake Powell on the Colorado River is about one km³ per year, or about 10 percent of the annual river flow. This

BOX 16.2   The Aral Sea Is Dying

The Aral Sea is a huge, shallow, saline lake hidden in the remote deserts of Uzbekistan and Kazakhstan in the south-central Soviet Union. Once the world's fourth largest lake in area (smaller only than the nearby Caspian Sea, America's Lake Superior, and East Africa's Lake Victoria), in 1960 the Aral Sea had a surface area of 68,000 km², and a volume of 1,090 km³. Its only water sources are two large rivers, the Amu Dar'ya and the Syr Dar'ya. Flowing northward from the Pamir mountains on the Afgan border, these rivers pick up salts as they cross the Kyzyl Kum and Kara Kum deserts. Evaporation from the landlocked sea's surface (it has no outlet) makes the water even saltier.

The Aral Sea's destruction began in 1918 when plans were initiated to draw off water to grow cotton, a badly needed cash crop for the newly formed Soviet Union. The amount of irrigated cropland was expanded greatly (from 2.9 to 7.6 million ha) in the 1950s and 1960s with the completion of the Kara Kum canal. Annual water flows in the Amu Dar'ya and Syr Dar'ya dropped from about 55 km³ (13 mi³) to less than 5 km³. In some years the rivers were completely dry when they reached the lake.

Scientists warned Soviet authorities that the sea would die without replenishment, but sacrificing a remote desert lake for the sake of economic development seemed an acceptable trade. Inefficient irrigation practices drained away the lifeblood of the lake. Dry years in the early 1970s and mid-1980s accelerated water shortages in the region. Now, in a disaster of unprecedented magnitude and

a.

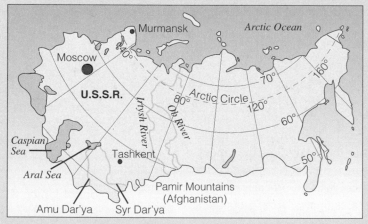

b.

BOX FIGURE 16.2   (a) Fishing boats lie marooned on the dry lake bed of the Aral Sea, which has lost two-thirds of its volume in 30 years due to excessive water withdrawals from its tributary rivers. (b) The Aral Sea lies in the Central Asian Soviet Republics of Uzbekistan and Kazakhstan.

amounts to nearly 4,500 l (1,200 gal) for each person in the United States per year. The salts left behind by evaporation nearly double the salinity of the river and make its water unusable when it reaches Mexico. To compensate, the United States is building a $350 million desalinization plant at Yuma, Arizona, to try to restore water quality.

As the turbulent Colorado River slows in the reservoirs created by Glen Canyon and Boulder Dams, it drops its load of suspended material. More than 10 million metric tons of silt per

year collect behind these dams. Imagine a line of twenty thousand dump trucks backed up to Lake Mead and Lake Powell every day, dumping dirt into the water. Within as little as one hundred years, these reservoirs will be full of silt and useless for either water storage or hydroelectric generation.

The accumulating sediments that clog reservoirs and make dams useless also represent a loss of valuable nutrients. The Aswan High Dam in Egypt was built to supply irrigation water to make agriculture more productive. Although thousands of

RESOURCES AND RESOURCE ECONOMICS

rapidity, the Aral Sea is disappearing as we watch.

Until 1960 the Aral Sea was fairly stable, but by 1990, it had lost 40 percent of its surface area and two-thirds of its volume. Surface levels dropped 14 m (42 ft.), turning 30,000 km² (about the size of Maryland) of former seabed into a salty, dusty, desert. Fishing villages that were once at the sea's edge are now 40 km (25 mi) from water. Boats trapped by falling water levels now lie abandoned in the sand. Salinity of the remaining water has tripled and almost no aquatic life remains. Commercial fishing that brought in 48,000 metric tons in 1957 was completely gone in 1990.

Winds whipping across the dried-up sea bed pick up salty dust poisoning crops and causing innumerable health problems for residents. An estimated 43 million metric tons of salt are blown onto nearby fields and cities each year. Eye irritations, intestinal diseases, skin infections, asthma, bronchitis, and a variety of other health problems have risen sharply in the past 20 years, especially among children. Infant mortality in the Kara-Kalpak Autonomous Republic adjacent to the Aral Sea is 60 per 1,000, twice as high as the rest of the Soviet Union.

Among adults, throat cancers have increased five fold in 30 years. Many physicians believe that heavy doses of pesticides used on the cotton fields and transported by runoff water to the lake sediments are now becoming air-borne in dust storms. Although officially banned, DDT and other persistent pesticides have been widely used in the area and are now found in mothers' milk. More than 35 million people are threatened by this disaster.

A report by the Institute of Geography of the Soviet Union Academy of Sciences predicts that without immediate action, the Aral Sea will vanish by 2010. It is astonishing that such a large body of water could dry up in such a short time. Philip P. Michlin of Western Michigan University, an authority on Soviet water issues says this may be the world's largest ecological disaster.

What can be done to avert this calamity? Clearly, one solution would be to stop withdrawing water for irrigation—but that would have disastrous social and economic effects. Making irrigation more efficient might save as much as half the water now lost without reducing crop yields. Restoring river flows to about 20 km³ per year would probably stabilize the sea at present levels. It would take perhaps twice as much to return it to 1960 conditions.

There is some talk of reviving grandiose plans to divert part of the northward flowing Ob' and Irtysh rivers in Western Siberia. In the early 1980s, a system of dams, pumping stations, and a 2,500 km (1500 mi) canal was proposed to move 25 km³ of water from Siberia to Uzbekistan and Kazakhstan. The cost of 100 billion rubles ($150 billion) and the potential adverse environmental effects led President Gorbechev to cancel this scheme in 1986.

Perhaps more than technological fixes, what we all need most is a little foresight and humility in dealing with nature. The words painted on the rusting hull of an abandoned fishing boat lying in the desert might express it best, "Forgive us Aral. Please come back."

---

hectares are being irrigated, the water available is only about half that anticipated because of evaporation in Lake Nasser behind the dam and seepage losses in unlined canals which deliver the water. Controlling the annual floods of the Nile also has stopped the deposition of nutrient-rich silt on which farmers depended for fertility of their fields. This silt is being replaced with commercial fertilizer costing more than $100 million each year. Furthermore, the nutrients carried by the river once supported a rich fishery in the Mediterranean that was a valuable food source for Egypt. After the dam was installed, sardine fishing declined 97 percent. To make matters worse, growth of snail populations in the shallow permanent canals that distribute water to fields has led to an epidemic of shistosomiasis. This debilitating disease is caused by blood flukes (parasitic flatworms) spread by snails living in permanent ponds and irrigation canals. In some areas, 80 percent of the residents are infected (chapter 21).

Loss of Free-Flowing Rivers

Among the environmental costs of many water projects are the loss of free-flowing rivers that are either drowned by reservoir impoundments or turned into linear, sterile irrigation canals.

Conservation history records show evidence of many battles between those who want to preserve wild rivers and those who would benefit from development (figure 16.19).

One of the first and most divisive of these battles was over the flooding of the Hetch-Hetchy Valley in Yosemite National Park. In the early 1900s, San Francisco wanted to dam the Tuolumne River to produce hydroelectric power and provide water for the city water system. This project was supported by many prominent San Francisco citizens because it represented an opportunity for both clean water and municipal power. Leader of the opposition was John Muir, founder of the Sierra Club and protector of Yosemite Park. Muir said that Hetch-Hetchy Valley rivaled Yosemite itself in beauty and grandeur and should be protected. After a prolonged and bitter fight, the developers won and the dam was built. Hard feelings from this controversy persisted for many years. In 1987, Interior Secretary Donald Hodel suggested that the Hetch-Hetchy Valley might be drained and restored to its former pristine condition. Supporters and opponents are lining up to fight the same battle again. Meanwhile, impoundment schemes for the remaining free stretches of the Tuolumne River are being proposed.

FIGURE 16.19 The recreational and aesthetic values of free-flowing wild rivers and wilderness lakes may be their greatest assets. Competition between *in situ* values and extractive uses can lead to bitter fights and difficult decisions.

WATER MANAGEMENT AND CONSERVATION

Watershed management and conservation are often more economical and environmentally sound ways to prevent flood damage and store water for future use than building huge dams and reservoirs.

Watershed Management

Rather than allowing residential, commercial, or industrial development on flood plains, these areas should be reserved for water storage, aquifer recharge, wildlife habitat, and agriculture. Sound farming and forestry practices can reduce runoff. Retaining crop residue on fields reduces flooding, and minimizing plowing and forest cutting on steep slopes protects watersheds. Wetlands conservation preserves natural water storage capacity and aquifer recharge zones. A river fed by marshes and wet meadows tends to run consistently clear and steady rather than in violent floods.

A series of small dams on tributary streams can hold back water before it becomes a great flood. Ponds formed by these dams provide useful wildlife habitat and stock-watering facilities. They also catch soil where it could be returned to the fields. Small dams can be built with simple equipment and local labor, eliminating the need for massive construction projects and huge dams.

Domestic Conservation

Similarly, we could save as much as half of the water we now use for domestic purposes without great sacrifice or serious changes in our life-styles (table 16.9). Simple steps, such as taking shorter showers, stopping leaks, and washing cars, dishes, and clothes as efficiently as possible, can go a long way toward forestalling the water shortages that many authorities predict (chapter 17). Isn't it better to adapt to more conservative uses now when we have a choice than to be forced to do it by scarcity in the future?

The use of conserving appliances, such as low-volume shower heads and efficient dishwashers and washing machines, can reduce water consumption greatly. If you live in an arid part of the country, you might consider whether you really need a lush green lawn that requires constant watering, feeding, and care. Planting native ground cover in a "natural lawn" or developing a rock garden or landscape in harmony with the surrounding ecosystem can be both ecologically sound and aesthetically pleasing. There are about 30 million ha (75 million acres) of cultivated lawns, golf courses, and parks in the United States. They receive more water, fertilizer, and pesticides per hectare than any other kind of land.

We dispose of relatively small volumes of waste with very large volumes of water. In many cases it is much better to treat or dispose of waste at its origin before it is diluted or mixed with other materials. For instance, each person in the United States uses about 50,000 l (13,000 gal) of drinking-quality water annually to flush toilets. This is more than one-third of the amount supplied to homes each year. There are now several types of waterless or low-volume toilets. The Swedish-made Clivus Multrum (figure 16.20) digests both human and kitchen wastes by aerobic bacterial action, producing a rich, nonoffensive compost that can be used as garden fertilizer. There are also low-volume toilets that use recirculating oil or aqueous chemicals to carry wastes to a holding tank, from which they are periodically taken to a treatment plant. Anaerobic digesters use bacterial or chemical processes to produce usable methane gas from domestic wastes. These systems provide valuable energy and save water, but are more difficult to operate than conventional toilets. Few cities are ready to mandate waterless toilets, but in 1988 a number of cities (including Los Angeles, California; Orlando, Florida; Austin, Texas; and Phoenix, Arizona) ordered that water-saving toilets, showers, and faucets be installed in all new buildings. The motivation was twofold: to relieve overburdened sewer systems and to conserve water.

An example of the savings that can be obtained through conservation was demonstrated in the 1977 drought in California. Through a combination of rationing, laws against nonessential uses, and public appeals for conservation, Marin County, north of San Francisco, was able to temporarily reduce water

TABLE 16.9 Annual U.S. household water use and potential savings with simple conservation measures*

Activity	Share of total indoor water use	Without conservation	With conservation	Savings
	(Percent)	(Thousand per capita)		(Percent)
Toilet flushing	38	34.5	16.4	52
Bathing	31	27.6	21.8	21
Laundry and dishes	20	18.0	13.1	27
Drinking and cooking	6	5.5	5.5	0
Brushing teeth, misc.	5	4.1	3.7	10
Total	100	89.7	60.5	33

*Estimates based on water use patterns for a typical U.S. household. European toilets, for example, often use less water than the figures given here would imply.

Source: U.S. Environmental Protection Agency, Office of Water Program Operations, Flow Reduction: Method, Analysis, Procedures, Examples, 1981.

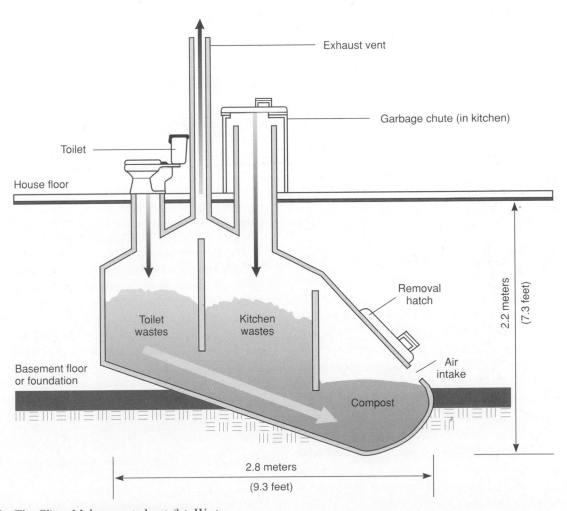

FIGURE 16.20 The Clivus Multrum waterless toilet. Wastes decompose and compost into an odorless, safe, rich, fertilizer as they slowly slide down to the bottom compartment.

Water Resources, Use, and Management

use by 65 percent. In 1988, San Francisco once again instituted water rationing and raised water prices to encourage conservation. Because of the severe droughts in the Midwest in 1988 and 1989, many communities ordered bans on all outdoor lawn sprinkling, car washing, and other nonessential uses. Whether we can make the attitudinal and behavioral changes necessary for conservation a permanent part of our lives remains to be seen.

Industrial and Agricultural Conservation

Perhaps half of all the agricultural water used is lost to leaks in irrigation canals, application to areas where plants don't grow, runoff, and evaporation. Better farming techniques, such as minimum tillage, leaving crop residue on fields and ground cover on drainage ways, intercropping, use of mulches, and trickle irrigation (chapter 11) could reduce these water losses dramatically.

Nearly half of all water use is for cooling of electric power plants and other industrial facilities. Some of this water use could be avoided by installing dry cooling systems similar to the radiator of your car. In many cases, cooling water could be reused for irrigation or other purposes in which water does not have to be drinking quality. The waste heat carried by this water could be a valuable resource if techniques were developed for using it.

Price Mechanisms

We have traditionally treated water as if there were an endless supply and as if the water itself had no intrinsic value. Federal water supply projects charge customers only for the immediate costs of delivery. The cost of building projects is usually subsidized, and the discount value of future supplies and foregone opportunities is ignored. A 1985 study by the National Resources Defense Council found that farmers in California's Central Valley were paying only $6.15 per acre-foot (85.8 cu m) for water that cost the government $72.99 to deliver. The subsidy represented by this underpriced water averaged almost $500,000 per farm per year in some areas.

Over the whole nation, the cost of federal water projects amounts to about $3.5 billion each year. Much of the water supplied by these projects is used to grow crops, such as corn, wheat, and cotton, that we have in embarrassing surplus. How ironic that we spend great sums of money to deliver water to arid lands in the Southwest to grow crops that we have paid farmers not to grow in the well-watered East and Midwest. Since water is so cheap in most western states, there is little incentive for farmers to practice conservation or efficient use. Agriculture accounts for 90 percent of the water consumed west of the 100th meridian. If farmers could save only 10 percent of the water they use, the amount available for all other uses would double.

Numerous municipalities also have unreasonably low prices for water. In New York City, for example, water was supplied to homes and businesses for many years at a flat rate. There were no meters because it was considered more expensive to install meters and read them than the water was worth. With no incentive to restrict water use or repair leaks, 750,000 cu m (200 million gal) of water were wasted each year from leaky faucets, toilets, and water pipes. The drought of 1988 convinced the city to begin a ten-year, $290 million program to install meters and reduce waste.

If water users were charged the real cost for environmental damage, future use, and public subsidies, conservation would be more attractive. One way to establish true water cost is to allow it to be marketed in interstate commerce. Laws intended to protect agriculture often prevent municipalities and industries from bidding on water supplies. This policy also protects inefficient and wasteful uses. Allowing the market to determine a price for water can encourage efficiency that makes more water available as if a new supply were being created. In 1982, the U.S. Supreme Court ruled that water is subject to the Interstate Commerce Clause of the Constitution so that state and local laws cannot interfere with its marketing.

It will be important, as water markets develop, to be sure that environmental, recreational, and wildlife values are not sacrificed to the lure of high-bidding industrial and domestic uses. The doctrine of public trust, which we discussed with reference to the California Water Plan, should help protect these diffuse public values.

S U M M A R Y

The hydrologic cycle constantly purifies and redistributes fresh water, providing an endlessly renewable resource. The physical processes that make this possible—evaporation, condensation, and precipitation—depend upon the unusual properties of water, especially its ability to absorb and store solar energy. Roughly 98 percent of all water in the world is salty ocean water. Of the 33,400 km³ that is fresh, 99 percent is locked up in ice or snow or buried in groundwater aquifers. Lakes, rivers, and other surface freshwater bodies make up only about 0.01 percent of all the water in the world, but they provide habitat and nourishment for aquatic ecosystems that play a vital role in the chain of life.

Water is essential for nearly every human endeavor. In the United States, direct personal use constitutes about one-tenth of the water we withdraw from our resources. Our two largest water uses are agricultural irrigation and industrial cooling. Altogether, the amount of water we withdraw from underground aquifers and surface reservoirs represents about 19 percent of the surplus runoff in the United States. Only about half the water we withdraw is consumed or degraded so that it is unsuitable for other purposes; much could be reused or recycled. Water conservation and recycling would have both economic and environmental benefits.

Water shortages in many parts of the world result from rising demand, unequal distribution, and increased contamination. Arid zones are especially vulnerable to the effects of natural droughts and land abuse by humans and domestic animals. Lakes, rivers, and groundwater reservoirs are being depleted at an

RESOURCES AND RESOURCE ECONOMICS

alarming rate, leading not only to water shortages, but also to subsidence, sinkhole formation, saltwater intrusion, and permanent loss of aquifers.

Water storage and transfer projects are a response to flooding and water shortages. Giant dams and diversion projects can have environmental and social costs that are not justified by the benefits they provide. Among the problems they pose are evaporation and infiltration losses, siltation of reservoirs, and loss of recreation and wildlife habitat. Watershed management and small dams are preferred by many conservationists as means of flood control and water storage.

There is much we can do to save water. Charging users the true cost of water is a good start towards conservation. We can each use less water in our personal lives, and society can encourage development of water-saving appliances, natural yards, recycling, and efficient water use. Perhaps the most important change we can make is to treat wastes at their sources rather than use precious water resources for waste disposal. Not everyone can live upstream.

◾ Review Questions

1. What is the difference between withdrawal, consumption, and degradation of water?
2. On average, how much water does each of us in the United States use each year? How is that water divided between domestic, agricultural, and industrial uses?
3. How do U.S. patterns of water use compare to those of other countries around the world?
4. Define desertification and describe how this cycle works.
5. What is subsidence? What are its results?
6. Describe some problems associated with dam building and water diversion projects.
7. Describe the path a molecule of water might follow through the hydrologic cycle from the ocean to land and back again.
8. Define evaporation, sublimation, condensation, precipitation, and infiltration. How do they work?
9. How do mountains affect rainfall distribution? Does this affect your part of the country?
10. What is the subtropical high-pressure belt and how does it cause deserts around the world?
11. What are the major water reservoirs of the world?
12. How much water is fresh (as opposed to saline) and where is it?
13. Define aquifer. How does water get into an aquifer?

◾ Questions for Critical Thinking

1. What changes would occur in the hydrologic cycle if our climate were to warm or cool significantly?
2. Why does it take so long for the deep ocean waters to circulate through the hydrologic cycle? What happens to substances that contaminate deep ocean water or deep aquifers in the ground?

3. Where would you most like to spend your vacations? Does availability of water play a role in your choice? Why?
4. Why do we use so much water? Do we need all that we use?
5. Are there ways you could use less water in your own personal life? Would that make any difference in the long run?
6. Should we use up underground water supplies now or save them for some future time?
7. How much should the United States invest to provide clean water to people in less developed countries?
8. How should we compare the values of free-flowing rivers and natural ecosystems with the benefits of flood control, water diversion projects, hydroelectric power, and dammed reservoirs?
9. Would it be feasible to change from flush toilets and using water as a medium for waste disposal to some other system? What might be the best way to accomplish this?
10. How does water differ from other natural liquids? How do the properties of water make the hydrologic cycle, and life, possible? (You may need to review chapter 2 to answer this question.)

◾ Key Terms

condensation (p. 302)
condensation nuclei (p. 302)
consumption (p. 311)
degradation (p. 311)
dew point (p. 302)
discharge (p. 307)
evaporation (p. 301)
groundwater (p. 305)
infiltration (p. 305)
public trust (p. 317)
rain shadow (p. 302)
recharge zones (p. 305)
relative humidity (p. 301)

residence time (p. 304)
runoff (p. 307)
saturation point (p. 301)
sinkholes (p. 315)
stable runoff (p. 307)
sublimation (p. 301)
subsidence (p. 314)
water table (p. 305)
withdrawal (p. 311)
zone of aeration (p. 305)
zone of saturation (p. 305)

SUGGESTED READINGS

◾ Biswas, A., et al. *Long-Distance Water Transfer: A Chinese Case Study and International Experiences*. New York: Unipub, 1985. Comparison of the Chang Jiang water diversion project with experiences in Egypt, India, the United States, and other countries.

◾ Carrier, J. 1991. "Water and the West: The Colorado River." *National Geographic* June 1991: 2. There isn't enough water in the Colorado River to meet all demands. How will this resource be managed? Great photos.

◾ El-Ashry, M. T., and D. C. Gibbons. 1988. *Water and Arid Lands of the Western United States*. Cambridge, U.K.: Cambridge University Press. An excellent study of agricultural and metropolitan water use in the western United States.

- Leopold, L. B. 1990. "Ethos, Equity and Water Resource." *Environment* 32 (2):16. A useful review of water resources and management by a distinguished hydrologist.
- Lewis, J. 1991. "The Ogallala Aquifer, an Underground Sea." *EPA Journal* 16 (6):42. A good description of the history of and possible solutions to overuse of this imperiled resource.
- Maurits la Riviere, J. W. 1989. "Threats to the World's Water." *Scientific American* 261 (3):80. Unless appropriate steps are taken soon, severe water shortages will occur.
- Micklin, P. P. 1988. "Dessication of the Aral Sea: A Water Management Disaster in the Soviet Union." *Science* 241:1170. An extensive and authoritative account of causes and effects of Aral Sea drying.
- Odell, R. September, 1975. "Silt, Cracks, Floods, and Other Dam Foolishness." *Audubon*. 77 (5):107. Descriptions of problems with dams in many different parts of the United States.
- Pearce, F. 1991. "Building a Disaster: The Monumental Folly of India's Tehri Dam." *Ecologist* 21 (3):123. If completed, the Tehri dam will be the largest in Asia. It's location in an earthquake-prone valley could be an ecological and social disaster.
- Plummer, C.C., and D. McGeary. *Physical Geology*. 2nd ed. Dubuque: Wm. C. Brown Publishers, 1982. Good overview of water resources, hydrologic cycle, river valley formation, etc.
- Postel, S. 1990. "Saving Water for Agriculture." *State of the World 1990*. Washington, D.C. Worldwatch Institute. Strategies for efficient water use in agriculture.
- Quammen, D. 1990. "A Long River with a Long History." *Audubon* March 1990: 68. A grand description of the Rio Grande and how it shapes the ecology and culture of the southwest.
- Quammen, D. September, 1983. "Bin of Water, River of Corn." *Audubon*. 85 (5):68. A wonderfully written article about a water diversion project for Nebraska's Niobara River.
- Reisner, M., and S. F. Bates. 1990. *Overtapped Oasis: Reform or Revolution for Western Water*. Covello, CA: Island Press. A stinging rebuke of business as usual in western water use.
- Sears, P. B. 1991. *Deserts on the March*. Covello, CA: Island Press. A re-issue of the classic 1935 study of human-caused desertification around the world.
- Stegner, W. 1954. *Beyond the Hundredth Meridian: John Wesley Powell and the Second Opening of the West*. Cambridge, MA: The Riverside Press. A classic history by the dean of western writers about the role of water in western settlement and politics.
- Tuan, Yi-Fu. *The Hydrologic Cycle and The Wisdom of God: A Theme in Geoteleology*. Toronto: University of Toronto Press, 1968. A historic and cultural review of attitudes towards nature with special emphasis on the hydrologic cycle.
- U.S. Environmental Protection Agency (EPA). 1980. *Primer for Wastewater Treatment*. Washington, D.C.: Government Printing Office. A brief, illustrated brochure describing strategies for municipal wastewater treatment.
- U.S. Geological Survey (USGS). 1984. *Estimated Use of Water in the United States in 1980*. Washington, D.C.: Government Printing Office. A good source of data on water consumption.
- Whittaker, R. *Communities and Ecosystems*. New York: Macmillan Co., 1975. A classic general ecology text with a good section on aquatic and marine ecosystems.
- Wilkinson, C. September/October 1989. "Water Rights and Wrongs." *Sierra* 74 (5):35. Water developments threaten wilderness values in the western states.
- Worster, D. 1985. *Rivers of Empire*. New York: Pantheon Press. A fascinating history of water use and abuse in the western United States.

C H A P T E R *17*

Air Resources: Atmosphere, Climate, and Weather

Major changes in the chemistry of the Earth's atmosphere are taking place with potentially calamitous effects for all.

Margaret Thatcher

C O N C E P T S

The atmosphere and living organisms have evolved together. Green plants, algae, and photosynthetic bacteria have added molecular oxygen to the air and removed carbon dioxide, methane, and ammonia. In doing so, they have created conditions that stabilize temperature and make life possible for oxidizing heterotrophs (like humans).

The atmosphere is layered in zones that differ greatly from each other in composition, temperature, and pressure. Upper layers absorb high-energy radiation from the sun and cosmic rays that would be lethal for most life-forms on the surface of Earth.

The peak energy transmission from the sun is in visible wavelengths, to which the atmosphere is relatively transparent. Energy is then absorbed by land or water on Earth's surface, and is re-emitted as longwave, infrared radiation, which is absorbed by the atmosphere. This trapping of heat helps keep the temperature of Earth's surface relatively uniform.

Solar energy is distributed over Earth's surface by large-scale circulation patterns created by wind and water currents. A large amount of the energy carried by these systems is transported by water or water vapor. Circulation patterns in the atmosphere create prevailing winds and long-range climatic patterns that play an important role in determining the biomes and ecosystems that exist in each area.

Topographic effects, Earth's rotation, and complex patterns of circulation in the atmosphere create large air masses that move over Earth's surface and control local weather. As these air masses push and shove against each other, they can cause violent storms. They also bring rain to replenish freshwater supplies and modify weather to avoid the great extremes that might otherwise exist.

Humans have tried many ways to modify the weather over the centuries, but we still have much to learn about this complex and massive global system. We may be altering the atmosphere in ways and with consequences that we don't yet fully understand.

INTRODUCTION

We notice the atmosphere mostly when the air is dirty or the weather is unpleasant; otherwise, we take it for granted. Earth, however, is the only known planet with an atmosphere capable of supporting life as we know it. We now are becoming aware that human activities are changing the atmosphere in ways that could have disastrous effects, not only on humans, but also on the basic life-support systems that make our planet habitable. In this chapter, we will consider the structure and composition of the atmosphere. We will look at the ways it affects and is, in turn, affected by living organisms, including humans.

Weather is a description of the physical conditions of the atmosphere (moisture, temperature, pressure, and wind), all of which play a vital role in shaping ecosystems. The dynamism of the atmosphere is maintained by a ceaseless flow of solar energy. Winds generated by pressure gradients push large air masses of differing temperature and moisture content around the globe. Our daily weather is created by the movement of these air masses (figure 17.1).

Climate is a description of the long-term pattern of weather in a particular area. Climates often undergo cyclic changes over decades, centuries, and millenia. Determining where we are in these cycles and predicting what may happen in the future is an important, but difficult, process. As human activities change the properties of the atmosphere, it becomes more difficult and more important to understand how the atmosphere works and what future weather and climate conditions may be.

Weather and climate are important, not only because they affect human activities, but because they are primary determinants of biomes and ecosystem distribution. Generalized, large-scale climate often is not as important in the life of an individual organism as is the microclimate of the atmosphere and soil in its specific habitat. Geographic boundaries that separate communities and ecosystems are established primarily by climatic boundaries created by temperature, moisture, and wind-distribution patterns (chapter 4). The movement and effects of pollutants also are strongly linked to weather and climatic conditions (chapter 23).

COMPOSITION AND STRUCTURE OF THE ATMOSPHERE

We live at the bottom of a virtual ocean of air. Extending upward about 1,600 km (1,000 mi), this vast, restless envelope of gases is far more turbulent and mobile than the oceans of water. Its currents and eddies are the winds.

Past and Present Composition

The composition of the atmosphere has changed drastically since it first formed as Earth cooled and condensed from interstellar gases. Geochemists believe that Earth's earliest atmosphere was

FIGURE 17.1 The atmospheric processes that purify and redistribute water, moderate temperatures, and balance the chemical composition of the air are essential in making life possible. Earth is the only planet we know with a habitable atmosphere. To a large extent, living organisms have created and help to maintain the atmosphere on which we all depend.

probably made up mainly of hydrogen and helium and was hundreds of times more massive than it is now. Over billions of years, much of that hydrogen and helium diffused into space or was swept away by passing asteroids or a stream of energized particles from the sun called the solar wind. At the same time, volcanic emissions have added carbon, nitrogen, oxygen, sulfur, and other elements to the atmosphere.

The current composition of Earth's atmosphere is unique in our solar system. Our planet alone has significant amounts of free oxygen and water vapor. We believe that virtually all the molecular oxygen in the air was produced by photosynthesis in blue-green bacteria, algae, and green plants. If that oxygen were not present, heterotrophs (like us) who oxidize organic compounds as an energy source could not exist. As chapter 3 discusses, producers and consumers create a balance between carbon dioxide and oxygen levels in the atmosphere. Thus, the composition of the atmosphere makes life possible, and living organisms, in turn, have modified the composition of the atmosphere. The **Gaia hypothesis,** first proposed by British chemist James Lovelock, suggests that the balance between atmospheric carbon dioxide and oxygen maintained by living organisms is responsible not only for creating a unique atmospheric chemical composition, but also for other environmental characteristics that make life possible. We will discuss how carbon dioxide levels regulate temperatures later in this chapter. Organisms—mainly humans—continue to modify the atmosphere, but at a dramatically accelerated rate, and in ways that may be disastrous for other forms of life.

Table 17.1 presents the main components of clean, dry air. Water vapor concentrations vary from near zero to 4 percent, depending on air temperature and available moisture (chapter 16). Small but important concentrations of minute particles and droplets of material—collectively called **aerosols**—also

TABLE 17.1 Present composition of the lower atmosphere*

Gas	Symbol or formula	Percent by volume
Nitrogen	N_2	78.08
Oxygen	O_2	20.94
Argon	Ar	0.934
Carbon dioxide	CO_2	0.033
Neon	Ne	0.00182
Helium	He	0.00052
Methane	CH_4	0.00015
Krypton	Kr	0.00011
Hydrogen	H_2	0.00005
Nitrous oxide	N_2O	0.00005
Xenon	Xe	0.000009

*Average composition of dry, clean air

are suspended in the air. The production of aerosols by human activities and their effects on human health and natural ecosystems are discussed in chapter 23.

A Layered Envelope

The atmosphere is layered into four distinct zones of contrasting temperature due to differential absorption of solar energy (figure 17.2). Understanding how these layers differ and what creates them helps us understand atmospheric functions.

The layer of air immediately adjacent to Earth's surface is called the **troposphere.** Ranging in depth from about 16 km (10 mi) over the equator to about 8 km (5 mi) over the poles, this zone is where most weather events occur. Due to the force of gravity and the compressibility of gases, the troposphere contains about 75 percent of the total mass of the atmosphere. The troposphere's composition is relatively uniform over the entire planet because this zone is strongly stirred by winds. Air temperature drops rapidly with increasing altitude in this layer, reaching about $-60°C$ ($-76°F$) at the top of the troposphere. A sudden reversal of this temperature gradient creates a sharp boundary, the tropopause, that limits mixing between the troposphere and upper zones.

The **stratosphere** extends from the tropopause up to about 50 km (31 mi). Air temperature in this zone is stable or even increases with higher altitude. Although more dilute than the troposphere, the stratosphere has a very similar composition except for two important components: water and ozone (O_3). The fractional volume of water vapor is about one thousand times lower, and ozone is nearly one thousand times higher than in the troposphere. Ozone is produced by lightning and solar irradiation of oxygen molecules and would not be present if photosynthetic organisms were not releasing oxygen. Ozone protects life on Earth's surface by absorbing most incoming solar ultraviolet radiation.

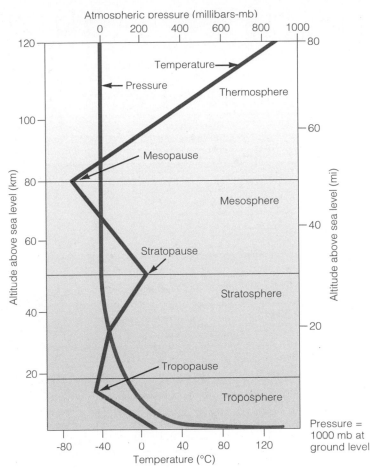

FIGURE 17.2 The atmosphere is layered into four distinct zones of contrasting temperature due to differential absorption of solar energy. Mixing between layers is inhibited by this temperature gradient. Note that pressure falls from 1,000 mb at ground level to near that of outer space in the stratosphere. Although temperatures rise in the mesosphere, there are too few molecules present to measure much pressure rise.

Recently discovered decreases in stratospheric ozone over Antarctica (and to a lesser extent over the whole planet) are of serious concern (see Box 23.1). If these trends continue, we could be exposed to increasing amounts of dangerous ultraviolet rays, resulting in higher rates of skin cancer, genetic mutations, crop failures, and disruption of important biological communities. Unlike the troposphere, the stratosphere is relatively calm. There is so little mixing in the stratosphere that volcanic ash or human-caused contaminants can remain in suspension there for many years (chapter 16).

Above the stratosphere, the temperature diminishes again, creating the **mesosphere** or middle layer. The minimum temperature reached in this region is about $-80°C$ ($-120°F$). At an altitude of 80 km, another abrupt temperature change occurs. This is the beginning of the **thermosphere,** a region of highly ionized gases, extending out to about 1,600 km (1,000 mi). Temperatures are very high in the thermosphere because molecules there are constantly bombarded by high-energy solar and cosmic radiation. There are so few molecules per unit area, how-

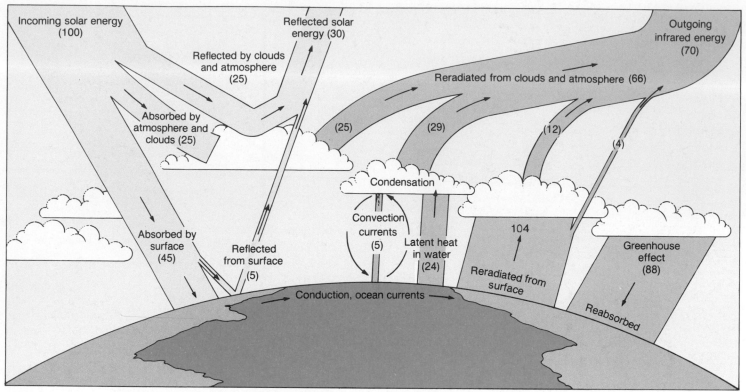

FIGURE 17.3 Energy balance between incoming and outgoing radiation. The atmosphere absorbs or reflects about half of the solar energy reaching Earth. Most of the energy reemitted from Earth's surface is long-wave, infrared energy. Most of this infrared energy is absorbed by aerosols and gases in the atmosphere and is reradiated towards the planet, keeping the surface much warmer than it would otherwise be. This is known as the greenhouse effect. The numbers shown are arbitrary units. Note that for 100 units of incoming solar energy, 100 units are reradiated to space, but more than 100 units are radiated from the earth's surface because of the greenhouse effect.

ever, that if you were cruising through in a spaceship, you wouldn't notice the temperature increase.

The lower part of the thermosphere is called the **iono-sphere.** This is where aurora borealis (northern lights) appears when showers of solar or cosmic energy cause ionized gases to emit visible light. There is no sharp boundary that marks the end of the atmosphere. Pressure and density decrease gradually as one travels away from Earth until they become indistinguishable from the near vacuum of intrastellar space. The composition of the thermosphere also gradually merges with that of intrastellar space, being made up mostly of helium and hydrogen.

THE GREAT WEATHER ENGINE

The atmosphere is a great weather engine in which a ceaseless flow of energy from the sun causes global cycling of air and water that creates our climate and distributes material through the environment.

Solar Radiation Heats the Atmosphere

The sun supplies Earth with an enormous amount of energy. Although it fluctuates from time to time, incoming solar energy at the top of the atmosphere averages about 340 watts per sq m.

About half of this energy is reflected or absorbed by the atmosphere, and half the Earth faces away from the Sun at any given time. The amount reaching the Earth's surface is still about 178,000 terawatts (trillion watts) per year. This is 15,000 times the commercial energy used by humans each year. Energy flows through space as photons (light units) of electromagnetic energy that seem simultaneously to be waves of pure energy and infinitesimal particles of matter. Figure 2.9 shows the frequencies and wavelengths of some members of this family.

On average, clouds and the atmosphere absorb or reflect about half the **insolation** (*in*coming *sol*ar rad*iation*) that reaches Earth. The absorption of solar energy by the atmosphere is selective. Visible light passes through almost undiminished, whereas ultraviolet light is absorbed mostly by ozone in the stratosphere. Infrared radiation is absorbed mostly by carbon dioxide (CO_2) and water (H_2O) in the troposphere. Scattering of light by water droplets, ice crystals, and dust in the air also is selective. Short wavelengths (blue) are scattered more strongly than long wavelengths (red). The blue of a clear sky or clean, deep water at midday, and the spectacular reds of sunrise and sunset are the result of this differential scattering. On a cloudy day, as much as 90 percent of insolation is absorbed or reflected by clouds. Figure 17.3 shows the average energy fluxes in the atmosphere.

Some solar energy is reflected from Earth's surfaces. **Albedo** is the term used to describe reflectivity. Fresh, clean snow can have an albedo of 90 percent, meaning that 90 percent of incident radiation falling on its surface is reflected. Dark surfaces, such as black topsoil or a dark forest canopy, absorb energy efficiently and might have an albedo of only 2 or 3 percent. The net average global albedo of Earth is about 30 percent. Clouds are responsible for most of that reflection. Earth's surface has a low average albedo (5 percent) due to the energy absorbency of the oceans that cover most of the globe.

Eventually, all the energy absorbed at Earth's surface is reradiated back into space. There is an important change in properties between incoming and outgoing radiation, however. Most of the solar energy reaching Earth is visible light, to which the atmosphere is relatively transparent; the energy reemitted by Earth is mainly infrared radiation (heat energy). These longer wavelengths are absorbed rather effectively in the lower levels of the atmosphere (mostly by carbon dioxide and water vapor), trapping much of the heat close to Earth's surface. If the atmosphere were as transparent to infrared radiation as it is to visible light, Earth's surface temperature would be about 35°C (63°F) colder than it is now.

This phenomenon is called the greenhouse effect because the atmosphere, like the glass of a greenhouse, transmits sunlight while trapping heat inside. (The analogy is not totally correct, however, because glass is much more transparent to infrared radiation than is air; greenhouses stay warm mainly because the glass blocks air movement.) Increasing atmospheric carbon dioxide due to forest clearing and fossil fuel burning appears to be causing a global warming that could cause major climatic changes. This may be *the* environmental concern of the next century (Box 17.1).

Because of recycling of infrared energy between the atmosphere and the planet, the amount of energy emitted from Earth's surface is about 30 percent greater than the total incoming solar radiation. The amount of energy reflected or reradiated from the top of the atmosphere must balance with the total insolation if Earth is to remain at a constant temperature.

Convection Currents and Latent Heat

Air currents, especially those carrying large amounts of water vapor, also play an important role in shaping our weather and climate. As the sun heats Earth's surface, some of that heat is transferred to adjacent air layers, causing them to expand and become less dense. This lighter air rises and is replaced by cooler, heavier air, resulting in vertical **convection currents** that stir the atmosphere and transport heat from one area to another.

Much of the solar energy absorbed by Earth is used to evaporate water. Because of the unique properties of water (chapter 2), it takes a significant amount of energy to change water from liquid to vapor state (chapter 16), and this energy is

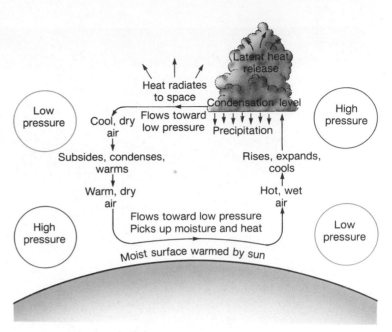

FIGURE 17.4 Convection currents and latent energy cause atmospheric circulation and redistribute heat and water around the globe.

stored in the water vapor as latent or potential energy. Latent energy is released as heat when the water condenses.

Water vapor carried into the atmosphere by rising convection currents transports large amounts of energy and plays an important role in the redistribution of heat from low to high altitudes, and from the oceans to the continental landmasses (figure 17.4). As warm, moist air rises, it expands (due to lower air pressures at higher altitudes) and cools. If condensation nuclei are present (chapter 16) or if temperatures are low enough, the water will condense to form water droplets or ice crystals and precipitation will occur. Release of latent heat causes the air to rise higher, cool more, and lose more water vapor. Rising, expanding air creates an area of relatively high pressure at the top of the convection column.

Air flows out of this high pressure zone towards areas of low pressure where cool, dry air is sinking (subsiding). This subsiding air is compressed (and therefore warmed) as it approaches Earth's surface, where it piles up and creates a region of relatively high pressure at the surface. Air flows out of this region back towards the area of low surface pressure caused by rising air, thus closing the cycle. These are the driving forces of the hydrologic cycle (chapter 16).

The convection currents we have just described can be as small and as localized as a narrow column of hot air rising over a sun-heated rock, or as large as the desert low-pressure cell that covers the U.S. Southwest most of the summer. The circulation patterns they create can be as mild as a gentle onshore breeze moving from the warm ocean toward the cooling shoreline as the sun goes down in the evening, or they can create monster cyclonic storms that drive hurricanes hundreds of kilometers wide

BOX 17.1   The Greenhouse Effect and Global Climate Change

Are we altering the atmosphere in ways that could lead to disastrous, worldwide climate change? Some climatologists point out that increasing concentrations of infrared-absorbing gases released into the atmosphere by human activities could trap heat and raise temperature with catastrophic effects. They picture an apocalyptic future in which summer heat is unbearable, farms are turned to deserts, famines sweep the globe, melting polar ice caps raise sea levels and flood coastal regions, and entire forests die along with thousands or even millions of species that can't migrate or adapt to sudden climatic changes.

About half of this predicted warming would be due to carbon dioxide (CO_2) released by burning fossil fuels, making cement, and cutting and burning forests (box figure 17.1a and 17.1b). Together, these activities release about 8.5 billion metric tons of CO_2 annually, causing atmospheric levels to rise about .4 percent each year. Two hundred years ago, at the beginning of the Industrial Revolution, the atmosphere contained about 280 parts per million (ppm) CO_2; now it contains about 350 ppm.

If current trends continue, pre-industrial CO_2 concentrations will have doubled by the year 2075. Computer models predict that doubling atmospheric CO_2 will cause global temperatures to rise 1.5 to 4.5°C (3° to 9°F). This may not sound like much change, but the difference between the temperature now and the last ice age about 10,000 years ago when glaciers covered much of North America was only about 5°C.

Global temperatures and CO_2 levels correlate throughout history. Evidence for this comes from air bubbles trapped in ice as glaciers form. Core samples from arctic glaciers provide a 160,000-year record (box figure 17.2). You can see that CO_2 concentrations follow temperatures. Both are higher now than at any time in the past 130,000 years. This correlation doesn't prove cause and effect, however.

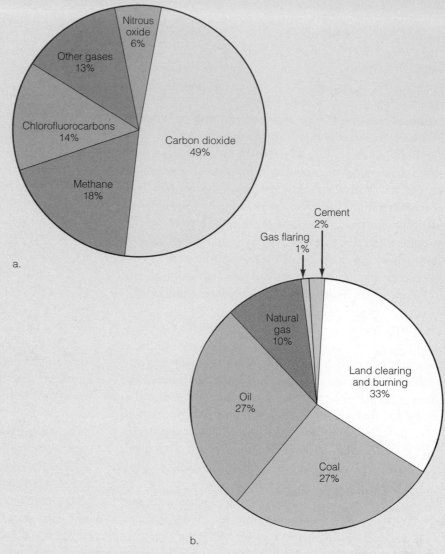

BOX FIGURE 17.1   (a) Greenhouse gas contributions to global warming—1980s, (b) sources of CO_2 from human activities.

Source: R. Morgenstern and D. Tirpack, Environmental Protection Agency, *EPA Journal*, 16 (2), March 1990.

Notice that both temperatures and CO_2 concentrations began to rise considerably before the start of industrialization.

If a greenhouse warming occurs, it probably will not be distributed evenly around the globe. Additional ocean evaporation and cloud cover will likely keep tropical coastal areas about as they are now. The greatest temperature changes are predicted to be at high latitudes and in the middle of continents (box figure 17.3). Siberia and the Canadian arctic might experience increases of 10° to 12°C (18° to 22°F). Chicago might go from an average of fifteen to forty-eight days each year above 32°C (90°F). Dallas may have 162 days per year that hot, rather than 100 as it does now. Calcutta, however, where the termperature is always hot, will not get much hotter.

Changing precipitation patterns might be one of the most serious consequences of the greenhouse effect. Some models predict that the mid-continents of

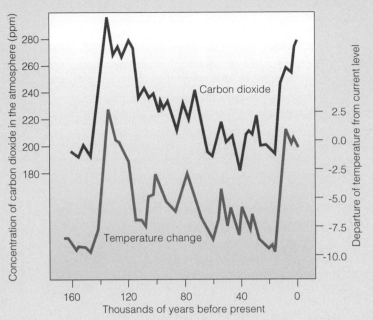

BOX FIGURE 17.2 Long-term variations in atmospheric carbon dioxide levels from arctic ice cores compared to global-mean temperature.

From J. M. Barnola, et al., "Vostok Ice Core Provides 160,000-year Record of Atmospheric CO_2" in *Nature*, Vol. 329, No. 6138, 1987, page 410.

North America, South America, and Asia will be significantly drier than they are now. This could have calamitous effects on world food supplies. The models don't agree on this point, however. While some show the American corn belt to be drier, others predict that it will be wetter than now. Model building is not yet an exact science.

Another disastrous effect could be rising sea levels. Some oceanographers calculate that thermal expansion alone could raise the sea level by one meter or more in the next century. An even worse scenario is that polar ice caps might melt. There is enough ice in the arctic and antarctic to raise the oceans by 30 meters (90 feet) if it all melted. Most of the world's major cities are on coasts only a few meters above sea level. The homes and businesses of about half the world's population would be threatened if the ice caps melt. A large amount of valuable farmland would also be lost. The United States Gulf Coast, for instance, would lose about 5,000 sq km, an area the size of Delaware.

Carbon dioxide is not the only gas that could cause climate warming. Methane, chlorofluorocarbons (CFCs), nitrous oxide, and other trace gases also absorb infrared radiation and warm the atmosphere (box figure 17.1a). Although these gases are more rare than CO_2, they absorb heat much more effectively. Methane, for instance, absorbs 20 to 30 times as much—molecule for molecule—as CO_2, and CFCs absorb approximately 20,000 times as much.

Methane is produced by intestinal bacteria in ruminant animals, anaerobic decomposition in wet-rice paddies, pipeline leaks, decaying wastes in landfills, and releases from coal mining (box figure 17.4). Atmospheric methane is increasing about 1 percent per year. Chlorofluorocarbons, used as spray propellents, degreasing agents, and refrigerants, have been accumulating at 5 percent each year. Nitrous oxide (N_2O), produced by burning organic material and soil denitrification, has been increasing at .2 percent annually.

Together, these minor greenhouse gases would have warming effects comparable to doubling CO_2 concentrations. That means that a 1.5° to 4.5° temperature rise could occur in 40 rather than 80 years if the models are accurate. The biggest share of greenhouse gases are produced by the developed countries of the world (box figure 17.5). Together, the United States, the Soviet Union, Europe, and Japan are responsible for about two-thirds of all potential global warming.

Some scientists are convinced that global climate change has already begun. By 1989, James Hansen of the Goddard Institute of Space Studies announced at a Congressional hearing in Washington that he was 99 percent confident that the greenhouse effect could already be detected. The unusually hot and dry summer of 1988 seemed to support his position. Many people were deeply worried and demanded that Congress and the Administration stop waffling around and do something to stop this ominous threat.

Not all of Hansen's scientific colleagues agreed, however. A vituperative exchange of letters and articles appeared in scientific journals. Charges of "voodoo science" were raised, and the motives and competence on both sides of the issue were questioned. Among the most notable supporters of the greenhouse effect were Steven Schneider of the National Center for Atmospheric Research, George Woodwell of the Woods Hole Oceanographic Laboratory, Michael Oppenheimer of the Environmental Defense Fund, and Dean Abrahamson of the University of Minnesota. Among the doubters were Richard Lindzen of MIT, Reid Bryson of the University of Wisconsin, Patrick Michaels of the University of Virginia, and Michael Schlesinger of the University of Illinois. All are distinguished and prominent scientists. How can we choose between these opposing views? Do observations support one side over the other?

Unfortunately, the data are equivocal. Some studies show increasing tem-

BOX 17.1   The Greenhouse Effect and Global Climate Change *(continued)*

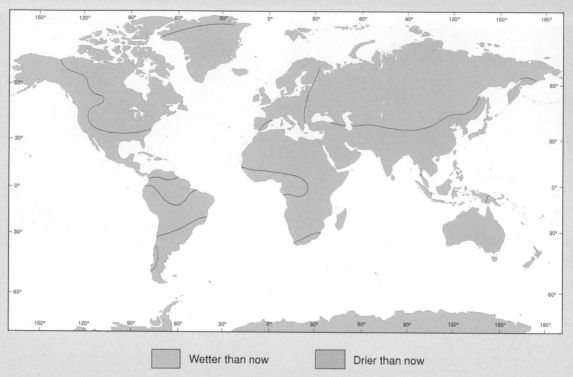

Wetter than now   Drier than now

BOX FIGURE 17.3   Predicted changes in precipitation patterns caused by global warming. High altitudes and mid-continent regions will probably be drier than now. Coastal areas and much of the tropics could be wetter. These changes could have severe effects on agriculture.

peratures; others show that temperatures are stable or even declining slightly. Among the confounding factors in these studies are dissimilar data bases and analytical methods, inaccurate thermometers, scarce and inconsistent readings in many areas, urbanization (sampling stations that were once in the country are now in cities that are known to trap heat), and unrepresentative sampling locations (most official weather stations in the United States are at airports, which are significantly warmer than surrounding areas).

Those who doubt climate change predictions often charge that current climate models do not adequately represent the effects of water vapor, clouds, air currents, ocean circulation, sulfate aerosols, decreasing solar radiation, biogeochemical ocean processes, or the possible growth stimulation of green plants and ocean plankton by higher CO_2. They point out that if the models' assumptions are changed slightly, the predictions

change from positive (warming) to negative (cooling) trends. If some of these factors are built into the models, they would show another ice age rather than a greenhouse effect.

Some scientists even claim that rising CO_2 levels could be beneficial. Sherwood B. Idso of the Department of Agriculture Water Conservation laboratory in Phoenix says that higher CO_2 will stimulate growth of many plants and result in more efficient water utilization. Lands now unsuitable for agriculture might come into production, and crop yields might increase rather than decrease. Growth of phytoplankton (single celled photosynthetic organisms) in the ocean might absorb most of the additional CO_2 and balance world temperatures. This could increase productivity of marine ecosystems and make more seafood available for human consumption.

What should we do about greenhouse gases in the face of such uncertainty? Should we take steps now to

reduce them or wait to see if their effects are as negative as some models suggest? Unfortunately, by the time we have evidence that a disaster is underway, it may be too late to make changes. Some authorities say it is too late already. Others argue that it would be wasteful to spend hundreds of billions of dollars and cause substantial disruptions to society to reduce these gases if they pose no threat.

There are, however, actions that could help reduce the greenhouse effect if it is real and that would be useful in their own right if it is not. These are steps that Steven Schneider calls a "no regrets" policy. For instance, the industrialized countries have agreed to phase out chlorofluorocarbons because they damage stratospheric ozone (chapter 23) as well as absorb heat. We could cut CO_2 emissions and save money by increasing energy efficiency in our homes, automobiles, factories, and offices. Increasing your automobile mileage from 25 to 50 mpg, for instance, would save about

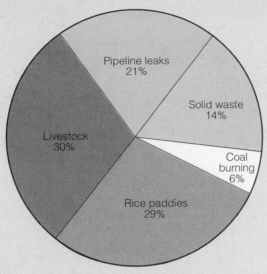

BOX FIGURE 17.4 Sources of methane from human activities.

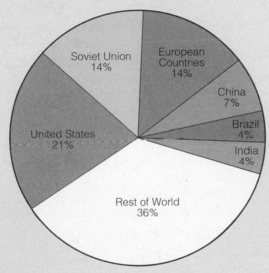

BOX FIGURE 17.5 Regional contributions to global warming.

Source: Environmental Protection Agency.

$3,000 and reduce CO_2 production by 20 metric tons over the life of the vehicle.

We could (and should) switch to alternative, renewable energy sources, such as solar, wind, tidal, and hydropower. This would reduce pollution, save fossil fuels, and reduce warming. We could stop destruction of tropical forests and plant more trees, which would save biological resources, shelter and feed wildlife, and provide a useful source of environmentally responsible wood products. Trap-

ping and burning methane from landfills and manure would provide useful energy while also reducing greenhouse effects. All of these actions are virtuous in their own right and deserve to be pursued, regardless of potential climate change. We might be able to reduce greenhouse emissions by 20 percent (enough to stabilize effects) over the next few decades without much sacrifice.

Achieving a 50 percent emission reduction (enough to reverse current

trends) would be much more difficult. Some of the actions proposed to accomplish this are controversial or risky. It might be prudent to wait and see how serious global climate change is before adopting them. For instance, some people advocate building many more nuclear power plants to reduce fossil fuel use. Others suggest spreading fertilizer in the ocean to stimulate phytoplankton growth. We don't know what ecological complications might arise from such actions. Some people advocate energy consumption taxes or rationing fuel supplies to promote conservation. Minimizing social and economic disruption and maximizing equity make these draconian responses unappealing unless they are absolutely necessary.

At a 1990 World Climate Conference in Geneva, Switzerland, the United States and the Soviet Union refused to commit themselves to specific reductions in CO_2 production because of scientific uncertainties and economic constraints. They were joined in this resolve by oil-producing countries like Saudi Arabia and Venezuela who fear a loss of revenues if fossil fuel consumption declines. Other developed countries—including Japan, Canada, Australia, New Zealand, and most of Europe—pledged a 20 percent reduction in greenhouse gases over 10 years. Developing countries demanded financial and technical assistance from their richer neighbors so they can reduce greenhouse emissions without abandoning their plans for economic development.

In many ways, your response to this issue comes down to a matter of philosophy and outlook. Are you pessimistic or optimistic? Do you believe it is more prudent to assume the worst-case scenario, or is it better to look on the bright side? Are draconian measures justified now, or should we be cautious until the science becomes clearer? This may well be the most difficult and the most important question in Environmental Science in the next few decades.

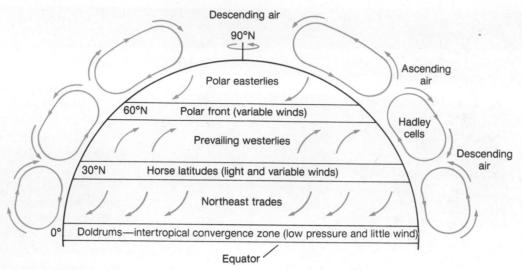

FIGURE 17.5 General circulation patterns over the Northern Hemisphere. The actual boundaries of the circulating Hadley cells vary from day to day and season to season, as do the local directions of surface winds. Surface topography also complicates circulation patterns, but within these broad regions, winds usually have a predominant and predictable direction.

across oceans and over continents. The force generated by such a storm (powered by latent heat released from condensing water vapor) can be equivalent to hundreds of megaton-sized nuclear bombs. These circulation systems—both large and small—are the driving force behind our weather, the moment-by-moment changes in the atmosphere.

WEATHER

Even though we have been watching the skies for thousands of years, trying to forecast what the weather will be, the science of meteorology (weather studies) is still rather imprecise and uncertain. Weather forces still are so important in our lives that it makes sense to learn as much as we can about how the atmosphere makes weather.

Energy Balance in the Atmosphere

Solar energy doesn't strike the whole globe equally. At the equator, the sun is almost directly overhead all year long. Its rays are very intense because it shines through a relatively short column of air (straight down), and energy flux (flow) is high. At the poles, however, sunlight comes in at an oblique angle. The long column of air through which light must pass before it reaches the surface causes much greater energy losses from absorption and scattering. Moreover, when the light does reach the ground, it is spread over a larger area because of its angle of incidence, reducing surface heating even more.

Furthermore, seasonal tilting of Earth's axis means that there is *no* sunlight at the poles during much of the winter. The equator, by contrast, has days about the same length all year long; thus, compared to the poles, the equatorial regions have an energy surplus. This energy imbalance is evened out by movement of air and water vapor in the atmosphere, and by liquid water in rivers and ocean currents. Warm, tropical air moving toward the poles and cold, polar air moving toward the equator account for about half of this energy transfer. Latent heat in water vapor (mainly from the oceans) makes up about 30 percent of the global energy redistribution. The remaining 20 percent is carried mainly by ocean currents.

Convection Currents and Hadley Cells

As air warms at the equator, rises, and moves northward, it doesn't go straight to the pole in a single convection current. Instead, this air sinks and rises in several intermediate bands, forming circulation patterns called **Hadley cells** (figure 17.5). Nor do the returning surface flows within these cells run straight north and south. Friction and drag cause air layers close to Earth's surface to be pulled in the direction of rotation. This deflection is called the **Coriolis effect.** In the Northern Hemisphere, the Coriolis effect deflects winds about 30° to the right of their expected path, creating clockwise or anticyclonic spiraling patterns in winds flowing out of a high-pressure center, and cyclonic or counterclockwise winds spiraling into a low-pressure area. In the Southern Hemisphere, winds and water movements shift in the opposite direction. Earth's rotation affects only large-scale movements, however. Contrary to popular beliefs, water draining out of a bathtub in the Southern Hemisphere is controlled by the shape of the drain, and will not necessarily swirl in the opposite direction from what it would in the north.

Prevailing Winds

A major zone of subsidence occurs at about 30° north latitude. Air flows into this region of low pressure both from the north and south. Air flowing back toward the equator is turned toward the west by the Coriolis effect, creating the steady northeast "trade winds" of subtropical oceans. Their name comes from the dependable trade routes they provided for sailing ships in earlier days. Where this dry, subsiding air falls on continents, it creates broad, subtropical desert regions (chapter 16). Air flowing north from this region of subsidence turns eastward, giving rise to the prevailing westerlies of middle latitudes. (Notice that an eastward flowing wind is called a west wind or a westerly, due to the direction from which it originates.)

Winds directly under regions of subsiding air often are light and variable. They create the so-called "horse latitudes" because sailing ships bringing livestock to the New World were often becalmed here and had to throw the bodies of dead horses overboard. Rising air at the equator creates doldrums where the winds may fail for weeks at a time. Another band of variable winds at about 60° north, called the polar front, tends to block the southward flow of cold polar air. As we will see in the next section, however, all these boundaries between major air flows wander back and forth, causing great instability in our weather patterns, especially in mid-continent areas. The Southern Hemisphere has more stable wind patterns because it has more ocean and less landmass than the Northern Hemisphere.

Rivers of Air

Superimposed on the major circulation patterns and prevailing surface winds are variations caused by large-scale upper air flows and shifting movements of the large air masses that they push and pull. The most massive of these rivers of air are the **jet streams,** powerful winds that circulate in shifting flows rivaling the oceanic currents in extent and effect. Generally following meandering paths from west to east, jet streams can be as much as 50 km wide and 5 km deep. The number, flowing speed, location, and size of jet streams all vary from day to day and place to place.

Wind speeds at the center of a jet stream are often 200 km/hr (124 mph) and may reach twice that speed at times. Located 6 to 12 km (3.7–7.5 mi) above Earth's surface, jet streams follow discontinuities in the tropopause (the boundary between the troposphere and the stratosphere), where they are broken into large, overlapping plates that fit together like shingles on a roof. The jet streams are probably generated by strong temperature contrasts where adjacent plates overlap.

There are usually two main jet streams over the Northern Hemisphere. The subtropical jet stream generally follows a sinuous path about 30° north latitude (the southern edge of the United States), while the northern jet stream follows a more irregular path along the edge of a huge cold air mass called the circumpolar vortex (figure 17.6) that covers Earth's top like a cap with scalloped edges. This whole polar vortex rotates from west to east slightly faster than the planet's rotation. As it moves, the lobes, or fingers, of cold air that protrude south from the vortex sweep across the United States. The clash between cold, dry arctic air masses pushing south against warm, wet air masses moving north from the Gulf of Mexico or the Pacific Ocean brings winds, rains, and storms to the middle of the continent.

One explanation of the 1987–1988 drought was that the circumpolar vortex slowed so that it was rotating at nearly the same speed as Earth, stalling the motion of the lobes or air masses, and locking a huge ridge of hot, dry air over mid-America for months at a time. What causes air flow to be stalled like this—or to resume normal circulation patterns—is unknown; but the amount of heat in the atmosphere surely plays a role.

During the winter, as the Northern Hemisphere tilts away from the sun and the atmosphere cools, the polar air masses become stronger and push further south, bringing snow and low temperatures across much of the United States. During the summer, as we tilt back toward the sun, warm air from the South pushes the polar jet stream back toward the pole.

Frontal Weather

The boundary between two air masses of different temperature and density is called a **front.** Fronts may be moving or stationary. When cooler air displaces warmer air, we call the moving boundary a **cold front.** Since cold air tends to be more dense than warm air, a cold front will hug the ground and push under warmer air as it advances. As warm air is forced upward, it cools adiabatically (without loss or gain of energy), and its cargo of water vapor condenses and precipitates. Upper layers of a moving cold air mass move faster than those in contact with the ground because of surface friction or drag, so the boundary profile assumes a curving, "bull-nose" appearance (figure 17.7). Notice that the region of cloud formation and precipitation is relatively narrow. Cold fronts generate strong convective currents and often are accompanied by violent surface winds and destructive storms. An approaching cold front generates towering clouds called thunderheads that reach into the stratosphere where the jet stream pushes the cloud tops into a characteristic anvil shape. The

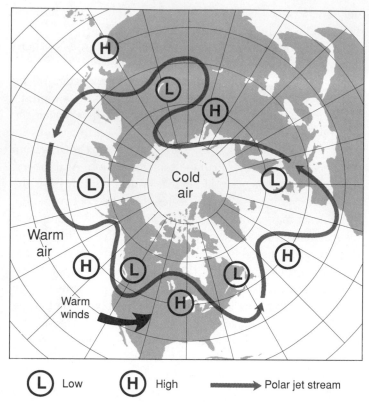

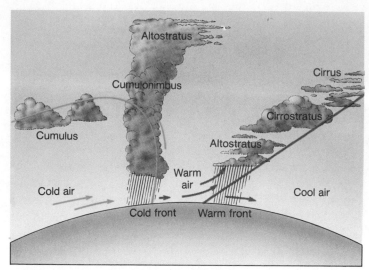

FIGURE 17.7   A cold front assumes a bulbous, "bull-nose" appearance because ground drag retards forward movement of surface air. As warm air is lifted up over the advancing cold front, it cools, producing precipitation. When warm air advances, it slides up over cooler air in front and produces a long, wedge-shaped zone of clouds and precipitation. The high cirrus clouds that mark the advancing edge of the warm air mass may be 1,000 km and forty-eight hours ahead of the front at ground level.

L Low   H High   ⟶ Polar jet stream

FIGURE 17.6   A typical pattern of the arctic circumpolar vortex. This large, circulating mass of cold air sends "fingers" or lobes across North America and Eurasia, spreading storms in their path. If the vortex becomes stalled, weather patterns stabilize, causing droughts in some areas and excess rain elsewhere.

weather after the cold front passes is usually clear, dry, and invigorating.

If the advancing air mass is warmer than local air, a **warm front** results. Since warm air is less dense than cool air, an advancing warm front will slide up over cool, neighboring air parcels, creating a long, wedge-shaped profile with a broad band of clouds and precipitation (figure 17.7). Gradual uplifting and cooling of air in the warm front avoids the violent updrafts and strong convection currents that accompany a cold front. A warm front will have many layers of clouds at different levels. The highest layers are often wispy cirrus (mare's tail) clouds that are composed mainly of ice crystals. They may extend 1,000 km (621 mi) ahead of the contact zone with the ground and appear as much as forty-eight hours before any precipitation. A moist warm front can bring days of drizzle and cloudy skies.

Cyclonic Storms

Few people experience a more powerful and dangerous natural force than cyclonic storms spawned by low-pressure cells. As we discussed earlier in this chapter, low pressure is generated by

rising warm air. Winds swirl into this low-pressure area, turning counterclockwise in the Northern Hemisphere due to the Coriolis effect. When rising air is laden with water vapor, the latent energy released by condensation intensifies convection currents and draws up more warm air and water vapor. As long as a temperature differential exists between air and ground and a supply of water vapor is available, the storm cell will continue to pump energy into the atmosphere. Winds near the center of these swirling air masses can reach hundreds of kilometers per hour and cause tremendous suffering and destruction. In 1970, a killer cyclone brought torrential rains, typhoon winds, and a tidal surge that flooded thousands of square kilometers of the flat coastal area of Bangladesh, drowning more than a half million people.

Storms over the land never have as much water vapor to pump energy into the atmosphere as those over the ocean, but cyclonic storms over the land can be terribly destructive in localized areas. Sometimes when a strong cold front pushes under a warm, moist air mass over the land, the updrafts create small cyclones that we call **tornadoes** (figure 17.8). These are most common in the spring and early summer, when temperature differentials are greatest. The southern plains states around Oklahoma are often referred to as "Tornado Alley" because of the frequency with which they experience tornadoes. Waterspouts over the ocean are also small cyclones caused by frontal weather patterns and temperature differentials.

FIGURE 17.8 Tornadoes are local cyclonic storms caused by rapid mixing of cold, dry air and warm, wet air. Wind speeds in the swirling funnel can be more than 160 km/hr (100 mph) and can be very destructive where that storm touches down on the ground.

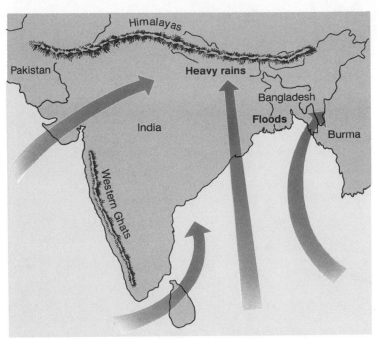

FIGURE 17.9 Summer monsoon air flows over the Indian subcontinent. Warming air rises over the plains of central India in the summer, creating a low-pressure cell that draws in warm, wet oceanic air. As this moist air rises over the Western Ghats or the Himalayas, it cools and heavy rains result. These monsoon rains flood the great rivers, bringing water for agriculture, but also causing much suffering.

Seasonal Winds

A **monsoon** is a seasonal reversal of wind patterns caused by different heating and cooling rates of the ocean and continents. The most dramatic examples of monsoon weather are found in tropical or subtropical countries where a large land area is cut off from continental air masses by mountain ranges and surrounded by a large volume of water. The Indian subcontinent is a good example (figure 17.9). As the summer sun heats the semiarid plains of India, a strong low-pressure system develops. Flow of cool air from the north is blocked by the Karakorum and Himalayan mountains. A continuous flow of moisture-laden air from the subtropical high-pressure area over the Arabian Sea sweeps around the tip of India and up into the Ganges Plain and the Bay of Bengal. As the onshore winds are driven against the mountains and rise, the air cools and drops its load of water. During the four months of the monsoon, which usually lasts from May until August, 1,000 cm (400 in.) of rain may fall on Nepal, Bangladesh, and western India. The highest rainfall ever recorded in a season was 2,500 cm (82 ft), which fell in about five months on the southern foothills of the Himalayas in 1970. These heavy rains result in high rates of erosion and enormous floods in the flat delta of the Ganges River, but they also irrigate the farmlands that feed the second most populous country in the world. In winter, the Indian landmass is cooler than the surrounding ocean and the wind flow reverses. The northeast winds pick up moisture as they blow over the Bay of Bengal and bring winter monsoon rains to Indonesia, Australia, and the Philippines.

Africa also experiences strong seasonal monsoon winds. Hot air rising over North Africa in the summer pulls in moist air from the Atlantic Ocean along the Gulf of Guinea. The torrential rains released by this summer circulation nourish the tropical forests of Central Africa and sometimes extend as far east as the Arabian Ocean. How far north into the Sahara Desert these rain clouds penetrate determines whether crops, livestock, and people live or die in the desert margin (figure 17.10). In the winter, the desert cools and air flows back out over the now warmer Atlantic. These shifting winds allowed Arabian sailors, like Sinbad, to sail from Africa to India in the summer and return in the winter.

There seems to be a connection between strong monsoon winds over West Africa and the frequency of killer hurricanes in the Carribean and along the Atlantic coast of North and Central America. In years when the Atlantic surface temperatures are high, a stronger than average monsoon trough forms over Africa. This pulls in moist maritime air that brings rain (and life)

FIGURE 17.10 Failure of monsoon rains brings drought, starvation, and death to both livestock and people in the Sahel desert margin of Africa. Although drought is a fact of life in Africa, many governments fail to plan for it, and human suffering is much worse than it needs to be.

to the Sahel (the margin of the Sahara). This trough gives rise to tropical depressions (low-pressure storms) that follow one another in regular waves across the Atlantic. The weak trade winds associated with "wet" years allow these storms to organize and gain strenth. Evaporation from the warm surface water provides energy.

During the 1970s and 1980s when the Sahel had devastating droughts, the weather was relatively quiet in North America. Only one killer hurricane (winds over 70 km/hr or 112 mph) reached the United States in two decades. By contrast, between 1947 and 1969 when rains were plentiful in the Sahel, thirteen killer hurricanes hit the United States. A climatic shift appeared to have started between 1988 and 1990. When ocean temperatures were up, rains returned to North Africa, and storms

were stronger and more frequent in the Carribean. What drives these climatic changes is unknown, but it may be related to long-term shifts in ocean circulation.

Weather Modification

As author Samuel Clemens (Mark Twain) said, "Everybody talks about the weather, but nobody does anything about it." People probably always have tried to influence local weather through religious ceremonies, dancing, or sacrifices. During the drought of the 1930s in the United States, "rainmakers" fleeced desperate farmers of thousands of dollars with claims of ability to bring rain.

Some recent developments appear to be effective in local weather modification, at least in some circumstances, but they are not without drawbacks and controversy. Seeding clouds with dry ice or ionized particles, such as iodine crystals, can initiate precipitation if water vapor is present and air temperatures are near the condensation point (chapter 16). Dry ice also is very effective at dispersing cold fog (where supercooled water droplets are present). Warm fog (air temperatures above freezing) and ice fog (ice crystals in the air) are not usually amenable to weather modification. Hail suppression by cloud seeding also can be effective, but dissipation of the clouds that generate hail diverts rain from areas that need it, as well. There are concerns that materials used in cloud seeding could cause air, ground, and water pollution. We also have inadvertently modified weather through desertification and pollution (chapter 23).

CLIMATE

If weather is a description of physical conditions in the atmosphere (humidity, temperature, pressure, wind, and precipitation), then climate is the *pattern* of weather in a region over long time periods. The interactions of atmospheric systems are so complex that climatic conditions are never exactly the same at any given location from one time to the next. While it is possible to discern patterns of average conditions over a season, year, decade, or century, complex fluctuations and cycles within cycles make generalizations difficult and forecasting hazardous. Do anomalies in local weather patterns represent normal variations, a unique abnormality, or the beginnings of a shift to a new regime?

Climatic Catastrophes

Major climatic changes, such as those of the Ice Ages, have drastic effects on biotic assemblages. When climatic change is gradual, species may have time to adapt or migrate to more suitable locations. Where climatic change is relatively abrupt, many organisms are unable to respond before conditions exceed their tolerance limits. Whole communities may be destroyed, and if the climatic change is widespread, many species may become extinct.

RESOURCES AND RESOURCE ECONOMICS

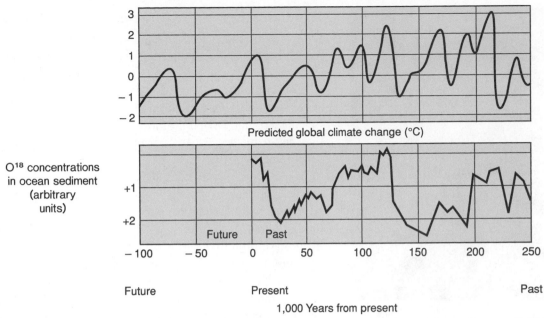

Predicted global climate change (°C)

O¹⁸ concentrations in ocean sediment (arbitrary units)

Future   Present   Past

1,000 Years from present

FIGURE 17.11   Milankovitch cycles (*top*) correlate roughly with concentration of a heavy isotope of oxygen (O¹⁸) in ocean sediment (*bottom*). As saturated air moves poleward, water molecules incorporating the heavy O¹⁸ isotope are preferentially removed. During glacial periods, when large amounts of water are tied up in polar ice caps, the concentration of O¹⁸ in ocean sediment rises, enabling scientists to infer estimates of total ice volume and global climate.

Perhaps the most well-studied example of this phenomenon is the great die-off that occurred about 65 million years ago at the end of the Cretaceous period. Most dinosaurs—along with 75 percent of all previously existing plant and animal species—became extinct, apparently as a result of sudden cooling of Earth's climate. Geologic evidence suggests that this catastrophe was not an isolated event. There appear to have been several great climatic changes, perhaps as many as a dozen, in which large numbers of species were exterminated.

Driving Forces and Patterns in Climatic Changes

What causes catastrophic climatic changes? As you can imagine, there are nearly as many theories as there are researchers studying this phenomenon. Some scientists believe that long-term climatic changes follow a purely random pattern brought about by chance interaction of unrelated events, such as asteroid impacts, cosmic radiation from exploding supernovas, massive volcanic eruptions, abrupt flooding of glacier meltwater into the ocean, and tectonic ocean spreading that changes patterns of ocean and wind circulation.

Other scientists discern periodic patterns in weather cycles. One explanation is that changes in solar energy associated with eleven-year sunspot cycles or twenty-two-year solar magnetic cycles might play a role. Another theory is that a regular 18.6-year cycle of shifts in the angle at which our moon orbits Earth alters tides and atmospheric circulation in a way that affects climate. A theory that has received a great deal of attention

in recent years is that orbital variations as Earth rotates around the sun might be responsible for cyclic weather changes. **Milankovitch cycles,** named after Serbian scientist Milutin Milankovitch, who first described them in the 1920s, are patterns caused by precession (wobbling) of Earth's axis, the direction of axial tilt, and eccentricity of Earth's orbit. Together, these variations change the amount and distribution of sunlight over Earth's surface and influence the timing of seasons and maximum and minimum temperatures (figure 17.11).

The frequency of the wobble (22,000 years), variations in tilt (44,000 years), and shape of the orbit (100,000 years) seem to match banding and isotopic changes in ocean sediments (figure 17.12). They also seem to match periodic cold spells and expansion of glaciers worldwide every 100,000 years or so and may be responsible for shorter-term cycles, although patterns are more difficult to discern in small-scale events.

A historical precedent for cyclic temperature changes that had disastrous effects on humans was the "little ice age" that occurred in the 1300s. Temperatures dropped so that crops failed repeatedly in parts of northern Europe that once were good farmland. Scandinavian settlements in Greenland founded by Eric the Red during the warmer period around 1000 A.D. lost contact with Iceland and Europe as ice blocked shipping lanes. It became too cold to grow crops, and fish that once migrated along the coast stayed further south. The settlers slowly died out, perhaps having been attacked by Inuit people who were driven south from the high Arctic by colder weather.

A more frequently occurring large-scale weather cycle is the El Niño-Southern Oscillation (ENSO) event, in which the

FIGURE 17.12 Milankovitch-induced periodicity? The banding of these sea sediments, now exposed along the coast of France, has a period of 20,000 to 40,000 years, fitting models of climate change caused by wobble, tilt, and changes in the shape of Earth's orbit.

large tropical low, usually centered over Indonesia, shifts eastward into the Central Pacific every few years. This reverses normal trade winds, diverts ocean currents, and blocks upwelling of cool, nutrient-rich, deep water that nourishes enormous schools of fish and flocks of seabirds along the west coast of South America. The lack of normal rainfall in Indonesia and Australia associated with the 1982–1983 ENSO caused crop failures and disastrous forest fires, including one that burned 3.3 million ha (8 million acres) in Kalimantan (Borneo), one of the worst fires in recorded history. Winter storms associated with an ENSO often bring torrential rains and mud slides to California, while the Midwest may enjoy an unusually mild winter as the polar front is pushed northward by warm Pacific air.

One explanation of this cyclic ENSO pattern is that ocean surface temperature, thickness of the ocean surface layers (and therefore depth of the thermocline), upwelling currents, ocean currents, and wind flow patterns form a dynamic, oscillating system driven by negative feedback mechanisms that reverses itself every few years. A complimentary series of events called La Niña pushes the tropical low back to the western Pacific, cools surface waters off the coast of South America, and brings drought to the western American deserts. Recent evidence suggests that high cirrus ice clouds formed by evaporation and strong updrafts during an El Niño event cool the ocean's surface contributing to La Niña and stabilizing world climate. Because there are so many interacting factors, the recurrence of this system is highly variable, as is the strength of the oscillations that drive it. Understanding the complex and highly variable interactions that regulate the ENSO system may give us some insights into the confusing patterns of other weather and climatic phenomena.

Human Effects on Climate

How are humans affecting global climate? This is a subject of much controversy and anxiety. Many scientists are concerned that we already have set in motion changes that could have disastrous consequences, such as destruction of the protective ozone layer and global warming through the greenhouse effect. Other scientists contend that the atmosphere and biosphere are resilient and self-regulating, so that changes we initiate may not be so important.

Although we don't fully understand the processes that create large-scale weather patterns and we can't predict with any assurance what the climate will be next year—let alone ten to one hundred years from now—it is clear that weather and climate have profound effects on natural ecosystems and human societies. Until now, Earth's atmosphere has been a remarkably stable, self-correcting system, providing just the right balance of gases, letting in just the right amount and type of solar energy, and providing just the right balance of moisture and temperature to sustain life. Ironically, just as we have begun to decipher the climatic rhythms that have existed for hundreds of millions of years, we also may have begun to change them irrevocably. Perhaps by working together, we can stabilize and protect this irreplaceable resource.

SUMMARY

The atmosphere and living organisms appear to have evolved together so that the present chemical composition of the air is both suitable for and largely the result of biological processes. Compression concentrates most gas molecules in a thin layer (the troposphere) near Earth's surface. The upper layers of the atmosphere, while too dilute for life, play an important role in protecting Earth's surface by intercepting dangerous, mutagenic ultraviolet radiation from the sun. The atmosphere is relatively transparent to visible light that warms Earth's surface and is captured by photosynthetic organisms and stored as potential energy in organic chemicals.

Heat is lost from Earth's surface as infrared radiation, but fortunately for us, carbon dioxide and water vapor that are naturally present in the air capture the radiation and keep the atmosphere warmer than it would otherwise be. When air is warmed by conduction or radiation of heat from Earth's surface, it expands and rises, creating convection currents. These vertical updrafts carry water vapor aloft and initiate circulation patterns that redistribute energy and water from areas of surplus to areas of deficit. Pressure gradients created by this circulation drive great air masses around the globe and generate winds that determine both immediate weather and long-term climate.

Earth's rotation causes wind deflection called the Coriolis effect, which makes air masses circulate in spiraling patterns. Strong cyclonic convection currents fueled by temperature and pressure gradients and latent energy in water vapor can create devastating storms. Another source of storms are the seasonal winds, or monsoons, generated by temperature differences

RESOURCES AND RESOURCE ECONOMICS

between the ocean and a landmass. Monsoons often bring torrential rains and disastrous floods, but they also bring needed moisture to farmlands that feed a majority of the world's population. When the rains fail, as they do in drought cycles, ecosystem disruption and human suffering can be severe.

Many procedures claiming to control the weather are ineffectual, but some human actions—both deliberate and inadvertent—may change local weather and long-term climate. Cloud seeding can induce rain or disperse fog under the right atmospheric conditions. Improving the local situation, however, often makes things worse somewhere else. It is not yet clear what the future of our weather and climate will be. Some scientists warn that the gaseous pollutants we release into the atmosphere may trap radiant energy and cause a global warming trend that could drastically disrupt human activities and natural ecosystems. Understanding and protecting this complex, vital aspect of our world is clearly essential.

Review Questions

1. What are the main constituents of air? What are their sources?
2. Name and describe the four layers of air in the atmosphere.
3. What is the greenhouse effect?
4. Describe the atmospheric heating and cooling cycle.
5. What are the "jet streams"? How do they influence weather patterns?
6. What are Hadley cells and how do they form?
7. Describe the Coriolis effect. What causes it?
8. What is the difference between weather and climate?
9. What are some theories explaining major climatic changes?
10. Would changes in Earth's orbit affect climate? How?

Questions for Critical Thinking

1. Discuss the environmental effects of a nuclear winter. How could such a catastrophe be avoided?
2. Should humans try to control the weather? What would be the positive effects? What would be the dangers?
3. Can we avoid great climatic changes, such as another ice age or a greenhouse effect? Will we have the gumption to change our ways?
4. Has there been a major change recently in the weather where you live? Can you propose any reasons for such changes?
5. What forces determine the climate in your locality? Are they the same for neighboring states?
6. What was the weather like when your parents were young? Do you believe the stories they tell you? Why or why not?
7. Have you ever experienced a tornado or hurricane? What was it like? What omens or warnings told you it was coming?
8. From which direction does bad weather come in your region? Why?
9. What would you do to adapt to a permanent drought? What effects would it have on your life?

10. What should we do about coastal cities threatened by rising oceans? Rebuild, enclose in dikes, or just move?
11. Would you favor building nuclear power plants to reduce CO_2 emissions?

Key Terms

aerosols (p. 326)
albedo (p. 329)
climate (p. 326)
cold front (p. 335)
convection currents (p. 329)
Coriolis effect (p. 334)
front (p. 335)
Gaia hypothesis (p. 326)
Hadley cells (p. 334)
insolation (p. 328)
ionosphere (p. 328)

jet streams (p. 335)
mesosphere (p. 327)
Milankovitch cycles (p. 339)
monsoon (p. 337)
stratosphere (p. 327)
thermosphere (p. 327)
tornadoes (p. 336)
troposphere (p. 327)
warm front (p. 336)
weather (p. 326)

SUGGESTED READINGS

■ Alverez, W. et al. 1990. "What Caused the Mass Extinction?" *Scientific American* 263 (4):76. The debate over asteroids or volcanoes as the cause of extinction.
■ Arthur, M. A., and R. E. Garrison, eds. 1986. "Milankovitch Cycles through Geologic Time." *Paleo-oceanography*. 1:369. A special section on global climate changes.
■ Abrahamson, D. E., ed. *The Challenge of Global Warming*. Washington, D. C.: Center for Resource Economics/Island Press, 1989. An anthology of articles by leading proponents of greenhouse theories.
■ Berner, R. A., and A. C. Lasaga. March 1989. "Modeling the Geochemical Carbon Cycle." *Scientific American* 260 (3):74. Follows carbon through the environment.
■ Broecker, W. S., and G. H. Denton. January 1990. "What Drives Glacial Cycles?" *Scientific American* 262 (1):49. A discussion of how changes in the Earth's orbit could lead to massive reorganization of ocean-atmosphere circulation patterns that determine climate.
■ Brookes, W. T. December 1989. "The Global Warming Panic." *Forbes* (12):96. An industry view (highly skeptical) of climate warming predictions.
■ Graedel, T. E., and P. J. Crutzen. September 1989. "The Changing Atmosphere." *Scientific American* 261 (3):58. A good overview of the atmosphere and how greenhouse gases might affect it.
■ Glantz, M. H. June 1987. "Drought." *Scientific American*. 256 (6):34. An excellent combination of meteorology and human ecology. If African leaders were to treat drought as a recurring phenomenon, they could deal with it in ways that would stabilize farm production.
■ Graham, N. E., and W. B. White. June 1988. "The ElNiño Cycle: A Natural Oscillation of the Pacific Ocean Atmospheric System." *Science*. 240 (4857):1293. Proposes that ENSO cycles are an oscillating system regulated by baronic, subsurface Kelvin and Rosby waves.

Jones, P. D., and T. M. L. Wigley. August 1990. "Global Warming Trends." *Scientific American* 263 (2):84. Recent results of climate changes and how to interpret them.

Kerr, R. A. February 1987. "Milankovitch Climate Cycles through the Ages." *Science*. 235 (4792):973. Brief review of climate cycles.

Kunzig, R. April 1991. "Earth and Ice." *Discover* 12 (4):55. Proposes how fertilizing the ocean could stimulate phytoplankton growth and reduce atmospheric CO_2.

Lindzen, R. S. March 1990. "A Skeptic Speaks Out." *EPA Journal* 16 (2):46. Criticism of climate models and climate change theories.

Lovelock, J. 1985. "Are We Destabilizing World Climate?" *The Ecologist*. 15 (1/2):52. A fascinating discussion of atmospheric stability by the author of the Gaia hypothesis.

Lyman, F. et al. 1990. *The Greenhouse Trap: What We're Doing to the Atmosphere and How We Can Slow Global Warming*. Washington D.C.: World Resources Institute. A good explanation of the greenhouse effect.

Mooney, H. A. et al. November 1987. "Exchange of Materials between Terrestrial Ecosystems and the Atmosphere." *Science*. 238 (4829):926. An excellent summary of the atmospheric balance of trace gases.

Navarra, J. *Atmosphere, Weather and Climate: An Introduction to Meteorology*. Philadelphia: W. B. Saunders Co., 1979. A well-written introductory textbook.

Oppenheimer, M. and R. H. Boyle. *Dead Heat: The Race Against the Greenhouse Effect*. New York: New Republic Books, 1990. A popular account of possible causes and effects of global climate change.

Ramage, C. S. June 1986. "El Niño." *Scientific American*. 254 (6):76. A good description of atmospheric pressure anomalies in the equatorial Pacific and their effects on large-scale climate changes.

Reifsnyder, W. *Weathering the Wilderness: The Sierra Club Guide to Practical Meteorology*. San Francisco: Sierra Club Books, 1980. A good practical weather guide.

Schaefer, V. and J. Day. *A Field Guide to the Atmosphere*. New York: Houghton Mifflin Co., 1981. Sponsored by the National Audubon Society and the National Wildlife Federation.

Stolarski, R. et al. 1986. "Stratospheric Ozone." *Geophysical Research Letters*. Vol. 13 (supplementary issue). More than forty pages by the world's leading atmospheric scientists.

Udall, J. R. July/August 1989. "Turning Down the Heat." *Sierra* 74 (4):26. Practical actions we can take to avert global climate change.

White, R. M. 1991. "Our Climatic Future: Science, Technology, and World Climate Negotiations." *Environment* 33 (2):18. A good discussion of the politics of protecting our atmosphere and climate.

CHAPTER *18*

Energy Use and Traditional Fuels

CONCEPTS

Energy is the capacity to do work. Power is the rate of energy delivery. Nearly 90 percent of all commercial energy worldwide is provided by fossil fuels: coal, oil, and natural gas. The proven reserves of oil represent only about thirty-five years' supply at present rates of usage. Proven reserves of natural gas will last about sixty years at current usage rates, while coal in proven reserves could last three hundred years if usage remains constant.

In the United States, industry consumes about 37 percent of the energy used, primarily in manufacturing and processing. Residential and commercial space and water heating account for about 35 percent of our energy use. Transportation consumes almost all of the remaining commercial energy.

Our present rates of energy use may have to be sharply curtailed because supplies of traditional fuels appear to be running out and because the environmental and health effects of fossil fuel combustion and nuclear power production are unacceptable.

There may be "unconventional" sources of petroleum and natural gas, as well as new technologies for preparing and using coal, that could greatly extend the useful life of these resources while also reducing adverse environmental impacts associated with current use patterns.

The United States is threatened far more by the hazards of too much energy, too soon, than by the hazards of too little, too late.

John Holdren

INTRODUCTION

Energy is, by now, a familiar concept to you. In previous chapters, we have talked about the energy needs of organisms and how they meet these needs. We looked at thermodynamics and the flow of energy through ecosystems. We studied ways the sun's energy heats Earth's atmosphere and surface, making life possible. In terms of life, it's no exaggeration to say that the sun is the ultimate source of energy and that life on Earth is totally dependent upon it. In terms of sustaining our complex industrial and technological energy needs, however, we clearly underutilize the sun's awesome power.

Using an external energy source to do useful work is one of the main features that distinguishes humans from most other animals. The ability to use external energy allows us to modify our environment so it is more hospitable. This ability is also a chief reason that humans have become a dominant life-form on the planet. We have become dependent on—some would say addicted to—external energy sources. Unfortunately, most of our energy now comes from nonrenewable sources (figure 18.1). The extraction and use of that energy is having environmental effects that soon may be intolerable. In the next two chapters, we'll explore some alternative sources of energy for the future.

In this chapter, we will examine the current primary sources of energy in the world, how the energy from those sources is extracted and used, and the history of energy technology and resource distribution that has led to current patterns of energy use. The relationship between economic development and energy use is an important aspect of energy policy. We will look briefly, therefore, at the disparities in energy use between the more developed and less developed countries of the world and ways in which economics and energy access are linked to environmental conditions, quality of life, and future development in the world. In the second half of this chapter, we will examine energy derived from fossil sources: coal, oil, and natural gas.

ENERGY UNITS

What is energy and how is it measured? **Energy** is the capacity to do work, changing the physical state or motion of an object. **Work** is the application of force through a distance. Energy is measured in joules, calories, ergs, or BTUs (table 18.1). **Power** is the rate of energy delivery and is measured in horsepower or watts.

A BRIEF HISTORY OF ENERGY USE

Humans tap external sources of energy to provide light and heat and to operate machines. Combustion of traditional fuels is the basis of most modern technology.

FIGURE 18.1 Fossil fuels provided the energy source for the Industrial Revolution and built the world in which we now live. We have used up much of the easily accessible supply of these traditional fuels, however, often through wasteful practices such as the flaring shown here. Adverse environmental effects of burning coal, oil, and natural gas may preclude further use of these fuels. We desperately need to find renewable, environmentally benign sources of energy.

Fire and Muscle Power

The first human energy technology, and certainly one of the turning points in human history, was the harnessing of fire. Charcoal from fires and evidence of cooking have been found along with remains of *Homo erectus* (Java man or Peking man) dating back to the beginning of the Pleistocene era, one million years ago. We don't know when animals first were used for riding and pulling burdens, but wheeled war chariots were in use by the beginning of recorded history, in cities of Sumer five thousand years ago. The muscle power of humans and domestic animals still supplies as much as half of all energy used in some less developed countries.

Wind, Water, Coal, and Steam

Wind and water power were major energy sources before the development of steam engines in the eighteenth century. The mechanical and engineering inventions needed to convert wind and water energy to useful power helped lay the groundwork for the Industrial Revolution. We will discuss these energy sources further in chapter 20.

The Industrial Revolution resulted to a large extent from a pair of energy shortages. The dense forests that once covered Britain were heavily depleted for fuelwood during the Middle Ages and Renaissance. People turned to coal for their fires in spite of the noxious fumes it produced because it was the only fuel available. As coal mines got deeper in the sixteenth and seventeenth centuries, groundwater seepage into the mines became a serious problem, and it appeared that only the coal immediately adjacent to the surface could be extracted. Thomas New-

TABLE 18.1 Some energy units*	
1 joule (J)	Work done when 1 kg is lifted 1 meter
1 gigajoule (GJ)	1 billion J
1 petajoule (PJ)	1 million GJ or 1×10^{15} J or 947.8×10^9 BTU
1 watt (w)	1 joule per second 1 watt hour = 3,600 J or 3.4 BTU
1 kilowatt hour (kWhr)	1,000 watts per hour = 3.6×10^6 J or 3,413 BTU
1 metric ton of oil equivalent (t.o.e.)	6.66 barrels crude oil
1 t.o.e.	Generates 4,000 kWhr or 13.6 million BTU
1 British thermal unit (BTU)	Energy needed to heat 1 pound of water 1°F = 246 cal or 0.293 watt hours
1 calorie (cal)	Energy needed to heat 1 gram of water 1°C
1 kilocalorie (kcal)	1,000 calories
1 Quad (Q)	1 quadrillion (1×10^{15}) BTU or 2.93 $\times 10^{11}$ kWhr or 1.055×10^3 PJ
1 million t.o.e.	13.65×10^{12} BTU or 0.0136 Quad or 41.87 PJ

*Energy (measured in joules and tons of oil equivalent or t.o.e.) is the capacity to do work.

Power (measured in watts) is the rate of energy delivery.

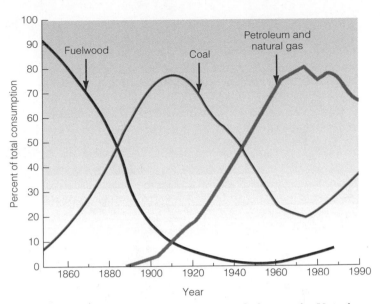

FIGURE 18.2 Changing patterns of major fuel use in the United States, 1850–1990.

comen's 1712 invention of a coal-fired steam pump to dewater mines and the improvement of the condensing steam engine by James Watt in 1769 are considered by some to be the beginning of the Industrial Revolution. Steam technology spread to spinning mills, eliminating dependence on water power, and made sweeping changes in society.

Figure 18.2 shows the changing production patterns of fuelwood, coal, and petroleum products over the past century in the United States. In the middle of the nineteenth century, wood and muscle power provided more than 90 percent of all energy used in this country, as they had in most places throughout history. Coal-fired steam boilers, however, were beginning to supply power to factories, railroads, and farms. During the Industrial Age of the United States and Europe (1775 to 1950), coal provided most of the energy for transportation, manufacturing, and space heating. As figure 18.2 shows, coal accounted for at least 80 percent of all energy used in the United States at the beginning of this century.

The Oil Age

The transition to petroleum also began with an energy shortage. Coal is a very efficient heat source but not a good light source.

During the eighteenth and early nineteenth centuries, whale oil was so popular for lamps that the great whales were quickly depleted (chapter 14). Strangely enough, the petroleum industry, which has dominated the energy mix in recent years, began by providing a poorly regarded, smelly substitute for whale oil.

Rapid Growth in Oil Dependence

The first commercial oil well in the world was drilled in Titusville, Pennsylvania, in 1859 (figure 18.3). The invention of the four-stroke internal combustion engine by Alphonse de Rochas the next year represented a new use for the gooey, ill-smelling crude oil. Refined petroleum products fed automobile engines as well as lanterns and heaters. By 1950, petroleum provided about 40 percent of all the energy used in this country. Twenty-five years later, oil and natural gas contributed nearly two-thirds of all energy consumed in the United States.

Price Shocks of the 1970s

World oil use hit its historic peak in 1979, when daily production passed 66 million barrels per day. At that point the Organization of Petroleum Exporting Countries (OPEC)—Saudi Arabia, Iran, Iraq, United Arab Emirates, Kuwait, Libya, Qatar, Abu Dhabi, Algeria, Nigeria, Ecuador, Gabon, Indonesia, and Venezuela—controlled more than 50 percent of the total world market.

Sharp price shocks in the oil market during the 1970s and again in 1990 rudely awakened the world to the cost of dependence on petroleum. In 1973, the Arab countries declared an embargo on oil shipments to the United States and other Western countries that had supported Israel in the war against Egypt. Oil shipments never declined very much, but average prices tripled,

FIGURE 18.3 The world's first commercial oil well near Titusville, Pennsylvania. On August 27, 1859, Edwin L. Drake (in top hat and frock coat) ushered in the age of petroleum with this well.

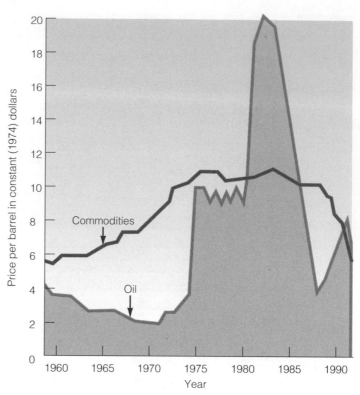

FIGURE 18.4 The price of a barrel of Saudi Arabian light crude oil 1960–1990 in constant 1973 dollars. The instability of oil prices has sent both positive and negative price shocks through the world economy. Particularly hard hit are Third World nations dependent on imported oil as their energy supply. Nonfuel commodities have not kept pace with rapidly changing oil prices.

Sources: P. Odell, *Draining the World of Energy*, (Center for International Energy Studies, Roggerdam, 1984) p. 8 and The World Bank, *The World Bank Annual Report 1985* (The World Bank, Washington, DC, 1985) p. 40.

from $3 to more than $9 per barrel. The Iranian revolution in 1979 reduced oil supplies again and raised prices to around $35 per barrel. Altogether, the price hikes between 1970 and 1980 amounted to nearly a tenfold increase in the real price of oil. The invasion of Kuwait by Iraq in 1990 again briefly pushed oil prices above $40 per barrel. In constant dollars, however, this was about equal to the 1973 price. Figure 18.4 shows the average price of Saudi Arabian light crude oil over the past 30 years.

The effects of rising oil prices are most devastating in the developing world. For industrialized countries, these shocks are lessened by accompanying price inflation in manufactured goods and decreasing prices of commodities, such as sugar, tin, rubber, and copper. The developing countries, however, pay more for *both* energy and manufactured goods and get less for the raw materials they sell abroad. The result is a growing trade deficit between more developed and less developed countries, and a spiraling international debt owed by the poorer countries of the world that threatens to keep them in a permanent underclass. They cannot afford to buy the energy necessary to develop economically, and they cannot afford to invest in conservation programs and modern equipment that will allow them to balance their import/export payment ratios.

Another striking effect of oil price increases is a sudden transfer of wealth to the oil-producing countries. In 1973, total

yearly OPEC oil income was $22.5 billion. Seven years later, this figure had risen to nearly $275 billion. The OPEC countries moved, in some cases, from feudal subsistence cultures to economic giants, practically overnight. Since most of these dollars eventually returned to the United States and other industrialized countries either for arms purchases or as investments, the economic effects of sending them abroad was not as severe as it might have seemed. The effects of draining purchasing power from the less developed countries was devastating, however.

Reverse Price Shocks of the 1980s

The rapid oil price increases in the 1970s resulted in a substantial decrease in worldwide oil consumption. As a result, the demand for oil decreased and production was cut back. From a high of 66 million barrels per day in 1979, oil production fell 15 percent to 56 million barrels per day in 1985, in spite of both population and economic growth. Some of this drop was due to successful conservation programs, some was the result of changing to other fuels, and some was caused by reduced consumption by marginal users who were simply forced out of the market by rising prices. The result was a market glut of oil, which caused prices to fall almost as sharply in the mid-1980s as they had risen a decade earlier.

By December 1986, the average price of a barrel of Saudi light crude oil had fallen below $10 per barrel. In constant dollars this is very close to the price in 1973 *before* prices started to rise (see figure 18.4). The price decreases were greeted joyfully in the consuming countries, but they caused hardships in the producing countries. Indonesia, for example, which depended on oil for about 60 percent of its foreign exchange, had to cut back on many planned projects. Mexico, already struggling under an enormous international debt, experienced strong inflationary pressures, economic recession, and political unrest. Unfortunately, as soon as oil prices fell back to bargain rates and surpluses made fuel plentifully (if temporarily) available, the public forgot about both conservation and the development of alternative and sustainable energy resources. In the United States, research programs in solar, wind, biomass, and other forms of renewable energy largely came to a standstill. People began buying big cars again and speed limits were raised to 65 mph in spite of evidence that 55 mph saved thousands of lives and millions of barrels of oil.

CURRENT PATTERNS OF ENERGY CONSUMPTION

Where are we now in regard to our energy sources? Figures 18.5 and 18.6 summarize current patterns of energy consumption by source, worldwide and for the United States. The data on which these figures are based are presented in table 18.2.

Major Energy Sources

Notice that oil is still the dominant energy source worldwide, although it has fallen to only 37 percent of the total in 1988. In the United States, coal and natural gas provide about equal shares; but worldwide, coal provides considerably more energy than natural gas. North America, Western Europe, and Asia consume more energy than they produce. Latin America, the Middle East, the Soviet Union, and Africa all produce much more energy than they consume.

Natural gas is a clean-burning, convenient fuel, but it is difficult to ship across oceans or to store. We in North Amercia are fortunate to have an abundant, easily available supply of gas and a pipeline network to deliver it to market. Europe is turning increasingly to natural gas as an energy source. In 1988, six Western European countries entered into a long-term contract to purchase gas from the Norwegian-controlled Troll Field located in the North Sea. Europe has also been negotiating with the Soviet Union for gas from fields in Siberia. Six huge pipelines are planned for completion by 1996 to bring gas from Yamal

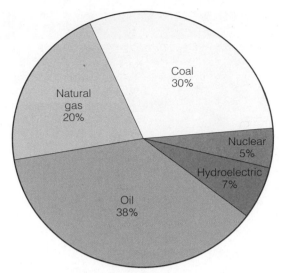

FIGURE 18.5 Worldwide commercial energy consumption, 1984. See table 18.2 for more detailed data.

Source: Data from World Resources Institute, 1990.

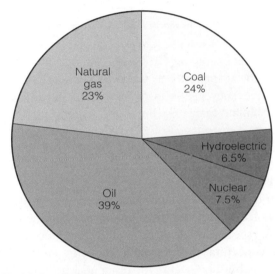

FIGURE 18.6 North American commercial energy consumption, 1988. Natural gas provides us with a larger share of our energy than is true for most of the world because we have a continent-wide pipeline system. Coal provides a much smaller energy share, not because we have less, but because of environmental constraints.

Source: Data from *World Resources 1990–1991*.

Peninsula in the Kara Sea. Environmental problems and challenges of this project will be as difficult as those in the United States and Canadian Arctic (chapter 15). Most countries cannot afford a pipeline network, and much of the natural gas that comes out of the ground in conjunction with oil pumping is simply burned (flared off), a terrible waste of a valuable resource (see figure 18.1).

Worldwide, potentially sustainable or renewable energy resources—such as solar, biomass, hydroelectric, and other less-developed types of power production—provide at least 18 per-

TABLE 18.2 Commercial energy production and consumption in petajoules, 1988

Region	Oil		Natural gas		Coal		Nuclear power	Hydropower	Total	
	Produc-tion	Consump-tion	Produc-tion	Consump-tion	Produc-tion	Consump-tion	Production and consumption	Production and consumption	Produc-tion	Consump-tion
North America	22,857	36,171	21,203	21,211	23,866	21,542	6,883	5,987	80,797	91,796
Latin America	14,278	9,551	3,601	3,308	888	959	96	3,902	22,765	17,816
Western Europe	8,290	24,875	6,297	8,332	7,821	11,033	6,155	4,509	33,073	54,904
Middle East	30,954	5,669	2,734	2,278	29	105	0	109	33,827	8,160
Africa	10,991	3,609	2,227	1,264	4,187	3,048	80	791	18,276	8,793
Asia and Australasia	6,816	19,507	4,539	4,091	9,877	12,779	2,759	2,491	26,483	41,627
Centrally Planned Economies										
Soviet Union	26,127	18,385	29,045	22,982	16,409	12,984	1,779	2,345	75,705	58,476
China	5,699	4,216	528	561	24,251	24,331	0	1,319	31,796	30,427
Other[a]	888	5,238	2,617	4,262	14,994	14,881	620	1,038	20,156	26,039
Total[b]	126,900	127,222	72,791	68,290	102,322	101,660	18,373	22,493	342,878	338,037

a. Albania, Bulgaria, Czechoslovakia, East Germany, Hungary, Kampuchea, Laos, Mongolia, North Korea, Poland, Romania, Vietnam, Yugoslavia.

b. Figures may not total because of rounding.

Source: Data from *World Resources 1990–1991*, page 142.

cent of the total energy use. In North America, by contrast, hydroelectric power supplies only 6.2 percent of our energy needs, and the other renewable sources provide 0.3 percent.

The United States has the world's largest nuclear power program, yet it provides just 7.5 percent of our energy needs. Worldwide, nuclear power provides about 5 percent of energy needs, led by western Europe with about 11 percent reliance on nuclear plants.

You may have noticed a conspicuous hole in the energy consumption data given for the United States and the world: use of biomass, especially fuelwood. Note that the preceding figures and tables concentrated on *commercial* energy sources, whereas most biomass burning occurs for residential use, mostly by individuals in the Third World where data collection is incomplete at best. Although statistics in this category may be "soft," the importance of biomass to billions of humans can't be neglected. Fuelwood is a significant means of meeting daily needs in almost all developing countries. Many countries use fuelwood (including charcoal) for more than 75 percent of their nonmuscle energy source. The poorest countries, such as Haiti, Mali, Malawi, and Burkina Faso, depend on biomass supplies for more than 90 percent of total energy for heating and cooking. This is a serious cause of forest destruction and soil loss (chapters 11 and 12). In middle-income countries, biomass use ranges from about 7 percent (Portugal) to about 40 percent (Philippines, Costa Rica) of total energy consumption.

Perhaps the most important facts about fossil fuel consumption are that the developed countries consume about 78 percent of the natural gas, 65 percent of the oil, and about 50 percent of the coal produced each year (table 18.2). Although they have less than *one-fifth* of the world's population, these countries use more than *two-thirds* of the commercial energy supply. North America, for instance, constitutes only 5 percent of the world's population, but consumes about 27 percent of the available energy. To be fair, however, some of that energy goes to manufacture goods that are later shipped to developing countries.

Japan is far more dependent on oil than most other industrialized countries because it has very few indigenous fossil fuel resources and has to import nearly 85 percent of its energy. Oil is easier to ship and handle than other fuels and, therefore, makes up 70 percent of Japanese energy imports and about 60 percent of its total fuel mix. As a result, Japan is very sensitive to oil prices and vulnerable to political disturbances that might disrupt supplies. Because of this energy dependence, Japan is working very hard to develop both nuclear and solar energy.

Hydroelectric power contributes nearly 6.6 percent of the total energy mix in the developing countries, a considerably higher fraction than in the rest of the world. This amount, large as it is, represents only a fraction of the potential for hydropower development in many Third World countries. We will discuss the pros and cons of further hydropower development in chapter 20.

Per Capita Energy Consumption

On average, each person in the United States uses about 280 GJ (equivalent to *7 tons* of oil) each year (figure 18.7). By contrast, the poorest countries of the world, such as Ethiopia, Kam-

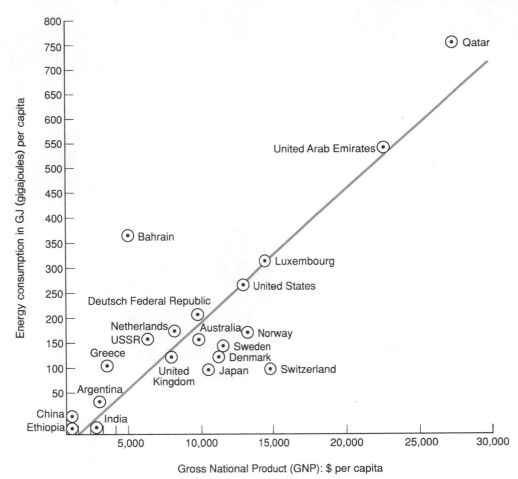

FIGURE 18.7 Per capita energy use and GNP. In general, higher energy use correlates with a higher standard of living. Scandinavia and Switzerland use about half as much energy as we do, yet have a higher standard of living by most measures. Oil-rich states, such as Qatar, Bahrain, and the United Arab Emirates (UAE), use two to three times as much energy as we do, but are still relatively backwards in most areas.

Source: World Resources Institue, 1990.

puchea, Nepal, and Bhutan, generally consume less than one GJ per person per year. Looking at it another way, each person in the United States consumes, on average, almost as much energy in a single day as a person in one of these countries would consume in a year. While most of the energy consumed by people in the less developed countries comes from biomass and muscle power, about 85 percent of our energy comes from fossil fuels, mostly oil and gas.

During the 1950s, when the United States had about two-thirds of all industrial capacity in the world, we consumed far more energy per person than any other nation. This is no longer true. Several of the oil-rich Middle Eastern countries consume much more energy per capita than the United States. Qatar, for instance, uses about three times as much per person as we do, mainly for air conditioning and water desalination. Oman and Bahrain also use a great deal of energy because they have a plentiful supply. Between 1970 and 1980, Oman's energy consumption rose 8,434 percent. Even some countries with an industrial base that is similar to ours consume more energy than

we do. Canada used 288 GJ per capita, and Luxembourg used 328 GJ in 1986 compared to our 280 GJ.

If you plot per capita energy consumption against gross national product (GNP) or average income per person, you find that most countries cluster more or less along a straight line, indicating a direct correlation between income and energy use. If people can afford energy and the goods and services that consume energy, they generally will buy as much as they can. Some important variations in per capita consumption show up, however, even among equally rich nations. Notice that Norway uses only about 60 percent as much energy per person as the United States, even though their GNP per person is about the same as ours. Switzerland has a higher income per person than we do, but uses only one-third as much energy.

Sweden and Denmark also use only half as much energy as we do, even though by most measures their standards of living rank higher than the United States. These countries have had effective energy conservation programs for many years. Japan, also, has a far lower energy consumption rate than might be ex-

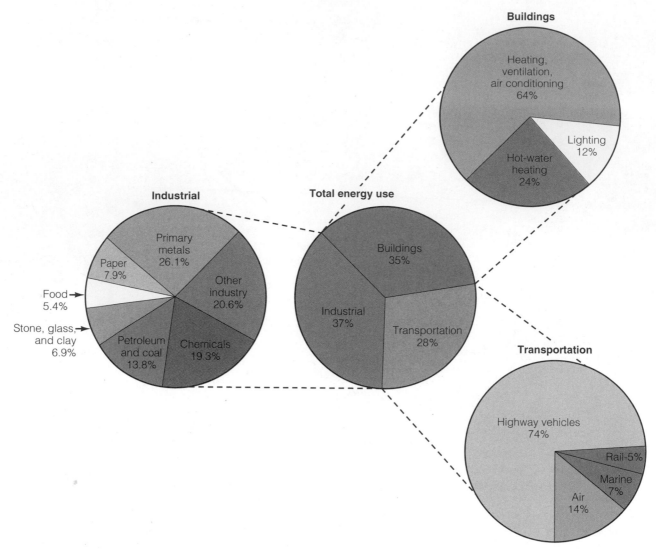

FIGURE 18.8 How energy is used in the United States.
Source: Data from U.S. Energy Agency.

How Energy Is Used

The largest share (37 percent) of the energy used in the United States is consumed by industry (figure 18.8). Mining, milling, smelting, and forging of primary metals consume about one-quarter of the industrial energy share. The chemical industry is the second largest industrial user of fossil fuels, but only half of its use is for energy generation. The remainder is raw material for plastics, fertilizers, solvents, lubricants, and hundreds of thousands of organic chemicals in commercial use. The manufacture of cement, glass, bricks, tile, paper, and processed foods also con-

pected for its industrial base and income level. Because Japan has few energy resources of its own, it has developed energy conservation measures that make it very efficient.

sumes large amounts of energy. Coal, oil, and natural gas each contribute about one-third of the energy used by industry. Residential and commercial buildings use some 35 percent of the primary energy consumed in the United States, mostly for space heating, air conditioning, lighting, and water heating. Small motors and electronic equipment take an increasing share of residential and commercial energy. In many office buildings, computers, copy machines, lights, and human bodies liberate enough waste heat so that a supplementary source is not required, even in the winter.

Transportation consumes 28 percent of all energy used in the United States each year. About 98 percent of that energy comes from petroleum products refined into liquid fuels, and the remaining 2 percent is provided by natural gas and electricity. Almost three-quarters of all transport energy is used by motor vehicles. Nearly 3 trillion passenger-miles and 600 billion ton-miles of freight are carried annually by motor vehicles in the United States. About 75 percent of all freight traffic in the United States is carried by trains, barges, ships, and pipelines, but because they are very efficient, they use only 12 percent of all transportation fuel.

Finally, analysis of how energy is used has to take into account waste and loss of potential energy. About *half* of all the energy in primary fuels is lost during conversion to more useful forms, while it is being shipped to the site of end use or during its use. Consider electrical power, for instance. It supplies about 35 percent of all energy used in the United States, but is not itself a primary energy source because we use other fuels to generate electricity. Coal-fired power plants produce 54 percent of our electricity, nuclear plants provide 17 percent, hydropower generates 15 percent, oil and gas together make 15 percent, and all other sources amount to only 1 percent of our electrical power.

Electricity is generally promoted as a clean, efficient source of energy because when it is used to run a resistance heater or an electrical appliance, almost 100 percent of its energy is converted to useful work and no pollution is given off. What happens before then, however? We often forget that huge amounts of pollution are released during mining and burning of the coal that fires power plants. Furthermore, nearly two-thirds of the energy in the coal that generated that electricity was lost because of inefficient conversion in the power plant. About 10 percent more is lost during transmission and stepping down to household voltages. Similarly, about 75 percent of the original energy in crude oil is lost during distillation into liquid fuels, transportation of that fuel to market, storage, marketing, and combustion in vehicles.

Natural gas is our most efficient fuel. Only 10 percent of its energy content is lost in shipping and processing, since it moves by pipelines and usually needs very little refining. Ordinary gas-burning furnaces are about 75 percent efficient, and high-economy furnaces can be as much as 95 percent efficient. Because natural gas is more efficient to produce and burn than oil or coal, it produces about half as much carbon dioxide—and therefore half as much contribution to global warming—per unit of energy.

COAL RESOURCES, USE, AND MANAGEMENT

Coal is fossilized plant material preserved by burial in sediments and altered by geological forces that compact and condense it into a carbon-rich fuel. Coal is found in every geologic system since the Silurian Age (Appendix A), but graphite deposits in very old rocks suggest that coal formation may date back to Pre-cambrian times. Most coal was formed during the Carboniferous period (286 to 360 million years ago) when Earth's climate was warmer and wetter than it is now.

Types of Coal

Coal varies widely, but can be grouped into four broad categories. **Lignite** is a soft brown to black coal of low BTU value, in which the original plant components are still discernible. It burns with little or no smoke. Its energy content ranges from 6,000 to 7,500 BTU per pound and its sulfur content is generally below 0.6 percent. **Subbituminous** coal is banded, black, and fairly soft, with woody layers commonly visible. It disintegrates in the air, although rather slowly, and it smokes when burned. Its energy content ranges from 9,000 to 14,000 BTU per pound and its sulfur content is usually below 0.6 percent. **Bituminous** coal is black, often banded, and can have a high content of volatile gases. It weathers very little and smokes when burned. When the volatile gases are driven off under reducing conditions it makes coke, a source of heat and carbon in steel production that often has high sulfur content (as much as 6 percent by weight). Its energy content ranges from 11,000 to 14,000 BTU per pound. Finally, **anthracite** is dense, black, and rocklike, with a glassy texture. It does not make coke and burns with a nonluminous flame. It is more rare and has a lower sulfur content (0.5 percent) and greater energy content (13,500 to 15,000 BTU per pound) than other coals. It is also very expensive.

Coal Resources and Reserves

World coal deposits are vast, ten times greater than conventional oil and gas resources combined. Coal seams can be 100 m thick and can extend across tens of thousands of square kilometers, created by vast swampy forests of prehistoric times. The total resource is estimated to be 10 trillion (10^{13}) metric tons. If all this coal could be extracted and coal consumption continued at present levels, this would amount to several thousand years' supply. Table 18.3 shows coal resources proven in place by region. At present rates of consumption, these reserves will last about 200 years. Remember that *proven in place* means the resource has been explored and mapped but may not be economically recoverable (chapter 8).

Where are these coal deposits located? They are not evenly distributed throughout the world (figure 18.9). Notice that, the countries with the largest land area have the most coal. North America has one-fourth of all proven reserves (mostly in the United States). The Soviet Union and China have nearly as much. Eastern and Western Europe have fairly large coal supplies in spite of their relatively small size. Latin America has very little coal. Antarctica is thought to have large coal deposits, but they would be difficult, expensive, and ecologically damaging to mine.

TABLE 18.3 Proven commercial energy resources (petajoules)

Region	Oil		Natural gas		Coal	
	Reserves	One year supply[a]	Reserves	Year supplies	Reserves	One year supply[a]
North America	230,285	10	310,720	50	5,622,779	286
Latin America	715,977	51	260,228	70	287,104	371
Western Europe	100,488	12	221,388	34	1,798,193	219
Middle East	3,236,551	100+	1,297,256	100+	x	x
Africa	314,025	29	275,764	100+	1,826,124	357
Asia	113,049	18	264,112	57	1,918,603	228
Soviet Union	334,960	13	1,650,700	55	4,895,548	x
China	129,797	23	34,956	64	4,504,393	x
Other centrally planned countries	8,374	11	31,072	12	1,506,751	x
Total	5,183,506	41	4,346,196	58	22,359,495	218

a. Proven reserves divided by annual consumption-years supply.

x—Data not available.

Source: *British Petroleum Statistical Review of World Energy 1988.*

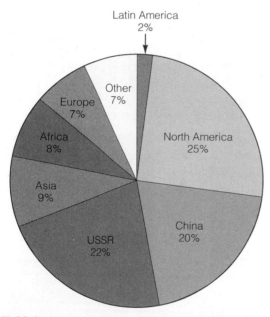

FIGURE 18.9 Proven-in-place coal reserves. The "other" category. Other includes eastern Europe (except the USSR) and Asia (except China).

Figure 18.10 shows locations of the major coal-bearing regions of the United States. Anthracite exists only in small deposits along the East Coast and in Mississippi and Alabama. Enormous bituminous deposits are found in the Appalachian Mountains and the Ohio and Tennessee River valleys. Other vast deposits underlie the Midwest from Illinois and Iowa to Texas. Rich deposits of subbituminous coal occur in the Rocky Mountains. Most of western North Dakota, eastern Montana, and east Texas are underlain by beds of lignite that are accessible to strip mining. Altogether, these deposits probably amount to 300 billion metric tons of coal, or one thousand years' supply at present rates of consumption.

It would seem that the abundance of coal deposits is a favorable situation. Do we really want to use all of the coal? What are the environmental and personal costs of coal extraction and use? In the next section, we will look at some of the disadvantages and dangers of mining and burning coal.

Environmental and Health Effects

It seems certain that extracting and using all the coal available with present technology would have disastrous environmental and human health effects. Figure 18.11 presents an overview of the coal production and use process, from mine to steam boiler.

Mining

Coal mining is a hot, dirty, and dangerous business. Underground mines are subject to cave-ins, fires, accidents, and accumulation of poisonous and/or explosive gases (carbon monoxide, carbon dioxide, methane, hydrogen sulfide). Between 1870 and 1950 more than 30,000 coal miners died of accidents and injuries in Pennsylvania alone, equivalent to one man per day for eighty years. Untold thousands have died of respiratory diseases. In some mines, nearly every miner who did not die early from some other cause was eventually disabled by **black lung disease,** inflammation and fibrosis caused by accumulation of coal dust in the lungs or airways (chapter 21). Few of these miners or their families were compensated for their illnesses by the companies for which they worked. The United States De-

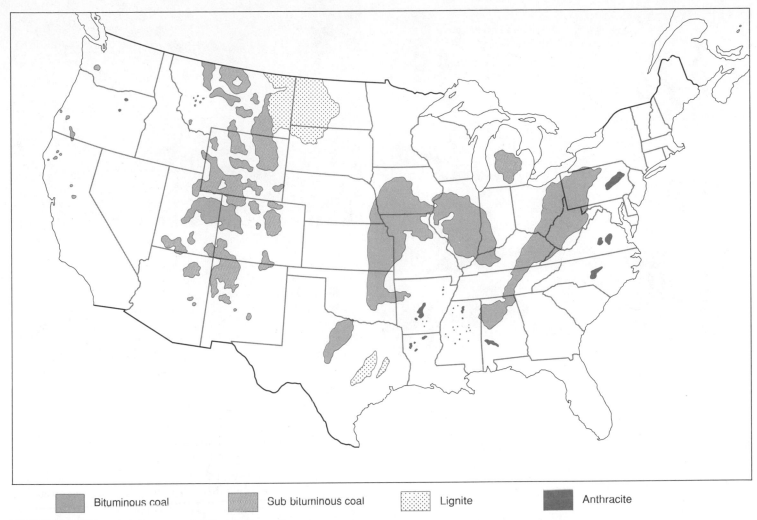

| | Bituminous coal | | Sub bituminous coal | | Lignite | | Anthracite |

FIGURE 18.10 Coal deposits in the United States. Bituminous coal and anthracite are found mainly in the East and Midwest. Subbituminous and lignite are found mainly in the Rocky Mountain states and the High Plains.

Source: 1976 *National Energy Outlook*, Federal Energy Administration.

partment of Labor now compensates miners and their dependents under the Black Lung Benefits Program, but workers often have difficulty proving that health problems are occupationally related. Even when compensation is provided, black lung remains a wretched occupational hazard of coal mining.

Coal mining also contributes to water pollution. Sulfur and other minerals in coal are often water soluble. Mine drainage and water leaching from coal piles and mine tailings are generally acidic and contaminated with highly toxic chemicals. Thousands of miles of streams in the United States have been poisoned by coal-mining operations (chapter 9).

Strip mining or **surface mining** involves removing overburden (overlying layers of noncommercial sediments) using giant machines (figure 18.12). Surface mining is cheaper and safer for workers than underground mining, but can be extremely damaging to natural ecosystems and soils if proper rec-

lamation techniques are not used. On relatively level land, it is possible to set aside topsoil when cutting down to the coal seam, and to recontour the land and replant native vegetation after the coal has been removed. In the United States, reclamation is now required by the Federal Surface Mine Reclamation Act. These procedures are expensive, however, and mining companies often do an inadequate job. On steep slopes, the overburden often slides downhill, smothering forests, filling streambeds, and burying farmland, roads, and even houses. Erosion on steep slopes makes reclamation especially difficult, if not impossible. Some areas simply should not be strip mined.

Air Pollution

Unless the sulfur in coal is removed by washing or flue-gas scrubbing, it is released during burning and oxidizes to sulfur dioxide (SO_2). The high temperatures and rich air mixtures or-

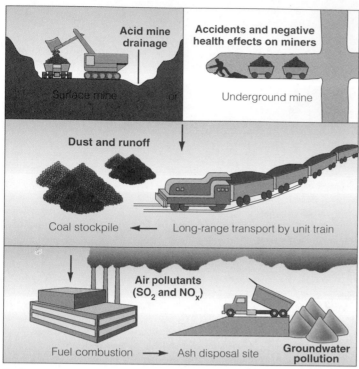

FIGURE 18.12 Strip mining or surface mining involves removal of soil and overburden to reach underlying coal seams. Federal requirements now stipulate that mine operations return the land surface to its original contours and replace cover vegetation but these programs are only marginally effective in many areas.

FIGURE 18.11 The coal story, from surface or underground mine to the power plant. Some associated environmental problems are also indicated.

dinarily used in coal-fired burners also oxidize nitrogen compounds (mostly from the air) to nitrogen monoxide, dioxide, and trioxide. Every year the 900 million tons of coal burned in the United States (83 percent for electric power generation) releases 18 million metric tons of SO_2, 5 million metric tons of nitrogen oxides (NO_x), 4 million metric tons of airborne particulates, 600,000 metric tons of hydrocarbons and carbon monoxide, and close to a trillion metric tons of CO_2. This is about three-quarters of the SO_2, one-third of the NO_x, and about half of the industrial CO_2 released in the United States each year.

The pollution-control equipment necessary to eliminate these emissions now typically accounts for 40 percent of the capital outlay and 35 percent of the operating costs of a coal-fired power plant. A 1,000 megawatt (MW) power plant burning coal containing 3 percent sulfur produces enough scrubber sludge in one year to cover 2.6 sq km (1 sq mi) 30.5 cm (1 ft) deep. Scrubbers often suffer from plugging and fouling of equipment, corrosion of ductwork, and unreliable operation. Pollution-control equipment can consume up to 10 percent of the electricity generated by a plant just to run pumps, blowers, and so forth.

What are the biological effects of these emissions? Sulfur and nitrogen oxides are oxidizing agents and strong tissue irritants. They interfere with plant metabolism, causing leaf damage and even death. In the air, SO_2 and NO_x oxidize further to SO_3 and NO_3, which combine with water vapor. The result is acid

deposition (commonly called acid rain) that damages buildings, aquatic ecosystems, and forests (chapter 17). Acid deposition in Canada caused by coal-burning power plants in the United States is a major issue (chapter 23). Hazardous particulate sulfates and nitrates also are produced, which not only damage crops, but also become lodged in the lungs of animals. Damage to crops, ornamental plants, and wild species from air pollutants has been estimated at $5 to $10 billion each year in the United States alone. Humans suffer respiratory damage and physiological stress, and it has been estimated that at least five thousand excess deaths per year can be attributed to coal burning.

Sulfur can be removed from coal before it is burned, or sulfur compounds can be removed from the flue gas after combustion. Formation of nitrogen oxides during combustion also can be minimized. Perhaps the ultimate limit to our use of coal as a fuel will be the release of carbon dioxide into the atmosphere. Carbon dioxide is generally a harmless gas in low concentrations, but it traps heat in the atmosphere and contributes to the greenhouse effect that threatens to cause major global climatic changes (chapter 17). It is difficult to imagine how we could trap and dispose of the 1 trillion tons of CO_2 generated each year in the United States by burning coal, not to mention that formed when we burn petroleum products. Alternative energy sources, such as solar and wind power (chapter 20), may be our only options.

Radiation and Heavy Metals

Few people know that radiation and heavy metals are among the contaminants released by coal burning. Most coal contains about 1 part per million uranium and 2 parts per million thorium. Similar levels of mercury, cadmium, chromium, lead, selenium, arsenic, copper, chlorine, nickel, and other elements also are often found in coal. A 1,000 MW coal-fired power plant burns about 20,000 tons of coal per day. Hundreds of kilograms of heavy metals, radioactive elements, and other toxic substances pass through the plant in the form of ashes, fine airborne particles, or gases each day (table 18.4).

Much of this material is removed in pollution-control systems, but mercury, selenium, fluorine, chlorine, bromine, iodine, and radon are ejected in the flue gas. Other toxic compounds attach to very fine particulate materials and can penetrate deeply into the lungs, where they damage sensitive respiratory epithelium. Weather conditions influence the distribution and concentration of these pollutants. When effluents from a power plant are trapped by inversion layers or concentrated in clouds and fogbanks, pollutants can build up to dangerous levels. It has been calculated that the radiation dose for people living near a coal-fired power plant may be significantly higher than for people living near a properly operated pressurized-water nuclear reactor.

New Coal Technology

What can be done to reduce pollution from coal burning? Through the 1980s the Reagan/Bush administrations invested billions of dollars for research on new coal-cleaning processes and coal-burning technologies to reduce acid rain and air pollution. This program is aimed primarily at reducing effluents so that consumption rates can remain high, rather than finding ways to reduce demand. The following paragraphs describe some promising new techniques being studied that may make coal a more acceptable fuel for the future.

Fuel switching is a change from high-sulfur eastern coal to low-sulfur western supplies. This is the cheapest solution, but it threatens to displace thousands of miners in already economically depressed states. Furthermore, western coals are lower in BTU potential and the surface-mined lands of the arid western states are very difficult to reclaim.

Coal washing involves crushing coal and washing out soluble sulfur compounds with water or other solvents. This is also relatively inexpensive, but it generally removes only about half of the sulfur and can result in water pollution and/or waste disposal problems.

Flue-gas scrubbing removes sulfur from flue gases in two ways. Sulfur is catalytically converted to water-soluble sulfuric acid, which is collected and sold to the chemical industry. Alternatively, a slurry of water and limestone is sprayed into the gas stream to convert SO_2 into an insoluble calcium sulfate (gypsum) sludge that is stored in landfill sites.

Fluidized bed combustion reduces both sulfur and nitrogen oxides. Coal is crushed and mixed with limestone and burned in suspension over an upwelling flow of air. The limestone removes most of the sulfur. Nitrogen oxide levels are reduced because burning temperatures are one-half to one-third lower than in most furnaces (see figure 23.18).

Element	Coal	Slag	Fly ash	Atmospheric discharge
Aluminum	14,000	10,400	3,600	400
Arsenic	5.3	0.5	1.8	0.2
Barium	96	42	55	0.3
Bromine	5.4	0.2	0.3	6
Cadmium	0.7	0.09	0.6	0.02
Chlorine	1,340	8	15	1,300
Cobalt	4.9	2.7	2.1	0.04
Chromium	27.5	21.2	22	0.3
Cesium	1.9	1.1	0.9	0.01
Mercury	0.18	0.002	0.004	0.1
Lanthanum	5.6	3.5	3	0.02
Lead	9.2	0.6	7.8	0.2
Rubidium	22.8	8.6	11.5	0.07
Antimony	0.74	0.05	0.9	0.2
Selenium	3.2	0.0	1.8	0.4
Thorium	3.1	1.3	1.5	0.01
Thallium	680	390	230	6
Uranium	3	2	1.1	0.02
Vanadium	46	34	23	0.8
Zinc	120	3	78	2

TABLE 18.4 Trace element flows through a coal-fired power plant (gms/min)

Total discharges for a given element do not always balance because of technical measuring errors.
To determine annual discharges, multiply these figures by 525,600 minutes per year.

Source: From J. P. McBride, et al., "Radiological Impact of Airborne Effluents of Coal and Nuclear Plants" in *Science*, 202:1045, December 8, 1978. Copyright 1978 by the American Association for the Advancement of Science, Washington, DC.

Coal gasification removes contaminants and produces a clean-burning fuel. Gas from coal usually is not economically competitive with natural gas, but when gasification is integrated into a combined-cycle electric generating plant, it can be very efficient. It also produces much less pollution and is more flexible in load dispatching (deciding which power plants will be used to meet changing load requirements). This system can be built in modular units by phased steps to meet growing utility needs without the difficulties in financing and planning that are inherent in large-scale boilers (Box 18.1).

Raprenox is a process to reduce nitrogen oxide emissions invented by Robert Perry of the United States Department of Energy's Sandia Labs. Combustion gases are mixed with isocyanic acid (HNCO) to convert NO_x to nitrogen gas. The isocyanic gas is generated by heating nontoxic cyanuric acid ($HNCO_3$) in the flue gas stack. NO_x removal rates in small-scale test systems have been 99 percent effective. It remains to be seen whether this system will work in giant power plants.

BOX 18.1 Gasification: New Coal Technology for Electric Generation

Coal is by far the most abundant fossil fuel in the United States, accounting for approximately half the electricity generated each year. Concerns about acid rain, "greenhouse" gases, toxic hydrocarbons, and heavy metals released by coal combustion have raised questions, however, about continued use of coal as an energy source. Some recent advances in coal technology may change this picture dramatically. One of the most promising new technologies is the integrated gasification combined-cycle (IGCC) system developed at Stanford University. A demonstration model of this system has been built by Southern California Edison Company near Barstow in the Mojave Desert.

Utilities initially were interested in the coal gasification process because it produces a clean-burning fuel that can be used directly in gas turbines, eliminating the need for costly and unreliable scrubbers. Sulfur and particulates are removed from the synthetic gas before it is burned, reducing emissions to less than one-tenth of existing federal air quality limits. The burning temperature of the gas is lower than in direct coal combustion so that formation of nitrogen oxides is minimized. In addition, the waste heat from

gas turbines can be captured to generate steam to run a conventional steam turbine, making this integrated system not only one of the cleanest but also one of the most efficient means available to generate electric power.

The system starts with a gasifier in which pulverized coal is mixed with water and partially burned at low temperatures (1,370°C) to release hydrogen and carbon monoxide. The remaining ash solidifies to a nontoxic glasslike slag that can be disposed of in a landfill. Captured heat is used to generate steam that drives electric generator turbines. The cool synthetic gas (syngas) passes through scrubbers that convert hydrogen sulfide gas (H_2S) to elemental (solid) sulfur. Clean syngas is burned in a gas turbine to generate more electricity, and still more heat is captured to make high-pressure steam to run turbogenerators (box figure 18.1). The overall efficiency of this integrated system can approach 40 percent.

Another advantage of this technology is its flexibility. An IGCC can be built in small modular units to match the load requirements of a utility system. A conventional boiler is very inefficient when run at low power loads and may suffer high maintenance costs and struc-

tural damage if it is started and stopped frequently. An IGCC can be run at one-third power by turning off two of its three gas turbines and running the remaining one at full power, thus maintaining high efficiency. Furthermore, since there is usually excess oxygen in the exhaust stream from the turbine, extra gas can be burned to boost steam generation and increase power a few percent to match load requirements more precisely. The modular units can be added to a plant one at a time in a phased construction plan that is more attractive to utilities during uncertain times than investing in a single billion dollar conventional boiler that may take five to seven years to build and not meet standards or be needed in the future.

There are also disadvantages in this new technology, however. There is no economical way to capture and dispose of the enormous amounts of CO_2 it will produce; thus, it will continue to contribute to the greenhouse effect. This process also consumes large amounts of water and may exacerbate water shortages in arid regions. Finally, without controls on mining, coal extraction will continue to pollute and scar the landscape.

OIL RESOURCES, USE, AND MANAGEMENT

Like coal, petroleum is derived from organic molecules formed by living organisms millions of years ago. Analysis of the hydrocarbons found in oil deposits suggests that they may have come primarily from marine bacteria and other microorganisms rather than higher plants and animals, however. When buried in sediments and subjected to high pressures and temperatures, these organic residues are concentrated and transformed into a variety of energy-rich compounds.

Petroleum Formation

It seems to take at least a million years to form oil deposits, since none have been found in sediments younger than that. In the early stages of petroleum formation, the deposit consists mainly

of larger "heavy" hydrocarbons that have the thick, nearly solid consistency of asphalt or paraffin waxes. As the petroleum matures and these long-chain molecules are broken down, successively "lighter" or more fluid compounds are formed. At temperatures above 200°C, all the organic material is transformed into methane and other gaseous hydrocarbons. Depending on its age and history, a deposit will have varying mixtures of oil, gas, and solid tarlike materials. Some very large deposits of heavy oils and tars are trapped in porous shales, sandstone, and sand deposits in the western areas of Canada and the United States.

Once the solid organic matter is converted to liquids and gases, the hydrocarbons can migrate out of the sediments in which they formed through cracks and pores in surrounding rock layers. Since oil and gas are lighter than water, they displace groundwater in porous sediments and flow upward as well as laterally.

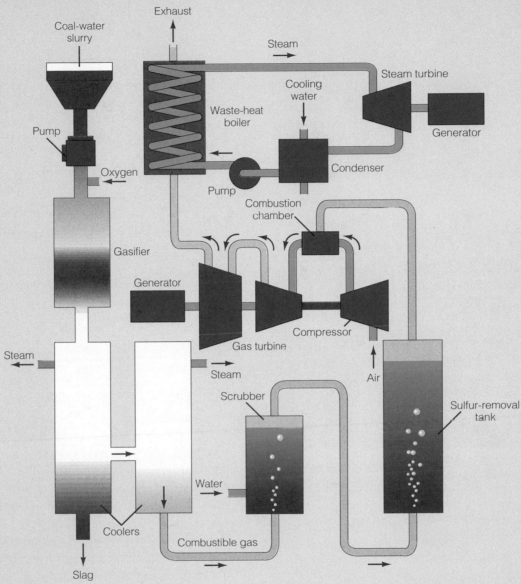

BOX FIGURE 18.1 Coal gasification involves heating a coal/water slurry in a low-oxygen atmosphere. The carbon monoxide and hydrogen produced are cleaned of contaminants and then burned to drive electric generators. Efficiency is improved by combining gas and steam turbines. This system emits far fewer air pollutants than conventional coal burners but is water intensive.

Unless stopped by impermeable layers, they tend to rise to the surface. Oil and gas seeps have been known since antiquity and small amounts of these materials have been used as medicines, fuels, and building materials for thousands of years.

Pools of oil and gas often accumulate under layers of shale or other impermeable sediments, especially where folding and deformation of systems create pockets that will trap upward-moving hydrocarbons (figure 18.13). Contrary to the image implied by its name, however, an oil pool is not usually a reservoir of liquid in an open cavern, but rather individual droplets or a thin film of liquid permeating spaces in a porous sandstone or limestone, much like water saturating a sponge.

Pumping oil out of a reservoir is much like sucking liquid out of a sponge. The first fraction comes out easily, but removing subsequent fractions requires increasing effort. We never recover all the oil in a deposit; in fact, a 30 to 40 percent yield is about average. There are ways of forcing water or steam into the oil-bearing formations to "strip" out more of the oil, but at

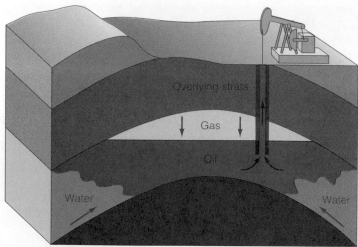

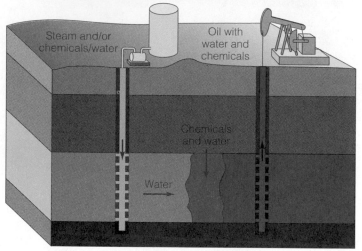

FIGURE 18.13 Recovery process for petroleum. In a "natural drive" well, water or gas pressure forces liquid petroleum into the well. In an enhanced recovery or "stripping" well, steam and/or water mixed with chemicals are pumped down a well behind the oil pool, forcing oil up the extraction well.

least half the total deposit usually remains in the ground at the point at which it is uneconomical to continue pumping. Methods for squeezing more oil from a reservoir are called **secondary recovery techniques.**

Oil Resources and Reserves

The total amount of oil in the world is estimated to be about 4 trillion barrels (600 billion metric tons), half of which is thought to be ultimately recoverable. Some 465 billion barrels of oil already have been consumed. In 1990 the proven reserves were roughly one trillion bbls, enough to last only 50 years at the current consumption rate of 20 billion bbls/yr. It is estimated that another 800 billion barrels either remain to be discovered or are not recoverable at current prices with present technology. As oil

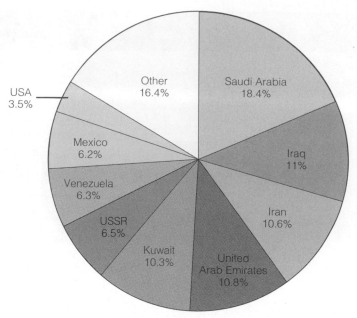

FIGURE 18.14 World proven recoverable oil reserves by country in 1991. Annual world consumption is about 20 billion barrels or 2.4% of the 823.6 billion barrel proven recoverable reserve.

Source: Data from World Resources Institute, 1991.

resources become depleted and prices rise, it probably will become economical to find and bring this oil to the market unless alternative energy sources are developed. This estimate of the resource does not take into account the very large potential from unconventional liquid hydrocarbon resources, such as shale oil and tar sands, which might double the total reserve if they can be extracted with reasonable social, economic, and environmental costs.

By far the largest supply of proven-in-place oil is in Saudi Arabia, which has 168.8 billion barrels, one-fourth of the total world reserve (figure 18.14). Kuwait had about 10 percent of the proven world oil reserves before Iraq invaded in 1990. Some 600 wells were blown up and set on fire by the retreating Iraqis. It may take several years to put out all the fires. Meanwhile, some 6 million gallons of oil per day are lost. Water seeping into depleted formations is mixing with the oil and many wells will be lost forever. Together, the Persian Gulf countries in the Middle East contain nearly two-thirds of the world's proven petroleum supplies. With our insatiable appetite for oil (some would say addiction), it is not difficult to see why this volatile region plays such an important role in world affairs.

Originally, the United States is estimated to have had about 200 billion barrels of recoverable oil, 10 percent of the total world oil resource. Our large supply of relatively accessible oil played an important part in developing our economic and industrial power during the twentieth century (figure 18.15). Until 1947, we were the leading oil export country in the world. By 1990, domestic production in the United States had fallen to 8 million

FIGURE 18.15   Major natural gas and oil deposits in the United
States.

Source: Council on Environmental Quality.

barrels per day, while imports had risen to 8.1 million barrels per day. In spite of the oil shocks of the 1970s, imports had climbed from 25 percent to over half of our total consumption, making the United States vulnerable to events in the Middle East.

Altogether, the United States has already used about 40 percent of its original recoverable resource of 200 billion barrels. Of the 120 billion barrels thought to remain, about 58 billion barrels are proven in place. If we stopped importing oil and depended exclusively on indigenous supplies, our proven reserve would last only *ten* years at current rates of consumption.

In the United States, it always has been government policy to keep fuel prices as low as possible, both to stimulate business and to make life comfortable for ordinary citizens. This policy has had beneficial effects in creating a strong economy and a high standard of living, but it also has encouraged wasteful practices and environmental degradation. Gasoline prices in the United States, for instance, have consistently been one-half to one-fourth as much as in most European countries, largely because favorable tax policies and government regulations keep prices low in the United States. Consequently, most people here

have come to depend on private automobiles rather than public transportation. As you well know, however, many social and environmental costs result from our infatuation with the automobile. Now that our energy supplies are running low, we may have to rethink policies that discourage conservation and suppress alternative and sustainable energy sources.

The 1970s price shocks resulted in a 20 percent reduction in oil consumption in the United States, partly from conservation and partly from switching to other fuels. New wells were also initiated. In the mid-1980s, however, lower oil prices made pumping from marginal "stripper" wells uneconomical and inhibited exploration for new fields so that domestic reserves fell again.

The success rate in finding new wells also has fallen as the easy oil has been used up. In the 1940s, the ratio was about one producer for every ten wells drilled; in recent years, it has been closer to one in fifty. Some industry spokesmen and government officials are predicting that discovery of new reserves will fall off in the next few years and that we will become even more dependent on foreign oil.

Energy Use and Traditional Fuels

The regions of the United States with the greatest potential for substantial new discoveries are portions of the continental shelf along the California coast, around the perimeter of Alaska, and in the Arctic National Wildlife Refuge near Prudhoe Bay on Alaska's North Slope. Proposals to do exploratory drilling in these areas have been strongly opposed by many people concerned about the potential for long-term environmental damage from drilling activities and oil spills like that from the *Exxon Valdez* in Prince William Sound in 1989 (see Box 22.1). Other people argue that energy independence, even if only for a few decades, is worth whatever environmental problems are encountered.

NATURAL GAS RESOURCES, USE, AND MANAGEMENT

Natural gas is the world's third largest commercial fuel (after oil and coal), making up 21 percent of global energy consumption. It is the most rapidly growing energy source because it is convenient, cheap, and clean burning. Natural gas could help reduce global warming (chapter 17) because it produces only half as much CO_2 as an equivalent amount of coal.

World Reserves

The Soviet Union has 44 percent of known natural gas reserves (much of it in Siberia and the Central Asian Republics that have declared independence from the Union) and accounts for 36.5 percent of all production. Both Eastern and Western Europe buy substantial quantities of gas from the Soviet Union. Figure 18.16 shows the distribution of proven natural gas reserves in the major producing countries of the world.

The total ultimately recoverable natural gas resources in the world are estimated to be 10,000 trillion cubic feet. One cubic foot of gas contains about 1,000 BTU of energy, so 6,000 cubic feet is equivalent to one barrel of oil. The recoverable gas resource, thus, is thought to be equivalent to about 1,666 billion barrels of oil (227 billion metric tons), or 80 percent as large as the recoverable reserves of crude oil. The proven world reserves of natural gas are 3,200 trillion cubic feet (73,000 million metric tons). Although natural gas reserves are not as large as oil reserves, gas consumption rates are only about half of those for oil, so that current gas reserves represent roughly a sixty-year supply at present usage rates. Proven reserves in the United States are about 200 trillion cubic feet, or 6 percent of the world total. This is a ten-year supply at current rates of consumption. Known reserves are more than twice as large.

Unconventional Gas Sources

Natural gas resources have not been as extensively investigated in most places as have oil resources. Until recently, gas has been

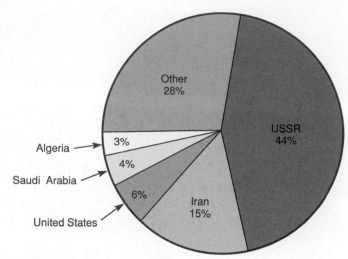

FIGURE 18.16 Percent of natural gas reserves by region in 1990. The total proven reserve is 3,200 trillion cubic feet. World annual production is approximately 54 trillion cubic feet or 1,409 million tons of oil equivalent.

Source: Data from World Resources Institute, 1990.

a by-product of crude oil drilling and production. In many areas where no convenient way exists to ship gas to market, the gas produced simultaneously with oil is simply flared off to get rid of it.

There may be many sources of "unconventional" gas, such as Devonian shales, "tight-sand" gas, and coal-seam methane, that have not been extensively explored but could provide large and valuable fuel sources. Some geologists theorize that natural methane may come from abiogenic chemical origins (inorganic carbon in rocks) deep within Earth and not, as generally supposed, primarily from degeneration of biological materials. Proponents of this point of view cite the presence of methane in formations that also contain helium and in undersea regions of tectonic plate spreading. Thus, they believe that the true resources of methane within Earth's crust may be vastly larger than current estimates predict.

Unlike coal and oil, natural gas is presently being formed as a by-product of human activities. Some municipalities heat their waste treatment plants with methane produced during sewage digestion. Methane is also collected from landfills. Some farmers in both China and the United States generate methane for heating, lighting, and cooking with animal waste digesters. Such domestic methane sources have potential for small-scale, low-cost, local applications (chapter 20).

Liquefied Natural Gas Shipments

Natural gas is quite easy to ship through pipelines as long as it is going from one place to another on the same continent. The problem is that much of the gas is in Russia or the Middle East, while the customers who want to buy it are in Europe, Japan, or North America. One way of shipping gas across oceans is to

FIGURE 18.17 Specialized tankers are required to carry liquefied natural gas at $-140°C$ ($-126.6°F$). Liquefied gas takes up only 1/600 the volume of its gaseous state, so these tankers can carry large amounts of gas.

liquefy it by cooling it below its condensation point ($-140°C$). Since liquefied natural gas (LNG) has only 1/600 the volume of the gaseous form, it is economical to transport by tanker ship (figure 18.17). Finding a deep-water port willing to serve as an unloading station for these tankers has been difficult, however. If a very large LNG tanker had an accident and blew up, it would release as much energy as several Hiroshima-sized atomic bombs.

HARD VERSUS SOFT PATHS IN OUR ENERGY FUTURE

As we have seen in this chapter, the future of fossil fuel use appears to be limited for a number of reasons. Known reserves of liquid petroleum and gas will be exhausted in a few decades at present rates of use. There are large supplies of shale oil, tar sands, and coal, but the environmental costs associated with their extraction and use may reduce or even preclude their use. It is possible that technological developments will reveal "unconventional" sources of oil and gas or new ways to process and use coal that will greatly extend the useful life of these energy sources, while also reducing the adverse effects of their use. Until that happens, however, it seems wise to plan for alternative energy.

Energy specialist Amory B. Lovins describes our current energy policy as based on "hard" technologies that depend on large-scale, highly complex, and very expensive systems. He points out that at least half of the cost of these systems is in fixed distribution and production facilities. Furthermore, most of the energy supplied by these systems is very high grade—that is, capable of carrying high-energy densities or producing high-temperature heat—such as liquid petroleum or electricity. The end uses of that energy, however, are low-grade applications, such as space heating or water heating, which are very wasteful. Once we have spent billions of dollars for a large electrical generating plant or a petroleum pipeline system, it is difficult to abandon that investment and change to a different fuel, even if supplies become limited or adverse effects mount. We are locked into a "hard" technology.

In chapter 19, we will look at what is probably the ultimate in "hard" technologies, nuclear power. The decision to embark on nuclear energy production has committed us (and our descendants) to thousands of years of guardianship over highly toxic radioactive wastes, and it will require significant social changes to prevent nuclear materials from getting into the hands of terrorists or from being used as weapons by other governments. The expense of building and maintaining a nuclear power system is likely to preempt many other energy options that might be more desirable.

An alternative advocated by Lovins and others is sometimes called a "soft" path. Low-technology, sustainable, small-scale systems, such as wood stoves, windmills, passive solar collectors, methane generation, and conservation, are not only cheaper and more benign, but they also offer more individual choice and more flexibility as times change. We will look at some alternative, renewable energy systems in chapter 20. Fortunately, shifting to these "soft" technologies doesn't require heroic sacrifices or great lifestyle changes because they are generally the least costly and least disruptive of all our options. They also offer attractive alternatives in terms of where and how individuals live.

SUMMARY

Energy is the capacity to do work. Power is the rate of energy delivery. The earliest human energy source was muscle power fueled by biomass. The first technological energy source was probably fire, also fueled by biomass. Over the centuries, humans have perfected a great variety of energy sources, ranging from wind and water power to nuclear power plants. Today, nearly 90 percent of all commercial energy is generated by fossil fuels, about 37 percent coming from petroleum. Next are coal, with 30 percent, and natural gas (methane), with 21 percent. Petroleum and natural gas were not used in large quantities until the beginning of this century, but supplies are already running low. Coal supplies will last several more centuries at present rates of usage, but it appears that the fossil fuel age will have been a rather short episode in the total history of humans. Nuclear power provides only about 5 percent of commercial energy worldwide.

Energy is essential for most activities of modern society. Its use generally correlates with standard of living, but there are striking differences between energy use per capita in countries with relatively equal standards of living. The United States, for instance, consumes about 2.5 times as much energy as does Switzerland, which is higher in many categories that measure quality of life. This difference is based partly on level of industrialization and partly on policies, attitudes, and traditions in the United States that encourage extravagant or wasteful energy use. Our wasteful energy use in the United States coupled with our heavy dependence on imported oil makes us highly vulnerable to events in the politically volatile Middle East.

The environmental damage caused by mining, shipping, processing, and using fossil fuels may necessitate cutting back on our use of these energy sources. Coal is an especially dirty and dangerous fuel, at least as we currently use it. Some new coal treatment methods remove contaminants, reduce emissions, and make its use more efficient (so less will be used). Coal combustion is a major source of acid precipitation that is suspected of being a significant cause of environmental damage in many areas. Coal burning releases heavy metals, toxic organic chemicals, radioactivity, and carbon dioxide. We now recognize that CO_2 buildup in the atmosphere has the potential to trap heat and raise Earth's temperature to catastrophic levels.

None of our current major energy sources appear to offer security in terms of stable supply or environmental considerations. Neither coal nor nuclear power is a good long-term energy source with our present level of technology. We urgently need to develop alternative sources of sustainable energy.

■ Review Questions

1. What is energy? What is power?
2. What are the major sources of commercial energy worldwide and in the United States? Why is data usually presented in terms of commercial energy?
3. How is energy used in the United States?
4. How does our energy use compare with that of people in other countries?
5. How far back in history can we trace our use of wind, water, and steam power?
6. How much coal, oil, and natural gas are in proven reserves worldwide? Where are those reserves located?
7. How is commercial energy used and by whom?
8. What are the most important health and environmental consequences of our use of fossil fuels?
9. What are "unconventional" gas and oil sources? Where are they located and what might they mean to our energy future?
10. Describe some new technologies that might make coal a more environmentally acceptable fuel in the future.

■ Questions for Critical Thinking

1. Why does the United States use so much more energy per person than most European countries? What effects (both positive and negative) does that level of energy use have on our lives?
2. Could you reduce your own personal energy use? Why or why not?

3. What changes in both your personal circumstances and in government policies would be necessary for you to save substantial amounts of energy?

4. Would the Industrial Revolution have been possible if steam power had not been invented? How might our society have been different if the internal combustion engine had not been invented?

5. What political consequences result from the unequal distribution of energy resources in the world?

6. Are the health and environmental costs of using fossil fuels worth the benefits we derive from these energy sources? What are our options?

7. What are "hard" energy technologies and "soft" energy paths? Which do you think we should adopt?

8. If your local utility company were going to build a new power plant in your community, what kind would you favor?

Key Terms

anthracite (p. 351)
bituminous (p. 351)
black lung disease (p. 352)
coal gasification (p. 355)
coal washing (p. 355)
energy (p. 344)
flue-gas scrubbing (p. 355)
fluidized bed combustion (p. 355)

fuel switching (p. 355)
lignite (p. 351)
power (p. 344)
raprenox (p. 355)
secondary recovery techniques (p. 358)
strip mining (surface mining) (p. 353)
subbituminous (p. 351)
work (p. 344)

SUGGESTED READINGS

- *Allar, B. March 1984. "No More Coal-Smoked Skies?" Environment.* 26(2):25. The end of King Coal?
- American Gas Association (AGA). *World Natural Gas Consumption Trends.* 1986. Arlington: AGA. How much gas is there?
- British Petroleum (BP). *BP Statistical Review of World Energy, 1986.* London: British Petroleum, 1986. Good data.
- Chadwick, M. et al. 1985. "Developing Coal in Developing Countries." *AMBIO.* 14(4–5). Problems and perspectives: a good overview.
- Corcoran, E. 1991. "Cleaning Up Coal." *Scientific American* 264(5): 106. Novel market-based approach to reducing air pollution from coal combustion are described.
- Davis, G. R. 1991. *Energy for Planet Earth.* New York: W. H. Freeman. Reading on energy from the *Scientific American.* Highly recommended.
- Edmonds, J., and J.M. Reilly. *Global Energy: Assessing the Future.* New York: Oxford University Press, 1985. How much do we have and where is it?
- Erbes, M.R. et al. July 1987. "Off-Design Performance of Power Plants: An Integrated Gasification Combined-Cycle Example." *Science.* 237(4813):379. Describes a promising new technology.
- Fisher, W.L. June 1987. "Can the U.S. Oil and Gas Resource Base Support Sustained Production?" *Science.* 236(4809):1631. What are our prospects for the future?
- Flavin, C. 1985. "World Oil: Coping With the Dangers of Success." *Worldwatch Paper #66.* Washington, D.C.: Worldwatch Institute. A thoughtful discussion of petroleum reserves and use.
- Fulkerson, W. et al. 1990. "Energy from Fossil Fuels." *Scientific American* 263(3):128. One of a series of articles in a special issue on energy. Also reprinted in book form. See Davis G. R. above.
- Gates, D.M. *Energy and Ecology.* Sunderland: Sinauer, 1985. An overview by a pioneer in this area.
- Hirsch, R.L. March 1987. "Impending United States Energy Crisis." *Science.* 235:1467. Argues that low energy prices have discouraged oil exploration and encouraged consumption so that we are headed toward another energy crisis.
- Masters, C.D. *World Petroleum Resources: A Perspective.* Open-file report 85–248. Reston: U.S. Department of Interior Geological Survey, 1985. How much oil is there?
- McBride, J.P. et al. December 1978. "Radiological Impact of Airborne Effluents of Coal and Nuclear Plants." *Science.* 202:1045. Startling conclusions that may make you think twice about nuclear power.
- Munson, R. *The Power Makers.* Emmaus: Rodale Press, 1985. Who controls our energy supplies?
- Osborne, D. 1985. "The Origin of Petroleum." *The Atlantic.* 257(2):39. Reviews theories of oil formations.
- Southern California Edison Company (SCE). *Cogeneration/Small Power Projects: Quarterly Report to the California Public Utilities Commission.* Rosemead: SCE, 1986. Similar reports from Pacific Gas and Electric Company and San Diego Gas and Electric Company.
- U.S. Environmental Protection Agency (EPA). *Environmental Perspective on the Emerging Oil Shale Industry.* Washington, D.C.: Government Printing Office, 1980. A massive potential source of energy, but also a massive potential source of environmental harm.
- World Bank. *The Energy Transition in Developing Countries.* Washington, D.C.: The World Bank, 1983. How will developing countries meet their energy needs?

CHAPTER 19

Nuclear Power: A Fading Dream?

You don't consult the frogs when you are planning to drain the swamp.

Remy Carle, Director, Electricite de France (commenting on the desirability of public input on nuclear power plant design and location)

When the United States led the world into the nuclear age during the 1950s, civilian nuclear power was expected to provide clean, safe electrical energy "too cheap to meter." Now, however, the nuclear age appears to be coming to a close. Few new nuclear plants are being ordered and an increasing number of countries have pledged to be "nuclear-free." Since 1975, only thirteen orders for nuclear power stations have been placed in the United States, and all of them subsequently were canceled.

Since the 1991 Persian Gulf war, energy companies have begun advertising nuclear energy again as a clean, safe alternative to fossil fuels. New reactor designs have been proposed to be "inherently safe." Whether nuclear power will once again become an important part of our energy sources remains to be seen.

Radioactivity is the emission of energy or subatomic particles (or both) from spontaneous decay (or alteration) of an unstable atom into a lighter element. There are many naturally occurring radioactive elements and there are also many synthetic or human-caused ones that do not occur in nature. People in the United States are exposed to an average annual radiation dose of about 360 millirem, four-fifths from natural sources (cosmic rays, radioactivity from soil, rocks, water, and air). Most of the rest comes from medical and dental X rays.

The basis of nuclear power is a chain reaction of nuclear fissions of unstable uranium, thorium, or plutonium isotopes sustained in an enclosed core or pile of nuclear fuel. Water, gases, or liquid metals circulating through the nuclear fuel absorb heat and generate steam that drives an electric turbine generator. Nuclear reactors are built with multiple layers of safeguards to prevent accidental release of radioactivity, but accidents have occurred, often because of operator error.

The worst possible accident at a large nuclear power station could cause thousands of immediate deaths, tens of thousands of cancer deaths and genetic abnormalities, and cost many billions of dollars in property damage and clean-up. It could contaminate a large geographic area so badly that it might have to be abandoned for hundreds of years. The 1986 explosion at Chernobyl in the Soviet Union was very close to this worst possible case. Under normal operating conditions, however, nuclear power probably causes far fewer deaths and much less pollution than generation of comparable amounts of energy in conventional coal-fired electric generating stations.

Some radioactive isotopes produced by nuclear power reactions will remain hazardous for hundreds of thousands of years. The United States has chosen Yucca Mountain, Nevada, as the first permanent storage site for these extremely dangerous materials, but many people still have questions about this site and how wastes there will be protected.

If fusion reactions could be contained and controlled in power plants, they might provide an essentially unlimited source of energy and should produce less waste and be less dangerous than fission reactors. No one has yet been able to contain or sustain a fusion reaction, and it is not clear that this energy source will ever be available.

Nuclear power may have been an expensive wild goose chase or it may be a valuable interim power source that will allow us to avoid the problems of fossil fuels while we develop environmentally benign, sustainable energy sources.

INTRODUCTION

In 1953, President Dwight Eisenhower presented his "Atoms for Peace" speech to the United Nations. He announced that the United States would build nuclear-powered electrical generators to provide clean, abundant energy. He predicted that nuclear energy would fill the deficit caused by predicted shortages of oil and natural gas. It would provide power "too cheap to meter" for continued industrial expansion of both the developed and the developing world. It would be a supreme example of "beating swords into plowshares." Technology and engineering would tame the evil genie of atomic energy and use its enormous power to do useful work.

Glowing predictions about the future of nuclear energy continued into the early 1970s. Between 1970 and 1974, American utilities ordered 140 new reactors for power plants. Some advocates predicted that by the end of the century there would be 1,500 reactors in the United States alone. The International Atomic Energy Agency (IAEA) projected worldwide electric generating capacities of at least 4.5 million megawatts (MW) by the year 2000, eighteen times more than the current nuclear capacity and twice as much as the total world electrical capacity from all sources.

The oil price shocks of the 1970s changed the energy picture dramatically. Rising energy prices, along with an emerging conservation ethic, created strong pressures to use less energy and to conserve fuel. The demand for electricity, which had been growing at a steady 4 percent per year since World War II, began to recede. As inflation drove up interest rates, the cost of building nuclear plants escalated from an average of $200 per kilowatt (kW) in 1970 to $3,500 per kW in 1987, making it increasingly difficult for utilities to finance further nuclear expansion.

Public opposition also held down the growth of the nuclear industry. The environmental movement that blossomed in 1970 made energy policy a priority, and people began to question the safety of nuclear energy (figure 19.1). New calculations about the numbers of cancer deaths that might occur if a full-scale nuclear program were implemented and details of the deaths and destruction that might result from a "worst-case" accident became widely known. Consumers objected to paying increased costs of nuclear power. Electricity from nuclear plants had been about half the price of coal in 1970, but increased to a rate that was twice as much (12 cents per kWhr vs. 6 cents) by the mid-1980s. The cost of nuclear-generated electricity is expected to double again by the mid-1990s and to be more costly than solar photovoltaic power (chapter 20).

The United States led the world into the nuclear age, and it appears that we are leading the world out of it. After 1975, only thirteen orders were placed for new nuclear reactors in the United States, and all of those orders subsequently were canceled (figure 19.2). In fact, 100 of the 140 reactors ordered before 1975 were canceled. It is beginning to look as if the much-acclaimed nuclear power industry may have been a very expensive wild goose chase that may never produce enough energy to compensate for the amount invested in research, development, mining, fuel preparation, and waste storage. In defense of nuclear power, it should be pointed out that it does not release carbon dioxide, it does not contribute to the greenhouse effect (chapter 17), and it may be an interim source of energy while we develop alternate sources.

After the 1991 Persian Gulf war, the nuclear industry began advertising again in the media promoting their reactors as a clean, safe "alternative to dependence on foreign fossil fuels." They clearly hope for a rebirth of a once moribund technology. At the end of 1990, there were 107 operating commercial reactors in the United States and 24 still under construction (figure 19.3). Some of the first plants built have reached the end of their useful life and are being dismantled and sent to nuclear waste repositories for storage. Table 19.1 describes the current status of nuclear power plants in the world.

In this chapter, we will review basic information about energy from atoms, how nuclear reactors work, their benefits, and the hazards and special problems associated with their use. Nuclear-generated power is a fact of life in our world, but one that is surrounded by strong controversy. Proponents and opponents have intense convictions about it. We hope this chapter will help you understand both the nature of nuclear power generation and the issues it involves.

ATOMIC THEORY AND ISOTOPES

To understand how nuclear energy is produced and why radiation is dangerous, we need to know about atoms and radioactivity. In this section, we will review some basic information about atomic structure, radioactivity, and the energy released from atoms when they split.

Atomic Structures

Atoms are composed of protons, neutrons, and electrons. Protons and neutrons are found in the nucleus, the center of the atom, where they are held together by strong atomic force. Electrons are extremely small, negatively charged packets of energy that rapidly circle the nucleus in defined orbitals. They are held in place by a balance between their kinetic energy, which tends to propel them out of their orbitals, and by their attraction to the positively charged protons in the nucleus. Atoms that have picked up extra electrons (and are therefore negatively charged) or that have lost electrons (and are therefore positively charged) are called **ions.** Nonionized atoms of a particular element always have a standard number of electrons and protons, giving them the same chemical characteristics.

The atomic number of an atom is determined by its number of protons. The lightest element, hydrogen (H), has 1

FIGURE 19.1   The explosion and fire at the Chernobyl Nuclear Power Plant in the Soviet Ukraine in 1986 was the world's worst civilian nuclear accident so far. It marked a major turning point in the history of nuclear power.

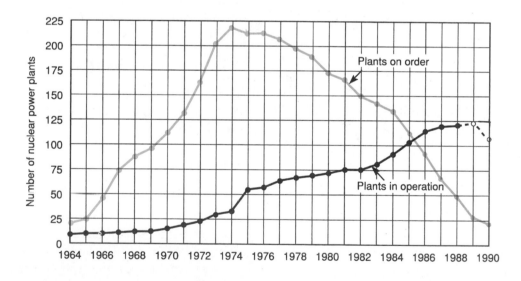

FIGURE 19.2   The changing fortunes of nuclear power in the United States are evident in this graph showing the number of nuclear plants on order (color) and plants in operation (black). Since the early 1970s, few utilities have ordered new plants and many previous orders have been canceled.

proton and 1 electron, so has an atomic number of 1. Uranium (U) has 92 protons and 92 electrons, so has an atomic number of 92. Atoms of the same element may, however, have different numbers of uncharged neutrons in their nuclei; thus they have different masses. These different forms of the same element are called **isotopes.** Isotopes are identified by appending the mass number (sum of protons and neutrons) to the name or symbol of the element; thus, an extra neutron added to hydrogen-1 becomes hydrogen-2 (deuterium, or H^2); if a second neutron is added, it makes hydrogen-3 (tritium, or H^3). Uranium has sev-

RESOURCES AND RESOURCE ECONOMICS

• Operating commercial reactors (97)

○ Commercial reactors under construction (27)

▲ Graphite reactors (1 commercial, 1 military)

■ Military reactors without containment (5)

FIGURE 19.3 Location and status of nuclear reactors in the United States as of 1990.

Radioactive and Stable Isotopes

eral isotopes and can change between them by adding or losing neutrons, making it extremely valuable as fuel for nuclear reactors and weapons.

An isotope can be stable (nonradioactive) or unstable (**radioactive**). What is the difference? The nuclei of radioactive isotopes spontaneously emit high-energy electromagnetic radiation or subatomic particles (or both), while gradually changing into another isotope or even a different element. We call this process radioactive decay, and we call the isotopes that exhibit this property **radionucleides.** As the unstable isotope disintegrates, it emits subatomic particles, such as beta particles (high-energy electrons), gamma rays (very short wavelength forms of the electromagnetic spectrum), and alpha particles (helium nuclei con-

sisting of two protons and two neutrons). The rate at which these reactions occur is characteristic for each isotope and is generally described in terms of **half-life,** the length of time it takes for half the nuclei in a sample to change into another isotope.

Some isotopes decay very rapidly, while others have extremely long half-lives. Table 19.2 shows radioactive isotopes commonly associated with nuclear power plants, some as fuels and some as by-products. Note that short-lived radioisotopes decay quickly when removed from the reactor. After twenty half-lives, only one-millionth of the original radioactivity still remains, for instance. For an isotope such as iodine[131], which has a half-life of only eight days, a year of storage (forty-six half-lives) reduces the radioactivity to insignificant levels. Isotopes such as plutonium[239], on the other hand, have half-lives in thousands of years and will continue to be radioactive hundreds of thousands of years from now. In fact, plutonium levels in nuclear wastes

TABLE 19.1 The state of commercial nuclear power around the world

Country	Operating reactors No. of units	Operating reactors Megawatts	Under construction No. of units	Under construction Megawatts	Cumulative operation years
Argentina	2	935	1	692	14
Belgium	8	5,486	0	0	64
Brazil	1	626	1	1,245	4
Bulgaria	4	1,632	2	1,906	30
Canada	16	9,776	6	4,789	152
China	0	0	2	1,500	0
Cuba	0	0	2	816	0
Czechoslovakia	5	1,980	11	6,284	22
Finland	4	2,310	0	0	27
France	43	37,533	20	25,017	338
Germany, East	5	1,694	6	3,432	57
Germany, West	19	16,413	6	6,585	215
Hungary	2	825	2	820	4
India	6	1,140	4	880	55
Iran	0	0	2	2,400	0
Italy	3	1,273	3	1,999	70
Japan	33	23,665	11	9,773	287
Korea, Republic of	4	2,720	5	4,692	15
Mexico	0	0	2	1,308	0
Netherlands	2	508	0	0	30
Pakistan	1	125	0	0	14
Poland	0	0	2	880	0
Romania	0	0	3	1,980	0
South Africa	2	1,840	0	0	2
Soviet Union	51	27,756	34	31,816	532
Spain	8	5,577	2	1,920	56
Sweden	12	9,455	0	0	99
Switzerland	5	2,882	0	0	54
Taiwan	6	4,918	0	0	26
United Kingdom	38	10,120	4	2,530	696
United States	94	77,804	26	29,258	955
Yugoslavia	1	632	0	0	4
Total	375	249,625	156	141,322	3,822

Source: From International Atomic Energy Agency, "Nuclear Power Reactors in the World," April 1986 Edition. Reprinted by permission.

increase over time as less stable radioactive elements, such as neptunium, decay to form plutonium.

How Nuclear Energy Is Released

In the process of radioactive decay, some isotopes split apart to create two smaller atoms. This is called **nuclear fission.** During fission, a part of the mass of the atomic nucleus of the isotope is converted into energy, mostly as infrared radiation (or heat). Subatomic particles, such as electrons or neutrons, also are usually emitted. If a neutron is released, it can hit another unstable atom and trigger another fission reaction. If this atom also releases a neutron, a **chain reaction** can occur that goes from atom to atom until the neutron is absorbed by a nonfissionable reaction. If each atom in the chain releases two or more neutrons, the reaction can accelerate exponentially. Since these reactions occur in billionths or trillionths of a second, enormous

numbers of fissioning reactions can occur very quickly if enough fissionable isotopes are present, and enormous amounts of energy can be released in a short period of time. This is the basis of both nuclear power plants and atomic bombs. The difference between them is that a bomb explosion is uncontrolled, while the reaction in a power plant proceeds more slowly in a controlled fashion. A mass of isotopes large enough to sustain a chain reaction is called a critical mass (figure 19.4).

Radiation Terms and Units

Table 19.3 shows some of the units and symbols used to describe radiation and its effects on organisms. Curies and Becquerels (named after pioneers in the study of radioactivity) are measures of the radiation produced by radioisotopes. Since different types of radiation vary greatly in penetrating power and effects on living tissues, however, we need other measures to describe the effects

RESOURCES AND RESOURCE ECONOMICS

TABLE 19.2 Some isotopes associated with nuclear power

Radioisotope	Half-life (years)	Radiation emitted
Americium^{241}	460	
Cesium^{134}	2.1	
Cesium^{135}	2,000,000	
Cesium^{137}	30	Beta and gamma
Curium^{243}	32	
Iodine^{131}	0.02	Beta and gamma
Krypton^{85}	10	Beta and gamma
Neptunium^{239}	2.4	
*Plutonium^{239}	24,000	Gamma and neutron
Radon^{226}	1,600	Alpha
Ruthenium^{106}	1	Beta and gamma
Strontium^{90}	28	Beta
Technetium^{99}	200,000	
*Thorium^{230}	76,000	
Tritium	13	Beta
*Uranium^{235}	713,000,000	Alpha and neutron
Xenon^{133}	0.01	
Zirconium^{93}	900,000	

*Used as fuel in nuclear reactions. All others are by-products.

TABLE 19.3 Common radiological units

Unit	Symbol	Description
Curie	Ci	3.7×10^{10} nuclear transformations per second
Becquerel	Bq	1 nuclear transformation per second
Radiation dose absorbed	rad	0.01 Joules absorbed per kg (= 100 erg/gm)
Gray	Gy	1 Joule/kg (= 100 rad)
Quality factor	Q	Biological effectiveness of radiation
Roentgen equivalent man	rem	Rad × Q × modifying factors
Seivert	Sv	Gy × Q × modifying factors (= 100 rem)
Millirem	mrem	1/1,000 of a rem

of radioactivity on organisms. A rad is the dose absorbed by the target. The quality factor (Q) describes the biological effectiveness of a particular type of radiation. Rem (roentgen equivalent man) describes the radiation dose times the quality factor times modifying factors, such as repair mechanisms, shielding, or behavior that might mitigate effects of radiation. Monitors worn by people who work in potentially radioactive environments record exposure in millirems (mrem) or thousandths of a rem.

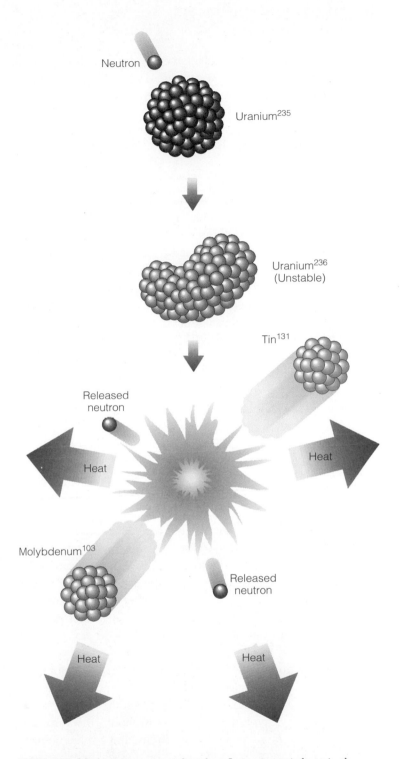

FIGURE 19.4 The process of nuclear fission is carried out in the core of a nuclear reactor. In the sequence shown here, the unstable isotope, Uranium-235, absorbs a neutron and splits to form tin-131 and molybdenum-103. Two or three neutrons are released per fission event and continue the chain reaction. The total mass of the reaction product is slightly less than the starting material. The residual mass is converted to energy (mostly heat).

RADIATION IN THE ENVIRONMENT

You are exposed to atomic radiation every day from both natural and human-originated sources (figure 19.5). There are many sources and types of radioactivity. Some sources are minor and some types are relatively innocuous; others are very dangerous. In this section, we will look at the major sources and types of radiation in the environment.

Sources of Radiation

For many years the average radiation dose received by residents of the United States was estimated to be about 160 mrem per year. Recent recognition of the widespread prevalence of radon gas produced by decay of naturally occurring uranium and thorium isotopes has caused the estimate to more than double to an average of 360 mrem per year. This doesn't mean that our environment has become more dangerous, but it indicates that we have a better understanding of radiation sources. Actual radiation exposures vary widely, depending on location and lifestyle. In some places, such as the radioactive black sand beaches of southern India and Brazil, natural radioisotopes can produce 2,000 mrem or more per year. There are no known health effects from living in these areas.

Cosmic rays from space provide from one-twentieth to one-half of your natural background radiation, depending mainly on the altitude at which you live. Much cosmic radiation is filtered out by the atmosphere. If you live at sea level, you probably receive only about 45 mrem per year from cosmic rays. By contrast, if you live in Boulder, Colorado, where the altitude is 1,600 m (5,200 ft), your cosmic ray dose is twice that amount. A person in Leadville, Colorado, at 3,200 m (10,500 ft), gets about 200 mrem from cosmic rays. The increase is nonlinear because the number of protective air molecules in the atmosphere decreases nonlinearly with altitude (chapter 17). Other natural sources of radiation include radioactive rocks and soil. Traces of radioactivity are stored in our bodies and found in food and drinking water.

Human sources now contribute as much radiation for most people in the United States as do Earth or cosmic rays. Medical and dental X rays are generally the largest source of anthropogenic radiation. A typical chest X ray is about 10 mrem, full mouth dental X rays are about 100 mrem, and a mammogram is about 1,000 mrem. In most cases, the medical benefits from these diagnostic and therapeutic procedures far outweigh the possible risk from radiation. It is wise, however, to keep these exposures as low as possible. Building materials usually contain small amounts of radioactivity. Houses made of stone or brick are about twice as radioactive as houses made of wood. Unvented natural gas burners release measurable quantities of radiation.

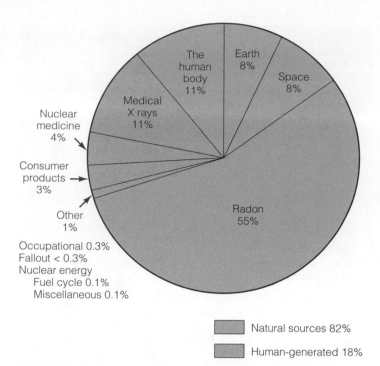

FIGURE 19.5 Sources of radiation. Naturally occurring radioactive elements in the Earth and in our bodies, cosmic radiation from space, and radon gas produced by the breakdown of uranium and thorium make up 82 percent of the average yearly radiation exposure. Human-caused exposures—mainly medical X rays and nuclear medicine—make up the other 18 percent. Recent discovery of the contributions of radon have doubled estimates of our total exposure.

Smoking is a major source of radioactivity because tobacco plants take up uranium and radon daughters from the soil. These radionucleides are concentrated in the smoke and carried deep into the lungs. The high incidence of lung cancer associated with smoking may be related to the combination of carcinogenic chemicals and radioactive isotopes in smoke. Frequent flyers get a significant amount of extra cosmic radiation because of the high altitudes at which jet airplanes fly. A round-trip flight from the United States to Europe gives you as much radiation as a full chest X ray.

The normal operation of nuclear power plants produces a very small fraction of our total radiation exposure. Altogether, nuclear power plants, nuclear weapons testing, uranium mining, and other nuclear power sources account for about 1 percent of our average radiation dose. Exposure is much higher for uranium miners, power plant workers, and others involved in the nuclear fuel cycle. Of course, a major accident like that in Chernobyl in the Soviet Ukraine in 1986 could expose many people to very high levels of radioactivity (Box 19.1).

Types of Radiation

The high-energy electromagnetic radiation or energetic subatomic particles released by nuclear decay are especially dangerous to living organisms because they are examples of **ionizing**

radiation. This means that they can tear electrons away from other atoms to form ions. Some of these ions are highly reactive and can disrupt biological structures, causing great damage. Ultraviolet (UV) light is another kind of ionizing radiation, as are the **X rays** given off when some of the electrons in orbit around the nucleus of an atom jump down from a higher to a lower energy state.

The penetrating power of these forms of ionizing radiation varies greatly. Gamma rays, X rays, and neutrons are extremely energetic; it takes a meter of concrete or several centimeters of lead shielding to block their passage. Beta particles, by contrast, usually just barely penetrate the skin. Alpha particles have the least penetrating power and can be stopped by a single sheet of paper. Because of their high energy and penetrating power, gamma rays, X rays, and neutrons are the most damaging to biological tissues (Box 19.2).

Radon gas released from uranium in the ground and ultraviolet (UV) radiation from the sun are probably much greater health risks to the average person than radiation from nuclear power (chapter 23). The American Cancer Society reported that in 1990 some 600,000 new cases of skin cancer were reported, most caused by UV radiation. Depletion of stratopheric ozone will exacerbate this situation. In 1990, it was predicted that two of three people in Australia will have skin cancer during their lifetime.

NUCLEAR REACTORS

How do nuclear reactors produce energy? In this section, we will look at the nuclear fuel cycle, reactor designs, and some of the concerns about accidents and security risks associated with nuclear power.

Reactor Fuel

The most commonly used fuel in nuclear power plants is the uranium nucleide U^{235}, a naturally occurring radioactive isotope of U^{238}, the more common form. Ordinarily, U^{235} makes up only about 0.7 percent of uranium ore, too little to sustain a chain reaction. Uranium-bearing ore is extracted from surface or underground mines and crushed in mills, usually located near the mines. Uranium is mechanically and chemically separated from the crushed ore, leaving surface piles of radioactive waste products, or **tailings.** The extracted concentrate is converted into **yellowcake,** which contains 70 to 90 percent uranium oxide (U_3O_8) (figure 19.6).

The yellowcake then is shipped to an enrichment plant for conversion to uranium hexafluoride gas (UF_6). The lighter U^{235} isotope is separated from the heavier, nonradioactive U^{238} either by gaseous diffusion or by centrifugation. When the concentration of U^{235} reaches about 3 percent, the UF_6 is converted to uranium dioxide (UO_2) powder and compacted into cylindrical pellets. Each pellet is slightly thicker than a pencil and about 1.5 cm long, but it packs an amazing amount of energy. A single 8.5 gram pellet has as much energy as a ton of coal or four barrels of crude oil.

The UO_2 pellets now are ready to become reactor fuel. They are stacked on end and slid into hollow steel rods approximately 4 m long. About one hundred of these rods are bundled together to make a **fuel assembly** (figure 19.7). Thousands of fuel assemblies containing millions of pellets (100 tons of uranium) are stacked in a heavy steel vessel, about 3 m in diameter and 4 m tall, that makes up the core of the reactor. Approximately one-third of the fuel assemblies must be replaced annually to keep the reaction going efficiently.

With such incredibly concentrated radioactivity, how are the nuclear reactors regulated? **Control rods** of neutron-absorbing material, such as carbon or boron, are inserted into spaces between fuel assemblies to slow the fission reaction and are withdrawn in increments to allow it to proceed.

Kinds of Reactors in Use

Seventy percent of the nuclear plants in the United States and in the world are Pressurized Water Reactors (PWR) (figure 19.8). Water is circulated through the core, absorbing heat as it cools the fuel rods. This primary cooling water is heated to 317°C and reaches a pressure of 2,235 psi. It then is pumped to a steam generator where it heats a secondary water-cooling loop. Steam from the secondary loop drives a high-speed turbine generator that produces electricity. Both the reactor vessel and the steam generator are contained in a thick-walled concrete and steel containment building that prevents radiation from escaping and is designed to withstand high pressures and temperatures in case of accidents. Engineers operate the plant from a complex, sophisticated control room containing many gauges and meters to tell them how the plant is running. Overlapping layers of safety mechanisms are designed to prevent accidents, but these fail-safe controls make reactors very expensive and very complex. A typical nuclear power plant has 40,000 valves compared to only 4,000 in a similar size fossil fuel-fired plant. In some cases, they are so complex that they confuse operators and cause accidents rather than prevent them. Under normal operating conditions, a PWR releases very little radioactivity and is probably less dangerous for nearby residents than a coal-fired power plant.

In the United States, the second most common reactor design (about 25 percent of all nuclear power plants) is called a Boiling Water Reactor (BWR) (figure 19.9). In this design, the primary cooling water expands to make steam that directly drives the turbine generator. The highly radioactive cooling water cannot be completely enclosed in this system because it would not condense after going through the turbine and a vapor lock

BOX 19.1   Chernobyl: The Worst Possible Accident?

In the early morning hours of April 26, 1986, residents of the Ukrainian village of Pripyat saw a spectacular and terrifying sight. A glowing fountain of molten nuclear fuel and burning graphite was spewing into the dark sky through a gaping hole in the roof of the Chernobyl Nuclear Power Plant only a few kilometers away. Although officials assured them that there was nothing to worry about in this "rapid fuel relocation," the villagers knew that something was terribly wrong. They were witnessing the worst possible nuclear power accident, a "meltdown" of the nuclear fuel and rupture of the containment facilities, releasing enormous amounts of radioactivity into the environment.

The accident was a result of a risky experiment undertaken by the plant engineers in violation of a number of safety rules and operational procedures. They were testing whether the residual energy of a spinning turbine could provide enough power to run the plant in an emergency shutdown if off-site power were lost. Reactor number four had been slowed down to only 6 percent of its normal operating level. To conserve the small amount of electricity being generated, they then disconnected the emergency core-cooling pumps and other safety devices, unaware that the reactor was dangerously unstable under these conditions.

The heat level in the core began to rise, slowly at first, and then faster and faster. The operators tried to push the control rods into the core to slow the reaction, but the graphite pile had been deformed by the heat so that the rods wouldn't go in. In 4.5 seconds, the power level rose two thousandfold, far above the rated capacity of the cooling system. Chemical explosions (probably hydrogen gas released from the expanding core) ripped open the fuel rods and cooling tubes. Cooling water flashed into steam and blew off the 1,000–ton concrete cap on top of the reactor. Molten uranium fuel puddled in the bottom of the reactor, creating a critical mass that accelerated the nuclear fission reactions. The metal superstructure of the containment building was ripped apart and a column of burning graphite, molten uranium, and radioactive ashes billowed 1,000 m (3,000 ft) into the air.

Panic and confusion ensued. Officials first denied that anything was wrong. The village of Pripyat was not evacuated for thirty-six hours. There was no public announcement for three days. The first international warning came, not from Soviet authorities, but from Swedish scientists 2,000 km away who detected unusually high levels of radioactive fallout and traced air flows back to the southern Soviet Union.

There were many acts of heroism during this emergency. Firemen climbed to the roof of the burning reactor building to pour water into the blazing inferno. Engineers dived into the suppression pool beneath the burning core to open a drain to prevent another steam explosion. Bus drivers made repeated trips into the contaminated area to evacuate nearby residents. Helicopter pilots hovered over the gaping maw of the ruined building to drop more than 7,000 tons of lead shot, sand, clay, limestone, and boron carbide onto the burning nuclear core to smother the fire and suppress the nuclear fission re-actions. The main fire was put out within a few hours, but the graphite core continued to smolder for weeks. It wasn't finally extinguished until tunnels were dug beneath the reactor building and liquid nitrogen was injected under the core to cool it.

A 10 km zone (6.25 mi) around the plant was evacuated first. Later, it was expanded to 30 km (18 mi). Altogether, more than 135,000 people were evacuated from seventy-one villages in the immediate area, and 250,000 children from Kiev, 80 km (50 mi) to the south, were sent on an "early summer holiday." Several hundred people were hospitalized for radiation sickness. Thirty-one people are officially reported to have died from direct effects of radiation. Critics claimed that the total was ten times higher.

The amount of radioactive fallout varied from area to area, depending on wind patterns and rainfall. Some places had heavy doses while neighboring regions had very little (box figure 19.1). One band of fallout spread across Yugoslavia, France, and Italy. Another crossed Germany and Scandinavia. Small amounts of radiation even reached North America. Altogether, about 7 tons of fuel containing 50 to 100 million curies were released, roughly 5 percent of the reactor fuel.

Assessments of the long-term effects of this accident differ widely. The U.S. Department of Energy estimates that between 400 and 28,000 cancer deaths will occur in the Northern Hemisphere in the next fifty years as a result of Chernobyl. With 30 million cancer deaths expected from all causes in the next fifty years, however, even 28,000 extra deaths prob-

would occur. As a result, a considerable amount of radio-activity escapes, mostly in the form of tritium (H^3 or ^3H) and iodine[125].

Canadian nuclear reactors use deuterium (H^2 or ^2H, the heavy, stable isotope of hydrogen) as both a cooling agent and a moderator. These Canadian Deuterium (CANDU) reactors operate with natural, unconcentrated uranium (0.7 percent U^{235}) for fuel, eliminating expensive enrichment processes. "Heavy water," deuterium-containing water (^2H_2O), is expensive, however. Also, these reactors have a *positive void coefficient*, which means that if cooling pumps fail, core temperatures rise very quickly. The cooling water turns to steam, which has very little

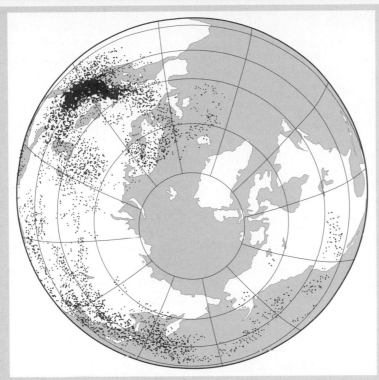

BOX FIGURE 19.1 Spread of radioactive fallout from the Chernobyl accident. Shifting winds and spotty rainfall deposited radioactivity in irregular patterns. Most of the fallout was in southern Europe, but significant amounts also fell on central and eastern Asia. Small amounts even reached North America.

(Lange, 1988).

sands of square meters of plastic film have been laid on the ground to contain radioactive dust. Both Chernobyl and Pripyat have been declared unsafe for habitation. Along with dozens of smaller towns and hundreds of farms, they have been abandoned and will be completely destroyed.

More than 600,000 people worked on decontamination and reconstruction. Those who worked in the most radioactive areas could stay there for only a few minutes before being rotated out. What health effects those workers will experience is not known. The immediate direct costs in the Soviet Union were roughly $3 billion; total costs might be one hundred times that much.

In 1990, Soviet officials revealed that contamination was much higher and more widespread than was previously admitted. A program is underway to resettle 250,000 people from the Ukraine and Byelorussia. Estimates are that 3 million people might eventually have to be moved. Also in 1990, the Soviet Parliament voted $26 billion for medical checkups, housing, and benefits for victims.

Several countries have suggested that the IAEA inspect all nuclear power plants. All IAEA members have signed an agreement to provide timely information to other countries in the event of a transboundary nuclear accident. Suits have been filed in international court by other countries claiming compensation for damages from this accident. Public opinion has swung against nuclear power in many countries (including the Soviet Union) since the Chernobyl accident, which may have been the death knell for the nuclear power industry.

ably will not be statistically significant or detectable.

Some scientists criticize these calculations, pointing out that iodine-131 and cesium-137 are taken up by plants and accumulated by animals (iodine in the thyroid, cesium in bones) so that effects may be greater than expected. Furthermore, they disagree with the cancer-risk coefficient used (the probable number of cancer fatalities per unit of radiation exposure). If these critics are right at least 100,000 deaths from cancer and genetic

defects could result from Chernobyl in the next fifty years. Some areas appear to be showing effects of Chernobyl already. Leukemia rates in Minsk have doubled from 41 per million in 1985 to 93 per million in 1990.

For the present, the damaged reactor has been entombed in a giant, steel-reinforced concrete sarcophagus where it is intended to remain for thousands of years. Forests have been cleared and topsoil removed over wide areas to clean up radioactivity. In the worst areas, thou-

heat-absorbing capacity and blocks introduction of additional cooling water. Unless steps are taken immediately, there is risk of a core meltdown.

In Britain, France, and the Soviet Union, a common reactor design uses graphite, both as a moderator and as the structural material for the reactor core. In the British MAGNOX

design (named after the magnesium alloy used for its fuel rods), gaseous carbon dioxide is blown through the core to cool the fuel assemblies and carry heat to the steam generators. In the Soviet design, called RBMK (the Russian initials for a graphite-moderated, water-cooled reactor), low-pressure cooling water circulates through the core in thousands of small metal tubes.

BOX 19.2 Why Is Radiation Dangerous?

A single whole-body exposure to 10,000 rems of high-energy radiation will almost certainly kill you within a few hours; 1,000 rems are generally lethal within a few days: and 600 whole-body rems will kill most people within a few weeks. Exposure to between 100 and 600 whole-body rems causes radiation sickness (nausea, skin burns, tiredness, hair loss, and disruption of the immune system), but most people recover within a few weeks or months. Exposure below 100 rems will probably not cause any outward symptoms, but damage to sensitive internal tissues and organs, such as bone marrow, spleen, lymph nodes, and circulating blood cells, can be detected with exposures as low as 10 rems over the whole body.

Even if you recover from radiation sickness and don't have any direct physical evidence of damage, however, other long-term problems can persist for many years after radiation exposure. Infertility, neurological damage, cancer, and premature death, as well as birth defects and genetic abnormalities in future offspring, have been associated with radiation exposures as low as 2 rems. These long-term effects appear to be cumulative; that is, a given dose seems to be equally dangerous whether it is received in a single exposure or over many years.

Most authorities now believe that *any* exposure to high-energy radiation carries some risk of long-term health problems. It is thought that as much as 5 to 10 percent of the cancers, birth defects, genetic abnormalities, spontaneous abortions, and premature deaths may be due to radiation exposure (either natural or human-caused). Some scientists believe there is a direct linear relationship between radiation dose and health risks; that is, any increase over zero exposure carries some additional risk. They argue, therefore, that people should keep their exposure to radiation as close to zero as possible.

Not everyone agrees with this point of view, however. It is difficult to measure the effects of very low levels of radiation because people vary considerably in their sensitivity to damage and because it is impossible to separate radiation damage from other kinds of environmental hazards, many of which affect the same tissues, organs, and functions that radiation does. Some scientists claim that there is a threshold below which the risks of radiation are negligible, either because any damage caused can be repaired by normal cellular mechanisms, or because at some point the risks are so low that they don't matter.

In 1988, results were reported from a 44-year study of atom bomb survivors in Hiroshima and Nagasaki, Japan. Genetic damage was lower than had been expected. It appears that humans are considerably less sensitive to radiation than are mice. It makes evolutionary sense that we would have better genetic repair mechanisms than mice since the time from birth to reproductive age is at least 25 times longer for humans than for mice. This does not mean, however, that radiation is not dangerous to humans, only that we must be careful in extrapolating data from laboratory animals to possible human risks.

Radiation damages tissues in two main ways: it can cause direct changes in organic molecules that make up cellular structures and carry out cellular functions, or it can create ions that subsequently bring about deleterious changes.

The most dangerous ions are called free radicals. They are powerful oxidizing or reducing agents that can break chemical bonds and disrupt the structure of organic molecules, such as proteins or nucleic acids, that make up the essential structural, functional, and genetic components of the cell. This sort of change may be expressed immediately, or it might be hidden for many years. If changes occur in gametes (sex cells), their effects may show up in future generations. Changes in the regulatory mechanisms that control cell growth and development can be especially dangerous because they can result in cancer.

At very low levels of exposure, the chances that an important molecule will be hit are slight; however, if the right molecule is hit, the results may be fatal. These results may not be expressed immediately, but as molecules and cells reproduce, there is an effective biological amplification so that what might have been a very simple change originally becomes of great importance at a later date. This is why we believe that any exposure to these hazards carries some risk. Your chances of injury might be one in a billion. However, if you are the one affected, your injury may not be just one-billionth of normal health; you could be 100 percent dead.

These designs were originally thought to be very safe because graphite has high capacity for both capturing neutrons and dissipating heat. Designers claimed that these reactors could not possibly run out of control. The fuel rods are dispersed in small groups rather than being packed in large bundles as in PWR and BWR designs. However, the small cooling tubes in the Soviet RBMK also have a positive void coefficient. That means if the pumps are turned off, the water turns to steam quickly and the reactor overheats.

Furthermore, graphite burns when heated and exposed to air. The two most disastrous reactor accidents in the world, so far, involved fires in graphite cores that allowed the nuclear fuel to melt and escape into the environment. In 1956, a fire at the Windscale Plutonium Reactor in England released roughly 100

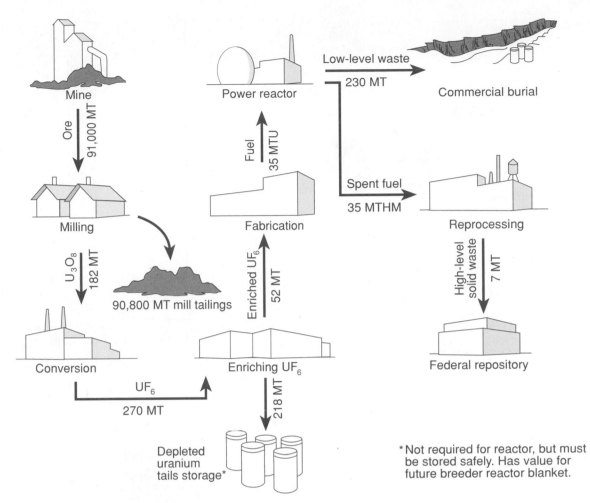

FIGURE 19.6 The nuclear fuel cycle. (MT = metric tons.)
Quantities represent the average annual fuel requirements for a typical
1,000 MW light water reactor.

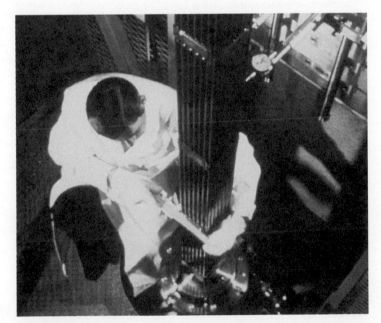

FIGURE 19.7 Inspecting uranium fuel bundles before installation in
the reactor. Each reactor contains hundreds of these bundles.

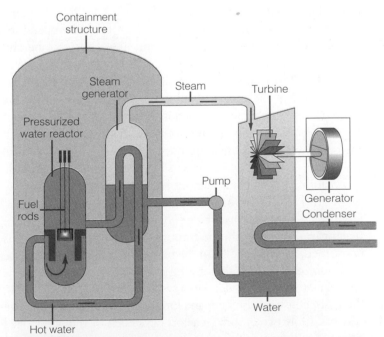

FIGURE 19.8 Pressurized water nuclear reactor. Water is
superheated and pressurized as it flows through the reactor core. Heat is
transferred to nonpressurized water in the steam generator. The steam
drives the turbogenerator to produce electricity.

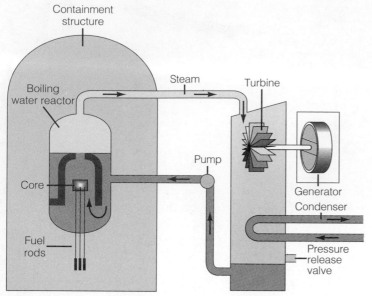

FIGURE 19.9 Boiling water reactor. Water flowing through the reactor core boils and produces steam that is piped out to the turbogenerators. Since it has passed through the reactor, this steam is radioactive. As it condenses, volatile gases, such as tritium and iodine, carry radioactivity into the environment.

million curies of radionucleides and contaminated hundreds of square kilometers of countryside. In 1986, a fire and explosion at the Chernobyl nuclear plant in Russia released several tons of nuclear fuel and contaminated most of Europe (Box 19.1).

Alternative Reactor Designs

Several other reactor designs are inherently safer than the ones we now use. Among these are the modular High-Temperature, Gas-Cooled Reactor (HTGCR) and the Process-Inherent-Ultimate-Safety (PIUS) reactor.

HTGCR is similar to the British MAGNOX design. Gaseous helium is the coolant, and graphite blocks form the core structure and moderate the reaction (figure 19.10). If the reactor core is kept small enough and the fuel is dispersed throughout the core, heat will not build up above 1,600°C, even if all coolant is lost; thus, a meltdown is impossible and operators could walk away during an accident without risk of a fire or radioactive release. An HTGCR is fueled by graphite or ceramic-coated pellets of UO_2 or uranium carbide (UC_2). They are loaded into the reactor from the top, shuffle through the core as the uranium is consumed, and emerge from the bottom as spent fuel. This type of reactor can be reloaded during operation. Since the reactors are small, they can be added to a system a few at a time, avoiding the costs, construction time, and long-range commitment of large reactors. Only two of these reactors have been tried in the United States: the Brown's Ferry reactor in Alabama and the Fort St. Vrain reactor near Loveland, Colorado. Both were continually plagued with problems (including fires in control buildings and

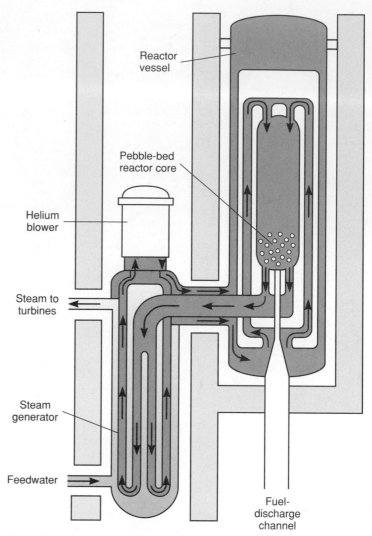

FIGURE 19.10 The modular, high-temperature, gas-cooled, graphite-core reactor shown here is cooled by pressurized helium, which carries heat to a steam generator. In this design, the fuel consists of thousands of tiny uranium pellets coated in layers of graphite and silicon. New pellets can be added to the top of the core while spent pebbles are withdrawn from the bottom. Since the helium does not boil, this design is theoretically safer than water-cooled reactors.

turbine-generators), and both were closed without producing much power.

A much more successful design has been built in Europe by General Atomic. In West German tests, a HTGCR was subjected to total coolant loss while running at full power. Temperatures remained well below the melting point of fuel pellets and no damage or radiation releases occured. These reactors might be built without expensive containment buildings, emergency cooling systems, or complex controls. They would be both cheaper and safer than current designs.

The PIUS design features a massive, 60 meters-high pressure vessel of concrete and steel, within which the reactor core is submerged in a very large pool of boron-containing water

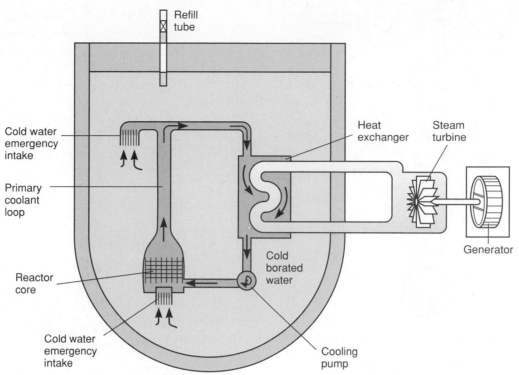

FIGURE 19.11 A PIUS reactor consists of a core and primary cooling system immersed in a very large pool of cold, borated water. As long as the reactor is operating, the cool water is excluded from the cooling circuit by the temperature and pressure differential. Any failure of the primary cooling system would allow cold water to flood the core and shut down the nuclear reaction. The large volume of the pool would cool the reactor for several days without any replenishment.

(figure 19.11). As long as the primary cooling water is flowing, it keeps the borated water away from the core. If the primary coolant pressure is lost, however, the surrounding water floods the core, and the boron poisons the fission reaction. There is enough secondary water in the pool to keep the core cool for at least a week without any external power or cooling. This should be enough time to resolve the problem. If not, operators can add more water and evaluate conditions further.

The Canadian "slow-poke" is a small-scale version of the PIUS design that doesn't produce electricity, but might be a useful substitute for coal, oil, or gas burners in district heating plants. The core of this "mini-nuke" is only about half the size of a kitchen stove and it generates only 1/10 to 1/100 as much power as a conventional reactor. The fuel sits in a large pool of ordinary water, which it heats to just below boiling and sends to a heat exchanger. A secondary flow of hot water from the exchanger is pumped to nearby buildings for space heating. Promoters claim that a runaway reaction is impossible in this design and that it makes an attractive and cost-efficient alternative to fossil fuels. Despite a widespread aversion to anything nuclear, Switzerland and West Germany are developing similar small nuclear heating plants.

If these reactors had been developed initially, the history of nuclear power might have been very different. Neither reactor type is suitable, however, for mobile power plants, such as nuclear submarines, which tells you something about the history of nuclear power and the motivation for its development. Aside from being inherently safer in case of coolant pressure loss, neither of these alternative reactor designs eliminates other problems that we will discuss later in this chapter. Perhaps nuclear power in one or more of these alternate forms will once again be an important energy source in the developed world.

Breeder Reactors

For more than thirty years, nuclear engineers have been proposing high-density, high-pressure fission reactors that would produce fuel rather than consume it. These reactors are called *breeders* because they create fissionable material from the abundant, but nonradioactive, uranium^{238}. Bombardment of U^{238} atoms with fast, high-energy neutrons from previous plutonium^{239} fissions converts them into more U^{239} (figure 19.12). For every 100 atoms of Pu^{239} consumed in the breeder reactor, 130 new atoms of Pu^{239} are produced from U^{238}. Similarly, thorium^{232} is converted into fissionable uranium^{233}. The starting material for this reaction would be Pu^{239} reclaimed from spent fuel from conventional fission reactors. After about ten years of operation, a breeder reactor would produce enough plutonium to start another reactor. Enough U^{238} currently is stockpiled in the United

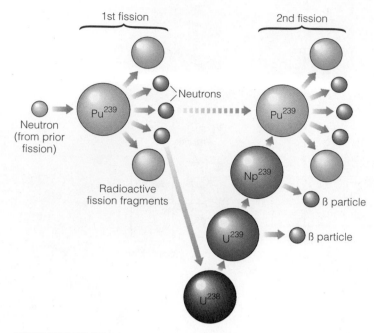

FIGURE 19.12 Reactions in a "breeder" fission process. Neutrons from a plutonium fission change U^{238} to U^{239} and then Pu^{239} so that the reactor creates more fuel than it uses.

States to produce electricity for one hundred years at present rates of consumption, if breeder reactors can be made to work safely and dependably.

Several problems have held back the breeder reactor program in the United States. One problem is the concern about safety. The reactor core of the breeder must be at a very high density for the breeding reaction to occur. Water doesn't have enough heat capacity to carry away the high heat flux in the core, so liquid sodium is used as a coolant. This design is called a **liquid metal fast breeder** (figure 19.13). Liquid sodium is very corrosive and difficult to handle. It burns with an intense flame if exposed to oxygen, and it explodes if it comes into contact with water. Because of its intense heat, a breeder reactor will melt down and self-destruct within a few seconds if the primary coolant is lost, as opposed to a few minutes for a normal fission reactor.

Another very serious concern about breeder reactors is that they produce excess plutonium that can be used for bombs. It is essential to have a spent-fuel reprocessing industry if breeders are used, but the existence of large amounts of weapons-grade plutonium in the world would surely be a dangerous and destabilizing development. If a breeder program were adopted by all of the countries that now have nuclear power, hundreds of planeloads and shiploads of spent fuel and starter plutonium could be shipped around the world every year. The chances of some of that material falling into the hands of terrorists or other troublemakers are very high.

A proposed $1.7 billion breeder-demonstration project in Clinch River, Tennessee, has been on and off for fifteen years. At last estimate, it would cost up to five times the original price if it is ever completed. The United Kingdom built a breeder reactor at Dounray, Scotland, but it didn't work well and was shut down in 1985 with great financial loss. In 1986, France put into operation a full-sized commercial breeder reactor, the SuperPhenix, near Lyons, which cost three times the original estimate to build and produces electricity at twice the cost per kW of conventional nuclear power. In 1987, a large crack was discovered in the inner containment vessel of the SuperPhenix, and it had to be shut down for repairs.

Nuclear Accidents

Most nuclear reactor accidents have resulted from operator errors. You can design an elegant machine with multiple layers of safeguards, but if those safety mechanisms are turned off by ignorant or malicious operators, "impossible" accidents can occur and do occur with disturbing frequency. In 1985, 3,000 mishaps and 746 emergency shutdowns occurred in United States nuclear power plants, 28 percent more than the previous year. Most of these mishaps and shutdowns were trivial, but 18 were serious enough that they could have led to core damage.

Even trivial accidents can lead to serious situations. In 1978, a technician at the Rancho Seco reactor in California dropped a light bulb he was changing into the control panel, shorting out vital safety controls. The steam generator dried out, and the thermal shock may have stressed and weakened the reactor vessel. In 1987, the Peach Bottom reactor near Delta, Pennsylvania, was closed by the Nuclear Regulatory Commission (NRC) when it was discovered that control room operators were regularly sleeping while on duty. The most serious nuclear accident in United States history occurred on March 28, 1979, when an operator error at the Three Mile Island reactor near Harrisburg, Pennsylvania, caused an out-of-control reaction that melted about 20 percent of the reactor core, releasing volatile radionucleides into the air. Further errors allowed radioactive water to escape into the Susquehanna River. Altogether, about a million curies of radionucleides were released. No excess deaths have been detected so far as a result of this accident, but local residents are understandably concerned about the long-term effects of radiation exposure.

In 1975, Norman Rasmussen, head of nuclear engineering at Massachusetts Institute of Technology, chaired a $4 million study for the NRC to assess the worst possible accident that could occur in a nuclear power plant. The conclusion of this study was that a complete core meltdown might cause 3,500 immediate deaths, 45,000 cancer deaths, and cost $14 billion for property damage and clean-up, but it was calculated that such an accident would happen only once in every 20,000 operating years. The Chernobyl accident did not cause this many imme-

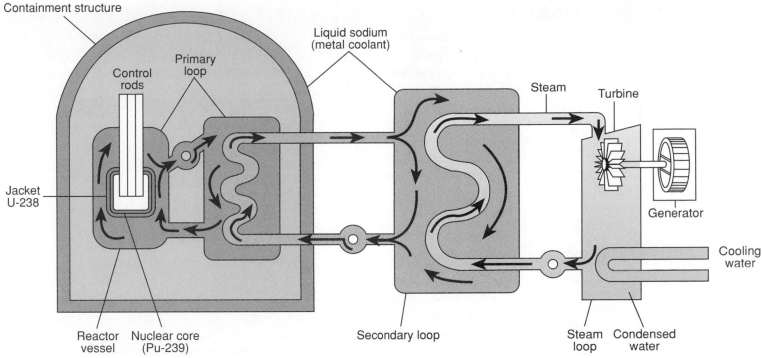

FIGURE 19.13 Liquid metal fast breeder nuclear reactor. The core of a breeder reactor is much hotter and more dense than an ordinary fission reactor. Water is not an adequate coolant, so molten liquid sodium is used as primary coolant. This corrosive material is very dangerous and difficult to handle.

diate deaths, but the reactor was in a fairly remote area and fortuitous winds and weather patterns carried most of the radio-activity out over the Arctic and Atlantic oceans. Some experts believe that a major nuclear accident comparable to Chernobyl is likely to occur every five to ten years somewhere in the world if current nuclear programs continue.

Security Concerns

Systems can be designed to minimize accidents, and better op-erating procedures can reduce mistakes, but there is probably no way to stop deliberate damage to nuclear reactors. No matter how rigid security is, a determined terrorist or undercover agent can gain access to a nuclear plant and its fissionable material and cause serious damage. Nuclear plants also are tempting targets for political retaliation.

With nearly four hundred plants now in operation around the world, thousands of tons of fissionable material are being shipped from fuel assembly plants to power stations and back to reprocessing plants. Enough plutonium and enriched uranium already has disappeared from the fuel cycle to make dozens of small atomic bombs. Some of that material may have been di-verted to such countries as South Africa, Israel, Pakistan, and India, which are thought to have nuclear bomb-making capa-bilities. With so much nuclear material being shipped back and forth, it seems inevitable that some of it will fall into the hands

of maniacs and terrorists. Bomb technology has advanced so that it would be possible to make a bomb small enough to be carried in a briefcase, but powerful enough to injure or kill millions of people if it were delivered to the center of one of the world's major cities. A good discussion of this topic is found in John McPhee's book, *The Curve of Binding Energy.*

Production of atomic bombs in the United States has always proceeded with great secrecy because of concerns about secu-rity. This has allowed operators of research and fabrication fa-cilities to be cavalier about production methods. In 1988, plants in Hanford, Washington; Rocky Flats, Colorado; Fernald, Ohio; Savannah, Georgia, and other areas released millions of curies of radioactive material into the environment through careless ac-cidents or through dumping of wastes. It may cost as much as $100 billion to clean and rebuild these facilities.

RADIOACTIVE WASTE MANAGEMENT

One of the most difficult problems associated with nuclear power is the disposal of wastes produced during mining, fuel produc-tion, and reactor operation. The United States currently has mil-lions of tons of mine tailings and thousands of tons of other radioactive wastes that must be stored and monitored. How these wastes are managed may ultimately be the overriding obstacle to nuclear power.

Uranium Mining and Producing

The dangers of uranium mining have been known for a long time. As early as 1546, miners of uranium-bearing ores in the Erz Mountains of central Europe were reported to have an unusually high frequency of fatal lung disease. Cases of lung cancer in uranium miners were first clinically and anatomically diagnosed in Germany in 1879. In 1913, it was reported that of 665 Schneeburg uranium miners dying between 1876 and 1912, 40 percent died of lung cancer. Because there was little silicosis in these cases, investigators concluded that the most probable cause of the tumors was radiation from radon and its daughters. They also suggested that a simple and cheap way to reduce the danger was adequate mine ventilation. In spite of this knowledge, the United States standards for allowable radiation doses are equivalent to between 3,000 and 15,000 chest X rays per year, and many mines in the 1950s and 1960s had radiation levels ten times the allowable standards. Consequently, lung cancer among uranium miners has been unacceptably high. Among Native American miners in the Four Corners area of the Southwest, as many as 70 percent of the workers in some mines have died from lung cancer.

Enormous piles of mine wastes and mill tailings that have accumulated and now are blowing in the wind and washing into streams in western America represent another problem with mining and fuel processing. Production of 1,000 tons of uranium fuel typically generates 100,000 tons of tailings and 3.5 million liters of liquid waste. There now are 191 million tons of radioactive waste in piles around mines and processing plants in the United States. Canada has even more radioactive mine waste on the surface than does the United States.

Much of this dangerous material has simply been abandoned to blow where it will. Radon gases, toxic metals, and water- or airborne radioactivity from these materials represent a significant health hazard. Mining companies are required by law to clean up the wastes they leave behind, but not until the operation is officially closed. Of the forty-six "active" plants in the United States, only five have produced any fuel in the last twenty years. There are also thousands of abandoned mines all over the Southwest. Where they all are, who dug them, and what is now coming out of them remain unknown.

Radioactive Waste Disposal

In addition to the leftovers from fuel production, there are about 100,000 tons of low-level waste (contaminated tools, clothing, building materials, etc.) and about 15,000 tons of high-level (very radioactive) wastes in the United States. The high-level wastes consist mainly of spent fuel rods from commercial nuclear power plants and assorted wastes from nuclear weapons production. By 1987, about 50,000 highly radioactive spent fuel assemblies were stored in deep water-filled pools at nuclear power plants. These

TABLE 19.4 Typical radioactive inventory of a 1,000 MW light-water reactor (millions of curies)

Isotope	At shutdown	After ten years	After 100 years
Cesium-134	7.5	5.0	0.6
Iodine-131	85	Negligible	Negligible
Neptunium-239	1,640	240.0	Negligible
Plutonium-239	0.051	0.055	0.6
Strontium-89	94	3.6	0.4
Ruthenium-106	25	0.01	Negligible
Tellurium-132	120	Negligible	Negligible
Xenon-133	170	Negligible	Negligible

pools were originally intended only for temporary storage until the wastes were shipped to reprocessing centers or permanent disposal sites. Since neither of these options now exists, utilities have been forced to hold highly radioactive materials for twenty years or more. Space in these pools is running out and fuel rods are being packed closer together than was intended. This problem is quickly approaching crisis proportions as some 14,000 additional spent fuel assemblies are added each year. High-level waste is especially dangerous since it contains high concentrations of very toxic radioactive elements, such as cesium[134], iodine[131], plutonium[239], strontium[89], tritium (H^3) and ruthenium[106], in addition to natural uranium, thorium, and radon daughters.

As table 19.4 shows, a typical 1,000 MW reactor has some 2 billion curies of these elements in its inventory at the end of a fuel cycle. Some of these isotopes have very long half-lives, requiring storage for at least 10,000 years before their radioactivity is reduced to harmless levels. No human civilization has ever lasted more than a few thousand years. How we can insure that these materials will be maintained and safeguarded if the time required to decay safely has not yet been determined?

The oceans have been and continue to be a repository for radioactive wastes. Until 1970, the United States and Britain dumped low-level radioactive wastes in the ocean. The United States dropped some 90,000 barrels offshore along the Atlantic and Pacific coasts between 1940 and 1970. Britain dumped about half as much in the Atlantic. Many of those barrels already have rusted through and released their toxic contents. Fish and other sealife are accumulating radioactivity near the dump sites. The United States and Britain stopped ocean dumping, spurred in part by adverse publicity generated by the activist organization Greenpeace. In the United States, these materials now are buried at thirteen sites operated by the U.S. Department of Energy (DOE) and three other sites run by private firms (figure 19.14). Switzerland, Japan, France, and Belgium continue to dump radioactive wastes in the ocean.

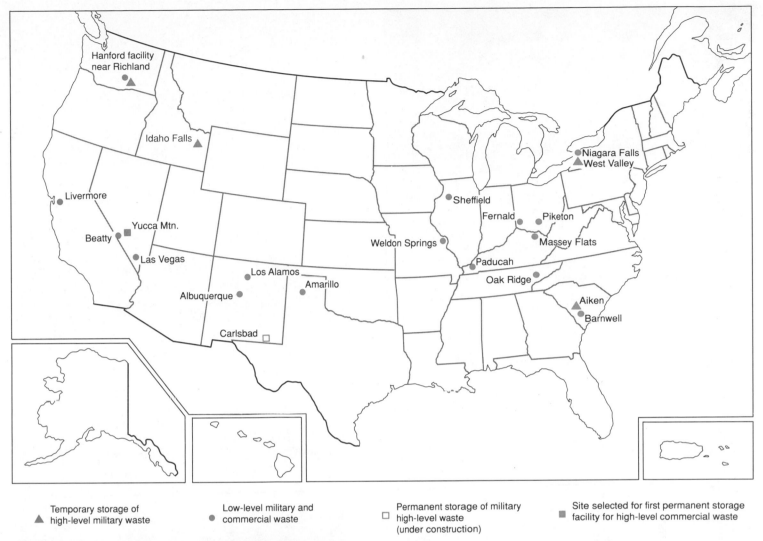

Temporary storage of
▲ high-level military waste

Low-level military and
● commercial waste

Permanent storage of military
☐ high-level waste
(under construction)

Site selected for first permanent storage
■ facility for high-level commercial waste

FIGURE 19.14 Nuclear waste disposal sites. Yucca Mountain, Nevada, has been selected as the first permanent underground storage site for high-level civilian wastes. Carlsbad, New Mexico, will store high-level military wastes. Other sites store low-level wastes.

In 1982, the United States Congress ordered the DOE to identify two sites for permanent radioactive waste disposal on land by 1998. In 1987, the DOE announced that it planned to build the first repository at a cost of $6 to $10 billion on a tuff formation (compacted volcanic ash) in Yucca Mountain, Nevada. Twelve sites in Georgia, Maine, Minnesota, Wisconsin, North Carolina, Virginia, and New Hampshire also have been identified as candidates for the second waste repository, but opposition has been strong in all of these states because of worries about transportation accidents and leaks from storage containers.

The first storage facility is intended to be a deep-mine burial facility that depends on geological confinement (figure 19.15). It is hoped that the volcanic tuff will remain stable, free of groundwater, and unaltered for many millenia. Glassified wastes in corrosion resistant metal canisters would be buried in vaults

at least 200 m (656 ft) below the surface. Some geologists question this approach, arguing that groundwater can percolate through fractures and seams into storage vaults, causing corrosion and leaking of canisters. Once the storage chambers are filled, the waste canisters will be inaccessible. If succeeding generations decide a thousand years from now that this kind of storage was a mistake, they will be unable to do much about it.

Some nuclear experts believe that monitored, retrievable storage would be a much better way to handle wastes. This method involves putting wastes in underground mines or secure surface facilities where they can be watched. In this system, the storage sites would be kept open and inspected regularly. If storage canisters begin to leak, they could be removed for repacking. This would be an expensive alternative, and it might be susceptible to wars or terrorist attacks. We might need a per-

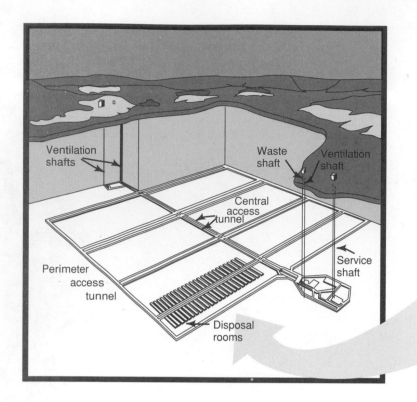

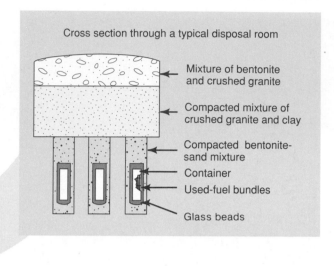

Cross section through a typical disposal room

Mixture of bentonite and crushed granite

Compacted mixture of crushed granite and clay

Compacted bentonite-sand mixture

Container

Used-fuel bundles

Glass beads

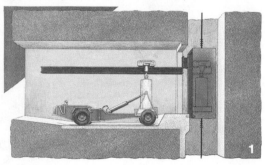

FIGURE 19.15 Proposal for the permanent disposal of used fuel is to bury the used fuel containers 500 to 1,000 m deep in the stable rock of the Canadian Shield. Spent fuel in the form of ceramic fuel pellets are sealed inside corrosion-resistant containers. Glass beads are compacted around the used fuel bundles and into the spaces between the fuel and the container shell. The containers are then buried in specially built vaults that are packed and sealed with special materials to further retard the migration of radioactive material.

petual priesthood of nuclear guardians to make sure that the wastes were never released into the environment.

Shooting wastes into space isn't usually considered seriously. Aside from ethical concerns about cluttering up space with toxic garbage, and the enormous costs of lifting materials into orbit, we do not have rockets reliable enough to handle such dangerous material. As the *Challenger* Space Shuttle disaster in 1986 showed, even very sophisticated systems sometimes fail. If a rocketload of radioactive waste were to blow up and shower its cargo into the atmosphere, it could contaminate the whole Earth.

Although none of these alternatives is very attractive, we have to do something with the wastes we already have generated.

Even if we had never developed civilian power plants, we still have vast quantities of nuclear wastes from atomic weapon production that have to be stored.

Decommissioning Nuclear Power Plants

Old power plants themselves eventually become waste when they have outlived their useful lives. Most plants are designed for a life of only thirty years. After that time, pipes become brittle and untrustworthy because of the corrosive materials and high radioactivity to which they are subjected. Plants built in the 1950s and early 1960s already are reaching the ends of their lives. You do not just lock the door and walk away from a nuclear power plant; it is much too dangerous. It has to be taken apart, and the

RESOURCES AND RESOURCE ECONOMICS

FIGURE 19.16 A specially built railroad car carrying a nuclear waste shipping cask is shown here slamming into a concrete wall at more than 130 km/hr (80 mph) to test its survivability. Even after this violent crash, the shipping cask remains intact.

most radioactive pieces have to be stored just like other wastes. This includes not only the reactor and pipes, but also the meter-thick, steel-reinforced concrete containment building. The pieces have to be cut apart by remote-control robots because they are too dangerous to be worked on directly.

Only a few plants have been decommissioned so far, but it has generally cost two to ten times as much to tear them down as to build them in the first place. That means that the one hundred reactors now in operation might cost somewhere between $200 billion and $1 trillion to decommission. No one knows how much it will cost to store the debris for thousands of years or how it will be done. However, we would face this problem, to some degree, even without nuclear electric power plants. Plutonium production plants and nuclear submarines also have to be decommissioned. Originally, the Navy proposed to just tow old submarines out to sea and sink them. The risk that the nuclear reactors would corrode away and release their radioactivity into the ocean makes this method of disposal unacceptable, however.

Shipping Nuclear Waste

Transporting nuclear fuels to reactors and nuclear wastes to storage sites is not a simple task. There has been a great deal of public concern about the possibility of a shipping accident that could cause disastrous contamination of the countryside. Special casks have been designed to withstand every conceivable accident. Even when crashed into an immovable concrete wall (figure 19.16) at more than 130 km/hour (80 mph), these casks remain intact. Shipping nuclear wastes still engenders a great deal of fear and protest, however.

OTHER PUBLIC ISSUES

Nuclear power has raised strong emotions among its supporters and opponents. Controversial topics include who should bear liability for accidents, how people will be evacuated from the danger zone around a runaway reactor, and when the public should be involved in nuclear planning. In this section, we will examine some of these issues.

Nuclear Accident Liability

When the "Atoms for Peace" program originally was proposed, utilities were unwilling to accept responsibility for a technology that they rightly guessed would be complex and dangerous. To make nuclear power more palatable, Congress limited utility liability under the Price-Anderson Act to no more than $700 million for any accident. This act also entirely shields contractors who work on federal projects from any liability, even if accidents are caused by negligence or malfeasance. The government would pay victims up to $500 million for damages on a contractor's project.

When this act was passed in 1957, it was intended to last only ten years or until the industry could get established technically and economically. It was extended, however, in 1965, 1966, 1975, and 1987. The most recent extension raised the liability limit for each accident to $7 billion, in spite of protests from industry that they could not afford the required insurance. An amendment to the act that would have made contractors on federal nuclear projects liable for accidents due to gross negligence or willful misconduct was defeated. Critics claim that the liability limits are still too low and are an unwarranted subsidy of nuclear power and an unreasonable risk for the public to bear. The estimated cost of the Chernobyl accident is $14.4 billion—twice what a United States company would pay. Many people feel that if the utilities that operate nuclear plants are not confident enough to assume responsibility for them, then the plants are too dangerous to continue operating. If we demand that utilities bear all liability for nuclear power, however, they will shut down existing plants. What do you think would be the consequences if that were to happen? What alternatives would you propose?

Emergency Evacuation Plans

After the 1979 Three Mile Island accident in Harrisburg, Pennsylvania, the NRC ordered all utility companies to prepare emergency evacuation plans for a 16 km (10 mi) zone around nuclear reactors. Utility companies have been trying to reduce the evacuation zone to a 3.2 km (2 mi) radius, arguing that a larger area is unnecessary. Public citizen groups have been trying to enlarge the zone, arguing that the Soviet Union found it necessary to evacuate a 30 km zone after the Chernobyl accident in 1986.

Failure to reach agreement on evacuation plans with local authorities delayed the licensing of a completely furnished, $5.2 million plant in Shoreham, New York for five years. Located about halfway out Long Island, this plant will represent a logistic nightmare in case of an emergency. Several million people live within the 16 km radius that would be evacuated. Narrow roads in this area are congested under normal conditions. Since the island is less than 16 km wide where the plant is located, people further out would either have to pass through the evacuation zone on their way to the mainland or jump into the ocean. In June, 1988, the Long Island Lighting Company agreed to close the Shoreham plant or convert it to a fossil fuel plant if most of the loss could be passed on to customers. The state legislature refused to approve the deal, and a few months later, the NRC eliminated the requirement that states agree with emergency evacuation plans and issued the plant an operating license.

Similar problems with evacuation plans delayed the licensing of the Seabrook Nuclear Reactor in Seabrook, New Hampshire, for nearly five years. This delay caused the bankruptcy of Public Service of New Hampshire, the first failure of a utility in United States history. In 1984, the Washington Public Power Supply Consortium (WPPSS—pronounced, aptly enough, woops) defaulted on about $20 billion in loans because of an overly ambitious nuclear building program. To compound the woes of the nuclear power industry, in 1989 the United States Supreme Court ruled that utilities cannot bill customers for planning and construction costs of nuclear power plants that are canceled before being put into operation.

Citizen Protests Against Nuclear Power

Opposing nuclear power has been a high priority for many conservation organizations in recent years. A federation of antinuclear alliances (e.g., Clamshell Alliance, Northern Sun Alliance) has rallied thousands of people to oppose nuclear power. Protests and civil disobedience at some sites have gone on for a decade or more, and several plants have been abandoned because there was so much opposition to them. Others feel that abandoning nuclear power—especially plants that are already finished—is an excessively high price to pay for elimination of risks that are not clearly understood and may be much lower than is commonly held (figure 19.17). What do you think?

Since the Chernobyl accident, public opinion in many countries has swung strongly against nuclear power. In France, more than 20,000 protesters battled police in 1985 at the site of the SuperPhenix near Lyons (figure 19.18); one person was killed. In Hong Kong, more than 1 million people (20 percent of the population) signed a petition opposing the 1,200 MW reactor being built by China at Daya Bay, only 50 km from the center of this densely populated city. Hong Kong is at the end of a narrow, rocky peninsula, so evacuation would be impossible in case of a nuclear emergency.

Even the Soviet Union has experienced public debate and demonstrations against nuclear power. In 1988, work was halted at nuclear plants under construction at Chigirin in the Ukraine and Krasnodar in southwest Russia (figure 19.19) because of public protests. More than 25 million rubles ($42 million) already had been spent at Krasnodar. Nobel prize winner Andrei Sakharov (chief designer of the Soviet hydrogen bomb) joined the debate, suggesting that all new nuclear reactors be built underground for safety.

FIGURE 19.17 The Seabrook nuclear reactor in New Hampshire has been the site of protests and counter-protests for twenty years. Here pipefitters march in support of the plant and their jobs.

FIGURE 19.18 Riot police fire tear gas to disperse protesters of SuperPhenix, the world's first large-scale fast breeder reactor, near Lyons, France. Many have been injured and one person was killed in confrontations over this plant.

FIGURE 19.19 Public concern about the safety of nuclear power stopped construction at plants in Chigirin in the Ukraine and Krasnodar in Russia.

Results of polls on nuclear power in selected countries before and after Chernobyl are summarized in table 19.5 and figure 19.20. Opposition ranged from 52 percent in France to 83 percent in England and West Germany after the Soviet accident. In the United States, public support for nuclear power fell from 64 percent in 1975 to 19 percent in 1986. Of the 81 percent who now oppose nuclear plants in the United States, half think that all plants should be shut down immediately, while the other half think existing plants should be phased out.

A few countries remain strongly committed to nuclear power. France generates 60 percent of its electricity with nuclear plants and plans to build more. The United Kingdom and Japan both depend on nuclear power for about 20 percent of their electricity. Many other countries, however, have decided to postpone building reactors, to phase out already built reactors, or simply to abandon reactors that have been built but not yet been put into service. In the Philippines, President Corazon Aquino decided to dismantle the billion dollar Baatan reactor (the only one in the country) before it was ever used. Austria has done the same thing. Finland and Sweden voted to phase out all nuclear energy by 2010. In Spain, five plants under construction have been mothballed or canceled. In 1990, Germany announced that it would close and dismantle five Soviet-built RBMK reactors in the former People's Democratic Republic (East Germany). Australia, Luxembourg, New Zealand, Denmark, Italy, and Greece have adopted "no-nuke" policies. Altogether, about half the nations with nuclear power have chosen to reduce or dismantle existing plants, and few new plants have been ordered in recent years. Opponents of nuclear power claim that the nuclear fission experiment may have been a very brief episode in our history, but one whose legacy will remain with us for a long time. Proponents of nuclear power argue that it may be our only near-term alternative to fossil fuels. What do you think?

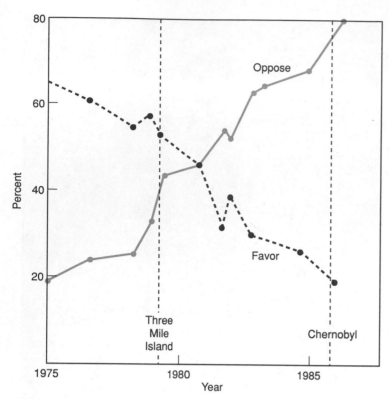

FIGURE 19.20 United States public opinion on building more nuclear power plants, 1975 to 1986.

TABLE 19.5	Public opposition in selected countries to building additional nuclear power plants	

Country	Before Chernobyl	After Chernobyl
United Kingdom	65%	83%
West Germany	46	83
Italy	n.a.	79
United States	67	78
Yugoslavia	40	74
Canada	60	70
Finland	33	64
France	n.a.	52

Source: Data from *Worldwatch Report #75*, 1987.

n.a. = no data available

NUCLEAR FUSION: A VISION OF THE FUTURE OR JUST MORE PROBLEMS?

Fusion energy is an alternative to fission energy that many scientists believe has virtually limitless potential. **Nuclear fusion** energy is the energy released when two smaller atomic nuclei fuse into one larger nucleus. Nuclear fusion reactions, the energy source for the sun and for hydrogen bombs, have not yet been harnessed by humans to produce useful net energy. The fuels for these reactions are deuterium and tritium, the heavy isotopes of hydrogen that are so plentiful in seawater that fusion often is cited as a possible ultimate energy source for the future.

It has been known for forty years that if temperatures in an appropriate fuel mixture are raised to 100 million degrees Celsius and pressures of several billion atmospheres are obtained, fusion of deuterium and tritium will occur. Under these conditions, the electrons are stripped away from atoms and the forces that normally keep nuclei apart are overcome. As nuclei fuse, some of their mass is converted into energy, which takes the form of kinetic energy of the reaction products. The reaction is

heat and
$$H^2 + H^3 \rightarrow \text{alpha particle (helium nucleus)} + \text{neutron}$$
pressure

There are two main schemes for creating these conditions: magnetic confinement and inertial confinement.

Magnetic confinement involves the containment and condensation of **plasma**, a hot, electrically neutral gas of ions and free electrons, in a powerful magnetic field inside a vacuum chamber. Compression of the plasma by the magnetic field should raise temperatures and pressures enough for fusion to occur. The most promising example of this approach, so far, has been a Russian design called *tokomak* (after the Russian initials for "tordial magnetic chamber"), in which the vacuum chamber is shaped like a large donut (figure 19.21).

Inertial confinement involves a small pellet (or a series of small pellets) bombarded from all sides at once with extremely high-intensity laser light. The sudden absorption of energy causes an implosion (an inward collapse of the material) that will increase densities by one thousand to two thousand times and raise temperatures above the critical minimum. These conditions will be maintained for a fraction of a second before the pellet flies apart in a miniature hydrogen explosion, but there should be enough reaction to release some harnessable energy. So far, no lasers powerful enough to create fusion conditions have been built, but researchers are hopeful that this may be achieved in a few years.

In both of these cases, high-energy neutrons escape from the reaction and are absorbed by molten lithium circulating in the walls of the reactor vessel. The lithium absorbs the neutrons and transfers heat to water via a heat exchanger, making steam that drives a turbine generator, as in any steam power plant.

The advantages of fusion reactions, if they are obtained, include production of fewer radioactive wastes, the elimination of fissionable products that could be made into bombs, and a fuel supply that is much larger and less hazardous than uranium. Even with breeder reactors that create fissionable isotopes (such

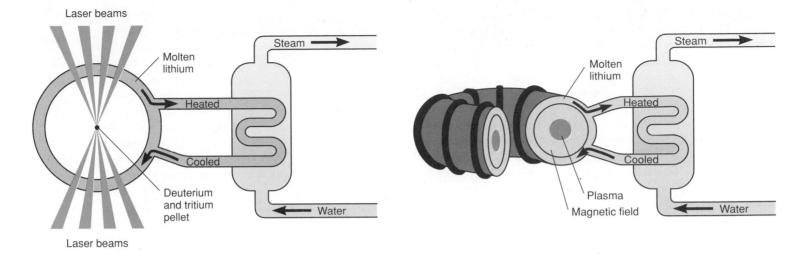

a.

b.

FIGURE 19.21 Nuclear fusion devices. (*a*) Inertial confinement is created by laser beams that bombard and ignite fuel pellets. Molten lithium absorbs energy from the fusion and transfers it as heat to a steam generator. (*b*) A powerful magnetic field confines the plasma and compresses it so that critical temperatures and pressures are reached. Again, molten lithium absorbs energy and transfers it to a steam generator.

as plutonium^{239}) from stable atoms (such as uranium^{238}) uranium and thorium supplies would last for approximately one hundred years, while hydrogen isotopes are plentiful enough to last for thousands of years. There would be some radioactivity and radioactive waste from fusion because the high neutron flux would irradiate the reactor vessel and create radionucleides, but nothing on the order of fission by-products.

After forty years and $20 billion of research, plasmas have been created with temperatures above 230 million degrees for a fraction of a second. Laser beams have achieved power densities of 10^{13} watts per sq cm, about one-tenth of what is needed. Proponents of fusion power predict that within ten years we will pass the break-even point at which a fusion reaction can be sustained long enough to produce more energy than it takes to start it. The discovery of new, high-temperature superconducting materials may greatly enhance progress in this field.

In 1989, scientists from the University of Utah announced they had achieved nuclear fusion reactions at room temperature using a palladium metal electrode immersed in "heavy" water containing deuterium rather than ordinary hydrogen. A great deal of excitement and media hoopla ensued. After a year of unsuccessful attempts to repeat or explain this phenomenon, most scientists concluded that it was a false hope and not worth pursuing.

One approach to cold fusion that might have some potential is muion-catalyzed fusion. Muions, subatomic particles that can be negatively charged, are much heavier than electrons and orbit closer to the atomic nucleus. Muions shared between two atoms can hold their nuclei close enough together to allow fusion to occur. Muions are expensive to produce and very short lived,

but recent advances in nuclear technology give hope that muion-catalyzed reactions might be feasible in the future.

Opponents of nuclear power, whether fission or fusion, point out that all the power we need is available from less expensive, less risky, better known, decentralized sources, such as biomass, hydropower, solar, wind, and geothermal energy. These alternatives are explored in chapter 20.

S U M M A R Y

Nuclear energy offers an alternative to many of the environmental and social costs of fossil fuels, but it introduces serious problems of its own. In the 1950s, there was great hope that these problems would be overcome and that nuclear power plants would provide energy "too cheap to meter." Recently, however, much of that optimism has been waning. Of the 140 new reactors ordered in the United States in the early 1970s, 100 have been canceled. Many countries are closing down existing nuclear power plants, and a growing number have pledged to remain or become "nuclear-free."

The greatest worry about nuclear power is the danger of accidents that might release the extremely hazardous radioactive materials produced in the nuclear fission reactions into the environment. Several accidents, most notably the "meltdown" at the Chernobyl plant in the Soviet Ukraine in 1986, have shown that such accidents are possible, even in the presence of elaborate safety controls. Proponents of nuclear power argue that accidents have not been common, in spite of nearly four thousand cumulative years of operating experience, worldwide. Furthermore, in the history of civilian nuclear power, only forty people have been killed (all at Chernobyl) directly by nuclear accidents. Opponents point out that the large number of nuclear power plants make it almost inevitable that more accidents will occur. The great danger represented by even small amounts of radioactivity means that many thousands of people could die from radiation exposure, cancer, and genetic defects as a result of a major nuclear power accident.

Other major worries about nuclear power include where to put the waste products of the nuclear fuel cycle and how to ensure that it will remain safely contained for the thousands of years required for "decay" of the radioisotopes to nonhazardous levels. Yucca Mountain, Nevada, was chosen for a high-level waste repository, but many experts believe that burying these toxic residues in nonretrievable storage is a mistake. There are serious concerns about evacuating people who live near nuclear plants in case of emergencies, the potential for terrorist attacks on nuclear plants, and the possibility of nuclear fuels being diverted for weapons production. How much liability utility companies and others should carry for nuclear accidents is another matter of debate.

Nuclear fusion occurs when atoms are forced together in order to create new, more massive atoms. These reactions provide the power of the sun and of hydrogen bombs. If fusion actions could be created under controlled conditions in a power plant, they might provide an essentially unlimited source of energy that would avoid many of the worst problems of both fossil fuels and fission-based nuclear power. So far, however, no one has been able to sustain controlled fusion reactions that produce more energy than they consume. It remains to be seen whether this will ever be possible, and whether other unforeseen problems will arise if it does become a reality.

■ Review Questions

1. Describe the process of nuclear fission. How is this reaction used differently in nuclear power plants and atomic bombs?
2. Name some sources of radiation. How harmful are they?
3. Why is radiation harmful?
4. What are the four most common reactor designs? How do they differ from each other?
5. Why are the HTGCR and PIUS reactors inherently safer than those in current use? Why are they not in widespread use?
6. What are the advantages and disadvantages of the breeder reactor?
7. Describe methods proposed for storing and disposing of nuclear wastes.
8. What would be the advantages of nuclear fusion?

■ Questions for Critical Thinking

1. Discuss the advantages and dangers of nuclear power. Are we overly concerned or not concerned enough about nuclear power plant accidents? Why?

2. Should more time, money, and research be spent on improving nuclear power capabilities and safety? If yes, what should be improved, and how could it be accomplished? If no, with what forms of energy should nuclear power be replaced? What are their advantages over nuclear power? What are their disadvantages?
3. Since the environmental impact of a nuclear power plant accident cannot be limited to national boundaries (as in the case of Chernobyl), should there be international regulations for power plants? Who would define these regulations? How could they be enforced?
4. Is it possible to keep nuclear material from falling into the hands of maniacs and terrorists? How?
5. What was the purpose of the Price-Anderson Act? Discuss the liability issue.
6. Discuss the advantages and disadvantages of nuclear waste disposal. What would be some alternatives?
7. Would you rather live next door to a nuclear power plant or a coal-burning plant? On what factors do you base your preference?

■ Key Terms

chain reaction (p. 368)
control rods (p. 371)
fuel assembly (p. 371)
half-life (p. 367)
inertial confinement (p. 386)
ionizing radiation (p. 370)
ions (p. 365)
isotopes (p. 366)
liquid metal fast breeder (p. 378)

magnetic confinement (p. 386)
nuclear fission (p. 368)
nuclear fusion (p. 386)
plasma (p. 386)
radioactive (p. 367)
radionucleides (p. 367)
tailings (p. 371)
X rays (p. 371)
yellowcake (p. 371)

SUGGESTED READINGS

- Beckmann, P. *The Health Hazards of Not Going Nuclear*. Boulder: Golem Press, 1976. A pronuclear viewpoint worth considering.
- Caldicott, H. *Nuclear Madness*. New York: Bantam Books, 1981. A frightening but highly readable discussion of the health risks of nuclear power by a physician and a leading antinuclear activist.
- Craxton, R. S. et al. August 1986. "Progress in Laser Fusion." *Scientific American* 255(2):68. Describes recent advances in fusion-energy research.
- Flavin, C. 1987. "Reassessing Nuclear Power: The Fallout from Chernobyl." *Worldwatch Paper #75*. Washington, D.C.: Worldwatch Institute. A useful analysis of the consequences of the Chernobyl accident on nuclear power.

Ford, D. *Three Mile Island: Thirty Minutes to Meltdown*. New York: Penguin Books, 1983. A description of the worst nuclear power accident in the United States and how it happened.

Gofman, J., and A. R. Tamplin. *Poisoned Power: The Case Against Nuclear Power Plants Before and After Three Mile Island*. Emmaus: Rodale Press, 1979. Two of the leading antinuclear scientists in the U.S. describe the health risks of nuclear power.

Golay, M. W., and N. E. Todreas. 1989. "Advanced Light-Water Reactors." *Scientific American* 262(4): 82. Discusses new passive safety features that can make nuclear energy safer and more attractive.

Goldsmith, E. et al. 1986. "Chernobyl: The End of Nuclear Power?" *The Economist* 16(4/5):138–209. Nuclear power from the European perspective. Useful articles on SuperPhenix, British nuclear power, ocean dumping of nuclear waste, contamination of the Irish Sea, the Windscale fire, and alternatives to nuclear energy.

Hammond, R. P. March/April 1979. "Nuclear Wastes and Public Acceptance." *American Scientist* 67(2):146–50. Describes how monitored containers in a controlled tunnel environment could prove more acceptable than the uncertainties of uncontrolled geological containment.

Hätele, W. 1990. "Energy from Nuclear Power." *Scientific American* 263(3). Nuclear power could supply much needed energy, this author claims, if safe reactors are designed and security and waste storage problems are solved.

Jungle, R. *The New Tyranny*. New York: Warner Books, 1979. Superb discussion of the dangers of nuclear power.

League of Women Voters Education Fund. 1985. *The Nuclear Waste Primer*. Washington, D.C.: League of Women Voters, 1985. An excellent, balanced summary of the options and problems of nuclear waste disposal.

Lester, R. K. March 1986. "Rethinking Nuclear Power." *Scientific American* 254(3):31. A good description of how a new generation of low-power, centrally fabricated nuclear reactors could be designed for inherent safety.

Lovins, A. B., and L. H. Lovins. *Energy/War: Breaking the Nuclear Link*. San Francisco: Friends of the Earth, 1980. Excellent discussion of the relationship between nuclear power and national security.

Manning, R. May 1985. "The Future of Nuclear Power." *Environment* 27(4):12. An interesting discussion of HTGCR and PIUS reactors.

Martin, D. H. "Nuclear Power." In *Introduction to Environment Toxicology*, edited by Frank Guthrie and J. J. Perry. New York: Elsevier, 1980. Pages 77–99. A useful survey of the nuclear fuel cycle and toxicology of nuclear products.

McPhee, J. *The Curve of Binding Energy*. New York: Farrar Straus Giroux, 1974. A fascinating story of Ted Taylor, bomb designer and antiwar activist. Some useful insight on how terrorists could use special nuclear materials. (*See also* "The Atlantic Generating Station," in *Giving Good Weight* by McPhee, 1975.)

Patterson, W. C. *The Plutonium Business and the Spread of the Bomb*. San Francisco: Sierra Club Books, 1984. Shows how nuclear fuel reprocessing creates a plutonium economy and provides fissionable material to countries or other groups that might want to make a bomb.

Pollack, C. 1986. "Decommissioning: Nuclear Power's Missing Link." *Worldwatch Paper #69*. Washington, D.C.: Worldwatch Institute. An excellent discussion of disposal of old nuclear power plants.

Rafelski, J., and S. E. Jones. 1987. "Cold Nuclear Fusion." *Scientific American* 257(1):84–89. An excellent article on muion-catalyzed nuclear fusion.

Weinburg, A. M. *Continuing the Nuclear Dialogue*. La Grange Park: American Nuclear Society, 1985. One of the leading nuclear scientists in the United States argues that we cannot afford to abandon this valuable energy source.

CHAPTER *20*

Sustainable Energy Resources

Each year the United States wastes more fuel than most of mankind uses . . . We could lead lives as rich, healthy, and fulfilling—with as much comfort, and with more employment—using less than half the energy now used.

Denis Hayes

C O N C E P T S

Renewable energy sources, like all energy sources, may have high initiation costs. The advantage of renewable energy is in elimination of economically and environmentally costly fuels, as well as elimination of pollution associated with conventional fuels. Investing in conservation is often the best and least expensive way to make more energy available while minimizing pollution and resource use.

Passive solar heat storage is an ancient and still valuable source of home heating—and cooling. Parabolic mirrors can focus solar energy on a central collector to produce high temperatures for both domestic and industrial uses. Direct solar-to-

electricity conversion is a promising technology that may be cost competitive with fossil fuels in a few years.

Many forms of domestic and industrial wind energy systems are available, some in use for thousands of years. Biomass also offers a huge potential for energy, especially in the forms of methane and ethanol, which burn more thoroughly and are cleaner than wood or dung. Geothermal, tidal, low-head hydropower, and many other developing energy technologies are beginning to contribute to world energy production and are potential sources of sustainable energy.

INTRODUCTION

Nonrenewable energy resources (fossil fuels and nuclear power) provide about 95 percent of the commercial energy used worldwide. Large supplies of fossil fuels remain in the earth, but we are rapidly using the easily retrievable deposits. Economic and environmental costs increase as the ease of resource recovery decreases. Maintaining our technological society consumes so much energy that we soon will run out of affordable supplies unless we radically change our consumption patterns or find some major renewable energy resources.

How serious is the problem? The developed countries of the world have only 20 percent of the total world population but consume two-thirds percent of all commercial energy. If the rest of the world were to use energy at the per capita rate that we do, there wouldn't be enough fossil fuel to support all of us. Furthermore, the environment couldn't withstand the assault. Energy consumption is rising rapidly in developing countries. If present trends continue, per capita consumption in 2020 will be four times that in 1980. Some people argue that the other four-fifths of the world cannot rise to the same standard of living we now enjoy.

The World Resources Institute calculates, however, that with some help from the industrialized countries, the developing world could apply energy-efficient technologies to boost living standards to levels approaching modern Europe by 2020 while increasing energy consumption per capita only 20 percent above the 1980 rate. To accomplish this goal, we need to develop renewable, environmentally benign energy resources as rapidly as possible. For our good and that of the planet, we should use these alternate sources ourselves and also make them available at low cost to the developing countries.

In this chapter, we will look at some energy conservation measures that can reduce energy consumption, as well as the outlook for promising sustainable energy resources, such as biomass, solar, wind, moving water, and geothermal energy (figure 20.1).

CONSERVATION

One of the best ways to avoid energy shortages and to relieve environmental and health effects of our current energy technologies is simply to use less. Conservation offers many benefits both to society and to the environment.

Utilization Efficiencies

Much of the energy we consume is wasted. This statement is not a simple admonishment to turn off lights and turn down furnace thermostats in winter; it is a technological challenge. Our energy conversion technologies are so inefficient that most potential energy in fuel is lost as waste heat, becoming a form of environmental pollution. Of the energy we do extract from pri-

FIGURE 20.1 Windfarm at Altamont Pass, California. This utility-scale use of wind energy has been especially successful in California, but it may spread across the nation in the coming decades.

mary resources, however, much is used for frankly trivial or extravagant purposes. As we saw in chapter 18, several European countries have higher standards of living than the United States, and yet use 30–50 percent less energy.

Many conservation techniques are relatively simple and highly cost effective. More efficient and less energy-intensive industry, transportation, and domestic practices could save large amounts of energy. Improved automobile efficiency, better mass transit, and increased railroad use for passenger and freight traffic are simple and readily available means of conserving transportation energy. In response to the 1970s oil price shocks, automobile gas-mileage averages in the United States rose from 13 mpg in 1975 to 28.8 mpg in 1988. The oil glut and falling fuel prices of the late 1980s discouraged further conservation. The 1991 average shipped only 28.2 mpg.

Much more could be done. Unfortunately, prototype high-efficiency automobiles that get about 100 mpg at highway speeds and over 65 mpg at normal city traffic speeds already have been road tested (figure 20.2). Amory B. Lovins of the Rocky Mountain Institute in Colorado estimated that raising the average fuel efficiency of the United States car and light truck fleet by one mile per gallon would cut oil consumption about 295,000 barrels per *day*. In one year, this would equal the total amount the Interior Department hopes to extract from the Arctic National Refuge in Alaska.

Similar improvements in domestic energy efficiency have occurred in the past decade. Today's average new home uses one-half the fuel required in a house built in 1974, but much more can be done. Figure 20.3 shows major routes of energy loss in a house in a cool climate. All these losses can be reduced by one-half to three-fourths through better insulation, double or triple glazing of windows, thermally efficient curtains or window coverings, and by sealing cracks and loose joints. Reducing air infiltration is usually the cheapest, quickest, and most effective way of saving energy because it is the largest source of losses in a typical house. It doesn't take much skill or investment to caulk

FIGURE 20.2 This experimental automobile with a superefficient turbo-charged diesel engine gets over 100 mpg at a constant 65 mph and 40 mpg in city driving conditions. In addition to standard diesel fuel, it can burn methanol, corn oil, kerosene, and a variety of other nontraditional fuels. Its acceleration, safety, and driving characteristics are comparable to standard autos.

around doors, windows, foundation joints, electrical outlets, and other sources of air leakage. For even greater savings, new houses can be built with extra thick super-insulated walls, air-to-air heat exchangers to warm incoming air, and even double-walled sections that create a "house within a house." Special double-glazed windows that have internal reflective coatings and that are filled with an inert gas (argon or xenon) have an insulation factor of R11, the same as a standard 4-inch thick wall or ten times as efficient as a single pane window. Superinsulated houses now being built in Sweden require 90 percent less energy for heating and cooling than the average American home.

Orienting homes so that living spaces have passive solar gain in the winter and are shaded by trees or roof overhang in the summer also helps conserve energy. Earth-sheltered homes built into the south-facing side of a slope or protected on three sides by an earth berm are exceptionally efficient energy savers because they maintain relatively constant subsurface temperatures (figure 20.4). In addition to building more energy-efficient homes, there are many personal actions we can take to conserve energy (box 20.1).

Industrial energy savings are another important part of our national energy budget. More efficient electric motors and pumps, new sensors and control devices, advanced heat-recovery sysems, and material recycling have reduced industrail energy requirements significantly. In the early 1980s, U.S. businesses saved $160 billion per year through conservation. When oil prices collapsed, however, we returned to wasteful ways.

Energy Conversion Efficiencies

Energy efficiency is a measure of energy produced compared to energy consumed. Table 20.1 shows the typical energy efficiencies of a variety of energy conversion devices. Thermal-conversion machines, such as steam turbines in coal-fired or nuclear power plants, can turn no more than 40 percent of the

TABLE 20.1 Typical net efficiencies of energy conversion devices

	Yield (Percent)
Electric power plants	
Hydroelectric (best case)	90
Combined-cycle steam	90
Fuel cell (hydrogen)	80
Coal-fired generator	38
Oil-burning generator	38
Nuclear generator	30
Photovoltaic generation	10
Transportation	
Pipeline (gas)	90
Pipeline (liquid)	70
Waterway (no current)	65
Diesel-electric train	40
Diesel-engine automobile	35
Gas-engine automobile	30
Jet-engine airplane	10
Space heating	
Electric resistance	99*
High-efficiency gas furnace	90
Typical gas furnace	70
Efficient wood stove	65
Typical wood stove	40
Open fireplace	−10
Lighting	
Sodium vapor light	60*
Fluorescent bulb	25*
Incandescent bulb	5*
Gas flame	1

*Note that 60 to 70% of the energy in the original fuel is lost in electric power generation.

Source: U.S. Department of Energy.

TABLE 20.2 Typical net useful energy yields

Energy source	Yield/Cost ratio
Nonrenewable Sources	
Coal (space or process heat)	20/1
Natural gas (as heat source)	10/1
Gasoline and fuel oil	7/1
Coal gasification (combined cycle)	5/1
Oil shale (as liquid fuel)	1/1
Nuclear (excluding waste disposal)	2/1*
Renewable Sources	
Hydroelectric (best case)	20/1**
Wind (electric generation)	2/1
Biomass methane	2/1
Solar electric (10% efficient)	1/1
Solar electric (20% efficient)	2/1

*Decommissioning of old nuclear plants and perpetual storage of wastes may consume more energy than nuclear power has produced.

**Hydropower yield depends on availability of water and life-expectancy of dam and reservoir. Most dams produce only about 40% of rated capacity due to lack of water. Some dams have failed or silted up without producing any net energy yield.

RESOURCES AND RESOURCE ECONOMICS

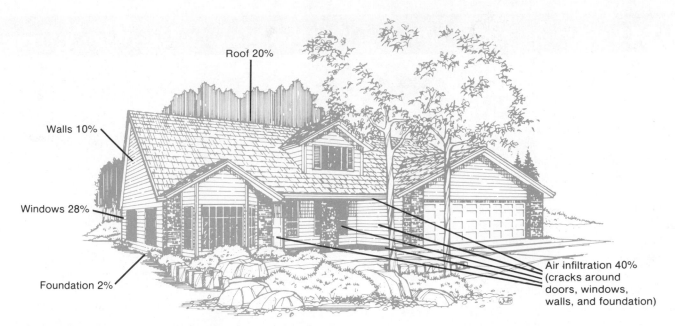

Roof 20%

Walls 10%

Windows 28%

Foundation 2%

Air infiltration 40%
(cracks around
doors, windows,
walls, and foundation)

FIGURE 20.3 Energy losses in a house in a cool climate. All these losses can be reduced by half to two-thirds through better insulation, double glazing, and sealing joints and cracks.

FIGURE 20.4 Earth-sheltered structures retain much of their heat during cold winters because the earth surrounding them remains considerably warmer than the outside air temperature. They likewise remain cool in the hot summer.

energy in their primary fuel into electricity or mechanical power because of the need to reject waste heat. Does this mean that we can never increase the efficiency of fossil fuel use? No. Some waste heat can be recaptured and used for space heating, raising the net yield to 80 or 90 percent. In another kind of process, fuel cells convert the chemical energy of a fuel directly into electricity without an intermediate combustion cycle. Since this process is not limited by waste heat elimination, its efficiencies can approach 80 percent with such fuel as hydrogen gas or methane.

Another way to look at energy efficiency is to consider the total **net energy yield** from energy conversion devices (table 20.2). Net energy yield is based on the total useful energy produced during the lifetime of an entire energy system minus the energy required to make useful energy available. To make comparisons between different energy systems easier, the net energy yield is often expressed as a *ratio* between the output of useful energy and the energy costs for construction, fuel extraction, energy conversion, transmission, waste disposal, etc.

For some people, home energy efficiency means designing and building a new house that incorporates methods and materials that have been proven to conserve energy. Most of us, however, have to live in houses or apartments that already exist. How can we practice home energy conservation?

Think about the major ways we use energy. The biggest factor is space heating. Heat conservation is among the simplest, cheapest, and most effective ways to save energy in the home (box table 20.1). Easy, surprisingly effective, and relatively inexpensive measures include weatherstripping, caulking, and adding layers of plastic to windows. Insulating walls, floors, and ceilings increases the energy conservation potential. Simply lowering your thermostat, especially at night, is a proven energy and money saver.

Water heating is the second major user of home energy, followed by large electric or gas appliances, such as stoves, refrigerators, washing machines, and dryers. Careful use of these appliances can save significant amounts of energy. Consider using less hot water, lowering the thermostat of your water heater, and buying an insulating blanket designed to wrap around it. Be sure that the refrigerator door doesn't stand open, and defrost it regularly (if it is not frostfree) to reduce the ice buildup that prevents the

heat exchange system from working well. Make sure your oven door has a tight seal, and plan ahead when you use it; bake several things at once rather than reheating it repeatedly. Wash your clothes in cold or cool water rather than hot. Air dry your laundry, especially in the summer when sun-dried clothes are so pleasant to wear.

Some energy-efficient appliances can save substantial amounts of energy. Air conditioners and refrigerators with better condensers and heat exchangers use one-half to one-fourth as much energy as older models. New, improved furnaces can offer 95 percent efficiency, compared

with conventional 70 percent efficiency in older models. The pay-back period may be as little as two to three years if you trade an old wasteful appliance for a newer, more efficient one. Some light bulbs are made to save energy (box figure 20.1). High-efficiency fluorescent lights emit the same amount of light but use only one-fourth as much energy as conventional incandescent bulbs. They cost ten times more than an ordinary bulb, but last ten times as long. Total life-time savings can be $30 to $50 per lamp. Almost all types of appliances, large and small, are available in efficient models and brands; we simply need to shop for them.

TABLE BOX 20.1 Energy saving potential

	Model average	Best on market	Best prototype	Saving (percent)
Automobile (mi/gal.)	18	55	107	83
Home (1,000 J/m²)	190	68	11	94
Refrigerator (Kw hr/day)	4	2	1	75
Gas furnace (million J/day)	210	140	110	48
Air conditioner (kw hr/day)	10	5	3	70

Source: U.S. Department of Energy.

Nuclear power is a good case for net energy yield studies. We get a large amount of electricity from a small amount of fuel in a nuclear plant, but it also takes a great deal of energy to extract and process nuclear fuel and to build, operate, and eventually dismantle power plants. As chapter 19 points out, we may never reach a break-even point where we get back more energy from nuclear plants than we put into them, especially considering the energy that may be required to decommission nuclear plants and guard their waste products in secure storage for thousands of years.

Net yields and overall conversion efficiencies are not the only considerations when we compare different energy sources. The yield/cost ratio and conversion-cycle efficiency is much higher for coal burning, for instance, than for photovoltaic elec-

trical production, making coal appear to be a better source of energy than solar radiation. Solar energy, however, is free, renewable, and nonpolluting. If we can use solar energy to get electrical energy, it doesn't matter how *efficient* the process is, as long as we get more out of it than we put in.

Conservation versus New Power Plants

Utility companies are finding it much less expensive to finance conservation projects than to build new power plants. Pacific Gas and Electric in California and Potomac Power and Light in Washington, D.C., both have instituted large conservation programs. They have found that conservation costs about $350 per kilowatt (kW) saved. By contrast, a new nuclear power plant costs between $3,000 and $8,000 per kW of installed capacity.

The easiest way to save energy is to keep an eye on energy-use habits. Turning off lights, televisions, and other devices when they are not in use is elementary. Line drying clothes and hand washing dishes are not difficult. In some cases, the most beneficial result of living a frugal, conservative life is not so much the total amount of energy you save, as the effect that this life-style has on you and the people around you. When you live conscientiously, you set a good example that could have a multiplying effect as it spreads through society. Making conscious ethical decisions about this one area of your life may stimulate you to make other positive decisions. You will feel better about yourself and more optimistic about the future when you make even small, symbolic gestures towards living as a good environmental citizen.

Some Things You Can Do to Save Energy

1. Drive less: make fewer trips, use telecommunications and mail instead of going places in person.
2. Use public transportation, walk, or ride a bicycle.
3. Use stairs instead of elevators.
4. Join a carpool or drive a smaller, more efficient car; reduce speeds.
5. Insulate your house or add more insulation to the existing amount.
6. Turn thermostats down in the winter and up in the summer.
7. Weatherstrip and caulk around windows and doors.
8. Add storm windows or plastic sheets over windows.
9. Create a windbreak on the north side of your house; plant deciduous trees or vines on the south side.
10. During the winter, close windows and drapes at night; during summer days, close windows and drapes if using air conditioning.
11. Turn off lights, television sets, and computers when not in use.
12. Stop faucet leaks, especially hot water.
13. Take shorter, cooler showers; install water-saving faucets and shower heads.
14. Recycle glass, metals, and paper; compost organic wastes.
15. Eat locally grown food in season.
16. Buy locally made, long-lasting materials.

New coal-burning plants with the latest air pollution-control equipment cost at least $1,000 per kW. By investing $200 to $300 million in public education, home improvement loans, and other efficiency measures, a utility can avoid building a new power plant that would cost a billion dollars or more. Furthermore, conservation measures don't consume expensive fuel or produce pollutants.

Can application of this approach help alleviate energy shortages in other countries as well? Yes. For example, Brazil could cut its electricity consumption an estimated 30 percent with an investment of $10 billion (U.S.). It would cost $44 billion to build new power plants to produce that much electricity. South Korea has instituted a comprehensive conservation program with energy-saving building standards, efficiency labels on new household appliances, depreciation allowances, reduced tariffs on energy-conserving equipment, and loans and tax breaks for upgrading homes and businesses. It also forbids some unnecessary uses, such as air conditioning and elevators between the first and third floors.

Cogeneration

One of the fastest growing sources of new energy is **cogeneration,** the simultaneous production of both electricity and steam or hot water in the same plant. By producing two kinds of useful energy in the same facility, the net energy yield from the primary fuel is increased from 30–35 percent to 80–90 percent. In 1900, half the electricity generated in the United States came from plants that also provided industrial steam or district heating. As power plants became larger, dirtier, and less acceptable as neighbors, they were forced to move away from their customers. Waste heat from the turbine generators became an unwanted pollutant to be disposed of in the environment. Furthermore, long transmission lines, which are unsightly and lose up to 20 percent of the electricity they carry, became necessary.

By the 1970s, cogeneration had fallen to less than 5 percent of our power supplies, but interest in this technology is being renewed. The capacity for cogeneration has more than doubled in the 1980s to about 30,000 megawatts (MW). District heating systems are being rejuvenated, and plants that burn municipal wastes are being studied. New combined-cycle coal-gasification plants (chapter 18) or "mini-nukes" (chapter 19) offer high efficiency and clean operation that may be compatible with urban locations. Small neighborhood- or apartment building-sized power-generating units are being built that burn methane (from biomass digestion), natural gas, diesel fuel, or coal. The Fiat Motor Company makes a small generator for about $10,000 that produces enough electricity and heat for four or five houses.

TAPPING SOLAR ENERGY

The sun serves as a giant nuclear furnace in space, constantly bathing our planet with a free energy supply. Solar heat drives winds and the hydrologic cycle. All biomass, as well as fossil fuels and our food (both of which are derived from biomass), results from conversion of light energy (photons) into chemical bond energy by photosynthetic bacteria, algae, and plants.

A Vast Resource

The average amount of solar energy arriving at the top of the atmosphere is 340 watts per square meter. About half of this energy is absorbed or reflected by the atmosphere (more at high latitudes than at the equator), but the amount reaching the Earth's surface is some 8,000 times all the commercial energy used each year. However, this tremendous infusion of energy comes in a form that, until this century, has been too diffuse and low in intensity to be used except for environmental heating and photosynthesis. Figure 20.5 shows solar energy levels over the United States for a typical summer and winter day.

Passive Solar Heat Collectors

Our simplest and oldest use of solar energy is **passive heat absorption,** using natural materials or absorptive structures with no moving parts to simply gather and hold heat. For thousands of years, people have built thick-walled stone and adobe dwellings that slowly collect heat during the day and gradually release that heat at night. After cooling at night, these massive building materials maintain a comfortable daytime temperature within the house, even as they absorb external warmth.

A modern adaptation of this principle is a glass-walled "sunspace" or greenhouse on the south side of a building. In-

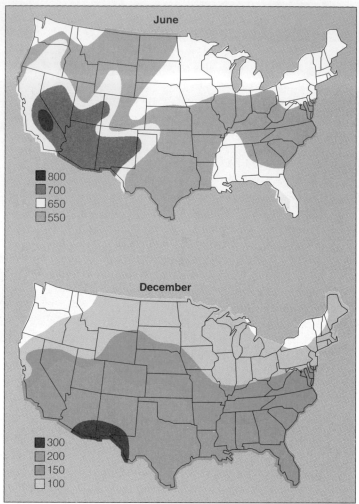

FIGURE 20.5 Average daily solar radiation in the United States in June and December. One langley, the unit for solar radiation, equals 1 calorie per square centimeter of Earth surface (3.69 Btu/square foot).

Source: National Weather Bureau, U.S. Department of Commerce.

corporating massive energy-storing materials, such as brick walls, stone floors, or barrels of heat-absorbing water into buildings also collects heat to be released slowly at night (figure 20.6). An interior, heat-absorbing wall called a trombe wall is an effective passive heat collector. Some trombe walls are built of glass blocks enclosing a water-filled space or water-filled circulation tubes so heat from solar rays can be absorbed and stored, while light passes through to inside rooms.

Active Solar Heat Systems

Active solar systems generally pump a heat-absorbing, fluid medium (air, water, or an antifreeze solution) through a relatively small collector, rather than passively collecting heat in a stationary medium like masonry. Active collectors can be located adjacent to or on top of buildings rather than being built into the structure. Because they are relatively small and structurally independent, active systems can be retrofitted to existing buildings.

South

FIGURE 20.6 Stone floors or barrels of water are simple, heat-absorbing devices. Insulated shutters keep heat inside at night.

A flat black surface sealed with a double layer of glass makes a good solar collector. A fan circulates air over the hot surface and into the house through ductwork of the type used in standard forced-air heating. Alternatively, water can be pumped through the collector to pick up heat for space heating or to provide hot water (figure 20.7). Water heating consumes 15 percent of the United States' domestic energy budget, so savings in this area alone can be significant. A simple flat panel with about 5 sq m of surface can reach 95°C (200°F) and can provide enough hot water for an average family of four almost anywhere in the United States. In California, 650,000 homes now heat water with solar collectors. In Greece, Italy, Israel, Asia, and Africa where fuels are more expensive, up to 70 percent of domestic hot water comes from solar collectors. Figure 20.8 shows a large solar water heater used by a restaurant in Santa Fe, New Mexico.

Sunshine doesn't reach us all the time, of course. How can solar energy be stored for times when it is needed? There are a number of options. In a climate where sunless days are rare and seasonal variations are small, a small, insulated water tank is a good solar energy storage system. For areas where clouds block the sun for days at a time or where energy must be stored for winter use, a large, insulated bin containing a heat-storing mass, such as stone, water, or clay, provides solar energy storage (figure 20.9). During the summer months, a fan blows the heated air from the collector into the storage medium. In the winter, a similar fan at the opposite end of the bin blows the warm air into the house. During the summer, the storage mass is cooler than the outside air, and it helps cool the house by absorbing heat. During the winter, it is warmer and acts as a heat source by radiating stored heat. Six or seven months' worth of thermal energy can be stored in 10,000 gallons of water or 40 tons of gravel, about the amount of water in a very small swimming pool or the gravel in two average-sized dump trucks.

Another heat storage system uses **eutectic** (phase-changing) **chemicals** to store a large amount of energy in a small volume. Heating melts these chemicals and cooling returns them to a solid state. This is an important way the hydrologic cycle stores solar energy in the natural environment, converting water from its solid phase to its liquid and gaseous phases. Most eutectic chemical systems (such as salts) do not swell when they solidify (as does water) and undergo phase changes at higher temperatures than water and ice, making them more convenient for storing heat. Since the latent heat of crystallization is usually much greater than the specific heat required to raise temperature, a eutectic system can store more energy in a smaller volume than can a simple, one-phase system.

Heat also can be captured in chemical reactions in the storage material. Some clays, for instance, undergo reactions that release or store heat when they absorb or release water. An insulated bin filled with a clay, such as bentonite, can be baked dry (absorbing heat) by blowing warm air into it in the summer. Sprinkling water on the clay to rewet it will release heat in winter.

High-Temperature Solar Energy Collection

Parabolic mirrors are curved mirrors that focus light from a large area onto a single, central point. Use of parabolic reflectors to focus intense heat on a central tube containing air, water, sodium, or antifreeze produces a much higher quality (high temperature) heat than does the basic flat panel collector. Temperatures in the collection medium can reach 500°C (1,000°F). The "Power Tower" in Barstow, California, (figure 20.10) focuses the reflected light from several thousand mirrors on a steam generator that drives turbines and produces 10 MW, enough for three thousand homes. Ideally, such collectors should reach 100 MW per 0.5 km² of reflectors.

The reliability and durability of large-scale active solar projects are issues of concern. Solar-powered electric motors must keep the mirrors properly aligned as the sun moves across the sky. We don't have enough experience with solar collectors to know what their life expectancy will be, but breakdowns and malfunctions have not presented a significant problem at the California station in more than ten years of operation. Another concern is that these projects occupy large land areas. If the entire present United States electrical output came from central tower solar steam generators, 2,000 km² (780 mi²) of collectors would be needed. This is less land, however, than would be strip-mined in a thirty-year period if all our energy came from coal or uranium. And we can put solar collectors wherever we choose (such as lands unsuited for agriculture, grazing, or habitation), whereas strip-mining occurs wherever coal or uranium exist, regardless of other values associated with the land.

Solar Cookers

Parabolic mirrors have been tested for home cooking in tropical countries where sunshine is plentiful and other fuels are scarce. They produce such high temperatures and intense light that they are dangerous, however. A much cheaper, simpler, and safer

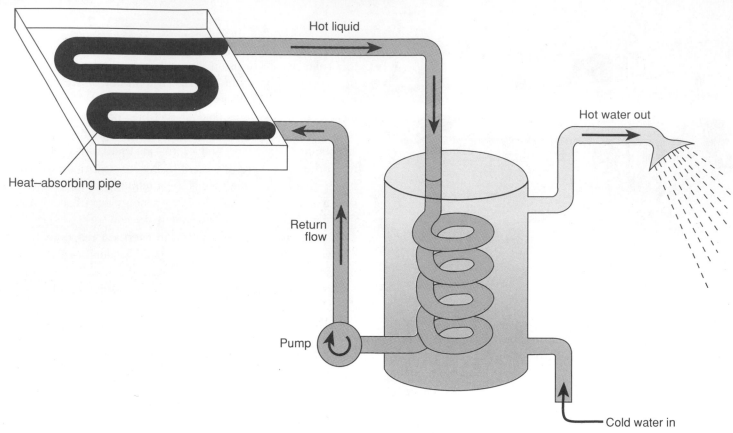

FIGURE 20.7 An active, closed-cycle solar water heating system. A cycling liquid, such as sodium or antifreeze, can reach hundreds of degrees Celsius in a parabolic reflector. Alternatively, a flat black panel enclosed with three layers of glass can absorb heat for the collecting pipes.

FIGURE 20.8 Active solar panels on a restaurant in Santa Fe, New Mexico, heat water for dish washing and space heating. Clean, cheap solar power could be an important part of the answer to our pollution and energy supply problems.

alternative is the solar box cooker (figure 20.11). An insulated box costing only a few dollars with a black interior and a glass lid serves as a passive solar collector. Several pots can be placed inside at the same time. Temperatures only reach about 120°C (250° F) so cooking takes longer than an ordinary oven. Fuel is free, however, and the family saves hours each day usually spent hunting for firewood or dung. These solar ovens help reduce tropical forest destruction and reduce the adverse health effects of smoky cooking fires.

Photovoltaic Solar Energy Conversion

The photovoltaic cell offers an exciting potential for capturing solar energy in a way that will provide clean, versatile, renewable energy. This simple device has no moving parts, negligible maintenance costs, produces no pollution, and has a lifetime equal to that of a conventional fossil fuel or nuclear power plant.

Photovoltaic cells capture solar energy and convert it directly to electrical current by separating electrons from their parent atoms and accelerating them across a one-way electrostatic barrier formed by the junction between two different types of semiconductor material (figure 20.12). The photovoltaic effect, which is the basis of these devices, was first observed in 1839 by French physicist Alexandre Edmond Becquerel, who also discovered radioactivity. He showed that shining a light on one of

RESOURCES AND RESOURCE ECONOMICS

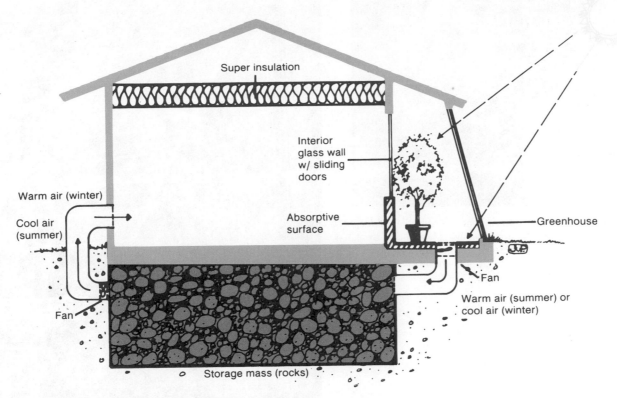

FIGURE 20.9 Underground massive heat storage unit. Heated air collected behind double- or triple-glazed windows is pumped down into a storage medium of rock, water, clay, or similar material, where it can be stored for a number of months.

a pair of plates immersed in a dilute acid solution increased the amount of electric current running through an attached circuit. His discovery didn't lead to any practical applications until 1954, when researchers at Bell Laboratories in New Jersey learned how to carefully introduce impurities into single crystals of silicon. They sliced these crystals into thin wafers that could be fashioned into cells capable of producing a steady supply of electrical current when exposed to sunlight.

These handcrafted single-crystal cells were much too expensive for any practical use until the advent of the United States space program. In 1958, when Vanguard I went into orbit, its radio was powered by six palm-sized photovoltaic cells that cost $2,000 per peak watt of output. This was more than two thousand times as much as conventional energy at the time, but it was the only way to provide a renewable source of energy in outer space. The experience gained by building photovoltaic cells for spacecraft and other specialized uses has reduced the price by a factor of ten each decade. In 1970, they cost $100 per watt; in 1990 they were about $5 per watt. This makes solar energy cost-competitive with other sources in some areas (figure 20.13).

By the mid-1990s, photovoltaic cells could be less than $1 per watt of generating capacity. With 15 percent efficiency and thirty-year life, they should be able to produce electricity for around 8¢ per kilowatt hour. At that time, coal-fired steam power

FIGURE 20.10 The Solar I facility at Barstow, California. Several similar projects have been built in the United States and France, each using hundreds of mirrors to focus the sun's energy on a central collector. This central collector reaches extremely high temperatures, which are useful for electricity production or industrial applications.

FIGURE 20.11 A simple box of cardboard and foil can help reduce tropical deforestation, improve women's lives, and avoid health risks from smoky fires in Third World countries. These inexpensive solar cookers could revolutionize energy use in developing tropical countries.

FIGURE 20.13 Photovoltaic cells provide electricity for this filling station. Recent improvements in this technology make it cost competitive with fossil fuels and nuclear power in many areas.

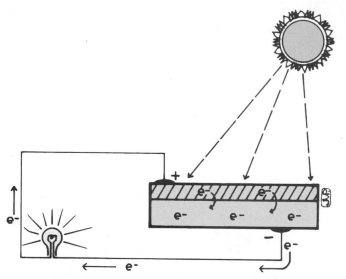

FIGURE 20.12 The operation of a photovoltaic cell. Boron impurities incorporated into the upper silicon crystal layers cause electrons (e-) to be released when solar radiation hits the cell. The released electrons move into the lower layer of the cell, thus creating a shortage of electrons, or a positive charge, in the upper layer and an oversupply of electrons, or negative charge, in the lower layer. The difference in charge creates an electric current in a wire connecting the two layers.

will probably cost about four times as much and nuclear power will likely cost ten times as much as photovoltaic cell energy.

During the last twenty-five years, the efficiency of energy capture by photovoltaic cells has increased from less than 1 percent of incident light to more than 10 percent under field conditions and over 30 percent in the laboratory. Promising experiments are underway using exotic metal alloys, such as gallium arsenide, cadmium telluride, and silicon germanium, which are more efficient in energy conversion than silicon crystals. In 1987, an experimental car built by General Motors won a solar-

powered 3,140 km (1,950 mi) race across Australia. Using gallium arsenide photovoltaic cells, superefficient electric motors, and a highly aerodynamic shape, the Sunraycer averaged nearly 80 km/hr (50 mph) and beat its nearest rival by almost twenty-four hours (figure 20.14).

One of the most promising developments in photovoltaic cell technology in recent years is the invention of amorphous silicon collectors. First described in 1968 by Stanford Ovshinky, a self-taught inventor from Detroit, these noncrystalline silicon semiconductors can be made into lightweight, paper-thin sheets that require much less material than conventional photovoltaic cells. They also are vastly cheaper to manufacture and can be made in a variety of shapes and sizes, permitting ingenious applications. Roof tiles with photovoltaic collectors layered on their surface already are available. Even flexible films can be coated with amorphous silicon collectors. Silicon collectors already are providing power to places where conventional power is unavail-

RESOURCES AND RESOURCE ECONOMICS

FIGURE 20.14 GM Sunraycer. Using gallium arsenide photovoltaic cells and superefficient electric motors, this experimental car raced 3,140 km (1,950 mi) across Australia at an average speed of nearly 80 km/hr.

able, such as lighthouses, mountain-top microwave repeater stations, villages on remote islands, and ranches in the Australian outback.

You probably already use amorphous silicon photovoltaic cells. They are being built into light-powered calculators, watches, toys, photosensitive switches, and a variety of other consumer products. Japanese electronic companies presently lead in this field, having foreseen the opportunity for developing a market for photovoltaic cells. This market is already more than $100 million per year. Within a few years, Japanese companies expect to have home-roof arrays capable of providing all the electricity needed for a typical home at prices competitive with power purchased from a utility. By the end of this century, Japan plans to meet a considerable part of its energy requirements with solar power, an important goal for a country lacking fossil and nuclear fuel resources. In 1990, the Sanyo Company sponsored the flight of a solar-powered airplane from San Diego, California, to near Kitty Hawk, North Carolina, where the Wright brothers made their first flight. The purpose was to announce a variety of products they intended to market utilizing amorphous silica, thin-film photovoltaic cells.

In 1988, Arco Solar and the Solar Energy Research Institute announced production of a multilayer, thin-film solar cell made of copper, indium, and diselenide that is more than 11 percent efficient. It seems feasible to mass produce these CIS films for about $50 per sq m and boost the efficiency to 15 percent, the point at which they will be competitive with other forms of power. An added benefit is that the CIS films do not degrade in sunlight as silicon cells do.

Think about how solar power could affect your future energy independence. Imagine the benefits of being able to build a house anywhere and having a cheap, reliable, clean, quiet source of energy with no moving parts to wear out, no fuel to purchase, and little equipment to maintain. You could have all the energy you need without commercial utility wires or monthly energy bills. Coupled with modern telecommunications and in-

formation technology, an independent energy source would make it possible to live in the countryside and yet have many of the employment and entertainment opportunities and modern conveniences available in a metropolitan area.

The possibility that we might all someday become independent of the commercial power grid worries large utility companies, and this concern has political and technological ramifications. Solar research in the United States has taken the opposite direction of the individualized, small-scale Japanese approach. Almost all the emphasis in the United States has been on building large-scale arrays suitable for operation by centralized utility companies, using their extensive power-grid distribution systems. In the late 1970s, the U.S. Department of Energy provided funds for nine solar-collection plants ranging in size from 10 kW to 100 kW. The largest photovoltaic solar plant now operating is the 7.2 MW Carrisa Plains facility in California, built by ARCO Solar and Pacific Gas and Electric. The European community is developing some twenty photovoltaic demonstration plants. Saudi Arabia, Israel, India, and several other countries also are building large-scale demonstration projects.

Storing Electrical Energy

Electrical energy is difficult and expensive to store. This is a problem for photovoltaic generation as well as other sources of electric power. Traditional lead-acid batteries are heavy and have low energy densities; that is, they can store only moderate amounts of energy per unit mass or volume. Lead from smelters and battery manufacturing is a serious health hazard in some areas, and discarded batteries contaminate the environment with both lead and caustic acid. A typical lead-acid battery array sufficient to store several days of electricity for an average home would cost about $5,000 and weigh 3 or 4 tons. All the components for an electric car are readily available except a cheap, light-weight battery. Lead acid batteries weigh more than the car itself and could only go 50 or 60 miles between recharges.

Other types of batteries also have drawbacks. Metal-gas batteries, such as the zinc-chloride cell, use inexpensive materials and have relatively high energy densities, but have shorter lives than other types. Sodium-sulfur batteries have considerable potential for large-scale storage. They store twice as much energy in half as much weight as lead-acid batteries. They require an operating temperature of about 300° C (572° F) and are expensive to manufacture. Alkali-metal batteries have a high storage capacity but are even more expensive. Lithium batteries have very long lives and store more energy than other types, but are the most expensive.

Another strategy is to store energy in a form that can be turned back into electricity when needed. Pumped-hydro storage involves pumping water to an elevated reservoir at times when excess electricity is available. The water is released to flow back down through turbine generators when extra energy is needed. Using a similar principle, pressurized air can be pumped into such reservoirs as natural caves, depleted oil and gas fields, abandoned mines, or special tanks. The electric motors that spin turbines to pressurize the air can run in reverse and become generators when energy is needed.

An even better way to use surplus electricity is in electrolytic decomposition of water to H_2 and O_2. These gases can be liquefied (like natural gas) at $-252°$ C ($-423°$ F), making them easier to store and ship than most forms of energy. They are highly explosive, however, and must be handled with great care. They can be burned in internal combustion engines, producing mechanical energy, or they can be used as fuel in fuel cells to produce more electrical energy. Liquid hydrogen cars are already being produced. They could be very attractive in smoggy cities because they produce no carbon monoxide, smog-forming hydrocarbons, carcinogenic chemicals, or soot.

Flywheels are the subject of current experimentation for energy storage. Massive, high-speed flywheels, spinning in a nearly friction-free environment, store large amounts of mechanical energy in a small area. This energy is convertible to electrical energy. It is difficult, however, to find materials strong enough to hold together when spinning at high speed. Flywheels have a disconcerting tendency to fail explosively and unexpectedly, sending shrapnel flying in all directions.

New high-temperature superconducting ceramics may offer yet another way to store electricity. Giant superconducting underground rings might be able to store massive amounts of energy as electricity circulates without resistance through this marvelous material.

ENERGY FROM BIOMASS

Photosynthetic organisms have been collecting and storing the sun's energy for more than two billion years. Plants capture about 0.1 percent of all solar energy that reaches Earth's surface. That kinetic energy is transformed, via photosynthesis, into chemical bonds in organic molecules (chapter 3). A little more than half of the energy that plants collect is spent in such metabolic activities as pumping water and ions, mechanical movement, maintenance of cells and tissues, and reproduction; the rest is stored in biomass.

The magnitude of this resource is difficult to measure. Most experts estimate useful biomass production at fifteen to twenty times the amount we currently get from all commercial energy sources. It would be ridiculous to consider consuming all green plants as fuel, but biomass has the potential to become a prime source of energy. It has many advantages over nuclear and fossil fuels because of its renewability and easy accessibility. Renewable energy resources account for about 18 percent of total world energy use, and biomass makes up three-quarters of that renewable energy supply. Biomass resources used as fuel include wood, wood chips, bark, branches, leaves, starchy roots, and other plant and animal materials.

Burning Biomass

Wood fires have been a primary source of heating and cooking for thousands of years. As recently as 1850, wood supplied 90 percent of the fuel used in the United States (chapter 18). Wood now provides less than 1 percent of the energy in the United States, but in many of the poorer countries of the world, wood and other biomass fuels provide up to 95 percent of all energy used. The 1,500 million cubic meters of fuelwood collected in the world each year is about half of all wood harvested.

In northern industrialized countries, wood burning has increased since 1975 in an effort to avoid rising oil, coal, and gas prices. Most of these northern areas have adequate wood supplies to meet demands at current levels, but problems associated with wood burning may limit further expansion in this use. Inefficient and incomplete burning of wood in open fireplaces and stoves produces smoke laden with fine ash and soot and hazardous amounts of carbon monoxide (CO) and hydrocarbons. In valleys where inversion layers trap air pollutants, the effluent from wood fires can present a major source of air quality degradation and health risk (chapter thirteen). Polycyclic aromatic compounds produced by burning are especially worrisome because they are carcinogenic (cancer-causing).

In Oregon's Willamette Valley or in the Colorado Rockies, where wood stoves are popular and topography concentrates contaminants, as much as 80 percent of air pollution on winter days is attributable to wood fires. Several resort towns, such as Vail, Aspen, and Telluride, Colorado, have banned installation of new wood stoves and fireplaces because of high pollution levels. Oregon, Colorado, and Vermont now have emission standards for new wood stoves. The Environmental Protection Agency ranks wood burners high on a list of health risks to the general population, and standards are being considered to regulate the use of wood stoves nationwide.

Highly efficient and clean-burning woodstoves are available but expensive. Running exhaust gases through heat exchangers recaptures heat that would escape through the chimney, but if flue temperatures drop too low, flammable, tarlike creosote deposits can build up, increasing fire risk. A better approach is to design combustion chambers that use fuel efficiently, capturing heat without producing waste products. Brick-lined fireboxes with after-burner chambers to burn gaseous hydrocarbons do an excellent job. Massive, honeycombed Scandinavian and Slavic stoves are appearing again. These traditional stone or ceramic units burn cleanly and hold heat for hours, producing little ash and emitting low levels of air pollution. The small ash residue results from extremely efficient burning; thus, fewer logs are needed each winter to heat a house. Catalytic combusters, similar to the catalytic converters on cars, also are placed inside stove pipes. They burn carbon monoxide and hydrocarbons to clean emissions and to recapture heat that otherwise would have escaped out the chimney in the chemical bonds of these molecules.

FIGURE 20.15 Harvesting marsh reeds (*Phragmities sp.*) in Sweden as a source of biomass fuel. In some places, biomass from wood chips, animal manure, food processing wastes, peat, marsh plants, shrubs, and other kinds of organic material make a valuable contribution to energy supplies. Care must be taken, however, to avoid environmental damage in sensitive areas.

FIGURE 20.16 The firewood shortage in less developed countries means that women and children must spend hours each day searching for fuel. Destruction of forests and removal of ground cover result in erosion and desertification.

Wood chips, sawdust, wood residue, and other plant materials (figure 20.15) are being used in some places in the United States and Europe as a substitute for coal and oil in industrial boilers. In Vermont, for instance, where fossil fuels are expensive and 76 percent of the land is covered by forest, 250,000 cords of unmarketable cull wood are burned annually to fuel a 50 MW power plant in Burlington. Michigan's Public Service Commission estimated that the state's surplus forest growth could provide 1,300 MW of electricity each year. Pollution-control equipment is easier to install and maintain in a central power plant than in individual home units. Wood burning also contributes less to acid precipitation than does coal. Because wood has little sulfur, it produces few sulfur gases and burns at lower temperatures than coal; thus, it produces fewer nitrogen oxides. Burning wood as a renewable crop doesn't produce any net increase in atmospheric carbon dioxide (and, therefore, doesn't add to global warming) because all the carbon released by burning biomass was taken up from the atmosphere when the biomass was grown.

Fuelwood Crisis in Less Developed Countries

Two billion people—about 40 percent of the total world population—depend on firewood and charcoal as their primary energy source. Of these people, three-quarters (1.5 billion) do not have an adequate, affordable supply. Most of these people are in the less developed countries where they face a daily struggle to find enough fuel to warm their homes and cook their food. The problem is intensifying because rapidly growing populations in many Third World countries create increasing demands for firewood and charcoal from a diminishing supply.

As firewood becomes increasingly scarce, women and children, who do most of the domestic labor in many cultures, spend more and more hours searching for fuel (figure 20.16). In some places, it now takes eight hours or more just to walk to the nearest fuelwood supply and even longer to walk back with a load of sticks and branches that will only last a few days.

For people who live in cities, the opportunity to scavenge firewood is generally nonexistent and fuel must be bought from merchants. This can be ruinously expensive. In Addis Ababa, Ethiopia, 25 percent of household income is spent on wood for cooking fires. A circle of deforestation has spread more than 160 km (100 mi) around some major cities in India, and firewood costs up to ten times the price paid in smaller towns.

What is the environmental impact of this increasing demand for a diminishing resource? As fallen wood becomes scarce, people lop off branches, fell trees, and even uproot seedlings, bushes, and herbaceous plants. Natural communities are destroyed, with predictable consequences. Removing ground cover robs soil of nutrients and exposes it to erosion. The denuded landscape becomes incapable of supporting either crops or new forest growth. The bare subsoil allows rain to run off without being absorbed. The climate becomes warmer and drier, and desert spreads over once-fertile land. Government policies exacerbate the problem by encouraging cash crop production for export, leaving less land available for domestic crop production. Forests are destroyed as new land is opened for farms. Mountain slopes, denuded of trees and bushes, are washed away by disastrous landslides. Increased albedo changes rainfall patterns, causing the spread of deserts into arid lands. In Africa, for instance, desertification is quickly expanding the already vast Sahara Desert across the once-fertile land of the Sahel.

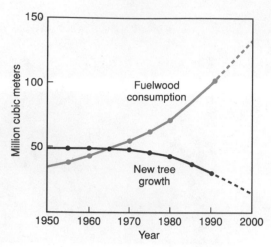

FIGURE 20.17 Fuelwood consumption and new tree growth in the Sudan, with projections to the year 2000. Firewood demand in the Sudan is already two to three times the sustainable yield of the forests. In Mauritania and Rwanda, demand is more than 10 times the sustainable yield.

This problem is expected to worsen unless steps are taken immediately to provide alternative energy sources. It is predicted that by the end of this century, some 2.9 billion people—about half the world's population—will find their minimum fuel needs unmet. The worst areas will continue to be the arid and semiarid regions of Africa, the populous regions of Southeast Asia, and the mountainous areas of Latin America, where demand is already as much as six times the available supply.

The 1,500 million cubic meters of fuelwood now harvested annually is about 500 million cubic meters short of the needed amount. By 2025, it is estimated that the worldwide demand for fuelwood will be 4,400 million cubic meters, while supplies will not expand much beyond present levels. This means that demand will be more than double the available supply. In many countries, the situation will be much worse than the world average. In some African countries, such as Mauritania, Rwanda, and the Sudan, firewood demand is already ten times the sustainable yield of remaining forests (figure 20.17).

Dung and Methane as Fuels

Where wood and other fuels are in short supply, people often dry and burn animal manure. This may seem like a logical use of waste biomass, but it can intensify food shortages in poorer countries. Not putting this manure back on the land as fertilizer reduces crop production and food supplies. In India, for example, where fuelwood supplies have been chronically short for many years, a limited manure supply must fertilize crops and provide household fuel. Cows in India produce more than 800 million tons of dung per year, more than half of which is dried and burned in cooking fires (figure 20.18). If that dung were applied to fields as fertilizer, it could boost crop production of edible grains by 20 million tons per year, enough to feed about 40 million people.

FIGURE 20.18 Animal dung is gathered by hand, dried, and used for fuel in many countries. This girl in Mali is carrying home fuel for cooking. Using dung as fuel deprives fields of nutrients and reduces crop production.

When cow dung is burned in open fires, more than 90 percent of the potential heat and most of the nutrients are lost. Compare that to the efficiency of using dung to produce methane gas, an excellent fuel. In the 1950s, simple, cheap methane digesters were designed for villages and homes, but they were not widely used. Customary uses and disposal of manure inhibit any changes in this area, and lack of funds and government support also have held back development. China has recently instituted a program of local methane production as an energy source. Perhaps other countries will follow China's lead.

Methane gas is a "natural" gas. It is produced by anaerobic decomposition (digestion by anaerobic bacteria) of any moist organic material. Many people are familiar with the fact that swamp gas is explosive. Swamps are simply large methane digesters, basins of wet plant and animal wastes sealed from the air by a layer of water. Under these conditions, organic materials are decomposed by anaerobic (oxygen-free) rather than aerobic (oxygen-using) bacteria, producing flammable gases instead of carbon dioxide. This same process may be reproduced artificially by placing organic wastes in a container and providing warmth and water (figure 20.19). Bacteria are ubiquitous enough to start the culture spontaneously.

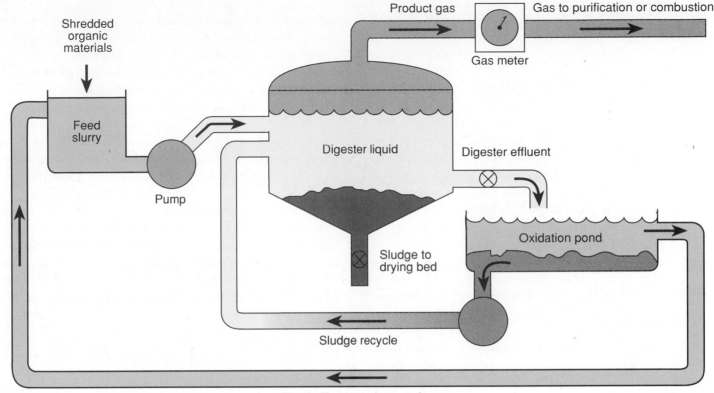

FIGURE 20.19 Continuous unit for converting organic material to methane by anaerobic fermentation. One kilogram of dry organic matter will produce 1–1.5 m³ of methane, or 2,500–3,600 million calories per metric ton.

Source: Solar Energy as a National Energy Resource, NSF/NASA Solar Energy Panel, National Science Foundation, December 1972.

Burning methane produced from manure provides more heat than burning the dung itself, and the sludge left over from bacterial digestion is a rich fertilizer, containing healthy bacteria as well as most of the nutrients originally in the dung. Whether the manure is of livestock or human origin, airtight digestion also eliminates some health hazards associated with direct use of dung, such as exposure to fecal pathogens and parasites.

All sizes and forms of containment have been used for methane production, from coffee cans to oil drums to specially designed tanks. Like other bacteria and yeast cultures, such as those used for bread, yogurt, and beer, methane-producing cultures need heat to thrive. The highest gas production is generally at 35°C (95°F), but once a culture is established, it produces enough of its own heat to continue. This warmth can be felt in all backyard compost heaps. It also accounts for early melting of snow over septic tanks.

How feasible is methane—from manure or from municipal sewage—as a fuel resource? Methane is a clean fuel that burns efficiently. It is produced in a low-technology, low-capital process from any kind of organic waste material: livestock manure, kitchen and garden scraps, and even municipal garbage and sewage. In fact, municipal landfills are active sites of methane production, contributing as much as 20 percent of the annual output of methane to the atmosphere. This is a waste of a valu-able resource and a threat to the environment because methane absorbs infrared radiation and contributes to the greenhouse effect (chapter 17). Some municipalities are drilling gas wells into landfills and garbage dumps. Cattle feedlots and chicken farms in the United States are a tremendous potential fuel source. Collectible crop residues and feedlot wastes each year contain 4.8 billion gigajoules (4.6 quadrillion BTUs) of energy, more than all the nation's farmers use. Municipal sewage treatment plants routinely use anaerobic digestion as a part of their treatment process, and many facilities collect the methane they produce and use it to generate heat and/or electricity for their operations. Although this technology is well-developed, its utilization could be much more widespread.

Alcohol from Biomass

Ethanol (grain alcohol) and methanol (wood alcohol) are produced by anaerobic digestion of plant materials with high sugar content, mainly grain and sugar cane. Ethanol can be burned directly in automobile engines adapted to use this fuel, or it can be mixed with gasoline (up to about 10 percent) to be used in any normal automobile engine. A mixture of gasoline and ethanol is often called **gasohol.** Ethanol in gasohol raises octane ratings and is a good substitute for lead antiknock agents, the major cause of lead pollution. Changing over to gasohol can cause problems in some cars because it loosens gasoline residues that have collected in gas tanks, gas lines, and the engine. Engines can be adjusted, however, to run just as well on gasohol or even straight ethanol as on other fuels.

Ethanol production could be a solution to grain surpluses and bring a higher price for grain crops than the food market offers. It also offers promise for reduced dependence on gasoline, which is refined from petroleum. Brazil has instituted an ambitious national program to substitute crop-based ethanol for imported petroleum. In 1985, the Brazilian sugar harvest produced 2.5 billion gallons of ethanol and plans are to produce 4 billion gallons in a few years. Brazil hopes to export much of its ethanol to the United States.

Researchers at the Solar Energy Research Institute in Golden, Colorado, have developed a new biomass-to-gasoline process that yields a high-octane, high-energy product from wood scraps, crop residues, and paper or municipal wastes. In this one-step process, organic materials are vaporized and passed over a zeolite (silica alumina) catalyst, which converts hydrocarbons to long-chain, gasolinelike fluids, carbon dioxide, and water. A ton of biomass will produce about 225 l (60 gal) of liquid fuel suitable for direct use in engines, or to be blended with other fuels to raise octane levels.

Methanol might have many advantages as a motor vehicle fuel. Since it burns at a lower temperature than gasoline or diesel, the bulky, heavy radiator might be eliminated, making automobiles sleeker and more efficient. Combined with flywheel-energy storage, intermittent burning engines, turbochargers, and other high-tech devices, we might be able to have personal transportation that is both energy efficient and less polluting.

Crop Residues, Energy Crops, and Peat

Crop residues, such as corn stalks, corn cobs, or wheat straw, can be used as a fuel source, but they are expensive to gather and often are better left on the ground as soil protection (chapter 11). The residue accumulated in food processing, however, can be a useful fuel. Hawaiian sugar growers burn bagasse, a fibrous sugar cane residue, to produce the state's second-largest energy source, surpassed only by petroleum.

Some crops are being raised specifically as an energy source. Fast-growing trees, such as *Leucaena* (chapter 12), and shrubs, such as alder and willow, are grown to provide wood for energy. Milkweeds, sedges, marsh grasses, cattails, and other biomass crops also might become useful energy sources, grown on land that is otherwise unsuitable for crops (figure 20.15). There, however, may be unacceptable environmental costs from disruption of wetlands and forests that are converted to energy crop plantations. Some plant species produce hydrocarbons that can be used directly as fuel in internal combustion engines without having to be fermented or transformed to hydrogen or methane gas. Sunflower oil, for instance, can be burned directly in diesel engines. In most cases, these oils and other high-molecular-weight

hydrocarbons are more valuable for other purposes, such as food, plastics, and industrial chemicals; they might be an attractive energy source in some circumstances, however.

Several other forms of biomass are actual or proposed sources of energy. People in northern climates have been digging up peat (partially decomposed plant residues) from bogs and marshes for centuries. There has been a resurgence of interest in exploiting the vast Minnesota and Canadian peat bogs, which some people regard as wastelands. After mining some or all of the peat, the land could be used for energy crops, such as cattails. Critics of these schemes point out the damaging effects of bog draining on ecosystems, watersheds, and wildlife. Burning of municipal garbage ("waste to watts") is a popular proposal; however, many people are worried about air pollution from these facilities (chapter 24). Garbage may contain hazardous, volatile, or explosive materials that are difficult or dangerous to sort out. Effects on air quality from all forms of direct combustion of biomass are a serious concern.

ENERGY FROM EARTH'S FORCES

The winds, waves, tides, ocean thermal gradients, and geothermal areas are renewable energy sources. Although available only in selected locations, these sources could make valuable contributions to our total energy supply.

Hydropower

Falling water has been used as an energy source nearly as long as wind power. In the fourteenth century, the fast-flowing mountain streams of Italy and the tributaries of the Rhone River in France and the Rhine River in Germany had clusters of water mills with power capacities that surpassed the steam boilers of most eighteenth-century industrial cities. Historian Louis Mumford has suggested that the modern industrial revolution would have progressed steadily based on water power alone, even if coal and oil had never been available as energy sources.

The invention of water turbines in the nineteenth century greatly increased the efficiency of hydropower dams (figure 20.20). By 1925, falling water generated 40 percent of the world's electric power. Since then, hydroelectric production capacity has grown fifteen-fold, but fossil fuel use has risen so rapidly that hydropower is now only one-quarter of total electrical generation. Still, many countries produce most of their electricity from falling water. Norway, for instance, depends on hydrogeneration for 99 percent of its electricity; Brazil, 87 percent; New Zealand and Switzerland, 75 percent; and Canada and Austria, 66 percent (figure 20.21).

The total world potential for hydropower is estimated to be about 3 million MW. If all of this capacity were put to use, the available water supply could provide between 8 and 10 terrawatt hours (10^{12} Whr) of electrical energy. The total installed capacity as of 1987 was only about 340,000 MW, or about 10

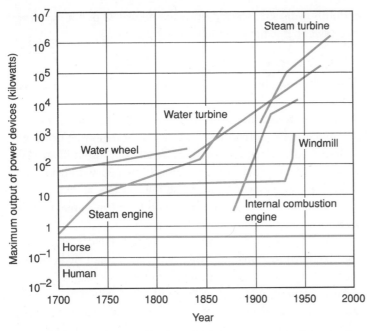

FIGURE 20.20 The development of power devices, 1700 to 1980. This graph shows improvements in output of state-of-the-art devices as they were invented and developed over recent history.

FIGURE 20.21 Hydropower dams like this one in New South Wales, Australia, can prevent floods and provide useful energy, but often they destroy valuable farmland, forests, villages, and historic areas when valleys are flooded for water storage. In addition, dams can also lose large amounts of water to seepage and evaporation, or they could fail and thus be a threat to downstream areas.

percent of the potential hydropower supply. The energy derived from this source in 1984 was equivalent to about 485 million tons of oil, or 6.7 percent of the total world commercial energy consumption.

Much of the hydropower development in recent years has been in enormous dams. There is a certain efficiency of scale in giant dams, and they bring pride and prestige to the countries that build them, but they can have unwanted social and environmental effects as well. The largest hydroelectric dam in the world at present is the Itaipu Dam on the Parana River between Brazil and Paraguay (figure 20.19). Designed to generate 12,600 MW of power, this dam should produce as much energy as thirteen large nuclear power plants when completed. The lake that it is creating already has flooded about 1,300 sq km (500 sq mi) of tropical rain forest and displaced many thousands of native people and millions of other creatures.

There are still other problems with big dams, besides human displacement, ecosystem destruction, and wildlife losses. Dam failure can cause catastrophic floods and thousands of deaths. Sedimentation often fills reservoirs rapidly and reduces the usefulness of the dam for either irrigation or hydropower. In China, the Sanmenxia Reservoir silted up in only four years, and the Laoying Reservoir filled with sediment before the dam was even finished. Rotting vegetation in artificial impoundments can have disastrous effects on water quality. When Lake Brokopondo in Suriname flooded a large region of uncut rain forest, underwater decomposition of the submerged vegetation produced hydrogen sulfide that killed fish and drove out villagers over a wide area.

Acidified water from this reservoir ruined the turbine blades, making the dam useless for power generation.

Floating water hyacinths (rare on free-flowing rivers) have already spread over reservoir surfaces behind the Tucurui Dam on the Amazon River in Brazil, impeding navigation and fouling machinery. Herbicides sprayed to remove aquatic vegetation have contaminated water supplies. Herbicides used to remove forests before dam gates closed cause similar pollution problems. Schistosomiasis, caused by parasitic flatworms called blood flukes (chapter 21), is transmitted to humans by snails that thrive in slow-moving, weedy tropical waters behind these dams. It is thought that 14 million Brazilians suffer from this debilitating disease.

As mentioned before, dams displace indigenous people. The Narmada Valley project in India will drown 150,000 ha of tropical forest and displace 1.5 million people, mostly tribal minorities and low-caste hill people. China's Three Gorges project, now under construction, will displace 1.4 million people. The Akosombo Dam built on the Volta River in Ghana nearly twenty years ago displaced 78,000 people from seven hundred towns. Few of these people ever found another place to settle, and those still living remain in refugee camps and temporary shelters.

In tropical climates, large reservoirs often suffer enormous water losses. Lake Nasser, formed by the Aswan High Dam in Egypt, loses 15 billion cubic meters each year to evaporation and seepage. Unlined canals lose another 1.5 billion cubic meters. Together, these losses represent one-half of the Nile River flow, or enough water to irrigate 2 million ha of land. The silt trapped by the Aswan Dam formerly fertilized farmland during seasonal flooding and provided nutrients that supported a rich fishery in the Delta region. Farmers now must buy expensive chemical fertilizers, and the fish catch has dropped almost to zero. As in South America, schistosomiasis is an increasingly serious problem.

In the United States, the only large rivers that haven't already been developed for hydropower production are those in the East with rich agricultural valleys, and a few wild rivers in the West that have been protected because of especially rich recreational, historical, or scenic values. Proposals to build dams on the Colorado River in the Grand Canyon in the 1950s were a major factor in the growth of the Sierra Club and other national environmental organizations. Continuing controversy surrounds proposals to dam the remaining wild stretches of the Stanislaus, Tuolumne, American, and Kings Rivers in California, where anglers, whitewater boaters, ecologists, and other river lovers are fighting to preserve free-flowing rivers.

If big dams—our traditional approach to hydropower—have so many problems, how can we continue to exploit the great potential of hydropower? Fortunately, there is an alternative to gigantic dams and destructive impoundment reservoirs. Small-scale, environmental **low-head hydropower** technology can extract energy from small headwater dams that cause much less damage. Some modern, high-efficiency turbines can even operate on **run-of-the-river flow.** Submerged directly in the stream and small enough to not impede navigation in most cases, these turbines don't require a dam or diversion structure and can generate useful power with a current of only a few kilometers per hour. They also cause minimal environmental damage and don't interfere with fish movements, including spawning migration. **Micro-hydro generators** operate on similar principles but are small enough to provide economical power for four to six homes. Where enough rain falls to keep rivers running all year around, small systems such as these enable a few families to form their own power cooperative, freeing them from dependence on large utilities and foreign energy supplies.

Small-scale hydropower systems also can cause abuses of water resources. The Public Utility Regulatory Policies Act of 1978 included economic incentives to encourage small-scale energy projects. As a result, thousands of applications were made to dam or divert small streams in the United States. Many of these projects have little merit. All too often, fish populations, aquatic habitat, recreational opportunities, and the scenic beauty of free-flowing streams and rivers are destroyed primarily to provide tax benefits for wealthy investors.

Wind Energy

The air surrounding Earth has been called a 20–billion-cubic-kilometer storage battery for solar energy. As chapter 17 shows, wind currents are the result of the uneven heating of air over land and water in equatorial zones and polar regions. The World Meteorological Organization has estimated that 20 million MW of wind power could be commercially tapped worldwide, not including contributions from windmill clusters at sea. This is about fifty times the total present world nuclear generating capacity.

Wind power has advantages and disadvantages, as do other nontraditional technologies. Like solar power and hydropower, wind power taps a natural physical force. Like solar power (its ultimate source), wind power is a virtually limitless resource, is nonpolluting, and causes minimal environmental disruption. Like solar power, however, it requires expensive storage during peak production times to offset nonwindy periods. Like the other nontraditional and renewable energy sources, wind power is limited by insufficient government allocations for research and development.

Power from the wind was first harnessed at least five thousand years ago when Egyptians began sailing on the Nile River. Windmills as a source of power to pump water and grind grain seem to have been invented in China and, independently, in Persia about three thousand years ago. The spread of Islam carried windmill technology throughout the Mediterranean region. By A.D. 1100, Crusaders and travelers took windmill knowledge back to Europe, where it became a valuable source of power in the Low Countries and other places where winds were steady and water power was lacking.

Windmills played a crucial role in the settling of the American West. The Great Plains had abundant water in underground aquifers, but little surface water. The strong, steady winds that blew across the prairies provided the energy to pump water that allowed agriculture to move west across the prairies. By the end of the nineteenth century, nearly every farm or ranch west of the Mississippi River had at least one windmill. The manufacture, installation, and repair of windmills was a major industry in America. Even today, some 150,000 windmills still spin productively in the Great Plains (figure 20.22).

Windmill technology was making great strides early in this century (see figure 20.20). Efficiencies increased ten times or more with the invention of aerodynamic propellers and affordable electric-wind generators in the 1920s. This promising technology was nipped in the bud, however, by passage of the Rural Electrification Act of 1935. This act brought many benefits to rural America, but it effectively killed wind-power development and encouraged the building of huge hydropower dams and coal-burning power plants. It is interesting to speculate what the course

FIGURE 20.22 Millions of traditional windmills once pumped water, generated electricity, and provided power for farms and ranches across America. Most have been displaced by internal combustion engines and rural electric power systems, but there is still potential for capturing large amounts of renewable energy from the wind.

for wind generators (average wind speeds in excess of 25 km/hr) and do not involve competing and conflicting uses, such as cities and national parks.

All wind conversion systems are composed of a support tower, a rotor, a mechanical energy transmission system, and an energy converter, usually an electrical generator or water pump. Most windmills also require a system to orient them toward the wind, a means of controlling rotor speed, and some form of energy storage. The standard modern wind turbine uses only two or three propeller blades. More blades on a windmill provide more torque in low-speed winds, so that the traditional Midwestern windmill, with twenty or thirty blades, was most appropriate for small-scale use in less reliable wind fields. Fewer blades operate better in high-speed winds, providing more energy for less material cost at wind speeds of 25–40 km/hr (15–25 mph). A two-blade propeller (figure 20.23) can extract most of the available energy from a large vertical area and has less metal or fiberglass that could weaken and break in a storm. Three-bladed propellers often are preferred because they are easier to balance and spin more smoothly.

In 1929, the French engineer G. J. M. Darrieus proposed an alternative to the traditional horizontal-axis windmill. Developed in Canada in the 1960s, the Darrieus rotor, a vertical-axis "egg beater" machine, is durable and inexpensive (figure 20.24). It needs no adjustment to wind direction and operates in any wind stronger than 19 km/hr (12 mph), compared to conventional turbines that generally require at least 24 km/hr (15 mph). Quantity manufacturing of Darrieus generators has reduced the price rapidly, bringing the current cost to around 5.4 cents per kWhr for a conservatively estimated fifteen-year life span.

Wind farms are large-scale public utility efforts to take advantage of wind power. The first major wind farm started in New Hampshire in 1981, but California quickly took the lead in wind farm development. By 1982, the first one hundred-turbine farm stretched across Altamont Pass in the rolling ranch country east of San Francisco. The windmills, mainly medium-sized, American- and Danish-built two- and three-bladed machines, stand along the ridge following a contour line of highest wind availability. By 1990, more than 7,000 windmills stood at Altamont. (see figure 20.1). Nine thousand more are installed at California's Tehachapi and San Gorgino Passes. The existing project already produces well over 1,500 MW, the equivalent of three large nuclear reactors. Altogether, California's 16,000 commercial turbines produce 95 percent of United States wind energy and 75 percent of the world's production.

Construction of wind farms is not limited to mountain ridges. The Great Plains are legendary for their strong, incessant winds. Sea coasts also provide excellent wind farm sites. Offshore wind farms have been proposed, with large (up to 8 MW each) generators to be anchored a short distance out at sea where competition for real estate development is low and winds are high year-round. According to such plans, collected energy would

of history might have been if we had not spent trillions of dollars on fossil fuel and nuclear energy, but instead had invested that money on small-scale, renewable energy systems.

As the world's conventional fuel prices rise, interest in wind energy is resurging. The United States and Denmark are presently the world's largest producers, but planners and manufacturers in other European countries also are studying wind power. Enthusiasm also is evident in Asia, India, and Australia, as well as the island countries of the Pacific and Caribbean, where energy costs are high.

Pilot projects already have demonstrated the economy of wind power. Theoretically 60 percent efficient, windmills typically produce at 35 percent efficiency under field conditions. General Electric estimates the wind could provide 14 percent of United States electrical demand by the year 2000. One thousand MW of installed wind power could meet the energy needs of 400,000 households at present use levels and save 3.7 million barrels of oil per year ($55.5 million each year at current world oil prices). A General Electric study also found more than 556,400 sq km of land in the United States alone that is suitable

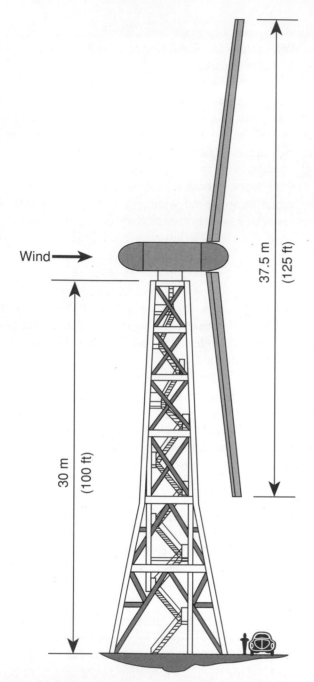

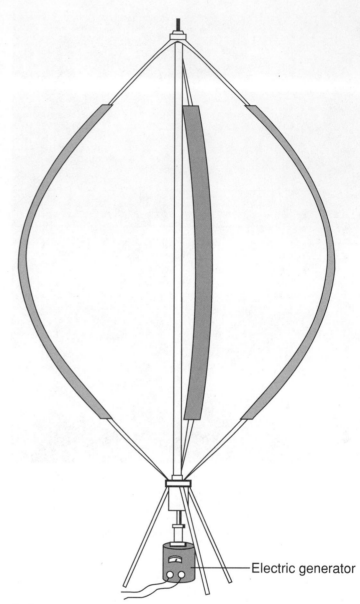

FIGURE 20.24 The Darrieus rotor. This "egg-beater" wind machine is a low-cost device capable of producing electricity at relatively low wind speeds. It has been in use since 1927 when it was patented by G. J. M. Darrieus in France.

FIGURE 20.23 A 100-kilowatt, two-bladed wind machine. A two-bladed generator requires stronger winds than do machines with more blades, but in high winds it extracts energy more efficiently. The blades are air-foils, similar in cross-section to an airplane propeller. Their pitch is adjustable to match wind speed. At maximum rpm the propeller tip speed approaches the speed of sound.

probably be stored in electrolyzed water (discussed earlier as storage for photovoltaic power) for transportation to shore.

Are there negative impacts of wind farms? They generally occupy places with wind and weather too severe for residential or other development. The wind speed at Altamont, for instance, averages 35 km/hr (22 mph) from March to September. Livestock graze undisturbed beneath the windmills. Wind farm opponents complain, however, of visual and noise pollution (the low, rhythmic whoosh of the blades); proponents claim these effects are negligible. Most wind farms are too far from residential areas to be heard or seen. But they do interrupt the view in remote isolated places and destroy the sense of isolation and natural beauty. They also can be a wildlife hazard. Hawks, eagles, and other birds fly into the whirling blades. If California condors are ever released again into the wild, they may be threatened by windmills on their favorite mountain passes.

The most visible and most heavily funded wind technology development has been in the large multimegawatt range. Federal research funding was given to NASA, the DOE, and large public corporations, including General Electric, Boeing, Bendix, Westinghouse, and United Technologies, all of which manufactured enormous machines with blades nearly 60 m (200 ft) from tip to tip, and outputs of up to 4 MW. These highly visible projects attracted much attention and applause, and consequently received increased funding. They were also frequently down for repair, had parts that were difficult to replace, and now are considered the dinosaurs of the field. Increasingly, purchasers of wind machines are turning to smaller Canadian and Danish models. By the end of 1990, the United States had imported five thousand wind generators from Denmark. Politics and the public's preference for high-visibility projects, however, are encouraging the industry's trend to take models that have demonstrated their dependability and simply make them bigger.

When a home owner or community invests independently in wind generation, the same question arises as with solar energy: What should be done about energy storage when electricity production exceeds use? Besides the storage methods mentioned earlier, many private electricity producers believe the best use for excess electricity is in cooperation with the public utility grid. When private generation is low, the public utility runs electricity through the meter and into the house or community. When the wind generator or photovoltaic systems overproduce, the electricity runs back into the grid and the meter runs backward. Ideally, the utility reimburses individuals for this electricity, for which other consumers pay the company. The 1978 Public Utilities Regulatory Policies Act required utilities to buy power generated by small hydro, wind, cogeneration, and other privately owned technologies at a fair price. Not all utilities yet comply, but some—notably in California, Oregon, Maine, and Vermont—are purchasing significant amounts of private energy.

Geothermal Energy

The earth's internal temperature can provide a useful source of energy in some places. High-pressure, high-temperature steam fields exist below Earth's surface. Around the edges of continental plates or where Earth's crust overlays magma (molten rock) pools close to the surface, this energy is expressed in the form of hot springs, geysers, and fumaroles (chapter 9). Yellowstone National Park is undoubtedly the most famous geothermal region in the United States. Iceland, Japan, and New Zealand also have high concentrations of geothermal springs and vents. Depending on the shape, heat content, and access to groundwater, these sources produce wet steam, dry steam, or hot water.

Until recently, the main use of this energy source was in baths built at hot springs. More recently, geothermal energy has been used in electric power production, industrial processing, space heating, agriculture, and aquaculture. In Iceland, most buildings are heated by geothermal steam. A few communities in the United States, including Boise, Idaho, use geothermal home heating. Recently, geothermal steam also has been developed for electrical generation.

California's Geysers project is the world's largest geothermal electric generating complex, with 200 steam wells that

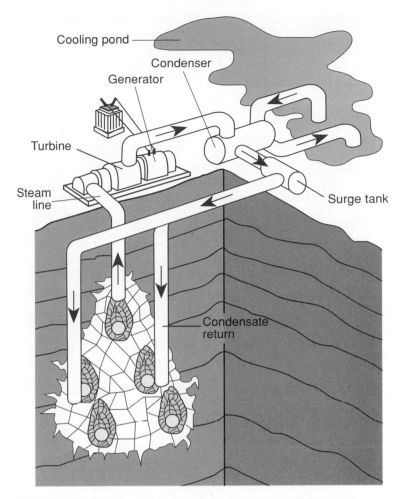

FIGURE 20.25 Geothermal steam extraction system for electricity production. Once the steam has spun the turbine, it is condensed, cooled, and pumped back into the underground steam field. The steam source for this relatively low-capital electric plant should last for several decades.

Sources: American Oil Shale Corporation and U.S. Atomic Energy Commission, *A Feasibility Study of a Plowshare of Geothermal Power Plant*, April 1971.

provide some 1,300 MW of power. This system functions much like other heat-driven generators. A shaft sunk into the subsurface steam reservoir brings pressurized steam up to a turbine at the surface. The steam spins the turbine to produce electricity and then is piped off to a condensing unit and then to a cooling pond (figure 20.25). The remaining brine (water heavily laden with dissolved minerals) is pumped back down to the reservoir. When the steam source has high mineral concentrations that would corrode delicate turbine blades, a heat exchanger is used to generate clean steam as is done in a nuclear power plant.

Among the advantages of geothermal generators are a reasonably long life span (at least several decades), no mining or transporting of fuels, and little waste disposal. The main disadvantages are the potential danger of noxious gases in the steam and noise problems from steam-pressure relief valves.

While few places have geothermal steam, the entire Earth is underlain by warm rocks that can be reached by drilling down

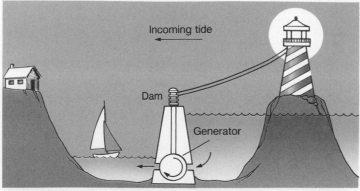

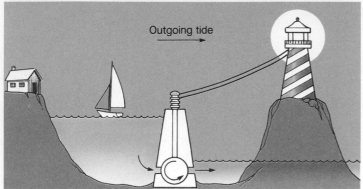

FIGURE 20.26 Tidal power station. Both incoming and outgoing tides are held back by a dam. The difference in water levels generates electricity in both directions as water runs through reversible turbogenerators.

far enough into the crust. Some heating systems pump groundwater from underground aquifers or inject surface water into deep strata and then withdraw it after it has been warmed, as a way of extracting geothermal energy. There have been experiments with pumping surplus heated water into underground reservoirs during the summer when the heat is not needed, and then withdrawing it in the winter as make-up water for steam-heating systems. A major concern with such systems, however, is to ensure that contaminants aren't injected into aquifers, which are important sources of drinking water.

Tidal and Wave Energy

Ocean tides and waves contain enormous amounts of energy that can be harnessed to do useful work. Tidal power exploitation is not new. The first known tide-powered mills were built in about 1100 in England. One built in Woodbridge, England, in 1170 functioned productively for eight hundred years. Through the seventeenth century, similar mills were built by the Dutch in the Netherlands and in New York.

The Rance River Power Station in France, in operation since 1966, was the first large tidal electric generation plant, producing 160 MW. A tidal station works like a hydropower dam, with its turbines spinning as the tide flows through them (figure 20.26). It requires a high-tide low-tide differential of several

meters to spin the turbines. Unfortunately, the tidal period of thirteen and one-half hours causes problems in integrating the plant into the electric utility grid, as it seldom coincides with peak use hours. Nevertheless, demand has kept the plant running for more than two decades.

The first North American tidal generator, producing 20 MW, was completed in 1984 at Annapolis Royal, Nova Scotia. A much larger project has been proposed to dam the Bay of Fundy and produce 5,000 MW of power on the Bay's 17-meter tides. The total flow at each tide through the Bay of Fundy theoretically could generate energy equivalent to the output of 250 large nuclear power plants. The environmental consequences of such a gargantuan project, however, may prevent its ever being built. The main worries are saltwater flooding of freshwater aquifers when seawater levels rise behind the dam and the flooding and destruction of rich shoals and salt flats, breeding grounds for aquatic species and a vital food source for millions of migrating shorebirds. There also would be heavy siltation, as well as scouring of the sea floor as water shoots through the dam. However, it appears that modest-sized plants like Annapolis Royal's will avoid most of these dangers. Three other medium-sized projects are under consideration for sites within the Bay of Fundy.

Ocean wave energy can easily be seen and felt on any seashore. The energy that waves expend as millions of tons of water are picked up and hurled against the land, over and over, day after day, can far exceed the combined energy budget for both insolation (solar energy) and wind power in localized areas. Captured and turned into useful forms, that energy could make a very substantial contribution to meeting local energy needs.

Numerous attempts have been made to use wave energy to drive electrical generators. Generally these take the form of a floating bar that moves up and down as the wave passes. When coupled to a dynamo, this mechanical energy can be converted to electricity in the same way that a water wheel or steam turbine works. England, with a long coastline facing the stormy North Sea, plans to build an extensive system of wave-energy platforms. Unfortunately for developers of this energy source, the stormy coasts where waves are strongest are usually far from major population centers that need the power. In addition, the storms that bring this energy destroy the equipment intended to exploit it.

An intriguing proposal has been developed for small rotating cylinders called ducks that would sit just off-shore, bobbing in the surf and generating electricity. There are claims that British energy officials deliberately suppressed this low-tech approach because it might compete with conventional sources.

Ocean Thermal Electric Conversion (OTEC)

Temperature differentials between upper and lower layers of the ocean's water also are a potential source of renewable energy. Energy can be extracted in either a closed-cycle or open-cycle system. In a closed cycle, heat from sun-warmed upper ocean

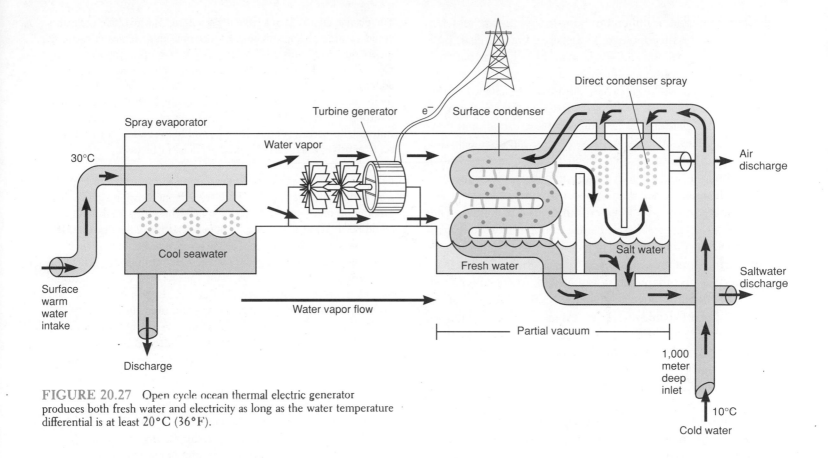

FIGURE 20.27 Open cycle ocean thermal electric generator produces both fresh water and electricity as long as the water temperature differential is at least 20°C (36°F).

layers is used to evaporate a working fluid, such as ammonia or Freon, which has a low boiling point. The pressure of the gas produced is high enough to spin turbines to generate electricity. Cold water then is pumped from the depths of the ocean to condense the gas.

An open-cycle system uses seawater itself as the working fluid (figure 20.27). Warm seawater is sprayed into a chamber, where low air pressure causes some water to evaporate. The resulting water vapor drives a massive, low-pressure turbogenerator and then passes through two condensers cooled by deep ocean water. The first one, a closed surface condenser, produces distilled (desalinized) water. The other is a direct contact condenser in which cold water is sprayed directly into the steam flow, condensing the remaining water vapor and creating the vacuum that drives the system. The outflow from this condenser is discharged into the sea. This process produces fresh water, as well as electricity, but it requires very large turbines and inlet pipes, making it expensive to install.

As long as a temperature difference of about 20°C (36°F) exists between the warm upper layers and cooling water, useful amounts of net power can, in principle, be generated with one of these systems. This differential corresponds, generally, to a depth of about 1,000 m in tropical seas. The places where this much temperature difference is likely to be found close to shore are islands that are the tops of volcanic seamounts, such as Hawaii,

or the edges of continental plates along subduction zones (chapter 9) where deep trenches lie just offshore. The west coast of Africa, the south coast of Java, and a number of South Pacific islands, such as Tahiti, have usable temperature differentials for OTEC power.

Disadvantages of OTEC include the energy cost of pumping up deep waters (which could eliminate most of the energy advantage of the system), saltwater corrosion of pipes and equipment, vulnerability to storm damage, and ecological destabilization from the upwelling of deep, nutrient-rich water and local alterations of ocean temperatures.

THE QUESTION OF SUSTAINABILITY

The rapid industrial growth and high standard of living enjoyed by the Western world for the past century has been based, to a large extent, on exploiting supplies of nonrenewable energy. The question is whether we will be able to shift to one or more of the renewable resources that we have discussed in this chapter without traumatic dislocations in society and the economy and before we run out of traditional fuels or are overwhelmed by their waste products.

Establishing new energy sources is not problem-free. Equipment that is manufactured on an experimental basis is costly and often unreliable or difficult to operate. People are reluctant to invest in untried technology, especially when familiar, well-tested, conventional technology is available. Traditional fuels benefit from the economic efficiency of a mass market, and the convenience of a well-established infrastructure. Many of the alternative energy sources we have discussed in this chapter suffer from being less glamorous than the complex, exotic, and powerful technologies, such as nuclear power, large-scale hydropower, or synthetic fuel plants. Politicians and investors often prefer to build a single facility that symbolizes power, prestige, and sophistication rather than support a widely dispersed and more ordinary project, such as retro-insulation of homes or individual photovoltaic collectors. Furthermore, utilities and agencies have little incentive to build an energy source that will supply essentially free, unlimited power to their former customers.

Alternatively, there are personal and environmental advantages to adopting renewable energy resources that offer the possibility of a sustainable future, are essentially free of pollution, and are much less damaging to the environment than those commonly used today. Renewable energy resources may enable people to live with more freedom and independence away from the utility power grid. If individuals or small communities were able to grow or generate their own energy sources, it might dramatically affect how and where we live in the future.

These alternative energy sources may not be suitable for every situation, however. The greatest security and flexibility undoubtedly will come from developing a wide range of options from which to choose. We clearly need innovative and courageous leadership from government, industry, and private citizens to bring these new energy options to the marketplace as quickly as possible.

S U M M A R Y

Several sustainable energy sources could reduce or eliminate our dependence on fossil fuels and nuclear energy. Some of these sources have been used for centuries but have been neglected since fossil fuels came into widespread use. Passive solar heat, fuelwood, windmills, and water wheels, for instance, once supplied a major part of the external energy for human activities. With increased concern about the dangers and costs associated with conventional commercial energy, these ancient energy sources are being reexamined as part of a more sustainable future for humankind.

Exciting new technologies have been invented to use renewable energy sources. Active solar air and water heating, for instance, require less material and function more quickly than passive solar collection. Parabolic mirrors can produce temperatures high enough to be used as process heat in manufacturing. Ocean thermal electric conversion, tidal and wave power stations, and geothermal steam sources can produce useful amounts of energy in some localities. One of the most promising technologies is direct electricity generation by photovoltaic cells.

Since solar energy is available everywhere, photovoltaic collectors could provide clean, inexpensive, nonpolluting, renewable energy, independent of central power grids or fuel-supply systems.

Biomass also may have some modern applications. In addition to direct combustion, biomass can be converted into methane or ethanol, which are clean-burning, easily storable, and transportable fuels. These alternative uses of biomass also allow nutrients to be returned to the soil and help reduce our reliance on expensive, energy-consuming artificial fertilizers.

Many of these sustainable energy sources depend on technology that is still experimental and too expensive to compete well with established energy industries. If the economies of mass production and marketing were applied, these new technologies could be made available more cheaply and more dependably. It may take special funding and other governmental incentives to make sustainable energy competitive. The subsidies for renewable energy sources have been especially meager in comparison to the billions of dollars spent on nuclear energy, large-scale hydropower, and fossil fuel extraction and utilization.

Although conventional and alternative energy sources offer many attractive possibilities, conservation often is the least expensive and easiest solution to energy shortages. Even basic conservation efforts, such as turning off lights, can save large amounts of energy when practiced by many people. More major conservation methods, such as home insulation and energy-efficient appliances and transportation, can drastically reduce energy consumption and similarly reduce energy expenses. Our natural resources, our environment, and our pocketbooks all benefit from careful and efficient energy consumption.

■ Review Questions

1. What is cogeneration and how does it save energy?
2. Explain the principle of net energy yield. Give some examples.
3. What is the difference between active and passive solar energy?
4. How do photovoltaic cells work?
5. Describe the advantages and disadvantages of a multiple-bladed, two-bladed, and Darieus windmill.
6. Describe some problems with wood burning in both industrialized nations and Third World nations.
7. How is methane made?
8. What are some advantages and disadvantages of large hydro-electric dams?
9. What are some examples of biomass fuel other than wood?
10. Describe how tidal power and ocean thermal conversion work.

■ Questions for Critical Thinking

1. What alternative energy sources are most useful in your region and climate?
2. What can you do to conserve energy where you live? In personal habits? In your home, dormitory, or workplace?

3. What massive heat storage materials can you think of that could be attractively incorporated into a home?

4. Do you think building wind farms in remote places, parks, or scenic wilderness areas would be damaging or unsightly?

5. If you were a government energy administrator or planner, how would you distribute energy research funds?

6. What are the advantages and disadvantages of being disconnected from central utility power?

7. Can you think of environmental consequences associated with tidal or geothermal energy? If so, how can they be mitigated?

8. You are offered a home solar energy system that costs $10,000 but saves you $1,000 a year. Will you take it at this rate? If the cost were higher and the payoff time longer, what is the threshold at which you would not buy the system?

Key Terms

active solar systems (p. 396)
cogeneration (p. 395)
energy efficiency (p. 392)
eutectic chemicals (p. 397)
gasohol (p. 405)
low-head hydropower (p. 408)
micro-hydro generators (p. 408)
net energy yield (p. 393)
parabolic mirrors (p. 397)
passive heat absorption (p. 396)
photovoltaic cell (p. 398)
run-of-the-river flow (p. 408)
wind farms (p. 409)

SUGGESTED READINGS

- Caufield, C. June 1983. "Dam the Amazon." *Natural History* 92(7):60. Discusses the economic importance and ecological troubles of several major dams on the Amazon River.

- Chiles, J. R. 1990. "Tomorrow's Energy Today." *Audubon* January, 59. A good overview of renewable energy in the United States.

- Davis, G. R. September, 1990. "Energy for Planet Earth." *Scientific American* 263(3):54. An excellent overview of our global energy resources and how we can achieve a sustainable relationship between energy use and the environment.

- Dorf, R. *The Energy Factbook*. New York: McGraw-Hill, 1981. Illustrations and tables describing alternative energy and conservation methods.

- Flavin, C. 1986. "Reforming the Electric Power Industry." *State of the World*. Washington, D.C.: Worldwatch Institute. Excellent discussion. See especially the sections on the cogeneration boom, renewable electricity, and electricity's future.

- Gray, C. L. Jr., and J. A. Alson 1989 "The Case for Methanol." *Scientific American* 261(5):108. The authors maintain that a move to pure methanol fuel would reduce vehicular emissions of hydrocarbons and greenhouse gases and could lessen U.S. dependence on foreign energy sources.

- Hamakawa, Y. April 1987. "Photovoltaic Power." *Scientific American* 256(4):86. An excellent discussion of photovoltaics and why they work.

- Hirst, E. March 1991. "Boosting Energy Efficiency through Federal Action." *Environment* 33(2):6. Describes some actions we could take to increase our energy efficiency.

- Jeffries, J. November/December 1983. "The New Alchemy of Photovoltaics." *Sierra* 68(6):45. An illustrated guide to domestic applications of photovoltaics.

- Keough, J. January/February 1986. "Sin of Emission." *Sierra* 71(1):22. The politics of trying to control wood stove emissions in the United States.

- Moretti, P., and L. Divone. June 1986. "Modern Windmills." *Scientific American* 254(6):110. Excellent introduction to modern wind technology and where it comes from.

- Palmer, T. July/August 1983. "What Price 'Free' Energy?" *Sierra* 68(4):40. Looks at the "staggering environmental costs" of hydroelectric projects on our rivers.

- Penney, T., and D. Bharathan. January 1987. "Power from the Sea." *Scientific American* 256(1):86. A good discussion of ocean thermal electric conversion.

- Reisner, M. Spring 1987. "The Rise and Fall and Rise of Energy Conservation. *Amicus Journal* 9(2):22. A very readable survey of American attitudes about energy conservation.

- Rosenfeld, A. H., and D. Hafernlester 1988. "Energy-Efficient Buildings." *Scientific American* 258(4):78. Highly efficient homes and offices will slash energy bills and avoid building new power plants.

- Schaeffer, J. et al. 1990. *Real Goods Alternate Energy Source Book*. Real Goods Trading Corp, Ukiah California. Products, graphics, instructions for setting up alternate energy and energy conservation systems.

- Weinberg, C. J., and R. H. Williams. September 1990. "Energy from the Sun." *Scientific American* 263(3):146. Describes how advances in wind, solar, thermal, and biomass technologies will soon render them cost-competitive with gasoline and coal-generated electricity.

- White, L., Jr. August 1970. "Medieval Uses of Air." *Scientific American* 223(12):92. A fascinating historical perspective on wind uses through time.

- Wilson, H. B., P. B. MacCready, and C. R. Kyle. 1989. "Lessons of Sunraycer." *Scientific American* 263(3):90. Building the world's first solar automobile.

PART FOUR

Environmental Pollution and Other Dilemmas

CHAPTER 21

Environmental Health and Toxicology

The best measure we have for designing our future technologies is human health. There is nothing that seems more immediate and important than personal health, our own and that of our loved ones. It is here that we feel the greatest urgency to solve problems of environmental pollution, and it is here that the consequences of our actions are most dramatically demonstrated.

Mike Samuels and Hal Zina Bennett

CONCEPTS

Health is a state of physical, mental, and social well-being, not merely the absence of disease or infirmity. The cause or development of nearly every human disease is at least partially related to environmental factors.

For most people in the world, the greatest environmental health threat is from pathogenic organisms. Gastrointestinal infections cause more deaths worldwide than any other group of diseases. About ten million children die each year from the effects of malnutrition and diarrhea. Most of these deaths could be prevented with simple oral rehydration therapy and good food.

There are numerous toxic and hazardous agents, both natural and human-made, in our environment. Some supertoxic materials are dangerous at such low levels that they are undetectable without sensitive analytic instruments. This makes it difficult for us to protect ourselves from them.

Carcinogens, teratogens, and mutagens are among the most sinister threats in the environment. It may be that a single chemical molecule or radioactive particle can cause damage that eventually leads to birth defects, genetic abnormalities, or cancer. For many of these agents, there seems to be no safe exposure; any dose brings some increased risk.

Stress, diet, and life-style are important health factors. Our social environment may be as important as our physical environment in determining the state of our health. In some areas of the world, people live exceptionally long and healthful lives. We might be able to learn from them how to live better ourselves.

Each chemical substance has a unique set of physical characteristics that determines its distribution and fate in the environment. Among the most important of these are solubility, bioreactivity, and persistence. Organisms have mechanisms to minimize the effects of environmental hazards. Among these are metabolic degradation, excretion, and avoidance of toxins.

Because of these defense and repair mechanisms, some materials have thresholds below which there are no measurable adverse effects. Regulations attempting to reduce emissions of such materials to zero may be a waste of resources, especially if these materials are already present naturally.

Our assessment of environmental risks can be skewed by misconceptions, personal bias, faulty perceptions, and unrepresentative media reporting. When risks are less than about one in a million, we tend to disregard them. We are willing to accept some risks that are much greater than one in a million if they accompany an activity that we consider desirable.

INTRODUCTION

If you read a newspaper regularly or watch television news, you undoubtedly have seen many stories about toxic and hazardous chemicals in the environment. There are scares about pesticide residues on fruits and vegetables; cancer-causing radon in our homes; neurotoxic heavy metals in fish; pathogenic bacteria in eggs, milk, and cheese; and dangerous synthetic chemicals released into the environment by industrial accidents or deliberate dumping (figure 21.1).

What are these toxic and hazardous agents? Why are they dangerous and how are we exposed to them? In this chapter, we will survey some principles of toxicology and environmental health that will help answer these questions. In subsequent chapters (22, 23, 24), we will look at how environmental contaminants are produced, released, and transported through the environment, and what we can do to reduce or eliminate these hazards.

TYPES OF ENVIRONMENTAL HEALTH HAZARDS

What is health? The World Health Organization defines **health** as a state of complete physical, mental, and social well-being, not merely the absence of disease or infirmity. By that definition, we all are ill to some extent. Likewise, we all can improve our health to live happier, longer, more productive, and more satisfying lives.

What is a disease? A **disease** is a deleterious change in the body's condition in response to an environmental factor that could be nutritional, chemical, biological, or psychological. The cause and development of nearly every human disease is somehow related to environmental factors. Diet and nutrition, infectious agents, toxic chemicals, physical factors, and psychological stress all play roles in the onset or progress of human diseases. To understand how these factors affect us, let's look at some of the major categories of environmental health hazards.

Infectious Organisms

For most people in the world, the greatest environmental health threat continues to be, as always, pathogenic (disease-causing) organisms (table 21.1). Although this chapter is primarily about toxic chemicals, we also should consider some of the biological hazards to which we are exposed. In a sense, we are engaged in a constant battle against infectious and parasitic organisms. It is surprising that we don't suffer more sickness than we do, considering the number of potential pathogens in our environment. In the less developed countries, where nearly 80 percent of the world population lives, infectious agents, parasites, and nutritional deficiencies still are the main cause of **morbidity** (illness) and mortality (death).

Gastrointestinal infections (diarrhea, dysentery, and cholera) probably cause more deaths worldwide than any other

FIGURE 21.1   The town of Times Beach, Missouri, is now closed and abandoned. Although the most heavily contaminated soil has been removed, it is still considered too dangerous to enter because of dioxin-contaminated oil spread on roads ten years ago.

TABLE 21.1   Some major environmental health problems

Disease	New cases each year	Yearly deaths
Diarrhea	1 Billion	10 Million
Malaria	800 Million	5–10 Million
Parasitic worms, including flukes	1 Billion	*
Anemia	375 Million	*
Respiratory diseases†	500 Million	5–6 Million
Trachoma	300 Million	*
Goiter and cretinism	200 Million	*
Tetanus	5 Million	800,000
Polio	2 Million	200,000

*Few people die directly from these diseases, but debilitation can be severe and can contribute to other diseases.

†Respiratory diseases include tuberculosis, influenza, pneumonia, and

group of diseases. Diarrhea can be caused by either bacteria or protozoans (figure 21.2). At least one billion new cases of diarrhea occur each year, mostly among young children, and around ten million deaths result from a combination of malnutrition and diarrhea. That means that one child dies every three seconds from these highly preventable diseases. In the time that it will take you to read this paragraph, ten children will die. In many of the less developed countries, one child in four dies before the age of five from these diseases.

Malnutrition and diarrhea create a vicious cycle. Malnutrition makes people more susceptible to infection, and infections, in turn, make it more difficult to obtain, absorb, and retain food. These diseases are both highly preventable and curable. Better sanitation and nutrition could prevent most, if not

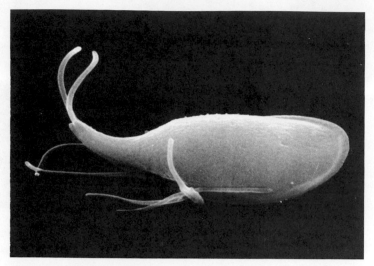

FIGURE 21.2 Giardia, a parasitic intestinal protozoan, is reported to be the largest single cause of diarrhea in the United States. It is spread from human feces through food and water. Even pristine wilderness areas have giardia outbreaks due to careless campers.

FIGURE 21.3 Malaria, spread by the *Anopheles* mosquito, may be the largest single cause of human disease and premature death in the world. Diarrhea and dysentery actually kill more people, but they represent a family of diseases rather than a single disease.

all, gastrointestinal infections. Simple oral rehydration therapy (ORT), in which patients are given an inexpensive mixture of sugar and salts in water, is highly effective in treating diarrhea, yet costs only a few cents per patient (Box 21.1).

Malaria is an infection of red blood cells by a parasitic protozoan (*Plasmodium* sp.). It is probably the second leading disease-caused source of mortality in the world and is especially common in the moist tropical countries of Africa, where the *Anopheles* mosquitoes that spread the disease thrive (figure 21.3). About 160 million people have malaria at any given time, and there are around 800 million new cases each year. Once the malaria parasite becomes established in the blood, the disease can recur every few months for many years. Insecticide use greatly reduced malaria incidence (as much as 90 percent in India and Sri Lanka) in the 1950s and 1960s, but pesticide-resistant mosquito populations have developed, and the disease is reappearing—in some cases at higher levels than before. Malaria has now become the most rapidly growing disease in the world, increasing at a much higher rate than AIDS in many areas.

Parasitic nematodes (roundworms) and flatworms (flukes and tapeworms) are very common in less developed countries where sanitation is primitive. People rarely die directly from the infections, but they can be extremely debilitated. Intestinal worms, such as tapeworms and some roundworms, are especially serious in persons who are malnourished in the first place. Worms that occupy other body sites produce some specific diseases.

Schistosomiasis is a disease caused by waterborne blood flukes. About 200 million people are infected worldwide, and about one million deaths are caused by complications of this disease. The adult flukes live in blood vessels of the human digestive tract where they cause dysentery, anemia, general weakness,

and greatly reduced resistance to other infections. They have a complex life cycle involving an aquatic snail as an intermediate host (figure 21.4). Rice paddy farming, the most common type of agriculture in many tropical countries, creates a nearly perfect environment for transmission of these flukes because human feces are used for fertilizer, the water is shallow and warm—just right for the snails—and people spend hours wading in the water tending rice plants. Large irrigation projects, such as those made possible by building the Aswan Dam in Egypt, bring about increased crop yields but also increase problems with schistosomiasis.

Nematodes are among the most numerous animals in the world, and several parasitic species cause serious human diseases. Onchocerciasis (river blindness) is caused by worms transmitted by the bite of black flies. Masses of dead nematodes accumulate in the eyeball, destroying vision. This disease affects 18 million people in the world and permanently blinds 500,000 each year. In some African villages, nearly every adult over thirty is blind from this disease (figure 21.5). The World Health Organization (WHO) undertook a massive pesticide spraying campaign in Africa during the early 1980s that reduced the incidence of river blindness in many areas, but the flies have developed pesticide resistance and many nontarget species were destroyed. The drugs, amocarzine and ivermectin, have been effective in controlling filariasis and a number of other parasitic infections in humans and domestic livestock. The WHO predicts that river blindness, and perhaps other parasitic infections, could be wiped out by the year 2000 with continued efforts of public education and early treatment.

Some other parasitic worms also cause dreadful diseases. Filariasis (one form of which is elephantiasis) is transmitted by

ENVIRONMENTAL POLLUTION AND OTHER DILEMMAS

BOX 21.1 The Child Survival Revolution

Every year in the developing countries of the world, some 14 million children under the age of five die of common infectious diseases (box figure 21.1). Most of these children could be saved by simple, inexpensive, preventative medicine. Many public health officials argue that it is as immoral and unethical to allow children to die of easily preventable diseases as it would be to allow them to starve to death or to be murdered. Preventing these deaths would not only eliminate tremendous pain and suffering; it would also preserve one of our most precious resources: children. As James P. Grant, head of the United Nations Childrens Fund (UNICEF), says, "children are our future." One measure of civilization is how we treat the most vulnerable members of society. The state of the world's children is clearly an indication of the level of our civilization. In 1986, the United Nations announced a worldwide campaign to prevent unnecessary child deaths. Called the "child survival revolution," this campaign is based on four principles designated by the acronym GOBI.

G is for growth monitoring. A healthy child is a growing child. Underweight children are much more susceptible to infectious diseases, retardation, and other medical problems than children who are better nourished. Regular growth monitoring is the first step in health maintenance.

O is for oral rehydration therapy (ORT). About one-third of all deaths under five years of age are caused by diarrheal diseases. A simple solution of salts, glucose or rice powder, and boiled water given orally is almost miraculously effective in preventing death from dehydration shock in these diseases. The cost of treatment is only a few cents per child. The British medical journal, *Lancet*, has called ORT "the most important medical advance of the century."

B is for breast-feeding. Babies who are breast-fed get natural immunity to diseases from antibodies in their mothers' milk, but infant formula companies have been persuading mothers in many developing countries that bottle-feeding is

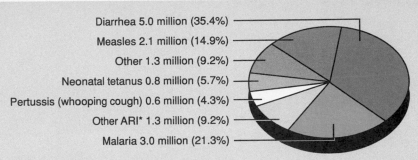

Diarrhea 5.0 million (35.4%)
Measles 2.1 million (14.9%)
Other 1.3 million (9.2%)
Neonatal tetanus 0.8 million (5.7%)
Pertussis (whooping cough) 0.6 million (4.3%)
Other ARI* 1.3 million (9.2%)
Malaria 3.0 million (21.3%)

Estimated total annual child deaths: 14.1 million

Notes: For purposes of this chart, one cause of death has been allocated for each child death when, in fact, children die of multiple causes.

*Other acute respiratory infections (ARI): Tuberculosis, diphtheria, pneumonia, influenza, pleurisy, acute bronchitis, otitis media and other respiratory tract diseases.

BOX FIGURE 21.1 Estimated annual deaths of children under five by cause. Most of these deaths could be prevented by better sanitation and inexpensive health care.

From WHO and UNICEF estimates, *State of the World's Children*. Copyright © 1987 Oxford University Press, Oxford, England. Reprinted by permission.

more modern and healthful than breast-feeding. Unfortunately, these mothers usually don't have access to clean water to combine with the formula and they can't afford enough expensive synthetic formula to nourish their babies adequately. Consequently, the mortality among bottle-fed babies is much higher than among breast-fed babies in developing countries.

I is for universal immunization against the six largest, preventable, communicable diseases of the world: measles, tetanus, tuberculosis, polio, diphtheria, and whooping cough. In 1975, less than 10 percent of the developing world's children had been immunized. By 1985, this number had risen to just under 50 percent. By 1990, the U. N. hopes to achieve full immunization, saving about five million children per year. In some countries, yellow fever, typhoid, meningitis, cholera, and other diseases also urgently need attention.

Burkina Faso provides an excellent example of how a successful immunization campaign can be carried out. Although this West African nation is one of the poorest in the world (annual GNP per capita of only $140), and roads, health care clinics, communication, and educational facilities are either nonexistent or woefully inadequate, a highly successful "vaccination commando" operation was undertaken in 1985. In a single three-week period, one million children

were immunized against three major diseases (measles, yellow fever, and meningitis) with only a single injection. This represents 60 percent of all children under age fourteen in the country. The cost was less than $1 per child.

In addition to being an issue of humanity and compassion, reducing child mortality may be one of the best ways to stabilize world population. There has never been a reduction in birth rates that was not preceded by a reduction in infant mortality. When parents are confident that their children will survive, they tend to have only the number of children they actually want, rather than "compensating" for likely deaths by extra births. In Bangladesh, where ORT was discovered, a children's health campaign in the slums of Dacca has reduced infant mortality rates 21 percent since 1983. In that same period, the use of birth control increased 45 percent and birth rates decreased 21 percent. Sri Lanka, China, Costa Rica, Thailand, and the Republic of Korea have reduced child deaths to a level comparable to those in many highly developed countries. This child survival revolution has been followed by low birth rates and stabilizing populations. The United Nations Children's Fund estimates that if all developing countries had been able to achieve similar birth and death rates, there would have been nine million fewer child deaths in 1987, and nearly 22 million fewer births.

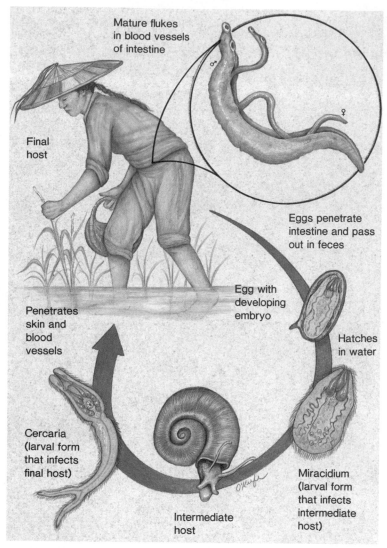

FIGURE 21.4 The blood fluke life cycle. Sexual forms (inset) live in human tissues. The mature female fits in a groove running the length of the larger male's body. Eggs pass out of the body in feces. They hatch in fresh water and infect the intermediate host (snails) where they reproduce asexually to produce free-swimming larvae that reinfect humans.

Labels in figure: Mature flukes in blood vessels of intestine; Final host; Eggs penetrate intestine and pass out in feces; Penetrates skin and blood vessels; Egg with developing embryo; Hatches in water; Cercaria (larval form that infects final host); Miracidium (larval form that infects intermediate host); Intermediate host

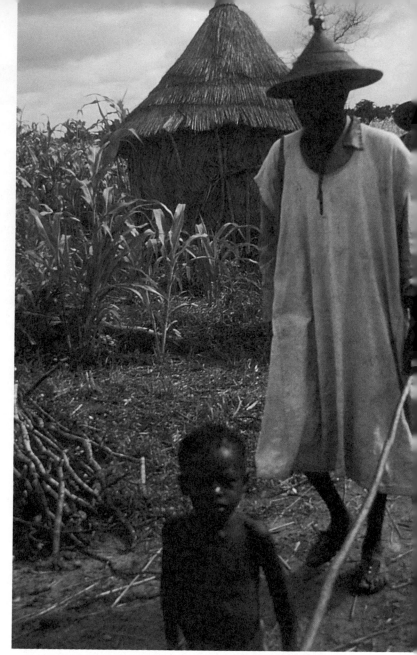

FIGURE 21.5 A child leads a blind adult in West Africa. River blindness, a parasitic disease transmitted by biting flies, affects 18 million people. In some African villages, nearly every adult over age thirty is blind due to this disease.

mosquitoes. The worms block the lymphatic system, causing large fluid accumulations and swellings in various body areas. Guinea worms (*Dracunculus*) live as larvae inside small aquatic crustaceans. They infect people who drink unfiltered water containing the crustaceans. Adult worms, can be up to 1 m (3 ft) long. After a year of growth, the worm emerges through the skin to lay its eggs. This painful ordeal may last several weeks, and the open sores it creates get infected, causing further illness.

Tuberculosis and other respiratory diseases (influenza and pneumonia) are the leading cause of death in many subtropical countries, especially Latin America. Until 1900, tuberculosis was the major cause of death in the United States—it is still endemic on some Indian reservations. Like diarrhea, tuberculosis rarely kills a well-fed individual, and it is highly preventable by good sanitation and inexpensive inoculations. Pneumonia and influenza now are the leading causes of deaths from infection in the United States, ranking sixth among all causes. This statistic reflects the fact that these diseases are often the last, fatal complication of a variety of other ailments.

Trachoma is another very widespread eye disease. It is a contagious inflammation of the inner eyelid, tear glands, and cornea caused by viruses. This disease is found where sanitation is poor. If not treated, it can cause blindness. Several hundred million people, mostly children, suffer from trachoma.

ENVIRONMENTAL POLLUTION AND OTHER DILEMMAS

Although sexually-transmitted diseases don't kill nearly the numbers that malaria, diarrhea, and tuberculosis do, there is a worry that they may become more serious problems in the future. Antibiotic-resistant "super strains" of the gonorrhea and syphilis bacteria have developed. The virus that causes acquired immune deficiency syndrome (AIDS) causes an invariably fatal disease because it attacks the immune system itself. Some epidemiologists warn that AIDS could become a pandemic equal to the bubonic plague (Black Death), which killed about one-third of the population of Europe in the fourteenth century.

In the more highly developed countries, improved sanitation, health care, and immunizations have greatly reduced the incidence of most life-threatening diseases. Life expectancy has doubled over the past century in these countries due largely to the reduction in communicable diseases.

Chemicals

Toxic chemicals in the environment are becoming a source of increasing concern to people in industrialized countries. Humans probably have always been subjected to a variety of toxic materials from natural sources. Now we are also exposed to an increasing variety and quantity of dangerous synthetic chemicals as well. In this section, we will look at some of these materials and why they are of concern.

Chemical agents are divided into two broad categories: those that are hazardous and those that are toxic. **Hazardous** means dangerous. This category includes flammables, explosives, irritants, sensitizers, acids, and caustics. Many chemicals that are hazardous in high concentrations are relatively harmless when dilute. **Toxins** are poisonous. That means they react with specific cellular components to kill cells. Because of this specificity, they often are harmful even in dilute concentrations. Toxins can be either general poisons that kill many kinds of cells, or they can be extremely specific in their target and mode of action. Ricin, for instance, is a protein found in castor beans. It is one of the most toxic organic compounds known. Three hundred picograms (trillionths of a gram) injected intravenously is enough to kill an average mouse. A single molecule can kill a cell. This is about two hundred times the acute toxic dose for dioxin. Table 21.2 shows some of the environmental toxins of greatest concern to the EPA. This group of chemicals includes heavy metals, inorganic chemicals, and both natural and synthetic organic compounds.

Active sites of cell division, such as bone marrow or the lining of the respiratory and gastrointestinal systems, often are the most sensitive to poisoning. Furthermore, malfunctioning of these tissues is more noticeable because they are so important to our health. Similarly, fetal growth, where cells are dividing rapidly and differentiating to form new tissues, is one of the most sensitive stages of life in terms of environmental toxins. Let's look more closely at some of the major types of hazardous and toxic chemicals and why they are dangerous.

TABLE 21.2	Toxic chemicals causing the greatest risk to human health (in order of relative importance: greatest to least)
Benzene	Methyl ethyl ketone
Cadmium	Methyl isobutyl ketone
Carbon tetrachloride	Nickel
Chloroform	Tetrachloroethylene
Chromium	Toluene
Cyanides	Trichloroethane
Dichloromethane	Trichloroethylene
Lead	Xylene(s)
Mercury	

Source: Environmental Protection Agency, 1991.

Irritants are corrosives (strong acids), caustics (alkaline reagents), and other substances that damage biological tissues on contact. Some examples are sulfuric and nitric acid, ammonia, sodium hydroxide, toxic metal fumes (e.g., beryllium or nickel), ozone, chlorine, sulfur or nitrogen oxides, formaldehyde, benzene hexachloride, and dioxin. These agents not only damage cells directly, but also make them susceptible to infections and can trigger transformations to a cancerous state. Skin diseases caused by irritants (dermatoses) are the most common occupational disease.

Respiratory fibrotic agents are a special class of irritants that damage the lungs, causing scar tissue formation that lowers respiratory capacity. This group includes both chemical reagents and particulate materials. Some conditions are common enough to be given specific names: silicosis (caused by silica dust), black lung (caused by coal dust), brown lung (caused by cotton fibers), asbestosis (caused by asbestos fibers), and farmer's lung (caused by moldy hay), among others. Some of these health problems are simply obstructive diseases in which the lungs fill with residue and tissue that interfere with breathing. Some also lead to cancer.

Asphyxiants are chemicals that exclude oxygen or actively interfere with oxygen uptake and distribution. Pure/or nitrogen oxides, methane, and carbon dioxide are representatives of passive asphyxiants. Under normal circumstances, they are relatively inert, but when they fill enclosed spaces like mines, caves, or farm silos, they can be deadly. By contrast, active asphyxiants react chemically with blood or lung tissue to prevent oxygen uptake. Some examples are carbon monoxide, hydrogen cyanide, hydrogen sulfide, and analine. The effects of these chemicals tend to be relatively irreversible and are toxic even in low concentrations.

Allergens are substances that activate the immune system. Some allergens act directly as **antigens**; that is, they are recognized as foreign by white blood cells and stimulate the production of specific antibodies. Other allergens act indirectly by binding to other materials and changing their structure or chemistry so they become antigenic and cause an immune response.

Formaldehyde is a good example of a widely used synthetic chemical that is a powerful sensitizer. It is both directly and indirectly allergenic. Some people who are exposed to formaldehyde in plastics, wood products, insulation, glue, fabric dyes, and a variety of other products become hypersensitive not only to formaldehyde itself, but also to many other materials in their environment. This is sometimes called "sick house" syndrome. Victims may have to go to great lengths to protect themselves from these allergenic substances.

Immune system depressants are pollutants that seem to suppress the immune system rather than activate it. Little is known about how this occurs or which chemicals are responsible. Immune system failure was thought to play a role, however, in the widespread die-off of seals in the North Atlantic in 1988 and of dolphins in the Mediterranean in 1989. In both cases, the dead animals contained high levels of pesticide residues, PCBs, and other contaminants. They seemed to have succumbed to a variety of opportunistic infections. Similarly, some humans with "sick-house" syndrome or other environmental illnesses seem to have defective immune responses. The evidence for a pollution link is mostly anecdotal, however, and little hard scientific data supports these claims.

Neurotoxins are a special class of metabolic poisons that specifically attack nerve cells (neurons). The nervous system is so important in regulating body activities that disruption of its activities is especially fast-acting and devastating. Different types of neurotoxins act in different ways. Anesthetics (ether, chloroform, halothane), chlorinated hydrocarbons (DDT, Dieldrin, Aldrin), and heavy metals (lead, mercury) disrupt the ion transport across cell membranes necessary for nerve action. Organophosphates (Malathion, Parathion) and carbamates (Sevin, Zeneb, Maneb) inhibit acetylcholinesterase, an enzyme that regulates nerve signal transmission between nerve cells and the tissues or organs they innervate (e.g., muscle). Most neurotoxins are both extremely toxic and fast-acting.

Mutagens are agents, such as chemicals and radiation, that damage or alter genetic material (DNA) in cells. This can lead to birth defects if the damage occurs during embryonic or fetal growth. Later in life, genetic damage may trigger neoplastic (tumor) growth. When damage occurs in reproductive cells, the results can be passed on to future generations. Cells have repair mechanisms to detect and restore damaged genetic material, but some changes may be hidden, and the repair process itself can be flawed. It is generally accepted that there is no "safe" threshold for exposure to mutagens. Any exposure has some possibility of causing damage.

Teratogens are chemicals or other factors that specifically cause abnormalities during embryonic growth and development. Some compounds that are not otherwise harmful can cause tragic

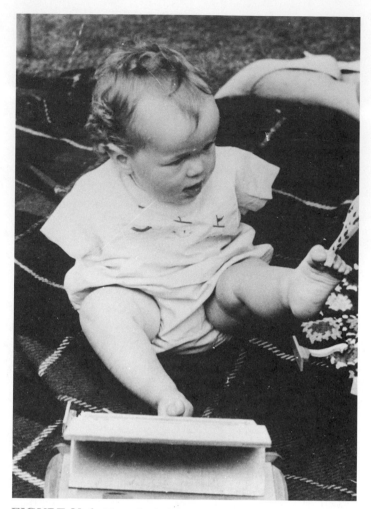

FIGURE 21.6   Tragic birth defects caused by the teratogenic sedative thalidomide. Fortunately, thalidomide was not approved for use in the United States.

problems in these sensitive stages of life. One of the most well-known examples of teratogenesis is that of the widely used sedative, thalidomide. In the 1960s, thalidomide (marketed under the trade name Cantergan) was the most widely used sleeping pill in Europe. It seemed to have no unwanted effects and was sold without prescription. When used by pregnant women, however, it caused abnormal fetal development resulting in phocomelia, meaning seal-like limbs, in which there is a hand or foot, but no arm or leg (figure 21.6). There is evidence that taking a single thalidomide pill in the first weeks of pregnancy is sufficient to cause these tragic birth defects. Altogether, about 10,000 children were affected before this drug was withdrawn from the market. Fortunately, thalidomide was not approved for sale in the United States because the Food and Drug Administration was not satisfied with the laboratory tests of its safety.

We don't know how many birth defects are due to mutagens, teratogens, or other environmental factors. Between 5 and 10 percent of all children born alive have birth defects that require medical attention in the first few years of life. It is thought

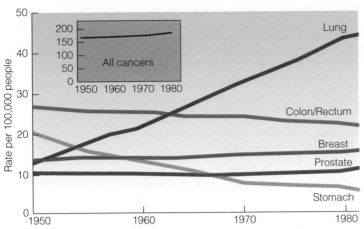

FIGURE 21.7 U.S. cancer death rates. Although the total cancer rate in the United States has increased in recent years, the age-adjusted rates have been constant or falling for all but respiratory cancers, most of which are caused by smoking.

that as many as 75 percent of all human embryos fail to implant in the uterus or undergo spontaneous abortions, often without the mother's knowledge that conception has occurred. Cell damage and errors in development caused by environmental factors may play a large role in preventing these pregnancies.

Carcinogens are substances that cause **cancer,** invasive, out-of-control cell growth that results in **malignant tumors.** Cancer rates have been rising rapidly in the United States and most other industrialized countries in recent years, and cancer is now the second leading cause of death in the United States, killing 510,000 people in 1990. Some investigators warn that we are entering an era of "cancer epidemic" due to the proliferation of toxic pollutants in the environment. They point out that 30 percent of Americans now living eventually will have cancer if present rates continue. A part of this rapid rise is due to increasing longevity. Since cancer often takes twenty or thirty years to develop, it is especially associated with old age. If we adjust for an aging population and better diagnosis, the rate for most cancers has remained steady or even declined. The only major types that have increased greatly are skin cancer and respiratory cancers, most of which are associated with excessive sun exposure and smoking. (figure 21.7). Still, in spite of improving preventive measures and increasingly successful therapy, cancer eventually will strike in approximately three of four families. Few of us will escape contact with this dreaded disease.

There are also many different kinds of cancer and probably many different environmental causes. Some viral infections may trigger cancer. Mutagenic agents, such as radiation, heavy metals, and organic chemicals, also initiate this process. It may be that every mutagen is a potential carcinogen. Crystals of asbestos can cause cancer when they are ingested or inhaled. Foreign ma-

terial, such as plastic in the body, also can trigger tumor formation. Repeated damage to cells by toxic agents like alcohol also can result in cancer, especially in the liver. Some agents are not carcinogenic themselves but assist in the progression and spread of tumors. These factors are called cocarcinogens or **promoters.** Table 21.2 lists some carcinogenic factors and promoters in the environment.

Natural Toxins and Carcinogens

Although technology is adding many new, synthetic chemicals to the environment, there probably are even more naturally occurring toxic chemicals to which we are exposed. Since most plants and many animal species can't escape from predators or defend themselves by fighting back, many of them have evolved a kind of chemical warfare, secreting or storing in their tissues a vast armamentarium of irritants, toxins, metabolic disrupters, and other chemicals that discourage competitors and predators. Some of the chemical defenses employed by organisms are very sophisticated and specific. Both plants and animals make chemicals similar to or even identical to neurotransmitters, hormones, or regulatory molecules of predators or potential enemies.

Humans are highly susceptible to poisoning by natural materials present in the environment. Some of the most potent toxins, mutagens, teratogens, and carcinogens known are natural products. Toxicologist Bruce Ames claims that there are 10,000 times more natural pesticides in our diets than synthetic ones. He argues that our fear of synthetic chemicals may divert our attention from more important issues. For instance, he has found natural pesticides in crops, such as potatoes, tomatoes, coffee, celery, and mushrooms, that are more carcinogenic than some commercial products. When plants are attacked by insects, for instance, they may synthesize natural toxins that are more dangerous than the residues left from protective treatment with synthetic pesticides. Similarly, treatment of crops with fungicides (which are carcinogenic) may prevent growth of molds that are even more carcinogenic. Simply because food is raised "organically" may not necessarily make it safer than food raised by current commercial practices.

Other environmental health specialists argue that there is a great difference between the effects of toxic chemicals in our diet when they are mixed with fiber and a multitude of other substances and their effects as pure chemicals in laboratory tests. There may be much less danger from natural foods than Professor Ames' results suggest.

Physical Agents, Trauma, and Stress

Physical agents, such as radiation, also are important environmental health hazards. The sources and effects of radiation are discussed in chapter 19. Noise is another important physical danger to health. Because this is an especially prominent factor in cities, it is discussed in chapter 25.

TABLE 21.3	Causes of death in the United States in 1984			
Cause of death	Years of potential life lost	(Rank)	Deaths per 100,000	(Rank)
All causes	11,761,000		866.7	
Unintentional injuries*	2,308,000	(1)	40.1	(4)
Malignant neoplasms	1,803,000	(2)	191.6	(2)
Heart diseases	1,563,000	(3)	324.4	(1)
Suicide, homicide*	1,247,000	(4)	20.6	(7)
Congenital anomalies	684,000	(5)	5.6	(10)
Premature birth	470,000	(6)	3.5	(11)
Sudden infant death syndrome	314,000	(7)	2.4	(12)
Cerebrovascular diseases	266,000	(8)	65.6	(3)
Chronic liver diseases and cirrhosis	233,000	(9)	11.3	(9)
Pneumonia and influenza	163,000	(10)	25.0	(6)
Chronic obstructive pulmonary diseases	123,000	(11)	29.8	(5)
Diabetes mellitus	119,000	(12)	15.6	(8)

*Trauma-related causes

(Ranked by estimated years of potential life lost before age sixty-five and by deaths per 100,000 population)
Source: From R. I. Glass, "New Prospects for Epidemiologic Investigations" in *Science*, 234:952, November 21, 1984. Copyright 1984 by the American Association for the Advancement of Science, Washington, DC. Reprinted by permission of the publisher and author.

Trauma, injury caused by accidents and violence, has surely always been a life-threatening environmental factor for humans. There is probably less danger from physical trauma in the more developed countries of the world now than ever before, even though modern media coverage might make it seem otherwise. The death rate from accidents in the United States is 40 per 100,000, about half what it was in 1900. Still, accidents, homicide, and suicide are the principal causes of death for people between the ages of one and thirty-eight in the United States, and trauma is the leading cause of years of life lost before age sixty-five in all industrialized countries (table 21.3). Every year there are about 100,000 premature deaths from trauma in the United States and twice as many cases of permanent disability. More than half of these deaths and injuries are caused by motor vehicle accidents. About 90 percent of the accidents involve private automobiles, trucks, or motorcycles, and at least half are caused by drivers under the influence of alcohol or other drugs.

The United States has nearly ten times as many intentional deaths as any other nation in the industrialized world. In fact, most of the trauma deaths not caused by motor vehicle accidents in the United States are the result of suicide (12.0 per 100,000) or homicide (8.6 per 100,000). The death rates for teenagers and young adults in the United States are 50 percent

higher than that for any of the other top twenty industrialized nations. These death rates are strongly related to race, sex, income, and social class. A young black male is four times as likely to be murdered as a young white male.

Stress and life-style once were considered the realm of psychologists and sociologists. We have learned, however, that there is a definite relationship between the level of stress in one's life and physical diseases, such as heart attack, stroke, and atherosclerosis, which are the leading causes of nontrauma-related death in the United States and most other industrialized countries (table 21.3). Since stress is an important aspect of our environment, it deserves a place in a discussion of environmental health.

What is **stress**? In the health sense, stress refers to a physical, chemical, or emotional factor that places a strain on an animal to which the animal does not make an adequate adaptation. As a result, physiological tensions produced may contribute to disease. Physical responses to stress are not unique to humans. When subdominant male animals are kept in cages next to especially aggressive males of the same species in laboratories or zoos, they often show many signs of anxiety and stress. Even though the animals are separated by glass walls so that no physical contact is possible, the stressed animals die prematurely of cardiovascular diseases. Gastrointestinal disturbances, such as ulcers, are common human responses to stress. Stress also contributes to susceptibility to infectious diseases, as many students realize at exam time.

Diet

Diet also has an important effect on health. For instance, there is a strong correlation between cardiovascular disease and the amount of salt and animal fat in one's diet (figure 21.8). Highly processed foods; fat; and smoke-cured, high-nitrate meats also are associated with cancer. Fruits, vegetables, whole grains, complex carbohydrates, and dietary fiber (plant cell walls), on the other hand, often have beneficial health effects. Certain dietary components, such as pectins; vitamins A, C, and E; substances produced in cruciferous vegetables (cabbage, broccoli, cauliflower, brussel sprouts); and selenium (in low levels) seem to have anticancer effects.

Eating too much food is a significant dietary health factor in developed countries and among the well-to-do everywhere. At least one-fourth of all Americans are considered overweight. Cutting back on the number of calories consumed reduces the strain on bones, muscles, and other organs, and has additional beneficial effects, including reduction of cardiovascular disease, diabetes, and—perhaps—cancer.

In some areas of the world, people seem to live exceptionally long lives. The Abkhasian people in the Caucasus Mountains of Soviet Georgia, the Hunzans in the mountains of Pakistan, and the Vilcabama villagers in Ecuador, for instance, are among the longest-lived people in the world. Many claim to

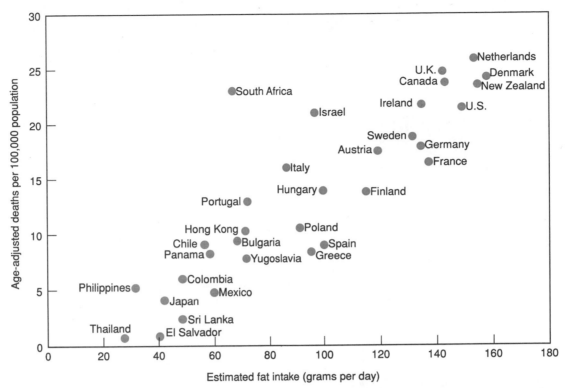

FIGURE 21.8 Strong linear correlation between dietary fat intake and deaths from breast cancer. The exact cause/effect relationship is unknown.

From "Diet and Cancer" by Leonard A. Cohen. Copyright © 1987 by Scientific American, Inc. All rights reserved.

be 120 to 140 years old, although it is difficult to substantiate when they were born. Still, they do appear to live longer and to be more physically active later in life than do most of us.

These people have several factors in common that probably contribute to longevity. They live at moderately high elevations where the climate is cool, dry, and sunny. They lead active, vigorous lives in nonindustrialized settings. In their small villages, life is uniform and predictable, and stress levels are low. All ages work together in the fields and in the home. The elderly are respected and live a useful, active life. The whole family shares in decision making, recreation, and religion. Their diet is usually simple, with low fat and salt, high fiber content, little meat, and lots of fruits and vegetables. They enjoy clean air and pure water. Their culture and uncomplicated life-styles reduce conflict and anxiety. It may be that we could benefit from incorporating some aspects of their lives into our own (table 21.4).

MOVEMENT, DISTRIBUTION, AND FATE OF TOXINS

There are many sources of toxic and hazardous chemicals in the environment and many factors related to the chemical itself, its route or method of exposure, and its persistence in the environ-

TABLE 21.4 National health recommendations and diet goals
Eat only enough calories to meet body needs (fewer if overweight).
Eat less fat and cholesterol.
Eat less salt.
Eat less sugar.
Eat more whole grains, cereals, fruits, and vegetables.
Eat more fish, poultry, beans, and peas.
Eat less red meat.
Eat less additives and processed foods.

Source: The Surgeon General's Report: *Healthy People*, 1980.

ment, as well as characteristics of the target organism (table 21.5), that determine the danger of each chemical. We can think of an ecosystem as a set of interacting compartments between which a chemical moves, based on its molecular size, solubility, stability, and reactivity (figure 21.9). The routes of entry of chemicals into the body also play an important role in determining how dangerous they are (figure 21.10). In this section, we will consider some of these characteristics and how they affect environmental health.

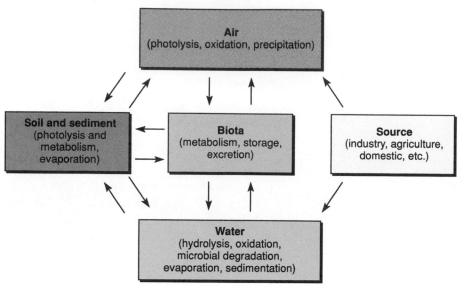

FIGURE 21.9 Movement and fate of chemicals in the environment. Mechanisms that modify, remove, or sequester compounds are shown in parentheses.

Factors related to the toxic agent
1. Chemical composition and reactivity
2. Physical characteristics (e.g., solubility, state)
3. Presence of impurities or contaminants
4. Stability and storage characteristics of toxic agent
5. Availability of vehicle (e.g., solvent) to carry agent
6. Movement of agent through environment and into cells

Factors related to exposure
1. Dose (concentration and volume of exposure)
2. Route, rate, and site of exposure
3. Duration and frequency of exposure
4. Time of exposure (time of day, season, year)

Factors related to organism
1. Resistance to uptake, storage, or cell permeability of agent
2. Ability to metabolize, inactivate, sequester, or eliminate agent
3. Tendency to activate or alter nontoxic substances so they become toxic
4. Concurrent infections or physical or chemical stress
5. Species and genetic characteristics of organism
6. Nutritional status of subject
7. Age, sex, body weight, immunological status, and maturity

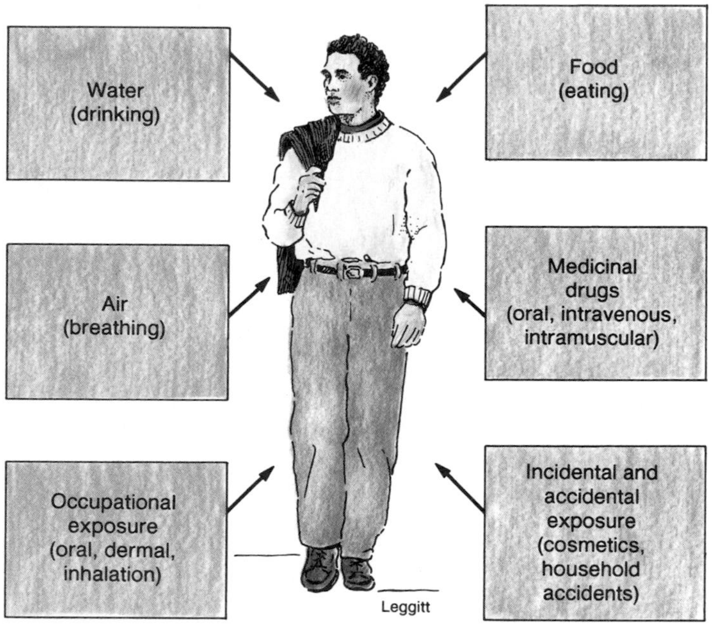

FIGURE 21.10 Routes of exposure to toxic and hazardous environmental factors.

Solubility

Solubility is one of the most important characteristics in determining how, where, and when a toxic material will move through the environment or through the body to its site of action. Chemicals can be divided into two major groups: those that dissolve more readily in water and those that dissolve more readily in oil. Water-soluble compounds move rapidly and widely through the environment because water is ubiquitous. They also tend to have ready access to most cells in the body because aqueous solutions bathe all our cells. Molecules that are oil- or fat-soluble (usually organic molecules) generally need a carrier to move through the environment, into, and within the body. Once inside the body, however, oil-soluble toxins penetrate readily into tissues and cells because the membranes that enclose cells are themselves made of similar oil-soluble chemicals. Once they get inside cells, oil-soluble materials are likely to be accumulated and stored in lipid deposits where they may be protected from metabolic breakdown and persist for many years.

ENVIRONMENTAL POLLUTION AND OTHER DILEMMAS

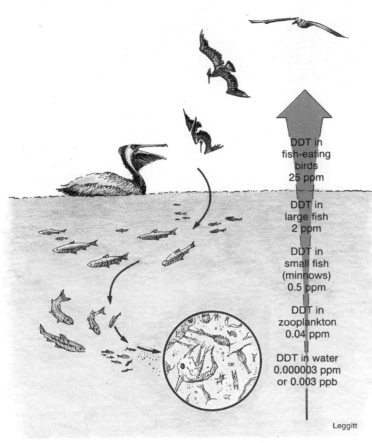

DDT in
fish-eating
birds
25 ppm

DDT in
large fish
2 ppm

DDT in
small fish
(minnows)
0.5 ppm

DDT in
zooplankton
0.04 ppm

DDT in water
0.000003 ppm
or 0.003 ppb

Leggitt

FIGURE 21.11 Bioaccumulation and biomagnification. Lower organisms take up and store toxins from the environment. They are eaten by larger predators, who are eaten, in turn, by even larger predators. The highest members of the food chain can accumulate very high levels of the toxin.

Bioaccumulation and Biomagnification

Cells have mechanisms for **bioaccumulation,** selectively absorbing and storing a great variety of molecules. This allows them to accumulate nutrients and essential minerals, but they also may absorb and store harmful substances through these same mechanisms. Toxins that are rather dilute in the environment can reach dangerous levels inside cells and tissues through this process of bioaccumulation.

The effects of toxins also are magnified in the environment through food chains. **Biomagnification** occurs when the toxic burden of a large number of organisms at a lower trophic level is accumulated and concentrated by a predator in a higher trophic level. Phytoplankton and bacteria in aquatic ecosystems, for instance, take up heavy metals or toxic organic molecules from water or sediments (figure 21.11). Their predators— zooplankton and small fish—collect and retain the toxins from many prey organisms, building up higher concentrations of toxins. The top carnivores in the food chain—game fish, fish-eating birds, and humans—can accumulate such high toxin levels that they suffer adverse health effects. One of the first known examples of bioaccumulation and biomagnification was DDT,

which accumulated through food chains so that by the 1960s it was shown to be interfering with reproduction of peregrine falcons, brown pelicans, and other predatory birds at the top of their food chains.

Persistence

Some chemical compounds are very unstable and degrade rapidly under most environmental conditions, so that their concentrations decline quickly after release. Some of the modern herbicides, for instance, quickly lose their toxicity. Other substances are more persistent and last for long times. Some of the most useful chemicals, such as chlorofluorocarbons, plastics, chlorinated hydrocarbons, and asbestos, are valuable because they are resistant to degradation. This stability also causes problems because these materials persist in the environment and have unexpected effects far from the sites of their original use. DDT, for instance, is a useful pesticide because it breaks down very slowly and doesn't have to be reapplied very often. Its toxic effects may spread to unintended victims, however, and it may be stored for long periods of time in organisms that lack mechanisms to destroy it.

MECHANISMS FOR AVOIDING OR MINIMIZING TOXIC EFFECTS

Organisms have ways to avoid toxins or minimize toxic effects. Sometimes, however, a mechanism that is beneficial with one type of toxin or at one stage in the life cycle becomes deleterious with another substance or in another stage of development. Let's look at how these processes help protect us from harmful substances, and how they can go awry.

Metabolic Degradation and Excretion

Most organisms have enzymes that process waste products and environmental poisons to reduce their toxicity and biological activity. In mammals many of these enzymes are located in the liver, the primary site of detoxification of both natural wastes and introduced poisons. Sometimes, however, these reactions work to our disadvantage. Compounds, such as benzepyrene, for example, that are not toxic in their original form, are processed by these same enzymes into cancer-causing carcinogens. Why would we have a system that makes a chemical more dangerous? The answer can only be speculative, but it seems likely that our defense mechanisms are "selected" by evolution to protect us from toxins and hazards early in life. Factors or conditions that affect postreproductive ages (like cancer or premature senility) don't affect reproductive success and don't exert selective pressure.

We also reduce the effects of waste products and environmental toxins by eliminating them from our body through excretion. Some volatile molecules, such as carbon dioxide,

TABLE 21.6 Toxicity rating chart

Class	Lethal dose*	Practical nonmetric equivalent
1. Practically nontoxic	15 g/kg	More than one quart
2. Slightly toxic	5–15 g/kg	Between one pint and one quart
3. Moderately toxic	0.5–5 g/kg	Between one ounce and one pint
4. Very toxic	50–500 mg/kg	Between a teaspoon and one ounce
5. Extremely toxic	5–50 mg/kg	Between 7 drops and a teaspoon
6. Supertoxic	less than 5 mg/kg	Less than 7 drops

*Probable lethal dose for average adult human.

hydrogen cyanide, and ketones are excreted via breathing. Some excess salts and other substances are excreted in sweat. Primarily, however, excretion is a function of the kidneys, which can eliminate significant amounts of soluble materials through urine formation. Accumulation of toxins in the urine can damage this vital system, however, and the kidneys and bladder often are subjected to harmful levels of toxic compounds. In the same way, the stomach, intestine, and colon often suffer damage from materials concentrated in the digestive system and may be afflicted by diseases and tumors.

Repair Mechanisms

In the same way that individual cells have enzymes to repair damage to DNA and protein at the molecular level (Box 19.2), tissues and organs that are exposed regularly to physical wear-and-tear or to toxic or hazardous materials often have mechanisms to repair damage to which they are subjected. Our skin and the epithelial linings of the gastrointestinal tract, blood vessels, lungs, and urogenital system have high cellular reproduction rates so injured cells can be replaced. With each reproduction cycle, however, there is a chance of error, and these and other active cell-replacement tissues are among the most likely in the body to develop cancers (table 21.6). This also may be the reason that ethyl alcohol is a powerful liver carcinogen.

Avoidance

Many organisms adopt protective behavior to avoid or mitigate the effects of toxins and other hazards. They learn to recognize sources of danger and protect themselves. Animals instinctively breathe shallowly when there are noxious gases in the air. They learn to avoid prey that contain poisonous chemicals. Even plants have protective behaviors, closing stomates (pores in their leaves) when exposed to ozone or sulfur oxide gases, for example. They also drop damaged leaves and seal off injured areas so that diseases and toxins will not spread further. There is evidence early humans knew many thousands of years ago about the dangers of toxins and about other threats to their health from certain environmental hazards, as well as ways to mitigate those hazards.

MEASURING TOXICITY

How toxic are the chemical substances to which we may be exposed? Determining and comparing the toxicity of various materials is difficult because of different routes of exposure, sensitivity, kinds of reactions, and effects of various chemicals in different species. Different individuals within a single species can have very different responses to a given exposure of toxic material. In this section, we will look at methods of toxicity testing and at how results are analyzed and reported.

Animal Testing

The most commonly used and most widely accepted toxicity test is to expose a population of laboratory animals to measured doses of a specific substance under controlled conditions. This procedure is expensive, time consuming, and often painful and debilitating to the animals being tested. It commonly takes hundreds—or even thousands—of animals, several years of hard work, and hundreds of thousands of dollars to thoroughly test the effects of a toxin at very low doses. More humane toxicity tests using computer simulation of model reactions, cell cultures, or other substitutes for whole living animals are being developed. However, conventional large-scale animal testing is the method in which we have the most confidence and on which most public policies about pollution and environmental or occupational health hazards are based.

In addition to humanitarian concerns, there are several problems in laboratory animal testing that trouble both toxicologists and policy makers. One problem is differences in sensitivity to a toxin of the members of a specific population. Figure 21.12 shows a typical dose/response curve for exposure to a hypothetical toxin. Some individuals are very sensitive to the toxin, while others are insensitive. Most, however, fall in a middle category forming a bell-shaped curve. The question for regulators and politicians is whether we should set pollution levels that will protect everyone, including the most sensitive people, or only aim to protect the average person. It might cost billions of extra dollars to protect a very small number of individuals at the extreme end of the curve. Is that a good use of resources?

Curves are not always uniformly symmetrical, making it difficult to compare toxicity of different chemicals or different species of organisms. A convenient way to describe toxicity of a chemical is to determine the dose to which 50 percent of the test population is sensitive. In the case of a lethal dose (LD), this is called the **LD50** (figure 21.13).

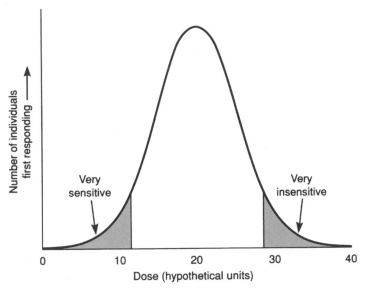

FIGURE 21.12 Variations in sensitivity to a toxin within a population. Some members of a population are very sensitive to a given toxin, while others are much less sensitive. The majority of the population falls somewhere between the two extremes.

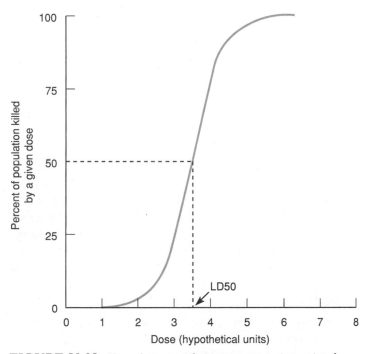

FIGURE 21.13 Cumulative population response to increasing doses of a toxin. The LD50 is the dose that is lethal to half the population.

Different species can react very differently to the same toxin, not only because body sizes vary but also because of variations in physiology and metabolism. Even closely related species can have very dissimilar sensitivities to a particular toxin. Hamsters, for instance, are nearly 5,000 times less sensitive to some dioxins than are guinea pigs. Of 226 chemicals found to

TABLE 21.7	Acute lethal dose for some toxic organic chemicals	
Chemical	**Exposure**	**LD50**
Ricin (castor bean)	Ivn-mus	3 ng/kg
	Orl-rat	100 mg/kg
Botulism toxin	Ipr-mus	160 ng/kg
Dioxin (tetrachlorodioxin)	Orl-gpg	600 ng/kg
	Orl-hmstr	3 mg/kg
Muscarine (mushroom poison)	Ivn-mus	250 µg/kg
Parathion (insecticide)	Ipr-rat	1.5 mg/kg
Aflatoxin (fungal toxin)	Orl-mky	1.75 mg/kg
Nicotine	Ivn-cat	2 mg/kg
	Orl-rat	53 mg/kg
DDT (dichlorodiphenyl-trichloroethane)	Orl-hum	50 mg/kg
Toxaphene	Orl-rat	60 mg/kg
2,4–D (dichlorophenoxyacetic acid)	Orl-hum	80 mg/kg

orl = oral, ivn = intravenous, ipr = intraperitoneal, mus = mouse, mky = monkey, hmstr = hamster, hum = human, gpg = guinea pig

nanogram (ng) = 1×10^{-9} gm
microgram (µg) = 1×10^{-6} gm
milligram (mg) = 1×10^{-3} gm

Source: *Registry of Toxic Effects of Chemical Substances*, National Institute for Occupational Safety and Health, 1985.

be carcinogenic in either rats or mice, 95 caused cancer in one species but not the other. These differences make it difficult to estimate the risks for humans since we can't perform controlled experiments in which we deliberately expose people to toxins.

Human Toxicity Ratings

Table 21.6 shows six toxicity categories based on the oral dose that is generally lethal for a human. The amount required for an average human is also shown. Supertoxic chemicals are extremely potent; many require far less than a single drop to make a lethal dose. Table 21.7 shows the LD50 for a variety of different animals and methods of exposure for some of the most directly toxic chemicals known. Note that these materials are not all synthetic (human-made). The most toxic of these chemicals is ricin, a protein found in castor bean seeds. It is so toxic that 0.3 billionths of a gram given intravenously will generally kill a mouse. If aspirin were this toxic, a single tablet, divided into small doses, could kill one million people.

This table doesn't consider carcinogenesis (cancer formation) since this is a different kind of health risk from direct toxicity. Many carcinogens are dangerous at levels far below their direct toxic effect because there is a biological amplification effect in cancer when they transform cells into a malignant state. Through cell growth and division a single molecular event can be multiplied into a deadly tumor. Just as there are different levels of direct toxicity, however, there are different degrees of carcinogenicity. Some chemicals, such as methanesulfonic acid, are highly carcinogenic while others, such as the sweetener saccharin, are suspected carcinogens, but their effects may be vanishingly small.

Acute versus Chronic Doses and Effects

Most of the toxic effects that we have discussed so far have been **acute effects.** That is, they are caused by a single exposure of the toxin and result in an immediate health crisis of some sort. Often, if the individual experiencing an acute reaction survives this immediate crisis, the effects are reversible. **Chronic effects,** on the other hand, are long-lasting, perhaps even permanent. A chronic effect can result from a single dose of a very toxic substance or it can be the result of a continuous or repeated sublethal exposure.

We also describe long-lasting *exposures* as chronic, although their effects may or may not persist after the toxin is removed. It usually is difficult to assess the specific health risks of chronic exposures because other factors, such as aging or normal diseases, act simultaneously with the factor you would like to study. It often requires very large populations of experimental animals to obtain statistically significant results for low-level chronic exposures. Toxicologists talk about "megarat" experiments in which it might take a million rats to determine the health risks of some supertoxic chemicals at very low doses. Such an experiment would be terribly expensive for even a single chemical, let alone for the thousands of chemicals and factors suspected of being dangerous.

An alternative to enormous studies involving millions of animals is to give massive doses of a toxin being studied to a smaller number of individuals and then to extrapolate what the effects of lower doses might have been. This is a controversial approach because it is not clear that responses to toxins are linear or uniform across a wide range of doses.

Figure 21.14 shows three possible results from low doses to a toxin. Curve a shows a baseline level of response in the population, even at zero dose of the toxin. This suggests that some other factor in the environment also causes this response. Curve b shows a straight-line response from the highest doses to zero exposure. Many carcinogens and mutagens show this kind of response. Any exposure to such agents, no matter how small, carries some risk. Curve c shows a threshold for the response where some minimal dose is necessary before any effect can be observed. This generally suggests the presence of some defense mechanism that prevents the toxin from reaching its target in an active form or repairs the damage that it causes. Low levels of exposure to the toxin in question may have no deleterious effects, and it might not be necessary to try to keep exposures to zero.

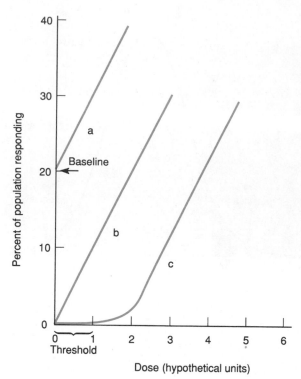

FIGURE 21.14 Three possible dose-response curves at low doses. (*a*) Threshold must be passed before any response is seen. (*b*) Response is linear down to the lowest possible dose. (*c*) Some individuals respond, even at zero dose, indicating that some other factor must be involved.

Whether there are thresholds for environmental health hazards is one of the most important questions in environmental science. The Delaney Clause to the U.S. Food and Drug Act forbids the addition of *any* amount of a known carcinogen to foods or drugs. This is based on the assumption that there is no threshold for these substances and that *any* exposure will result in more cancer. This may not be true in every case. Holding exposures to absolute zero may be impossible and unnecessary; however, attempting to do so seems to be a prudent precaution until we learn more.

Detection Limits

You may have seen or heard dire warnings about toxic materials detected in samples of air, water, or food. A typical headline announced recently that 23 pesticides were found in 16 food samples. What does that mean? The implication seems to be that the mere presence of dangerous materials is equivalent to risk and that counting the numbers of compounds detected is a reliable way to establish danger. We have seen, however, that the dose makes the poison. It matters not only what is there, but how much, where it is located, how accessible it is, and who is exposed. At some level, the mere presence of a substance is insignificant.

Toxins and pollutants may seem to be more widespread now than in the past, and this is surely a valid perception for many substances. The daily reports we hear of new materials found in new places, however, are also due in part, to our more sensitive measuring techniques. Twenty years ago parts per million were generally the limits of detection for most chemicals. Anything below that amount was often reported as zero or absent. A decade ago, new machines and techniques were developed to measure parts per billion. Suddenly, chemicals were found where none had been suspected. Now we can detect parts per trillion or even parts per quadrillion in some cases. Increasingly sophisticated measuring capabilities may lead us to believe that toxic materials have become more prevalent. In fact, our environment may be no more dangerous; we are just better at finding trace amounts.

RISK ASSESSMENT AND ACCEPTANCE

Even if we know with some certainty how toxic a specific chemical is in laboratory tests, it still is difficult to determine how dangerous that chemical will be if it is released into the environment. As you already have seen, many factors complicate the movement and fate of chemicals both around us and within our bodies. Furthermore, public perception of relative dangers from environmental hazards can be skewed so that some risks seem much more important than others.

Table 21.8 shows estimates of the relative risks of some common activities and technologies by three very different groups of people. As you can see, each group had quite different ideas about how dangerous each of these activities is.

Assessing Risk

A number of factors influence how we perceive relative risks associated with different situations.

1. Social, political, or economic interests tend to downplay certain risks and emphasize others that suit their own agendas. We do this individually, as well, building up the dangers of things that don't benefit us, while diminishing or ignoring the negative aspects of activities we enjoy or profit from.

2. Most people have difficulty understanding and believing probabilities. We feel that there must be patterns and connections in events, even though statistical theory says otherwise. If the coin turned up heads last time, we feel certain that it will turn up tails next time. In the same way, it is difficult to understand the meaning of a 1-in-10,000 risk of being poisoned by a chemical.

Activity or Technology	League of women voters	College students	Risk analysis experts
Nuclear power	1	1	20
Motor vehicles	2	5	1
Handguns	3	2	4
Smoking	4	3	2
Motorcycles	5	6	6
Alcoholic beverages	6	7	3
General (private) aviation	7	15	12
Police work	8	8	17
Pesticides	9	4	8
Surgery	10	11	5
Fire fighting	11	10	18
Large construction	12	14	13
Hunting	13	18	23
Spray cans	14	13	26
Mountain climbing	15	22	29
Bicycles	16	24	15
Commercial aviation	17	16	16
Electric power (non-nuclear)	18	19	9
Swimming	19	30	10
Contraceptives	20	9	11
Skiing	21	25	30
X rays	22	17	7
High school and college football	23	26	27
Railroads	24	23	19
Food preservatives	25	12	14
Food coloring	26	20	21
Power mowers	27	28	28
Prescription antibiotics	28	21	24
Home appliances	29	27	22
Vaccination	30	29	25

TABLE 21.8 Ordering of perceived risk for 30 activities and technologies*

*Ordering is based on the geometric mean risk ratings within each group. Rank 1 represents the perceived most risky activity or technology.

From Paul Slovic, "Perception of Risk" in *Science*, vol. 236, April 1987. Copyright © 1987 by the American Association for the Advancement of Science, Washington, D.C.
NutraSweet® is a registered trademark of the NutraSweet Company.

3. Our personal experiences often are misleading. When we have not personally experienced a bad outcome, we feel it is more rare and unlikely to occur than it actually may be. Furthermore, the anxieties generated by life's gambles make us want to deny uncertainty and to misjudge many risks.

4. We have an exaggerated view of our own abilities to control our fate. We generally consider ourselves above-average drivers, safer than most when using appliances or power tools, and less likely than others to suffer medical problems, such as heart attacks. People often feel they can avoid hazards because they are wiser or luckier than others.

5. News media give us a biased perspective on the frequency of certain kinds of health hazards, overreporting the frequency of some accidents or diseases, while downplaying or underreporting others. Sensational, gory, or especially frightful causes of death like murders, plane crashes, fires, or terrible accidents occupy a disproportionate amount of attention in the public media. Heart disease, cancer, and stroke kill nearly fifteen times as many people in the United States as do accidents and seventy-five times as many people as do homicides, but the emphasis placed on accidents and homicides in the media are nearly inversely proportional to their relative frequency compared to either cardiovascular disease or cancer. This gives us an inaccurate picture of the real risks to which we are exposed.

6. We tend to have an irrational fear or distrust of certain technologies or activities that leads us to overestimate their dangers. Nuclear power, for instance, is viewed as very risky, while coal-burning power plants seem to be familiar and relatively benign, yet coal mining, shipping, and combustion cause around 10,000 known deaths each year in the United States, compared to none known so far for nuclear. An old, familiar technology seems safer and more acceptable than does a new, unknown one.

Accepting Risks

How much risk is acceptable? How much is it worth to minimize and avoid exposure to certain risks? Most people will tolerate a higher probability of occurrence of an event if the harm caused by that event is low. Conversely, harm of greater severity is acceptable only at low levels of frequency. A 1-in-10,000 chance of being killed might be of more concern to you than a 1-in-100 chance of being injured. For most people, a 1-in-100,000 chance of being killed by some event or some factor is a threshold for changing one's exposure to that event or factor. That is, if the chance of death is less than 1 in 100,000, we are not likely to be worried enough to change our ways. If the risk is greater, we will probably do something about it. The Environmental Protection Agency generally assumes that a risk of 1 in 1 million is acceptable for most environmental hazards. Critics of this policy ask acceptable to whom?

For activities that we enjoy however, or find profitable, we are often willing to accept far greater risks than this general threshold. Conversely, for risks that benefit someone else we demand far higher protection. For instance, your chance of dying in a motor vehicle accident in any given year is about 1 in 5,000, but that doesn't deter many people from riding in motor vehicles. Your chances of dying from lung cancer if you smoke one pack of cigarettes per day is about 1 in 1,000. By comparison,

Activity	Resulting death risk
Smoking 1.4 cigarettes	Cancer, heart disease
Drinking 0.5 liter of wine	Cirrhosis of the liver
Spending 1 hour in a coal mine	Black lung disease
Living 2 days in New York or Boston	Air pollution
Traveling 6 minutes by canoe	Accident
Traveling 10 miles by bicycle	Accident
Traveling 150 miles by car	Accident
Flying 1,000 miles by jet	Accident
Flying 6,000 miles by jet	Cancer caused by cosmic radiation
Living 2 months in Denver	Cancer caused by cosmic radiation
Living 2 months in a stone or brick building	Cancer caused by natural radioactivity
One chest X ray	Cancer caused by radiation
Living 2 months with a cigarette smoker	Cancer, heart disease
Eating 40 tablespoons of peanut butter	Cancer from aflatoxin
Living 5 years at the site boundary of a typical nuclear power plant	Cancer caused by radiation from routine leaks
Living 50 years 5 miles from a nuclear power plant	Cancer caused by accidental radiation release
Eating 100 charcoal-broiled steaks	Cancer from benzopyrene

TABLE 21.9 Activities estimated to increase your chances of dying in any given year by 1 in 1 million*

*From William Allman, "Staying Alive in the Twentieth Century" in *Science 85,* 5(6):31, October 1985. Copyright 1985 by the Association for the Advancement of Science, Washington, D.C. Reprinted by permission of the author.

the risk from drinking water with the EPA limit of trichloroethylene is about 2 in 1 billion. Strangely, many people demand water with zero levels of trichloroethylene, while continuing to smoke cigarettes.

Table 21.9 lists some activities estimated to increase your chances of dying in any given year by 1 in 1 million. These are statistical averages, of course, and there clearly are differences in where one lives or how one rides a bicycle that affect the danger level of these activities. Still, it is interesting how we readily accept some risks while shunning others.

Our perception of relative risks is strongly affected by whether risks are known or unknown, whether we feel in control of the outcome, and how dreadful the results are. Risks that are unknown or unpredictable and results that are particularly gruesome or disgusting seem far worse than those that are familiar and socially acceptable. Figure 21.15 shows the relative acceptability of a variety of technologies and activities judged from their

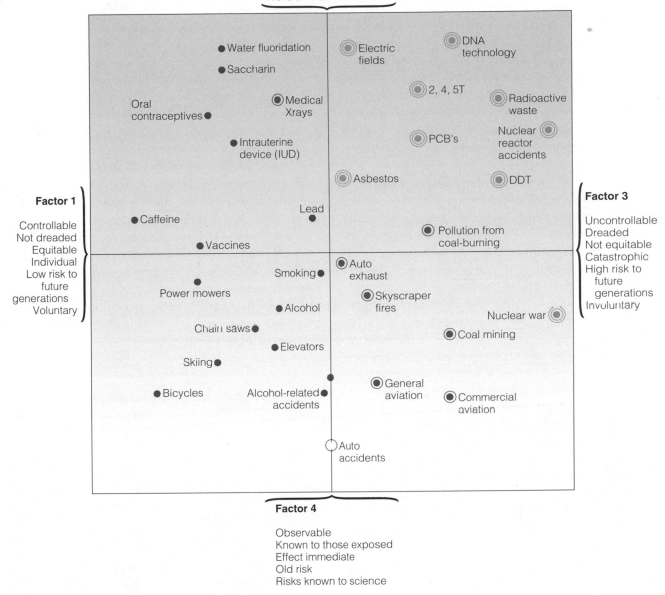

Factor 2

Not observable
Unknown to those exposed
Effect delayed
New risk
Risks unknown to science

● Water fluoridation
● Saccharin

Oral
contraceptives ●

◎ Medical
Xrays

● Intrauterine
device (IUD)

Factor 1

Controllable
Not dreaded
Equitable
Individual
Low risk to
future
generations
Voluntary

● Caffeine

Lead
●

● Vaccines

●
Power mowers

Smoking ●

● Alcohol

Chain saws ●

● Elevators

Skiing ●

● Bicycles

Alcohol-related ●
accidents

◎ Electric
fields

◎ DNA
technology

◎ 2, 4, 5T

◎ Radioactive
waste

◎ PCB's

Nuclear ◎
reactor
accidents

◎ Asbestos

◎ DDT

◎ Pollution from
coal-burning

◎ Auto
exhaust

◎ Skyscraper
fires

Nuclear war ◎

◎ Coal mining

◎ General
aviation

◎ Commercial
aviation

○ Auto
accidents

Factor 3

Uncontrollable
Dreaded
Not equitable
Catastrophic
High risk to
future
generations
Involuntary

Factor 4

Observable
Known to those exposed
Effect immediate
Old risk
Risks known to science

FIGURE 21.15 Relative perception of various risks. Size of circle indicates perception of risk; location on graph indicates contribution of different factors in apprehension of risk. Each factor is derived from combinations of characteristics.

familiarity, controllability, and consequences. The relative undesirability of these risks is indicated by the location of the circle that marks its position. Note that factors in the upper right quadrant tend to be much more feared than those in the lower left quadrant, even though the actual numbers of deaths or disease in some cases might be fairly comparable.

ESTABLISHING PUBLIC POLICY

A problem in setting environmental standards is that we are dealing with many sources of harm to which we are exposed simultaneously or sequentially. It is difficult to separate the effects of all these different hazards and to evaluate their risks accurately, especially when the exposures are near the threshold of measurement and response. In spite of often vague and contradictory data, public policy makers must make decisions.

The case of the sweetener saccharin is a good example of the complexities and uncertainties of risk assessment in public health. Studies in the 1970s at the University of Wisconsin and the Canadian Health Protection Branch suggested a link between saccharin and bladder cancer in male rats. Critics of these studies pointed out that humans would have to drink eight hundred cans of diet soda *per day* to get a saccharin dose equivalent to that given to the rats. Furthermore, they argued that people are not just very large rats. Although the Food and Drug Act forbids the addition of any substance that causes cancer in any amount in any animal, Congress has repeatedly exempted saccharin from being banned from food because of the uncertainty about its risks.

Experiments testing the toxicity of saccharin in rats merely give a range of probable toxicities in humans. The lower end of this range indicates that only 1 person in the United States would die from using saccharin every 1,000 years. That is clearly inconsequential. The higher estimate, however, indicates that 3,640 people would die each year from the same exposure. Is that too high a cost for the benefits of having saccharin available to people who must restrict sugar intake? How does the cancer risk compare to the dangers of obesity, cardiovascular disease, etc.? What other alternatives might there be to saccharin? A popular but more expensive alternative (aspartame, derived from the amino acid aspartic acid) which bears the trade name Nutrasweet®* also is controversial because of uncertainties about its safety.

In 1989, a near-panic swept the United States after the Natural Resources Defense Council issued a report claiming that children were being exposed to dangerous chemicals in fruits and vegetables. The worst of these was believed to be Alar (Daminozoide), used to treat apples to promote even ripening and reduce surface blemishes. The apple industry claimed that you would have to eat 28,000 pounds of treated apples every day for 70 years to approach the exposures that caused cancer in laboratory animals.

In setting standards for environmental toxins, we need to consider (1) combined effects of exposure to many different sources of damage, (2) different sensitivities of members of the population, and (3) effects of chronic as well as acute exposures. Some people argue that pollution levels should be set at the highest amount that does *not* cause measurable effects. Others demand that pollution be reduced to zero if possible, or as low as is technologically feasible. It may not be reasonable to demand that we be protected from every potentially harmful contaminant in our environment, no matter how small the risk. As we have seen, our bodies have mechanisms that enable us to avoid or repair many kinds of damage so that most of us can withstand some minimal level of exposure without harm.

On the other hand, each challenge to our cells by toxic substances represents stress on our bodies. Although each individual stress may not be life-threatening, the cumulative effects of all the environmental stresses, both natural and human-caused, to which we are exposed may seriously shorten or restrict our lives. Furthermore, some individuals in any population are more susceptible to those stresses than others. Should we set pollution standards so that no one is adversely affected, even the most sensitive individuals, or should the acceptable level of risk be based on the average member of the population?

Finally, policy decisions about hazardous and toxic materials need to be based also on information about how such materials affect the plants, animals, and other organisms that define and maintain our environment. In some cases, pollution can harm or destroy whole ecosystems (chapters 22 and 23) with devastating effects on the life-supporting cycles on which we depend. In other cases, only the most sensitive species are threatened. The 1992 budget of the Environmental Protection Agency reflects a concern that our exclusive focus on reducing pollution to protect human health has neglected risks to natural ecological systems. Administrator William K. Reilly points out that while there have been many benefits from a case-by-case approach in which we evaluate the health risks of individual chemicals, we have often missed broader ecological problems that may be of greater ultimate importance.

SUMMARY

Health is a state of physical, mental, and social well-being, not merely the absence of disease or infirmity. The cause or development of nearly every human disease is at least partly related to environmental factors. For most people in the world, the greatest health threat in the environment is now, as always, from pathogenic organisms. Bacteria, viruses, protozoans, parasitic worms, and other infectious agents probably kill more people each year than any other cause of death.

Stress, diet, and life-style also are important health factors. Our social environment may be as important as our physical environment in determining the state of our health. There are some areas in the world where people live exceptionally long and healthful lives. We might be able to learn from them how to live healthier lives.

Estimating the potential health risk from exposure to specific environmental factors is difficult because information on the precise dose, length, method of exposure, and possible interactions between the chemical in question and other potential toxins to which the population may have been exposed is often lacking. In addition, individuals have different levels of sensitivity and response to a particular toxin and are further affected by general health condition, age, and sex.

The distribution and fate of materials in the environment depend on their physical characteristics and the processes that transport, alter, destroy, or immobilize them. Uptake of toxins into organisms can result in accumulation in tissues and transfer from one organism to another.

Estimates of health risks for large, diverse populations exposed to very low doses of extremely toxic materials are inexact because of biological variation, experimental error, and the necessity of extrapolating from results with small numbers of laboratory animals. In the end, we are left with unanswered questions. Which are the most dangerous environmental factors that we face? How can we evaluate the hazards of all the natural and synthetic chemicals that now exist? What risks are acceptable? We have not yet solved these problems or answered all the questions raised in this chapter, but it is important that these issues be discussed and considered seriously.

5. Should pollution levels be set to protect the average person in the population or the most sensitive? Why not have zero exposure to all hazards?

6. What level of risk is acceptable to you? Are there some things for which you would accept more risk than others?

7. Do you believe that testing requirements for new chemicals inhibit invention and introduction of new products?

8. If you were setting environmental standards, would you ban production and use of any chemicals? Which ones and why?

9. Why do you suppose we have enzymes in our livers that turn relatively harmless materials into dangerous carcinogens?

10. What do you think of the claim that plants make thousands of times more pesticides and other toxins than does human industry? Is there a fundamental difference between natural products and synthetic materials?

■ Review Questions

1. What is the difference between toxic and hazardous? Give some examples of materials in each category.

2. What are some of the most important infectious diseases in the world? How are they transmitted?

3. How do stress, diet, and life-style affect environmental health? What diseases are most clearly related to these factors?

4. How do the physical and chemical characteristics of materials affect their movement, persistence, distribution, and fate in the environment?

5. Define LD50. Why is it more accurate than simply reporting toxic dose?

6. What is the difference between acute and chronic toxicity?

7. Define carcinogenic, mutagenic, teratogenic, and neurotoxic.

8. What are irritants, sensitizers, allergens, caustics, acids, and fibrotic agents?

9. How do organisms reduce or avoid the damaging effects of environmental hazards?

10. What are the relative risks of smoking, driving a car, and drinking water with the maximum permissible levels of trichloroethylene? Are these relatively equal risks?

■ Questions for Critical Thinking

1. Do you think that the environment is more hazardous now than it was a century ago? Why?

2. List some of the benefits you enjoy of synthetic chemicals. Do you believe that the benefits outweigh the risks?

3. If there are thresholds for some chemicals or levels below which we can't measure any effects, do you think that we should set pollution levels up to those thresholds?

4. What might be the costs and benefits of setting zero pollution levels (i.e., no pollution at all)? Is it feasible?

■ Key Terms

acute effects (p. 432)
allergens (p. 423)
antigens (p. 423)
asphyxiants (p. 423)
bioaccumulation (p. 129)
biomagnification (p. 429)
cancer (p. 425)
carcinogens (p. 425)
chronic effects (p. 432)
disease (p. 419)
hazardous (p. 423)
health (p. 419)

irritants (p. 423)
LD50 (p. 430)
malignant tumor (p. 425)
morbidity (p. 419)
mutagens (p. 424)
neurotoxins (p. 424)
promoters (p. 425)
respiratory fibrotic agents (p. 423)
stress (p. 426)
teratogens (p. 424)
toxins (p. 423)
trauma (p. 426)

SUGGESTED READINGS

■ Albert, A. *Selective Toxicity: The Physico-Chemical Basis of Therapy*. 5th ed. London: Chapman and Hall, 1985.

■ Allman, W. October 1985. "Staying Alive in the Twentieth Century." *Science 85* 5(6):31. A highly readable summary of risks and risk perception.

■ Ames, B., R. Magaw, and L. Gold. April 17, 1987. "Ranking Possible Carcinogenic Hazards." *Science* 236(4799):271. Startling new conclusions about natural versus synthetic products.

■ Campt, D. 1990. "Reducing Dietary Risk." *E.P.A. Journal* 16(3):18 There is a crisis in public confidence about the safety of food. How can food be made more safe?

■ Castleman, M. March/April 1985. "Toxics and Male Infertility." *Sierra* 70(2):49. Is there a connection?

■ Efron, E. 1984. *The Apocalyptics: Cancer and the Big Lie—How Environmental Politics Controls What We Know about Cancer*. New York: Simon & Schuster, 1984. A blistering attack on the cancer establishment.

Guthrie, F., and J. Perry. *Introduction to Environmental Toxicology*. New York: Elsevier, 1980. A good introductory textbook in toxicology.

Hirschhorn, N., and W. B. Greenough III. 1991. "Progress in Oral Rehydration Therapy." *Scientific American* 264(5): 50. Treatment of diarrhea-induced dehydration with simple electrolyte solutions now saves one million children a year from death.

Lave, L. April 17, 1987. "Health and Safety Risk Analysis: Information for Better Decisions." *Science* 236(4799):291. A thoughtful discussion of risk analysis.

Moriarity, F. *Ecotoxicology: The Study of Pollutants in Ecosystems*. Orlando, Florida: Academic Press, 1983. What is dangerous and why?

NIOSH. 1985. *Registry of Toxic Effects of Chemical Substances*. Lewis, R., and D. Sweet, eds. Washington, D.C.: Government Printing Office. The official toxic substances list.

Regenstein, L. *America the Poisoned*. Washington, D.C.: Acropolis Books Ltd., 1982. A critical look at pesticide use.

Reilly, W. K. 1991. "Why I Propose a National Debate on Risk." *E.P.A. Journal* 17(2):2. The EPA proposes to weigh the relative importance of different risks in setting environmental policies.

Russell, M., and M. Gruber. April 17, 1987. "Risk Assessment in Environmental Policy-Making." *Science* 236(4799):286. How should we measure environmental hazards?

Slovic, P. April 17, 1987. "Perception of Risk." *Science* 236(4799):280. An excellent discussion of risk assessment.

Trunkey, D. August 1983. "Trauma." *Scientific American* 249(2):28. An excellent overview of accidents and other hazards.

Tschirley, F. February 1986. "Dioxin." *Scientific American* 254(2):29. Argues that dioxins may be less dangerous to humans than previously thought.

CHAPTER 22

Water Pollution

CONCEPTS

Water pollution is any physical, biological, or chemical change in water quality that adversely affects living organisms or makes water unsuitable for desired uses. For regulatory purposes, it is useful to distinguish between highly concentrated point sources and diffuse nonpoint sources.

The major categories of human-caused water pollution are infectious agents, toxic organic and inorganic chemicals, radioactive wastes, sediment or suspended solids, plant nutrients, oxygen-demanding wastes, and thermal pollution. Plant nutrients and sediments hasten the natural process of eutrophication, changing oligotrophic lakes, rivers, and streams to eutrophic ones.

The United States has made encouraging progress in recent years controlling the major point sources of pollution, but nonpoint sources remain difficult, serious pollution problems. Surface water quality has been significantly improved. Some 75 percent of streams, rivers, and lakes monitored by the EPA now "fully support their designated uses." Surface waters in less developed countries often are incredibly polluted and dangerous, even lethal, to all living organisms.

Groundwater reservoirs in the developed countries are being contaminated by agricultural and industrial chemicals. There are thousands of hazardous waste dumps, millions of leaking underground storage tanks, billions of liters of wastes injected into deep wells, and unknown numbers of other sources of groundwater pollution in the United States. It will be very difficult to clean up this valuable resource.

Oil spills, garbage and incinerator sludge dumping, discarded plastic litter, and discharge from rivers and land runoff all threaten ocean water quality. Some marine biologists warn that the oceans are "dying," but others feel that the oceans are more resilient and dynamic than we may have thought.

There are many techniques for preventing pollution or reducing its effects. Septic systems and drain fields are effective under the right conditions. Municipal sewage treatment plants can remove essentially all contaminants if they are designed and operated properly.

INTRODUCTION

In the 1950s and 1960s, the Cuyahoga River running through Akron and Cleveland, Ohio, was little better than an open sewer. Chemical and steel factories discharged up to 155 tons a day of toxic chemicals, oil, solvents, and sludge into the river. Poor land-use practices upstream added sediment, animal manure, pesticides, and fertilizer. Raw or inadequately treated municipal sewage contributed disease-causing bacteria and viruses. Oxygen levels in the water fell so low that only the most resistant life-forms could survive. Blobs of discolored foam, dead fish, and rotting refuse circled slowly in the scum on the river's surface; the stench was nauseating. Lake Erie, into which the Cuyahoga and numerous other rivers dumped their noxious loads, was becoming increasingly polluted. There were warnings that the lake was dying.

In 1959, oil slicks on the Cuyahoga's surface caught fire and burned for eight days, destroying several bridges. Fireboats were sent to battle the fire, but the water they sprayed was from the river itself and only spread the blaze. Ten years later, the river burned again. The image of this foul, burning river caught public attention and became a symbol of what we were doing to our environment (figure 22.1). We began to realize that many, perhaps most, of our surface waters were threatened by pollution. Even the oceans were considered in peril. Among the many concerns that initiated the United States' "environmental decade" of the 1970s, perhaps none was stronger than protecting and restoring water quality.

What are the sources of this water pollution? How does it affect ecosystems and human health? Most important of all, what can we do as a society and as individuals to reduce pollution levels and restore the quality of our vital water resources? These are some of the questions we will address in this chapter.

WHAT IS WATER POLLUTION?

Any physical, biological, or chemical change in water quality that adversely affects living organisms or makes water unsuitable for desired uses can be considered pollution. Often, however, a change that adversely affects one organism may be advantageous to another. Nutrients that stimulate oxygen consumption by bacteria and other decomposers in a river or lake, for instance, may be lethal to fish, but will stimulate a flourishing community of decomposers. Whether the quality of the water has suffered depends on your perspective. There are natural sources of water contamination, such as poison springs, oil seeps, and sedimentation from erosion, but in this chapter we will focus primarily on human-caused changes that affect water quality or usability.

Pollution control standards and regulations usually distinguish between point and nonpoint pollution sources. Factories, power plants, sewage treatment plants, underground coal mines, and oil wells are classified as **point sources** because they dis-

charge pollution from specific locations, such as drain pipes, ditches, or sewer outfalls (figure 22.2). These sources are discrete and identifiable, so they are relatively easy to monitor and regulate. It is generally possible to divert effluent from the waste streams of these sources and treat it before it enters the environment.

In contrast, **nonpoint sources** of water pollution are scattered or diffuse, having no specific location where they discharge into a particular body of water. Nonpoint sources include runoff from farm fields, golf courses, lawns and gardens, construction sites, logging areas, roads, streets, and parking lots. Whereas point sources may be fairly uniform and predictable throughout the year, nonpoint sources are often highly episodic. The first heavy rainfall after a dry period may flush high concentrations of gasoline, lead, oil, and rubber residues off city streets, for instance, while subsequent runoff may have much less of these pollutants. Spring snowmelt carries high levels of atmospheric acid deposition into streams and lakes in some areas. The irregular timing of these events, as well as their multiple sources and scattered location, makes them much more difficult to monitor, regulate, and treat than point sources.

The details of human sewage treatment are discussed later in the chapter, but you need to be introduced to a few definitions at this time. Untreated sewage is simply dumped on the ground or into the water. **Primary treatment** removes solids from the waste stream. **Secondary treatment** is bacterial decomposition of remaining suspended solids. **Tertiary treatment** removes inorganic minerals and plant nutrients.

MAJOR TYPES AND EFFECTS OF WATER POLLUTANTS

Although the types, sources, and effects of water pollutants are often interrelated, it is convenient to divide them into major categories for discussion (table 22.1). Let's look more closely at some of the important sources and effects of each type of pollutant.

Infectious Agents

The most serious water pollutants in terms of human health worldwide are pathogenic organisms (chapter 21). Among the most important waterborne diseases are typhoid, cholera, bacterial and amoebic dysentery, enteritis, polio, infectious hepatitis, and schistosomiasis (chapter 21). Malaria, yellow fever, and filariasis are transmitted by insects that have aquatic larvae. Altogether, at least 25 million deaths each year are blamed on these water-related diseases. Nearly two-thirds of the mortalities of children under five years old are associated with waterborne diseases (Box 21.1 and figure 22.3).

The main source of these pathogens is from untreated or improperly treated human wastes. Animal wastes from feedlots or fields near waterways and food processing factories with in-

FIGURE 22.1 Cleanup after the burning of the Cuyahoga River in Cleveland in 1969. This infamous episode was a major factor in passage of the Clean Water Act.

TABLE 22.1 Major categories of water pollutants

Category	Examples	Sources
A. *Health problems*		
1. Infectious agents	Bacteria, viruses, parasites	Human excreta
2. Organic chemicals	Pesticides, plastics, detergents, oil, and gasoline	Industrial and farm use
3. Inorganic chemicals	Acids, caustics, salts, metals	Industrial effluents, household cleansers, surface runoff
4. Radioactive materials	Uranium, thorium, cesium, iodine, radon	Mining and processing of ores, power plants, weapons production, natural sources
B. *Ecosystem disruption*		
1. Sediment	Soil, silt	Land erosion
2. Plant nutrients	Nitrates, phosphates	Agricultural and urban fertilizers, sewage, manure
3. Oxygen-demanding wastes	Plant and animal manure and residues	Sewage, agricultural runoff, paper mills, food processing
4. Thermal	Heat	Power plants, industrial cooling

adequate waste treatment facilities also are sources of disease-causing organisms.

In the more developed countries of the world, sewage treatment plants and other pollution-control techniques have reduced or eliminated most of the worst sources of pathogens in inland surface waters. Furthermore, drinking water is generally disinfected by chlorination so epidemics of waterborne diseases have become rare in Europe, North America, and Japan. The United Nations estimates that 90 percent of the people in the more developed countries have adequate (safe) sewage disposal, and 95 percent have clean drinking water.

The situation is quite different in the less developed countries of the world. The United Nations estimates that less than 25 percent of the people in these countries have adequate sanitation, and that only 43 percent have access to clean drinking water. Conditions are even worse in remote, rural areas where sewage treatment is usually primitive or nonexistent, and purified water is either unavailable or too expensive to obtain. The World Health Organization estimates that 80 percent of all sickness and disease in less developed countries can be attributed to waterborne infectious agents.

Detecting specific pathogens in water is difficult, time-consuming, and costly: thus, water quality control personnel usually analyze water for the presence of **coliform bacteria,**

FIGURE 22.2

Drawing by Stevenson; © 1970 The New Yorker Magazine, Inc.

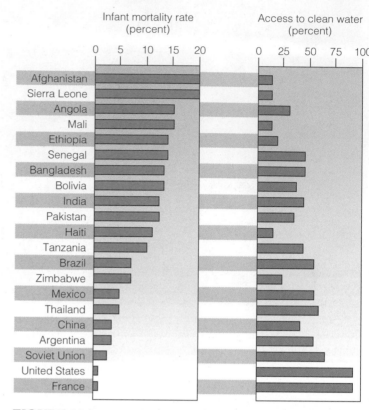

FIGURE 22.3 Infant mortality and percent of population with access to clean drinking water in selected countries 1982.

Source: Derived from UNICEF, *State of the World's Children*, 1984, as published in William U. Chandler, *Investing in Children*, *Worldwatch Paper 64* (Washington, DC: Worldmatch Institute, 1985), p. 32.

any of the many types that live in the intestines of humans and other animals. If large numbers of these organisms are found in a water sample, recent contamination by untreated feces is indicated. Exposure to an alien strain of coliform bacteria is usually the cause of upset stomach and diarrhea that strike tourists. It is usually assumed that if coliform bacteria are present in a water sample, infectious pathogens are present also.

To test for coliform bacteria, a 100 ml (4 oz) sample of water is passed through a filter that removes bacterial cells. The filter is placed in a dish containing a liquid nutrient medium that supports bacterial growth. After twenty-four hours at the appropriate temperature, each living cell will have produced a small colony of cells on the filter (figure 22.4). If more than one colony per sample is found in a drinking water sample, the U.S. Environmental Protection Agency (EPA) considers the water unsafe and requiring chlorination. The EPA-

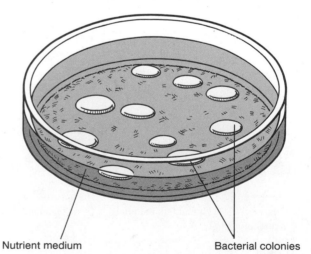

Nutrient medium   Bacterial colonies

FIGURE 22.4 Coliform bacteria growing in a petri dish. More than one colony per sample indicates water is unsafe for drinking.

ENVIRONMENTAL POLLUTION AND OTHER DILEMMAS

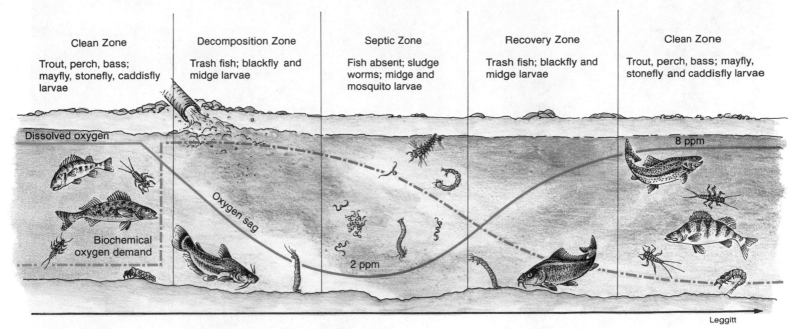

Clean Zone

Trout, perch, bass; mayfly, stonefly, caddisfly larvae

Decomposition Zone

Trash fish; blackfly and midge larvae

Septic Zone

Fish absent; sludge worms; midge and mosquito larvae

Recovery Zone

Trash fish; blackfly and midge larvae

Clean Zone

Trout, perch, bass; mayfly, stonefly and caddisfly larvae

Dissolved oxygen

8 ppm

Oxygen sag

Biochemical oxygen demand

2 ppm

Leggitt

FIGURE 22.5 Oxygen sag downstream of an organic source. A great deal of time and distance may be required for the stream and its inhabitants to recover.

recommended maximum coliform count for swimming water is 200 colonies per 100 ml, but some cities and states allow higher levels. If the limit is exceeded, the contaminated pool, river, or lake usually is closed to swimming.

Oxygen-Demanding Wastes

The amount of oxygen dissolved in water is a good indicator of water quality and of the kinds of life it will support. Water with an oxygen content above 8 parts per million (ppm) will support game fish and other desirable forms of aquatic life. Water with less than 2 ppm oxygen will support only worms, bacteria, fungi, and other decomposers. Oxygen is added to water by diffusion from the air, especially when turbulence and mixing rates are high, and by photosynthesis of green plants, algae, and cyanobacteria. Oxygen is removed from water by respiration and chemical processes that consume oxygen.

The addition of certain organic materials, such as sewage, paper pulp, or food processing wastes, to water stimulates oxygen consumption by decomposers. The impact of these materials on water quality can be expressed in terms of **biological oxygen demand (BOD)**: a standard test of the amount of dissolved oxygen utilized by aquatic microorganisms over a five-day period. A new method, called the chemical oxygen demand (COD), uses a strong oxidizing agent (dichromate ion in 50 percent sulfuric acid) to completely break down all organic matter in a water sample. This method is much faster than the BOD test, but normally gives much higher results because it oxidizes compounds not ordinarily metabolized by bacteria. A third method of as-

saying pollution levels is to measure **dissolved oxygen** (DO) **content** directly, using an oxygen electrode. The DO content of water depends on factors other than pollution (e.g., temperature and aeration), but it is usually more directly related to whether aquatic organisms survive than is BOD.

The effects of oxygen-demanding wastes on rivers depends to a great extent on the volume, flow, and temperature of the river water. Aeration occurs readily in a turbulent, rapidly flowing river, which is, therefore, often able to recover quickly from oxygen-depleting processes. Downstream from a point, such as a municipal sewage plant discharge, a characteristic decline and restoration of water quality can be detected either by measuring dissolved oxygen content or by observing the flora and fauna that live in successive sections of the river.

The oxygen decline and rise downstream are called the **oxygen sag** (figure 22.5). Above the pollution source, oxygen levels support normal populations of clean-water organisms. Immediately below the source of pollution, oxygen levels begin to fall as decomposers metabolize waste materials. Rough fish, such as carp, bullheads, and gar, are able to survive in this oxygen-poor environment where they eat both decomposer organisms and the waste itself. Further downstream, the water may become anaerobic (without oxygen) so that only the most resistant microorganisms and invertebrates can survive. Eventually most of the nutrients are used up, decomposer populations are smaller, and the water becomes oxygenated once again. Depending on the volumes and flow rates of the effluent plume and the river receiving it, normal communities may not appear for several miles downstream.

FIGURE 22.6   Eutrophic lake. Nutrients from agriculture and domestic sources have stimulated growth of algae and aquatic plants in this lake. This reduces water quality, alters species composition, and lowers recreational and aesthetic values of the lake.

Plant Nutrients and Cultural Eutrophication

Water clarity (transparency) is affected by the abundance of plankton organisms and is a useful measure of water quality and water pollution. Rivers and lakes that have clear water and low biological productivity are said to be **oligotrophic** (oligo = little + trophic = nutrition). By contrast, **eutrophic** (eu + trophic = truly nourished) waters are rich in organisms and organic materials. Eutrophication, an increase in nutrient levels and biological productivity, is a normal part of successional changes (chapter 5) in most lakes. Tributary streams bring in sediments and nutrients that stimulate plant growth. Over time, the pond or lake tends to fill in, eventually becoming a marsh and then a terrestrial biome (figure 22.6). The rate of eutrophication and succession depends on water chemistry and depth, volume of inflow, mineral content of the surrounding watershed, and the biota of the lake itself.

Human activities can greatly accelerate eutrophication. An increase in biological productivity and ecosystem succession caused by human activities is called **cultural eutrophication.** Cultural eutrophication can be brought about by increased nu-

trient flows, higher temperatures, more sunlight reaching the water surface, or a number of other changes. Increased productivity in an aquatic system sometimes can be beneficial. Fish and other desirable species may grow faster, providing a welcome food source. Often, however, eutrophication has undesirable results. An oligotrophic lake or river usually has aesthetic qualities and species of organisms that we value.

The high biological productivity of eutrophic systems is often expressed as "blooms" of algae or thick growths of aquatic plants and high levels of sediment accumulation. Bacterial populations also increase, fed by larger amounts of organic matter. The water often becomes opaque and has unpleasant tastes and odors. The deposition of silt and organic sediment caused by cultural eutrophication can accelerate the "aging" of a water body enormously over natural rates. Lakes and reservoirs that normally might exist for hundreds or thousands of years can be filled in a matter of decades.

Eutrophication also occurs in marine ecosystems, especially in near-shore waters and partially enclosed bays or estuaries. Blooms of minute organisms called dinoflagellates produce toxic **red tides** that kill fish. The Mediterranean Sea is in es-

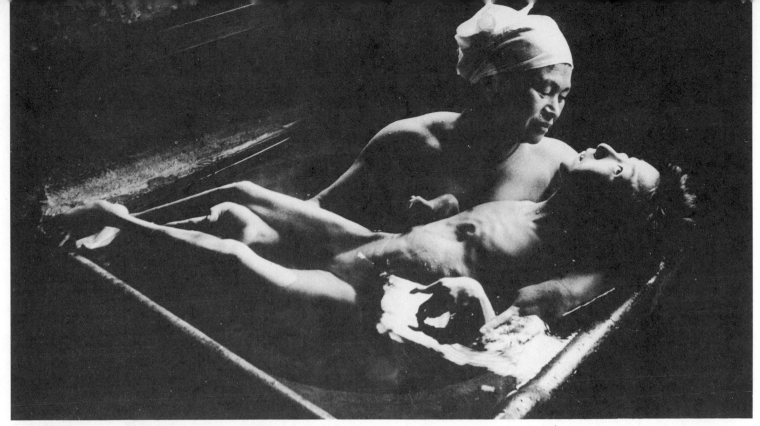

FIGURE 22.7 A mother from Minamata, Japan, bathes her daughter, who suffered permanent brain damage and birth defects from mercury-contaminated seafood the mother ate while pregnant. This kind of poisoning is now known as Minamata Disease.

pecially critical condition. During the tourist season, the coastal population swells to 200 million people. Eighty-five percent of the effluents from large cities go untreated into the sea. Extensive beach pollution, fish kills, and contaminated shellfish result. Massive algal blooms suffocate the Venice lagoons. The stench is unbearable. Swarms of flies feeding on dead plants infest the city. The lagoons and the sea along much of the adjacent Adriatic coast is essentially devoid of oxygen during most of the summer.

Toxic Inorganic Water Pollutants

Some toxic inorganic chemicals are released from rocks by weathering, carried by runoff into lakes or rivers, or percolate into groundwater aquifers. This pattern is part of natural mineral cycles (chapter 3). Humans often accelerate the transfer rates in these cycles thousands of times above natural background levels through the mining, processing, using, and discarding of minerals.

In many areas, toxic, inorganic chemicals introduced into water as a result of human activities have become the most serious form of water pollution. Among the chemicals of greatest concern are heavy metals, such as mercury, lead, tin, and cadmium. Supertoxic elements, such as selenium and arsenic, also have reached hazardous levels in some waters. Other inorganic

materials, such as acids, salts, nitrates, and chlorine, that normally are not toxic in low concentrations may become concentrated enough to lower water quality or adversely affect biological communities.

Heavy Metals

In the early 1950s, people in the small coastal village of Minamata, Japan, experienced an alarming incidence of nervous disorders including numbness, tingling sensations, headaches, blurred vision, and slurred speech. For an unlucky few, these milder symptoms were followed by violent trembling, paralysis, and even death. An abnormally high rate of birth defects also occurred. Children were born with tragic deformities, paralysis, and mental deficiency. Lengthy investigations indicated that these symptoms were caused by mercury poisoning (figure 22.7).

For years the Chisso Chemical Plant had been releasing residues containing mercury into Minamata Bay. Since elemental mercury is not water soluble, it was assumed that it would sink into the bottom sediments and remain inert. Scientists discovered, however, that bacteria living in the sediments were able to convert metallic mercury into soluble methyl mercury, which was absorbed from the water and concentrated in the tissues of aquatic organisms. People who ate fish and shellfish from the bay were exposed to dangerously high levels of this toxic chemical. Altogether, more than 3,500 people were affected and about

50 died of what became known as Minamata Disease. Since then, the mercury-bearing sediments have been dredged up, and Minamata Bay is now considered safe for fishing.

Another mercury-poisoning disaster appears to be in process in South America. Since the mid-1980s, a gold rush has been under way in Brazil, Ecuador, and Bolivia. Forty thousand *Garimperios* or prospectors have invaded the jungles on the Amazon River and its tributaries to pan for gold. They use mercury to trap the gold and separate it from sediments. Then, the mercury is burned off with a blow torch. Miners and their families suffer nerve damage from breathing the toxic fumes. Estimates are that 130 tons of mercury per year are deposited in the Amazon. It will probably be impossible to clean up this huge river system.

We have come to realize that other heavy metals released as a result of human activities also are concentrated by hydrological and biological processes so that they become hazardous to both natural ecosystems and human health. A condition known as Itai-Itai (literally ouch-ouch) disease that developed in Japanese living near the Jintsu River was traced to cadmium poisoning. Bacteria from methylate tin have been found in sediments in Chesapeake Bay, leading to worries that this toxic metal also may be causing unsuspected health effects. The use of tin compounds as antifouling agents on ship bottoms has been criticized as a potential source of dangerous pollution.

Lead poisoning has been known since Roman times to be dangerous to human health. Lead pipes are a serious source of drinking water pollution, especially in older homes or in areas where water is acidic and, therefore, leaches more lead from pipes. Even lead solder in pipe joints and metal containers can be hazardous. In 1988, the EPA set the maximum limit for lead in public drinking water at 50 parts per billion (ppb). Suppliers of water with 20 to 50 ppb lead must notify customers of possible hazards. Some public health officials argue that lead is neurotoxic at any level, and the limits should be less than 10 ppb.

Mine drainage and leaching of mining wastes are serious sources of metal pollution in water. A recent survey of water quality in eastern Tennessee found that 43 percent of all surface streams and lakes and more than half of all groundwater used for drinking supplies was contaminated by acids and metals from mine drainage. In some cases, metal levels were two hundred times higher than what is considered safe for drinking water.

Toxic Nonmetals and Salts

Desert soils often contain high concentrations of soluble salts, including toxic selenium and arsenic. You have probably heard of poison springs and seeps in the desert where these compounds are brought to the surface by percolating groundwater. When the water evaporates, they are left in increasing concentrations. Irrigation and drainage of desert soils mobilize these materials on a larger scale and can result in serious pollution problems, as in Kesterson Marsh in California where selenium poisoning killed thousands of migratory birds in the 1980s.

Such salts as sodium chloride (table salt), that are nontoxic at low concentrations also can be mobilized by irrigation and concentrated by evaporation, reaching levels that are toxic for plants and animals. Salt levels in the San Joaquin river in central California rose from 0.28 gm/l in 1930 to 0.45 gm/l in 1970 as a result of agricultural runoff. Salinity levels in the Colorado River and surrounding farm fields have become so high in recent years that millions of hectares of valuable croplands have had to be abandoned. The United States is building a huge desalinization plant at Yuma, Arizona, to reduce salinity in the river. In northern states, millions of tons of sodium chloride and calcium chloride are used to melt road ice in the winter. The corrosive damage to highways and automobiles and the toxic effects on vegetation are enormous. Leaching of road salts into surface waters has a similarly devastating effect on aquatic ecosystems.

Acids and Bases

Acids are released as by-products of industrial processes, such as leather tanning, metal smelting and plating, petroleum distillation, and organic chemical synthesis. Coal mining is an especially important source of acid water pollution. Sulfides in coal are solubilized to make sulfuric acid. Thousands of kilometers of streams in the United States have been acidified by acid mine drainage, some so severely that they are essentially lifeless.

Coal and oil combustion also leads to formation of atmospheric sulfuric and nitric acid (chapter 23), which are disseminated by long-range transport processes and deposited via precipitation (acid rain, acidic snow, acid fog, or dry deposition) in surface waters. Where soils are rich in such alkaline material as limestone, these atmospheric acids have little effect because they are neutralized. In high mountain areas or recently glaciated regions where crystalline bedrock is close to the surface and lakes are oligotrophic, however, there is little buffering capacity (ability to neutralize acids) and aquatic ecosystems can be severely disrupted. These effects were first recognized in the mountains of Northern England and Scandinavia about thirty years ago. In recent years, aquatic damage due to acid precipitation has been reported in about two hundred lakes in the Adirondack Mountains of New York State and in several thousand lakes in Eastern Quebec, Canada. Game fish, amphibians, and sensitive aquatic insects are generally the first to be killed by increased acid levels in the water. If acidification is severe enough, aquatic life is limited to a few resistant species of mosses and fungi. Increased acidity may result in leaching of toxic metals, especially aluminum, from soil and rocks, making water unfit for drinking or irrigation, as well.

Toxic Organic Chemicals

Thousands of different natural and synthetic organic chemicals are used in the chemical industry to make pesticides, plastics, pharmaceuticals, pigments, and other products that we use in everyday life. Many of these chemicals are highly toxic (chapter 21). Exposure to very low concentrations (perhaps even parts per quadrillion in the case of dioxins) can cause birth defects, genetic disorders, and cancer (chapter 21). They also can persist in the environment because they are resistant to degradation and toxic to organisms that ingest them. Contamination of surface waters and groundwater by these chemicals is a serious threat to human health.

The two most important sources of toxic organic chemicals in water are improper disposal of industrial and household wastes and runoff of pesticides from farm fields, forests, roadsides, golf courses, and other places where they are used in large quantities. The EPA estimates that about 410,000 metric tons of pesticides are used in the United States each year. Much of this material washes into the nearest waterway, where it passes through ecosystems and may accumulate in high levels in certain nontarget organisms. The bioaccumulation of DDT in aquatic ecosystems was one of the first of these pathways to be understood (chapter 21). Polychlorinated biphenyls, dioxins, and other chlorinated hydrocarbons (hydrocarbon molecules that contain chlorine atoms) also have been shown to accumulate to dangerous levels in the fat of salmon, fish-eating birds, and humans.

Hundreds of millions of tons of hazardous organic wastes are thought to be stored in dumps, landfills, lagoons, and underground tanks in the United States. Many, perhaps most, of these sites are leaking toxic chemicals into surface waters or groundwater or both. The EPA estimates that about 26,000 hazardous waste sites will require cleanup because they pose an imminent threat to public health, mostly through water pollution.

Sediment

Sediment and suspended solids make up the largest volume of water pollution in the United States and most other parts of the world. Rivers have always carried sediment to the oceans, but erosion rates in many areas have been greatly accelerated by human activities. As chapter 11 shows, some rivers carry astounding loads of sediment. Erosion and runoff from croplands contribute about 25 billion metric tons of soil, sediment, and suspended solids to world surface waters each year. Forests, grazing lands, urban construction sites, and other sources of erosion and runoff add at least 50 billion additional tons (chapter 11). This sediment fills lakes and reservoirs, obstructs shipping channels, clogs hydroelectric turbines, and makes purification of drinking water more costly. Water with high levels of suspended

FIGURE 22.8 This dam is now useless because its reservoir has filled with silt and sediment.

solids is less suitable for fish and other forms of aquatic life. Sunlight is blocked so that plants cannot carry out photosynthesis and oxygen levels decline. Murky, cloudy water also is less attractive for swimming, boating, fishing, and other recreational uses (figure 22.8).

Sediment can be beneficial. Mud carried by rivers nourishes floodplain farm fields. Sediment deposited in the ocean at river mouths creates valuable deltas and islands. The Ganges River, for instance, builds up islands in the Bay of Bengal that are eagerly colonized by land-hungry people of Bangladesh. In Louisiana, lack of sediment in the Mississippi River (it is being trapped by locks and dams upstream) is causing biologically rich coastal wetlands to waste away. Sediment also can be harmful. Excess sediment deposits can fill estuaries and smother aquatic life on coral reefs and shoals near shore. As with many natural environmental processes, acceleration as a result of human intervention generally diminishes the benefits and accentuates the disadvantages of the process.

Thermal Pollution and Thermal Shocks

Raising or lowering water temperatures from normal levels can adversely affect water quality and aquatic life. Water temperatures are usually much more stable than air temperatures, so aquatic organisms tend to be poorly adapted to rapid temperature changes. Lowering the temperature of tropical oceans by even one degree can be lethal to some corals and other reef species. Raising water temperatures can have similar devastating effects on sensitive organisms. Oxygen solubility in water decreases as temperatures increase, so species requiring high oxygen levels are adversely affected by warming water.

Humans cause thermal pollution by altering vegetation cover and runoff patterns, as well as by discharging heated water

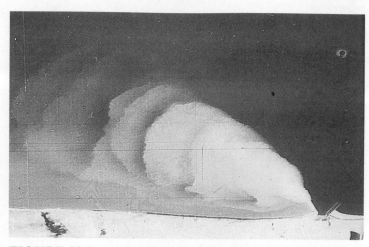

FIGURE 22.9 An infrared photo shows a thermal plume spreading 1,600 m (1 mi) from its source at a power plant discharge on Lake Michigan.

directly into rivers and lakes. As chapter 16 shows, nearly half the water we withdraw is used for industrial cooling. Metal smelters and processing mills use and release large amounts of cooling water, as do petroleum refineries, paper mills, food-processing factories, and chemical manufacturing plants. The electric power industry uses about three-quarters of all cooling water in the United States. Steam-driven turbogenerators of both nuclear and fossil fuel power plants (chapters 18 and 19) are only about 40 percent efficient. That means that the other 60 percent of the energy released from the fuel—almost entirely heat energy—must be gotten rid of in some way.

The cheapest way to remove heat from an industrial facility is to draw cool water from an ocean, river, lake, or aquifer, run it through a heat-exchanger to extract excess heat and then dump the heated water back into the original source. Figure 22.9 shows the **thermal plume** of heated water being discharged into a lake. Raised temperatures can disrupt many processes in natural ecosystems and drive out sensitive organisms. To minimize these effects, power companies frequently are required to construct artificial cooling ponds or wet- or dry-cooling towers in which heat is released into the atmosphere and water is cooled before being released into natural water bodies. Wet cooling towers are cheaper to build and operate than dry systems, but lose large quantities of water to evaporation.

In some circumstances, introducing heated water into a water body is beneficial. Warming catfish-rearing ponds, for instance, can increase yields significantly. Warm water plumes from power plants often attract fish, birds, and marine mammals that find food and refuge there, especially in cold weather. This artificial environment can be a fatal trap, however. Organisms dependent on the warmth may die if they leave the plume or if the flow of warm water is interrupted by a plant shutdown. The manatee, for example, is an endangered marine mammal species that lives in Florida (chapter 15). Manatees are attracted to the abundant food supply and warm water in power plant thermal plumes and are enticed into spending the winter much further north than they normally would. On several occasions, a mid-winter power plant breakdown has exposed a dozen or more of these rare animals to a sudden thermal shock that they could not survive.

CURRENT WATER QUALITY CONDITIONS

In 1989, the EPA announced that 17,365 segments of surface water in the United States and its territories were contaminated by toxic chemicals, sewage, or other pollutants. This contamination affects about 10 percent of the river, stream, coastal water, lake, and estuary mileage in the country. In addition, between 1 to 2 percent of the groundwater near the surface is also polluted. How does this situation compare to past pollution levels? How does the United States compare to other countries? In the next section, we will look at areas of progress and at remaining problems in water pollution control.

Surface Waters in the United States

Water pollution problems in surface waters are often both highly visible and a direct threat to environmental quality. Consequently, more has been done to eliminate surface water pollution than any other type. This is probably the greatest success story in our anti-pollution efforts. Much remains to be done, however.

Areas of Progress

Like most developed countries, the United States has made encouraging progress in protecting and restoring water quality in rivers and lakes over the past forty years. In 1948, only about one-third of Americans were served by municipal sewage systems, and most of those systems discharged sewage without any treatment or with only primary treatment (the bigger lumps of waste are removed). Most people depended on cesspools and septic systems to dispose of domestic wastes.

The 1972 Clean Water Act established a National Pollution Discharge Elimination System (NPDES), which requires an easily revoked permit for any industry, municipality or other entity dumping wastes in surface waters. The permit requires disclosure of what is being dumped and gives regulators valuable data and evidence for litigation. As a consequence, only about 10 percent of our water pollution now comes from industrial or municipal point sources. One of the biggest improvements has been in sewage treatment.

Since the passage of the Clean Water Act in 1972, the United States has spent more than $100 billion in public funds and much more in private investments to control these "conventional" point sources of water pollution. About one-third of the public expenditures have been used to build or upgrade

FIGURE 22.10 Our national goal of making all surface waters in the United States "fishable and swimmable" has not been fully met, but scenes like this have been reduced by pollution control efforts.

thousands of municipal sewage treatment plants. As a result, by 1990, 70 percent of the U.S. population was served by municipal sewage systems. Four-fifths of these plants had secondary or tertiary treatment, and no major city was discharging raw sewage into a river or lake except as overflow during heavy rainstorms. Some coastal cities, such as New York and Boston, still dump sewage sludge or partially treated effluent into the ocean. Their wastes come back to haunt them, however (figure 21.10).

Widespread and significant decreases in fecal coliform bacterial count were reported in 1987 by the U.S. Geological Survey, sampling 380 stations in a nationwide river monitoring network. Municipal BOD loads were reported to have decreased 46 percent. Industrial BOD discharges were down an impressive 71 percent since 1972, in spite of an 11 percent population growth and a 25 percent increase in GNP. The national goal of making all U.S. surface waters "fishable and swimmable" by 1985 was not met, but 75 percent of the streams, rivers, and lakes monitored by the EPA fully supported their designated uses by that date. Thirteen percent of the monitored streams and rivers on which the EPA has information have improved in recent years in terms of reduced BOD, increased dissolved oxygen content, and greater diversity of aquatic life. Only 3 percent of the monitored stations have recorded a decrease in water quality, while 84 percent have remained about the same since 1972.

Remaining Problems

The greatest impediments to achieving our national goals in water quality are nonpoint discharges of pollutants. These sources are harder to identify and to reduce or treat than are specific point sources. The EPA estimates that about three-fourths of the water pollution in the United States comes from soil erosion, fall-out of air pollutants, and surface runoff from urban areas, farm fields, and feedlots. In the United States, as much as 25 percent of the 46,800,000 metric tons (52 million tons) of fertilizer spread on farmland each year is carried away by runoff.

Cattle in feedlots produce some 129,600,000 metric tons (144 million tons) of manure each year, and the runoff from these sites is rich in viruses, bacteria, nitrates, phosphates, and other contaminants. A single cow produces about 30 kg (14 lb) of manure per day, or about as much as that produced by ten people. Some feedlots have 100,000 animals with no provision for capturing or treating runoff water. Imagine drawing your drinking water downstream from such a facility. Pets also can be a problem. It is estimated that the wastes from about a half million dogs in New York City are disposed of primarily through storm sewers, and therefore, do not go through sewage treatment.

Loading of both nitrates and phosphates in surface water have decreased from point sources but have increased about fourfold since 1972 from nonpoint sources. "Toxic fallout" from the atmosphere is a major nonpoint source of water pollutants. Fossil fuel combustion has become a major source of nitrates, sulfates, arsenic, cadmium, mercury, and other toxic pollutants that find their way into water. Carried to remote areas by atmospheric transport, these combustion products now are found nearly everywhere in the world. Toxic organic compounds, such as DDT, PCBs, and dioxins, also are transported long distances by wind currents. More than nine hundred synthetic chemicals and metals have been identified in the Great Lakes, for instance. The distribution and concentration of these materials cannot be accounted for by local sources alone. Atmospheric transport is thought to be the single largest source of toxic pollution in the Great Lakes. Perhaps as much as 13,500 kilograms (30,000 lb) of carcinogenic combustion compounds (polycyclic aromatic hydrocarbons) are deposited into the Great Lakes yearly. A study by the Canadian Government revealed that cancer rates in people living along the lower Great Lakes and Niagara River are twice as high as those in the rest of the country. Children raised in communities in this area have lower IQs, higher incidence of birth defects, and more frequent illnesses than normal. All these problems are thought to be directly related to water pollution.

Surface Waters in Other Countries

Japan, Australia, and most of Western Europe also have improved surface water quality in recent years. Sewage treatment in the wealthier countries of Europe generally equals or surpasses that in the United States. Sweden, for instance, serves 98 percent of its population with at least secondary sewage treatment (compared with 70 percent in the United States), and the other 2 percent have primary treatment. Denmark and Germany have municipal sewage treatment for 90 percent and 84 percent of their populations, respectively. The poorer countries have much less to spend on sanitation. Spain serves only 18 percent of its population with even primary sewage treatment. In Ireland, it is only 11 percent, and in Greece, less than 1 percent of the people have even primary treatment. Most of the sewage, both domestic and industrial, is dumped directly into the ocean.

This lack of pollution control is reflected in inland water quality as well. In Poland, 95 percent of all surface water is unfit to drink. The Vistula River, which winds through the country's most heavily industrialized region, was so badly polluted in 1978 that only 432 of its 1,068 km were suitable even for industrial use. It was reported to be "utterly devoid of life." In 1980, however, the Polish government instituted an ambitious program to build domestic and industrial waste treatment plants and to clean up the river. In the Soviet Union, the lower Volga River is reported to be on the brink of disaster due to the 300 million tons of solid waste and 20 trillion liters (5 trillion gal) of liquid effluent dumped into it annually

There are also some encouraging pollution control stories. One of the most outstanding is the Thames River in London. Since the beginning of the Industrial Revolution, the Thames had been little more than an open sewer, full of vile and toxic waste products from domestic and industrial sewers. In the 1950s, however, England undertook a massive cleanup of the Thames. More than $250 million in public funds plus millions more from industry were spent to curb pollution. By the early 1980s, the river was showing remarkable signs of rejuvenation. Some ninety-five species of fish had returned, including pollution-sensitive salmon, which had not been seen in London for three hundred years.

The less developed countries of South America, Africa, and Asia have even worse water quality than do the poorer countries of Europe. Sewage treatment is usually either totally lacking or woefully inadequate. Low technological capabilities and little money for pollution control are made even worse by burgeoning populations (chapter 7), rapid urbanization (chapter 25), and the shift of much heavy industry (especially the dirtier ones) from developed countries where pollution laws are strict to less developed countries where regulations are more lenient.

Appalling environmental conditions often result from these combined factors (figure 22.11). Two-thirds of India's surface waters are contaminated sufficiently to be considered dangerous to human health. The Yamuna River in New Delhi has 7,500 coliform bacteria per 100 ml (thirty-seven times the level considered safe for swimming in the U.S.) *before* entering the city. The coliform count increases to an incredible 24 *million* cells per 100 ml as the river leaves the city! At the same time, the river picks up some 20 million l of industrial effluents every day from New Delhi. It's no wonder that disease rates are high, and life expectancy is low in this area. Only 217 of India's 3,119 towns and cities have any sewage treatment, and only 8 cities have anything beyond primary treatment.

FIGURE 22.11 Severe contamination of this tidal canal in Jakarta exposes nearby residents to a variety of health hazards. It also contributes to the pollution of Jakarta Bay and the Java Sea.

In Malaysia, forty-two of fifty major rivers are reported to be "ecological disasters." Residues from palm oil and rubber manufacturing, along with heavy erosion from logging of tropical rain forests, have destroyed all higher forms of life in most of these rivers. In the Philippines, domestic sewage makes up 60 to 70 percent of the total volume of Manila's Pasig River. Thousands of people use the river not only for bathing and washing clothes, but also as their source of drinking and cooking water. China treats only 2 percent of its sewage. Of seventy-eight monitored rivers in China, fifty-four are reported to be seriously polluted. Of forty-four major cities in China, forty-one use "contaminated" water supplies, and few do more than rudimentary treatment before it is delivered to the public.

Groundwater

About half the people in the United States, including 95 percent of those in rural areas, depend on underground aquifers for their drinking water. This vital resource is threatened in many areas by overuse and pollution. In even more places, the precious remaining reserves are being made unfit for use by a wide variety of industrial, agricultural, and domestic contaminants. For decades it was widely assumed that groundwater was impervious to pollution because soil would bind chemicals and cleanse water as it percolated through. Springwater or artesian well water was considered to be the definitive standard of water purity, but that is no longer true in many areas.

The Office of Technology Assessment estimates that every day some 4.5 trillion l (1.2 trillion gal) of contaminated water seep into the ground in the United States from septic tanks, cesspools, municipal and industrial landfills and waste disposal sites, surface impoundments, agricultural fields, forests, and wells (figure 22.12). The most important of these in terms of toxicity are probably waste disposal sites. The most important in terms of total volume of pollutants and area affected are agricultural

ENVIRONMENTAL POLLUTION AND OTHER DILEMMAS

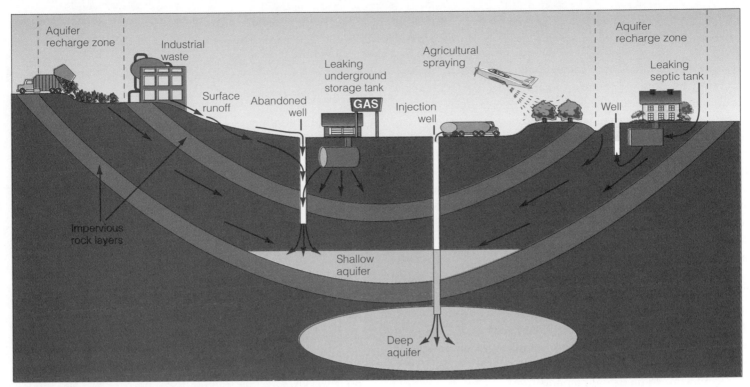

FIGURE 22.12 Sources of groundwater pollution. Septic systems, landfills, and industrial activities on aquifer recharge zones leach contaminants into aquifers. Wells provide a direct route for injection of pollutants into aquifers.

chemicals. It usually takes hundreds to thousands of years for most deep aquifers to turn over their water content, and many contaminants are extremely stable once underground. It is possible, but expensive, to pump water out of aquifers, clean it, and then pump it back. For very large aquifers, pollution may be essentially irreversible.

We don't know exactly how contaminated our aquifers already are because we have access to few aquifers, and testing is expensive. Results of recent groundwater studies have been alarming. A 1979 survey of groundwater quality in Massachusetts found that 351 communities were drawing drinking water from contaminated wells. Concentrations of industrial solvents and chemical wastes that were as much as one hundred times accepted levels were discovered. In Woburn, two of eight municipal wells were found to contain suspected carcinogens. A Harvard study found that childhood leukemia in neighborhoods served by these wells was twice as high as expected. In 1986, families of eight of these leukemia victims were awarded $1 million each in an out-of-court settlement of a lawsuit against the company that had been accused of dumping potentially carcinogenic chemicals.

In a 1982 survey of large public water systems served by groundwater, the EPA found that 45 percent were contaminated with industrial solvents, agricultural fertilizers or pesticides, or other toxic synthetic chemicals. In a 1986 survey in Iowa, pesticides and other synthetic chemicals were detected in half of all wells tested. One-fifth of these wells had nitrate levels from fertilizer infiltration that exceeded federal standards. These high nitrate levels are dangerous to infants (nitrate combines with hemoglobin in the blood and results in "blue-baby" syndrome). They also are transformed into cancer-causing nitrosamines in the human gut. In Florida, one thousand drinking water wells were shut down by state authorities in 1986 because of excessive levels of toxic chemicals, mostly ethylene dibromide (EDB), a pesticide used to kill nematodes (roundworms) that damage plant roots.

The United States has at least 2.5 million underground chemical storage tanks. Most of the tanks are reaching the end of their life expectancy, and one-third of them are believed to be leaking. Many of these tanks were left behind at abandoned gasoline stations or industrial sites, their exact whereabouts and contents unknown. The EPA estimates that about 42 million l (11 million gal) of gasoline are lost each year from leaking underground storage tanks (LUST). Considering that a single gallon of gasoline can make an aquifer unsuitable for drinking, these old, rusting, forgotten tanks represent a problem of tremendous proportions. In 1986, Congress placed a tax on motor fuels to generate a $500 million fund for cleaning up abandoned tanks.

The EPA now requires that all new tanks have double walls or be placed in concrete vaults to help prevent leaks into groundwater.

Aquifers in the United States also are threatened by direct injection of wastes. Every year, 38 billion l (10 billion gal) of liquid wastes, such as oil field brine, effluents from chemical plants, and treated sewage, are pumped down deep wells as an alternative to incineration or other treatment. The EPA estimates that 58 percent of all hazardous wastes generated are injected into deep wells. No permits are required, nor are there any limits on where or how wastes are pumped. Opponents of this disposal method argue that we don't know exactly how underground aquifers are connected or how toxic substances might flow through them. Some three hundred deep injection wells are in use, and many billions of liters of wastes already have been pumped into them. It seems likely that some of those wastes have made or will make their way into aquifers used for domestic and municipal water supplies. How do you feel about this situation, and what do you think should be done?

Abandoned wells represent another major source of groundwater contamination. Most domestic wells have no casings to prevent surface contaminants from leaking directly into aquifers that they penetrate. When these wells are no longer in use, they are rarely capped adequately, and people forget where they are. They become direct routes for drainage of surface contaminants into aquifers. Oil wells and municipal water wells do have casings to prevent leakage into aquifers, but these casings corrode and crack as they age.

Aquifer recharge zones are the normal route for replenishing groundwater (figure 22.12). When surface waters that percolate through the recharge zone are contaminated by runoff from farm fields, feedlots, city streets, or industrial sites, the underlying aquifers are threatened. If pollution levels are low and the filtering capacity of surface sediments is high, the aquifer may remain clean; but when pollution levels are high, aquifers become contaminated. Sound land-use practices exclude polluting activities from aquifer recharge zones (chapter 13). When well-built and properly maintained, septic systems are appropriate for domestic wastes in rural areas, but they should not be built on recharge zones.

Relatively little information is available about groundwater quality in most countries, especially in the less developed countries, because (1) it is expensive to drill test wells and to monitor pollutants and (2) this is not yet a major priority. In Europe, where fertilizer use is even more intensive than in the United States (chapter 11), nitrate levels in groundwater are reported to be alarmingly high in many areas. Britain, for instance, calculates that half of its underground reservoirs have been contaminated by fertilizer nitrates. In 1987, China reported that forty-one of forty-four large cities suffer from polluted groundwater. As ag-

ricultural modernization increases the use of fertilizer and pesticides in less developed countries, we may see more groundwater pollution there as well.

Ocean Pollution

During the summer of 1988, bathers from New Jersey to Massachusetts experienced unwelcome firsthand evidence of increasing levels of ocean pollution. They found floating garbage, ranging from untreated sewage to used drug paraphernalia and medical wastes, washing up on their favorite beaches. One author reported that the experience was as safe and appealing as bathing in an unflushed toilet.

This distressing situation is only one aspect of a global problem. Near-shore zones around the world, especially bays, estuaries, shoals, and reefs near large cities or the mouths of major rivers, are being overwhelmed by human-caused contamination. Suffocating and sometimes poisonous blooms of algae regularly deplete ocean waters of oxygen and kill enormous numbers of fish and other marine life. High levels of toxic chemicals, heavy metals, disease-causing organisms, oil, sediment, and plastic refuse are adversely affecting some of the most attractive and productive ocean regions. The potential losses caused by this pollution amount to billions of dollars each year. In terms of quality of life, the costs are incalculable. Oceanographer Jacques Cousteau warns that the oceans are dying and that our own survival is threatened.

One of the most massive and least understood sources of this pollution is agricultural and urban runoff. Fertilizers, manure, pesticides, and crop residues from farm fields combine with oil, rubber, metals, salts, and other urban contaminants and are carried by rivers to the ocean. Industrial wastes and municipal sewage effluents are also chronic pollution sources of near-shore ocean zones. In the United States, 1,300 major industrial and 600 municipal facilities dump untreated wastewater directly into estuaries and coastal regions. Thousands of other facilities discharge a variety of toxic wastes into rivers that run into the oceans.

The United States and England are the only countries that use the ocean for sludge disposal. Each year England dumps about 10 million metric tons and the United States dumps about 7 million metric tons of sewer sludge and industrial waste contaminated with heavy metals, such as mercury, cadmium, and lead, as well as an abundance of toxic organic and inorganic matter. The environmental effects of sludge dumping are well known. The fine sludge covers the bottom with a thick, toxic goo that kills organisms and slowly releases harmful chemicals that bioaccumulate in marine food chains. In 1988, fishermen reported an alarming increase in deformed and diseased fish, crabs, and lobsters around sites of sewer sludge dumping along the Atlantic coast.

FIGURE 22.13 Plastic litter covers this beach in Niihau.

FIGURE 22.14 Deadly necklace. Marine biologists estimate that castoff nets, plastic beverage yokes, and other packing residue kill hundreds of thousands of birds, mammals, and fish each year.

Years of dumping have made the harbors of such American cities as Boston, Baltimore, and New York so toxic that virtually no marine life can survive there. EPA regulations now require cities to dump their wastes further offshore in deeper water where the effluent will be more dispersed, presumably will have less harmful effects on bottom dwellers, and will be less likely to contaminate coastal regions. Critics of this program argue that careless loading and transportation allow spills that contaminate swimming beaches of New York and New Jersey and that uncharted ocean currents bring wastes back inshore where they poison fish and other marine life.

Discarded plastic flotsam and jetsam are becoming an ubiquitous mark of human impact on the oceans (figure 22.13). Since plastic is lightweight and nonbiodegradable, it is carried thousands of miles on ocean currents and lasts for years. Even the most remote beaches of distant islands are likely to have bits of styrofoam containers or polyethylene packing material that were discarded half a world away by some careless person. It has been estimated that some 6 million metric tons of plastic bottles, packaging material, and other litter are tossed from ships every year into the ocean where they ensnare and choke seabirds, mammals (figure 22.14), and even fish. Sixteen states now require that six-pack yokes be made of biodegradable or photodegradable plastic, limiting their longevity as potential killers. The amount of municipal and industrial plastic that finds its way to the ocean is unknown but immense. In one day, volunteers in Texas gathered more than three hundred tons of plastic refuse from gulf coast beaches.

Fishing boats lose or discard 136 million tons of plastic fishing gear each year, much of it in the form of nonbiodegradable fishing nets. Among the worst of these nets are the enormous drift and gill nets that float unattended just below the ocean surface. Up to 50 km (35 mi) long and 10 m (30 ft) deep, these spidery, monofilament nets are practically invisible once in the water. Ostensibly set to catch tuna, squid, and swordfish, these "walls of death" sweep through the water trapping and drowning everything that becomes entangled in their web. Millions of whales, dolphins, sharks, seals, and sea birds are killed.

In the North Pacific, a fleet of 1,500 fishing boats, mainly from Japan, Taiwan, and South Korea, set out an estimated 20,000 nautical miles of drift nets every day. This practice is severely depleting fishing stocks. Both young and adult fish are caught and killed; consequently, no breeding stock is left to replenish the species.

Even worse is the problem of abandoned or lost nets. When a net becomes snagged or tangled, it is cut loose. These "ghost nets" drift until the dead fish and sea mammals they have caught cause them to sink. After the organic material decays, the net may rise again to resume fishing. This deadly cycle may continue for years.

An estimated 600 nautical miles of drift nets are lost every year. Altogether, discarded fishing gear and plastic waste kill at least 1 million sea birds and 150,000 marine mammals each year. Many nations outlaw the use or disposal of drift nets in coastal waters, but their use is difficult to control since most of this type of fishing is in international waters.

Few coastlines in the world remain uncontaminated by oil or oil products. Tar granules and sticky crude oil droplets stick to feet on beaches everywhere. Because oil floats on water, it is easily detected on the open ocean. Figure 22.15 shows locations where sailors reported visible oil slicks in the early 1980s. Oceanographers estimate that somewhere between 3 million and 6 million metric tons of oil are discharged into the world's oceans each year from both land- and sea-based operations. About half of this amount is due to maritime transport. Most of the 40 million liters of discharge (nearly 11 million gal) is not from dramatic, headline-making accidents, such as the 1989 oil spill from the *Exxon Valdez* in Prince William Sound, Alaska (Box 22.1), but

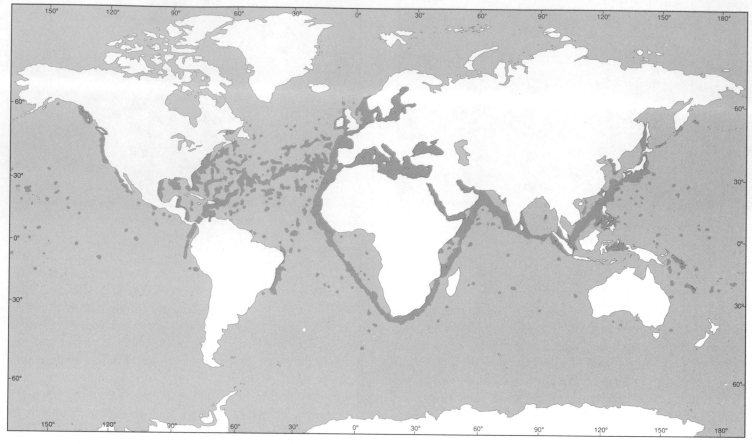

FIGURE 22.15 Oil slicks are visible in every ocean and along the coasts of most continents. Note the prevalence of oil spills along major shipping routes.

from routine open-sea bilge pumping and tank cleaning, which are illegal but, nonetheless, are carried out once ships are beyond sight of land. Much of the rest comes from land-based municipal and industrial runoff or from atmospheric deposition of residues from refining and combustion of fuels.

The transport of huge quantities of oil creates opportunities for major pollution episodes through a combination of human and natural hazards. Military conflict in the Middle East and increasing amounts of oil being pumped and shipped from offshore drilling areas in inhospitable places, such as the notoriously rough North Sea and the Arctic Ocean, make it likely that more oil spills will occur. Plans to drill for oil along the seismically active California and Alaska coasts have been controversial because of the damage that oil spills could cause to these biologically rich coastal ecosystems.

The toxic chemicals of all sorts that we dump into the ocean are having deadly effects on marine life. In 1987, more than 750 dolphins died mysteriously along the Atlantic coast. Some had snouts, flippers, and tails pocked with blisters and craters. Others had huge patches of skin sloughed off. The most likely cause of these distressing conditions is viral infections. It is thought that exposure to pesticides and other water pollutants may have weakened the mammals' immune systems and made them susceptible to infections. Harbor seals in the Gulf of Maine have the highest pesticide levels of any U.S. mammals on land or sea. Fishermen bring up lobsters and crabs with gaping holes in their shells, and fish have rotted fins and ulcerous lesions. In Louisiana, 35 percent of the oyster beds have been closed because of sewage contamination. Japan's heavily used inland sea has more than two hundred toxic red tides each year. One of these episodes killed more than 1 million yellowtail that would have been worth $15 million in the market.

We often don't know exactly which pollutant is causing these distressing biological effects. In many cases, there may not be a single pollutant, but a complex series of interactions in marine ecosystems that have many manifestations. Sometimes symptoms of shifting balances may not be noticeable until a disastrous population crash occurs. We have always assumed that the ocean is so vast that its capacity to absorb and neutralize contaminants would be inexhaustible. Marine ecosystems do have an enormous ability to recover from pollution episodes and to regenerate biological communities. Some enormous oil spills, such as

ENVIRONMENTAL POLLUTION AND OTHER DILEMMAS

the wreck of the *Amoco Cadiz* on the coast of France in 1978 or the blowout of a Mexican oil well in the Gulf of Mexico in 1979, had far less disastrous consequences than had been feared. Many scientists feel that localized, short-term pollution episodes in warm tropical waters may not cause serious long-term damage. Spills in cold arctic waters, like that of the *Exxon Valdez* in 1989, may take much longer to dissipate and may cause much more damage. Attention is now turning more to the chronic, land-based pollution from industrial, municipal, and agricultural wastes that slowly build up until they overwhelm natural systems and destroy the ocean's regenerative capacity.

WATER POLLUTION CONTROL

Appropriate land-use practices and careful disposal of industrial, domestic, and agricultural wastes are essential for control of water pollution.

Source Reduction

In many cases, the cheapest and most effective way to reduce pollution is to avoid producing or releasing it to the environment in the first place. Better land-use regulations and improved forestry, agriculture, and construction practices can greatly reduce surface runoff (chapters 11 and 12). Prohibiting logging or cultivation on steep slopes, leaving undisturbed plant cover or crop residues on the ground, protecting vegetation in and along watercourses, and other soil conservation techniques can prevent erosion and hold back both water and pollutants. Applying precisely determined amounts of fertilizer, irrigation water, and pesticides also reduces soil salinization and contamination of runoff.

Nonagricultural, nonpoint sources of water pollution also can be controlled through source reduction. Elimination of lead from gasoline has resulted in a widespread and significant decrease in the amount of lead in surface waters in the United States. Studies have shown that as much as 90 percent less road deicing salt can be used in many areas without significantly affecting the safety of winter roads. Careful handling of oil and petroleum products can greatly reduce the amount of water pollution caused by these materials.

Industry can modify manufacturing processes so fewer wastes are created. Recycling or reclaiming materials that otherwise might be discarded in the waste stream also reduces pollution. Both of these approaches usually have economic as well as environmental benefits. In some states, "sewering" of heavy metals by industry has been outlawed. Producers are required instead to separate their wastes. It turns out that a variety of valuable metals can be recovered from wastes and reused or sold for other purposes. The company benefits by having a product to sell, and the municipal sewage treatment plant benefits by not having to deal with highly toxic materials mixed in with millions of gallons of other types of wastes.

Human Waste Disposal

As we have already seen, human and animal wastes usually create the most serious health-related water pollution problems. More than 500 types of disease-causing (pathogenic) bacteria, viruses, and parasites can travel from human or animal excrement through water. In this section, we will look at how to prevent the spread of these diseases.

Natural Processes

In the poorer countries of the world, most rural people simply go out into the fields and forests to relieve themselves as they have always done. Where population densities are low, natural processes eliminate wastes quickly, making this an effective method of sanitation. The high population densities of cities make this practice unworkable, however. Even major cities of many less developed countries are often littered with human waste which has been left for rains to wash away or for pigs, dogs, flies, beetles, or other scavengers to consume. This is a major cause of disease, as well as being extremely unpleasant. Studies have shown that a significant portion of the airborne dust in Mexico City is actually dried, pulverized human feces.

The first examples of municipal sanitation appear to have been in the cities of the Indus Valley (Harappa and Mohenjo-Daro), where covered sewers were built five thousand years ago. The Romans built huge sewers two thousand years ago that are still in use. The *Cloaca maxima* of Rome has enough capacity to serve a city of 1 million people. The wastes collected by these sewers were simply discharged into the nearest lake, river, or ocean. This made the city more habitable, but transferred the pollution problem from one place to another. As we saw earlier in this chapter, dumping raw sewage into bodies of water is still a common practice in many parts of the world and is a major source of water pollution.

Where intensive agriculture is practiced—especially in wet rice paddy farming in Asia—it has long been customary to collect "night soil" (human and animal waste) to be spread on the fields as fertilizer. This is a valuable source of plant nutrients, but it is also a source of disease-causing pathogens in the food supply. It is the main reason that travelers in less developed countries must be careful to surface sterilize or cook any fruits and vegetables they eat. Collecting night soil for use on farm fields was common in Europe and America until about one hundred years ago when the association between pathogens and disease was recognized.

Until about fifty years ago, most rural American families and quite a few residents of towns and small cities depended on a pit toilet or "outhouse" for waste disposal. Untreated wastes tended to seep into the ground, however, and pathogens sometimes contaminated drinking water supplies. The development of septic tanks and properly constructed drain fields represented a considerable improvement in public health (figure 22.16). In

BOX 22.1 The Wreck of the *Exxon Valdez*

At 9 P.M. on March 23, 1989, the tanker *Exxon Valdez* cast off from the Alyeska oil terminal in Valdez, Alaska, and began moving slowly down the narrow, twisting fjord toward Prince William Sound and the Gulf of Alaska. Loaded with 1.2 million barrels of crude oil (50 million gal), the huge ship moved ponderously through the cold night. More than 300 m (987 ft) from stem to stern, this behemoth is as long as a 50 story building is tall. If your city has a building that big, go look at it and imagine laying it on its side and driving it through high waves and shifting currents down a narrow, rocky channel on a dark night.

About two hours after leaving the terminal, Captain Joseph Hazelwood radioed the Coast Guard to say that he was moving the ship from the normal outbound shipping lane to the inbound lane to avoid floating ice. A few minutes later, Hazelwood, whose license was later revoked for drinking while on duty and negligence in leaving the bridge, turned over control of the vessel to Third Mate Gregory Cousins and went to his cabin.

Nine minutes later, at 12:04 A.M., the *Valdez* ran aground on the well-marked Bligh Reef, ripping eight huge gashes in its hull. Gooey, foul-smelling crude oil gushed from the ruptured hull at a rate of 76,000 l (20,000 gal) per hr. Calls for help went out immediately, but it took nearly twenty-four hours for the Alyeska Oil Spill Team to assemble its gear and reach the site (box figure 22.1). The barge that was supposed to be loaded and ready to go at all times was damaged and barely seaworthy. Hoses, lights, and containment booms were missing or inoperable. By the time the flow was stanched many days later, about 250,000 barrels (11 million gal) of oil had hemorrhaged into the ocean. Promises that an accident of this size could not possibly happen and that a response would be mounted within three hours if it did, obviously were empty rhetoric.

Hopelessly unprepared and understaffed, there was little that the Oil Spill

BOX FIGURE 22.1 The *Exxon Valdez* aground on Bligh Reef in March, 1989. Nearly 200,000 barrels (11 million gallons) of crude oil spilled by the wreck spread over 4,600 km² (1,800 mi²) of the pristine waters of Prince William Sound. The one million barrels (42 million gallons) remaining in the Exxon *Valdez* are being pumped into the *Baton Rouge*.

Team could do to scoop up the viscous, black crude oil. Only 10 precent was ultimately recovered. The rest evaporated or spread over 4,600 km² (1,800 mi²) of Prince William Sound. More than 2,500 km (1,560 mi) of pristine beaches and rocky islands were smeared with a filthy, smelly, oily sludge that stuck to everything it touched. More and more workers were rushed to the scene as the massive oil slick spread inexorably across the sound, along the coast, and into Kenai Fjords and Katmai National Parks.

As the oil slick spread, workers began bringing in oil-soaked birds and mammals. Emergency treatment centers were overwhelmed. Wildlife experts estimate that between 100,000 and 300,000 sea birds were killed by the oil. Some were poisoned from injesting toxic chemicals, some died from starvation or exhaustion trying to swim through the sticky crude, but most died from hypothermia when oil-soaked feathers lost their insulating capacity. Bald eagles were poisoned or fouled by eating contaminated fish and birds. A 1989 survey found 60 dead eagles and no surviving chicks in any of the nests around the sound.

Alaska has the largest remaining sea otter population in the world, and Prince William Sound is the center of their range. At least 1,000 otters were contaminated and about 80 percent of them died from hypothermia and poisoning. Teams of dedicated volunteers and wildlife professionals worked to save otters that were brought into rescue centers (box figure 22.2). In spite of the care and attention, however, mortality rates were very high. Oil is much more toxic to marine mammals than had been supposed. More than two-thirds of the dead animals had lung damage from oil fumes. Many also had liver, kidney, or spleen abnormalities. The $18.3 million paid by Exxon for otter recovery works out to $51,260 apiece for each of the 225 surviving animals.

Fortunately, fish seem to have been less adversely affected by the oil than was first feared. Prince William Sound is one of the richest and most productive fisheries in the world. Every year millions of

BOX FIGURE 22.2 A tranquilized sea otter is washed by volunteers at the Valdez Otter Rescue Center. Even though the detoxification process is stressful for the otters, 225 were saved. However, about five times that number were brought in dead or died while being treated.

BOX FIGURE 22.3 Workers on Naked Island in Prince William Sound use high-pressure hot water hoses to wash oil from the beach and into the boomed-off bay where skimmers wait. This process cleaned the surface, but it also drove oil deep into the sand and rocks. Wind and waves reoiled the beaches almost immediately. Furthermore, this treatment killed plants and animals that lived on and among the rocks. Critics claim that natural processes are less destructive and more effective than this brute-force approach.

pink salmon pass through the sound on their way to spawn in the cold, clear rivers that tumble down the mountain valleys. The fish catch is worth about $90 million per year, and most of the villages around the sound live entirely on fishing and tourism. In 1989 the harvest was a disaster, at least in part because many crews were making more money working on the cleanup than fishing.

Some 43 million salmon were caught in 1990, however, far more than the previous record of 29 million in 1987. Pink salmon have a two-year life cycle. Many of the fish harvested in 1990 were either hatched in waters fouled by the *Exxon Valdez* or swam through the area on their way to the open ocean. Apparently, neither the young salmon nor the plankton, their main food at that stage, were irreparably damaged by the oil.

During the summer of 1989, more than 10,000 people worked scrubbing rocks and spraying oil-fouled beaches (box figure 22.3). It was mostly a futile exercise: as quickly as the surface oil was re-

moved, more washed in from offshore or oozed up from beneath the rocks. In the first work season, only 3 percent of the shoreline was cleaned satisfactorily. Exxon spent about $2 billion in 1989 and another $200 million in 1990 on this effort.

Interestingly, the most effective cleanup was done by nature. Over the winter, fierce storms and pounding waves scoured at least 75 percent of the surface oil and half of the buried oil from the beaches. At the start of the 1990 season, only 160 km (100 mi) of shore remained badly contaminated. By September, when work stopped again for the winter, only 6 km (4 mi) was still considered heavily oiled.

Often the best way to clean oil contaminated beaches is to just let nature take its course. A technique called bioremediation works on this principle. Instead of blasting the beach with high-pressure steam that washes off surface oil but also kills all living organisms, bioremediation encourages growth of creatures that me-

tabolize the oil. Spreading fertilizers and straw stimulates growth and recovery of natural ecosystems. In some cases, microbes genetically engineered to consume oil are spread over the area to speed up natural processes. The Alaskan beaches treated by bioremediation looked considerably healthier more rapidly than those scrubbed and steamed. This is also much cheaper than the brute-force approach. One of the heartening lessons learned from this terrible episode is that nature may be more resilient and less fragile than we generally suppose.

Perhaps the most shocking fact about the wreck of the *Exxon Valdez* is that the 11 million gal it spilled represent only about 5 percent of the total oil spilled or dumped worldwide in 1989. With 20 *billion* barrels of oil pumped and shipped around the world every year, oceans and land inevitably will be fouled by oil and its by-products in the future. Unless we curb our insatiable appetite for oil and our wasteful habits in using it, similar accidents are bound to occur.

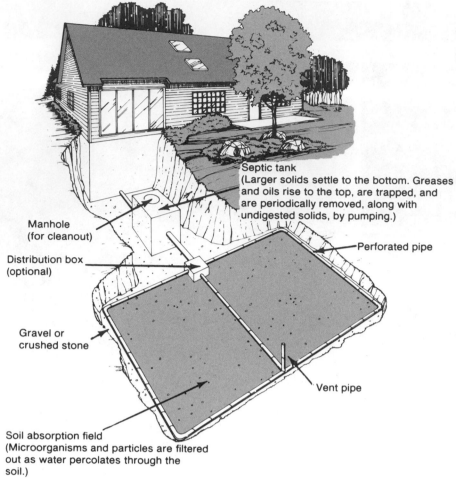

Manhole
(for cleanout)

Distribution box
(optional)

Gravel or
crushed stone

Septic tank
(Larger solids settle to the bottom. Greases
and oils rise to the top, are trapped, and
are periodically removed, along with
undigested solids, by pumping.)

Perforated pipe

Vent pipe

Soil absorption field
(Microorganisms and particles are filtered
out as water percolates through the
soil.)

FIGURE 22.16 A domestic septic tank and drain field system for
sewage and wastewater disposal.

a typical septic system, wastewater is first drained into a septic
tank. Grease and oils rise to the top and solids settle to the bottom,
where they are subject to bacterial decomposition. The clarified
effluent from the septic tank is channeled out through a drainfield of small perforated pipes embedded in gravel just below the
surface of the soil. The rate of aeration is high in this drainfield
so that pathogens (most of which are anaerobic) will be killed,
and soil microorganisms can metabolize any nutrients carried by
the water. Excess water percolates up through the gravel and
evaporates. Periodically, the solids in the septic tank are pumped
out into a tank truck and taken to a treatment plant for disposal.

Where land is available and population densities are not
too high, this can be an effective method of waste disposal. It is
widely used in suburban areas, but as suburban densities grow,
groundwater pollution often becomes a problem, indicating the
need to shift to a municipal sewer system. It doesn't work well
in cold, rainy climates where the drainfield may be too wet for
proper evaporation, or where the water table is close to the surface. Private septic systems have been banned recently from many
areas because of septic seepage into surface or groundwaters.

Residents have been required to install holding tanks that collect
wastewater, which then must be pumped and removed by a septic
service. Although this is an extra nuisance and expense for residents, it is regarded as a necessary tradeoff to maintain (or restore) water quality. How would you respond to such a regulation
if it affected your property?

Municipal Sewage Treatment

Over the past one hundred years, sanitary engineers have developed ingenious and effective municipal wastewater treatment
systems to protect human health, ecosystem stability, and water
quality. This topic is an important part of pollution control, and
is a central focus of every municipal government; therefore, let's
look more closely at how a typical municipal sewage treatment
facility works.

Primary treatment is the first step in municipal waste treatment. It physically separates large solids from the waste stream.
As raw sewage enters the treatment plant, it passes through a
metal grating that removes branches, tires, dead animals, and
other large debris (figure 22.17a). A moving screen then filters

ENVIRONMENTAL POLLUTION AND OTHER DILEMMAS

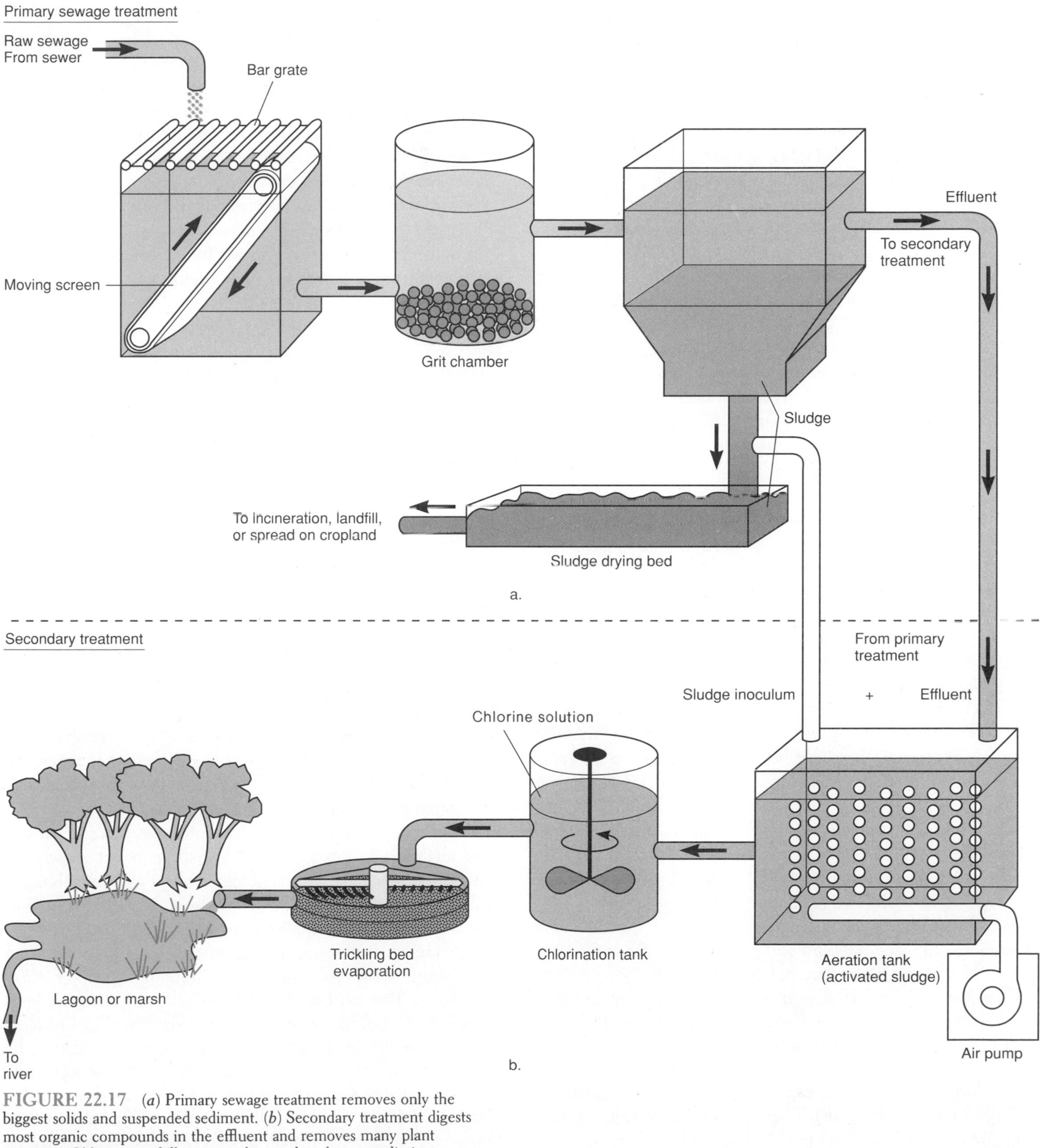

Primary sewage treatment

Raw sewage
From sewer

Bar grate

Moving screen

Grit chamber

Effluent

To secondary
treatment

Sludge

To Incineration, landfill,
or spread on cropland

Sludge drying bed

a.

Secondary treatment

From primary
treatment

Sludge inoculum   +   Effluent

Chlorine solution

Lagoon or marsh

To
river

Trickling bed
evaporation

Chlorination tank

Aeration tank
(activated sludge)

Air pump

b.

FIGURE 22.17 (*a*) Primary sewage treatment removes only the
biggest solids and suspended sediment. (*b*) Secondary treatment digests
most organic compounds in the effluent and removes many plant
nutrients. Chlorination kills most pathogens but does not eliminate
adverse ecological effects.

out diapers, tampons, bottles, and similar smaller items. Brief residence in a grit tank allows sand and gravel to settle. The waste stream then is pumped into the primary sedimentation tank where about half the suspended, organic solids settle to the bottom as sludge. Many pathogens remain in the effluent and it is not yet safe to discharge into waterways or onto the ground.

Secondary treatment consists of biological degradation of the remaining suspended solids. The effluent from primary treatment is pumped into a trickling filter bed, an aeration tank, or a sewage lagoon. The trickling filter is simply a bed of stones or corrugated plastic sheets through which water drips from a system of perforated pipes or a sweeping overhead sprayer (figure 22.17*b*). Bacteria and other microorganisms in the bed catch organic material as it trickles past and aerobically decompose it.

Aeration tank digestion is also called the activated sludge process. Effluent from primary treatment is pumped into the tank and mixed with a bacteria-rich slurry. Air or pure oxygen pumped through the mixture encourages bacterial growth and decomposition of the organic material. Water is siphoned off the top of the tank and sludge is removed from the bottom. Some of the sludge is used as an inoculum for incoming primary effluent. The remainder would be valuable fertilizer if it were not contaminated by metals, toxic chemicals, and pathogenic organisms. The toxic content of most sewer sludge necessitates disposal by burial in a landfill or incineration. Sludge disposal is a major cost in most municipal sewer budgets.

Where space is available for sewage lagoons, the exposure to sunlight, algae, aquatic organisms, and air does the same job more slowly but with less energy costs. Effluent from secondary treatment processes is usually treated with chlorine to kill harmful bacteria before it is released.

Tertiary treatment removes plant nutrients, especially nitrates and phosphates, from the secondary effluent. Although wastewater is usually free of pathogens and organic material after secondary treatment, it still contains high levels of inorganic nutrients, such as nitrates, phosphates, iron, potassium, and calcium. When discharged into surface waters, these nutrients stimulate algal blooms and eutrophication. To preserve water quality, these nutrients also must be removed. Chemicals often are used to bind and precipitate nutrients.

Onland disposal uses secondary treatment effluent for crop irrigation, removing nutrients and conserving water at the same time. This is safe if crops are not consumed directly by people. Wetland disposal works well in some areas. Marshes and bogs have a great capacity for absorbing nutrients and other pollutants (chapter 5). In some cases, artificial marshes have been constructed as a cheap but effective way to remove water pollution. The biomass they produce can be used to generate methane, a valuable, clean burning fuel. They also serve as a home for wildlife and a pleasant addition to the urban landscape.

In most American cities, sanitary sewers are connected to storm sewers, which carry runoff from streets and parking lots.

A large line called an interceptor delivers the combined stream of storm runoff and domestic and industrial waste to the municipal treatment plant. Storm sewers are routed to the treatment plant rather than discharged into surface waters because runoff from streets, yards, and industrial sites generally contains a variety of refuse, fertilizers, pesticides, oils, rubber, tars, lead (from gasoline), and other undesirable chemicals. During dry weather, this plan works well. Heavy storms often overload the system, however, causing bypass dumping of large volumes of raw sewage and toxic surface runoff directly into receiving waters. To prevent this overflow, cities are spending hundreds of millions of dollars to separate storm and sanitary sewers. These are huge, disruptive projects. When they are finished, surface runoff will go directly into the river and cause another pollution problem.

Heavy Metals in Sewage Sludge

Sewage wastes are biodegradable and could be a potential source of fertilizer if they were less contaminated with industrial wastes. What can be done? Metropolitan sewer plants often receive hundreds of tons of toxic metals, such as cadmium, chromium, copper, lead, nickel, and zinc in the sewer stream each year. The largest single contributor is usually lead residue from gasoline. Fortunately, as leaded gasoline is phased out, this source is decreasing. Other important sources are electroplating, metal finishing, and printed circuit board manufacturing. The presence of these metals in the sludge makes it unsafe for land disposal and for use as fertilizer. It would be much cheaper to prevent these metals from entering the waste stream in the first place than to try to remove them after they are mixed with massive quantities of other refuse.

There are several possible solutions to this problem. One option is to change industrial processes so these residues are not created, or metals could be trapped in the effluent on ion-exchange resins, which work much the same way as water softeners do. A filter made of an absorbent resin binds metal ions and releases a less toxic ion, such as sodium, in exchange. This makes it possible to recycle and reuse industrial chemicals rather than dump them. Recycling also reduces the need for mining, processing, and transportation of new materials. The building of pretreatment plants is, however, an expensive proposition for most industries and/or municipalities wanting to keep or attract industries. The ash from sewage sludge incineration has been considered a source of valuable minerals, but recovery techniques have not been proven on a large scale. Sewer sludge also makes an excellent additive in brick making. It produces bricks of better quality, lighter weight, better absorbancy, and higher strength than the normal clay-shale mixture. There is such an enormous volume of sludge, however, that it would make far more bricks than we can use (figure 22.18).

"WELL, IF YOU CAN'T USE IT, DO YOU KNOW ANYONE WHO CAN USE 3000 TONS OF SLUDGE EVERY DAY?"

FIGURE 22.18

© 1976 by Sidney Harris—*American Scientist* Magazine.

WATER LEGISLATION

Water pollution legislation has been among the most popular and most effective of all environmental legislation. In this section, we will look at some major water quality laws and their provisions (table 22.2).

The Clean Water Act of 1972 is the farthest reaching of all federal water legislation to date. It regulates release of "conventional pollutants" (dirt, organic wastes, and sewage), toxic chemicals (from the Toxic Substances Control Act), and "nonconventional pollutants" (those whose toxicity is suspected but not yet determined). For such point sources as industries, municipal facilities, and other discrete sources of conventional pollutants, the act requires discharge permits and **best practical control technology (BPT)** to reduce pollution. It sets national goals of **best available, economically achievable technology (BAT)** for toxic substances and zero discharge for 129 priority toxic pollutants. An important provision of the act provided $52 billion in federal grants to help states and communities construct sewage treatment plants. By 1988, all municipalities were required to have at least secondary sewage treatment.

Nonpoint sources of pollution (runoff from city streets, construction sites, farmland, mines, and waste dumps) are more difficult to control. When the U.S. Senate voted in 1985 to renew and extend the Clean Water Act, one of the more significant provisions of this bill was a program to require and help states control nonpoint sources. Another important section of the act protects wetlands and aquifer recharge zones. A powerful clause of the act allows environmental groups to sue polluters for exceeding NPEDES permits. Using data that polluters themselves are required to submit, groups like the Natural Resources Defense Council and the Sierra Club have won million dollar settlements under this act. This helps clean up the environment while deterring pollution.

The 1990 London Dumping Convention calls for phasing out all ocean dumping of industrial waste, tank-washing effluent, and plastic trash by 1995. The sixty-four nations that have signed the Law of the Sea Treaty are bound by this agreement. The United States has already passed legislation to support its provisions. Whether this can be enforced remains to be seen, however.

Is it safe to assume that we are well on our way to solving our water pollution problems? We are better off in terms of legislation, policy, and practice than we were in 1960. Laws, however, are only as good as (1) the degree to which they are not weakened by subsequent amendments and exceptions and (2) the degree to which they are funded for research and enforcement. Economic interests cause continued pressure on both of these points, so that the importance of an overriding national attitude to maintain the intent of protective legislation must continually be stressed and retaught.

S U M M A R Y

Worldwide, the most serious water pollutants, in terms of human health, are pathogenic organisms from human and animal wastes. We have traditionally taken advantage of the capacity of ecosystems to destroy these organisms, but as population density has grown, these systems have become overloaded and ineffective. Effective sewage treatment systems are needed that purify wastewater before it is released to the environment.

In industrialized nations, toxic chemical wastes have become an increasing problem. Agricultural and industrial chemicals have been released or spilled into surface waters and are seeping into groundwater supplies. The extent of this problem is probably not yet fully appreciated.

Ultimately, all water ends up in the ocean. The ocean is so large that it would seem impossible for human activities to have a significant impact on it, but pollution levels in the ocean are increasing. Major causes of ocean pollution are oil spills from tanker bilge pumping or accidents and oil well blowouts. Surface runoff and sewage outfalls discharge fertilizers, pesticides, organic nutrients, and toxic chemicals that have a variety of deleterious effects on marine ecosystems. We usually think of eutrophication (increased productivity due to nutrient addition) as a process of inland waterways, but this can occur in oceans as well.

The major water pollutants in terms of quantity are silt and sediments. Biomass production by aquatic organisms, land erosion, and refuse discharge all contribute to this problem. Addition of salts and metals from highway and farm runoff and industrial activities also damage water quality. In some areas, drainage from mines and tailings piles deliver sediment and toxic materials to rivers and lakes. Water pollution is a major source of human health problems. As much as 80 percent of all disease and some 25 million deaths each year may be attributable to water contamination.

Appropriate land-use practices and careful disposal of industrial, domestic, and agricultural wastes are essential for control of water pollution. Natural processes and living organisms have a high capacity to remove or destroy water pollutants, but these systems become overloaded and ineffective when pollution levels are too high. Municipal sewage treatment is effective in removing organic material from wastewater, but the sewage sludge is often contaminated with metals and other toxic industrial materials. Reducing the sources of these materials is often the best solution to our pollution problems.

■ Review Questions

1. Define water pollution.
2. List eight major categories of water pollutants and give an example for each category.
3. Describe ten major sources of water pollution in the United States. What pollution problems are associated with each source?
4. Name some waterborne diseases. How do they spread?
5. What is eutrophication? What causes it?
6. What are the origins and effects of siltation?
7. Describe primary, secondary, and tertiary processes for sewage treatment. What is the quality of the effluent from each of these processes?
8. Why do combined storm and sanitary sewers cause water quality problems? Why does separating them also cause problems?
9. What pollutants are regulated by the 1985 Clean Water Act? What goals does this act set for abatement technology?
10. What is the difference between best practical control technology and best available technology? Which do you think should be required? Why does the Clean Water Act start with BPT?

■ Questions for Critical Thinking

1. Do you believe that a definition of pollution should be limited to human-caused changes? What natural changes affect desired human uses?
2. Do you think that water pollution is worse now than it was in the past? In what ways is the situation better or worse now?
3. Many rural Americans use cesspools or septic systems. What would be the effects if these systems were prohibited? Is such prohibition feasible?
4. What sorts of chemicals are flushed down the drains in your house? What else could you do with these chemicals?
5. Should farmers be required to catch and treat feedlot runoff? What level of treatment should be required?
6. Proponents of deep well injection of hazardous wastes argue that it will never be economically feasible to pump water out of aquifers more than one kilometer below the surface. Therefore, they say, we might as well use those aquifers for hazardous waste storage. What do you think? Why?
7. What limits could be applied to very large crude oil carriers to insure safety of marine ecosystems? How can marine protection be enforced in international waters?
8. Under what conditions might cultural eutrophication be beneficial? How should we weigh beneficial and negative effects?
9. Suppose your city were going to separate storm and sanitary sewers. What problems might occur? Would it be possible?
10. Should all urban areas be required to install tertiary sewage treatment? Why or why not?

ENVIRONMENTAL POLLUTION AND OTHER DILEMMAS

Key Terms

best available, economically
achievable technology
(BAT) (p. 461)
best practical control
technology (BPT)
(p. 461)
biological oxygen demand
(BOD) (p. 443)
coliform bacteria (p. 441)
cultural eutrophication
(p. 444)
dissolved oxygen (DO)
content (p. 443)

eutrophic (p. 444)
nonpoint sources (p. 440)
oligotrophic (p. 444)
oxygen sag (p. 443)
point sources (p. 440)
primary treatment (p. 440)
red tides (p. 444)
secondary treatment
(p. 440)
tertiary treatment (p. 440)
thermal plume (p. 448)

SUGGESTED READINGS

■ Batisse, M. June 1990. "Probing the Future of the
Mediterranean Basin." *Environment* 32(5):4. Describes the
Mediterranean Blue Plan for international cooperation in
cleaning up this sea.

■ Brodie, P. September/October 1983. "The Clean Water
Act." *Sierra* 68(5):39. Traces the history, provisions, and
politics of the Clean Water Act. Contains a good history of
the Cuyahoga River.

■ Colburn, T., and R. Liroff. November/December 1990.
"Toxics in the Great Lakes." *EPA Journal* 16(6):5.
Increasing levels of toxins in the Great Lakes make clean-
up more urgent and more complicated.

■ Dorreboom, L., and R. Grace. November/December 1990.
"Deadly Secret of the Deep." *Greenpeace* 15(6):17.
Driftnets are snaring and killing marine wildlife in tragic
numbers. Images from the Rainbow Warrior's expedition to
the North Pacific.

■ Fischhoff, B. March 1991. "Report from Poland: Science
and Politics in the Midst of Environmental Disaster."
Environment 33(2):12. Air and water pollution were
ignored and information was suppressed during forty years
of communist rule. Now an enormous task of
environmental cleanup faces the country.

■ Glomb, S. November/December 1990. "Measuring
Environmental Success." *EPA Journal* 16(6):57. How do
we know whether efforts to clean up the major bodies of
water are really succeeding?

■ Goldstein, B. 1991. "The Folly of the *Exxon Valdez*
Cleanup." *Earth Island Journal* 6(1):30. Much of the
cleanup of Prince William Sound was done by nature. The
expensive and highly publicized steam-cleaning of beaches
may have done more harm than good.

■ Goldstein, J. *Sensible Sludge*. Emmaus: Rodale Press,
1977. A description of practical and useful methods of
sludge disposal.

■ Laycock, G. 1989. "Baptism of Prince William Sound."
Audubon September 1989. Lead article in a special report
on the wreck of the *Exxon Valdez* and its effect on the
ecosystem.

■ Leighton, T. June/July 1984. "Water, Water Everywhere
(and Hardly a Drop above Suspicion)." *Harrowsmith
Magazine*. A good survey of drinking water contamination
problems and household devices for water purification.

■ National Academy of Science. *Drinking Water and Health*,
Vol. 5. Washington, D.C.: National Academy Press, 1983.
Review of human health hazards of water pollutants.

■ Siegel, L. November/December 1984. "High-Tech
Pollution." *Sierra* 69(6):58. Groundwater contamination
from computer chip companies have polluted wells in
California's "silicon valley."

■ Sierra Club Legal Defense Fund. 1990. *The Poisoned
Well: New Strategies for Groundwater Protection*. San
Francisco: Sierra Club Press. An excellent source of
information about sources of and protection against
groundwater pollution.

■ Smith, R. A., et al. March 1987. "Water-Quality Trends in
the Nation's Rivers." *Science* 235:1607. Results of two
nationwide surveys of water quality are presented.
Improvements have been made, but much remains to be
done.

■ Woodwell, G. M. 1977. "Recycling Sewage through Plant
Communities." *American Scientist* 65:556. Removal of
sewage effluent nutrients and toxins by processing through
managed ecosystems.

C H A P T E R *23*

Air Pollution

It was a town of machinery and tall chimneys out of which interminable serpents of smoke trailed themselves forever and ever, and never got uncoiled. It had a black canal in it, and a river that ran purple with ill-smelling dye, and vast piles of buildings full of windows, where there was a rattling and trembling all day long, and where the piston of the steam engine worked monotonously up and down, like the head of an elephant in a state of melancholy madness. . . . The attributes of Coketown were in the main inseparable from the work by which it was sustained.

Charles Dickens *Hard Times*

Air pollution is everywhere. Even the most remote and pristine areas have traces of industrial pollutants that are transported worldwide by winds. Humans probably have caused air pollution ever since we began building fires, but growing populations and industrialization have created vast quantities of contaminants, including some highly toxic synthetic chemicals. Although there are natural sources for most of the large-volume "conventional" air pollutants, in most urban areas—and rural areas in which such industrial facilities as mines are located—human-caused pollutants usually far outweigh those from natural sources.

Primary pollutants are released directly into the air in a harmful form. Secondary pollutants are modified to a hazardous form or are created by physical and chemical reactions in the atmosphere. Photochemical oxidants produced by reactions driven by solar energy are among the most harmful kinds of air pollutants.

Indoor air can be more polluted and dangerous than outdoor air. Concentrations of toxic materials in our homes can be thousands of times higher than would be legal in the workplace or outside. Smoke from cooking and heating fires in unventilated rooms adversely affects health of hundreds of millions of people in less developed countries. Tobacco smoking is the largest single environmental health threat in the United States and the largest cause of preventable deaths.

Topography, climate, air stability, wind currents, and physical processes, such as diffusion, precipitation, adsorption, and absorption, play important roles in transport, concentration, dispersal, or removal of air pollutants from the atmosphere. In earlier times, it was thought that "dilution was the solution to pollution," but we realize now that not creating pollutants and removing them from effluents before release are better approaches.

Ecosystems and building materials also are sensitive to air pollution. Air pollutants have killed aquatic life in millions of lakes and are thought to play a role in the widespread rapid death of forests in many areas. Crop losses resulting from air pollution in the United States may reach $10 billion per year, and the losses from damage to building materials and property values probably are similar. Destruction of the protective stratospheric ozone shield by air pollutants could expose us to damaging ultraviolet rays that would increase cancers and damage crops.

There are several effective techniques for preventing air pollution, including precipitators, filters, scrubbers, and catalytic converters on effluent streams. We have made significant progress in reducing air pollution in most of the United States in the past two or three decades, but the biggest cities have not yet met the air quality standards mandated by the Clean Air Act, and may have to restrict automobile use or industrial growth in order to clean up their air. Some of the worst pollution problems in the world are in developing countries where control technology, regulatory procedures, and environmental ethics are decades behind those in the more developed countries.

INTRODUCTION

How does the air taste, feel, smell, and look in your home or your neighborhood? Chances are that wherever you live, the air is contaminated to some degree. Smoke, haze, dust, odors, corrosive gases, noise, and toxic compounds are present nearly everywhere, even in the most remote, pristine wilderness. Air pollution is generally the most widespread and obvious kind of environmental damage. According to the Environmental Protection Agency (EPA), some 147 million metric tons of air pollution (not counting carbon dioxide or wind-blown soil) are released into the atmosphere each year in the United States by human activities (figure 23.1). Total worldwide emissions of these pollutants are around 2 billion metric tons per year. The air in a typical industrial city can contain unhealthy concentrations of hundreds of different toxic substances; indoor air can be even worse.

Over the past twenty years, air quality has improved appreciably in most cities in Western Europe, North America, and Japan. At the same time, however, we have discovered dangers from air pollutants that did not exist or were not recognized in the past. Manufacturing, shipping, use, and disposal of thousands of new supertoxic chemicals have introduced a great variety of new hazardous materials into the air we breathe. Exposure to parts per billion or even parts per trillion of these chemicals may be dangerous to sensitive members of the community.

In this chapter, we will examine the major types and sources of air pollution. We will study how they enter and move through the atmosphere and how they are changed into new forms, concentrated or dispersed, and removed from the air by physical and chemical processes. We also will look at some of the major effects of air pollution on human health, ecosystems, and materials. Finally, we will survey some of the control methods available to reduce air pollution or mitigate its effects, and the results of air pollution control efforts on ambient air quality in the United States and elsewhere.

NATURAL SOURCES OF AIR POLLUTION

It is difficult to give a simple, comprehensive definition of pollution. The word comes from the Latin *pollutus*, which means made foul, unclean, or dirty. Some authors limit the use of the term to damaging materials that are released into the environment by human activities. There are, however, many natural sources of air quality degradation. Volcanoes spew out ash, acid mists, hydrogen sulfide, and other toxic gases. Sea spray and decaying vegetation are major sources of reactive sulfur compounds in the air. Forest fires create clouds of smoke that blanket whole continents. Trees and bushes emit millions of tons of volatile organic compounds (terpenes and isoprenes), creating, for example, the blue haze that gave the Blue Ridge Mountains

their name. Pollen, spores, viruses, bacteria, and other small bits of organic material in the air cause widespread suffering from allergies and airborne infections. Storms in arid regions raise dust clouds that transport millions of tons of soil and can be detected half a world away. Bacterial metabolism of decaying vegetation in swamps and of cellulose in the guts of termites and ruminant animals is responsible for as much as two-thirds of the methane (natural gas) in the air.

In many cases, the chemical compositions of pollutants from natural and human-related sources are identical, and their effects are inseparable. Sometimes, however, materials in the atmosphere are considered innocuous at naturally occurring levels, but when humans add to these levels, overloading of natural cycles or disruption of finely tuned balances in the environment can occur. It may be the total amount of a substance in the air that causes trouble, or it may be that the location or timing of human-caused emissions are such that sensitive organisms or materials are adversely affected. While the natural sources of suspended particulate material in the air outweigh human sources at least tenfold worldwide, in many cities more than 90 percent of the airborne particulate matter is anthropogenic (human-caused).

TYPES AND SOURCES OF AIR POLLUTION

What are the major types of air pollutants and where do they come from? In this section, we will define some general categories and sources of air pollution and survey the characteristics and emission levels of the seven conventional pollutants regulated by the Clean Air Act.

Primary and Secondary Pollutants

Primary pollutants are those released directly into the air in a harmful form. **Secondary pollutants,** by contrast, are modified to a hazardous form after they enter the air or are formed by chemical reactions as components of the air mix and interact. Solar radiation often provides the energy for these reactions. Photochemical oxidants and atmospheric acids formed by these mechanisms are probably the most important secondary pollutants in terms of human health and ecosystem damage. We will discuss several important examples of such pollutants in this chapter.

Fugitive emissions are those that do not go through a smokestack. By far the most massive example of this category is dust from soil erosion, strip mining, rock crushing, and building construction (and destruction). In the United States, natural and anthropogenic sources of fugitive dust add up to some 100 million metric tons per year. The amount of CO_2 released by burning fossil fuels and biomass is nearly equal in mass to fugitive dust. Fugitive industrial emissions are also an important source

FIGURE 23.1 Many human activities cause air pollution that affects ecosystems, wildlife, building materials, and human health. Air pollution controls have alleviated some of the worst problems in developed countries, but many serious problems remain, especially in developing countries.

of air pollution as well. Leaks around valves and pipe joints contribute as much as 90 percent of the hydrocarbons and volatile organic chemicals emitted from oil refineries and chemical plants.

Conventional or "Criteria" Pollutants

The Clean Air Act of 1970 designated seven major pollutants (sulfur dioxide, carbon monoxide, particulates, hydrocarbons, nitrogen oxides, photochemical oxidants, and lead) for which maximum **ambient air** (that all around us) levels are mandated. These seven **conventional** or **criteria pollutants** contribute the largest volume of air quality degradation and also are considered the most serious threat of all air pollutants to human health and welfare. Figure 23.2 shows the major sources of the first five criteria pollutants. Table 23.1 shows an estimate of the total annual worldwide emissions of some important air pollutants, and table 23.2 shows estimated annual emissions for some selected countries. Now let's look more closely at the sources and characteristics of each of these major pollutants.

Sulfur Compounds

Natural sources of sulfur in the atmosphere include evaporation of sea spray, erosion of sulfate-containing dust from arid soils, fumes from volcanoes and fumaroles, and biogenic emissions of hydrogen sulfide (H_2S) and organic sulfur-containing compounds, such as dimethylsulfide, methyl mercaptan, carbon disulfide, and carbonyl sulfide. Total yearly emissions of sulfur from all sources amounts to some 182 million metric tons (figure 23.3).

Although anthropogenic sources represent only one-fourth of the total sulfur flux worldwide, in most urban areas they contribute as much as 90 percent of the sulfur in the air. The predominant form of anthropogenic sulfur is sulfur dioxide (SO_2) from combustion of sulfur-containing fuel (coal and oil), purification of sour (sulfur-containing) natural gas or oil, and industrial processes, such as smelting of sulfide ores. China and the United States are the largest sources of anthropogenic sulfur, primarily from coal burning (table 23.2).

Sulfur dioxide is a colorless corrosive gas, which is directly damaging to both plants and animals. Once in the atmosphere, it can be further oxidized to sulfur trioxide (SO_3), which reacts with water vapor or dissolves in water droplets to form sulfuric acid (H_2SO_4). Very small solid particles or liquid droplets can transport the acidic sulfate ion (SO_4^-) long distances through the air or deep into the lungs where it is very damaging. Sulfur dioxide and sulfate ions are probably second only to smoking as causes of air-pollution-related health damage. Sulfate particles and droplets reduce visibility in the United States as much as 80 percent.

Nitrogen Compounds

Nitrogen oxides are highly reactive gases formed when nitrogen in fuel or combustion air is heated to temperatures above 650°C (1,200°F) in the presence of oxygen, or when bacteria in soil or water oxidize nitrogen-containing compounds. The initial product, nitric oxide (NO), oxidizes further in the atmo-

ENVIRONMENTAL POLLUTION AND OTHER DILEMMAS

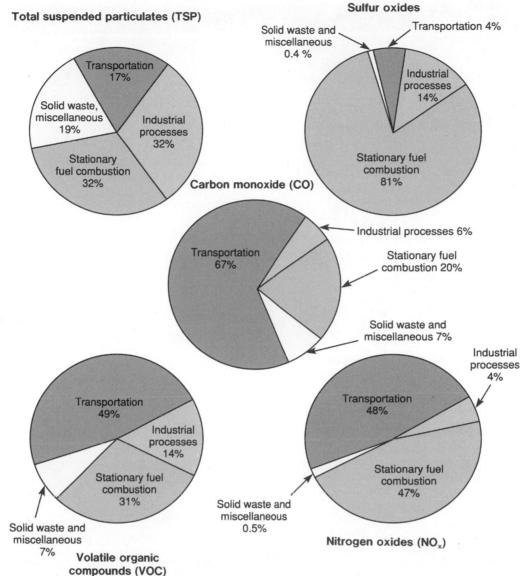

FIGURE 23.2 Sources of five major pollutants in the United States. Yearly emissions are 7.5 megatons TSP, 21.5 megatons SO_2 + SO_4, 79 megatons CO, 20.3 megatons NO_x, and 18.6 megatons VOC from anthropogenic sources. Total sum is not percent due to rounding of figures.

sphere to nitrogen dioxide (NO_2), a reddish brown gas that gives photochemical smog its distinctive color. Because of their interconvertibility, the general term NO_x is used to describe these gases. Nitrogen oxides combine with water to make nitric acid (HNO_3), which is also a major component of atmospheric acidification.

The total annual emissions of reactive nitrogen compounds into the air are about 210 million metric tons worldwide (table 23.1). Anthropogenic sources account for 45 percent of these emissions. About 95 percent of all human-caused NO_x in the United States is produced by fuel combustion in transportation and electric power generation (figure 23.4). Ammonia from fertilizer and decaying organic material is oxidized to NO_x

and is an important source of nitrogen loading in rural areas. Nitrous oxide (N_2O) is an intermediate in soil denitrification that absorbs ultraviolet light and plays an important role in climate modification (chapter 17).

Carbon Oxides

The predominant form of carbon in the air is carbon dioxide (CO_2). It is usually considered nontoxic and innocuous, but increasing atmospheric levels (about 0.4 percent per year) due to human activities appear to be causing a global climate warming that may have disastrous effects. As table 23.1 and figure 23.5 show, 90 percent of the CO_2 emitted each year is from respiration (oxidation of organic compounds by plant and animal cells).

TABLE 23.1 Estimated fluxes of pollutants and trace gases to the atmosphere

Species	Major sources	Approximate annual flux (millions of tons)	Average half life in atmosphere (days)
CO_2 (Carbon dioxide)	Respiration	100,000	2,500
CO_2 (Carbon dioxide)	Biomass, fossil fuel burning	10,000	2,500
CO (Carbon monoxide)	Biomass, fossil fuel burning	1,000	75
CH_4 (Methane)	Wetlands, rice paddies, termites, ruminant animals	400	3,600
VOC[a]	Human-made	100	1–1,000
VOC	Isoprene, terpenes from plants	800	<1
NO_x (Nitrogen oxides)	Soils, burning biomass, fossil fuel	100	4
N_2O (Nitrous oxide)	Fertilizer, tropical forests	10	60,000
NH_3 (Ammonia)	Industrial and biological Nitrogen-fixation	100	9
SO_2 and SO_4 (Sulfur dioxide and sulfate)	Sea spray, fossil fuels, smelting	90	1–4
H_2S and organic sulfur[b] (Hydrogen sulfide)	Biogenic, anthropogenic	90	1–900
Metals[c]	Leaded gasoline, coal, industrial waste	3	1–30
SPM[d]	Wind erosion, fires, volcanoes, human sources	10,000	1–1,000

[a]Volatile organic compounds (VOC) (other than methane) include benzene, formaldehyde, vinyl chloride, phenol, chloroform, trichloroethylene, gasoline ingredients, and chlorofluorocarbons.
[b]Organic sulfur includes methyl mercaptan, carbon disulfide, carbonyl sulfide, and dimethyl sulfide, among others.
[c]Metals include lead, cadmium, nickel, beryllium, mercury, and arsenic.
[d]Suspended particulate material (SPM) includes soot, dust, ash, pollen, and algae.

From: Mooney, et al., *Science*, 288:926, 1987. Copyright © 1987 by the American Association for the Advancement of Science, Washington, D.C.

TABLE 23.2 Estimated emission of pollutants in selected countries

Country	Sulfur oxides	Particulates	Nitrogen oxides (Thousands of metric tons)	Carbon monoxide
China	12,920	13,740	4,130	X
France	1,460	210	1,730	6,330
Italy	2,230	410	1,530	5,420
Japan	1,610	X	1,420	420
Poland	3,700	3,350	1,770	3,300
Sweden	285	40	295	1,600
United Kingdom	3,750	230	1,770	5,180
United States	21,100	6,900	19,500	69,230

Source: Data from Global Environmental Monitoring System, 1984.
X = data not available

These releases are usually balanced by an equal uptake by photosynthesis in green plants. Burning of fossil fuels and biomass each contribute about 5 billion metric tons per year to the air.

Carbon monoxide (CO) is a colorless, odorless, nonirritating but highly toxic gas produced by incomplete combustion of fuel (coal, oil, charcoal, or gas), incineration of biomass or solid waste, or partially anaerobic decomposition of organic material. CO inhibits respiration in animals by binding irreversibly to hemoglobin. About 1 billion metric tons of CO are released to the atmosphere each year, half of that from human activities. In the United States, two-thirds of the CO emissions are created by internal combustion engines in transportation. Land-clearing fires and cooking fires also are major sources. About 90 percent of the CO in the air is consumed in photochemical reactions that produce ozone.

ENVIRONMENTAL POLLUTION AND OTHER DILEMMAS

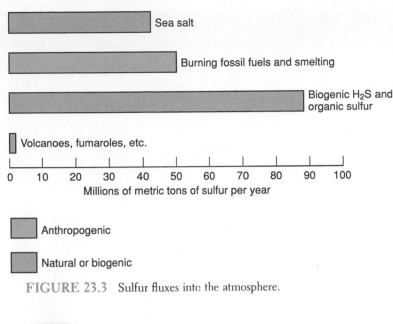

FIGURE 23.3 Sulfur fluxes into the atmosphere.

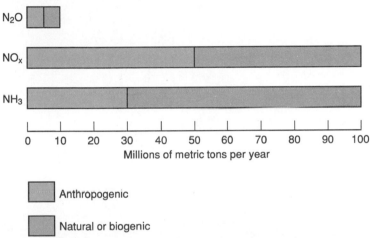

FIGURE 23.4 Fluxes of reactive nitrogen gases into the atmosphere.

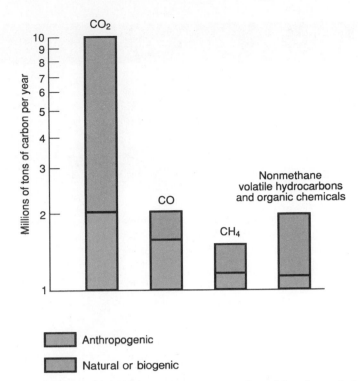

FIGURE 23.5 Carbon fluxes into the atmosphere. Note that scale is logarithmic: CO_2 emissions from natural sources are four times as great as those from human-caused releases.

Metals and Halogens

Many toxic metals are mined and used in manufacturing processes or occur as trace elements in fuels, especially coal. These metals are released to the air in the form of metal fumes or suspended particulates by fuel combustion, ore smelting, and disposal of wastes. Worldwide lead emissions amount to about 2 million metric tons per year, or two-thirds of all metallic air pollution. Most of this lead is from leaded gasoline. Lead is a metabolic poison and a neurotoxin that binds to essential enzymes and cellular components and inactivates them. An estimated 20 percent of all inner city children suffer some degree of mental retardation from high environmental lead levels.

Mercury is another dangerous neurotoxin that is widespread in the environment. The two largest sources of atmospheric mercury appear to be coal-burning power plants and mercuric fungicides in house paint. Mercury batteries discarded in domestic waste make garbage incinerators another dangerous source of mercury vapors.

Other toxic metals of concern are nickel, beryllium, cadmium, thallium, uranium, cesium, and plutonium. Some 780,000 tons of arsenic, a highly toxic metalloid, are released from metal smelters, coal combustion, and pesticide use each year. Halogens (fluorine, chlorine, bromine, and iodine) are highly reactive and generally toxic in their elemental form. About 600 million tons of highly persistent chlorofluorocarbons (CFCs) are used annually worldwide in spray propellants, refrigeration compressors, and for foam blowing. They diffuse into the stratosphere where they release chlorine and fluorine ions that destroy the ozone shield that protects Earth from ultraviolet radiation (Box 23.1).

Particulate Material

An **aerosol** is any system of solid particles or liquid droplets suspended in a gaseous medium. For convenience, we generally describe all atmospheric aerosols, whether solid or liquid, as **particulate material.** This includes dust, ash, soot, lint, smoke, pollen, spores, algal cells, and many other suspended materials. Anthropogenic particulate emissions amount to about 100 million metric tons per year worldwide. Wind-blown dust, volcanic ash, and other natural materials may contribute one hundred times that much.

BOX 23.1 A Hole in the Ozone Shield

In 1985, the British Antarctic Atmospheric Survey announced a startling and disturbing discovery: ozone levels in the stratosphere over the South Pole drop precipitously during September and October as the sun returns at the end of the long polar spring. This ozone depletion has been occurring since the late 1960s (box figure 23.1) but was not recognized because researchers programmed their computers to ignore changes in ozone levels that were presumed to be erroneous. In 1987 and 1989, ozone levels were 50 percent less than normal and in 1990 stratospheric ozone was depleted by 70 percent over an area about the size of the continental United States. The cause of the ozone depletion and its possible consequences for life on Earth have been the subject of intense debate and concern. Research bases that once were empty and quiet in the paralyzing cold of the polar winter are now buzzing with scientists who have rushed to Antarctica to study this phenomenon.

Ozone is a harmful pollutant in the lower atmosphere, damaging plants, building materials, and human health, but it is an irreplaceable resource in the upper atmosphere, where it screens out more than 99 percent of the dangerous ultraviolet (UV) rays from the sun. Without this shield, organisms on Earth's surface would be subjected to life-threatening radiation burns and genetic damage; thus, it is urgent that we learn what is depleting the ozone in Antarctic skies and whether it is a general phenomenon or limited to the unique conditions of the polar atmosphere.

The exceptionally cold temperatures (−85 to −90°C) of the Antarctic winter play a role in ozone destruction. During the long, dark months of the polar winter, isolation of the air mass within the

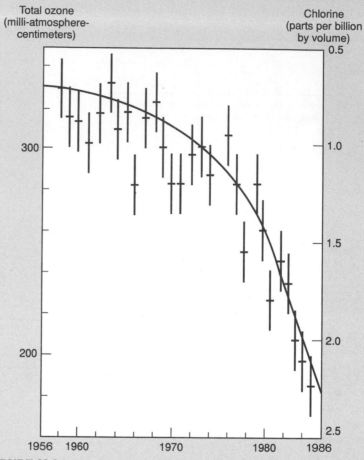

BOX FIGURE 23.1 Monthly means of total ozone at Halley Bay, Antarctica, October 1956–1986. Bars represent standard deviation above and below mean.

Source: F. Sherwood Rowland. University of California at Irvine (unpublished October 1986).

polar vortex allows stratospheric temperatures to drop low enough for water vapor to freeze, creating a dense haze of ice particles at high altitudes. Molecules containing ozone and chlorine are adsorbed on the surface of the ice crystals. When the sun returns in the spring and provides UV energy to liberate active chlorine atoms from precursor molecules, they already are in close contact with ozone, and destructive chemical reactions proceed quickly.

The most likely sources of chlorine in the stratosphere appear to be chlorine monoxide (ClO) and hydrochloric acid (HCl) derived from chlorofluorocarbons (CFCs) and halon gases. CFCs were first synthesized in 1928 by scientists at General Motors as propellants for spray paint. Known commonly by the trade name, Freon, these nontoxic, nonflammable, chemically inert, cheaply produced, and versatile compounds soon found a wide variety of applications.

Particulates often are the most apparent form of air pollution since they reduce visibility and leave dirty deposits on windows, painted surfaces, and textiles. Respirable particles smaller than 2.5 micrometers are among the most dangerous of this group because they can be drawn into the lungs, where they damage respiratory tissues. Asbestos fibers and cigarette smoke are among the most dangerous respirable particles in urban and indoor air because they are carcinogenic. The World Health Organization estimates that 70 percent of the global urban population, primarily in developing countries, breathes air that has unhealthy particulate concentrations.

$$CCl_3F \xrightarrow{\text{UV radiation}} CCl_2F + Cl^-$$
(Freon)

$$Cl^- + O_3 \longrightarrow ClO + O_2$$
(Chlorine ion)　(Ozone)　　　(Chlorine monoxide)　(Oxygen)

$$ClO + O_3 \longrightarrow ClO_2 + O_2$$
(Chlorine dioxide)

$$ClO_2 \xrightarrow{\text{UV radiation}} Cl^- + O_2$$

BOX FIGURE 23.2
Photodissociation of Freon II. When struck by UV radiation, chloroflurocarbons release chlorine ions. These chlorine ions combine with ozone to make chlorine monoxide plus oxygen. The chlorine monoxide reacts with another ozone molecule to make chlorine dioxide, which dissociates into oxygen and a reactive chlorine ion that starts the cycle over again.

Until 1978, aerosol spray cans used more CFCs than any other application (box figure 23.2), but concerns about a possible role in ozone destruction prompted the United States, Canada, and the Scandinavian countries to ban non-essential propellant uses. Other countries continue to use CFCs as spray propellants, however, and this use still represents nearly one-third of the 320,000 metric tons of CFC consumed annually worldwide. CFCs are mainly used now as refrigerants and solvents for cleaning printed circuits and electronic parts. They also are used extensively as blowing agents for making styrofoam cups, fast-food containers, and other foam products of modern society. Halons are used mainly as fire extinguishers.

We don't know whether the polar ozone reductions are an isolated phenomenon or an ominous warning signal of a slowly progressing worldwide ozone depletion. Data indicate that the total amount of ozone in the atmosphere has declined 4 to 5 percent over the past decade, but sunspot activity and other sources of normal fluctuation could account for most of that amount. Calculations based on atmospheric chemistry modeling suggest that continuing emissions of CFCs at 1987 levels could reduce ozone levels another 4 to 5 percent by the middle of the next century.

The extra UV light allowed to reach Earth's surface by this depletion of the ozone shield could cause a million extra cases of skin cancer (about 30,000 would probably be fatal) each year, as well as cataracts, suppression of the immune system, severe sunburns, and accelerated skin aging. It also could damage plants (including vital crops) and degrade plastics and other polymers, causing losses amounting to billions of dollars each year.

Some of this cancer increase seems already to have begun. Southern Australia is experiencing a skin cancer epidemic that may be related to ozone loss. Some oncologists predict that two out of three Australians will have skin cancer before age 70. While much of this is due to fair skin and a love of sun bathing, ozone-depleted air spreading from polar regions has increased UV irradiation over Australia.

In response to this threat, eighty-one nations signed a treaty in 1989 in Helsinki, Finland, agreeing to phase out production of CFCs by the end of the century. This treaty does not cover carbon tetrachloride or methyl chloroform. In addition, China, India, and a group of developing nations asked for help in acquiring new technology so they can obtain refrigeration. Fortunately, an alternative to CFCs already exists. The DuPont Company, which developed CFCs thirty years ago, announced that it has developed a family of hydrofluorocarbons that can replace CFCs with only a fraction of the damage to the atmosphere. However, these hydrofluorocarbons are more expensive and their health safety has not yet been established.

Many observers are amazed and encouraged that nations could respond so quickly to the threat of ozone depletion. Others doubt that the reductions in CFC use will be quick enough to protect us from the danger of UV irradiation. This issue, like so many in environmental studies, raises difficult questions about how much disruption and expense is justified when scientific evidence is equivocal but the potential risks are very great.

Volatile Organic Compounds

Volatile organic compounds (VOC) are organic chemicals that exist as gases in the air. Plants are the largest source of VOC, releasing an estimated 350 million tons of isoprene (C_5H_8) and 450 million tons of terpenes ($C_{10}H_{15}$) each year. About 400 million tons of methane (CH_4) are produced by natural wetlands and rice paddies and by bacteria in the guts of termites and ruminant animals. These volatile hydrocarbons are generally oxidized to CO and CO_2 in the atmosphere.

In addition to these natural VOCs, a large number of other synthetic organic chemicals, such as benzene, toluene, formaldehyde, vinyl chloride, phenols, chloroform, and trichloroethy-

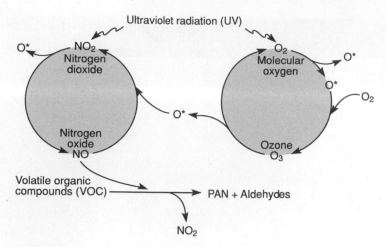

Summary of reactions shown in diagram:

1. $O_2 + UV \longrightarrow O^* + O^*$ (oxygen ion or free radical)
2. $NO_2 + UV \longrightarrow NO + O^*$
3. $O^* + O_2 \longrightarrow O_3$
4. $O_3 + NO \longrightarrow O_2 + NO_2$
5. $NO + VOC \longrightarrow NO_2 + PAN$ (peroxyacetylnitrate) + aldehydes[a]

Net results:

$NO_2 + UV + VOC + O_2 \longrightarrow NO_2 + O_3 + PAN + $ aldehydes

[a]Examples of aldehydes are formaldehyde, acetylaldehyde, and benzaldehyde.

FIGURE 23.6 Some photochemical atmospheric reactions that contribute to smog formation.

lene, are released into the air by human activities. About 28 million tons of these compounds are emitted each year in the United States, mainly unburned or partially burned hydrocarbons from transportation, power plants, chemical plants, and petroleum refineries (figure 23.2). These chemicals play an important role in the formation of photochemical oxidants.

In 1987, the EPA began requiring industries to report releases of some 332 toxic organic chemicals into the air. The expectation was that these emissions would amount to about 36,000 metric tons per year. To everyone's surprise, the reports totaled 2 *million* metric tons or 5 billion pounds. The largest carcinogen emission was 52,000 tons (115 million lbs) of dichloromethane, which is used as an industrial solvent and paint stripper. This startling revelation was instrumental in the regulation of these hazardous organic compounds in the 1990 Clean Air Act revisions.

Photochemical Oxidants

Photochemical oxidants are products of secondary atmospheric reactions driven by solar energy (figure 23.6). One of the most important of these reactions involves formation of singlet (atomic) oxygen by splitting either molecular oxygen (O_2) or nitrogen dioxide (NO_2). This singlet oxygen then reacts with another molecule of O_2 to make **ozone** (O_3). Ozone formed in

the stratosphere provides a valuable shield for the biosphere by absorbing incoming ultraviolet radiation. In ambient air, however, O_3 is a strong oxidizing reagent and damages vegetation, building materials (such as paint, rubber, and plastics), and sensitive tissues (such as eyes and lungs). Ozone has an acrid, biting odor that is a distinctive characteristic of photochemical smog. Hydrocarbons in the air contribute to accumulation of ozone by removing NO in the formation of compounds, such as peroxyacetyl nitrate (PAN), which is another damaging photochemical oxidant. Figure 23.6 also shows some of the chemical reactions in the ozone cycle.

Unconventional or "Noncriteria" Pollutants

The EPA has authority under the Clean Air Act to set **emission standards** (regulating the amount released) for certain **unconventional** or **noncriteria pollutants** that are considered especially toxic or hazardous. Among the materials regulated by emission standards are asbestos, benzene, beryllium, mercury, polychlorinated biphenyls, PCBs, and vinyl chloride. Most of these materials have no natural source in the environment (to any great extent) and are therefore only anthropogenic in origin.

In addition to these toxic air pollutants, some other unconventional forms of air pollution deserve mention. **Aesthetic degradation** includes any undesirable changes in the physical characteristics or chemistry of the atmosphere. Noise, odors, and light pollution are examples of atmospheric degradation that may not be life-threatening but reduce the quality of our lives. This is a very subjective category. Odors and noise (such as loud music) that are offensive to some may be attractive to others. Often the most sensitive device for odor detection is the human nose. We can smell styrene, for example, at 44 parts per billion (ppb). Trained panels of odor testers often are used to evaluate air samples. Factories that emit noxious chemicals sometimes spray "odor maskants" or perfumes that are sprayed into smokestacks to cover up objectionable odors.

In most urban areas, it is difficult or impossible to see stars in the sky at night because of dust in the air and stray light from buildings, outdoor advertising, and streetlights. This light pollution has become a serious problem for astronomers.

Indoor Air Pollution

We have spent a considerable amount of effort and money to control the major outdoor air pollutants, but we have only recently become aware of the dangers of indoor air pollutants. In many cases, the air indoors is much more dangerous than that outdoors because pollutants can be trapped and concentrated indoors. Furthermore, people generally spend more time inside than out, and therefore, are exposed to higher doses of these pollutants.

Smoking is without doubt the most important air pollutant in the United States in terms of human health. The Surgeon

General estimates that 350,000 people die each year from emphysema, heart attacks, strokes, lung cancer, or other diseases caused by smoking. Banning smoking probably would save more lives than any other pollution-control measure.

Other major indoor air pollution health hazards include asbestos, formaldehyde, vinyl chloride, radon, and combustion gases. Asbestos was widely used in floor and ceiling tiles, plaster, cement, insulation, and soundproofing. It is a serious concern in indoor air because of its carcinogenicity. Formaldehyde is used in more than three thousand products, including such building materials as particle board, waferboard, and urea-formaldehyde foam insulation. Vinyl chloride is used in plastic plumbing pipe, floor and wall coverings, and countertops. New carpets and drapes typically contain two dozen chemicals designed to kill bacteria and molds, resist stains, bind fibers, and retain colors.

In many cases, indoor air in homes has concentrations of chemicals that would be illegal outside or in the workplace. The EPA has found that concentrations of such compounds as chloroform, benzene, carbon tetrachloride, formaldehyde, and styrene can be 70 times higher in indoor air than in outdoor air. Many people are highly sensitive to these chemicals, and it is not uncommon to trace illness to a "sick house syndrome" caused by polluted indoor air. Next to smoking, the most serious indoor air pollutant in the United States is probably radon gas that leaks into houses from surrounding soil and rock (Box 23.2).

In the less developed countries of Africa, Asia, and Latin America where such organic fuels as firewood, charcoal, dried dung, and agricultural wastes make up the majority of household energy, smoky, poorly ventilated heating and cooking fires represent the greatest source of indoor air pollution (figure 23.7). The World Health Organization (WHO) estimates that some 400 million people, mostly women, are adversely affected by pollution from this source. Many women spend long hours each day cooking over open fires or unventilated stoves in enclosed spaces. The levels of carbon monoxide, particulates, aldehydes, and other toxic chemicals can be one hundred times higher than would be legal for outdoor ambient concentrations in the United States. Designing and building cheap, efficient, nonpolluting energy sources for the developing countries would not only save shrinking forests, but would make a major impact on health as well.

INFLUENCES OF CLIMATE, TOPOGRAPHY, AND ATMOSPHERIC PROCESSES

Topography, climate, and physical processes in the atmosphere play an important role in transport, concentration, dispersal, and removal of many air pollutants. Wind speed, mixing between air layers, precipitation, and atmospheric chemistry all determine whether pollutants will remain in the locality where they are pro-

FIGURE 23.7 Smoky cooking and heating fires may cause more ill health effects than any other source of air pollution except tobacco smoking. Levels of carbon monoxide, particulates, and cancer-causing hydrocarbons can be one thousand times higher indoors than outdoors. Some 400 million people, mostly women and children, spend hours each day in poorly ventilated kitchens and living spaces.

duced or go elsewhere. In this next section, we will survey some environmental factors that affect air pollution levels.

Inversions

Temperature inversions occur when a stable layer of warmer air overlays cooler air, reversing the normal temperature decline with increasing height (chapter 17) and preventing convection currents from dispersing pollutants. Several mechanisms create inversions. When a cold front slides under an adjacent warmer air mass or when cool air subsides down a mountain slope to displace warmer air in the valley below, an inverted temperature gradient is established. These inversions are usually not stable,

BOX 23.2 Radon in Indoor Air

The Environmental Protection Agency (EPA) estimates that one home in ten in the United States has excessive indoor radon levels, and that 5,000 to 20,000 of the 136,000 deaths from lung cancer can be attributed to radiation from indoor radon. This makes indoor radon second only to smoking as an air pollution health hazard. It appears that radon causes ten times more deaths each year than all other regulated toxic air pollutants combined.

In surveys of areas suspected to be at high risk for radon, one home in four exceeds the "action level" of 4 picocuries per liter (pCi/l). The highest individual concentrations were over 200 pCi/l in Pennsylvania, Ohio, and on a South Dakota Indian reservation. Iowa has the highest percentage of homes with excess radon. Nearly three quarters of Iowa homes exceeded 4 pCi/l.

Breathing air with 4 pCi/l of radon and its daughters continuously gives a radiation dose of about 2,000 mrem per year. This is equivalent to two hundred chest X rays each year and carries about a 1-in-100 lifetime risk of dying from lung cancer. In other words, breathing 4 pCi/l continuously is equivalent to smoking about half a pack of cigarettes per day. About 1 percent of the homes surveyed had more than 20 pCi/l, which would produce an annual exposure equal to or exceeding that received by underground uranium miners.

Radon is a colorless, odorless, and tasteless gas that is produced from the radioactive decay of naturally occurring uranium in rocks and soil. It is found almost everywhere, although some areas have geological structures that are especially high in radon.

How is radon formed? Uranium-238 undergoes fourteen conversion steps as it decays to stable, nonradioactive lead-206. Among the first intermediates formed are immobile metals: thorium, protactinium, and radium. Radium decays slowly (half-life of 1,660 years) to radon-222, a gas with a half-life of 3.8 days. Radon can seep through cracks in rock formations and filter through soil and cracks in home foundations, where it gives rise to its "daughters," polonium-218, lead-214, bismuth-214, polonium-214, and lead-210.

BOX FIGURE 23.3 Areas with potentially high levels of uranium in soil and rocks. Radon levels in air and water may be high in these areas.

Source: U.S. Geological Survey.

Although radon gas is chemically inert, the daughters are reactive and can bind to dust particles, clothing, walls of houses, and—if inhaled—to lung tissues. The mean life of the five daughters between radon-222 and lead-210 is one hour. During that time, they give rise to a combination of alpha, beta, and gamma radiation that can be very damaging to lungs. Polonium-218 and polonium-214 produce more than 95 percent of the radiation exposure from radon daughters in indoor air.

The amount of radon entering a house depends not only on the underlying soil and rock type (box figure 23.3), but also on how the house is built and how much ventilation it has. Radon filters up through the ground and seeps into the house through porous basement walls, cracks in the foundation, loose pipe fittings, sump pumps, and enclosed crawl spaces (box figure 23.4). The more cracks and other openings in the basement, the more radon can enter. Radon concentrations in basements are generally about twice those on the first floor. Except for some basement apartments, multistory apartment buildings rarely have a radon problem, and most concern is focused on single-family houses.

If houses are well ventilated, radon is quickly diluted to harmless levels. Outdoor air typically contains 0.1 to 0.2 pCi/l of radon. Houses that are tightly sealed to prevent energy losses tend to trap radon inside, along with formaldehyde and other toxic air pollutants that may have synergistic effects. There often is a conflict between the need to conserve energy and the dangers of indoor air pollution. In cold climates, winter levels of indoor radon are usually twice as high as summer levels when windows are open and air flow is greater. In the south, where air conditioning is common, just the opposite may be true.

Radon also accumulates in groundwater. A survey of groundwater-based public drinking water supplies revealed that 20,000 (roughly 40 percent) of such systems in the United States contain more than the 500 pCi/l considered to be a health risk. The danger is not so much from drinking the water as it is from inhaling radon evaporated during showers, bathing, cooking, and toilet flushing. At 500 pCi/l, the maximum limit for drinking water proposed by the EPA, radon would carry a 1–in–10,000 risk of causing a fatal cancer. This is one hundred to one thousand times higher than the risk considered acceptable for any other toxic substance in water, and makes radon the leading water pollutant, as well as the leading air pollutant, in terms of health effects.

How can you find out if your house exceeds safe levels of radon? Your State Public Health Department or local branch of the American Lung Association probably has radon detectors or can tell you where to obtain one. There are two main types of detectors: charcoal canisters and alpha-track. The charcoal canister is slightly cheaper, but measures radon levels over only three to seven days. The alpha-track gives an average over one to three months, and is usually a better representation of actual levels.

What can you do if you find that your house has excess radon? The first step is to seal cracks in the foundation and openings around pipes. Sealing walls with a plastic lining or epoxy paint may be desirable. Gravel or other porous fill around foundations can allow radon to escape outside rather than seep into the house. Drilling holes in the basement floor and installing sealed pipes (perhaps with a small fan to improve air flow) to vent radon from subfloor spaces to the outside can reduce radon infiltration. Finally, you can improve ventilation of indoor spaces to remove radon. Where heat loss is a problem, it may be worthwhile to install an air-to-air heat exchanger. Various combinations of these techniques, depending on particular house design and local conditions, have been able to reduce indoor radon levels by 95 percent or more. Carrying out all of these steps might cost thousands of dollars, but, in many cases, a few hundred dollars invested wisely is sufficient to make most houses relatively safe.

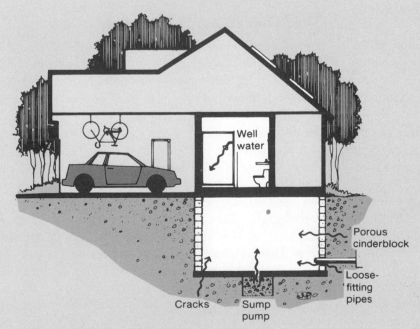

BOX FIGURE 23.4 Five common ways radon can enter your house.

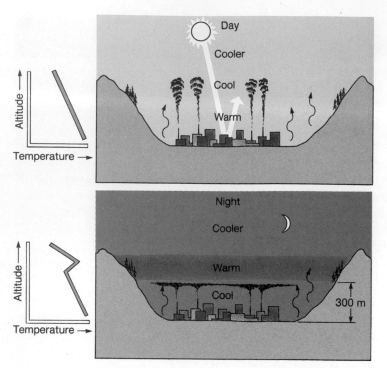

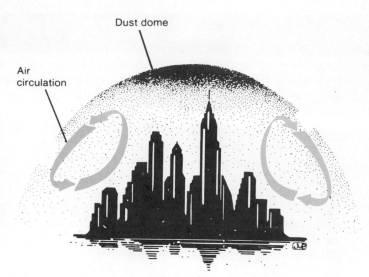

FIGURE 23.9 Streets and buildings trap solar energy, creating convective updrafts that sweep pollutants into a dust dome over the city. This heat island changes the climate of the city and its surrounding region.

FIGURE 23.8 The Los Angeles Basin is a classic example of conditions that generate an atmospheric temperature inversion. During the day, the sun heats the ground, warming the air near the surface and carrying dust and pollution aloft. Contaminated air is trapped by encircling mountains, forming a thick layer over the city. At night, the bare desert ground and paved streets radiate heat quickly into the cloudless sky. A layer of air immediately adjacent to the ground cools as well. This cooling is accelerated by cool, humid, on-shore breezes from the coast. The high particulate and pollutant levels in the upper air retain heat and form a cap that holds in and concentrates contaminants.

however, because winds accompanying these air exchanges tend to break up the temperature gradient fairly quickly and mix air layers.

The most stable inversion conditions are usually created by rapid nighttime cooling in a valley or basin where air movement is restricted. Los Angeles is a classic example of the conditions that create temperature inversions and photochemical smog (figure 23.8). The city is surrounded by mountains on three sides and the climate is dry and sunny. Extensive automobile use creates high pollution levels. Skies are generally clear at night, allowing rapid radiant heat loss, and the ground cools quickly. Surface air layers are cooled by conduction, while upper layers remain relatively warm. Density differences retard vertical mixing. During the night, cool, humid, onshore breezes slide in under the contaminated air, squeezing it up against the cap of warmer air above and concentrating the pollutants accumulated during the day.

Morning sunlight is absorbed by the concentrated aerosols and gaseous chemicals of the inversion layer. This complex mixture quickly cooks up a toxic brew of hazardous compounds. As the ground warms later in the day, convection currents break up the temperature gradient and pollutants are carried back down to the surface where more contaminants are added. Nitric oxide (NO) from automobile exhaust is oxidized to nitrogen dioxide. As nitrogen oxides are used up in reactions with unburned hydrocarbons, the ozone levels begin to rise. By early afternoon, an acrid brown haze fills the air, making eyes water and throats burn. On a typical summer day, ozone concentrations in the Los Angeles basin reach 0.34 ppm or more by late afternoon and the pollution index will be 300, the stage considered a health hazard.

Dust Domes and Heat Islands

Even without mountains to block winds and stabilize air layers, many large cities create an atmospheric environment quite different from the surrounding conditions. Sparse vegetation and high levels of concrete and glass in urban areas allow rainfall to run off quickly and create high rates of heat absorption during the day and radiation at night. Tall buildings create convective updrafts that sweep pollutants into the air. Temperatures in the center of large cities are frequently 3 to 5°C (5 to 9°F) higher than the surrounding countryside. Stable air masses created by this "heat island" over the city concentrate pollutants in a "dust dome" (figure 23.9). Rural areas downwind from major industrial areas often have significantly decreased visibility and increased rainfall (due to increased condensation nuclei in the dust plume) compared to neighboring areas with cleaner air. In the late 1960s, for instance, areas downwind from Chicago and St. Louis reported up to 30 percent more rainfall than upwind regions.

Long-Range Transport

Air pollutants can be carried long distances by wind currents. In 1971, scientists at the Nagoya Water Research Institute observed dust passing over Japan from Asia. A few days later, the same dust was collected at Hawaii, some 10,000 km across the Pacific Ocean. Mineralogic content, timing of events, and calculated air paths suggest that this dust came as a single surge from a storm in the Gobi Desert. Similar dust storms in the Algerian Sahara have been traced to islands in the Caribbean. Industrial pollutants also are transported great distances by wind currents.

Increasingly sensitive tools of atmospheric chemistry have revealed contaminants in places considered to be among the cleanest in the world. Samoa, Greenland, and Antarctica all have concentrations of heavy metals, pesticides, radioactive elements, and acidity that cannot be explained by local, natural sources. These contaminants must have been transported many thousands of kilometers to reach these remote sites. A polluted air mass often has a specific chemical composition that serves as a fingerprint in tracing its origin and path of transport through the atmosphere. An aerosol of sulfates, organic matter, and heavy metals, such as vanadium, manganese, and lead, creates a dense winter haze in the high Arctic that is traceable from the industrialized countries of western Europe, over Scandinavia, to Greenland, the Canadian Arctic, and to Point Barrow, Alaska.

Some of the most toxic and corrosive materials delivered by long-range transport are secondary pollutants (such as sulfuric and nitric acids or ozone), which are produced by chemical and physical reactions as contaminants mix and interact in an air mass. This makes it difficult to identify the sources of secondary pollutants or to devise simple control strategies. Furthermore, controversies are bound to result when the region that produces the pollution and bears the costs of its reduction is distant from the region where benefits of pollution control will be experienced. It is estimated, for instance, that 80 percent of the pollution in Lake Superior comes via long-range air transport from distant parts of the United States or even Central America. Persuading those regions to give up economically beneficial activities for the sake of people thousands of kilometers away is difficult.

EFFECTS OF AIR POLLUTION

So far we have looked at the major types and sources of air pollutants. Now we will turn our attention to the effects of those pollutants on human health, physical materials, ecosystems, and global climate.

Human Health

The primary human health effects of most air pollutants seems to be injury of delicate tissues, usually by damaging cellular membranes. This often sets in motion an **inflammatory response,** a complex series of interactions between damaged cells, surrounding tissues, and the immune system. One of the first symptoms of inflammation is leakage of fluid (plasma) from blood vessels. Exposure of respiratory tissues to severe irritants can result in so much edema (fluid accumulation) in the lungs that one effectively drowns.

Bronchitis is a persistent inflammation of bronchi and bronchioles (large and small airways in the lung) that causes a painful cough, copious production of sputum (mucus and dead cells), and involuntary muscle spasms that constrict airways. Acute bronchitis can obstruct airways so severely that death results. Smoking is undoubtedly the largest cause of chronic bronchitis in most countries. Persistent smog and acid aerosols also can cause this disease.

Severe bronchitis can lead to **emphysema,** an irreversible obstructive lung disease in which airways become permanently constricted and alveoli are damaged or even destroyed. Stagnant air trapped in blocked airways swells the tiny air sacs in the lung (alveoli), blocking blood circulation. As cells die from lack of oxygen and nutrients, the walls of the alveoli break down, creating large empty spaces incapable of gas exchange (figure 23.10). Thickened walls of the bronchioles lose elasticity and breathing becomes more difficult. Victims of emphysema make a characteristic whistling sound when they breathe. Often they need supplementary oxygen to make up for reduced respiratory capacity.

Cardiovascular stress from lack of oxygen in the blood is a common complication of all obstructive lung diseases. About twice as many people die of heart failure associated with smoking as die of lung cancer.

Irritants in the air are so widespread that about half of all lungs examined at autopsy in the United States have some degree of alveolar deterioration. The Office of Technology Assessment (OTA) estimates that 250,000 people suffer from pollution-related bronchitis and emphysema in the United States, and some 50,000 excess deaths each year are attributable to complications of these diseases, which are probably second only to heart attack as a cause of death.

Asthma is a distressing disease characterized by unpredictable and disabling shortness of breath caused by sudden episodes of muscle spasms in the bronchial walls. These attacks are often triggered by inhaling allergens, such as dust, pollen, animal dander, or corrosive gases. In some cases, there is no apparent external factor, and internal release of triggering agents is suspected. It isn't known whether asthma is genetic, environmental, or a combination of the two.

Fibrosis is the general name for accumulation of scar tissue in the lung. Among the materials that cause fibrosis are silica or coal dust, asbestos, glass fibers, beryllium and aluminum whiskers, metal fumes, cotton lint, and irritating chemicals, such as

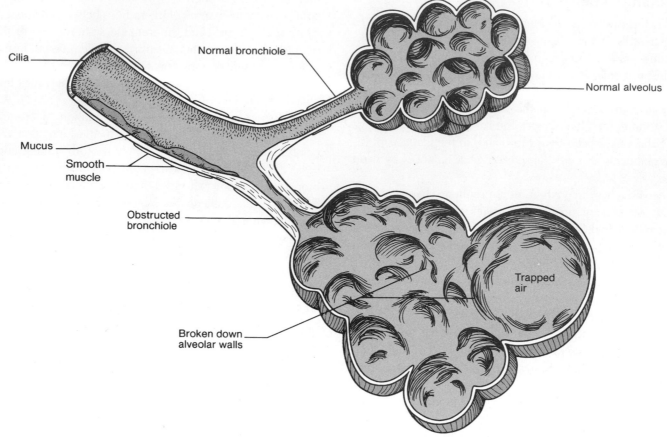

Cilia

Mucus

Smooth
muscle

Normal bronchiole

Normal alveolus

Obstructed
bronchiole

Trapped
air

Broken down
alveolar walls

FIGURE 23.10 Emphysema results from chronic irritation and obstruction of airways and alveoli.

the herbicide paraquat. We give each of these diseases an individual name (silicosis, black lung, asbestosis, beryllium lung disease, brown lung, or paraquat lung), but they really are very similar in development and effect. Cells respond to irritants and foreign material in the lungs by sealing off damaged areas with scar tissue (produced either by interstitial cells in the walls of airways or by the epithelial linings). As the lung fills up with fibrotic tissue, respiration is blocked and one slowly suffocates. In some cases, cell growth stimulated by the presence of foreign material in the lung results in tumor formation. Lung cancers are often lethal.

Plant Pathology

In the early days of industrialization, fumes from furnaces, smelters, refineries, and chemical plants often destroyed vegetation and created desolate, barren landscapes around mining and manufacturing centers. The copper-nickel smelter at Sudbury, Ontario, is a spectacular and notorious example of air pollution effects on vegetation and ecosystems. In 1886, the corporate ancestors of the International Nickel Company (INCO) began open-bed roasting of sulfide ores at Sudbury. Sulfur dioxide and sulfuric acid released by this process caused massive destruction

of the plant community within about 30 km (18.6 mi) of the smelter. Rains washed away the exposed soil, leaving a barren moonscape of blackened bedrock. Super-tall, 400 m smokestacks were installed in the 1950s and sulfur scrubbers were added 20 years later. Emissions were reduced by 90 percent and the surrounding ecosystem is beginning to recover. The area near the factory is still a grim, empty wasteland, however, (figure 23.11). Similar destruction occurred at many other sites during the nineteenth century. Copperhill, Tennessee, Butte, Montana, and the Ruhr Valley in Germany are some well-known examples, but these areas also are showing signs of recovery since corrective measures were taken.

There are two probable ways that air pollutants damage plants. They can be directly toxic, damaging sensitive cell membranes much as irritants do in human lungs. Within a few days of exposure to toxic levels of oxidants, mottling (discoloration) occurs in leaves due to chlorosis (bleaching of chlorophyll), and then necrotic (dead) spots develop (figure 23.12). If injury is severe, the whole plant may be killed. Sometimes these symptoms are so distinctive that positive identification of the source of damage is possible. Often, however, the symptoms are vague and difficult to separate from diseases or insect damage.

ENVIRONMENTAL POLLUTION AND OTHER DILEMMAS

FIGURE 23.11 Sulfur dioxide emissions and acid precipitation from the International Nickel Company copper smelter (*background*) killed all vegetation over a large area near Sudbury, Ontario. Even the pink granite bedrock has burned black. Recently installed scrubbers have dramatically reduced sulfur emissions. The ecosystem farther away from the smelter is slowly beginning to recover.

FIGURE 23.12 Soybean leaves exposed to 0.8 parts per million sulfur dioxide for 24 hours show extensive chlorosis (chlorophyll destruction) in white areas between leaf veins.

Another mechanism of action is exhibited by chemicals, such as ethylene, that act as metabolic regulators or plant hormones and disrupt normal patterns of growth and development. Ethylene is a component of automobile exhaust and is released from petroleum refineries and chemical plants. The concentration of ethylene around highways and industrial areas is often high enough to cause injury to sensitive plants. Some scientists

believe that the devastating forest destruction in Europe and North America may be partly due to volatile organic compounds.

Certain combinations of environmental factors have **synergistic effects** in which the injury caused by exposure to two factors together is more than the sum of exposure to each factor individually. For instance, when white pine seedlings are exposed to subthreshold concentrations of ozone and sulfur dioxide individually, no visible injury occurs. If the same concentrations of pollutants are given together, however, visible damage occurs. In alfalfa, SO_2 and O_3 together cause less damage than either one alone. These complex interactions point out the unpredictability of future effects of pollutants. Outcomes might be either more or less severe than previous experience indicates.

Pollutant levels too low to produce visible symptoms of damage may still have important effects. Field studies using open-top chambers and charcoal-filtered air show that yields in some sensitive crops, such as soybeans and alfalfa, may be reduced as much as 50 percent by currently existing levels of oxidants in ambient air. Some plant pathologists suggest that ozone and photochemical oxidants are responsible for as much as 90 percent of agricultural, ornamental, and forest losses from air pollution. The total costs of this damage may be as much as $10 billion per year.

Acid Deposition

Most people in the United States became aware of problems associated with **acid precipitation** (the deposition of wet acidic

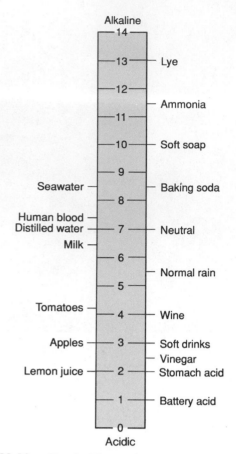

FIGURE 23.13 pH scale. The numbers represent the negative logarithm of the hydrogen ion concentration in water.

solutions or dry acidic particles from the air) within the last decade or so, but English scientist Robert Angus Smith coined the term "acid rain" in his studies of air chemistry in Manchester, England, in the 1850s. By the 1940s, it was known that pollutants, including atmospheric acids, could be transported long distances by wind currents. This was thought to be only an academic curiosity until it was shown that precipitation of these acids can have far-reaching ecological effects.

pH and Atmospheric Acidity

We describe acidity in terms of pH (the negative logarithm of the hydrogen ion concentration in a solution). The pH scale ranges from 0 to 14, with 7, the midpoint, being neutral (figure 23.13). Values below 7 indicate progressively greater acidity, while those above 7 are progressively more alkaline. Since the scale is logarithmic, there is a tenfold difference in hydrogen ion concentration for each pH unit. For instance, pH 6 is ten times more acidic than pH 7; likewise, pH 5 is one hundred times more acidic, and pH 4 is one thousand times more acidic than pH 7.

Normal, unpolluted rain generally has a pH of about 5.6 due to carbonic acid created by CO_2 in air. Volcanic emissions, biological decomposition, and chlorine and sulfates from ocean spray can drop the pH of rain well below 5.6, while alkaline dust can raise it above 7. In industrialized areas, anthropogenic acids in the air usually far outweigh those from natural sources. Acid rain is only one form in which acid deposition occurs. Fog, snow, mist, and dew also trap and deposit atmospheric contaminants. Furthermore, fallout of dry sulfate, nitrate, and chloride particles can account for as much as half of the acidic deposition in some areas. These particles are converted to acids when they dissolve in surface water or contact moist tissues (e.g., in the lungs). We have considerable evidence that acid aerosols are a human health hazard. The EPA is considering listing acid aerosols as a criteria pollutant. This would be the first new criteria pollutant added since the list was established in 1970.

Aquatic Effects

It has been known for about thirty years that acids—principally H_2SO_4 and HNO_3—generated by industrial and automobile emissions in northwestern Europe are carried by prevailing winds to Scandinavia where they are deposited in rain, snow, and dry precipitation. The thin, acidic soils and oligotrophic lakes and streams in the mountains of southern Norway and Sweden have been severely affected by this acid deposition. Some 18,000 lakes in Sweden are now so acidic that they will no longer support game fish or other sensitive aquatic organisms.

There has been a great deal of research on the mechanisms of damage by acidification. Generally, reproduction is the most sensitive stage in the life cycle. Eggs and fry of many fish species are killed when the pH drops to about 5.0. This level of acidification also can disrupt the food chain by killing aquatic plants, insects, and invertebrates on which fish depend for food. At pH levels below 5.0, adult fish die as well. Trout, salmon, and other game fish are usually the most sensitive. Carp, gar, suckers, and other less desirable fish are more resistant. There are several ways acids kill fish. Acidity alters body chemistry, destroys gills and prevents oxygen uptake, causes bone decalcification, and disrupts muscle contraction. Another dangerous effect (for us as well as fish) is that acid water leaches toxic metals, such as aluminum, out of soil and rocks. Which of these mechanisms is the most important is open to debate, but it is clear that acid deposition has had disastrous effects on sensitive aquatic ecosystems.

In the early 1970s, evidence began to accumulate suggesting that air pollutants are acidifying many lakes in North America. Studies in the Adirondack Mountains of New York revealed that about half of the high altitude lakes (above 1,000 m or 3,300 ft) are acidified and have no fish. Precipitation records show that the average pH of rain and snow has dropped significantly over a large area of northeastern United States and Canada in the past two decades. Some 48,000 lakes in Ontario are endangered and nearly all of Quebec's surface waters, including about 1 million lakes, are believed to be highly sensitive to acid deposition. Figure 23.14 shows the location of acid-

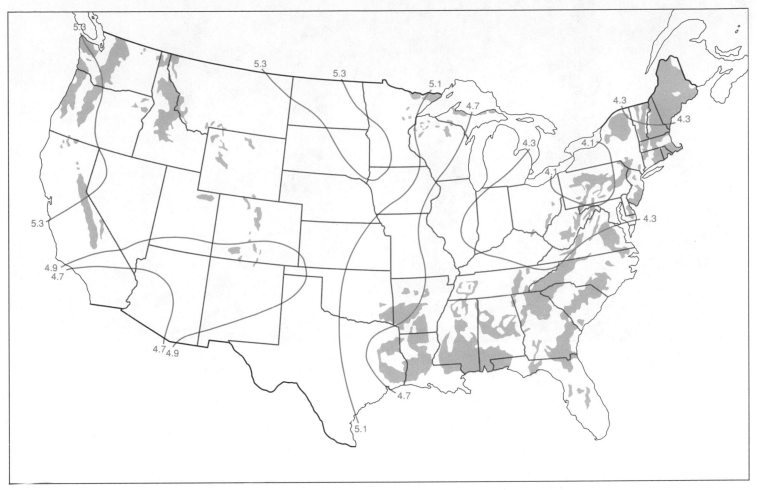

FIGURE 23.14 Acid-rain sensitive regions of the United States. Dark areas contain lakes likely to be acidified by acid deposition. Contour lines chart the average pH of precipitation. Values below pH 5.6 (now nearly the entire United States) are considered abnormally acidic.

sensitive regions of the United States and regions with sensitive lakes. About 50 percent of the acid deposition in Canada comes from the United States, while only 10 percent of United States pollution comes from Canada. Canadians are understandably upset about this imbalance.

Sulfates account for about two-thirds of the acid deposition in eastern North America and most of Europe, while nitrates contribute most of the remaining one-third. In urban areas, where transportation is the major source of pollution, nitric acid is equal to or slightly greater than sulfuric acids in the air. A vigorous program of pollution control has been undertaken by Canada, with promises of 50 percent reduction of SO_2 emissions and significant lowering of NO_x production.

Much of the western United States has relatively alkaline bedrock and carbonate-rich soil, which counterbalance acids from the atmosphere. Recent surveys of the Rocky Mountains, the Sierra Nevadas in California, and the Cascades in Washington, however, have shown that many high mountain lakes and streams have very low buffering capacity (ability to resist pH change) and are susceptible to acidification.

Damage to Forests

In the early 1980s, disturbing reports appeared of rapid forest declines in both Europe and North America. One of the earliest was a detailed ecosystem inventory on Camel's Hump Mountain in Vermont. A 1980 survey showed that seedling production, tree density, and viability of spruce-fir forests at high elevations had declined about 50 percent in fifteen years. By 1990, almost all the red spruce, once the dominant species on the upper part of the mountain, were dead or dying. A similar situation was found on Mount Mitchell in North Carolina where almost all red spruce and Frasier fir above two thousand meters are in a severe decline. Nearly all the trees are losing needles and about half of them are dead.

European forests also are dying at an alarming rate. West German foresters estimated in 1982 only 8 percent of their forests showed air pollution damage. By 1983, some 34 percent of the forest was affected, and in 1985, more than four million hectares (about half the total forest) was reported to be in a state of decline (figure 23.15). The loss to the forest industry is estimated to be about one billion DM (Deutsche marks) per year.

FIGURE 23.15 Waldsterben: The forests are dying! Crosses painted by environmentalists on spruce in Germany's Black Forest mark dead and dying trees. Large areas of forest in Europe, Canada, and the eastern United States have begun to show signs of air pollution damage. Various stresses, both natural and human-caused, probably play a role.

Similar damage is reported in Czechoslovakia, Poland, Austria, and Switzerland. Again, high elevation forests are most severely affected. This is a disaster for mountain villages in the Alps that depend on forests to prevent avalanches in the winter. Sweden, Norway, the Netherlands, Romania, China, and the Soviet Union also have evidence of growth reduction, defoliation, root necrosis, lack of seedling growth, and premature tree death. The species afflicted vary from place to place, but the overall picture is of widespread forest destruction.

This complex phenomenon probably has many contributing factors, but air pollution and deposition of atmospheric acids are thought to be leading causes of forest destruction in many areas. Considerable research has shown that acids are directly toxic to tender shoots and roots. High-altitude forests are subjected to especially intense doses of these acids because clouds saturated with pollutants tend to hang on mountain tops, bathing forests in a toxic soup for days or even weeks at a time.

Scientists have suggested that other mechanisms may play a role in forest decline. Over-fertilization by nitrogen compounds may make trees sensitive to early frost. Essential minerals, such as magnesium, may be washed out of foliage or soil by acidic precipitation. Toxic metals, such as aluminum, may be solubilized by acidic groundwater. Plant pathogens and insect pests may damage trees or attack trees debilitated by air pollution. Microrhizzae (fungi) that form essential mutual associations with trees may be damaged by acid rain. Other air pollutants, such as sulfur dioxide, ozone, or toxic organic compounds may damage trees. Repeated harvesting cycles in commercial forests may remove nutrients and damage ecological relationships essential for healthy tree growth. Perhaps the most likely scenario is that all these environmental factors act cumulatively but in different combinations in the deteriorating health of individual trees and entire forests.

Damage to Buildings and Monuments

In cities throughout the world, some of the oldest and most glorious buildings and works of art are being destroyed by air pollution. Smoke and soot coat buildings, paintings, and textiles. Limestone and marble are destroyed by atmospheric acids at an alarming rate. The Parthenon in Athens, the Taj Mahal in Agra, the Colosseum in Rome, frescos and statues in Florence, medieval cathedrals in Europe (figure 23.16), and the Lincoln Memorial and Washington Monument in Washington, D.C., are slowly dissolving and flaking away because of acidic fumes in the air. Medieval stained glass windows in Cologne's gothic cathedral are so porous from etching by atmospheric acids that pigments disappear and the glass literally crumbles away. Restoration costs for this one building alone are estimated at three to four billion German marks ($1.5 to $2 billion).

On a more mundane level, air pollution also damages ordinary buildings and structures. Corroding steel in reinforced concrete weakens buildings, roads, and bridges. Paint and rubber deteriorate due to oxidization. Limestone, marble, and some kinds of sandstone flake and crumble. The Council on Environmental

FIGURE 23.16 Atmospheric acids, especially sulfuric and nitric acids, have almost completely eaten away the face of this medieval statue. Each year, the total losses from air pollution damage to buildings and materials amounts to billions of dollars.

Quality estimates that U.S. economic losses from architectural damage caused by air pollution amount to about $4.8 billion in direct costs and $5.2 billion in property value losses each year.

Atmospheric Effects of Pollution

We are beginning to appreciate the effects of atmospheric pollutants on global climate and atmospheric structure. Several recent discoveries, such as the springtime ozone "hole" over the Antarctic (Box 23.1) and the global warming caused by infrared-absorbing trace gases (Box 17.1) suggest that anthropogenic air pollutants may cause catastrophic changes in the atmosphere that could far outweigh all other effects described so far.

AIR POLLUTION CONTROL

What can we do about air pollution? In this section we will look at some of the techniques that can be used to avoid creating pollutants or to clean up effluents before they are released. We also will look at some legislation that regulates pollutant emissions and ambient air quality.

Moving Pollution to Remote Areas

Among the earliest techniques for improving local air quality was moving pollution sources to remote locations and/or dispersing emissions with smokestacks. These approaches exemplify the attitude that "dilution is the solution to pollution." One electric utility, for example, ran newspaper and magazine ads in the early 1970s, claiming to be a "pioneer" in the use of tall smokestacks on its power plants to "disperse gaseous emissions widely in the atmosphere so that ground level concentrations would not be harmful to human health or property." The company claimed that their smoke would be "dissipated over a wide area and come down finally in harmless traces." Far from being harmless, how-

ever, those "traces" are the main source of many of our current problems. We are finding that there is no "away" to which we can throw our unwanted products. A far better solution to pollution is to prevent its release. We will now turn our attention to emission-control technology.

Particulate Removal Techniques

Filters remove particles physically by trapping them in a porous mesh of cotton cloth, spun glass fibers, or asbestos-cellulose, which allows air to pass through but holds back solids. Collection efficiency is relatively insensitive to fuel type, fly ash composition, particle size, or electrical properties. Filters are generally shaped into giant bags 10 to 15 meters long and 2 or 3 meters wide. Effluent gas is blown into the bottom of the bag and escapes through the sides much like the bag on a vacuum cleaner (figure 23.17a). Every few days or weeks, the bags are opened to remove the dust cake. Thousands of these bags may be lined up in a "baghouse." These filters are usually much cheaper to install and operate than electrostatic filters (described in the following paragraphs).

Cyclone collectors use gravitational settling to remove heavy particles from an effluent stream. Spiral fins or grooves in the collector cause the gas stream to swirl, creating a centrifugal force that spins out large particles (figure 23.17b). This is the most primitive and least effective of the three methods, but when combined with other methods, it can be useful.

Electrostatic precipitators (figure 23.17c) are the most common particulate controls in power plants. Fly ash particles pick up an electrostatic surface charge as they pass between large electrodes in the effluent stream. This causes the particle to migrate to and accumulate on a collecting plate (the oppositely charged electrode). These precipitators consume a large amount of electricity, but maintenance is relatively simple and collection efficiency can be as high as 99 percent. Performance depends on particle size and chemistry, strength of the electric field, and flue gas velocity.

The ash collected by all of these techniques is a solid waste (often hazardous due to the heavy metals and other trace components of coal or other ash source) and must be buried in landfills or other solid waste disposal sites.

Sulfur Removal

As we have seen earlier in this chapter, sulfur oxides are among the most damaging of all air pollutants in terms of human health and ecosystem damage. It is important to reduce sulfur loading. This can be done either by using low-sulfur fuel or by removing sulfur from effluents.

Fuel Switching and Fuel Cleaning

Switching from soft coal with a high sulfur content to low-sulfur coal can greatly reduce sulfur emissions. This may eliminate jobs, however, in such areas as Appalachia that are already econom-

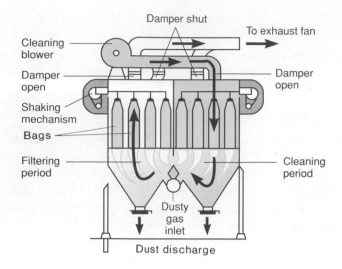

a. Typical bag filter

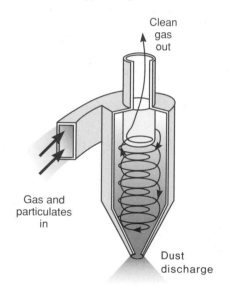

b. Basic cyclone collector

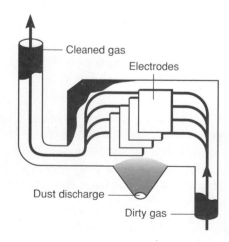

c. Electrostatic precipitator

ically depressed. Changing to another fuel, such as natural gas or nuclear energy, can eliminate all sulfur emissions as well as those of particulates and heavy metals. Natural gas is more expensive and more difficult to ship and store than coal, however, (chapter 18) and many people prefer the sure dangers of coal pollution to the uncertain dangers of nuclear power (chapter 19). Alternative energy sources, such as solar power, would be preferable to either fossil fuel or nuclear power, but are not yet economically competitive (chapter 20) in most areas. In the interim, coal can be crushed and washed to remove sulfur and metals before combustion. This improves heat content and firing properties but may replace air pollution with solid waste and water pollution problems. Coal gasification can also greatly reduce sulfur emissions (Box 18.1).

Limestone Injection and Fluidized Bed Combustion

Sulfur emissions can be reduced as much as 90 percent by mixing crushed limestone with coal before it is fed into a boiler. Calcium in the limestone reacts with sulfur to make calcium sulfite ($CaSO_3$), calcium sulfate ($CaSO_4$), or gypsum ($CaSO_4 \cdot 2H_2O$). In ordinary furnaces, this procedure creates slag, which fouls burner grates and reduces combustion efficiency.

A relatively new technique for burning called fluidized bed combustion offers several advantages in pollution control. In this procedure, a mixture of crushed coal and limestone particles about a meter deep (3 ft) is spread on a perforated distribution grid in the combustion chamber (figure 23.18). When high-pressure air is forced through the bed, the surface of the fuel rises as much as one meter and resembles a boiling fluid as particles hop up and down. Oil is sprayed into the suspended mass to start the fire. During operation, fresh coal and limestone are fed continuously into the top of the bed, while ash and slag are drawn off from below. The rich air supply and constant motion in the bed make burning efficient and prevent buildup of large slag clinkers. Steam generator pipes are submerged directly into the fluidized bed, and heat exchange is more efficient than in the water walls of a conventional boiler. More than 90 percent of SO_2 is captured by the limestone particles, and NO formation is reduced by holding temperatures around 800°C (1,500°F) instead of twice that figure in other boilers. These low temperatures also preclude slag formation, which aids in maintenance. The efficient burning of this process makes it possible to use cheaper fuel, such as lignite or unwashed subbituminous coal, rather than higher priced hard coal.

FIGURE 23.17 Typical emission-control devices: (*a*) bag filter, (*b*) cyclone precipitator, and (*c*) electrostatic precipitator. Note that two stages in the operational cycle are shown in (*a*) the filtering period on the left and the period of cleaning the filter bag on the right.

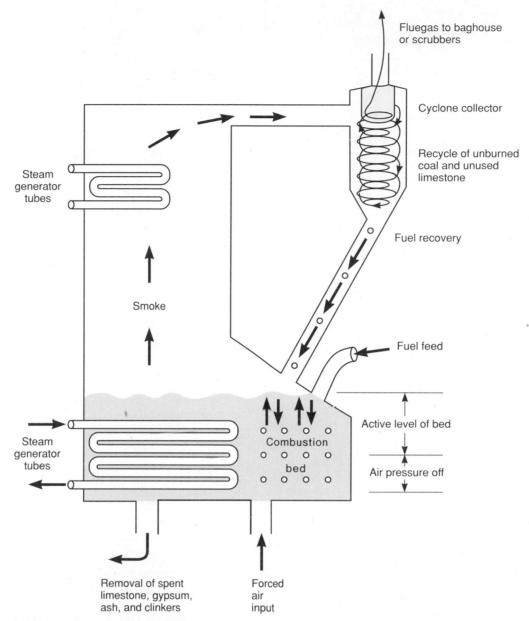

Fluegas to baghouse
or scrubbers

Cyclone collector

Recycle of unburned
coal and unused
limestone

Fuel recovery

Steam
generator
tubes

Smoke

Fuel feed

Active level of bed

Combustion
bed

Air pressure off

Steam
generator
tubes

Removal of spent
limestone, gypsum,
ash, and clinkers

Forced
air
input

FIGURE 23.18 Fluidized bed combustion. Fuel is lifted by strong air jets from underneath the bed. Efficiency is good with a wide variety of fuels, and SO_2, NO_x, and CO emissions are much lower than with conventional burners.

Flue Gas Desulfurization

Crushed limestone, lime slurry, or alkali (sodium carbonate or bicarbonate) can be injected into a stack gas stream to remove sulfur after combustion. These processes are often called flue gas scrubbing. Spraying wet alkali solutions or limestone slurry is relatively inexpensive and effective, but maintenance can be difficult. Rock-hard plaster and ash layers coat the spray chamber and have to be chipped off regularly. Corrosive solutions of sulfates, chlorides, and fluorides erode metal surfaces. Electrostatic precipitators don't work well because of fouling and shorting of electrodes after wet scrubbing.

Dry alkali injection (spraying dry sodium bicarbonate into the flue gas) avoids many of the problems of wet scrubbing, but the expense of appropriate reagents is prohibitive in most areas. A hybrid procedure called spray drying has been tested successfully in pilot plant experiments. In this process, a slurry of pulverized limestone or slaked lime is atomized in the stack gas stream. The spray rate and droplet size are carefully controlled so that the water flash evaporates and a dry granular precipitate is produced. Passage through a baghouse filter removes both ash and sulfur very effectively.

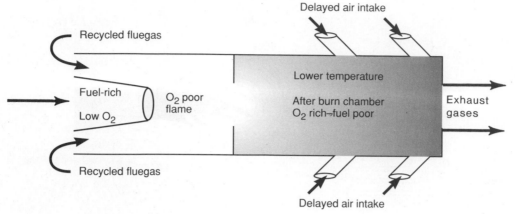

FIGURE 23.19 A staged burner reduces NO_x formation by burning fuel first in an oxygen-poor environment and then burning residual gas at lower temperatures with delayed air intake. This is somewhat similar to combined cycle coal gasification (see Box 18.1).

As with coal washing, scrubbing often results in a trade-off of an air pollution problem for a solid waste disposal problem. Sulfur slag, gypsum, and other products of these processes can amount to three or four times as much volume as fly ash. A large power plant can produce millions of tons of waste per year.

Sulfur Recovery Processes

Instead of making a throwaway product that becomes a waste disposal problem, sulfur can be removed from effluent gases by processes that yield a usable product, such as elemental sulfur, sulfuric acid, or ammonium sulfate. Catalytic converters are used in these recovery processes to oxidize or reduce sulfur and to create chemical compounds that can be collected and sold. Markets have to be reasonably close for economic feasibility, and fly ash contamination must be reduced as much as possible.

Nitrogen Oxide Control

Undoubtedly the best way to prevent nitrogen oxide pollution is to avoid creating it. A substantial portion of the emissions associated with mining, manufacturing, and energy production could be eliminated through conservation (chapter 20).

Staged burners, in which the flow of air and fuel are carefully controlled, can reduce nitrogen oxide formation by as much as 50 percent (figure 23.19). This is true for both internal combustion engines and industrial boilers. Fuel is first burned at high temperatures in an oxygen-poor environment where NO_x cannot form. The residual gases then pass into an afterburner where more air is added and final combustion takes place in an air-rich, fuel-poor, low-temperature environment that also reduces NO_x formation. Stratified-charge engines and new orbital automobile engines use this principle to meet emission standards without catalytic converters.

The approach adopted by United States auto makers for NO_x reductions has been to use selective catalysts to change pollutants to harmless substances. Three-way catalytic converters use platinum-palladium and rhodium catalysts to remove up to 90 percent of NO_x, hydrocarbons, and carbon monoxide at the same time (figure 23.20). Unfortunately, this approach doesn't work on diesel engines, power plants, smelters, and other pollution sources because of problems with back pressure, catalyst life, corrosion, and production of unwanted by-products, such as ammonium sulfate (NH_4SO_4), that foul the system.

Raprenox (*rap*id *r*emoval of *n*itrogen *ox*ides) is a new technique for removing nitrogen oxides that was developed by the U.S. Department of Energy Sandia Laboratory in Livermore, California. Exhaust gases are passed through a container of common, nonpoisonous cyanuric acid. When heated to 350°C (662°F), cyanuric acid releases isocyanic acid gas, which reacts with NO_x to produce CO_2, CO, H_2O, and N_2. In small-scale diesel engine tests, this system eliminated 99 percent of the NO_x. Whether it will work in full-scale applications, especially in flue-gases contaminated with fly ash, remains to be seen.

Hydrocarbon Emission Controls

Hydrocarbons and volatile organic compounds are produced by incomplete combustion of fuels or solvent evaporation from chemical factories, painting, dry cleaning, plastic manufacturing, printing, and other industrial processes that use a variety of volatile organic chemicals. Closed systems that prevent escape of fugitive gases can reduce many of these emissions. In automobiles, for instance, positive crankcase ventilation (PCV) systems collect oil that escapes from around the pistons and unburned fuel and channels it back to the engine for combustion. Modification of carburetor and fuel systems prevents evaporation of gasoline (figure 23.20). In the same way, controls on fugitive losses from valves, pipes, and storage tanks in industry can have a significant impact on air quality. Afterburners are often the best method for destroying volatile organic chemicals in industrial exhaust stacks. High air-fuel ratios in automobile engines and other

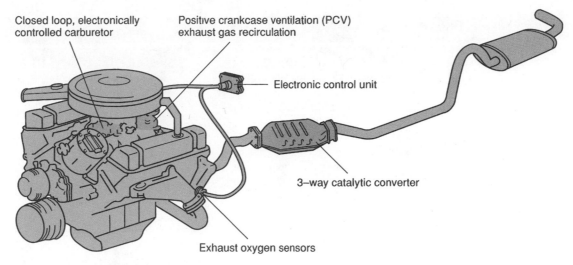

FIGURE 23.20 Elements of a modern automobile emission-control system. A closed-loop, electronically controlled carburetor or fuel-injector carefully meters fuel/air ratios to optimize combustion. Exhaust oxygen sensors measure completeness of fuel burning. Positive crankcase ventilation captures oil "blow-by" and unburned exhaust gases and recycles them to the cylinder.

burners minimize hydrocarbon and carbon monoxide emissions, but also cause excess nitrogen oxide production. Careful monitoring of air-fuel inputs and oxygen levels in exhaust gases can minimize all these pollutants.

CLEAN AIR LEGISLATION

Throughout history there have been countless ordinances prohibiting emission of objectionable smoke, odors, and noise. Air pollution traditionally has been treated as a local problem, however, to be regulated by local authorities. The Clean Air Act of 1963 was the first national legislation in the United States aimed at air pollution control. This act called for research to be carried out by the U.S. Public Health Service on the sources and effects of air pollution. Federal grants were provided to States to combat pollution, but the Act was careful to preserve States' rights to set and enforce air quality regulations. It soon became obvious that some pollution problems cannot be solved on a local basis. In 1965, amendments to the Clean Air Act called for national standards for automobile carbon monoxide and hydrocarbon exhausts.

On December 31, 1970, President Nixon signed an extensive set of amendments that essentially rewrote the Clean Air Act. These amendments identified the "criteria pollutants" discussed earlier in this chapter, and established national ambient air quality standards. These standards are divided into two categories. **Primary standards** (table 23.3) are intended to protect human health while **secondary standards** are set to protect materials, crops, climate, visibility, and personal comfort. Primary and secondary standards are the same for all pollutants except total suspended particulates (TSP), which have a maximum annual geometric mean of 60 $\mu g/m^3$. Ambient standards assume that pollutants have no adverse effects beyond certain thresholds.

TABLE 23.3 National Ambient Air Quality Standards (NAAQS)

Pollutant	Primary (health-based) standard	
	Averaging time	*Concentration*
TSP[a]	Annual geometric mean[b]	75 $\mu g/m^3$
	24 hours	260 $\mu g/m^3$
SO$_2$	Annual arithmetic mean[c]	80$\mu g/m^3$ (0.03 ppm)
	24 hours	365 $\mu g/m^3$ (0.14 ppm)
	3 hours	1,300 $\mu g/m^3$ (0.5 ppm)
CO	8 hours	10 mg/m^3 (9 ppm)
	1 hour	40 mg/m^3 (35 ppm)
NO$_2$	Annual arithmetic mean	100 $\mu g/m^3$ (0.05 ppm)
O$_3$	Daily max 1 hour avg	235 $\mu g/m^3$ (0.12 ppm)
Lead	Maximum quarterly avg	1.5 $\mu g/m^3$
Hydrocarbons	3 hours	160 $\mu g/m^3$ (0.24 ppm)

[a]Total suspended particulates
[b]The geometric mean is obtained by taking the nth root of the product of n numbers. This tends to reduce the impact of a few very large numbers in a set.
[c]An arithmetic mean is the average determined by dividing the sum of a group of data points by the number of points.

They also assume that pollutants arising from numerous diverse sources are more reasonably and effectively regulated by setting maximum total levels in the atmosphere than by regulating individual emissions. Some environmentalists disagree with both of these assumptions. These standards are the basis of a warning system called the Air Pollution Standards Index (table 23.4).

In 1990, after many years of acrimonious debate and political maneuvering, Congress finally passed another set of

TABLE 23.4 Air Pollutant Standards Index

Rating	Description	Health effects	Suggested actions
500	Disaster	Very hazardous to all; serious injury and excess deaths especially in sensitive persons	Stay inside with doors and windows closed; avoid all physical activity
400	Emergency	Hazardous to general population; grave injury possible	Avoid outdoor exercise; young, elderly, and ill should reduce all activity
300	Warning	Very unhealthy for all; serious threat to young, elderly, or ill	Elderly or those with heart or lung disease stay indoors
200	Alert	Irritation of eyes and lungs; aggravation of existing disease	Sensitive persons stay indoors
100	Moderate	NAAQS maximum permissible levels	Avoid traffic and congestion
50	Good	No known short-term effects	No restrictions

amendments to the Clean Air Act to protect public health, property, and the environment. Among the most important provisions of this legislation are the following:

- *Acid rain.* Sulfur dioxide releases will be cut from 24 million tons in 1990 to 10 million tons in 2000 by requiring the 111 largest sulfur emitters to meet strict standards. Nitrogen oxide emissions will be reduced from 6 million tons in 1990 to 4 million tons in 1992. This is the first specific limit set by Congress on nitrogen emissions.

- *Urban smog.* Motor vehicle tailpipe emissions of hydrocarbons and nitrogen oxides will be reduced 35 percent and 60 percent respectively in all new cars by 1996. Beginning in 1998, pollution control equipment on new cars must last ten years or 100,000 miles. Oil companies will be required to offer alternative fuels, such as methanol or ethanol (sometimes called oxygenated fuels), hydrogen, or compressed natural gas (methane) in cities with the worst pollution problems by 1992. Automobile manufacturers will be required to produce 300,000 alternative-fuel cars per year by 1998. Cities not meeting air quality standards for ozone and smog are divided into five categories (marginal, moderate, serious, severe, and extreme); deadlines for attaining standards are set for three, six, nine, fifteen, and twenty years respectively. Pollution sources emitting 100 tons per year in marginal and moderate areas, 50 tons per year in serious areas, 25 tons per year in severe areas, and 10 tons per year in extreme areas are regulated.

- *Toxic air pollutants.* Although the EPA has had authority to set emission standards for air toxics since 1970, only seven (beryllium, mercury, asbestos, lead, vinyl chloride, benzene, and PCBs) have been regulated. Now 189 chemicals are listed, and the EPA is required to establish about 250 source categories

(chemical factories, dry cleaners, coke furnaces, printing plants, etc.) to be regulated. The largest polluters will be required to install the best available technology to reduced emissions 90 percent by 2003. The safety standard for acceptable cancer risk to nearby residents is set at 1 in 10,000, a rate much too high according to most environmentalists.

- *Ozone protection.* Chlorofluorocarbons and carbon tetrachloride will be phased out by the year 2000. Recovery and recycling programs for existing CFCs will be instituted. Methyl chloroform will be outlawed by 2002. Hydrochlorofluorocarbons in aerosol cans and insulation will be phased out by 2030.

- *Marketing pollution rights.* Corporations are allowed to offset emissions by buying, selling, and "banking" pollution rights from other factories at an expected savings of $2 billion to $3 billion per year. This is a controversial free-market approach (chapter 8) that may make economic sense for industry and environmental sense on average but may be disastrous for some localities.

- *Workers compensation.* A $250 million fund is set up to retrain and compensate workers displaced by provisions of this law. This is intended primarily for high-sulfur coal miners in Appalachia who will lose jobs due to fuel switching.

Industrial economists calculate that these regulations will double the current $30 billion per year cost of pollution control. They warn that consumers will pay more for electricity, space heating, transportation, and consumer goods. Some claim health and ecosystem benefits will amount to only about half the $30 billion price of controls. They also warn that these costs will reduce our competitiveness in foreign markets and damage our economy. Congress calculates that the cost of air pollution control will be between $4 billion and

$10 billion per year. Proponents of these provisions argue that pollution control will be a lucrative new business that other countries will seek eagerly in coming years.

Although these revisions are a dramatic step toward cleaner skies, environmentalists didn't get everything they wanted. Electric utilities fought off regulations that would have reduced emissions of mercury and other toxic materials from coal-fired power plants. Steelmakers pleaded financial hardships and were given until the year 2020 to eliminate cancer-causing emissions from coke ovens, providing they take interim steps to reduce pollution.

California has gone further than the federal government in making specific plans for air pollution control. In 1990, the South Coast Air Quality Management District adopted 160 rules to clean the air in the Los Angeles Basin. If these measures are successful, smog-causing emissions could be reduced by 70 percent. By the year 2000, visibility would increase from a 10 mi current average to 60 mi. The number of days when the air is considered hazardous to breathe would decrease from 150 per year to 0 per year.

Reaching these goals will require substantial life-style changes for most Californians. Aerosol hair sprays, deodorants, charcoal lighter fluid, gasoline-powered lawnmowers, and drive-through burger stands could be banned. Paints and cleaning solutions would have to emit fewer volatile solvents. Radial tires and more stringent emission controls would be mandated for automobiles. Clean-burning oxygenated fuels or electric motors would be required for all vehicles. Car pooling would be encouraged, parking lots would be restricted, and limits would be placed on the number of cars a family could have.

California's land-use zones and housing codes might have to be changed to accommodate new commuting patterns. Substantial relocations could result. The cost is estimated to be about 60 cents per person per day. Opponents argue that the price tag could be as high as $15 billion a year and 30,000 lost jobs. Whether Californians, or any of us, care enough about health and the environment to make these changes and pay these costs remains to be seen.

CURRENT CONDITIONS AND PROSPECTS FOR THE FUTURE

Although we have not yet achieved the Clean Air Act goals in many parts of the United States, air quality has improved dramatically in the last decade in terms of the major large-volume pollutants. For twenty-three of the largest U.S. cities, the number of days in which air quality reached the hazardous level (PSI greater than 300) is down 93 percent from an average of 1.8 days/year a decade ago to 0.13 days/year now. In those same cities, the average number of days above PSI 200 (very unhealthful) was 11.3 days in 1982, down 65 percent from the average of 32.2 days/year ten years earlier. Unhealthful days (above PSI 100) still average about thirty-three days per year, but that is only one-half the level of the 1970s.

The EPA estimates that emissions of particulate materials are down 60 percent, lead is down 80 percent, SO_2 and CO are down 30 percent, and O_3 is down 18 percent over the past two decades (figure 23.21). Industrial cities, such as Chicago, Pittsburgh, and Philadelphia, that suffered "smokestack" pollution have had 90 percent reductions in number of days exceeding NAAQS maxima. Filters, scrubbers, and precipitators on power plants and other large stationary sources are responsible for most of the particulate and SO_2 reductions. Catalytic converters on automobiles are responsible for most of the CO and O_3 reductions. The only conventional "criteria" pollutant that has not dropped significantly is NO_x, which has risen 300 percent since 1940 and dropped only slightly over the last ten years.

Because automobiles are the main source of NO_x, cities where pollution is largely from traffic have had increased PSI levels in recent years. Los Angeles, Anaheim, and Riverside, California, are the only cities in the country in the extreme urban smog category, exceeding the ozone standards an average of 137.5 days per year between 1987 and 1989. Baltimore, New York City, Chicago, Gary, Houston, Milwaukee, Muskegan, Philadelphia, and San Diego are all in the severe category. Eighty-five other urban areas are still considered nonattainment regions. In spite of these local failures, however, 80 percent of the United States now meets the NAAQS goals. This improvement in air quality is perhaps the greatest environmental success story in our history (figure 23.21).

The outlook is not so encouraging in other parts of the world, however. The major metropolitan areas of many developing countries are growing at explosive rates to incredible sizes (chapter 25), and environmental quality is abysmal in many of them. The composite average annual levels of SO_2 in São Paulo, Brazil, for instance, are more than 100 $\mu g/m^3$, and peak levels can be up to ten times higher. Mexico City remains notorious for bad air. Its 131,000 industries and 2 million cars and buses spew out more than 5,500 tons of air pollutants daily. Santiago, Chile, averages 299 days per year on which suspended particulates exceed WHO standards of 150$\mu g/m^3$. Tehran, Iran, is nearly as bad.

While there are few statistics on China's pollution situation, it is known that many of China's 400,000 factories have no air pollution controls. Experts estimate that home coal burners and factories emit 10 million tons of soot and 15 million tons of sulfur dioxide annually. Sheyang, an industrial city in northern China, is thought to have the world's worst particulate problem with peak winter concentrations of 690 $\mu g/m^3$ (nine times U.S. maximum standards). Airborne particulates in Sheyang exceed

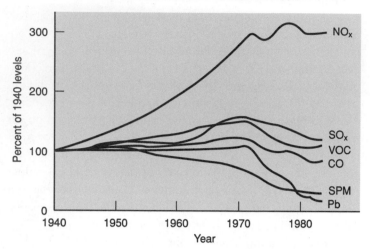

NO$_x$ = nitrogen oxides (1940 level: 6.8 x 10^6 metric tons/yr)
SO$_x$ = sulfur oxides (1940 level: 18.0 x 10^6 metric tons/yr)
VOC = volatile organic compounds (1940 level: 18.5 x 10^6 metric tons/yr)
SPM = suspended particulate matter (1940 level: 22.8 x 10^6 metric tons/yr)
Pb = lead (1940 level: c. 230 x 10^3 metric tons/yr)
CO = carbon monoxide (1940 level: 81.6 x 10^6 metric tons/yr)

FIGURE 23.21 Air pollution trends in the United States, 1940–1984. Emissions of six "criteria" pollutants as percent of 1940 levels. Source: Data from the Environmental Protection Agency (EPA).

WHO standards on 347 days per year. Beijing, Xian, and Guangzhou are nearly as bad. The high incidence of cancer in Shanghai is thought to be linked to air pollution.

As political walls came down across Eastern Europe and the Soviet Union at the end of the 1980s, horrifying environmental conditions in these centrally-planned economies were revealed. Inept industrial managers, a rigid bureaucracy, and lack of democracy have created ecological disasters. Where government owns, operates, and regulates industry, there are few checks and balances or incentives to clean up pollution. Much of the Eastern Bloc depends heavily on soft brown coal for its energy. Pollution controls are absent or highly inadequate.

Southern Poland and northern Czechoslovakia are covered most of the time by a permanent cloud of smog from factories and power plants. Acid rain is eating away historic buildings and damaging already inadequate infrastructures. The haze is so dark that drivers must turn on their headlights during the day. Residents complain that washed clothes turn dirty before they can dry. Zabrze, near Katowice in southern Poland, has particulate emissions of 3,600 metric tons per square kilometer. This is more than seven times the emissions in Baltimore, Maryland, or Birmingham, Alabama, the dirtiest cities (for particulates) in the United States. Home gardening in Katowice has been banned because vegetables raised there have unsafe levels of lead and cadmium.

For miles around the infamous Romanian "black town" of Copsa Mica, the countryside is so stained by soot that it looks as if someone had poured black ink over everything. Birth defects afflict 10 percent of infants in northern Bohemia. Workers in factories there get extra hazard pay; burial money, they call it. Life expectancy in these industrial towns is as much as ten years less than the national average. Espenhain, in the industrial belt of former East Germany, has one of the world's highest rates of sulfur dioxide pollution. One of every two children has lung problems, one of every three has heart problems. Brass doorknobs and name plates have been eaten away by the acidic air in a few months.

Not all is pessimistic, however. There have been some spectacular successes in air pollution control. Sweden and West Germany (countries affected by forest losses due to acid precipitation) cut their sulfur emissions by two-thirds between 1970 and 1985. Austria and Switzerland have gone even further. They even regulate motorcycle emissions. The Global Environmental Monitoring System (GEMS) reports declines in particulate levels in 26 of 37 cities worldwide. Sulfur dioxide and sulfate particles, which cause acid rain and respiratory disease, have declined in 20 of these cities.

In the first edition of this book, we described Cubato, Brazil as the "Valley of Death," one of the most dangerously polluted places in the world. In the mid-1980s, a steel plant, a huge oil refinery, and fertilizer and chemical factories churned out thousands of tons of air pollutants every year. Trees died on the surrounding hills. Birth defects and respiratory diseases were alarmingly high.

Since that time, however, the citizens of Cubato have made remarkable progress in cleaning up their environment. The end of military rule and restoration of democracy allowed residents to publicize their complaints. The environment became an important political issue. The state of São Paulo invested about $100 million, and the private sector spent twice as much to clean up most pollution sources in the valley. Particulate pollution was reduced 92 percent. Ammonia emissions were reduced 97 percent, other hydrocarbons that cause ozone and smog were cut 78 percent, and sulfur dioxide production fell 84 percent. Fish are returning to the rivers, and forests are regrowing on the mountains. Progress is possible! We hope that similar success stories will be obtainable elsewhere.

SUMMARY

In this chapter, we have looked at major categories, types, and sources of air pollution. We have defined air pollution as chemical or physical changes brought about by either natural processes or human activities, resulting in air-quality degradation. Air pollution has existed as long as there has been an atmosphere. Perhaps the first major human source of air pollution was fire. Burning fossil fuels, biomass, and wastes continues to be the largest source of anthropogenic (human-caused) air pollution. The six "conventional" large-volume pollutants are NO_x, SO_2, CO, lead, particulates, and volatile organic compounds. The major sources of air pollution are transportation, industrial processes, stationary fuel combustion, and solid waste disposal.

We also looked at some unconventional pollutants. Indoor air pollutants, including formaldehyde, asbestos, toxic organic chemicals, radon, and tobacco smoke may pose a greater hazard to human health than all of the conventional pollutants combined. Odors, visibility losses, and noise generally are not life threatening but serve as indicators of our treatment of the environment. Some atmospheric processes play a role in distribution, concentration, chemical modification, and elimination of pollutants. Among the most important of these processes are long-range transport of pollutants and photochemical reactions in trapped inversion layers over urban areas.

Encouraging improvements have been made in ambient outdoor air quality over most of the United States in the last decade. We have made considerable progress in designing and installing pollution-control equipment to reduce the major conventional pollutants. There are many types of scrubbers, filters, catalysts, fuel modification processes, and new burning techniques for controlling pollution. The Clean Air Act regulates air quality in the United States through both ambient standards and emission limits, and its 1990 amendments promise a dramatic improvement in our atmosphere. There is much yet to be done, especially in developing countries and in Eastern Europe, but air-pollution control is, perhaps, our greatest success in environmental protection and an encouraging example of what can be accomplished in this field.

■ Review Questions

1. What is the difference between bronchitis, emphysema, and fibrosis? What causes these diseases?
2. What are the most important causes of human death from air pollution?
3. What is acid deposition? What causes it?
4. What have been the effects of acid deposition on aquatic and terrestrial ecosystems?
5. How do electrostatic precipitators, baghouse filters, flue gas scrubbers, and catalytic converters work?
6. What is the difference between primary and secondary standards in air quality?
7. What is the difference between ambient standards and emission limits?
8. What are some of the major toxic air pollutants, and what are their sources?
9. Describe the health effects and suggested actions for each of the levels of the pollution standards index (PSI).
10. Which of the conventional pollutants has decreased most in the recent past and which has decreased least?

■ Questions for Critical Thinking

1. Which air pollution-related lung disease worries you most?
2. What might be done to improve indoor air quality?
3. Why do you suppose that air pollution is so much worse in Eastern Europe than the West?
4. Suppose air pollution causes a billion dollars in crop losses each year but controlling the pollution would also cost a billion dollars. Should we insist on controls?
5. In 1984, David Stockman, Director of the Office of Management and Budget for President Reagan, said that it would cost $1,000 per fish to control acid precipitation in the Adirondack lakes and that it would be cheaper to buy fish for anglers than to put scrubbers on power plants in Ohio. Do you agree?
6. What will the ban on fluorocarbon production do to your life? Will it be worth it to save the ozone layer?
7. Is it possible to have zero emissions of pollutants?
8. If there are thresholds for pollution effects (at least as far as we know now), is it reasonable or wise to depend on environmental processes to disperse, assimilate, or inactivate waste products?
9. Catalytic converters on automobiles definitely improve air quality, but up to one-fourth of car owners disable the converters on their cars by using leaded gasoline. What should we do about this?
10. Do you think that we should continue to use ambient air-quality standards or change to absolute emission standards for all pollutants?

Key Terms

acid precipitation (p. 479)
aerosol (p. 469)
aesthetic degradation
  (p. 472)
ambient air (p. 466)
asthma (p. 477)
bronchitis (p. 477)
carbon monoxide (p. 468)
conventional or criteria
  pollutants (p. 466)
cyclone collectors (p. 483)
dry alkali injection (p. 485)
electrostatic precipitators
  (p. 483)
emission standards (p. 472)
emphysema (p. 477)
fibrosis (p. 477)
filters (p. 483)
fugitive emissions (p. 465)

inflammatory response
  (p. 477)
nitrogen oxides (p. 466)
ozone (p. 472)
particulate material (p. 469)
photochemical oxidants
  (p. 472)
primary pollutants (p. 465)
primary standards (p. 487)
secondary pollutants
  (p. 465)
secondary standards (p. 487)
sulfur dioxide (p. 466)
synergistic effects (p. 479)
unconventional or
  noncriteria pollutants
  (p. 472)
volatile organic compounds
  (p. 471)

SUGGESTED READINGS

■ Brasseur, G. January/February, 1987. "The Endangered Ozone Layer." *Environment* 29 (1):6. Excellent summary of the status of research on stratospheric ozone.

■ Cicerone, R. J. July, 1987. "Changes in Stratospheric Ozone." *Science* 237 (4810):35. A useful overview of atmospheric chemistry and human effects on stratospheric ozone over Antarctica.

■ Committee on Global Change. *Toward an Understanding of Global Change*. Washington, D.C.: National Academy Press, 1988. A report of the United States' committee for the International Geosphere-Biosphere Program.

■ Fischhoff, B. 1991. "Report from Poland: Science and Politics in the Midst of Environmental Disaster." *Environment* 33 (2):21. Decades of neglect and oppression have created an environmental disaster in Eastern Europe. The new democratic government has little money to clean up the mess they inherited.

■ French, H. F. 1991. "Restoring Eastern European and Soviet Environments." *State of the World 1991*. Washington, D.C. World Watch Institute p. 93. Environmental reconstruction is imperative in Eastern Europe and the Soviet Union. The Green ecological movement has been instrumental in both political change and environmental restoration.

■ Frenzel, G. May, 1985. "The Restoration of Medieval Stained Glass." *Scientific American* 252 (5):126. Discusses air pollution's effect on building materials and artworks.

■ Graedel, T. E., and P. J. Crutzen. 1989. "The Changing Atmosphere." *Scientific American* 261 (3): 58. Describes how human activities are polluting the atmosphere.

■ Krahl-Urban, B. et al. 1988. *Forest Decline*. U.S. Environmental Protection Agency. Corvallis, OR. A thorough and beautifully illustrated report on the effects of air pollution on forests in North America and Germany. Present data and theories on several mechanisms of forest ecosystem damage.

■ Reilly, W. K. 1991. "The New Clean Air Act: An Environmental Milestone." *EPA Journal* 17 (1):2. The head of the EPA explains the significance of the new act not only as a tool for management, but also as an environmental precedent in a special volume entirely devoted to this issue.

■ Spengler, J., and K. Sexton. 1983. "Indoor Air Pollution: A Public Health Perspective." *Science* 221 (4605):9. What are the dangers of this most widespread form of pollution?

■ Torrens, I. M. 1990. "Developing Clean Coal Technologies." *Environment* 32 (6): 10. Coal combustion is the single largest source of air pollution in North America. Billions of dollars are being spent to find ways to reduce this pollution.

■ Turiel, I. *Indoor Air Quality and Human Health*. Palo Alto, California: Stanford University Press, 1984. More on indoor air pollution.

■ United Nations Environment Programme. 1989. "Monitoring the Global Environment: An assessment of Urban Air Quality." *Environment* 31 (8): 6. The first assessment of world-wide urban air quality since 1974.

CHAPTER *24*

Solid, Toxic, and Hazardous Wastes

Disposing of hazardous waste is expensive and risky. It is better to stop producing waste in the first place.

Joel Hirschhorn, *Cutting Production of Hazardous Wastes*

C O N C E P T S

Enormous amounts of solid wastes are being produced in the industrialized countries, and disposing of those wastes in an environmentally safe manner is an increasing problem. Land that is suitable for waste disposal near major urban centers is becoming more scarce and expensive. Costs for solid waste disposal in the U.S., totalling nearly $10 billion in 1990, are expected to climb to more than $100 billion per year by the end of this century.

Landfill burial, incineration, recycling, reusing, and producing less waste are ways of dealing with solid wastes. Water pollution, smoke from garbage fires, and fostering of pest populations are environmental hazards associated with landfills that may be eliminated through new landfill construction techniques.

Incineration of wastes is being explored by many communities as a way of getting rid of wastes and producing valuable energy. Incineration can be combined with resource recovery, a mechanized process for sorting and recovering useful materials from the waste stream. Incinerators may release unacceptable amounts of toxins into the air, however.

Most of the materials we use can be recycled, either to reproduce the original product or to make new and different products. Recycling saves energy, resources, and money, and lowers pollution emissions. Reusing materials is even more efficient and cost-effective than recycling.

About 60 million metric tons of hazardous and toxic wastes enter the waste stream each year in the United States. More than 90 percent of this material is now disposed of by environmentally unsound practices. Some of the materials that cause the most concern are heavy metals, synthetic organic chemicals (pesticides and solvents), toxic inorganic materials (asbestos and arsenic), and radioactive substances.

In addition to the materials now being produced, it is estimated that some 5 trillion tons of toxic and hazardous wastes were previously disposed of or accidentally released into the environment and now contaminate groundwater, soil, surface waters, and air. How to remove or decontaminate this residue of toxic material is one of the greatest challenges that we face today.

There are several attractive alternatives for handling hazardous wastes. Reducing the production of hazardous wastes, recycling, detoxifying, processing, immobilizing, or storing wastes in secure, permanent storage can reduce the release of these materials into the environment in the future, and may help clean up past mistakes.

INTRODUCTION

On August 31, 1986, the cargo ship *Khian Sea* loaded one month's production of ash (14,000 tons) from the Philadelphia municipal incinerator and set off on an odyssey that symbolizes a predicament we all share. The ship's first port of call in search of a dumping place for its noxious cargo was the poor Caribbean nation of Haiti. Four thousand tons of ash had been carried ashore by the time representatives of the environmental group Greenpeace alerted local residents to the potentially dangerous levels of arsenic, mercury, dioxins, and other toxins in the wastes. Authorities ordered the ship to take its objectionable goods elsewhere.

For twenty-four months this pariah ship wandered from port to port in the Caribbean, across to West Africa, around the Mediterranean, through the Suez Cannal, past India, and over to Singapore looking for a place to dump its toxic materials. Its name was changed from *Khian Sea* to *Felicia* to *Pelacano*. Its registration was transferred from Liberia to the Bahamas to Hondurus in an attempt to hide its true identity, but nobody wanted it or its contents. Like Coleridge's ancient mariner, it seemed cursed to wander the oceans forever. Two years, three names, four continents, and eleven countries later, the onerous load was still aboard.

Then, somewhere on the Indian Ocean between Columbo, Sri Lanka, and Singapore, 14,000 tons of toxic ash disappeared. When questioned about this remarkable occurrence, the crew had no comment except that it was all gone. Everyone assumes, of course, that once the ship was out of sight of land, the ash was dumped into the ocean.

If this were just an isolated incident, perhaps it wouldn't matter too much. However, some 3 million tons of toxic and hazardous waste goes to sea every year. How much ends up in the ocean and how much is deposited in poor countries is unknown.

The problem is that we generate vast amounts of nasty stuff every year (figure 24.1), and places to put it are becoming more and more scarce as the waste is becoming increasingly unpleasant and dangerous. No one wants this waste in their backyards. Wealthy communities send their garbage to poorer ones; rich nations send it to their improverished neighbors. What can we do to stop this process? In this chapter, we will look at the kinds of waste we produce, who makes them, what problems their disposal causes, and how we might reduce our waste production and dispose of our refuse in environmentally safe ways. Notice that our presentation begins with the least desirable historic approaches and proceeds to the more desirable methods to avoid producing wastes in the first place.

SOLID WASTE

Waste is everyone's business. We all produce wastes in nearly everything we do. According to the EPA, the United States pro-

FIGURE 24.1 Bulldozers pack down trash at a municipal landfill near Niagara Falls, Canada. Rainwater percolating through unsealed landfills carries toxic pollutants into the groundwater and contaminates drinking supplies. Landfills are being closed all over the United States, but other forms of waste disposal are not being developed or are equally controversial.

duces 11 billion tons of solid waste each year (figure 24.2). Nearly half of that amount consists of agricultural waste, such as crop residues and animal manure. Most agricultural wastes are recycled into the soil on the farms where they are produced. They represent a valuable resource as ground cover to reduce erosion and fertilizer to nourish new crops, but they also constitute the single largest source of nonpoint air and water pollution in the country. About one-third of all solid wastes are mine tailings, overburden from strip mines, smelter slag, and other residues produced by mining and primary metal processing. Most of this material is stored in or near its source of production and isn't mixed with other kinds of wastes. When properly stored and protected, these wastes have little impact beyond their immediate location. Improper disposal practices, however, can result in serious and widespread pollution (chapter 9).

Industrial waste—other than mining and mineral production—amounts to some 400 million metric tons per year in the United States. Most of this material is recycled, converted to other forms, destroyed, or disposed of in private landfills or deep injection wells (chapter 22). About 60 million metric tons of

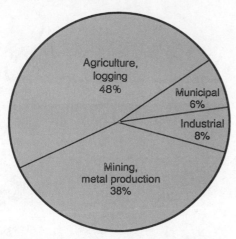

FIGURE 24.2 Sources of solid wastes in the United States. Most mining, logging, and agricultural wastes are disposed in place and never enter the waste stream.

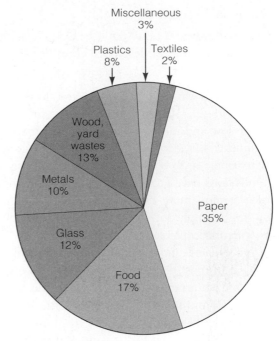

FIGURE 24.3 Composition of domestic waste in the United States, by weight. Plastics make up only 5 percent of the weight but up to 15 percent of the volume of our waste.

industrial waste falls in a special category of hazardous and toxic waste, which we will discuss later in this chapter.

Municipal waste—a combination of household and commercial refuse—amounts to about 300 million metric tons per year in the United States. That's approximately 1.24 tons for each man, woman, and child every year—twice as much per capita as Europe or Japan, and five to ten times as much as most developing countries.

The Waste Stream

Does it surprise you to learn that you generate that much garbage? Think for a moment about how much we discard every year. There are organic materials, such as yard and garden wastes, food wastes, and sewage sludge from treatment plants; junked cars, worn out furniture, and consumer products of all types. Newspapers, magazines, advertisements, and office refuse make paper one of our major wastes (figure 24.3). In spite of recycling programs, a majority of the 200 *billion* metal, glass, and plastic food and beverage containers that we use every year in the United States ends up in the trash. Wood, concrete, bricks, and glass come from construction and demolition sites, as do dust and rubble from landscaping and road building. All of this varied and voluminous waste has to arrive at a final resting place somewhere (figure 24.4).

The **waste stream** is a term that describes the steady flow of varied wastes that we all produce, from domestic garbage and yard wastes to industrial, commercial, and construction refuse. Many of the materials in our waste stream would be valuable resources if they were not mixed with other garbage. Unfortunately, our collecting and dumping processes mix and crush everything together, making separation an expensive and sometimes impossible task. In a dump or incinerator, much of the value of recyclable materials is lost.

Another problem with refuse mixing is that hazardous materials in the waste stream get dispersed through thousands of

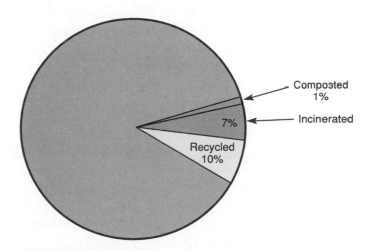

FIGURE 24.4 Fate of municipal (commercial plus domestic) solid waste in the United States.

tons of miscellaneous garbage. This mixing makes the disposal or burning of what might have been rather innocuous stuff a difficult, expensive, and risky business. Spray paint cans, pesticides, batteries (zinc, lead, or mercury), cleaning solvents, smoke detectors containing radioactive material, and plastics that produce dioxins and PCBs when burned are mixed willy-nilly with paper, table scraps, and other nontoxic materials. The best thing to do with household toxic and hazardous materials is to separate them for safe disposal or recycling, as we will see later in this chapter.

Waste Disposal Methods

Where do we put our wastes now? Sanitary secured landfills are replacing open dumps in most developed countries.

Open Dumps

For many people, the way to dispose of waste is to simply drop it someplace (figure 24.5). In primitive cultures, where population levels are low and most wastes are organic material that decomposes quickly, this technique may work relatively well, especially when people are nomadic and don't stay long in one place. As populations have grown larger and become more stationary, however, and more nonbiodegradable material is included in our trash, simply dumping it in the nearest convenient spot has become unacceptable.

Open, unregulated dumps are still the predominant method of waste disposal in most developing countries, however. The giant megacities have an enormous garbage problem. Mexico City, the largest city in the world, generates some 10,000 tons of trash *each day*. Until recently, most of this torrent of waste was left in giant piles, exposed to the wind and rain, as well as rats, flies, and other vermin. Manila, in the Philippines, has at least ten huge open dumps. The most notorious is called Smoky Mountain because of its constant smoldering fires (figure 24.6). Thousands of people live and work on this 30 m (90 ft) high heap of refuse. They spend their days sorting through the garbage for edible or recyclable materials. Health conditions are abysmal, but these people have nowhere else to go. The government would like to close these dumps, but how will the residents be housed and fed? Where else will the city put its garbage?

Most developed countries forbid open dumping, at least in metropolitan areas, but illegal dumping is still a problem. You have undoubtedly seen trash accumulating along roadsides and in vacant, weedy lots in the poorer sections of town. Is this just a question of aesthetics? Consider the problem of waste oil and solvents. It is estimated that 200 million liters of waste motor oil is poured into the sewers or allowed to soak into the ground every year in the United States. No one knows the volume of other solvents disposed of by similar methods.

Increasingly, these toxic chemicals are showing up in the groundwater supplies on which nearly half the people in America depend for drinking (chapter 22). An alarmingly small amount of oil or other solvents can pollute large quantities of drinking or irrigation water. One liter of gasoline, for instance, can make a million liters of water undrinkable. The problem of illegal dumping is likely to become worse as acceptable sites for waste disposal become more scarce and costs for legal dumping escalate. We clearly need better enforcement of antilittering laws as well as a change in our attitudes and behavior.

Landfills

Over the past fifty years most American and European cities have recognized the health and environmental hazards of open dumps. Increasingly, cities have turned to **landfills,** where solid

FIGURE 24.5 For many people, the way to dispose of waste is simply to dump it someplace. This roadside dump spreads disease, encourages vermin, and pollutes the environment with toxic and hazardous materials, as well as being unsightly.

waste disposal is regulated and controlled. To decrease smells and litter and to discourage insect and rodent populations, landfill operators are required to compact the refuse and cover it every day with a layer of dirt (figure 24.7). This method helps control pollution, but the dirt fill also takes up to 20 percent of landfill space. New methods of landfill construction are being developed to control such hazardous substances as oil, chemical compounds, toxic metals, and contaminated rainwater that seeps through piles of waste. An impermeable clay or plastic lining underlies and encloses the storage area. Drainage systems are installed in and around the liner to catch drainage and to help monitor any chemicals that may be leaking.

More careful attention is now paid to the siting of new landfills. Sites located on highly permeable or faulted rock formations are passed over in favor of sites with less leaky geologic foundations. Landfills are being built away from rivers, lakes, floodplains, and aquifer recharge zones rather than near them, as was often done in the past. More care is being given to a landfill's long-term effects so that costly cleanups and rehabilitation can be avoided.

Historically, landfills have been a convenient and relatively inexpensive waste-disposal option in most places, but this situation is changing rapidly. Rising land prices and shipping costs, as well as increasingly demanding landfill construction and maintenance requirements, are making this a more expensive disposal method. The cost of disposing a ton of solid waste in Philadelphia went from $20 in 1980 to more than $100 in 1990. Union

ENVIRONMENTAL POLLUTION AND OTHER DILEMMAS

FIGURE 24.6 Scavengers sort through the trash at "Smoky Mountain," one of the huge metropolitan dumps in Manila, Philippines. Some 20,000 people live and work on these enormous garbage dumps. The health effects are tragic.

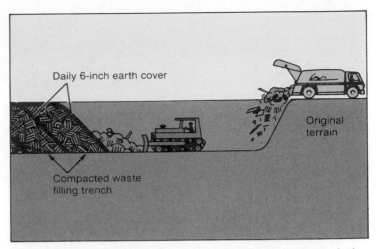

FIGURE 24.7 In a sanitary landfill, trash and garbage are crushed and covered each day to prevent accumulation of vermin and spread of disease. Traditionally, most landfills have not been enclosed in a waterproof lining to prevent leaching of chemicals into underground aquifers.

County, New York, experienced an even steeper price rise. In 1987, it paid $70 to get rid of a ton of waste; a year later, that same ton cost $420, or about $10 for a typical garbage bag. In 1990, the United States spent about $10 billion to dispose of trash. By the year 2000, it may cost us $100 billion per year to dispose of our trash and garbage.

Suitable places for waste disposal are becoming scarce in many areas. Other uses compete for open space. Citizens have become more concerned and vocal about health hazards, as well as aesthetics. It is difficult to find a neighborhood or community willing to accept a new landfill. More than half of all U.S. cities will exhaust their landfills by 1995. Since 1984, when stricter financial and environmental protection requirements for landfills took effect, more than 1,200 of the 1,500 existing landfills in the United States have closed. Many major cities already have no more dumping space. They export their trash, at enormous expense, to neighboring communities and even other states. More than half the solid waste from New Jersey goes out of state, some of it up to 800 km (500 mi) away.

Exporting Waste

As disposal costs rise and restrictions on what can be dumped become more stringent, many European and American cities and industries are sending their wastes abroad to less developed countries as the story of the *Khian Sea* illustrates. In 1988, the environmental organization Greenpeace identified more than fifty plans to send wastes from the United States and Europe to Africa, Latin America, and the Middle East. This is becoming a touchy political issue. Local people usually aren't told what is in the waste being dumped on their land. Provincial officials don't have the resources or the knowledge to test or regulate toxic materials. In the West African nation of Guinea, children were found playing on a huge pile of incinerator ash from Philadelphia. Nigerian officials discovered 8,000 leaking drums of radioactive waste stored in an open yard in the port city of Koko.

In 1985, two Americans were jailed in the United States for exporting 6,000 l (1,500 gal) of toxic waste to Zimbabwe under the guise of cleaning fluid. In 1988, West African countries arrested sixty-two corrupt local officials and foreigners for illegally dumping dangerous materials. In 1989, a treaty regulating international shipping of toxics was signed by 105 nations. This treaty requires that the receiving country give its permission for dumping. This doesn't stop exports completely, but it gives local authorities a chance to find out what is being dumped and to object before it happens.

It's not surprising that exports are attractive. It commonly costs $800 per barrel to dispose of toxic waste in the United States, but some African countries will take that barrel for $50. Toxic trash doesn't only move from north to south. East Germany earned $600 million in the 1970s and 1980s by accepting 5.5 million tons of household rubbish and 800,000 tons of toxic waste from its more prosperous western cousins. Other Eastern

Solid, Toxic, and Hazardous Wastes

TABLE 24.1 Annual hazardous waste trade, selected countries (mid-1980s)		
Country	Imports	Exports
	(Thousands of metric tons)	
Belgium	914.0	13.2
Brazil	40.0	x
Canada	130.0	65.0
France	95.9	25.0
Germany (East)	814.3	x
Germany (West)	75.0	1,695.6
Guinea	15.0	x
Iceland	28.6	x
Mexico	7.0	x
Netherlands	320.0	250.0
Switzerland	7.1	68.0
United States	45.3	203.4

x = no data or negligible

Source: Data from U.N. Global Environmental Monitoring System, 1990.

Bloc countries are waste repositories, while western countries generally are exporters (table 24.1).

Incineration and Resource Recovery

Landfilling is still the disposal method for the majority of municipal waste in the United States (table 24.2) Faced with growing piles of garbage and a lack of available landfills at any price, however, public officials are investigating other disposal methods. The method to which they frequently turn is burning. Another term commonly used for this technology is **energy recovery** or waste-to-energy because the heat derived from incinerated refuse is a useful resource. Burning garbage can produce steam used directly for heating buildings or generating electricity. Internationally, well over 1,000 waste-to-energy plants operate in Brazil, Japan, the Soviet Union, and Western Europe. In the United States, more than 110 waste incinerators burn 45,000 tons of garbage daily. Some of these are simple incinerators; others produce energy.

Incineration Processes

Municipal incinerators are specially designed burning plants capable of burning thousands of tons of waste per day. Two different approaches to garbage incineration are used. In some plants, refuse is sorted as it comes in. Most unburnable or recyclable materials are removed before they go to the combustion chamber. This is called **refuse-derived fuel** because the enriched burnable fraction has a higher energy content than the raw trash. The other approach, called **mass burn,** is to shred everything into small pieces and then burn as much as possible (figure 24.8). This technique avoids the expensive and unpleasant job of sorting through the garbage for nonburnable ma-

TABLE 24.2 Comparison of solid waste management in selected countries (percent of total waste stream)			
	United States	Japan	West Germany
Recycled or Reused	11	50	15
Waste-to-Energy	6	23	30
Landfilled or Other	83	27	55

Source: Data from *The Global Ecology Sourcebook*, 1990.

terials, but it often causes greater problems with air pollution and corrosion of burner grates and chimneys.

In either case, residual ash and unburnable residues representing 10 to 20 percent of the original volume are taken to a landfill for disposal. Because the volume of burned garbage is reduced by 80 to 90 percent, disposal is a smaller task. However, the residual ash usually contains a variety of toxic components that make it an environmental hazard if not disposed of properly. Ironically, one worry about incinerators is whether enough garbage will be available to feed them. Some communities in which recycling has been really successful have had to buy garbage from neighbors to meet contractual obligations to waste-to-energy facilities. In other places, fears that this might happen have discouraged recycling efforts.

Cost and Safety Considerations

The cost-effectiveness of garbage incinerators is the subject of heated debates. Initial construction costs are high—usually between $100 million and $300 million for a typical municipal facility. Tipping fees at an incinerator, the fee charged to haulers for each ton of garbage dumped, are often much higher than those at a landfill. As landfill space near metropolitan areas becomes more scarce and more expensive, however, landfill rates are certain to rise. It may pay in the long run to incinerate refuse so that the lifetime of existing landfills will be extended.

Environmental safety of incinerators is another point of concern. In 1988, the EPA released a report of alarmingly high levels of dioxins, furans, lead, and cadmium in incinerator ash. These toxic materials were more concentrated in the fly ash (lighter, airborne particles capable of penetrating deep into the lungs) than in heavy bottom ash. Dioxin levels can be as high as 780 parts per billion. One part per billion of TCDD, the most toxic dioxin, is considered a health concern. All of the incinerators studied exceeded cadmium standards, and 80 percent exceeded lead standards. Proponents of incineration argue that if they are run properly and equipped with appropriate pollution-control devices, incinerators are safe to the general public. Opponents counter that neither public officials nor pollution control equipment can be trusted to keep the air clean. They argue that recycling and source reduction efforts are better ways to deal with waste problems.

ENVIRONMENTAL POLLUTION AND OTHER DILEMMAS

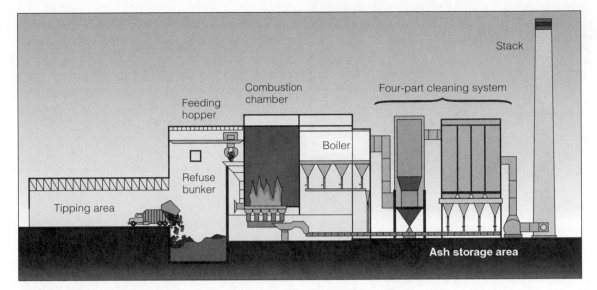

FIGURE 24.8 A diagram of a municipal "mass-burn" garbage incinerator. Steam produced in the boiler can be used to generate electricity or to heat nearby buildings.

The EPA, which has supported incineration in the past, acknowledges the health threat of incinerator emissions but holds that the danger is very slight. The EPA estimates that dioxin emissions from a typical municipal incinerator may cause one death per million people in seventy years of operation. Critics of incineration claim that a more accurate estimate is 250 deaths per million in seventy years.

One way to reduce these dangerous emissions is to remove heavy metal-containing batteries and chlorine-containing plastics before wastes are burned. Bremen, West Germany, is one of several European cities now trying to control dioxin and PCB emissions by keeping all plastics out of incinerator waste. Bremen is requiring households to separate plastics from other garbage. This is expected to eliminate nearly all dioxins and PCBs and prevent the expense of installing costly pollution-control equipment that otherwise would be necessary to keep the burners operating. Minneapolis has initiated a recycling program for the small "button" hearing aid batteries in an attempt to lower mercury emissions from its incinerator.

Recycling

The term "recycling" has two meanings in common usage. Sometimes we say we are *recycling* when we really are *reusing* something, such as refillable beverage containers. In terms of solid waste management, however, **recycling** is the reprocessing of discarded materials into new, useful products. Some recycling processes reuse materials for the same purposes; for instance, old aluminum cans and glass bottles are usually melted and recast into new cans and bottles. Other recycling processes turn old materials into entirely new products. Old tires, for instance, are shredded and turned into rubberized road surfacing. Newspapers become cellulose insulation, kitchen wastes become fuel pellets, and steel cans become automobiles and construction ma-

TABLE 24.3 Benefits from using recycled materials (percent reductions)

Benefit	Aluminum	Steel	Paper	Glass
Reduction of:				
Energy use	90–97%	47–74%	23–74%	4–32%
Air pollution	95	85	74	20
Water pollution	97	76	35	—
Mining wastes	—	97	—	80
Water use	—	40	58	50

Source: *Worldwatch Paper 76*, 1987

terials. The United States generally lags behind Japan and many European countries in recycling (table 24.2).

Benefits of Recycling

Recycling is usually a better alternative to either dumping or burning wastes. It saves money, energy, and resources and reduces pollution (table 24.3). It also encourages individual awareness and responsibility for the refuse produced. Curbside pickup of recyclables costs around $35 per ton, as opposed to the $80 paid to dispose of them at an average metropolitan landfill. Some recycling programs cost nothing; they cover their own expenses with materials sales and may even bring revenue to the community (figure 24.9).

Another benefit of recycling is that it could cut our waste volumes by 50 percent or more, drastically reducing the pressure on disposal systems. Philadelphia is investing in neighborhood collection centers that will recycle 600 tons a day, enough to eliminate the need for a previously planned, high-priced incinerator. New York City, down to one available landfill but still producing 27,000 tons of garbage a day, has set a target of 50 percent waste reduction to be accomplished by recycling office

FIGURE 24.9 Recycling is the cheapest way to dispose of our waste, as well as being the cleanest and most environmentally sound.

paper and household and commercial waste. New York's curbside collection service, projected to be the nation's largest, should more than pay for itself simply in avoided disposal costs.

Japan probably has the most successful recycling program in the world. About half of all household and commercial wastes in Japan are recycled. About one-quarter is incinerated, while some 27 percent is landfilled (table 24.2). By comparison, the United States landfills more than 80 percent of all solid waste. Japanese families diligently separate wastes into as many as seven categories, each picked up on a different day. Would we do the same? Some authors say that Americans are too lazy to recycle. North Stonington, Connecticut, however, faced with escalating disposal costs, reduced its waste volume by two-thirds in just two years.

Recycling also reduces our demand on raw resources. In the United States, we cut down 2 million trees every day to produce newsprint and paper products, a heavy drain on our forests. Recycling the print run of a single Sunday issue of the New York Times would spare 75,000 trees. Every piece of plastic we make reduces the reserve supply of petroleum and makes us more dependent on foreign oil. Recycling 1 ton of aluminum saves 4 tons of bauxite (aluminum ore) and 700 kg (1,540 lb) of petroleum coke and pitch, as well as keeping 35 kg (77 lb) of aluminum fluoride out of the air.

Recycling also reduces energy and air pollution. Plastic bottle recycling could save 50 to 60 percent of the energy needed to make new ones. Making new steel from old scrap offers up to 75 percent energy savings. Producing aluminum from scrap instead of bauxite ore cuts energy use by 95 percent, yet we still throw away more than a million tons of aluminum every year. If aluminum recovery were doubled worldwide, more than a million tons of air pollutants would be eliminated every year.

Reducing litter is an important benefit of recycling. Ever since disposable paper, glass, metal, foam, and plastic packaging began to accompany nearly everything we buy, these discarded wrappings have collected on our roadsides and in our lakes, rivers, and oceans. Without incentives to properly dispose of beverage cans, bottles, and papers, it often seems easier to just toss them aside when we have finished using them. Litter is a costly as well as unsightly problem. We pay an estimated thirty-two cents for each piece of litter picked up by crews along state highways, which adds up to $500 million every year. A deposit on bottles and cans has had a measurable beneficial impact on reducing littering.

Recycling Specific Materials

Nearly all our refuse materials are reusable or recyclable. Paper, aluminum, and glass are most commonly recycled, but rubber, organic waste, and plastics also are valuable material resources. Let's look more closely at some specific advantages and problems of the major categories of recyclable materials.

Paper. Paper is probably our most familiar recyclable material, accounting for about 25 percent of the paper pulp in the United States. Notice that this book is printed on recycled paper. Roundwood (trees cut specifically as pulpwood) makes up only half of our paper supply, with the remaining a quarter coming from sawdust, wood chips, branches, and other logging refuse. Used paper can be reworked into new paper of any weight, but much of it is made into cardboard boxes and packing material. Newsprint and high-quality office paper are easiest to recycle. The clay coating of slick magazine paper makes it difficult to repulp. Wastepaper is also shredded, fluffed, and coated with fire retardant to make cellulose home insulation. Paper can be repulped only a limited number of times before the fibers become so broken and short that the paper weakens. The solution is to mix some new virgin pulp in with recycled material.

Metals. Every year more than 100 *billion* cans of food and drink are sold in the United States. Two-thirds of these are beverage cans, half of which are recycled. Many industries have been recycling scrap metal for years. Auto scrap yards shred junked cars into small pieces and separate different kinds of metal. Copper radiator cores, electrical wiring, chrome bumpers, aluminum engine blocks, and steel body parts are all valuable sources of materials. Photo-finishing factories reclaim silver from developing solutions and plating companies reduce pollution and recover valuable metals from solutions that were once dumped into the sewer. The ash from coal-fired power plants and sewage sludge incinerators also contains small, but valuable, concentrations of such metals as gold, silver, vanadium, gallium, and germanium. The enormous volumes of ash generated each day could become an important source of these valuable minerals.

Glass. Remelting glass uses only 40 to 60 percent as much energy as melting sand to make new glass. Sand is also easier to melt when mixed with old glass. For this reason, it was customary to add 15 or 20 percent of crushed glass to the furnace even before energy prices jumped in the 1970s (chapter 18). Many newer systems are designed to run almost exclusively on crushed glass. In addition, remelted glass cuts air pollution by

ENVIRONMENTAL POLLUTION AND OTHER DILEMMAS

one-fifth, which brings used glass into higher demand in Japan, Europe, and the United States where emissions standards are tightening. The United States is behind other industrialized nations in glass recovery, however. We recycle about 10 percent of our glass, while recovery rates in Northern Europe run between 30 and 50 percent.

Most communities have opportunities for voluntary glass recycling. What can you do to make glass processing easier for your recycler? Remove paper and plastic labels, metal lids, and metal rings from screw-top caps. Sort glass by color. Glass pigments are difficult to remove, so once colored items are added, the whole melt is stained. Since glass is more suitable for recycling than plastic, you probably should buy more glass than plastic packaging.

Rubber. Automobile tire disposal is a serious problem everywhere in the United States. We discard some 200 million tires each year, but they are banned from most landfills because they "float," working their way back to the surface where they trap water and become breeding habitats for mosquitoes and biting gnats. In many places, mountains of discarded tires have accumulated in junkyards and dumps (figure 24.10). These tire mountains are not only unsightly and unsanitary, they are also a fire hazard. The filthy, black, noxious smoke from a burning tire dump can contaminate large areas and is a serious environmental and health hazard.

What else can we do with old tires? Recently, powerful new shredding machines have been developed that can grind up old tires or melt them to extract their rubber. When mixed with asphalt, recycled tire material makes durable, long-lasting surfacing for roads and airport runways. Rubber-asphalt surfaces do not crack quickly like other surfaces, eliminating millions of dollars worth of repaving and patching work.

Organic Wastes. Every year we throw away the energy equivalent of 80 million barrels of oil in organic waste in the United States. In developing countries, up to 85 percent of the waste stream is food, textiles, vegetable matter and other biodegradable materials. Worldwide, at least one-fifth of municipal waste is organic kitchen and garden refuse. Much of this matter is decomposed by microorganisms generating billions of cubic meters of methane ("natural gas"), much of which escapes into the atmosphere where it contributes to the "greenhouse" problem (chapter 17). To use this organic waste instead of burying it, many cities are **co-composting;** that is, they separate organic domestic wastes and compost them with sewage sludge or feedlot manure. Methane released by decomposition in enclosed landfills can be collected and sold to local energy utilities. The organic compost resulting from bacterial decomposition makes a nutrient-rich soil amendment that aids water retention, slows soil erosion, and improves crop yields.

Sixteen billion soiled disposable diapers are discarded in the United States every year. Next to newspapers and beverage

FIGURE 24.10 Tire dumps accumulate millions of used tires. They can be a valuable resource, but they also can be an environmental hazard. Water in tires serves as a breeding habitat for flies and mosquitoes. Fires in a tire dump can contaminate hundreds of square kilometers with noxious black smoke.

containers, diapers are the third biggest source of domestic solid waste. In 1989, the Procter and Gamble Company announced it would fund several projects for composting disposable diapers. After removing the plastic cover, the absorbent padding will be sterilized and composted to make a soil additive. Critics of this process argue that it would be better to use cloth diapers or to develop biodegradable ones.

Most large composting projects are still in Europe. The Netherlands' Waste Treatment Company (figure 24.11) generates 70,000 tons of compost annually for sale to farmers and gardeners, as well as separating and selling 20,000 tons of paper, 3,000 tons of iron, and 6,000 tons of plastics. France has more than 100 composting plants producing 800,000 tons of compost each year. Vineyards using this compost report 14 percent yield increases. European compost is so valuable for food production that it is even being exported to the Arabian Peninsula to build up that region's dry, sandy soil. Composting also can be used on a small scale. Millions of household methane generators provide fuel for cooking and lighting for homes in China and India (chapter 20). In the United States, some farmers use animal manure "digesters" to generate methane for heating.

Plastics. Plastics are tremendously popular because they are relatively lightweight, unbreakable, and rarely attacked by biological or chemical decomposition. Plastics make up an ever-increasing portion of our waste stream; each year we produce over 10 million tons of plastic, little of which is reused or recycled. The problem with recycling plastics lies in the wide variety of plastics we use (table 24.4). Hundreds of different types of plastics are used every day. Many are chemically incompatible so they can't simply be melted together to produce a material of any strength or structural reliability.

There are two broad categories of plastics with very different recycling potential. **Thermoplastics** are single-chain, unlinked polymers, such as polyethylene, polypropylene, polyvinylchloride, polystyrene, and polyester. These materials

FIGURE 24.11 The world's largest compost pile, the Netherland's waste treatment facility in Wijister, receives about one million tons of refuse per year. After microbial degradation of organic material and removal of glass and metal, about 100,000 tons of compost are sold for farm and garden use.

Type	Percent of weight	Uses
Low-density polyethylene	24	Squeezable bottles, dry cleaning, grocery and garbage bags, paper coating, detergent bottles
High-density polyethylene	19	Six-pack rings, milk jugs, ice-cube trays, garbage cans, laundry baskets, base cups for solf-drink bottles
Polystyrene	14.5	Foam cups, plates, plastic utensils, hamburger containers, disposable razors, battery cases, refrigerator parts, radio and T.V. cabinets
Polyvinyl chloride	6.5	Plastic wrap, shrink wrap, clear containers, plastic pipe, phonograph records, toys, upholstery, shower curtains
Polyethylene terephthalate (PET)	5	Solf-drink bottles
Polyurethane	5	Foam insulation, furniture
Polypropylene	14.5	Valves, packaging films, wire insulation, soft tubing, bottles for hot-filled food such as syrup
Other	11.5	

TABLE 24.4 Plastics in municipal waste

generally make soft plastics (squeezable bottles, foam cups, shower curtains, etc.) or plastic sheets and films. If separated into specific types, these polymers can be remelted and reformed into useful products. Polyethylene terephthalate (PET), for example, from soft-drink bottles is recycled into fiberfill for pillows, ski jackets, and sleeping bags. Even when mixed together, thermoplastics can be granulated and used to make fence posts, drain gutters, paving blocks, and cargo skids. Health regulations prohibit making new food or beverage containers from recycled plastic because sterilization cannot be guaranteed at the low temperature at which they melt. In response to both public opinion against plastics and a growing market for recycled stock, several of the largest manufacturers have recently instituted recycling programs for polystyrene. It is thought that we might supply about half of our annual demand for plastic with recycled material.

The other type of plastic is **thermoset polymers.** This category includes cross-linked molecular networks making hard plastics that cannot be remelted. Some examples are urea polymers, acrylics, phenolics, and epoxies. These materials are found in such items as fiberglass, hard casings, mechanical parts, and plexiglass. They are neither easily burnable nor recyclable. Landfilling is about the only way to dispose of them.

Encouraging Recycling

Restructuring fee systems for waste collection can offer a direct financial incentive for people to recycle. Garbage haulers usually charge a flat fee for collection from each household, regardless of the volume collected. A volume-based fee system, on the other hand, charges less to those who generate less waste. When Seattle, Washington, adopted a volume-based fee system, recycling doubled.

Many cities offer free curbside pickup of recyclables to encourage public participation in recycling programs. More than five hundred curbside pickup programs now operate in the United States, with compartmentalized trucks circulating once or twice a month. Some programs provide residents with attractive, uniform, color-coded containers in which to put their various recyclables (figure 24.12). Program operators find that this

FIGURE 24.12 Source separation in the kitchen—the first step in a strong recycling program.

encourages participation, either because people feel good about using attractive receptacles, or because of peer pressure. When recycling day rolls around, nobody wants to be the only one on the block whose recycling cans are missing from the curb!

Fee systems and taxes could be used to stabilize prices and create markets for recycled materials. The demand for the most easily recycled materials—paper, glass, scrap metal—are notoriously volatile. When recycling programs are successful, they tend to flood the market and drive down prices. This instability is deadly for marginally capitalized programs. Guaranteed prices and policies that encourage use of recycled material would do a great deal toward making recycling successful.

Our present public policies tend to favor extraction of new raw materials. Energy, water, and raw materials are often sold to industries below their real cost to create jobs and stimulate the economy. Setting the prices of natural resources at their real cost would tend to encourage efficiency and recycling and would help create a market for used materials.

Public education is an important tool in waste management. For many people, out of sight is out of mind with their garbage. They have no idea about where their wastes go or what problems they cause. If people understood the real costs of waste disposal and what the options are for reusing, recycling, and reducing our production of waste, they would probably participate. Public education may well be one of the most powerful tools in reducing our problems of waste disposal. Perhaps we should have guided tours of landfills and incinerators to make people aware of the magnitude of our problems.

Shrinking the Waste Stream

An even better resource and money saver than recycling is to clean and reuse materials in their present form, saving the energy of remelting and shaping. We do this already with some specialized items. Auto parts are regularly sold from junkyards, especially for older car models. In some areas, stained glass windows, brass fittings, fine woodwork, and bricks salvaged from old houses bring high prices. Some communities sort and reuse a variety of materials received in their dumps (figure 24.13).

In many cities, glass and plastic bottles are routinely returned to beverage producers for washing and refilling. The reusable, refillable bottle is the most efficient beverage container we have. It is much more efficient than remelting and more profitable for local communities. A reusable glass container makes an average of fifteen round-trips between factory and customer before it becomes so scratched and chipped that it has to be recycled. Reusable containers also favor local bottling companies and help preserve regional differences. Since the advent of cheap, lightweight, disposable food and beverage containers, many small, local breweries, canneries, and bottling companies have been forced out of business by huge, national conglomerates. These big companies can afford to ship food and beverages great distances as long as it is a one-way trip. If they had to collect their containers and reuse them, canning and bottling factories serving large regions would be uneconomical. Consequently, the national companies favor recycling rather than refilling because they prefer fewer, larger plants and don't want to be responsible for collecting and reusing containers.

FIGURE 24.13 Reusing discarded products is a creative and efficient way to reduce wastes. This dump in San Francisco has become a recycling center, a valuable source of secondary materials and a money saver for the whole community.

In less affluent nations, reuse of all sorts of manufactured goods is an established tradition. Where most manufactured products are expensive and labor is cheap, it pays to salvage, clean, and repair products. Cairo, Manila, Mexico City, and many other cities have large populations of poor people who make a living by scavenging. Entire ethnic populations may survive on scavenging, sorting, and reprocessing scraps from city dumps.

Producing Less Waste

What is even better than reusing materials? Generating less waste in the first place. Table 24.5 lists some contributions you can make to reducing the volume of our waste stream.

Industry can play an important role in source reduction. The 3M Company saved over $500 million since 1975 by changing manufacturing processes, finding uses for waste products, and listening to employees' suggestions. What is one division's waste is a treasure to another.

One of our greatest sources of unnecessary waste is in the excess packaging of food and consumer products. Paper, plastic, glass, and metal packaging material make up 50 percent of our domestic trash by volume. Much of that packaging is primarily for marketing and has little to do with product protection. Manufacturers and retailers could do a great deal to reduce these wasteful practices. Consumers can choose products without excess packaging. Manufacturers can avoid unnecessary wrappings. Communities also can have an impact. Berkeley, California; Portland, Oregon; Minneapolis and St. Paul, Minnesota; and about 25 other cities have passed ordinances requiring that fast-food restaurants package food in paper or other biodegradable wrappings, both to reduce litter and to protect the atmosphere. In 1990 responding to nation-wide pressure, Burger King and McDonald's announced they would stop using plastic foam hamburger boxes everywhere.

Where disposable packaging is necessary, we still can reduce the volume of waste in our landfills by using biodegradable materials. Usually this means no plastics. Recently, however, plastics have become available that do break down in the environment under ideal circumstances. **Photodegradable plastics** break down in ultraviolet radiation. **Biodegradable plastics** incorporate such materials as cornstarch that can be decomposed by microorganisms. Several states have introduced legislation requiring biodegradable or photodegradable six-pack beverage yokes, fast-food packaging, and disposable diapers. Critics claim that these degradable plastics don't decompose completely but only break down to small particles. In doing so, they can release toxic chemicals into the environment. Furthermore, they make recycling less feasable and may cause people to believe that littering is acceptable.

HAZARDOUS AND TOXIC WASTES

The most dangerous aspect of the waste stream we have described is that it often contains highly toxic and hazardous materials that are injurious to both human health and environmental quality. We now produce and use a vast array of flammable, explosive, caustic, acidic, and highly toxic chemical substances for industrial, agricultural, and domestic purposes (figure 24.14). According to the EPA, industries in the United States generate about 265 million metric tons of *officially* classified hazardous wastes each year, slightly more than 1 ton for each person in the country. In addition, considerably more toxic and hazardous waste material is generated by industries or processes not regulated by the EPA. Shockingly, at least 40 million metric tons (22 billion lbs) of toxic and hazardous wastes are released into the air, water, and land in the United States each year.

ENVIRONMENTAL POLLUTION AND OTHER DILEMMAS

FIGURE 24.14 The Office of Management and Budget estimates that there may be 100,000 abandoned toxic and hazardous waste dumps in the United States. Rusted, leaking drums in old hazardous waste landfills and abandoned dumps seep wastes that evaporate into the air or percolate through the soil to groundwater. Cleanup after the wastes escape and mix together in the environment is often hundreds of times more difficult and expensive than proper disposal.

What Is Hazardous Waste?

Legally, a **hazardous waste** is any discarded material, liquid or solid, that contains substances known to be (1) fatal to humans or laboratory animals in low doses, (2) toxic, carcinogenic, mutagenic, or teratogenic to humans or other life-forms, (3) ignitable with a flash point less than 60°C, (4) corrosive, or (5) explosive or highly reactive (undergoes violent chemical reactions either by itself or when mixed with other materials). Notice that this definition includes both toxic and hazardous materials as defined in chapter 21. Certain compounds are exempt from classification as hazardous waste if they are accumulated in less than 1 kg (2.2 lb) of commercial chemicals or 100 kg of contaminated soil, water, or debris. Even larger amounts (up to 1,000 kg) are exempt when stored at an approved waste treatment facility for the purpose of being beneficially used, recycled, reclaimed, detoxified, or destroyed. Table 24.6 shows some of the risks associated with hazardous wastes.

Although most households don't accumulate large quantities of hazardous wastes, we all have materials in our homes that are both toxic and hazardous. Many solvents, cleaners, pesticides, fuels, and other everyday household products are just as dangerous as their industrial counterparts. When we discard these materials through the sewer system, they combine with the wastes of our neighbors to make truly significant quantities. Figure 24.15 shows some common household hazardous wastes and table 24.7 suggests some ways to manage them.

As of 1985, the National Institute of Occupational Health and Safety listed 99,585 different kinds of chemical and physical substances as potentially toxic or hazardous, some of which are common in our daily lives (table 24.8). However, since only about 20 percent of the more than 700,000 different chemicals in commercial use have been subjected to thorough toxicity testing

TABLE 24.6 Economic and environmental risks of hazardous wastes

1. **Groundwater/Water supplies.** Contamination of water supplies is probably the greatest threat of hazardous and toxic chemicals. We generally think of "spring water" as being synonymous with "pure," but this is not the case in much of the United States.
2. **Soil contamination.** While most landfills and industrial sites have some degree of contamination, there are some areas in which persistent, toxic wastes have percolated hundreds of feet into the soil, making it unfit for any use, threatening groundwater, and creating a massive cleanup problem.
3. **Habitat destruction.** There have been incidents in which major ecosystems, such as lakes, estuaries, and whole river drainages have been contaminated and made unsuitable for indigenous species. This can mean extinction for some rare and endangered species.
4. **Human health.** In chapter 8, we reviewed some of the health effects related to exposure to hazardous and toxic materials. We find traces of persistent compounds, such as DDT and PCBs, in the most remote parts of the world. All people in the world are probably affected, to some extent, by human-generated radioactive materials and toxic chemicals.
5. **Fires, explosions, property damage.** The cost of corrosion repair (repainting, replacing, refinishing) has been estimated to be about $70 billion dollars per year in the United States. Much of this corrosion has natural origins, but the release of caustics, acids, solvents, oxidizers, and highly reactive compounds adds to these losses.

and at least one-third of all known chemicals have never been tested at all for toxicity, this list would undoubtedly be much larger if the hazards of everything we use were thoroughly understood.

Who Makes Hazardous Waste?

As figure 24.16 shows, by far the greatest volume of hazardous waste is produced by the chemical and petroleum industries. Together they account for 71 percent of all our hazardous wastes. Within this category, some companies stand out as especially notorious generators of toxic substances. The companies that make chlorinated hydrocarbons for pesticides and other products have probably contributed more toxic materials to our environment than any other group in the United States. The metal mining, smelting, forming, and plating industries account for some 22 percent of all hazardous wastes. All other businesses and factories together make up the remaining seven percent.

Hazardous Waste Disposal

Most hazardous waste is recycled, converted to nonhazardous forms, stored, or otherwise disposed of on site by the generators—chemical companies, petroleum refiners, and other large industrial facilities—so that it doesn't become a public problem. The 60 million tons per year that does enter the waste stream or the environment represent one of our most serious environ-

WHERE ARE HAZARDOUS WASTES IN MY HOME?

Home checklist

Where to look
- Basement
- Kitchen
- Garage
- Utility room
- Storage shed
- Laundry room
- Bathroom

Hobby and health care products
- Artist's paint and inks
- Waterproofers
- Photographic chemicals
- Glues and cements

Cleaning products
- Drain, toilet, and window cleaners
- Disinfectants
- Septic tank cleaners
- Bleach and ammonia
- Cleaning solvents and spot removers
- Oven cleaners

Automotive products
- Antifreeze
- Solvents
- Battery acid
- Gasoline
- Rust inhibitor, remover
- Used motor oil
- Brake and transmission fluid

Paint/building products
- Paint thinners, strippers and solvents
- Spray cans
- Lacquers, stains and varnishes
- Wood preservatives
- Acids for etching
- Asphalt and roof tar
- Latex and oil-based paints

Gardening/pest control products
- Sprays and dusts
- Ant and rodent killers
- Flea powder
- Weed killers
- Banned pesticides

FIGURE 24.15 How many different kinds of hazardous materials can you find in your home?

mental problems. Much of this material can be a serious threat to both environmental quality and human health.

Finding ways to safely ship, handle, and dispose of this enormous volume of dangerous material is a great challenge. For years, little attention was paid to this material. Wastes stored on private property, buried, or allowed to soak into the ground were considered of little concern to the public. An estimated 5 billion metric tons of highly poisonous chemicals were improperly disposed of in the United States between 1950 and 1975 before regulatory controls became more stringent.

Old Waste Sites

The EPA estimates that there are at least 26,000 abandoned hazardous waste disposal sites in the United States. The General Accounting Office (GAO) places the number much higher, perhaps 425,000 sites when all are identified. By 1990, some 1,226 sites had been placed on the National Priority List (NPL) for cleanup with financing from the 1980 federal Superfund Act (figure 24.17). The EPA expects that at least 2,500 sites eventually will be included on the NPL and that the total cost for cleanup will be between $8 billion and $16 billion. The GAO estimates the cost to be as high as $100 billion.

TABLE 24.7 Guidelines for disposal of hazardous wastes in the home

Do:

1. Use up any product that you buy, or give it to someone who can. Make sure that any product you give away is in its original container with label intact and any use and disposal instructions included. Give leftover paint to a local community or theater group; donate leftover pesticides to a local garden club, church, or neighborhood center. Small amounts of paint can be left to dry in the can.

2. Take used or contaminated motor oil, transmission fluid, kerosene, and diesel fuel to an automotive service center, oil recycling station, or authorized collection site.

3. Turn in your old car battery; do not take it to the dump.

4. Dispose of a container once it is empty. Follow label instructions about rinsing the container or wrapping it in newspaper. Before disposing of cans, always wrap in newspaper those that contained substances that could harm the skin.

5. Empty all aerosol cans by depressing the button until no more product comes out before wrapping the container in newspaper and disposing of it with the trash. Never throw empty aerosol containers into an incinerator or trash compactor.

6. Follow *all* label directions.

7. Call your local environmental or public health agency with questions about any material you think may pose a disposal problem.

8. Contact your local government agency to find out what kind of disposal systems your community has and whether there are any materials that should not go through normal municipal disposal.

Do Not Do:

1. Do not dispose of liquid chemicals, banned pesticides, batteries, or motor oil in the trash.

2. Do not dispose of any liquid chemicals by pouring on the ground or into a storm sewer.

3. Do not bury any chemical containers, full or empty, in your yard.

4. Do not burn containers of leftover chemicals.

5. Never reuse pesticide or chemical containers for other purposes. Residues remain in the container and will contaminate other materials subsequently placed in the container.

6. Avoid using aerosol spray cans. Purchase pump dispensers instead.

7. Do not mix chemical wastes.

8. Do not dispose of any chemicals along roadsides.

Source: Minnesota Pollution Control Agency, 1987.

What qualifies a site for placement on the NPL? These sites are considered to be especially hazardous to human health and environmental quality because they are known to be leaking or have a potential for leaking supertoxic, carcinogenic, teratogenic, or mutagenic materials (chapter 21). So far, 444 toxic pollutants have been identified at these sites. The ten most commonly found substances are lead, trichloroethylene, toluene, benzene, PCBs, chloroform, phenol, arsenic, cadmium, and chromium. These or other hazardous materials are known to have contaminated groundwater at 75 percent of the sites now

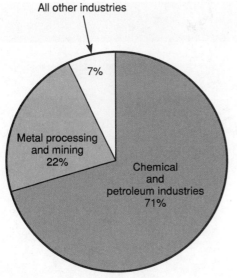

All other industries 7%

Metal processing and mining 22%

Chemical and petroleum industries 71%

FIGURE 24.16 Producers of hazardous wastes.

TABLE 24.8 Hazardous materials in products we use

Product	Potentially hazardous wastes
Plastics	Organic chlorine compounds
Pesticides	Organic chlorides and phosphates
Medicines	Solvents, residues, heavy metals
Paints	Heavy metals, pigments, solvents
Oil, gasoline	Phenols, benzene, lead, solvents
Metals	Fluorides, cyanides, plating salts
Leather	Chromium, aldehydes, solvents
Textiles	Metals, dyes, solvents, fibers
Insulation	Asbestos, urethane foams, fiberglass
Paper	Hydrogen sulfide, mercaptans, mercury, coatings, fibers

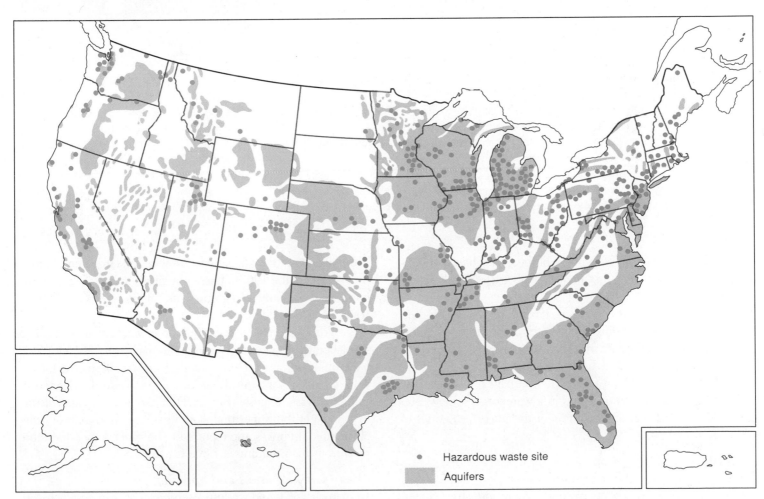

• Hazardous waste site

Aquifers

FIGURE 24.17 Some of the hazardous waste sites on the EPA priority cleanup list. Sites located on aquifer recharge zones represent an especially serious threat. Once groundwater is contaminated, cleanup is difficult and expensive. In some cases, it may not be possible.

Source: Data from the Environmental Protection Agency.

Solid, Toxic, and Hazardous Wastes

BOX 24.1   The Forgotten Wastes of Love Canal

Love Canal is a 16–acre landfill in a residential neighborhood of Niagara Falls, New York. It was intended to be the center of a budding nineteenth-century industrial empire, but the canal never was finished and finally became a dump for industrial waste. This was the beginning of a story that came back decades later to haunt the people of Niagara Falls. It cost them their homes, their health, and hundreds of millions of dollars in rescue efforts.

Early in the century, Love Canal stood empty. It was used mostly as a local swimming hole. In 1942, the city of Niagara Falls began dumping garbage there. The Hooker Chemical and Plastics Corporation, a local industry that had been in the neighborhood since the early days of the canal, also dumped chemical wastes into it. Few people lived in the area, and there was little opposition to the dumping. In April 1945, a Hooker engineer wrote in an internal memo that Love Canal was a "Guagmire which will be a potential source of lawsuits." A year later, the company purchased the canal and turned it into a large-scale industrial landfill. Over the next six years, more than 20,000 metric tons of chemical wastes, including

BOX FIGURE 24.1   Love Canal, Niagara Falls, N.Y. A housing development and a school were built on top of this unsealed hazardous waste dump. Residents' health problems forced the state to purchase and destroy homes and played a catalytic role in passage of the Superfund Act.

highly toxic pesticides, herbicides, and other chemicals—some of them contaminated with dioxins—were dumped into the canal.

In the 1950s, the city of Niagara Falls was growing rapidly and had surrounded the canal. Neighbors complained of foul

odors and rats in the dump. It was thought that a solution to both land shortage and pollution problems would be to fill in the site and develop it for housing. In 1953, Hooker sold the site to the Board of Education and the City of Niagara for $1.00 on the condition that the company be re-

---

on the NPL. In addition, 56 percent of these sites have contaminated surface waters, and airborne materials are found at 20 percent of the sites.

Where are these thousands of hazardous waste sites, and how did they get contaminated? Wastes come from countless sources and are disposed of in just as many ways. Chances are that one or more are near where you live. Some are large and well known; others are long forgotten—perhaps only a few barrels buried in the ground or stored in a warehouse somewhere. In this section, we will review a few of the major categories of waste sites.

Landfills

The most common method of waste disposal, for hazardous as well as solid waste, has been landfills. Many industries buried metal drums full of a variety of toxic chemicals on their own or leased property (figure 24.14). Sometimes the volumes were truly staggering. The Velsicol Chemical Company, for instance, buried at least 250,000 barrels containing about 60 million l

(16 million gal) of toxic pesticide residue in Hardeman County, Tennessee, in the 1960s. These chemicals have leaked from the barrels and contaminated groundwater aquifers used for drinking water by local residents.

Toxic landfills are frequently in or near urban areas. The revelation that homes and a school in Niagara Falls, New York, were built over a waste dump containing 20,000 metric tons of toxic chemical waste awakened many people to the dangers of improper waste disposal (Box 24.1). It also played a catalytic role in passage of the Superfund Act. There are thousands of other abandoned dumps, some reported to be much larger and more dangerous than the one at Love Canal.

Waste Lagoons and Land Spreading

In many areas, liquid wastes weren't even put in barrels before they were dumped; they were simply pumped into lagoons and holding ponds or sprayed on fields where they could evaporate and/or soak into the ground. The Stringfellow Acid Pit in California is a notorious example. Between 1956 and 1972, more

leased from any liability for injury or damage caused by the dump's contents. Homes were built on the land adjacent to the canal, and in 1954, a school and playground were built on top of the chemical dump itself (Box figure 24.1).

Much of the abandoned canal was a swampy, weedy gully with pools of stagnant, oily water. Children played in this wasteland, poking sticks in the black sludge that accumulated on the water and throwing rocks at the drums that floated to the surface. Parents complained that their children were burned by chemicals in the canal; dogs that roamed there developed skin diseases, and their hair fell out in clumps. Clearly, something was wrong.

In 1977, an engineering firm was hired to inspect the site and determine why basements in the area were filled with dark, smelly seepage after every rain. They discovered that the groundwater was contaminated with a variety of toxic organic chemicals. Several mothers, concerned about the health of their children, circulated a petition to close the school and adjacent playing fields. As they went from door to door, they became aware that many families had children with birth defects or chronic medical problems, such as asthma, bronchitis, continuing infections, and hyperactivity. There seemed to be an unusually high rate of miscarriages and stillbirths in the area as well. These informal surveys were dismissed by authorities as "housewife research," but on August 2, 1978, New York State ordered the emergency evacuation of 240 families living within two blocks of the canal.

Those people whose houses were not purchased by the State watched with mixed feelings as their neighbors departed. Suppose the house across the street from you had been condemned but you were just outside the quarantine area and had to stay. How safe would you feel? Residents traced old streambeds that crossed the canal and showed that chemical residues came up in wet areas, sometimes blocks from the dump site. Disputes, public rallies, lawsuits, and negotiations continued during the next six years.

In 1988, an agreement was reached between Occidental Petroleum (the parent company of Hooker Chemical and Plastics) and 1,345 members of the Love Canal Homeowners Association to pay some $250 million in damages. Whether this is an adequate compensation for possible future effects is not clear. We don't really know the dangers of long-term exposure to low levels of extremely toxic materials.

In 1990, the Love Canal Revitalization Agency began selling 236 abandoned houses adjacent to the dumpsite at bargain prices. Prospective buyers, many of whom otherwise could not afford a home, lined up eagerly. The government claims that pollution levels have been reduced enough to make the houses safe. Critics argue that the area is still dangerous and that people should not be allowed to live there. What do you think? Would you move into one of those houses? Should others be allowed to do so?

Love Canal has become a symbol of the dangers and uncertainties of toxic industrial chemicals in the environment. The tragedy is that there probably are many Love Canals, some much worse than the original one. No one knows what the total cost of our carelessness in disposing of these chemical wastes ultimately may be.

than 120 million l (32 million gal) of toxic chemicals were poured into shallow ponds at this site in Riverside County on the eastern edge of Los Angeles. A chemical plume is now creeping through the aquifer that supplies drinking water to 500,000 people in the Los Angeles Basin.

Unsafe Waste Storage

Large quantities of hazardous wastes have simply been stacked in old warehouses and abandoned. The high concentrations of flammable materials and highly poisonous chemicals at some of these sites create a lethal combination. On April 21, 1980, a fire broke out in an old warehouse at an inactive waste treatment facility in Elizabeth, New Jersey. More than 20,000 leaking and corroded drums containing pesticides, explosives, radioactive wastes, acids, and other hazardous substances were piled in and around the building (figure 24.18). A cloud of toxic gas drifted over heavily populated areas close to the site. Contaminated water from fighting the fire ran off into the Elizabeth River.

The site had been licensed as an incinerator facility for hazardous waste, but, in fact, the operators—some of whom have been linked with organized crime in New York and New Jersey—were simply collecting disposal fees and stacking the barrels in warehouses and on parking lots. A former employee of the company was quoted as saying: "There's millions to be made in this business. The overhead is nothing. You find a place—don't buy it, don't buy anything; just rent it and fill it up. When they catch up to you, just declare bankruptcy and get out."

Illegal Dumping

The EPA estimates that as much as 90 percent of the hazardous wastes handled by private contractors is disposed of by unapproved methods. So-called "midnight dumpers" use a variety of illegal and highly dangerous methods to get rid of unwanted toxic wastes. Sometimes liquid wastes are emptied at night into a storm sewer or river. Some drivers pick up a tanker full of wastes and

FIGURE 24.18 The morning after. Twenty thousand burned hazardous waste drums remain after a midnight fire at their "storage" site. Much of the drums' contents has washed into the Elizabeth River. Notice New York City across the river in the distance.

simply drive down a deserted country road with the discharge valve open. "About 60 miles is all it takes to get rid of a load," boasted one driver, "and the only way I can get caught is if the windshield wipers or the tires of the car behind me start melting."

Options for Hazardous Waste Management

We already have seen some notorious examples of how hazardous wastes have been managed—or mismanaged. Unfortunately, unsafe and unacceptable disposal methods have been the rule rather than the exception. Figure 24.19 shows estimates from the Congressional Budget Office of how hazardous wastes are handled once they leave the site where they were generated. About 25 percent of our wastes, including radioactive wastes, are injected into deep wells or stored in salt mines. Landfills account for about 23 percent of all hazardous wastes; discharge into sewers, rivers, lakes, and streams accounts for another 22 percent. Warehouses, wastepits, and surface lagoons provide temporary storage for another 19 percent. Only 11 percent of our waste is incinerated, detoxified, recycled, or recovered, and

Technique	Estimated costs/ metric ton
Land spreading	$2 – 25
Chemical processing	$5 – 500
Surface impoundment	$14 – 180
Secure landfill	$50 – 400
Incineration	$75 – 200
Physical and biological treatments	Variable

TABLE 24.9 Costs of hazardous waste disposal

most of the incineration is done in ways that do not completely destroy toxins.

We have three major options for controlling and managing hazardous wastes: (1) produce less waste, (2) convert wastes to less hazardous or nonhazardous substances, or (3) put them in safe, permanent storage. The costs of disposal methods vary widely, as you can see in table 24.9. Let's look more closely at some of the most promising options for hazardous waste management.

Produce Less Waste

As with other wastes, the safest and least expensive way to avoid hazardous waste problems is to avoid creating the wastes in the first place. Manufacturing processes can be modified to reduce or eliminate waste production. The Minnesota Mining and Manufacturing Company (3M) of St. Paul, Minnesota, reformulated products and redesigned manufacturing processes between 1975 and 1985 to eliminate more than 140,000 metric tons of solid and hazardous wastes, 4 billion l (1 billion gal) of wastewater, and 80,000 metric tons of air pollution each year. They frequently found that these new processes not only spared the environment, but also saved money by using less energy and fewer raw materials.

Recycling and reuse of materials also eliminates hazardous wastes and pollution. Many waste products of one process or industry are valuable commodities in another. In 1987, about 10 percent of the wastes that would otherwise have entered the waste stream in the United States were sent to regional waste exchanges where they were sold as raw materials for recycling or reuse by other industries. This figure could probably be raised substantially with better waste management. In Europe, at least one-third of all industrial wastes are exchanged in clearinghouses where beneficial uses are found. This represents a double savings: the generator doesn't have to pay for disposal, and the consumer pays little, if anything, for raw materials.

Convert to Less Hazardous Substances

Several processes are available to make hazardous materials less toxic. *Physical treatments* tie up or isolate substances. Charcoal or resin filters are used to absorb toxins. Distillation is used to separate hazardous components from aqueous solutions. Precipitation and immobilization in ceramics, glass, or cement isolate

ENVIRONMENTAL POLLUTION AND OTHER DILEMMAS

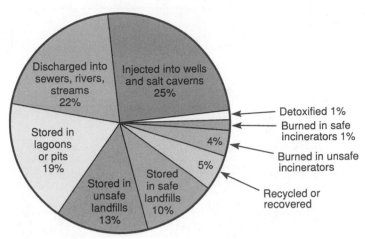

FIGURE 24.19 Fate of hazardous wastes in the United States. Estimated percentages of volumes of waste disposal in various legal and illegal manners.

Source: Data from the Congressional Budget Office, 1989.

toxins from the environment so that they become essentially nonhazardous. One of the few ways to dispose of metals and radioactive substances is to fuse them in silica at high temperatures to make a stable, impermeable glass that is suitable for long-term storage.

Chemical processing can restructure materials so they become nontoxic. Included in this category are neutralization, ion exchange to remove metals or halogens (chlorine, bromine, etc.), oxidation (removal of electrons), and reduction (addition of electrons). These are essentially the reactions carried out in our cells (chapter 21) to detoxify the waste products of metabolism or to protect us from the effects of environmental toxins. Industrial processes employ these reactions on a large scale to take care of massive quantities of pollutants. The Sunohio Corporation of Canton, Ohio, for instance, has developed a process called PCBx in which chlorine in such molecules as PCBs is replaced with other ions that render the compounds less toxic. A portable unit can be moved to the location of the hazardous wastes, eliminating the need for shipping them.

Biological waste treatment taps the great capacity of biological organisms to absorb, accumulate, and detoxify a variety of toxic compounds. Bacteria in activated sludge ponds, aquatic plants (such as water hyacinths or cattails), soil microorganisms, and other species remove toxic materials and purify effluents. If carefully managed, land farming (spreading toxic wastes on aerated soil), waste lagoons, and spray ponds can safely and effectively dispose of wastes. The Koch Refining Company of Rosemount, Minnesota, operates a thirty–acre "farm" to dispose of refinery sludge and petroleum-based solvents. These biodegradable chemicals are mixed into the soil in fenced and diked plots where microbes degrade and detoxify them. Tractors with discs and harrows till the soil and aerate it to speed the purification process. Rehabilitation of the wastes generally occurs within one season.

Biotechnology offers exciting possibilities for finding or creating organisms to eliminate specific kinds of hazardous or toxic wastes. PCBs, for instance, are usually highly resistant to microbial degradation both because they are toxic and because the chlorine atoms attached to their hydrocarbon rings make them impervious to attack by most enzymes. By using a combination of classic genetic selection techniques and high-technology gene-transfer techniques, however, scientists have recently been able to generate bacterial strains that are highly successful at metabolizing PCBs.

Bioengineered organisms could be used in two ways. The cheapest method would be simply to inoculate a culture of the appropriate bacterial strain into contaminated soil or water and let them destroy the pollutants in place. Many people worry about the ecological effects of releasing such exotic organisms into the environment, for fear they might become a worse problem than the one they were intended to solve. A more expensive, but probably safer method would be to take the degrading organisms to the site in a portable reaction vessel in which they would remain confined while they did their job. Contaminated soil or water would be fed through the reaction vessel, then sterilized to kill the bacteria before being returned to the environment. When the site was cleaned, the reaction vessel and the organisms would be removed and taken to the next job.

Incineration is really a physical method of processing wastes to a less hazardous form, but it is such an important one that it deserves separate discussion here. Incineration has several advantages. It is applicable to mixtures of wastes. It is a permanent solution to many problems. It is quick and relatively easy but not necessarily cheap—nor always clean—unless it is done correctly. Wastes must be heated to over 1,000°C (2,000°F) for a sufficient period of time to complete destruction. The ash resulting from thorough incineration is reduced in volume by at least 90 percent and is generally safer to store in a landfill or other disposal site than the original wastes.

Several sophisticated features of modern incinerators improve their effectiveness. Liquid injection nozzles atomize liquids and mix air into the wastes so they burn thoroughly. Fluidized bed burners pump air from the bottom up through burning solid waste as it travels on a metal chain grate through the furnace. The air velocity is sufficient to keep the burning waste partially suspended. Plenty of oxygen is available, and burning is quick and complete. Afterburners add to the completeness of burning by igniting gaseous hydrocarbons not consumed in the incinerator. Scrubbers and precipitators remove minerals, particulates, and other pollutants from the stack gases.

Store Permanently

Retrievable Storage. One of the main sources of hazardous waste problems is that we try to throw them away so we won't have to take care of them anymore. When we dump our wastes in the ocean or bury them in the ground, however, we have lost control of them. If we learn later that our disposal techniques were faulty, it is impossible to go back and do it over. For many supertoxic materials, the best way to store them may be in **permanent retrievable storage.** This means placing storage containers in a secure building, salt mine, or bedrock cavern where they can be inspected periodically and retrieved, if necessary, for repacking or for transfer if a better means of disposal is developed. This technique is more expensive than burial in a landfill because the storage area must be guarded and monitored continuously to prevent leakage, vandalism, or other dispersal of toxic materials. Remedial measures are much cheaper with this technique, however, and it may be the best system in the long run.

Secure Landfills. One of the most popular solutions for hazardous wastes disposal has been landfilling. As we saw earlier in this chapter, many such landfills have been environmental disasters. They need not have been so terrible. There are techniques for creating safe, modern **secure landfills** that are acceptable methods of disposing of wastes in many cases. The first line of defense in a secure landfill is a thick bottom cushion of compacted clay, which surrounds the pit like a bathtub (figure 24.20). Moist clay is flexible and resists cracking if the ground shifts. It is impermeable to groundwater and will safely contain wastes. A layer of gravel is spread over the clay liner and perforated drain pipes are laid in a grid to collect any seepage that escapes from the stored material. A thick polyethylene liner, protected from punctures by soft padding materials, covers the gravel bed. A layer of soil or absorbent sand cushions the inner liner and the wastes are packed in drums, which then are placed into the pit, separated into small units by thick berms of soil or packing material.

When the landfill has reached its maximum capacity, a cover much like the bottom sandwich of clay, plastic, and soil caps the site. Vegetation stabilizes the surface and improves its appearance. Sump pumps collect any liquids that filter through the landfill, either from rainwater or leakage of the drums. This water is treated and purified before being released. Monitoring wells check groundwater around the site to ensure that no toxins have escaped.

Most landfills are buried below ground level to be less conspicuous; however, in areas where the groundwater table is close to the surface, it is safer to build above-ground storage. The same protective construction techniques are used as a buried pit. An advantage to such a facility is that leakage is easier to monitor because the bottom is at ground level.

Several nontechnical problems in the siting and management of secure landfills that have yet to be solved are as follows:

1. *Transportation to the site.* Some states ban shipping of hazardous materials across their territory because of the dangers of accidental spills. There is just cause for alarm because our safety record for shipping is not very reassuring. If we concentrate wastes on a few sites where

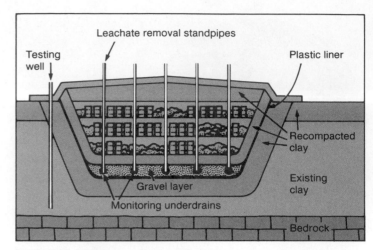

FIGURE 24.20 A secure landfill. Contents are enclosed by a thick plastic liner and two or more layers of impervious compacted clay. A gravel bed between the clay layers collects any lactrate, which can then be pumped out and treated. Testing wells sample for escaping contaminants.

they can be monitored and handled safely, it will mean that a few major transportation corridors will handle enormous volumes of toxic materials.

2. *Citizen opposition to sites.* Proposals for siting of hazardous waste facilities almost invariably elicit what planners call the NIMBY response: "Not In My Back Yard! Put it any place but here." This also is understandable, given the irresponsible and untrustworthy way that many companies have done business in the past. Where can we put it? Even if we stopped generating toxic materials right now, we already have vast quantities with which we must do something. We cannot put it in rockets and shoot it into space. The costs would be astronomical (no pun intended), and what would we do about the rockets that misfire or explode? Would you live near the launch pad?

3. *Financial responsibility for abandoned sites.* As we have seen, some unscrupulous operators of hazardous waste facilities took in large quantities of wastes and then declared bankruptcy or simply disappeared. Who is responsible for those wastes now? The public? The generators of the waste? What about waste generators who paid for what they thought would be adequate disposal? Some companies are being held liable for cleanup of dumps to which they were minor, but identifiable, contributors. Is that fair? Can a business continue to operate with that liability? What will we do about generators or handlers who followed all the legal requirements when they dumped, using state-of-the-art techniques for that time, but now have caused a problem? Is it reasonable to penalize them for doing something that neither we, nor they, knew at the time

was wrong? Can laws be made retroactive? These are difficult questions that require carefully negotiated answers.

Government Regulations

Ever since the National Environmental Policy Act was passed in 1969, the federal government has been trying to legislate controls on hazardous wastes in the environment. Its most practical and direct efforts include the 1976 Toxic Substances Control Act (TOSCA), the 1976 Resource Conservation and Recovery Act (RCRA), and the so-called Superfund, established by the 1980 Comprehensive Environmental Response, Compensation, and Liability Act (CERCLA).

TOSCA was the first law to require premarket testing of toxic substances. It allows the EPA to ban the manufacture, sale, and use of any chemical presenting an unreasonable risk of injury to health or the environment. This act specifically prohibited most uses of polychlorinated biphenols (PCBs), one of our most dangerous groups of substances.

RCRA established a regulatory system to track hazardous substances from the time of generation to disposal, a unique approach to pollution control that requires "cradle to grave" record keeping and accountability. The resulting information base allows remedial action after disposal, an option that would otherwise be impossible. RCRA also requires that safe and secure procedures be used in treating, storing, and disposing of hazardous substances. It is intended to prevent tragedies such as Love Canal, but it does not allow direct federal response to problems at waste sites already in existence.

CERCLA was passed to permit direct federal response to a broad range of environmental dangers and disasters. Superfund is a multibillion dollar cash pool instituted to finance government actions. Money is raised from public funds and taxes on designated hazardous chemicals, and all money spent from the fund is to be repaid to the federal government by the parties responsible for the hazardous waste release.

Superfund was first established as a $1.6 billion pool, but in 1986 it was raised to $9.6 billion, far less than the $100 billion that eventually may be needed. Cleanup of individual sites has been more expensive than anticipated, and the work has progressed slowly. After a decade of work and $9 billion in expenditures, the EPA has cleaned up only 27 out of 1,226 sites on the NPL. Critics charge that 80 percent of the money has been used for administration and high-priced consultants while only a pittance has been available for actual cleanup.

CERCLA has both strong points and weak points. Its flexibility in remedial action means that a wide variety of problems can be dealt with individually and appropriately. It provides immediate funding for cleanup in emergencies that might otherwise have to wait for years of litigation. CERCLA also encourages voluntary cleanup by the producer, carrier, or storage company responsible for a hazardous waste release. Such private parties handle about 90 percent of all releases that would otherwise require Superfund action. Unfortunately, quality and reliability of private responses vary greatly, and federal officials have difficulty ensuring a thorough cleanup. CERCLA establishes no standards for allowable levels of pollutants, and levels allowed by states differ widely. Superfund is administered by the EPA, which sometimes has been criticized for its inability to use its important regulatory powers.

The Danish Model

One of the most successful systems in the world for handling hazardous wastes has been developed in Denmark. This small, heavily populated country has no "away" to which it can put wastes. They have to be in someone's backyard. The Danes have developed a joint government/industry system for collecting and disposing wastes that combines penalties for waste production with incentives for waste reduction. A nonprofit corporation has a monopoly over off-site treatment, recycling, and waste disposal facilities. There are no midnight haulers or cheap landfills. Wastes are gathered, sorted, and pretreated at a network of collection facilities. Everything is recycled, detoxified, or incinerated. Nothing is landfilled except boiler slag and some nontoxic chemical residues.

The fees charged for this service are high. It's not a competitive market. They do things right and generators pay the cost. This creates a very strong incentive for generators to reduce, recycle, or treat wastes in-house. It also makes consumer goods more expensive as increased manufacturing costs are passed on. A firm that misrepresents the contents of its wastes is charged 30 percent more than the normal fee for the first offense. The second offense brings a 60 percent increase in charges plus the costs of thorough chemical analysis of the wastes. There has never been a second offense.

The system is not perfect. There is no incentive for the waste corporation to make its operation cost-effective, since it has a monopoly and its customers have no alternative to its service. There also is little reason to develop or adopt new technologies, such as advanced plasma-gas incineration, since criteria for operation are legislatively mandated. Many people in the United States would object to the level of social control and the loss of personal freedom represented by a nationwide agency with broad powers and rigid regulations. If the alternative is to be poisoned, however, we may have to change our traditional patterns of behavior.

S U M M A R Y

We produce enormous volumes of solid waste in industrialized societies, and there is an increasing problem of how to dispose of those wastes in an environmentally safe manner. In this chapter, we have looked at the character of our solid and hazardous wastes. We have surveyed the ways we dispose of our wastes and the environmental problems associated with waste disposal.

Solid wastes are domestic, commercial, industrial, agricultural, and mining wastes that are primarily nontoxic. About 80 percent of our domestic and industrial wastes are deposited in landfills; most of the rest is incinerated or recycled. Landfills are often messy and leaky, but they can be improved with impermeable clay or plastic linings, drainage, and careful siting. Incineration can destroy organic compounds, but whether incinerators can or will be operated satisfactorily is a matter of debate. Recycling is growing nationwide, encouraged by the economic and environmental benefits it brings. City leaders tend to doubt the viability of recycling programs, but successful programs have been sustained in other countries and in some American cities.

Near major urban centers, land suitable for waste disposal is becoming increasingly scarce and expensive. Costs for solid waste disposal totalled nearly $10 billion in the United States in 1990 and are expected to climb to $100 billion per year by the end of this century. A few cities now ship their refuse to other states or even other countries, but worries about toxic and hazardous material in the waste are leading to increasing resistance to shipping or storing it.

Hazardous and toxic wastes, when released into the environment, cause such health problems as birth defects, neurological disorders, reduced resistance to infection, and cancer. Environmental losses include contamination of water supplies, poisoning of the soil, and destruction of habitat. The major categories of hazardous wastes are ignitable, corrosive, reactive, explosive, and toxic. About 60 million metric tons of hazardous waste enter the waste stream each year in the United States, more than 90 percent of which is disposed of by environmentally unsound practices. Some materials that cause the most concern are heavy metals; synthetic organic chemicals, such as halogenated hydrocarbons, organophosphates, and phenoxy herbicides; and aerosols of toxic materials, such as asbestos, glass fibers, and beryllium whiskers.

Disposal practices for solid and hazardous wastes have often been unsatisfactory. Thousands of abandoned, often unknown waste disposal sites are leaking toxic materials into the environment. Some alternative techniques for treating or disposing of hazardous wastes include not making the material in the first place, incineration, secure landfill, and physical, chemical, or biological treatment to detoxify or immobilize wastes. People are often unwilling to have transfer facilities, storage sites, disposal operations, or transportation of hazardous or toxic materials in or through their cities. Questions of safety and liability remain unanswered in solid and hazardous waste disposal.

■ Review Questions

1. What are solid wastes and hazardous wastes? What is the difference?
2. How much solid and hazardous waste do we produce each year in the United States? How do we dispose of the waste?
3. Why are landfill sites becoming rare around most major urban centers in the United States? What steps are being taken to solve this problem?

4. Describe the concerns about waste incineration.
5. List some benefits and drawbacks of recycling wastes. What are the major types of materials recycled from municipal waste and how are they used?
6. What is composting, and how does it fit into solid waste disposal?
7. Describe some ways that we can reduce the waste stream to avoid or reduce disposal problems.
8. List ten toxic substances in your home and how you would dispose of them.
9. What are some illegal hazardous waste disposal methods, and why are they dangerous?
10. What societal problems are associated with waste disposal? Why do people object to waste handling in their neighborhoods?

■ Questions for Critical Thinking

1. Do you think the chemical companies that dumped wastes into unlined pits fifty years ago, when that was a legal disposal method, should be held responsible today for the damages those wastes are causing now?
2. Why do you suppose we create so much more solid waste than do other countries? Does this reveal anything about our history or national character?
3. What social, economic, and life-style factors influence people to participate or not to participate in recycling programs?
4. A 1,000-ton-per-day incinerator is planned to be built down the street from your house. You worry about the health of your young child, but the city desperately needs the incinerator, and your own disposal costs are rising. Do you support the plan or not?
5. If you were a city planner assigned the task of developing a city recycling program, how would you do it? What are the pros and cons of different collection systems, and how would you encourage participation?
6. Find out how your community disposes of its wastes and the costs of disposal, both monetary and environmental.
7. Should companies be held responsible for contamination created by wastes that they paid licensed trash haulers and waste disposers to remove from their property?
8. What products do you use that are hazardous in themselves, or that create hazardous wastes in their manufacture?
9. In your opinion, how serious is the problem of hazardous wastes? How would you rate it against other environmental and social problems?
10. What do you think can and should be done about the problem of "midnight dumpers"? Who should be penalized—the dumpers or the people who hire them?
11. If you were a lawmaker, would you favor establishment of a system like Denmark's? What are its advantages and disadvantages in your opinion?

Key Terms

biodegradable plastics
    (p. 504)
co-composting (p. 501)
energy recovery (p. 498)
hazardous waste (p. 505)
landfill (p. 496)
mass burn (p. 498)
permanent retrievable
    storage (p. 512)

photodegradable plastics
    (p. 504)
recycling (p. 499)
refuse-derived fuel (p. 498)
secure landfills (p. 512)
thermoplastics (p. 501)
thermoset polymers (p. 502)
waste stream (p. 495)

SUGGESTED READINGS

Block, A. A., and F. R. Scarpatti. *Poisoning for Profit: The Mafia and Toxic Waste in America*. New York: William Morrow, 1984. Excellent reporting on a dangerous business.

Blumberg, L. and R. Gottlieb. 1989. *War on Waste*. Washington D.C.: Island Press. An excellent book on all aspects of waste from generation, to shipping, to incineration, and landfilling. Includes case studies and examples.

Elder, J., and M. R. Wetzel. March/April 1985. "A Fine Kettle of Fish." *Sierra* 70(2):33. Chemical toxins make Great Lakes fish unsafe to eat.

Grieder, W. 1990. "Hazardous Waste Exports: Changes in Sight." *EPA Journal* 16(4): 46. Although progress is being made in controlling hazardous waste exports, much dangerous material still crosses boundaries. The U.S. Congress and the United Nations are considering measures to curb toxic exports.

Hall, B., J. Karliner, and P. Whitney. Fall 1987. "Garbage Imperialism." *Earth Island Journal* 2(4):10. An excellent, if radical, report on our garbage export to developing nations.

Hall, R. H. 1986. "Poisoning the Lower Great Lakes: The Failure of U.S. Environmental Legislation." *The Ecologist* 16(2/3):118. Exposes the broader threats behind and beyond Love Canal.

Holmes, H. 1991. "Recycling Plastics." *Garbage* Jan/Feb 1991: 32. A clear and comprehensive discussion of how recycling works, when it works.

Institute for Local Self-Reliance. *Environmental Review of Waste Incineration*. Washington, D.C.: Institute for Local Self Reliance, 1986. A good review of this controversial issue with some alternative suggestions.

Kamrin, M. A., and P. W. Rodgers. *Dioxins in the Environment*. New York: Harper & Row, 1985. A report on dioxins and furans in the environment with articles by forty authors.

Kunreuther, H., and R. Patrick. 1991. "Managing the Risks of Hazardous Waste." *Environment* 33 (3): 12. This article suggests that we need better information about the risks of hazardous waste before effective management strategies can be adopted.

Laurence, D., and B. Wynne. 1989. "Transporting Waste in the European Community: A Free Market?" *Environment* 31(6): 12. As the European Community moves toward free trade, some groups fear the implications for hazardous waste transport between member nations.

Levenson, H. 1990. "Wasting Away: Policies to Reduce Trash Toxicity and Quantity." *Environment* 32(2): 10. Describes policies that could reduce the quantity and toxicity of the waste we generate.

Luoma, J. 1990. "Trash Can Realities." *Audubon* March 1990: Garbage is rife with symbols of wickedness and virtue, but conventional environmental wisdom may not be so wise.

Newsday. *Rush to Burn: Solving America's Garbage Crisis*. 1989. Washington DC: Island Press. Written by a team of reporters from Newsday, this book is an excellent review of our throwaway society and the affects of incineration. One of a series of excellent books on this subject published by Island Press.

Packard, V. *Waste Makers*. New York: David McKay, 1960. A classic but still highly readable account of how we have become a throw-away society.

Postel, S. 1987. "Defusing the Toxics Threat: Controlling Pesticides and Industrial Waste." *Worldwatch Paper #79*. Washington, D.C.: Worldwatch Institute. A fine analysis of the risks of mismanagement and the potential for reducing waste production and improving disposal practices.

Rathje, W. R. 1991. "Once and Future Landfills." *National Geographic* 179(5): 116. Great photographs and good introduction to garbology, the science of analyzing garbage.

Stigliani, W. M. et al. 1991. "Chemical Time Bombs: Predicting the Unpredictable." *Environment* 33(4): 4. We have treated our environment as a "sink" for waste disposal. Now some areas have become poisonous wastelands. How can these chemical time bombs be detected and defused?

Starr, D. 1990. "Shoppers, Shoppers, Everywhere." *Audubon* March 1990: 98. Italians use 7.5 billion plastic shopping bags a year and most of them wind up littering the landscape. This article tells how one outraged mayor took precipitous action.

Thorp, L. January/February 1991. "How to Stop Incinerators: A Community Owner's Manual." *Greenpeace* 3:20. Lessons from the grassroots campaign to ban the burners.

Young, J. E. 1991. "Reducing Waste, Saving Materials." *State of the World*. Washington, D.C.: Worldwatch Institute 39. Describes how we can reduce waste and save resources at the same time.

CHAPTER 25

Urbanization

C O N C E P T S

Ten men are too few for a city; 100,000 are too many. A man is not fully human unless he is a citizen. Men come together in cities in order to live; they remain together in order to live the 'good' life—a common life for noble ends. The polis population should be self-sufficient for living the good life as a realizable community, but not so large that a sense of conscious unity is lost. The real difficulty is in putting theory into practice; the planner must always apply ingenuity to gain a maximum advantage from site conditions.

Aristotle, *On Civics*, 350 B. C.

Throughout history, most humans have lived in rural areas where they were supported by hunting, fishing, gathering or farming. People have congregated in family groups, tribes, or other communities that are the basis for rural villages.

Cities have long been centers of religion, education, the arts, commerce, and political power. The city provides the markets, capital, labor force, information flow, and materials that allow efficiency, experimentation, and speculation that lead to new developments. In many ways, the city has been both the focal point and the driving force for civilization.

The world is rapidly becoming urbanized. By the end of this century, more than half the world's population will live in cities. Most urban growth will occur in the less developed countries and especially in the giant metropolitan centers or urban cores. People move into cities because they are *pushed* out of the countryside by a lack of land and jobs or because they are *pulled* by the attractions of the city.

Most large Third World cities are being overwhelmed by their rapid growth and enormous size. They have terrible traffic problems, severe air and water pollution, high disease rates, housing and food shortages, and inadequate waste disposal systems. Millions of people live under desperate conditions, yet many of them have hope for the future and are working to better their lives and those of their children. The inner-city slums of American cities often are as bad as any Third World city.

INTRODUCTION

For more than 6,000 years, cities have been powerful sources of technological developments and social change, reflecting both the best and worst of human nature and human history. In many ways, the growth of cities and the emergence of civilization have been interdependent. Cities are cultural and racial melting pots in which information and technology are exchanged and resources are mobilized. Advances in art, science, education, architecture, and ethical concepts that are the human legacy for the future have come from the great cities of the world. Cities have been sources of energy, vitality, and progress; they also have been the source of pollution, crowding, disease, misery, and oppression.

Nearly half the people in the world now live in urban areas. It is predicted that by the end of the next century 80 or 90 percent of all humans will live in cities and that some giant interconnecting metropolitan areas could have hundreds of millions of residents. In this chapter, we will look at how cities came into existence, why people live there, and what the environmental conditions of cities have been, are now, and might be in the future. Some of the most severe urban problems in the world are found in the giant megacities of the developing countries. Far more lives may be threatened by the desperate environmental conditions in these cities than any other issue we have studied. In this chapter, we will look at a few of those problems and possible solutions that could improve the urban environment.

TRENDS AND TERMS OF URBANIZATION

Since their earliest origins, cities have been centers of education, religion, commerce, recordkeeping, communication, and political power (figure 25.1). Because of their importance as centers of civilization, cities have exerted an influence on culture and society that has greatly exceeded their proportion of the total population. Until recently, only a small percentage of the world's people have lived permanently in cities; the vast majority have always lived in rural areas where they were supported by farming, fishing, hunting, gathering timber harvest, animal herding, mining, or other natural resource-based occupations. Population sizes and distributions have, therefore, been limited mainly by productivity of the land. Even the greatest cities of antiquity were small by modern standards.

Since the beginning of the Industrial Revolution some three hundred years ago, however, cities have been growing rapidly in both size and power. In every developing country, the transition from an agrarian society to an industrial one has been accompanied by **urbanization,** an increasing concentration of the population in cities and a transformation of land use and society to a metropolitan pattern of organization. Urbanization now affects nearly every part of the world.

FIGURE 25.1   Since their earliest origins, cities have been centers of education, religion, commerce, politics, and culture. Cities, however, have also been the source of pollution, crowding, disease, and misery. The Parthenon, one of the most beautiful buildings in the world, is being eaten away by air pollution, vibrations, and other urban ills of Athens, Greece.

Degrees of Urbanization

Just what makes up an urban area or a city? Definitions differ. Some countries classify any collection of 100 occupied houses as an urban area. Others insist that it takes 10,000, 20,000, or 50,000 residents within a prescribed area to make it urban. The United States Census Bureau considers any incorporated community to be a city, regardless of size, and defines any city with more than 2,500 residents as urban. More meaningful definitions are based on *functions*. In a **rural area,** most residents depend on agriculture or other ways of harvesting natural resources for their livelihood. In an **urban area,** by contrast, a majority of the people are not directly dependent on natural resource-based occupations.

Additional terms reflect increasingly complex levels of urbanization (table 25.1). A **village** is a collection of rural households linked by culture, custom, and association with the land. Economic relationships in villages are based more on family ties and communal obligations than on rights and privileges; tradition generally dictates how things are done. This can make a village resistant to change, but it also gives members of the community a sense of security and connectedness that is often missing in the city. By contrast, a **city** is a differentiated community with a large enough population and resource base to allow residents to specialize in arts, crafts, services, or professions rather than rural-based occupations (figure 25.2). To support this specialization, the city needs a higher level of organization, services, and planning than does a rural village. Because of its size and resources, the city can mobilize economic resources (capital), skilled labor, and social organization necessary for projects that would be impossible in rural villages.

A **Standard Metropolitan Statistical Area (SMSA)** is defined by the Census Bureau as an urbanized region with at least 100,000 inhabitants having strong economic and social ties

FIGURE 25.2 A city is a differentiated community with a large enough population and resource base to allow specialization in arts, crafts, services, and professions. Although there are many disadvantages in living so closely together, there are also many advantages. Cities are growing rapidly and most of the world's population will live in urban areas if present trends continue.

TABLE 25.1   Definitions related to urbanization

Rural (from Latin *ruralis* = country). An area in which most people are supported by agriculture, fishing, logging, gathering or mining.

Urban (from Latin *urbanus* = city). An area in which the majority of people are not directly dependent on harvesting natural resources. It has sufficient population density and economic activity to create the characteristics of a city.

Village (from Latin *villa* = country house). A small community where houses are clustered together for cooperation and social interaction. Villages are usually relatively homogeneous and lack the services and specializations of the city.

City (from Latin *civitas* = citizen). A larger community differentiated to carry out specialized functions, such as education, religion, manufacturing, or commerce.

Metropolis (from Greek *metris* = mother + *polis* = state). A major city or capital that exerts its influence over a broad region and requires more complex systems of transportation, services, and administration than an ordinary city. The threshold for this transition is generally about 100,000 inhabitants.

Supercity, Megacity or **Megalopolis.** A metropolitan area so large (at least ten million people) that it experiences another quantum leap in economic power, environmental problems, and administrative complexity.

Core Region. A dominant metropolitan area (usually the capital and/or major port) that expands to occupy a broad area containing a major share of a country's population and economic activity. As a core region grows it often incorporates nearby metropolitan areas into a giant complex called a conurbation. This process is called hyperurbanization.

---

to the central city of at least 50,000 people. At this size, the complexity of a city seems to enter a new realm.

A **Consolidated Metropolitan Statistical Area (CMSA)** or **conurbation** is an area with more than one million people that is formed by the merger of two or more population centers and meets certain other specified requirements. Beyond about ten million inhabitants, an urban area is considered a **supercity, megacity,** or **megalopolis.**

Very large cities have a dominating influence on their states and countries. A **core region** is the primary industrial region of a country. It is usually located around the capital or largest port, and has both the greatest population density and the greatest economic activity of the country. Business is drawn to the core region because of proximity to government offices, availability of transportation, capital, and skilled labor, access to markets, and well-developed communication systems.

Core regions of industrialized countries have grown to enormous size. In the United States, urban areas between Boston and Washington, D.C., have merged into a nearly continuous megacity (sometimes called Bos-Wash) containing about 35 million people. The Tokyo-Yokohama-Osaka-Kobe corridor contains nearly 50 million people. The Paris Basin (or Ille de France) encompasses nearly half of that country, and the Antwerp-Brussels-Hamburg complex of northern Europe spreads across four countries (figure 25.3). Architect and city planner C. A. Dioxiadis predicts that if current trends continue, the sea coasts

ENVIRONMENTAL POLLUTION AND OTHER DILEMMAS

FIGURE 25.3 Map of core regions or major metropolitan conglomerations of Europe. Europe is so highly urbanized and densely populated that the Paris-Brussels-Amsterdam-Berlin-Prague-Cracow-Warsaw axis is on the verge of becoming a single giant megacity.

From "Migration Between the Core and the Periphery" by Daniel R. Vining, Jr. Copyright © 1982 by Scientific American, Inc. All rights reserved.

and major river valleys of most continents will be covered with continuous strip cities, which he calls Ecumenopolises, each containing billions of people.

Urbanization in the United States

The United States has changed from a predominantly rural to a largely urban country over the past two centuries (figure 25.4). At the beginning of the eighteenth century, only 5 percent of the population lived in cities, while 95 percent lived on farms or in rural villages. Now, about 75 percent of us live in cities. Of the 25 percent who live in rural areas, most work in small towns, commute to work in the city, or are retired or unemployed. Only 2 percent of the population are still supported full time by farming.

Industrialization is the primary underlying cause of this rural to urban shift. As mechanization of agriculture decreased the rural demand for labor, manufacturing and commerce created a demand for it in the cities. Many U.S. cities grew at a phenomenal rate during this transition era. In 1840, Chicago was a muddy pioneer village of only 4,000 people. Then the railroads arrived. By 1890, Chicago had 1 million residents (2,500 percent growth in fifty years), and by 1910, it had grown to more than 2 million. New York City experienced similar growth during this period. In 1850, the population of Manhattan was 200,000; in 1860, it was 900,000. This explosive growth and the conditions it created were much like what is happening now in many developing countries.

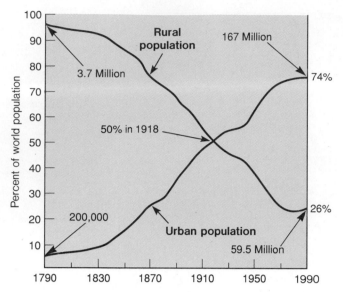

FIGURE 25.4 U.S. rural-to-urban shift. Percentages of rural and urban populations, 1790 to 1990.

Source: U.S. Bureau of the Census.

Region	1950	1990	2000 (estimated)
TABLE 25.2 Urban share of total population (percent)			
North America	64	74	78
Europe	56	73	79
Soviet Union	39	68	74
East Asia	43	65	79
Latin America	41	65	79
Oceania	61	73	73
China	12	32	40
Africa	15	34	42
South Asia	15	25	35
World	29	43	48

Source: Data from *World Resources 1990–1991*.

World Urbanization

In 1850, only about 2 percent of the world's population lived in cities. By 1900, urban areas had grown to include 14 percent of the total population. In 1990, 43 percent of the people in the world were urban. By the end of this century, for the first time in history, more people will live in cities than in the country. Only Africa and South Asia remain predominantly rural, but people there are swarming into cities in ever-increasing numbers. About three-fourths of the people in the developed countries and about two-thirds of those in Latin America and East Asia already live in cities (table 25.2). Some urbanologists predict that by 2100 the whole world will be urbanized to the levels now seen in developed countries.

As figure 25.5 shows, 90 percent of the population growth over the next fifty years is expected to occur in the less developed countries of the world. Three-quarters of that growth will be in the already overcrowded cities of the least affluent countries, such as India, China, Mexico, and Brazil. The combined population of these cities is expected to jump from its present 1 billion to nearly 4 billion in 2025. Meanwhile, rural populations in these countries are expected to decline somewhat as rural people migrate into the cities.

Recent urban growth has been particularly dramatic in the largest cities, especially those of the developing world. In 1900, thirteen cities had populations over 1 million; all of them except Tokyo were in Europe or North America. By 1990, there were 235 metropolitan areas of more than 1 million people—an eighteenfold increase. Presently, only one of the ten largest cities (New York) is in North America or Europe (table 25.3). In 1900,

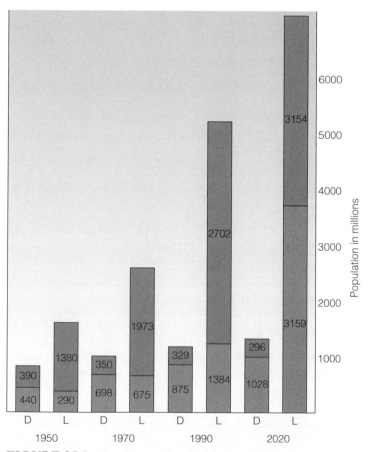

FIGURE 25.5 Distribution of rural and urban population in developed (D) and less developed (L) countries. Over the next fifty years, more than 90 percent of all urban growth will take place in the less developed countries.

ENVIRONMENTAL POLLUTION AND OTHER DILEMMAS

TABLE 25.3 Urban areas with populations of 2 million or more in 1985 and predicted average rate of growth 1970–2000

Rank in 1985	Agglomeration	Population (in millions)			Average annual rate of growth	
		1970	1985	2000	1970–1985	1985–2000
1	Tokyo/Yokohama (Japan)	14.87	19.04	21.32	1.65	0.75
2	Mexico City (Mexico)	8.74	16.65	24.44	4.30	2.56
3	New York (United States)	16.19	15.62	16.10	−0.24	0.20
4	Sao Paulo (Brazil)	8.06	15.54	23.60	4.38	2.79
5	Shanghai (China)	11.41	12.06	14.69	0.37	1.32
6	Buenos Aires (Argentina)	8.31	10.76	13.05	1.72	1.29
7	London (United Kingdom)	10.55	10.49	10.79	−0.04	0.19
8	Calcutta (India)	6.91	10.29	15.94	2.65	2.92
9	Rio de Janeiro (Brazil)	7.04	10.14	13.00	2.43	1.66
10	Seoul (Korea, Rep.)	5.31	10.07	12.97	4.27	1.69
11	Los Angeles (United States)	8.38	10.04	10.91	1.20	0.55
12	Osaka/Kobe (Japan)	7.60	9.56	11.18	1.53	1.04
13	Greater Bombay (India)	5.81	9.47	15.43	3.26	3.25
14	Beijing (China)	8.29	9.33	11.47	0.79	1.38
15	Moscow (Soviet Union)	7.11	8.91	10.11	1.50	0.84
16	Paris (France)	8.33	8.75	8.76	0.33	0.01
17	Tianjin (China)	6.87	7.96	9.96	0.98	1.49
18	Cairo/Giza (Egypt)	5.33	7.92	11.77	2.64	2.64
19	Jakarta (Indonesia)	4.32	7.79	13.23	3.93	3.53
20	Milan (Italy)	5.53	7.50	8.74	2.03	1.02
21	Teheran (Iran, Islamic Rep.)	3.29	7.21	13.73	5.23	4.29
22	Metro Manila/Quezon City (Philippines)	3.53	7.09	11.48	4.65	3.21
23	Delhi (India)	3.53	6.95	12.77	4.52	4.06
24	Chicago (United States)	6.72	6.84	6.98	0.12	0.14
25	Karachi (Pakistan)	3.13	6.16	11.57	4.51	4.20
26	Bangkok (Thailand)	3.11	5.86	10.26	4.22	3.73
27	Lagos (Nigeria)	2.02	5.84	12.45	7.08	5.05
28	Lima/Callao (Peru)	2.84	5.44	8.78	4.33	3.19
29	Hong Kong (Hong Kong)	3.40	5.16	6.09	2.78	1.10
30	Leningrad (Soviet Union)	3.98	5.11	5.84	1.67	0.89
31	Madras (India)	3.03	4.87	7.85	3.16	3.18
32	Madrid (Spain)	3.37	4.83	5.42	2.40	0.77
33	Dacca (Bangladesh)	1.50	4.76	11.26	7.70	5.74
34	Bogota (Colombia)	2.37	4.74	6.94	4.62	2.54
35	Baghdad (Iraq)	2.11	4.39	7.66	4.88	3.71

Source: From *Prospects of World Urbanization, 1988* (United Nations publication, Sales No. E.89.XIII.8), p.19.

London was the only city with more than 5 million people; now nineteen cities have populations above 5 million. Some futurists predict that by 2025 at least four hundred cities will have populations of 1 million or more, and ninety-three supercities each will have 5 million or more residents. Three-fourths of those cities will be in the developing nations (figure 25.6).

The rate of growth of the most rapidly expanding cities and the sizes they are predicted to reach are truly astounding (table 25.4). Many cities are growing at rates above 5 percent per year, which means they double in less than fifteen years. Mexico City, for instance, had a population of about 2.45 mil-lion in 1950. By 1990, it was the largest city in the world with 19.4 million residents. It is now growing at a rate of about two thousand people per day (figure 25.7). By the end of this century it is expected to pass 26 million, an elevenfold increase in fifty years. If that growth trend continues, Mexico City could have 100 million inhabitants by the middle of the next century.

Can a city function with that many people? Can it supply food, water, sanitation, transportation, jobs, housing, police protection, fire control, electricity, education, and other public services necessary to sustain a civilized way of life? The 750,000 new people added each year to Mexico City amount to building

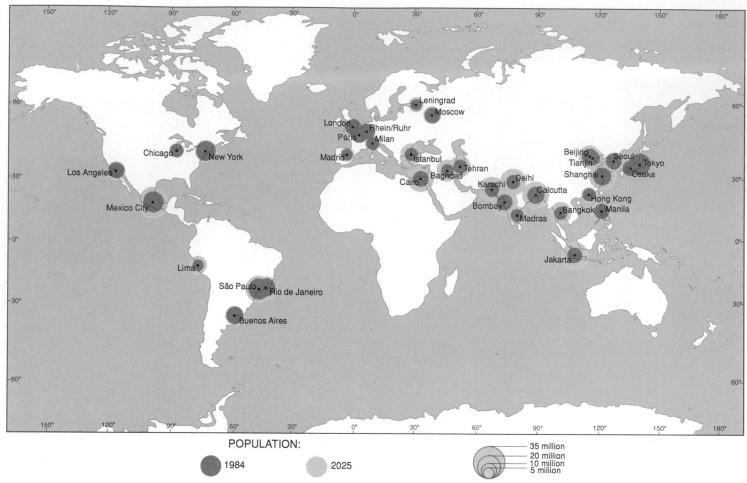

POPULATION:

● 1984   ○ 2025

35 million
20 million
10 million
5 million

FIGURE 25.6 By 2025 at least four hundred cities will have populations of one million or more and 93 supercities will have populations above five million. Three-fourths of the world's largest cities will be in developing countries that already have trouble housing, feeding, and employing their people.

TABLE 25.4 Cities that have experienced rapid population growth (population in millions)				
City	1950	1985	2000 (est.)	% increase 1950–2000
Nouakchott, Mauritania	0.006	0.25	1.1	18,333
Amman, Jordan	0.03	0.8	1.5	5,000
Lagos, Nigeria	0.27	4.0	10.5	3,888
Nairobi, Kenya	0.14	0.8	5.3	3,790
Baghdad, Iraq	0.65	7.2	12.8	1,969
Santa Cruz, Bolivia	0.06	0.26	1.0	1,670
Bogota, Colombia	0.61	3.5	9.6	1,570
Mexico City, Mexico	2.45	16.0	26.0	1,061
Manaus, Brazil	0.11	0.51	1.1	1,000
Sao Paulo, Brazil	2.45	12.5	24.0	980
Jakarta, Indonesia	1.45	6.2	13.0	897
Delhi, India	1.74	5.2	13.0	747
Cairo, Egypt	2.45	8.5	12.9	516
Bombay, India	3.34	8.0	16.8	494

Source: *Global Possible*, 1986.

a new city the size of Baltimore or San Francisco every year. This growth is occurring, as is most urban growth in the world, in a country with a sagging economy, an unstable government, and a high foreign debt load. The environmental costs (air and water pollution, soil erosion, solid and hazardous waste accumulation, noise and congestion) as well as the human costs (lack of housing, jobs, transportation, health care, social services, and living space) are tragic.

CAUSES OF URBAN GROWTH

Urban populations grow in two ways: by natural increase (more births than deaths) and by immigration. Natural increase is fueled by improved food supplies, better sanitation, and advances in medical care that reduce death rates and cause populations to grow both within cities and in the rural areas around them (chapter 7). In Latin America and East Asia, natural increase is responsible for two-thirds of urban population growth. In Africa and West Asia, immigration is the largest source of urban growth. Immigration to cities can be caused both by **push factors** that

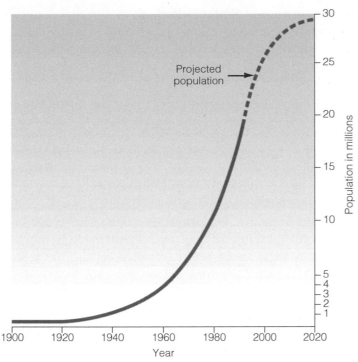

FIGURE 25.7 Growth of Mexico City metropolitan area, 1900–2020.

Source: U.N. Population Council.

force people out of the country and by **pull factors** that draw them into the city.

Immigration Push Factors

People migrate to cities for many reasons. In some areas, the countryside is overpopulated and simply can't support more people. The "surplus" population is forced to migrate to cities in search of jobs, food, and housing. Not all rural-to-urban shifts are caused by overcrowding in the country, however. In some places, economic forces or political, racial, or religious conflicts drive people out of their homes. The countryside may actually be depopulated by such demographic shifts. Of the 14.4 million refugees in the world in 1990, many were forced to live under harsh conditions in the already overcrowded megacities of the developing world. The largest groups of refugees were 5.9 million Afghans, 2.1 million Palestinians, 1.1 million Mozambiqueans, and 1.0 million Ethiopians.

Land tenure patterns and changes in agriculture also play a role in pushing people into cities. The same pattern of agricultural mechanization that made farm labor obsolete in the United States early in this century is spreading now to developing countries. Furthermore, where land ownership is concentrated in the hands of a wealthy elite, subsistence farmers are often forced off the land so it can be converted to grazing lands or monoculture cash crops. In some countries, speculators and absentee landlords let good farm land sit idle that otherwise might house and feed rural families.

Immigration Pull Factors

Even in the largest and most hectic cities, many people are there because they choose to be. People are attracted by excitement, vitality, and the chance to meet others like themselves. The city offers jobs, housing, entertainment, and freedom from the constraints of village traditions. The city has possibilities for upward social mobility, prestige, and power not ordinarily available in the country. Urban areas offer opportunities for specialization in arts, crafts, and professions for which there is not a sufficient market elsewhere.

Modern communications also draw people to cities by broadcasting images of luxury and opportunity. An estimated 90 percent of the people in Egypt, for instance, have access to a television set. The immediacy of television makes city life seem more familiar and attainable than ever before. When we see pictures of beggars and homeless people on the streets of teeming Third World cities, we generally assume that they have no other choice; but many of these people want to be in the city. However bad conditions are, many people prefer the city to their previous life in the country.

Government Policies

Government policies often favor urban over rural areas in ways that both push and pull people into the cities. Developing countries commonly spend most of their budgets on improving urban areas (especially around the capital city where leaders live), even though only a small percentage of the population lives there or benefits directly from the investment. This gives the major cities a virtual monopoly on new jobs, housing, education, and opportunities, all of which bring in rural people searching for a better life. In Peru, for example, Lima accounts for 20 percent of the country's population but has 50 percent of the national wealth, 60 percent of the manufacturing, 65 percent of the retail trade, 73 percent of the industrial wages, and 90 percent of all banking in the country. Similar statistics pertain to Sao Paulo, Mexico City, Manila, Cairo, Lagos, Bogota, and a host of other cities.

Governments often manipulate exchange rates and food prices for the benefit of more politically powerful urban populations at the expense of rural people. Importing lower-priced food holds prices down, which pleases city residents, but it makes it uneconomical for farmers to grow crops. This increases the number of people leaving rural areas and maintains a large work force in the city, keeping wages down and industrial production high. Zambia, for instance, sets maize prices below the cost of local production to discourage farming and to maintain a large pool of workers for the mines. Keeping the currency exchange rate high stimulates export trade but makes it difficult for small farmers to buy the fuels, machinery, fertilizers, and seeds that they need. This depresses rural employment and rural income while stimulating the urban economy. The effect is to transfer wealth from the country to the city.

CURRENT URBAN PROBLEMS

Large cities in both developed and developing countries face similar challenges in accommodating the needs and by-products of dense populations. The problems are most intense, however, in rapidly growing cities of developing nations.

The Developing World

As we mentioned earlier, demographers predict that 90 percent of the human population growth in the next century will occur in the underdeveloped and overcrowded countries of Africa, Asia, and South America. Almost all of that growth will occur in cities—especially the largest cities—which already have trouble supplying food, water, housing, jobs, and basic services for their residents. The unplanned and uncontrollable growth of those cities causes serious urban environmental problems. Let's examine some of these problems. Keep in mind that they are not limited to Third World cities, but that their intensity is magnified there compared to wealthy cities of the developed countries.

Traffic and Congestion

A first-time visitor to a supercity—particularly in a less developed country—is often overwhelmed by the immense crush of pedestrians and vehicles of all sorts that clog the streets. The noise, congestion, and confusion of traffic make it seem suicidal to venture onto the street (figure 25.8). Cairo, for instance, is one of the most densely populated cities in the world. Traffic is chaotic almost all the time. People commonly spend three or four hours each way commuting to work from outlying areas. Calcutta also has monumental traffic problems. The Howrah Bridge over the Hooghly River carries 30,000 motor vehicles and 500,000 pedestrians daily. Gigantic traffic jams occur day and night and there is almost always a wait to cross over.

Air Pollution

The dense traffic (commonly old, poorly maintained vehicles), smoky factories, and use of wood or coal fires for cooking and heating often create a thick pall of air pollution in the world's supercities. Lenient pollution laws, corrupt officials, inadequate testing equipment, ignorance about the sources and effects of pollution, and lack of funds to correct dangerous situations usually exacerbate the problem. What is its human toll? Cubatão, near Sao Paulo, Brazil, has been called the "valley of death" because of its high levels of pollution. An estimated 60 percent of Calcutta's residents are thought to suffer from respiratory diseases linked to air pollution. Lung cancer mortality in Shanghai is reported to be four to seven times higher than rates in the countryside. Mexico City, which sits in a high mountain bowl with abundant sunshine, little rain, high traffic levels, and frequent air stagnation, has one of the highest levels of photochemical smog (chapter 23) in the world.

Sewer Systems and Water Pollution

Few cities in developing countries can afford to build modern waste treatment systems for their rapidly growing cities. The World Bank estimates that only 35 percent of urban residents in developing countries have satisfactory sanitation services. The situation is especially desperate in Latin America, where only 2 percent of urban sewage receives any treatment. In Egypt, Cairo's sewer system was built about fifty years ago to serve a population of 2 million people. It is now being overwhelmed by more than 11 million people. Only 217 of India's 3,119 towns and cities have even partial sewage systems and water treatment facilities. These systems serve less than 16 percent of India's 200 million urban residents. In Colombia, the Bogota River, 200 km (125 mi) downstream from Bogota's 5 million residents, still has an average fecal bacterial count of 7.3 million cells per liter, more than 7,000 times the safe drinking level and 3,500 times higher than the limit for swimming.

Some 400 million people or about one-third of the population in developing world cities do not have safe drinking water, according to the World Bank. Where people have to buy water from merchants, it often costs one hundred times as much as piped city water and may not be safe to drink after all. Many rivers and streams in Third World countries are little more than open sewers, and yet they are all that poor people have for washing clothes, bathing, cooking, and, in the worst cases, for drinking (chapter 22). Diarrhea, dysentery, typhoid, and cholera are widespread diseases in these countries, and infant mortality is tragically high.

Housing

The United Nations estimates that at least 1 billion people—20 percent of the world's population—live in crowded, unsanitary slums of the central cities and in the vast shantytowns and squatter settlements that ring the outskirts of most Third World cities. Around 100 million people have no home at all. In Bombay, for example, it is thought that half a million people sleep on the streets, sidewalks, and traffic circles because they can find no other place to live (figure 25.9). In Sao Paulo, at least 3 million "street kids" who have run away from home or have been abandoned by their parents live however and wherever they can. This is surely a symptom of a tragic failure of social systems.

Slums are generally legal but inadequate multifamily tenements or rooming houses, either custom built for rent to poor people or converted from some other use. The chals of Bombay, for example, are high-rise tenements built in the 1950s to house immigrant workers. Never very safe or sturdy, these dingy, airless buildings are already crumbling and often collapse without warning. Eighty-four percent of the families in these tenements live in a single room; half of those families consist of six or more people. Typically, they have less than 2 square meters of floor

FIGURE 25.8 Motorized rickshaws, motor scooters, bicycles, street vendors, and pedestrians all vie for space on the crowded streets of Delhi. The heat, noise, smells, and sights are overpowering. In spite of the difficulties, however, many people here lead reasonably happy lives.

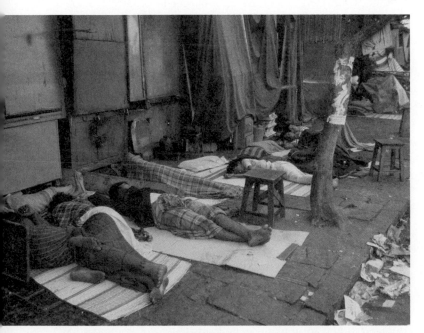

FIGURE 25.9 In Bombay, as many as half a million people sleep on the streets because they have no other place to live. Ten times as many live in crowded, dangerous slums and shantytowns throughout the city.

space per person and only one or two beds for the whole family. They may share kitchen and bathroom facilities down the hall with fifty to seventy-five other people. Even more crowded are the rooming houses for mill workers where up to twenty-five men sleep in a single room only 7 meters square. Because of this crowding, household accidents are a common cause of injuries and deaths in Third World cities, especially to children. Charcoal braziers or kerosene stoves used in crowded homes are a routine source of fires and injuries. With no place to store dangerous objects beyond the reach of children, accidental poisonings and other mishaps are a constant hazard.

Shantytowns are settlements created when people move onto undeveloped lands and build their own houses. Shacks are built of corrugated metal, discarded packing crates, brush, plastic sheets, or whatever building materials people can scavenge. Some shantytowns are simply illegal subdivisions where the landowner rents land without city approval. Others are spontaneous or popular settlements or **squatter towns** where people occupy land without the owner's permission. Sometimes this occupation involves thousands of people who move onto unused land in a highly organized, overnight land invasion, building huts and laying out streets, markets, and schools before authorities can root them out (figure 25.10). In other cases, shantytowns just gradually "happen."

Called barriads, barrios, favelas, or turgios in Latin America, bidonvillas in Africa, or bustees in India, shantytowns surround every megacity in the developing world. They are not an exclusive feature of poor countries, however. Some 200,00 immigrants live in the "colonies" along the southern Rio Grande in Texas. Only 2 percent have access to adequate sanitation. Most live in conditions as awful as you would see in any Third World city. Smaller enclaves of the poor and dispossessed can be found in most American cities.

The problem is magnified in less developed countries. Nouakchott, Mauritania, the fastest growing city in the world, is almost entirely squatter settlements and shantytowns. It has been called "the world's largest refugee camp." About 80 percent of the people in Addis Ababa, Ethiopia, and 70 percent of those in Luanda, Angola, live in these squalid refugee camps. Two-thirds of the population of Calcutta live in shantytowns or squatter settlements and nearly half of the 19.4 million people in Mexico City live in uncontrolled, unauthorized shantytowns and squatter settlements. Many governments try to clean out illegal settlements by bulldozing the huts and sending riot policemen to drive out the settlers, but the people either move back in or relocate in another shantytown elsewhere.

These popular but unauthorized settlements usually lack sewers, clean water supplies, electricity, and roads. Often the land on which they are built was not previously used because it is unsafe or unsuitable for habitation. In Bhopal, India, and Mexico City, for example, squatter settlements were built next to deadly industrial sites. In Rio de Janeiro, La Paz (Bolivia), Guatemala City, and Caracas (Venezuela), they are perched on landslide-prone hills. In Bangkok, thousands of people live in shacks built over a fetid tidal swamp. In Lima (Peru), Khartoum (Sudan), and Nouakchott, shantytowns have spread onto sandy deserts. In Manila, 20,000 people live in huts built on towering mounds of garbage and burning industrial waste in city dumps.

As desperate and inhumane as conditions are in these slums and shantytowns, many people do more than merely survive there. They keep themselves clean, raise families, educate their children, find jobs, and save a little money to send home to their parents. They learn to live in a dangerous, confusing, and rapidly changing world and have hope for the future. The people have parties; they sing and laugh and cry. They are amazingly adaptable and resilient. In many ways, their lives are no worse than those in the early industrial cities of Europe and America a century ago. Perhaps continuing development will bring better conditions to cities of the Third World as well.

Informal Economy

A striking aspect of most Third World cities is the number of people selling goods and services of all types on the streets or

FIGURE 25.10 A shantytown or spontaneous settlement on the outskirts of Mexico City. Millions of people live with no water, power, or sanitation in *colonias* such as this.

from small stands in informal markets. Food vendors push carts through crowded streets; children dart between cars selling papers or cigarettes; curb-side mechanics do repairs using primitive tools and great ingenuity. Nearly everything city residents need is available on the streets. These individual entrepreneurs are part of the **informal economy;** small-scale family business in temporary locations outside the control of normal regulatory agencies. In many developing countries, this informal sector accounts for 60 to 80 percent of the economy.

Governments often consider these independent businesses to be backward and embarrassing, a barrier to orderly development. It is difficult to collect taxes or to control these activities. In many cities, police drive food vendors, beggars, peddlers, and private taxis off the streets at the same time that they destroy shanties and squatter settlements.

Recent studies, however, have shown that this informal economy is a vital, dynamic force that is more often positive than negative. The sheer size and vigor of this sector means that it can no longer be ignored or neglected. The informal economy is often the only feasible source of new housing, jobs, food distribution, trash removal, transportation, or recycling for the city. Small businesses and individual entrepeneurs provide services that people can afford and the city cannot or will not provide.

The businesses common to the informal sector are ideal in a rapidly changing world. They tend to be small, flexible, and labor intensive. They are highly competitive and dynamic, avoiding much of the corruption and sclerosis of Third World bureaucracies. Government leaders are beginning to recognize that the informal sector should be encouraged rather than discouraged. They are making microloans (Box 8.2) and assisting communities with self-help projects. When people own their houses or business, they put more time, energy, and money into improving and upgrading them. Land tenure and methods of improving cities are discussed further in chapter 26.

ENVIRONMENTAL POLLUTION AND OTHER DILEMMAS

The Developed World

For the most part, the rapid growth of central cities that accompanied industrialization in nineteenth- and early twentieth-century Europe and North America has now slowed or even reversed. London, for instance, once the most populous city in the world, has lost nearly 2 million people, dropping from its high of 8.6 million in 1939 to about 6.7 million now. While the greater metropolitan area surrounding London has been expanding to about 10 million inhabitants, it is now only the twelfth largest city in the world.

Many of the worst urban environmental problems of industrialized countries have been substantially reduced in recent years. Air and water quality have improved greatly. Working conditions and housing are better for most people. Improved sanitation and medical care have reduced or even completely eliminated many of the worst communicable diseases that once were the scourge of cities. Dispersal of the people and businesses to the suburbs has greatly reduced crowding and congestion. Still, city life is generally regarded as being more stressful than life in the country or suburbs (Box 25.1). Part of the stress is related to the high noise levels in busy cities (Box 25.2).

The major problems now facing cities of the United States tend to be associated with decay and blight. Businesses and jobs have fled to the suburbs along with the middle class. Tax revenues are down. The city infrastructure—streets, sewers, schools, and housing—is getting old and dilapidated. Drug dealers, pawn shops, bars, sleazy rooming houses, graffiti, broken windows, trash, and abandoned cars have become symbols of the worst parts of inner cities. Freeways make it much easier for suburbanites to get into and out of the city, but they also destroy housing, cause noise, increase air pollution, and create vast concrete landscapes that divide neighborhoods and make travel difficult for people without cars.

Many of the social problems of central cities are a result of the entrapment of concentrated minority populations—the poor, elderly, handicapped, unemployed, homeless, or otherwise socially depressed members of society who were left behind when the middle and upper classes moved to the suburbs. In some inner-city areas, unemployment is routinely 50 percent or higher, and jobs that are available generally are menial, minimum-wage work with little prospect for advancement. Homelessness is an increasing problem that is full of depressing statistics. Welfare agencies estimate that 3.5 million people are homeless in the United States and ten to twenty times that number live in substandard housing. Most of the homeless are in large cities, and about one-third are now families. Of the single females living on the street, perhaps three-fourths are mentally disturbed and would formerly have been cared for in hospitals or institutions. Of the single, homeless men, two-thirds are elderly poor, disabled, drug addicts, or alcoholics.

The increasing concentration of poor and disadvantaged people in cities occurred at a time when federal assistance was

FIGURE 25.11 A "bag lady" carrying all her possessions on her back looks for shelter, food, and safety in an affluent American city. Many people fall through the safety net of social services designed to help them.

being cut by 80 percent and city tax revenues were declining. The rising statistics of violence, drug abuse, child neglect, and unwanted pregnancies are symptoms of poverty and decay in the cities (figure 25.11).

How do we break the cycle of poverty, ignorance, and lack of opportunity in which inner-city residents often seem to be trapped? How do we deal with the homeless people, especially those who are unable or barely able to take care of themselves? Are the negative aspects of large cities doomed to increase, or can they be dealt with constructively? Consider Detroit and Watts; they not only survived their crises, but have been improved. Chapter 26 deals further with potential ways of handling problems that accompany increasing urbanization.

TRANSPORTATION AND CITY GROWTH

Transportation has probably always played an essential role in city development. Most cities are located at crossroads, river fords, seaports, junctions of major rivers, agricultural centers, or other sites where travelers were brought together and merchandise could be traded, bought, or sold. Transportation is needed to bring building materials, energy sources, and food into the city and to remove wastes. Most older cities have been remodeled and reshaped repeatedly by the changing transportation systems that provide their life-giving circulation.

Before the Industrial Revolution, most cities were compact and densely populated (figure 25.12a). Except for a few broad

BOX 25.1   Crowding, Stress, and Crime in Cities

Large cities in America clearly have more crimes per capita than do smaller towns and rural areas. In 1985, according to the Federal Bureau of Investigation, cities with populations above 250,000 had an average murder rate more than twice the national average of 8.0 per 100,000. Box table 25.1 shows that larger cities had approximately three times the murder rate of either rural areas or cities under 100,000 residents, and they had six times as many murders per capita as towns under 10,000 people.

Some people believe that the sheer size of big cities makes them violent and inhumane, that constantly being surrounded by too many people (Box figure 25.1) causes nervous overload and leads to mental illness, deviant behavior, and cultural breakdown. It is argued that the anonymity and stress of the urban environment underlie much of the hostility, paranoia, criminality, drug abuse, and other abnormal or antisocial behaviors that plague modern urban societies.

One line of evidence linking crowding to abnormal behavior comes from studies of laboratory animals living in very high population densities. When rats or mice are given unlimited food and water and allowed to reproduce to very high numbers in a confined space, they frequently exhibit a condition called "stress-related disease" (chapter 6).

BOX FIGURE 25.1   Crowds of people jam the streets in Calcutta. Although prices are high, living conditions are crowded, and air quality is bad, many people still want to come here because of the job opportunities, entertainment, education, and vitality of the city.

Whether this says anything about human biology and human behavior, however, remains to be seen. Humans are much more complex than rodents, and are able to adapt to or change their environment in ways that rats cannot. People can live successfully in very high densities if that is considered normal in their culture, if they have a choice of where and how they live, and if they have a sense of hope for the future. Boats carrying refugees from German concentration camps to Israel in 1945 were so crowded that disease and conflict would seem to be inevitable. However, these people were going from much worse to much better conditions, and they survived very well.

In some societies, people always are in direct physical contact with others. They find privacy simply by withdrawing mentally and emotionally from what is going on around them. Cultures that are adjusted to crowded conditions develop behavioral traits, such as soft voices, polite speech, and confrontation avoidance, that minimize conflict. Hong Kong, for instance, is one of the most densely populated cities in the world. More than 5 million people live in approximately 65

ceremonial boulevards, streets were narrow and twisting, and often barely passable for wheeled traffic. Transportation was by sailboat, horse-drawn vehicles, or on foot. The wealthiest citizens usually lived in the middle of the city close to the centers of power and prestige while the poorer people lived on the outskirts. Housing was mainly multistoried apartment complexes over street-level shops.

The introduction of steam engines in the first decades of the nineteenth century brought many benefits, but also expanded the problems of cities. Railroads made it possible to move freight and passengers quickly and inexpensively, but they greatly increased noise, congestion, and pollution. Tracks were generally laid through river valleys and right into the already congested core of the city (figure 25.12b), much as freeways would be built a century later, lowering property values and adversely affecting

the environment of existing neighborhoods. The central railroad station (e.g., Grand Central Station in New York) was ordinarily the main focal point of the city.

Automobiles brought many changes to cities. The middle class moved to single-family housing developments that filled in open spaces between transit lines (figure 25.12c). The wealthy moved out of town to rural estates and satellite cities. Business offices remained downtown, but shopping centers sprang up at the nodes where major streets intersected transit lines. In the country, small towns and crossroads stores that had been built at about five-mile intervals to accommodate horse-and-buggy travelers dwindled when people were able to travel to the county seat or the nearest big city to do their shopping. The development of chain stores and brand-name merchandise was largely due to this new mobility of shoppers.

City population	Murder	Rape	Robbery	Assault	Property crimes
Above 250,000	18.6	74	670	526	8,596
50,000–100,000	5.9	38	185	320	5,898
Under 10,000	2.9	17	39	192	3,946
Rural areas	5.2	18	15	160	1,900

BOX TABLE 25.1 Annual crime rate per 100,000 residents in United States cities

sq km (220 sq mile), yet they have one of the lowest crime rates in the world. By contrast, metropolitan Los Angeles has 1/100 the density of Hong Kong, but has ten times the murder rate and one hundred times the violent crime rate.

It is often suggested that western, urban, technological societies are more violent and crime-prone than others, but statistics do not bear this out. Many rural, sparsely populated countries are just as violent as industrial, densely crowded countries. The three highest murder rates in the world (number per 100,000 people per year) are in Lesotho (140), the Bahamas (23), and Guyana (22). None of these countries is highly urbanized, but all have a history of foreign domination, high poverty levels, and cultural and racial conflict.

By contrast, some of the lowest murder rates in the world are in the more densely populated and highly urbanized countries. Indonesia, South Korea, and Japan have less than two murders per 100,000 people; Denmark, England, Malaysia, Hong Kong, the Philippines, and Singapore each have between two and three murders per 100,000. Clearly, the residents of the slums and shanty-towns of Manila, Jakarta, Hong Kong, and Kuala Lumpur are subjected to crowding, poverty, pollution, noise, squalor, and other stressful aspects of urban living that are at least as bad as the worst parts of American cities, and yet they don't seem to have nearly as much violence as western countries have. There must be other factors that we need to take into account.

Some sociologists argue that it is neither the absolute size nor density of cities, but rather the concentration of poverty and social problems that cause most crime and violence. They point out that it was mostly people at the bottom of the socio-economic spectrum who migrated from other countries or moved into the city during the rural-to-urban shift at the beginning of this century. Furthermore, it was the least successful and, therefore, least mobile people who remained in the city when the upper and middle classes fled to the suburbs after the Second World War. The result has been an accumulation of people who are trapped in a "culture of poverty" from a continuing cycle of family violence, drug abuse, low self-esteem, hopelessness, undereducation, unemployment, and neglect.

What can we do about crime and violence in the city? Clearly, we all would benefit from reducing the stresses in urban environments that are caused by pollution, noise, litter, squalor, and congestion. We also should try to make decent housing, jobs, and education available to all. Perhaps we also need to find ways to divide the city into smaller neighborhoods, to mix the kinds of people and housing, or to reduce the absolute number or density of people, as well as provide community support and opportunities for advancement.

The decision to build freeways has been called the most important land-use decision ever made in the United States. Freeways have profoundly reshaped where we live, work, and shop, and how we get from place to place. Shopping malls located at freeway interchanges have largely replaced downtown department stores and have become the urban centers of expanding rings of suburbs around major cities (figure 25.12d). Freeways allow us to travel with greater privacy, freedom, convenience, and speed (usually) than a mass transit system.

But freeways also bring many problems. They have torn through neighborhoods, choked cities with traffic, enormously increased energy consumption, and caused pollution, noise, and urban sprawl (figure 25.13). Some of the most contentious U.S. environmental battles of the last two decades concerned freeways planned to cut across residential neighborhoods, parks, scenic and historic areas, or farmlands. Many people were initiated into environmental activism through their opposition to a particular stretch of freeway that threatened an area they cared about.

Los Angeles epitomizes the modern postindustrial city. Its multiple suburban centers are linked by a 600-mile-long network of freeways. It has been estimated that two-thirds of downtown Los Angeles and one-third of the total metropolitan area is devoted to roads, parking lots, service stations, and other automobile-related uses. Five million vehicles crowd onto the roads and highways each day, causing about 85 percent of both the air pollution and urban noise in the metropolitan area. Most Los Angeles freeways no longer have morning rush in one direction and evening rush the opposite way. Traffic comes to a standstill morning and evening in both directions on most freeways, while

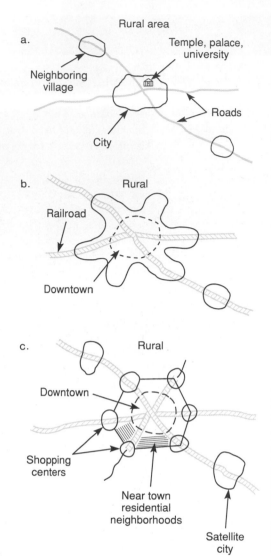

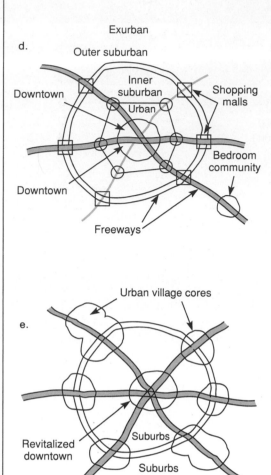

FIGURE 25.12 Patterns of city development from preindustrial (*a*) the polycentric postindustrial (*e*). Note how housing areas follow rail lines in early industrial stage (*b*). Introduction of the automobile allows people to move away from the rail lines. (*c*). Freeways (*d*) divert business to shopping malls and encourage growth of suburban rings beyond the central city. Finally, (*e*) these urban village cores develop a full range of services and amenities that rival the central core. Removal of heavy manufacturing from downtown allows gentrification and revitalization.

busy intersections have slowdowns throughout the day and into the night. It can take three hours to drive 20 miles at peak traffic periods. Between commuting to work and driving to beaches, the mountains, or for shopping, many people spend a substantial portion of their lives on the freeways.

What's next now that we seem to have saturated our urban areas with automobiles? Some cities are using modern mass transit systems to redesign where people live and work. Shopping centers are developing into urban villages with a complete set of services and facilities (figure 25.12*e*). The old downtown, freed of the congestion and pollution of heavy industry, is undergoing revitalization as a cultural center. Chapter 26 further discusses transportation and urban design.

REGIONAL IMPACTS OF CITIES

Critics of urbanization regard expanding urban areas as huge, shapeless organisms that crawl across the land consuming open space, farmland, forests, wetlands, and rural villages. Metropolitan New York City, for instance, now extends over a circle of newly suburbanized towns, villages, and hamlets more than 80 km (50 mi) from Manhattan. Spending three or four hours each day commuting to and from jobs in the city center is not uncommon for suburban residents. Even people who work in the suburbs may have long commutes because home prices in areas near their offices have risen so high that only people with high incomes can afford to live there. Those with more modest incomes are forced to move even further out—to suburbs beyond

FIGURE 25.13 The freeway makes it easy for suburban residents to get into and out of the city, but it consumes land, brings pollution into the city, creates noise and visual blight, and destroys neighborhoods.

suburbs! The suburbs themselves have become vast seas of characterless tract houses that lack both the close community spirit of the rural village and the vibrant vitality and diversity of the city (figure 25.14). In chapter 26, we will discuss some alternatives to typical suburban development.

The rapid spread of a city over former rural areas creates problems for those already there. Narrow country roads designed for light traffic are suddenly crowded with thousands of cars trying to commute to the city. Small-town services are overburdened. Many new residents spend so much time commuting to distant jobs that they have little time to meet neighbors or get involved in civic affairs. Without roots in the community, newcomers are unaware of its history and have little interest in local events. Growing populations overload water and sewer systems, runoff from construction pollutes streams and lakes, and increased traffic pollutes the air. Characteristically, noise, crime, and congestion increase while environmental quality deteriorates.

Rural townships and villages have to raise taxes for new roads, schools, and public services required by growing popu-

FIGURE 25.14 Suburbs provide low-cost housing, open space, and privacy for middle-income families, but they often lack the close community and vitality of the city. They also make us dependent on automobiles for transportation. New housing developments, built for economy rather than aesthetics, can be bleak and sterile environments.

BOX 25.2 Noise

Every year since 1973, the U.S. Department of Housing and Urban Development has conducted a survey to find out what city residents dislike about their environment. And every year the same factor has been named most objectionable. It is not crime, pollution, or congestion; it is noise—something that reaches every part of the city every day.

We have known for a long time that prolonged exposure to noises, such as loud music or the roar of machinery, can result in hearing loss. Evidence now suggests that noise-related stress also causes a wide range of psychological and physiological problems ranging from irritability to heart disease. An increasing number of people are affected by noise in their environment. By age forty, nearly everyone in America has suffered hearing deterioration in the higher frequencies. An estimated 10 percent of Americans (24 million people) suffer serious hearing loss, and the lives of another 80 million people are significantly disrupted by noise.

What is noise? There are many definitions, some technical and some philosophical. What is music to your ears might be noise to someone else. Simply defined, noise pollution is any unwanted sound or any sound that interferes with hearing, causes stress, or disrupts our lives. Sound is measured either in dynes, watts, or decibels (box table 25.2). Note that decibels (db) are logarithmic; that is, a 10 db increase represents a doubling of sound energy.

City noises come from many sources. Traffic is generally the most omnipresent noise. Cars, trucks, and buses create a roar that permeates nearly everywhere in the city. Around airports, jets thunder overhead, stopping conversation, rattling dishes, sometimes even cracking walls. Jackhammers rattle in the streets; sirens pierce the air; motorcycles, lawnmowers, snowblowers, and chain saws create an infernal din; and music from radios, TVs, and loudspeakers fills the air everywhere.

We detect sound by means of a marvelous set of sensory cells in the inner ear. These cells have tiny projections (called microvilli and kinocilia) on their surface. As sound waves pass through the fluid-filled chamber within which these cells are suspended, the microvilli rub against a flexible membrane lying on top of them. Bending of fibers inside the microvilli sets off a mechanico-chemical process that results in a nerve signal being sent through the auditory nerve to the brain where the signal is analyzed and interpreted.

The sensitivity and discrimination of our hearing is remarkable. Normally, humans can hear sounds from 16 to 20,000 cycles per second. A young child whose hearing has not yet been damaged by excess noise can hear the whine of a mosquito's wings at the window when less than one quadrillionth (1×10^{-15}) of a watt per cm^2 is reaching the eardrum.

The sensory cell's microvilli are flexible and resilient, but only up to a point. They can bend and then spring back up, but they die if they are smashed down too hard or too often. Prolonged exposure to sounds above about 90 decibels can flatten some of the microvilli permanently and their function will be lost. By age thirty, most Americans have lost 5 db of sensitivity and can't hear anything above 16,000 Hertz (Hz); by age sixty-five, the sensitivity reduction is 40 db for most people, and all sounds above 8,000 Hz are lost. By contrast, in the Sudan, where the environment is very quiet, even seventy–year-olds have no significant hearing loss.

Extremely loud sounds—above 130 db, the level of a loud rock band or music heard through earphones at a high setting—actually can rip out the sensory microvilli, causing aberrant nerve signals that the brain interprets as a high-pitched whine or whistle. You may have experienced ringing ears after exposure to very loud noises. Coffee, aspirin, certain antibiotics, and fever also can cause ringing sensations, but they usually are temporary.

A persistent ringing is called tinnitus. It has been estimated that 94 percent of the people in the United States suffer some degree of tinnitus. For most

---

lations. Land prices escalate rapidly as new areas are settled. In the worst cases, rising prices, higher taxes, and the changing character of the community squeeze out long-time residents. Farmers sell agricultural land to developers. More housing developments and shopping centers bring in more people, more business, and more congestion. Urbanization moves ever outward in a dominolike chain reaction. Is this scenario inevitable? No. Some communities have undergone planned, coordinated growth based on socially enlightened and environmentally sound principles.

Balanced against the many drawbacks of "creeping suburbanism," however, is the fact that it enables people to have private homes in smaller communities with more access to nature and outdoor recreation than they had in the city. There are some signs that the mass exodus from the cities is slowing. Perhaps urban expansion is reaching an equilibrium and we now can look at ways of consolidating city centers and creating communities within metropolitan regions. Time, higher energy and land costs, and zoning restrictions are some factors dealt with further in chapter 26.

BOX TABLE 25.2 Sources and effects of noise

Source	Sound pressure (dynes/cm^2)	Decibels (db)	Power at ear (watts/cm^2)	Effects
Shotgun blast (1 m)		150	10^{-1}	Instantaneous damage
Stereo headphones (full volume)	2,000	140		Hearing damage in 30 sec
50 hp siren (at 100 m)		130	10^{-3}	Pain threshold
Jet takeoff (at 200 m)	200	120		Hearing damage in 7.5 min
Heavy metal rock band		110	10^{-5}	Hearing damage in 30 min
Power mower, motorcycle	20	100		Damage in 2 hr
Heavy city traffic		90	10^{-7}	Damage in 8 hr
Loud classical music	2	80		OSHA 8-hour standard
Vacuum cleaner		70	10^{-9}	Concentration disrupted
Normal conversation	0.2	60		Speech disrupted
Background music		50	10^{-11}	
Bedroom	0.02	40		Quiet
Library		30	10^{-13}	
Soft whisper	0.002	20		Very quiet
Leaves rustling in the wind		10	10^{-15}	Barely audible
Mosquito wings at 4 m	0.0002	0		Hearing threshold youth 1,000–4,000 Hz

people, the ringing is noticeable only in a very quiet environment, and we rarely are in a place that is quiet enough to hear it. About thirty five out of one thousand people have tinnitus severely enough to interfere with their lives. Sometimes the ringing becomes so loud that it is unendurable, like shrieking brakes on a subway train. Unfortunately, there is not yet a treatment for this distressing disorder.

One of the first charges to the EPA when it was founded in 1970 was to study noise pollution and to recommend ways to reduce the noise in our environment. Standards have since been promulgated for noise reduction in automobiles, trucks, buses, motorcycles, mopeds, refrigeration units, power lawnmowers, construction equipment, and airplanes. The EPA is considering ordering that warnings be placed on power tools, radios, chain saws, and other household equipment. The Occupational Safety and Health Agency also has set standards for noise in the workplace that have considerably reduced noise-related hearing losses.

Noise is still all around us. In many cases, the most dangerous noise is that to which we voluntarily subject ourselves. Perhaps if people understood the dangers of noise and the permanence of hearing loss, we would have a quieter environment.

SUMMARY

A rural area is one in which a majority of residents are supported by methods of harvesting natural resources. An urban area is one in which a majority of residents are supported by manufacturing, commerce, or services. A village is a rural community. A city is an urban community with sufficient size and complexity to support economic specialization and to require a higher level of organization and opportunity than is found in a village. Metropolitan areas, conurbations, megacities, and core regions are even higher levels of urbanization and urban growth.

Urbanization in the United States over the past two hundred years has caused a dramatic demographic shift. In 1800, only 5 percent of the population lived in cities; now, 75 percent live in the city and only 2 percent are full-time farmers. A similar shift is occurring in most parts of the world. Only Africa and South Asia remain predominantly rural, and cities are growing rapidly there as well. By the end of this century, we expect that more than half the world's people will live in urban areas. By the middle of the next century, the urban population could exceed 5 billion and will probably encompass two-thirds of all humans. Most of that urban growth will be in the supercities of the Third

World. A century ago only thirteen cities had populations above 1 million; now there are 235 such cities. In the next century, that number will probably double again, and three-fourths of those cities will be in the Third World.

Cities grow by natural increase (births) and migration. People move into the city because they are "pushed" out of rural areas or because they are "pulled" in by the advantages and opportunities of the city. Huge, rapidly growing cities in the developing world often have appalling environmental conditions. Among the worst problems faced in these cities are traffic congestion, air pollution, inadequate or nonexistent sewers and waste disposal systems, water pollution, and housing shortages. Millions of people live in slums and shantytowns where conditions would crush any but the strongest spirit yet, these people raise families, educate their children, learn new jobs and new ways of living, and have hope for the future.

The problems of developed world cities tend to be associated with decay and blight. Over the past fifty years, many in the urban middle- and upper-class moved to the suburbs, leaving the old, very poor, handicapped, and economically marginal people in the inner city. Increasing joblessness and poverty in the inner city create a cycle of poverty from which it is difficult to escape. One of the most momentous land-use decisions ever made in the United States was building of the Interstate Freeway System. This system insured the domination of the automobile over our lives and profoundly reshaped our urban environment.

◼ Review Questions

1. What is the difference between a city and a village and between rural and urban?
2. How many people now live in cities, and how many live in rural areas worldwide?
3. What changes in urbanization are predicted to occur in the next fifty years and where will that change occur?
4. Locate the ten largest cities in the world. Has the list changed in the past fifty years? Why?
5. What is a core region? Why are core regions growing and spreading?
6. Describe the current conditions in Mexico City. What forces contribute to its growth?
7. Describe the difference between slums and shantytowns.
8. Why are urban areas in U.S. cities decaying?
9. How has transportation affected the development of cities? What have been the benefits and disadvantages of freeways?
10. Describe some of the regional impacts of growing urban areas.

◼ Questions for Critical Thinking

1. What are the advantages and disadvantages of living in a city as compared to a rural area? Given a choice, which would you prefer?

2. Why are Third World cities growing so large? What is different about their history, economy, and society from ours?
3. Is there an optimum size for a city? What size city would you prefer to live in?
4. If you were the mayor of Mexico City, what would you do to control growth and improve the quality of life?
5. Do you suppose that "push" factors and "pull" factors result in different kinds of city growth and city environment?
6. Imagine yourself in a Third World city. What would your impressions be? What aspects might you enjoy? What would you not enjoy?
7. In many cases, people who live in shantytowns and squatter settlements are more hopeful about the future than are slum residents. Why do you suppose this may be so?
8. Do you think that Third World governments should recognize and encourage squatter settlements?
9. Compare conditions in the developing world now to conditions in Europe or North America one hundred years ago. What factors might change the situation now?
10. Do you believe that crime and crowding are linked? Why are crime rates so high in American cities?

◼ Key Terms

city (p. 517)
Consolidated Metropolitan Statistical Area (CMSA) (p. 518)
conurbation (p. 518)
core region (p. 518)
informal economy (p. 526)
megacity (p. 518)
megalopolis (p. 518)
pull factors (p. 523)
push factors (p. 522)

rural area (p. 517)
shantytowns (p. 525)
slums (p. 524)
squatter towns (p. 525)
Standard Metropolitan Statistical Area (SMSA) (p. 517)
supercity (p. 518)
urban area (p. 517)
urbanization (p. 517)
village (p. 517)

SUGGESTED READINGS

◼ Brown, L. R., and J. Jacobson. 1987. "Assessing the Future of Urbanization." *State of the World 1987*. Worldwatch Institute. A good survey of the state of the world's cities.

◼ Chandler, T., and G. Fox. *3000 Years of Urban Growth*. New York: Academic Press, 1974. A monumental compilation of city history. Fascinating material.

◼ de Soto, H. *The Other Path: The Invisible Revolution in the Third World*. New York: Harper and Row, 1989. Describes the benefits of the informal economy in Third World cities.

◼ Dunkle, T. April, 1982. "The Sound of Silence." *Science* 3(3):30. A highly readable account of tinnitus, ringing in the ears from noise damage.

◼ Herbes, J. *The New Heartlands: America's Flight Beyond the Suburbs*. New York: Time Books, 1986. Good analysis of the growth of regional cities.

ENVIRONMENTAL POLLUTION AND OTHER DILEMMAS

- Huth, M. J. *The Urban Habitat: Past, Present and Future*. Chicago: Nelson-Hall, 1970. A review of the urbanization phenomenon from antiquity to present, urban planning in Europe and America, and a formula for a more humane urban America.
- Kozol, Jonathan. *Rachel and Her Children*. New York: Crown Publishers, 1988. A vivid, personal account of the plight of homeless people in New York.
- Leepson, M. February 22, 1980. "Noise Control." *Congressional Quarterly Editorial Research Report* 1:7. Useful data.
- Livermash, R. "Human Settlements." *World Resources 1990–1991*. Washington, D.C. 1990. A good overview of urban problems and solutions in developing countries.
- Mumford, Lewis. *The City in History: Its Origins, Its Transformations, and Its Prospects*. New York: Harcourt, Brace & World, Inc., 1961. A comprehensive, classic review of city planning through history.
- Steffens, L. *The Shame of the Cities*. New York: Hill & Wang, 1904. Reprinted by Hill & Wang in 1969. A hard-hitting account of urban conditions and how they evolve that still is valid today.
- Steinhart, P. March, 1986. "Personal Boundaries." *Audubon* 88(2):8. An insightful essay on attitudes towards crowding in different cultures.
- Teaford, J. C. *The Twentieth-Century American City*. Baltimore: John Hopkins Press, 1984. Puts urban problems into historical context.
- Tolstoy, L. N. 1881. *What Then Shall We Do?* An account of the census of 1880 and of conditions in Moscow during the age of industrialism. One of the seminal books in social philosophy.
- Vininy, D. R., Jr. April 1985. "The Growth of the Core Regions of the Third World." *Scientific American* 252(4):42. A good discussion of the causes and effects of megalopolis growth.
- The World Bank. *The Urban Edge*. Vol. 14. Washington, D.C., 1990 Annual review of urban development in the Third World.
- Wachs, M., and M. Crawford (eds.). 1990. *The Car and the City: The Automobile, the Built Environment, and Daily Urban Life*. Ann Arbor, Michigan: University of Michigan Press. An interdisciplinary perspective on urban development and change detailing the impact of the automobile on the form and functioning of the city.

PART FIVE

Global

Future

CHAPTER *26*

Building Sustainable Cities

. . . Through sanitary and remedial action in the houses that we have; and then the building of more, strongly, beautifully, and in groups of limited extent, kept in proportion to their streams and walled round, so that there may be no festering and wretched suburb anywhere, but clean and busy streets within and the open country without, with a belt of beautiful garden and orchard round the walls, so that from any part of the city perfectly fresh air and grass and sight of far horizon might be reachable in a few minutes walk.

John Ruskin, *Sesame and Lillies*, 1862

People have lived in cities for over 10,000 years. Many early cities were well planned and built with remarkable levels of comfort and convenience. By rediscovering and incorporating some of the best aspects of small towns and historic cities, we can create human scale and interest in our modern cities.

Our attitudes towards the city have varied greatly over the years. Some societies, such as the early Greeks, viewed the city as the only place where one can be truly human. Since the Industrial Revolution, some people have viewed the city as a source of evil, corruption, and human degradation.

Many utopian communities have been established in efforts to reform societies, individuals, and the environment in which we live. In order to develop sustainable cities in the future, we must consider options, such as urban renewal, suburban redesign, new land-use plans, and new modes of transportation.

Development and reform of Third World cities presents a host of unique problems related to economic, political, and social circumstances in these countries.

INTRODUCTION

As cities grow, we are faced with new challenges for managing and improving the urban environment. We saw in chapter 25 that most people in the developed countries already live in urban areas and the rest of the world is quickly becoming urbanized. Many social and environmental problems have arisen from the rapid pace of urbanization, the size and density of modern cities, and the fact that city development has tended to be a sprawling, unplanned phenomenon. What can we do about these problems? Can cities (and urban societies) be redesigned to create healthier, happier, more civilized environments? Are the rise of technology and separation from nature in cities causes of, or solutions to, urban problems? What role does transportation play in creating urban problems and in offering solutions to them? In this chapter, we will look at some answers to these questions and some options for managing urban environments. We will begin by exploring the history of cities and city planning to give you an overview of how humans have coped with modernization.

CITY PLANNING THROUGH HISTORY

City planning has a long history. Many of the earliest cities in Mesopotamia, India, and Egypt were built in an orderly and civilized fashion. The Greek city-states of the first millenium B.C. reached a level of architectural beauty and rational city planning that has rarely been surpassed. Miletus, Priene, and Pergamum on the Ionian coast of Turkey, for instance, and Alexandria on the Nile delta in Egypt, were splendid examples of Greek city planning. They were beautifully situated, with an orderly, unified design, magnificent buildings, and a pleasant mix of public and private spaces.

Residents of these cities regarded them as oases of comfort, convenience, safety, and delight in an otherwise cruel and dangerous world. Cities were seen as elysian refuges protected by a sacred wall from the chaos of evil spirits, wild animals, and dangerous people in the profane world outside. City temples were regarded as homes for the gods; their gardens were sanctuaries of beauty and sacred order. As centers of education, religion, political power, commerce, communication, and technology, cities offered many advantages over life in the country.

The preference for city over country is implicit in our language. The Latin word for city (*civitas*) is also the root for civil, civilization, civility, citizen, and civics. In ancient cities, the wealthy persons (called citizens) lived in the center near the temple, palace, marketplace, and forum. Serfs and villeins (now spelled villains) lived outside the city wall. The city, as home of the gods, was sacred; wilderness, where evil spirits lived, was profane. Rural areas (*pagus*) were the homes of pagans; the remote country (*heath*) was the land of heathens. No person was fully civilized unless he or she was part of a *polis* or body of citizens. Banishment from the polis was the worst punishment that could befall a citizen.

The Greeks considered optimum city size to be 30,000 to 50,000 people. This size was large enough to support a theater, university, library, temple, lively marketplace, and an active political and intellectual life, yet small enough to draw food from the surrounding region and compact enough so that a person could walk out into the country within a few minutes from anywhere in the city. Although such planned cities weren't called by the term, they represent the concept of cities as ecosystems where the inhabitants participated in exchanges of matter and energy with their environment.

In the sixteenth and seventeenth centuries, Rome, London, Paris, Berlin, Vienna, and Brussels all were rebuilt with spacious boulevards, parks, and imposing public buildings following the rediscovery of Greek and Roman principles of city planning. The design of these European capitals is reflected in Pierre L'Enfant's plan for Washington, D.C., which is one of the most notable city plans in the United States (figure 26.1). Washington's rectangular street grid is cut by diagonal avenues converging on the Capitol and Executive Mansion, forming numerous squares, circles and triangles that make interesting spaces for pocket parks. The broad, tree-shaded streets and impressive vistas created by this plan make Washington a grand setting for the national capital.

The Industrial Revolution brought new levels of social, economic, and environmental problems to European and American cities in the eighteenth and nineteenth centuries. Conditions in which the poorest people lived were worse than most of us can imagine. Filth, poverty, drunkenness, disease, noise, pollution, crowding, appalling working conditions, and wretched housing were standard. Life was short, brutal, and cheap. For the poorest urban residents, conditions in the major European and American cities of the eighteenth and nineteenth centuries were at least as bad as those in Third World cities today.

Many people came to believe that industrial cities—and the society and technology that spawned them—were inherently evil. Ned Ludd and his followers tried to smash the machines that were destroying their way of life. The simplicity of agrarian villages began to seem both wholesome and endangered, hemmed in by evil forces in the wilderness and the city. The United States' founding fathers had deep suspicions of large cities and urban people. They strongly favored a rural Jeffersonian democracy over an urbanized society. The belief that cities are evil and dangerous and that rural life is wholesome and pure still is a powerful political force.

URBAN REFORM

The unplanned growth and often dirty and sordid conditions of industrial cities have given rise to a variety of utopian communes

FIGURE 26.1 Pierre L'Enfant's plan for Washington, D.C. is based on neoclassical principles of city planning, with broad avenues cutting across a rectangular street pattern. Its many pocket parks, squares, and traffic circles create a pleasant urban environment. Although the ceremonial areas are grand, poverty and neglect in nearby residential areas cast a pall on our nation's capital.

and experimental new towns. Some notable examples include the Shaker communities of Kentucky and Massachusetts, the Oneida community in New York, and the Amana colonies in Iowa. New Harmony, Indiana, founded in 1815 as a religious community and expanded in 1825 by visionary industrialist Robert Owen, was one of the most important city planning experiments in America, combining scientific farming, communal living, and progressive social principles.

Experimentations with alternative life-styles and communal living continue today. During the 1970s, thousands of communes and experimental living arrangements sprang up in the United States. Most lasted only a short time, but others have become important centers for exploring a variety of alternative economic, religious, educational, and social systems. These "intentional communities" may have valuable lessons to transfer to mass society.

Garden Cities and New Towns

The twentieth century also has seen numerous experiments in building **new towns** for society at large that try to combine the best features of the rural village and the modern city. One of the most influential of all urban planners was Ebenezer Howard (1850–1929), who not only wrote about ideal urban environments, but also built real cities to test his theories. In his *Garden Cities of Tomorrow*, written in 1898, Howard proposed that the congestion of London could be relieved by moving whole neighborhoods to **garden cities** separated from the central city by a greenbelt of forests and fields. Howard was the first modern urban planner to advocate comprehensive land-use planning and to reintroduce the Greek concept of organic growth and human-scale measure to the city.

In the early 1900s, Howard worked with architect Raymond Unwin to build Letchworth and Welwyn Garden just outside of London. Interurban rail transportation provided access to these cities. Houses were clustered in "superblocks" surrounded by parks, gardens, and sports grounds (figure 26.2). Streets were curved. Safe and convenient walking paths and overpasses protected pedestrians from traffic. Businesses and industries were screened from housing areas by vegetation. Each city was limited to about 30,000 people to facilitate social inter-

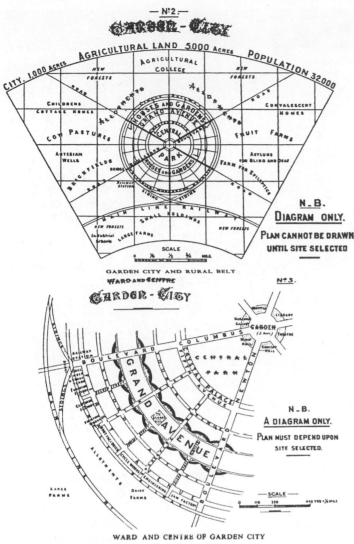

FIGURE 26.2 A plan for a model garden city, designed by Ebenezer Howard in 1902, combines comprehensive land-use planning with housing, jobs, transportation, recreation, nature, and civic life. Houses are clustered on curving streets, interspersed with landscaped boulevards, walking paths, and gardens. The surrounding countryside is zoned to produce food for the town and to protect environmental values.

action. Housing and jobs were designed to create a mix of different kinds of people and to integrate work, social activities, and civic life. Trees and natural amenities were carefully preserved and the towns were laid out to maximize social interactions and healthful living. Care was taken to meet residents' psychological needs for security, identity, and stimulation.

Letchworth and Welwyn Garden each have seventy to one hundred people per acre. This is a true urban density, about the same as New York City in the early 1800s and five times as many people as most suburbs today. By planning the ultimate size in advance and choosing the optimum locations for housing, shopping centers, industry, transportation, and recreation, Howard believed he could create a hospitable and satisfying urban setting while protecting open space and the natural environ-

ment. He intended to create parklike surroundings that would preserve small-town values and encourage community spirit in neighborhoods.

The Scandinavian countries have been especially successful in building new towns (Box 26.1). Other countries also have built garden cities. The Soviet Union has built about two thousand new towns. Some are satellites of existing cities, and some are entirely new wilderness communities. As of 1976, Great Britain had about 1 million people living in thirty-two new towns.

Planned communities also have been built in the United States following the theories of Ebenezer Howard, but most plans have been based on personal automobiles rather than public transit. In the 1920s, Lewis Mumford, Clarence Stein, and Henry Wright drew up plans that led to the establishment of Radburn, New Jersey, and Chatham Village (near Pittsburgh). Reston, Virginia, and Columbia, Maryland, both were founded in the early 1960s and are widely regarded as the most successful attempts to build new towns of their era (figure 26.3). Another movement to build new towns according to Howard's principles has sprung up in the 1990s. Towns, such as Seaside in northern Florida, Kentlands in the Maryland suburbs of Washington, and Laguna West near Sacramento, have cluster houses to save open space and create a sense of community. Commercial centers are located within a few minutes walk of most houses and streets are designed to encourage pedestrians and to provide places to gather and visit.

American Suburbs

As we saw in chapter 25, better transportation—first by streetcar or trolley and then by private automobile—made it possible to move away from the congestion and pollution of the industrial city. The resulting "bedroom" communities generally have fallen far short of the better features of urban planning, however, and vast tracts of monotonously uniform housing developments were built that lack many of the benefits of either urban or rural living. This urban sprawl was fueled by a pressure for new housing after the Second World War. The new prosperity of the working classes and the "baby-boom" (chapter 7) created a huge demand for houses. Young families wanted to move into the country where there was open space and where land was less expensive. Automobiles and cheap gasoline made it possible to commute from suburbs to jobs in town. Builders saw the advantages of large-scale developments where houses were put up in assembly-line fashion.

Among the most famous of these mass housing developments were the Levittowns built on Long Island in 1947, outside Philadelphia in 1951, and in New Jersey in 1958. They featured small, three-bedroom, expandable homes for young working-class and middle-class families. A few styles were repeated over and over again. Similar housing developments sprang up around every major American city. Often built on treeless farm fields to reduce construction costs, these housing devel-

BOX 26.1   Tapiola and Farsta

Tapiola, Finland, and Farsta, Sweden, are outstanding examples of successful new towns built according to Ebenezer Howard's ideals for garden cities. Tapiola, built in 1951 as a suburb of Helsinki, occupies a beautiful setting along the shores of the Gulf of Finland. It is divided into three distinct neighborhoods separated by greenbelts. About 20 percent of the housing is high-rise apartments, and the rest is mostly single-family terrace and row houses nestled among rocky hills and lush evergreen forests (box figure 26.1). Building designs harmonize. Housing is socially integrated and cooperatively managed. You can't detect from the outside what the income of the occupants might be.

Parks and playgrounds radiate from a central plaza where a striking town center is located. Walkways lead to carefully landscaped residential areas. Pedestrians need never cross streets at grade level. Industrial centers are unobtrusively situated and screened by vegetation so that people can work close to where they live and yet not suffer impacts of noise, pollution, and heavy traffic.

Farsta, Sweden, was built in 1954, 9 mi south of Stockholm and beyond the greenbelt enclosing the metropolitan area. Farsta is centered around a pedestrian shopping mall with an underground parking garage and a rapid railroad station with express service to the central station in downtown Stockholm. About half of all adult residents work in Farsta. Most of the rest commute by train to Stockholm. Housing is primarily in high-rise apartment buildings close to the central plaza, but single-family houses are also

BOX FIGURE 26.1 Tapiola, Finland, occupies a beautiful setting on the Baltic Ocean near Helsinki. High-rise apartments and cluster housing sit in parklike open spaces. Mass transit provides easy access to the city.

available. Parks and recreation areas surround the village center. No one walks far to school, work, shopping areas, or playgrounds. About 36 percent of all trips are by foot, 27 percent are by public transit, and only 30 percent are by car—even though most families own at least one automobile. The town has a boat harbor and is located in a nature reserve. Within a few minutes of leaving home, one can be in a pristine spruce-fir-birch forest.

Farsta uses less than half as much energy per person as a typical American city. It gets electricity and steam from the Agesta Atomic Power Plant, one of the

first civilian nuclear reactors built in the world. The reactor is only 3 mi from the town center (which makes district heating possible) but is built in a deep cavern excavated into the solid granite bedrock.

Both Tapiola and Farsta have international reputations for beauty and a high quality of life. Because housing uses much less space in these cities than in a typical American suburb, more land is available for open space, parks, and recreation areas. Their higher density also supports neighborhood shops, mass transit, and other urban community services not commonly available in American suburbs.

opments had little to break the monotony of long rows of identical houses (figure 26.4). Such suburbs offered little in the way of jobs, recreation, social services, shopping, or community involvement.

What are some of the difficulties in building a planned suburban community? First, it's hard to find enough open land for a complete town at an affordable price. There is a conflict between being far enough away from an urban area to find

cheaper land and yet near enough for commuting. New town builders compete with conventional housing developments that offer low prices by virtue of high density, mass-assembly construction, uniform design, and minimal public services. The market for new housing is largely young families that can afford only the most basic housing.

A planned town needs many features to make it work; housing, businesses, shopping, transportation, and government

FIGURE 26.3 Reston, Virginia, is among the most successful urban planning experiments in the United States. A mix of people live and work amidst lakes, woods, play-grounds and convenient shopping centers. Housing is clustered in villages, each of which has both multifamily and single family units. Walking and bike trails make nonmotorized travel safe and convenient.

FIGURE 26.4 This housing development south of San Francisco inspired Malvina Reynolds to write the song "Little houses on the hillside—and they're all made out of ticky tacky, and they all look just the same."

all need to be in place right away. Since most developers are undercapitalized, however, it's hard to get everything going at once.

Polycentric Cities

The recent movement of businesses and people from the central cities to small, remote cities and rural areas might have been even larger except that most people want access to the opportunities and amenities of urban areas. The suburbs are often a good compromise between city and country. In many suburbs, urban cores have developed that rival the downtown "loop" in business activity and urban benefits. Some cities are changing from a concentric set of suburban housing developments around a single central core to a **polycentric complex** of several urban cores linked by freeways (figure 25.12e). People and goods now move more commonly between suburban centers than between suburbs and the main urban center.

Our postindustrial economy has changed from primary resource extraction and manufacturing to information processing, communications, finance, and administration. Few of the original reasons for business to locate in the central city still apply. Service jobs are replacing the "smoke-stack industries" that originally built our big cities. Businesses no longer need to be in the center of big cities to be close to transportation, customers, and markets. The needs of a computer manufacturing company, an insurance office, or a biotechnology research laboratory in the 1990s will be very different from those of a steel mill, shipyard, or automobile factory of the 1920s. By moving out of the cities, businesses find cheaper land prices, lower taxes, parking spaces, and freeway access. Suburban locations are closer to executive and white-collar housing areas, airports, restaurants, hotels, and other amenities that attract business managers. It may be easier

to start over with new buildings specifically designed for the modern electronic office than to remodel space in old buildings that were designed to meet the needs of an earlier era. But this flow of jobs, taxes, and amenities to the suburbs strands the elderly and other low-income groups in the inner city, compounding problems there.

Regional Relocation

In the process of moving from the older central business district to the suburbs, many businesses decide to relocate in a new region of the country, as well. Both industry and the work force have migrated from the "Rust Belt" of the Northeast and Great Lakes states to the "Sun Belt" of the South and Southwest. In 1980, for the first time in our history, most (53 percent) of the United States population was in the South and West. Instead of movement from one big city to another, however, much of the migration has been to satellite cities on the outskirts of metropolitan areas or to autonomous cities that are environmentally and culturally more attractive. Places like Chapel Hill, North Carolina; Boulder, Colorado; San Jose, California; and Orlando, Florida, have grown rapidly because of the amenities and opportunities they offer. Access to higher education and research communities is a major attraction of these growing cities.

Cities of the Future

Is the expansion of cities—especially onto agricultural land—a waste of land and resources? Some people feel that growing cities decrease local agricultural resources and corrupt once pristine countryside and quaint rural towns. Other people see this as a healthy opportunity to decongest central cities and build smaller, more pleasant, more livable communities that have the benefits of both technological cities and rural villages. Perhaps the demographic shifts of the last forty years have merely been a transition between nineteenth-century industrial cities and garden cities of tomorrow. The polycentric network of urban villages in surrounding metropolitan areas may be a step towards more human-scale communities.

An alternative to spreading the population across a wide area of the countryside is to build upwards. This model, which depends strongly on technology, has been called the **technopolis, vertical city,** or **city of the future.** Its form has been a science fiction image since the 1930s (figure 26.5). The central hub of most big American cities now is dominated by skyscrapers and a highly technological environment. The emerging supercities of the Third World are also moving toward this style, in part because of its association with wealth, power, and progress.

There have been many proposals for gigantic, vertical supercities based on skyscrapers and completely artificial environments. An example was Buckminster Fuller's proposal for a mile-high geodesic dome to cover most of Manhattan. He also proposed a hollow, tetrahedronal building two hundred stories high that would float on the ocean and could house 1 million people. It could be built in stages, he suggested, like a beehive, with trailerlike modules inserted in successive layers into a mountainous, open-truss framework. Some architects have talked seriously about the eventual possibility of erecting 500-story, mile-high buildings! No one knows what the psychological and physiological effects might be of living 5,000 feet above street level.

Japan is now building eighteen new high-technology cities intended to be centers for economic and scientific growth in the next century. With names like Teletopia, Agripolis, and New Media City, these regional research, education, and marketing centers will have innovative housing, enclosed shopping malls, and high-technology communication and transportation systems. They will concentrate on leading-edge research and industries, such as fifth-generation supercomputers, biotechnology, lasers, ceramics, and bioelectronics. Some such high-tech cities may be giant, floating structures not unlike Buckminster Fuller's visionary proposals. Others may consist of a maze of tunnels and chambers entirely underground, not unlike a giant ant nest, opening onto huge twenty-story-deep air wells. This plan would conserve energy and would preserve scarce surface air; however, many technical and psychological problems need to be overcome.

PLANNING SUSTAINABLE CITIES AND SUBURBS

For many people, the idea of a completely enclosed, artificial environment sounds too much like science fiction—or technology fiction—and not a very good way of living. What are the alternatives? How can we develop a humane, aesthetic, efficient, and ecologically balanced urban environment that will be suitable for everyone?

FIGURE 26.5 The future city has grown overwhelmingly large and artificial in this illustration by Frank Paul for *Amazing Stories of 1928.* Although the architecture is Victorian, the skyways, skyscrapers, and helicopters are prescient of the vertical city of the future.

Urban Renewal

The United States began a massive program for renewal of deteriorating cities in the 1950s. Billions of dollars were spent to revitalize blighted areas of the inner city, but for some residents the solutions were worse than the problems they were intended to solve. Old neighborhoods, slums, and decaying industrial areas were cleared or renovated but usually to create luxury housing or glamorous projects, such as sports arenas, concert halls, high-rise office towers, or "festival" shopping malls. Often little attention was devoted to relocating former residents or solving the underlying social and economic problems that created the slums.

Who are the people being displaced by urban renewal projects? We tend to think of stereotyped images of homeless people in "Skid Row" flophouses, drunks, and junkies, as well as businesses that cater to the down-and-out, such as pawn shops and liquor stores. With such images in mind, it's easy to justify sweeping, civic-minded changes. Usually, however, such people and places simply shift to the next nearest rundown area, which quickly degenerates into the same slum conditions.

FIGURE 26.6 The Pruitt-Igoe public housing project in St. Louis was a dismal failure in urban planning. All 33 high-rise buildings were demolished in 1976.

In fact, inner cities also have cohesive neighborhoods that reflect the earlier settlement patterns of a city and are, therefore, often ethnic clusters with strong traditions, and cultural identities. These established communities also are disrupted and displaced by urban renewal. The residents are forced to leave their friends and cultural support relationships to seek new housing because their existing homes and businesses are torn down or because renovations increase the property values, creating unaffordable rents and/or taxes. Furthermore, such neighborhoods often have indigenous shops and other small businesses that provide income and services for residents. The emotional and cultural shock of moving families into unfamiliar neighborhoods or public housing projects can be devastating, especially when it is combined with the need to seek new means of economic support.

Much of the public housing built during the 1950s and 1960s consisted of massive blocks of high-rise apartment buildings that tended to concentrate and aggravate the problems of the poor, predominantly racial minority people who occupied them. The physical plans of the buildings tended to isolate residents rather than promote a spirit of community identity. There were few jobs nearby and few opportunities to escape from the poverty and violence. Vandalism, drug problems, and crime rates were high. Police often refused to respond to calls from these projects because they were so dangerous. One of the most infamous public housing projects was the Pruitt-Igoe project in Saint Louis, Missouri. Built in 1958 to house about 10,000 people, it won an architectural award for modern design; however, it lacked secure space, quality construction materials, and attention to residents' needs. By 1970, it was so badly run down

and dangerous that most residents had left. The city concluded that all thirty-three buildings were beyond repair, and demolished the entire project in 1976. (figure 26.6).

Not all public housing has been so disastrous. In the Cochran project, also in Saint Louis and not far from Pruitt-Igoe, residents have formed an association to clean up and manage the buildings. They hire their own security guards, set standards for cleanliness and orderliness, and evict incorrigible troublemakers. The association also mediates disputes, organizes day care, and helps residents find jobs. By taking charge of their own lives and building a sense of community, these people have created a better place to live.

Suburban Redesign

What are the drawbacks of suburban living? Most suburbs are too spread out for people to meet and interact, except with their immediate neighbors. Residential streets are empty during much of the day. Suburbs often have no sense of community and can be places of alienation and loneliness, just as cities often are. Although many urban problems have been eliminated, so have many of the finer aspects of city living. Suburbs, for instance, lack the activities, energy, and diversity that make cities exciting and dynamic. They have very limited artistic, cultural, and educational opportunities, compared to cities.

The low population density of the suburb makes it impossible to support an efficient public transportation system, so private automobiles are essential. Carpools for work, school, and extracurricular activities help fill group transportation needs. The uniform, single-family, detached houses of the suburbs offer few options to those who don't fit into the traditional, middle-class, nuclear family that is the foundation of suburban neighborhoods. As children grow up and move away from home, parents find that they don't need a big house anymore. Single-parent families, households of single adults, the elderly, and the poor tend to have difficulty finding affordable suburban housing that fits their needs.

What can or should be done to create ideal suburban environments? Obviously not everyone in America will want to leave established suburban communities for new garden communities or wilderness utopias. Nor should they. It would be a terrible waste to abandon existing buildings and civic infrastructures. We need, instead, to find ways to remodel and revitalize existing suburban cities, to reduce their problems, and adapt to the changing needs of their residents.

How can suburbs be redesigned to make them more diverse, flexible, and energy efficient? Ten proposals are listed below that might give them some of the better aspects of both the rural village and the big city.

1. Limit city size or organize them in modules of 30,000 to 50,000 people. This size is big enough to be a complete city but small enough to be a community. A greenbelt of agricultural and recreational land around the city sets limits to growth while ensuring the most efficient use of land within the city. By careful planning and cooperation with neighboring regions, a city of 50,000 people can have real urban amenities such as museums, performing arts centers, schools, hospitals, etc.

2. Establish a city plan to determine in advance where development will take place and to ensure rational land use. This protects property values and prevents chaotic development in which the lowest uses drive out the better ones. It also takes into account historical and cultural values, agricultural resources, and such ecological factors as impact on wetlands, soil types, groundwater replenishment and protection, and preservation of aesthetically and ecologically valuable sites.

3. Turn shopping malls into real city centers that invite people to stroll, meet friends, or listen to a debate or a street musician (figure 26.7). It has been said that if there aren't one hundred places for an impromptu celebration, then a place isn't a real city. Another test of a city is a vital night life. Design city spaces with sidewalk cafes, pocket parks, courtyards, balconies, and porticoes that shelter pedestrians, bring people together, and add life and security to the street. Restaurants, theaters, shopping areas, and public entertainment that draw people to the streets generate a sense of spontaneity, excitement, energy, and fun.

4. Locate everyday shopping and services so people can meet daily needs with greater convenience, less stress, less automobile dependency, and less use of time and

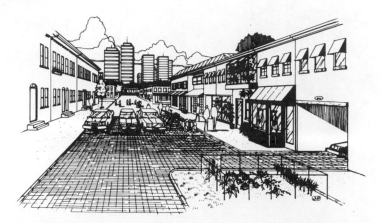

FIGURE 26.7 Limited-access residential streets mix pedestrians and autos while creating room for play areas, gardens, greenhouses, and small shops. Such streets, called "Woonerf" in Holland, have been used with great success in Europe.

energy. This might be accomplished by encouraging small-scale commercial development in or close to residential areas. Perhaps we should once again have "mom and pop" stores on street corners or in homes.

5. Increase jobs in the community by locating offices, light industry, and commercial centers in or close to suburbs, or by enabling work at home via computer terminals. Such alternatives save commuting time and energy and provide local jobs. On the other hand, there are concerns about work-at-home employees being exploited in low-paying "sweat-shop" conditions if cottage industries are revived. This might have to be handled carefully.

6. Make it possible to use small, low-speed, energy-efficient vehicles (microcars, motorized tricycles, bicycles, etc.) or to walk for many local trips now performed by full-size automobiles. Creating special traffic lanes, reducing the number or size of parking spaces, or closing shopping streets to big cars might encourage such alternatives. Clusters of high-density housing also might help a community support a viable public-transit system.

7. Encourage more diverse, flexible housing to provide alternatives to conventional, detached, single-family houses. "In-fill" building between existing houses saves energy, reduces land costs, and might help provide a variety of living arrangements. Building duplexes and allowing owners to turn basements, attics, garages and additions into rental units provides space for those who can't afford a house and brings income to retired people, for instance, who don't need a whole house themselves. Allowing single-parent families or groups of unrelated adults to share housing and to use facilities cooperatively also provides alternatives to those not living in a traditional nuclear family. One of the great "discoveries" of urban planning is that mixing various types of

housing—individual homes, townhouses, and high-rise apartments—can be attractive if buildings are aesthetically arranged in relation to one another.

8. Create housing "superblocks" that use space more efficiently and foster a sense of security and community (figure 26.8). Widen peripheral arterial streets and provide pedestrian overpasses so traffic flows smoothly around housing areas; then reduce interior streets within blocks to narrow access lanes with speed bumps and barriers to through traffic so children can play more safely. The land released from streets can be used for gardens, linear parks, playgrounds, and other public areas that will foster community spirit and encourage people to get out and walk. Cars can be parked in remote lots or parking ramps, especially where people have access to public transit and can walk to work or shopping.

9. Make cities more self-sustainable by growing food locally, recycling wastes and water, using renewable energy sources, reducing noise and pollution, and creating a cleaner, safer environment. A greenbelt of agricultural and forest land around the city not only provides food and open space as well as such valuable ecological services as purifying air, supplying clean water, and protecting wildlife habitat and recreation land.

10. Encourage public participation in decision making. Emphasize local history, culture, and environment to create a sense of community and identity. Create local networks in which residents take responsibility for crime prevention, fire protection, and home care of children, the elderly, sick, and disabled. Coordinate regional planning through metropolitan boards that cooperate with but do not supplant local governments.

Zoning and Land-Use Plans

Zoning and comprehensive, long-range planning are among the most important tools of city planners in accomplishing these goals of creating sustainable, hospitable communities. The first zoning laws in America were passed in New York City in 1916 to limit the size of new skyscrapers that were casting shadows on nearby streets. The concept is not new, however; Nero established similar rules in first century A.D. Rome. Building codes aimed at eliminating dangerous and fraudulent building practices also originated in Rome at about the same time because shoddy high-rise apartment buildings were collapsing.

Civic planners now use zoning to improve and protect urban and suburban areas by reserving open space and preventing conflicting uses from being adjacent to each other. Some of the most common zoning categories are presented in table 26.1. There are dangers in both too much and too little zoning. Overzoning causes undue restrictions and stifles innovation and adaptation to changing needs in the city. Underzoning will not be strict enough to prevent disadvantageous land uses. It is im-

TABLE 26.1 Common zoning districts

Category	Use allowed
Residential	
R-1	One-family houses
R-2	One- and two-family houses
R-3	Low-density, multiple family townhouses
R-4	Medium-density, multiple-family apartment buildings
R-5	High-density, multiple-family apartment buildings
Business	
B-1	Limited business and office buildings
B-2	Retail business
Industry	
I-1	Light industry and manufacturing
I-2	Heavy industry
Special purpose	
P-1	Public land, schools, parks
P-1/R-1	Present public, if reused R-1
P-1/B-1	Present public, if reused B-1

TABLE 26.2 Land characteristics and recommended uses

Land type	Recommended uses
Surface waters and riparian lands	Ports, harbors, marinas, parks and open space, industry requiring water access.
Marshes and wetlands	Water storage, recreation, wildlife habitat.
Fifty-year flood plain	Agriculture, forestry, recreation, some development not sensitive to flooding.
Scenic, historic, and cultural areas	Parks, recreation, greenbelts, open space, wildlife habitat.
Aquifer recharge zone	Agriculture, forestry, open space, low-density housing, industry that produces no toxic effluents; restricted fertilizer pesticide, solvent, and herbicide use.
Steep, erodable land	Forestry, recreation, open space; maximum housing density: one house per 3 acres with careful erosion control.
Prime agricultural land	Agriculture; housing density not more than one house per 25 acres.

From Ian McHarg, *Design with Nature*. Copyright © 1969 Ian L. McHarg. Reprinted by permission of the author.

portant to set up comprehensive, long-range, land-use plans because it is difficult, if not impossible, to do retroactive zoning.

In recent years, we have come to understand that zoning can and should be used to protect environmental values as well as property values. Table 26.2 describes some categories of particularly sensitive lands that might be damaged by urban development. This table, which is adapted from *Design with Nature* by Ian McHarg, also provides some compatible uses for these

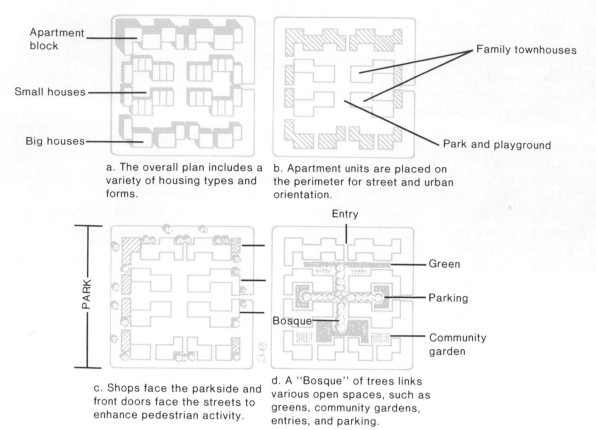

a. The overall plan includes a variety of housing types and forms.

b. Apartment units are placed on the perimeter for street and urban orientation.

c. Shops face the parkside and front doors face the streets to enhance pedestrian activity.

d. A "Bosque" of trees links various open spaces, such as greens, community gardens, entries, and parking.

FIGURE 26.8 (*a*) Design for a superblock of mixed housing options. (*b*) and (*c*) Apartments and shops are placed on the perimeter for street orientation. Family townhouses face playgrounds and parks in the center. (*d*) A "Bosque" of trees links various open spaces, greens, community gardens, and promenades. Underground parking keeps cars off streets.

lands. McHarg and other land-use planners have developed sophisticated techniques for taking environmental, social, and economic values into account in comprehensive, land-use planning.

To provide flexibility to their land-use plans, some communities allow trade-offs and marketing of use permits. In Montgomery County, Maryland, the total number of acres that can be developed is strictly controlled. Farmers who do not want to develop their land sell "rural development rights" to builders in nearby Washington, D.C., suburbs such as Bethesda and Silver Spring, Maryland. Since the farmer then can never sell the land for housing, taxes are based on farm values rather than subdivision rates, and tax pressure to sell the land for development is eliminated. In San Francisco, zoning regulations require developers to balance losses in housing or open space in one area by building new housing or establishing a new park somewhere else.

One of the best ways to control and direct city growth is to regulate the extension of city services to suburban and exurban areas. Boulder, Colorado, for example, has stabilized its population and preserved a greenbelt that separates it from Denver's sprawl by establishing a "blueline" around the city, beyond which no water, sewer, or services will be provided. Boulder also has purchased 14,000 acres of land in its greenbelt to control uses and has established cooperative planning with the county and surrounding cities to control growth. Within the city, Boulder

limits new housing developments to two percent of the existing housing stock each year.

Transportation Options

Transportation is one of the most critical needs of any urban area. Food, fuel, water, and building materials must be delivered to consumers. Workers need to get to their jobs. People place a high priority on being able to travel to visit friends and family, and for shopping, church, recreation, and entertainment. We will put up with a great deal of trouble, expense, discomfort, and even danger, to be able to travel. As we saw in chapter 25, transportation systems have been major factors in shaping cities. A big problem in redesigning cities is to find cheap, safe, reliable, convenient, and environmentally benign transportation. We need to relieve congestion and crowding in cities and make transportation available to those who do not drive or cannot afford a car.

Present Preferences

We make some 127 billion trips each year in the United States, traveling nearly 1.6 trillion km, or 3 trillion passenger mi in total. About three-quarters of all passenger trips and total distance traveled is interurban transport (figure 26.9). Private automobiles are used for 93 percent of all trips and account for 90 percent of total passenger miles. We have about 18 million au-

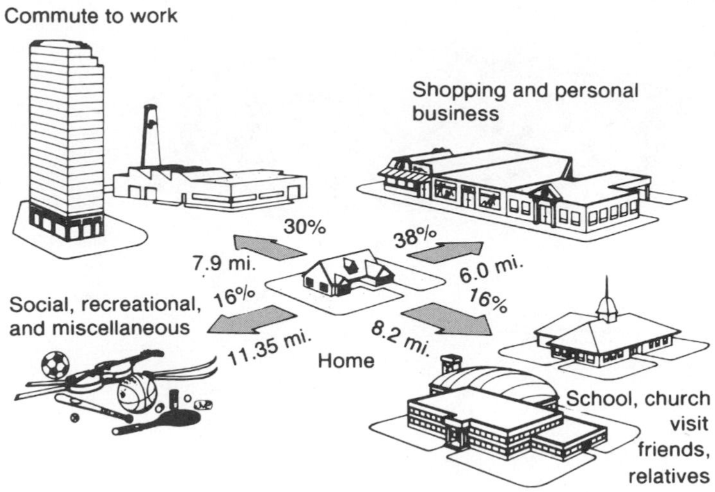

FIGURE 26.9 Destination percentages and average distances in miles of daily trips in the United States. Although the average trip is about eight miles, the most frequent trip is less than three miles and involves only one or two persons.

tomobile accidents each year in the United States, resulting in 2 million serious injuries and 45,000 deaths in 1990. Automobiles consume about 60 percent of our oil imports and cause about 60 percent of air pollution from nitrogen oxides, carbon monoxide, and lead. Public transportation accounts for 20 percent of commuting to work, but only 1 percent of other trips, so altogether it makes up only 7 percent of all passenger transportation. Airlines are the most popular form of interurban mass transit, providing 8.6 percent of total passenger miles. Buses and trains make up slightly more than 1 percent of all passenger transportation in the United States.

The supremacy of the automobile in our society cannot easily be overturned, even if we assume it would be advantageous in terms of energy use, pollution, public expense, and safety. We enjoy the privacy, comfort, convenience, and speed of our private automobiles. In some cases, automobiles are the most efficient way of moving people from one place to another, especially when we live in very low densities and travel long distances. Furthermore, automobiles are improving. Their fuel-use efficiencies have nearly doubled in the past two decades, from an average of 13 mpg in 1970 to 28.1 mpg in 1990 and much more could be done to improve mileage.

We shouldn't assume that any new mass-transit system will be superior in cost and energy efficiency to automobiles. If one takes into account the lifetime costs of both building and operating new transportation systems, they often turn out to be worse than existing systems, especially if automobiles are used for carpools or in combination with express buses or other mass-transit systems on congested routes. However, if cities were redesigned to increase housing densities and decrease trip distances, other options could substitute for automobiles for many trips.

Individual Transit

The invention of the safety bicycle in 1885 was hailed as the "nineteenth century's greatest invention," a "boon to all mankind," and "a force for social revolution." The reason for this enthusiasm was that the bicycle is cheap, fast, easy to operate and store, and simple to maintain. Where only the rich could afford a horse and buggy, even ordinary people could buy a bicycle. Thousands of people joined bicycle clubs, and touring became a popular sport. It was the Society of American Wheelmen that first campaigned for good roads and paved the way (literally) for automobiles.

With the introduction of cheap automobiles in 1908, all progress in bicycle design came to an abrupt halt, and bicycles were considered only children's toys. They presently are enjoying renewed popularity and would be suitable for many urban trips if streets were designed to accommodate them. New models have been designed that are safer, faster, and easier to ride than previous versions. Small trailers for carrying children or cargo are available. Recumbent self-propelled vehicles enclosed in a fairing (cover), some with three or four wheels and places for two or more riders, have reached speeds of 62 mph on level ground (figure 26.10).

For those who want or need motorized vehicles, motorcycles, mopeds, enclosed tricycles, or electric cars can be used for many short, urban trips. In 1990, General Motors invented an electric car suitable for freeway use. The *Impact* can go 55 mph for 120 mi on one battery charge. Its maximum speed is 100 mph, and it can accelerate from zero to 60 mph in 8 sec. The 400 kg (880 lb) lead-acid battery will last only 2 years and costs $1,500 (figure 26.11). At present, this electric car costs about twice as much to operate as a conventional gasoline engine. If new, long-lasting batteries are invented or gasoline prices go up, however, electric cars may become more feasible. Although cars themselves create no pollution (and thus are suitable for smoggy cities such as Los Angeles), generating the electricity to run them with coal-fired or nuclear plants does. An experimental solar car park in Basil allows commuters to recharge their batteries with photovoltaic energy (Chapter 20) while they work.

Personal rapid transit (PRT) systems have been proposed in which small, individual, computer-directed cars would travel along a monorail network nonstop from wherever you are to wherever you want to go. Prototype systems have been built, but this technology has never been demonstrated in a full-scale system. In 1990, Chicago awarded a contract to the Taxi 2000 Corporation for preliminary design of a PRT system.

Mass Transit

Mass transit systems are usually divided into heavy rail systems (railroads and subways), light rail systems (trolleys, streetcars, and monorails), and buses (either on ordinary streets or on special roadways). Each of these systems has advantages and disadvantages, depending on the situation. Fixed-rail systems, for instance, can work only when trip origins and destinations are close

Zzipper

Vector single

Avatar 2000

Schöndorf all–weather vehicle

FIGURE 26.10 For many trips, human-powered vehicles provide an inexpensive, energy-conserving, nonpolluting, healthful alternative to the automobile. Addition of a clear plastic faring or cover (upper left) on the front of a bicycle boosts efficiency about 20 percent. Recumbent bicycles both reduce air drag and are more comfortable to ride than conventional machines. The Vector single (top right) has a full faring and has reached 100 km/hr (62 mi/hr). Other recumbents (lower left and lower right) are designed for commuting, shopping, and other short trips.

to rail lines. If people have to drive to the rail station, they are likely to continue to drive all the way to their destination. One way to encourage mass transit use is to create convenient park-and-ride lots at the beginning of the rail system and to limit street widths and parking spaces within the city.

In general, the higher the population density, the more necessary and the more feasible is mass transport. It takes about ten families per acre to support regular bus service, although half that density can support a minibus or dial-a-ride system. Inner-city densities exceed two hundred persons per acre. A fixed rail system requires a population of about 500,000 people concentrated around a few major trip origin and destination points.

One of the most successful new subway systems in the world is in Hong Kong (figure 26.12). It was finished on or ahead of schedule and well within its budget of $4.2 billion. Revenues are 50 percent greater than operating costs, even though fares are only twenty-five to sixty cents for most trips. Stations are roomy, spotless, and quiet. Trains are fast, frequent, and clean. Several factors contribute to this success. The 5.3 million people in Hong Kong are densely packed between the ocean and the mountains, and it only takes three intersecting lines to reach most parts of the city. Almost everyone lives in high-rise apartment

buildings, and half of them live within a five-minute walk of a subway station. Only one person in thirty owns a car, so most residents are dependent on public transportation. Finally, the population is predominantly homogeneous, respectful, and well-behaved.

The Washington, D.C., Metro is generally considered to be the premier subway system in the United States. Originally estimated to cost $2.5 billion for a 100-mile system, it has already cost $6 billion and is only half finished. The trains are safe, clean, comfortable, and well run, but they don't yet reach far enough into the suburbs to catch many potential riders. Most people still drive to work, and streets are congested. The $250 million per year subsidy that Congress gives to the Metro is nearly half the mass-transit budget for the whole country. Without the subsidy, however, rides would cost twice as much as they do now, and the system might well collapse.

Streetcars, trolleys, and electric interurbans are the oldest motorized urban transit systems. Many cities tore up tracks of profitable systems in the 1950s because they were considered old-fashioned. Automobile manufacturers, oil companies, and others with vested interests supported this trend by buying up streetcar systems and then scrapping them.

GLOBAL FUTURE

FIGURE 26.11 The Hong Kong subway comes above ground when it reaches the end of the line on the Kowloon Peninsula. This is one of the best subway lines in the world. It is clean, fast, quiet, and it makes money. More than 5 million people live within walking distance of a subway station and only 1 percent owns an automobile.

FIGURE 26.12 The General Motors *Impact* uses conventional lead-acid batteries and efficient electric motors to achieve freeway speeds. It can travel 120 miles at 55 mph before needing recharging.

Now many planners are suggesting that light rail may be the best urban transit system after all. Light rail systems cost only $8 million to $10 million per mile for construction, compared to about $100 million per mile for urban freeways and subway systems. By coupling several cars together, a light rail system can carry three hundred to four hundred people with one driver versus only forty to fifty passengers on a bus. An electric system may be more compatible with future energy sources than internal combustion engines and doesn't produce combustion exhaust emissions.

Contrary to what you might think, buses are generally cheaper, quieter, more energy-efficient, provide better service, and have less overhead in construction and operation than any other mass-transit motor vehicle system. Buses carry nearly as many people per mile of transit lane as trains. They can fan out in residential areas to gather passengers and they can take riders more directly to dispersed destinations than any other system. Many cities have reserved exclusive freeway lanes for express buses, car pools, and passenger vans. When coupled with park-and-ride lots and minibus or dial-a-ride collecting systems, express buses can make a very satisfactory transit system.

THIRD WORLD CITIES

What can be done to improve conditions in Third World cities? While the advantages of garden cities and advanced transportation systems would greatly improve conditions in the developing world, the overcrowded cities of Third World countries have many more basic problems to solve first. Among the immediate needs are housing, clean water, sanitation, food, education, health care, and transportation for their residents.

A number of programs have been proposed to improve urban conditions in the Third World. Programs to reduce the costs of building materials and to provide small, flexible, easily obtainable loans have been of great assistance in many countries. Building codes, however, have often functioned mainly to protect building trades, landlords, speculators, and other special interests. They should be replaced with flexible, realistic guidelines that promote real public health and safety, and that allow poor people to build their own houses.

Perhaps more than anything else, people need land to live on. Thousands of hectares of land lie vacant in and around cities, such as Manila, Bombay, Delhi, Bangkok, and Sao Paulo, where housing is scarce. It's possible that Sao Paulo, for instance, could accommodate a 60 percent increase in population without expanding its boundaries by in-fill housing on vacant land between existing structures. Speculators, however, often hold land hoping for further price escalation. Meanwhile, poor people live crowded together in miserable shantytowns unable or afraid to put much time or effort into improving their housing because they could be forced out at any time.

Some countries are recognizing the need to use vacant land in their cities and are redistributing land held unproductively or closing their eyes to illegal land invasions. Indonesia, Peru, Tanzania, Zambia, and Pakistan recognize that squatter settlements make a valuable contribution to meeting national housing needs. Squatters' rights are being upheld in some cases, and such services as water, sewers, schools, and electricity are being provided to the settlements. Some countries intervene directly in land distribution and land prices. Tunisia, for instance, has a "rolling land bank" to buy and sell land. This strong and effective program controls urban land prices and reduces speculation and unproductive land ownership.

Nicaragua, Brazil, Cuba, and the Philippines also have undertaken effective land reform programs. Chile was redistributing land under the government of Salvadore Allende, but progress was stifled when General Pinochet seized control of the government. Since the 1960s, the percentage of Cuba's population living in Havana has remained essentially constant due to agrarian reform, government efforts to stimulate economic growth and social development in the countryside, and deliberate policies to restrict housing, jobs, and infrastructure development in the metropolitan area.

Some sociologists believe that inadequate and unstable income is the main reason for the poverty in which so many people are trapped. They argue that if people have enough money, they can and will buy better housing, adequate food, clean water, sanitation, and other things they need for a decent life. There seems to be a positive correlation between improved economic conditions, slowing population growth, and reduced rural-to-city movement. For most industrialized countries, the rate of population growth has slowed somewhat and migration to the cities has reversed in recent years. The threshold for this shift appears to be around $3,000 per person per year average income. Perhaps a more important measure of progress may be institution of a social welfare "safety net" that guarantees that people will not be abandoned and alone when they are old or sick. Some countries have accomplished these goals even without industrialization and high incomes.

Stabilizing society and sharing the benefits of progress with rural areas are components in slowing or even stopping the flood of migrants pouring into cities in the Third World. Sri Lanka, for instance, has lessened the disparity between the core and periphery of the country. By giving all people equal access to food, shelter, education, and health care, the incentives for interregional migration have been eliminated. Both population growth and city growth have been stabilized, even though the per capita income is only $800 per year. China has done something similar on a per capita income around $300 per year. A high level of government control over family planning and lower national economic growth have played important roles in these successes.

One of the many difficulties that challenge the economic progress of developing countries is that change is occurring so rapidly. The industrialization of Europe took place over several hundred years. There was time for adaptation. European countries had colonies to supply raw materials and to absorb surplus industrial products and surplus population. Countries developing today don't usually have that cushion of time, resources, or space. They are competing against the industrial giants of the world on a playing field of complex banking rules, international markets, and political systems designed by the developed countries for their own benefit. They are trying to get into a game already in progress, one in which some teams already have a large lead. On the other hand, they have the opportunity to learn from our experience and to avoid our mistakes. Will they be able to do so?

Some political scientists and economists believe that the richer nations of the Northern Hemisphere have created a state of economic servitude for poorer nations of the Southern Hemisphere. They argue that the international debt crisis severely limits the ability of developing countries to respond to urban problems. Even if they want to redistribute land or spend money on internal social services, they may be prevented from doing so by international banks and monetary funds. The best hope for these countries may be to "delink" themselves from the established international markets and develop direct south-south trade based on local self-sufficiency, regional cooperation, barter, and other

552

forms of nontraditional exchange that are not infused with lingering colonial attitudes and policies.

SUMMARY

This chapter has followed city planning from its earliest origins in antiquity to the most modern garden cities and high-rise technopolises of the developed world. Attitudes toward the city have varied over the past 10,000 years, from the sacred home of gods to the dens of evil. Many utopian communities have been established to try to reform society, individuals, and urban living conditions, but few have persisted for long. One of the most influential urban planners was Ebenezer Howard, who designed and built garden cities in Britain early in this century. These innovative cities have been the models for much of the suburban growth in developed countries, although satellite cities and suburbs rarely include the amenities and urban features that Howard envisioned.

In recent years, suburbs around American cities have been urbanizing so that most cities are now polycentric networks of urban villages rather than a series of single-purpose concentric rings. This transition may prove to be a step toward more humane postindustrial cities that will offer the advantages of urban life without the disadvantages of big industrial cities. In order to transform our cities into ideal living and working environments, we must consider new options, such as urban renewal, suburban redesign, new land-use plans, and new modes of transportation.

Third World cities are plagued by challenges unique to developing countries. Before planning and building more sustainable cities, basic problems of employment, clean water, health care, housing, and food resources need to be addressed.

■ Review Questions

1. Why is Washington, D.C., considered a well-planned city? What are some design features that make it livable? Why is it also having problems?
2. What are intentional communities? What are some of the principles they are exploring?
3. What did the ancient Greeks think about living in the city? What city planning principles did they introduce?
4. What is a garden city? Give an example of such a city.
5. Cities with a single urban core are being replaced by polycentric cities. Why?
6. What was the Pruitt-Igoe project? Why did it fail?
7. How has Boulder, Colorado, restricted urban sprawl?
8. Our society is shaped by the automobile. What are some alternative forms of transportation?
9. Why is the Hong Kong subway more successful than the Washington, D.C., subway?
10. Would you prefer to live in a high-rise building surrounded by a park or a row of single-family houses, each with a small private yard? What are the advantages and disadvantages of each?

■ Questions for Critical Thinking

1. List ten problems faced by someone living in New York. List ten advantages.
2. Do you prefer an urban, suburban, or rural living environment? Why? Describe what you think would be an ideal living environment.
3. What did early Romans consider the optimum size for a city? What changes (e.g., transportation) have allowed cities to grow bigger?
4. How did conditions in such cities as Liverpool during the Industrial Revolution change our conception of the city?
5. Do you think that zoning and strict land planning are important? Since we cherish ideals such as freedom and independence, why not determine land use with factors of supply and demand and cost and desirability instead of central planning?
6. Do you think cities should continue to expand into agricultural land? If cities continue to grow, should they move up or move out?
7. How do you think countries should deal with squatter settlements? Should people have the right to live on land that they don't rent or own?
8. How can developing countries build better cities? What aspects of our cities should they adopt and what should they avoid?
9. Our cities are changing from single urban cores to polycentric complexes. This change is largely due to our change from an industry-based economy to an information-based economy. Do you have any advice for Third World countries that are taking over our role in manufacturing and other industries?

■ Key Terms

garden cities (p. 540)
new towns (p. 540)
polycentric complex (p. 543)
technopolis, vertical city, or city of the future (p. 544)

SUGGESTED READINGS

■ Bookchin, M. *The Limits of the City*. Montreal: Black Rose, 1986. What makes a livable city?
■ Creese, W. L. *The Search for Environment: The Garden City Before and After*. New Haven: Yale University Press, 1966. A summary of the works of Howard, Parker, and Uniwin.
■ Doxiadis, C. A. *Anthropolis*: A *City for Human Development* and the proceedings of a symposium with Rene Dubos, Erik Erikson, Margaret Mead, et al. New York: W.W. Norton & Co., 1974. An idealistic plan for reorganizing cities into urban villages.
■ Geddes, P. *Cities: Being an Introduction to the Study of Civics*. University of London Extension Lectures Syllabus,

1907. The origins of the idea of organic growth in cities. See also *Cities in Evolution*. London, 1915. Most popular and available of his writings on cities.

■ Gross, A. C., et al. 1983. "The Aerodynamics of Human-Powered Land Vehicles." *Scientific American* 249(6):142. A fascinating look at modern efficient bicycles. See also *Scientific American* 253(5):(1985) and 255(6):(1986).

■ Howard, E. *Garden Cities of Tomorrow*. London: Farber and Farber, 1902. A classic in city planning that sparked the new town movement. Many of the features of modern suburbs feature designs of Howard and his chief architect, Raymond Uniwin. Unfortunately, they do not usually include the total community design envisioned by these pioneers.

■ Kropotkin, P. *Fields, Factories and Workshops or Industry Combined with Agriculture and Brainwork with Manual Work*. First published in Boston, 1899. Revised edition London: G. P. Putnam's Sons, 1913. Sociological and economic intelligence of the first order founded on Kropotkin's specialized competence as a geographer and his passion as a communist anarchist. Especially emphasizes planning for undeveloped areas.

■ Le Corbusier. *Urbanisme*. Paris, 1924. Translation: *The City of Tomorrow and its Planning*. New York: Dover Press, 1930. Suggestions for a mechanical metropolis with widely-spaced skyscrapers and multiple-decked traffic ways. One of the most influential books of its generation.

■ Lowe, M. D. 1991. "Rethinking Urban Transport." *State of the World* Washington D.C.: Worldwatch Institute. Describes approaches to mass transit and environmentally friendly transport systems.

■ McHarg, I. *Design with Nature*. Philadelphia: Natural History Press, 1969. A classic of landscape architecture that pioneered a holistic approach and consideration of natural features in urban design.

■ Park, R. *Human Communities: The City and Human Ecology*. Glenco: Free Press, 1952. The origin of the idea of human ecology.

■ Register, R, et al. 1990. *International Ecocity Conference Report* Berkeley CA.: Urban Ecology Institute. More than 150 experts on ecological city planning met in Berkeley, CA. to discuss the building of healthy and humane cities.

■ Sargent, F. O., et al. 1991. *New Rural Environmental Planning for Sustainable Communities*. Washington, D.C.: Island Press. A new edition of a classic guide to development and planning in harmony with nature.

■ Todd, N. J., and J. Todd. *Bioshelters, Ocean Arks, City Farming*. San Francisco: Sierra Club Books, 1985. An excellent discussion of appropriate technology and ecological design from New Alchemy Institute in Massachusetts.

■ Van der Ryn, S., and P. Calthorpe. *Sustainable Communities*. San Francisco: Sierra Club Books, 1986. Proceedings of 1980 Westerbeke Ranch Conference (near Sonoma, California) sponsored by Solar Energy Research Institute (SERI). An excellent description of principles for building (and rebuilding) ecologically sound cities.

■ Willis, C. Fall 1986. "The Titan City." *Invention and Technology* 2(2):44. An interesting historical survey of visions of what the city of the future would be like.

■ Wright, F. L. *The Living City*. New York: Horizon Press, 1938. A plan for Broadacre City, Usonionan House, Civic Center, car filling station, ceremonial stadium, roadside market, etc.

C H A P T E R *27*

Planning for a Sustainable Future: What Then Shall We Do?

C O N C E P T S

The choices we make now determine what our lives and those of our children will be like in the future. There is cause to be pessimistic about the state of our environment, but there is also reason to be optimistic.

Population growth must be controlled if we are to stay within the carrying capacity of the biosphere. Sustainable development calls for nondestructive economic growth and real progress in alleviating poverty, suffering, and oppression. The four bases of sustainable development are scientific information, ethical principles, hope for the future, and sharing the benefits of growth.

Environmental ethics is concerned with moral rules that govern relations between humans and the natural world. One of the most important and difficult questions in this field is whether other species and nonliving components of the environment have rights and values independent of their usefulness to humans. Some ethicists claim that human claims and interests always have priority in cases of moral conflict; others argue that humans have no more rights than any other species. Some view these questions as a matter of rights and obligations to be weighed and adjudicated; others see stewardship, reciprocity, care, and kinship as the solutions to environmental problems.

Laws and regulations are the foundation of most environmental protection policies. The global nature of resources and pollution make international laws and conventions essential. A wide variety of nongovernmental organizations, political parties, environmental clubs and societies, professional bodies, and other groups play important roles in environmental protection. They apply many different tactics and approaches. Some work within existing social and political systems; others practice civil disobedience or direct revolutionary action.

Since the earliest origins of the conservation movement, altruistic preservationists have debated utilitarian conservationists over the purposes of preserving resources and the tactics to be used. The latest version of this debate is between the "deep ecologists" and the political pragmatists. Which group you support depends on your beliefs, temperament, philosophical position, and political persuasion.

Each of us can individually do much to help preserve our environment and conserve resources. Our actions not only help society in general, but they often yield direct personal benefits as well.

If present trends continue, the world in 2000 will be more crowded, more polluted, less stable ecologically and more vulnerable to disruption. Serious stresses involving population, resources, and environment are clearly visible ahead. Despite greater material output, the world's people will be poorer in many ways than they are today.

Council on Environmental Quality, *Global 2000 Report*

We must mobilize now to achieve the global possible. If we do, the future can be bright. We have sufficient knowledge, skill, and resources— if we use them. If we remain inactive, whether through pessimism or complacency, we shall only make certain of the darkness that many fear.

R. Repetto, *World Enough and Time*, 1986

INTRODUCTION

What will the world's future be? Are we headed for crisis and disaster, or will we enjoy a happier, more comfortable, more affluent life than is possible now? In many ways, the choices we make now in managing our resources and our societies are determining what our lives—and those of our children—will be like in the future. Some people warn that we are on a downward spiral of environmental, economic, and social decay that will make the world of the future a bleak and dismal place. Others feel that we are on the threshold of a healthier, more just, more satisfying, and more meaningful life for everyone. Most people agree, however, that current trends are not predictions of what *must* be, but only indications of what *might* be, depending on the courses we choose to follow. In this chapter, we will review some of the major problems covered earlier in this book and examine some of our options in building a better society and managing our resources to create a sustainable future.

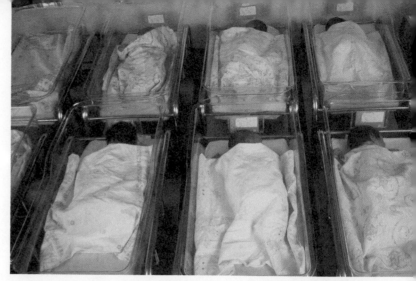

FIGURE 27.1 Are these babies our hope for the future or a threat to life on Earth? There is general agreement that we must stabilize our population, but there are differing opinions about how to accomplish this goal. Some favor aggressive birth-control programs. Others feel that if we eliminate poverty and infant mortality, population will stabilize automatically.

WHERE DO WE GO FROM HERE?

There are good reasons to be both optimistic and pessimistic about what the future holds. To some extent, the choice between these positions depends on your worldview and what you think we might or might not accomplish.

A Pessimistic View

It is not difficult to find evidence that supports a pessimistic view of the future. Throughout this book, we have seen examples of ways that humans are abusing environmental resources and threatening the basic support systems on which all life depends. If population growth rates continue as they have in recent years in many of the developing countries of the world, the sheer number of people in many places could easily overwhelm society's ability to provide jobs, shelter, food, and other necessities of life (figure 27.1). In both developed and less developed countries, consumption of resources and destructive forms of technology threaten to overwhelm supplies of energy, water, and clean air, as well as the environment's ability to absorb wastes and resist ecological disturbance. There are already ominous warnings that we may be irrevocably losing arable soil, destroying forests, altering atmospheric chemistry and global climate, and reducing biological diversity—all changes that could make the world less livable in years to come.

An Optimistic View

There also are trends that give hope for the future. In many areas, human ingenuity, resourcefulness, enterprise, and cooperation have dramatically improved the quality of life and the state of the environment. If the gains in population stabilization experienced in China, Sri Lanka, Cuba, and Singapore can be extended to other countries, it is possible that world population

could stabilize early in the twenty-first century at levels that appear to be within the carrying capacity of the planet. If energy conservation, material substitution and recycling, and solar energy reach the potential they seem to offer—and if new technologies are made available to developing countries at affordable prices—the disastrous consequences of obtaining and using energy and materials by conventional techniques may be avoided. If recent gains in food production and health care can be sustained, the famine and disease that now kill or incapacitate hundreds of millions of people each year may be eliminated (figure 27.2).

Many of the advances needed to direct our course away from crisis and toward a positive outcome already exist. Innovators have shown the way; it is now a matter of making the necessary knowledge and technology available *and* persuading people and governments that it is in their best interest to evaluate traditional, environmentally exploitative ways of doing things. Although we seem to be moving toward crisis points at an ever-increasing rate, there is hope that we can make the necessary changes in time. With the amazing developments in transportation, communication, and information technologies in recent years, advances can be transferred from country to country almost instantly. Successful experiences can be emulated, and progress can become synergistic.

Economic Growth and Ecojustice

Nearly everyone now accepts that human population growth must be controlled. The question of whether further economic growth and development are possible remains controversial. It seems clear that growth as it has occurred in the past cannot continue indefinitely. Our wasteful, throw-away approach to resources will have to be changed. Instead of constantly extracting new materials to produce goods designed to wear out quickly and be discarded along with the waste products of manufac-

FIGURE 27.2 New crop varieties, such as this high-yielding wheat developed by the CGIAR Institute in Mexico, have greatly increased food production land reform. Changes in the general agreement on tariffs and trade and new monetary policies will be needed to extend the benefits of these new crops to everyone.

TABLE 27.1 Principles for sustainable development
1. A demographic transition to a stable world population of low birth and death rates
2. An energy transition to high efficiency in production and use and increasing reliance on renewable sources
3. A resource transition to reliance on nature's "income" without depletion of its "capital"
4. An economic transition to a system based on human resources and intangible benefits with a broader sharing of power and benefits both between nations and individuals
5. A political transition to a global bargain, grounded in complementary interests between North and South, East and West

From Robert Repetto, *World Enough and Time*. Copyright © 1986 Yale University Press, New Haven, CT. Reprinted by permission.

turing, we need a conservation ethic that emphasizes long-lasting products, reduced consumption, repair, reuse, and recycling. Future economic systems will probably emulate steady-state natural systems in which materials are used over and over again and energy recovery is maximized wherever possible. We need to regard reused and recycled materials as our basic stock and native resources as a backup, rather than the other way around.

How we get to a more nearly steady-state system is a matter of great concern. Some people argue that the resource depletion and environmental crises caused by our once-through system are so grave that we must slow growth immediately. Others argue that some kinds of growth—at least for a while—are essential to pay for pollution control, to bring the poor of the world up to a more equitable standard of living, and to finance new technology that is needed to build a sustainable society.

The question of how and when growth should be controlled was the cause of a stormy confrontation at the landmark United Nations Conference on the Human Environment in Stockholm in 1972. On one side were environmentalists from the industrialized nations who argued that growth must slow immediately. On the other were representatives of less developed nations, along with a variety of advocates for poor and oppressed people. This group pointed out that under existing structures of power, patterns of distribution, and operating procedures in the economic order, a cessation of economic growth would halt badly needed development in the Third World and freeze in place the existing patterns of inequality.

They further stressed that cutting off growth abruptly would actually be worse for the environment because people who are struggling for survival have few options for protecting nature. You can't expect people to preserve wildlife when they are starving or to protect the forest when they are freezing. It may be that the richer countries of the world will have to reduce con-

sumption in order to allow the poorer countries to rise to a relative level of affluence that will give them the means to build a sustainable society. Justice in the social order, together with integrity in the natural order, is sometimes called **ecojustice.**

Sustainable Development

A path toward the twin goals of social justice and environmental protection also is described as sustainable development (table 27.1). This approach calls for sound economic development and real progress in alleviating hunger and poverty everywhere. It aims toward reducing income disparities and increasing access to health care, shelter, education, jobs, and other essentials of life. It requires management of all assets and natural and human resources to increase long-term wealth and well-being for all. Sustainable development rejects policies and practices that require depletion of the productive base and that leave future generations with poorer prospects and greater risks than our own. John Locke's seventeenth-century criterion for natural resource appropriation is a good guide for sustainable development. He said that property claims are valid only when they leave "as much and as good for others." This principle applies to intergenerational justice, as well as fairness between contemporaries.

Many kinds of growth are compatible with and contribute to sustainable development. They include advances in art, education, physical fitness, religion, health, aesthetics, and cultural diversity. Among the technologies and businesses that contribute to sustained development are scientific research, information management, pollution control, renewable energy production, recycling and resource recovery, resource management, and a multitude of services. All of these fields have the potential to improve general well-being while preserving resources and protecting the environment.

The four bases for sustainable development are (1) reliable scientific information, (2) consensus on ethical principles, (3) hope for the future, and (4) consideration of personal interest and incentives. We can manage our resources only if we

TABLE 27.2 World conservation strategy

1. To maintain essential ecological processes and life-support systems on which human survival and development depend
2. To preserve genetic diversity
3. To ensure the sustainable utilization of species and ecosystems

From United Nations Environment Program and World Wildlife Fund, 1987. Reprinted by permission.

know what we have and what we are doing to them. We need to agree on the reasons for preserving and distributing resources. We also need assurance that progress is possible and that we or our descendents will benefit from that progress. We have learned through history that people don't work efficiently and effectively toward a goal unless they know why they are working and see some future benefits for themselves or their family (table 27.2). Giving people some control over their own health care, family planning, urban development, soil conservation, watershed protection, community forestry, fisheries management, and other resource programs often has been the key to the success of these programs.

Some Challenges and Opportunities

The Chinese character for crisis (figure 27.3) is actually a combination of characters for danger and opportunity. This reminds us that crises have both negative and positive aspects. They can represent a threat to the status quo, but they also can represent a challenge to break through our limitations and improve our ways of doing things. In many ways, we are at a turning point in history, where we have a chance—perhaps an imperative—to make a major change. In the next section, we will review some important environmental problems that face us and some solutions that offer hope for the future.

ENVIRONMENTAL ETHICS

Ethics is a branch of philosophy concerned with morals (the distinction between right and wrong) and values (the ultimate worth of actions or things). It considers the relationships, rules, principles, or codes that require or forbid certain conduct. The famous questions posed by Socrates in the fifth century B.C. remain the focus of moral philosophy: "What is the good life? How ought we, as moral beings, to behave?"

Environmental ethics asks about the moral relationships between humans and the natural world. Do humans have duties, obligations, or responsibilities to Earth's natural environment? Are there ethical principles that subject humans to moral constraints with respect to the natural world? If so, what are those principles and how do they differ from principles governing our relations to other humans? How are these principles justified? How are our obligations and responsibilities toward the natural

FIGURE 27.3 Crisis or opportunity? The Chinese triple character *Ji* (*right*) can mean opportunity or a new direction. When combined with the double character *Zuan* (*left*), which means a turning point or critical change, it describes a crisis or bad situation that can be turned into an advantage if swift, determined action is taken. This could apply to our environmental condition.

world weighed against human values and interests? Do some interests supersede others?

Are There Universal Ethical Principles?

The first question to be considered in serious examination of ethics is whether *any* moral laws are objectively valid, independent of history, cultural context, or situation. **Universalists,** such as Plato and Kant, assert that some fundamental ethical principles are universal, unchanging, and eternal. These rules have validity regardless of interests, attitudes, desires, or tastes and can be discovered through reason and knowledge.

Relativists, such as Plato's opponents, the Sophists, assert that moral principles are always relative to a particular person, society, or situation. They argue that there are no ahistorical, transcendent, absolute principles that *always* apply regardless of circumstances. In this view, ethical values are always contextual.

Nihilists, such as Schopenhauer, claim that the world makes no sense at all. Everything is completely arbitrary, and there is no meaning or purpose in life other than the dark, instinctive, unceasing struggle for existence. According to this view, there is no reason to behave morally. Only power, strength, and sheer survival matter. There is no such thing as a "good" life; humans live in a world of uncertainty, dreariness, pain, and despair. Both Schopenhauer and Nietzsche, however, enjoyed living in a comfortable, civilized society where rules of normative behavior and good conduct prevailed.

Utilitarianism

Many branches of moral and ethical philosophy have developed in the 2,500 years since Socrates. Philosophers have argued about *a priori* versus empirical knowledge, about the nature of truth, about whether humans have real freedom and choice in our actions, about the goals (if any) of life, and about the ultimate worth of the goals we seek. Often these discussions lead to abstract arguments that are difficult to follow. One school of ethics that has persistent appeal because of its simple rule for solving ethical dilemmas is **utilitarianism**. It holds (among other things) that an action is right if it produces the greatest good for the greatest number of people.

Because he named it, the utilitarian philosophy is usually associated with the English philosopher Jeremy Bentham (1748–1832); however, something very similar had been suggested by Socrates, Plato, and Aristotle. Bentham was an eccentric genius and a hedonist who equated goodness with happiness and happiness with pleasure. He claimed that pleasure was the only thing worth having in its own right, and therefore, the good life was a life of maximum pleasure. Insofar as people are moral animals, they should act to produce the greatest pleasure for the greatest number. To do so is good; not to do so is wrong.

Utilitarianism was modified and made less hedonistic by Bentham's brilliant protege (and godson), John Stuart Mill (1806–1873). Mill believed that pleasures of the intellect were superior to pleasures of the body. He held that the greatest pleasure was to be educated and to act according to enlightened, humanitarian principles. This empirical, scientific form of utilitarianism inspired Gifford Pinchot and the early conservationists (chapter 13). They argued that the purpose of conservation was the protection of resources to produce the "greatest good for the greatest number *for the longest time.*"

Although utilitarianism is often practical and fairly easy to apply, it has drawbacks. It can, for instance, be used to justify reprehensible acts. If ten thousand Romans greatly enjoyed watching a few Christians being eaten by lions, did that make it the right thing to do? Does the pleasure of tormentors outweigh the suffering of victims? Most would conclude that it does not and that some pleasures are evil in and of themselves and should be avoided. Furthermore, some considerations are more important than pleasure. Justice and freedom take precedence over pleasure or even happiness, although it could be argued that furthering justice, freedom, and right action ought to bring the greatest happiness.

Moral Agents and Moral Subjects

The arguments presented so far in the utilitarian philosophy are phrased in terms of human values and interests. This is the **anthropocentric** (human-oriented) view. Many philosophers argue that only humans can be **moral agents,** beings who are capable of acting morally or immorally and who accept responsibility for their actions. Capacities that enable humans to form moral judgments include moral deliberation, the exercise of resolve and willpower to carry out decisions, and the responsibility to hold themselves answerable for failing to do what is right.

Of course, not all humans have all these capacities at all times. Children, the mentally retarded, mentally ill, and others who lack full use of reason are not regarded as moral agents. However, they still have rights. They are considered **moral subjects,** beings who are not moral agents themselves but who have moral interests of their own and can be treated rightly or wrongly

by others. Moral agents have certain duties and responsibilities towards moral subjects with whom they interact.

Means and Ends

In this view, moral subjects have inherent or intrinsic worth independent of the values or interests of others. A being has a right to respect; its treatment should reflect that it is an end in itself (has a purpose for existence) and not merely a means to some other purpose or interest. For instance, it is self-evident to most of us now that all humans have certain inalienable rights: life, liberty, and the pursuit of happiness. No one has a right to treat another being as a mere object for their own pleasure, gratification, or convenience.

Although we recognize that nonliving objects can have value, concepts of moral interests or inherent worth are not usually extended to nonliving objects, such as rocks, rivers, or machines. Instead, their value may be monetary (an expensive machine is worth a lot of money, for instance), aesthetic, historic, or cultural (due to the class of object, uniqueness, size, age, or beauty). The crucial distinction is that these objects are useful, beautiful, or inspiring to some person.

Philosophers call this **instrumental value**—these objects are instruments for the satisfaction of some other moral agent. They are not ends in themselves; they are only means for some other ends. They are valuable because someone values them. We don't have duties or responsibilities towards the instruments themselves, only to those who intend to use them. This philosophy has been applied in many ways that seem distasteful, inappropriate, and even cruel to us now. Rene Descartes (1596–1650) argued that animals, such as dogs and horses, are incapable of reason and feelings. He claimed that they are mere automata (machines) and can be operated on without anesthetics because they feel no pain. We continue to treat many domestic animals as if they are merely means to some desirable ends (figure 27.4).

Moral Extensionism

In his famous essay on environmental ethics, Aldo Leopold pointed out that the concepts of inherent worth and intrinsic rights have not always been extended to all humans. Children, women, foreigners, and some indigenous peoples were once regarded as less than persons. They were objects, personal property of an owner who could do with them as he (only adult males were considered citizens and moral agents) wished. Most civilized countries now recognize that all humans have inherent worth and intrinsic rights. These rights are also extended to inanimate entities—such as trusts, corporations, ships, joint ventures, municipalities, and nations—that can own property, sue each other, and seek protection of the law.

Many environmental philosophers believe we must extend our recognition of inherent worth to other organisms or nonliving components of the natural world. They argue that our

FIGURE 27.4 Do other organisms have inherent values and rights? These chickens are treated as if they are merely egg-laying machines. Many people argue that we should treat them more humanely.

anthropocentric view, which regards other organisms and natural objects as mere means rather than ends, is a major cause of environmental degradation. The alternative they propose is **biocentrism**: a belief that other organisms have inherent worth that gives them rights in and of themselves regardless of their potential for human use.

In his book *Respect for Nature*, Paul Taylor proposes four tenets of biocentrism:

1. Humans are members of Earth's community of life in the same sense and on the same terms as all other living things.

2. Humans and other species are interdependent; our survival and chances of doing well are determined not only by physical conditions of the environment but by our relations with other living things.

3. All organisms are teleological (purposeful) centers of life in the sense that each is a unique individual pursuing its own good in its own way.

4. Humans are not inherently superior to other living things. These tenets, known as *deep ecology*, will be discussed later in this chapter.

Professor Taylor derived three principles of ethical conduct from the biocentric outlook. The first principle is **nonmalfeasance**: Do no harm to any natural entity that has a good of its own. The second principle is **noninterference**: Do not try to manipulate, control, modify, manage, or otherwise interfere with the normal functioning of natural ecosystems, biotic communities, or individual wild organisms. The fact that organisms suffer and die does not itself call for corrective action when humans have nothing to do with the cause of death or suffering. The third principle is **fidelity**: Do not deceive or mislead any animal capable of being deceived or misled. In hunting, fishing,

and trapping, for instance, humans deceive prey into thinking they aren't present or don't intend to harm them.

These principles led Professor Taylor to advocate vegetarianism and to call for an end to hunting, fishing, and trapping and for the exclusion of industry or agriculture from wilderness areas. However, he doesn't suggest abandoning the ecosystems (farm fields, cities, commercial forests, etc.) that humans already manage.

Furthermore, Taylor extends only limited moral rights to plants and domestic animals, and no moral rights at all to non-living natural objects. He argues that a rock or a pile of sand has no rights or interests of its own. It is only useful as a means to an end; thus, humans may do to it anything they desire. Plants may have moral interests of their own in that they are living, but since plants cannot feel pain, humans have greater freedom to harvest and eat them. Domestic animals, on the other hand, clearly feel pain, but since their existence depends entirely on humans, we can own them and use them as property as long as we treat them kindly.

Not everyone agrees with these limits on the extension of moral rights. Some people argue that all natural objects, living or not, have rights. Priests of the Jain religion in India hold the most extreme respect for nature. They believe it is wrong to kill anything for any reason. The holiest among them never walk because they might step on an invisible insect and kill it. They wear a mask over their nose and mouth so they won't inadvertently inhale a pollen grain or bacterial cell. Some refuse to eat anything because all organic material comes from some living organism. Since it is impossible to live this way, the priests who hold this view starve to death.

Stewardship

Obviously, we must use the natural environment if we are to live. We eat other organisms, and we compete with them for living space, water, sunlight, and other essentials of life. We control or eliminate species of plants or animals that are harmful to our survival. How can we justify this intervention in the natural order?

One answer to this question lies in the principle of **stewardship,** which holds that humans have a unique responsibility to manage and care for domestic plants and animals as well as the larger natural world. This theory argues that humans are neither external nor superfluous to the natural world. We and our machines and knowledge are a part of the evolutionary process and an essential part of natural history. This philosophy calls for respect for and cooperation with nature to bring forth the greatest good.

The idea of stewardship is usually based on a belief that humans have a responsibility to care for nature because of their superior intellect and wisdom. It holds that humans can improve the world and make it a better place, both for themselves and for other organisms. Rene Dubos, in *So Human an Animal*, calls this the "wooing of the Earth," bringing out the best in it through love, care, and attention. The foremost symbol of stewardship is the garden, an ancient image of order, comfort, peace, and tranquility. The original meaning of Eden was a garden oasis. Humans are to be gardeners, actively caring for the Earth and its inhabitants. As Voltaire said in *Candide*, "This may be the best of all possible worlds, but we must tend our garden."

Competing Claims and Ethical Dilemmas

An ethical dilemma is a situation in which two courses of action each have valid ethical claims. It arises when whichever action we take will bring harm to ourselves or others. It may also be an ethical dilemma when two parties each have equally valid but incompatible ethical interests. How can we act morally and justly when ethical interests are in conflict?

Professor Taylor suggests the following principles for fair resolution of competing claims:

1. *Self-defense*. It is generally permissible for moral agents to defend themselves.
2. *Proportionality*. Basic interests (those necessary for survival) have priority over nonbasic interests.
3. *Minimum wrong*. Where basic interests are in conflict, the least wrong should be done to all agents involved. This is not merely a quantitative calculation; it also includes qualitative differences in interests and agents.
4. *Distributive justice*. If possible, the disadvantages resulting from competing claims should be borne equally by all participants.
5. *Restitutive justice*. The greater the harm done to a moral agent, the greater the compensation required. Restitution should restore harmony and justice.

Ecofeminism

Many feminists argue that neither utilitarianism, biocentrism, nor stewardship are sufficient to solve environmental problems or to tell us how we ought to behave as moral agents. They argue that all these philosophies have come out of a patriarchial system based on domination and duality. This framework assigns prestige and importance. It puts men "up" and women "down," minds "up" and bodies "down," culture "up" and nature "down." These feminists assert that there are important historic, symbolic, and theoretical interconnections between the domination, exploitation, and mistreatment of women, children, minorities, and nature.

Their alternative is **ecofeminism,** a pluralistic, relationship-oriented philosophy that suggests how humans could reconceive themselves and their relationships to nature in nondominating ways. Ecofeminism is concerned not so much with rights, obligations, ownership, and responsibilities as with care, appropriate reciprocity, and kinship. It promotes a richly textured understanding or sense of what human life is and how this understanding can shape people's encounters with the natural world.

When people see themselves as related to others and to nature, they will see life as bounty rather than scarcity, as cooperation rather than competition, and as a network of personal relationships rather than isolated egos. Whether people can let go of their feelings of superiority, privilege, and importance remains to be seen. Whether we truly can have what theologian Martin Buber called an I-thou relationship with nature and with other humans rather than an I-it relationship remains a challenge.

ENVIRONMENTAL LAWS

Local, state, and national laws are the foundation of most environmental protection. Generally, these laws have been designed to protect property rights and personal comfort and safety. In recent years, we have realized that the interconnectedness of everything in the biosphere means that we have to be concerned about even obscure aspects of our world and things that seem far removed from our daily lives. We have begun to establish a body of law concerned with broad questions of environmental quality and stability. We also have begun to recognize that the global nature of environmental problems requires international, as well as local, laws and conventions.

Throughout the history of pollution control and resource management, there has been a struggle over whether power for environmental protection should rest in a central authority or with local government. A central government usually has more resources for research, management, and enforcement than does a local unit of government. Central control is more uniform in its protection than are "hit or miss" local regulations. It can be more democratic and less likely to be captured by special interests than local power structures. Since pollution and environmental problems cross boundaries, it is important to have nationwide or even worldwide standards to prevent industry from shopping around for the least stringent pollution regulations. It also prevents communities from getting into bidding wars over who can offer the most lenient regulatory climate to attract business. Central government—in theory at least—should be more cost effective than many independent local regulatory agencies.

Many people distrust and dislike centralization, however, preferring local autonomy and individual action instead. They point out that central planners don't understand local conditions, customs, or preferences. Bureaucracies tend to be unwieldy and unresponsive, running roughshod over local sensibilities. Centralization also tends to diminish local initiative, incentives, and rights. Many examples of environmental disaster in Eastern Europe resulting from failed central planning are coming to light as Communist rule there collapses.

Perhaps both sides in this debate are partly right. Since we are citizens of a single global system, we need comprehensive environmental controls that protect the system as a whole. Our history for thousands of years has been as tribal people, however, and we still seem to need personal contact, involvement, and rewards to be inspired to act. Maybe as telecommunications and transportation break down national barriers, we will begin to feel more like members of a global village in which all humans (and even other life-forms) are part of our extended family.

HOW GOVERNMENT WORKS

The United States Government is based on checks and balances among three branches: the legislative, the judicial, and the executive. All three branches establish laws, legal opinions, and regulations that are important tools in protecting our environment. **Statutory law** is passed by a state or national legislature. **Common law** is a body of unwritten principles based on custom and the precedents of previous legal decisions in the courts. **Regulations** are rules promulgated by administrative agencies. All three branches of government are vital, and all three provide a forum for public concerns about resource use and the state of the environment. Every citizen ought to be familiar with how these systems function and should participate in the political process to make his or her opinions known. There isn't space here for a complete civics course, but we will briefly survey how the U.S. government works and how you might be involved.

Congress

At the national level, the legislative branch is made up of two houses: the House of Representatives and the Senate (figure 27.5). The 435 congressmen and congresswomen in the House are elected for two-year terms from congressional districts of about five hundred thousand people each. California has forty-five Representatives and Alaska has one. Two senators are elected at large from each state for six-year terms. These short terms make our representatives and senators responsive to the people. If you don't like what elected officials are doing, you can vote them out without much delay; however, the short terms may distract them from taking a long-term view in resource allocation and environmental protection. The large size of the Congress tends to be expensive and inefficient, but it means that people are more likely to know their representative personally. It also is harder for a single-interest group to take over Congress, since there are so many members to influence.

Figure 27.6 shows the pathway that a piece of legislation (usually called a **bill** while being considered by Congress) follows, from inception to being signed into law by the President. The idea for a law can originate from an ordinary citizen, as well as from an elected official. If you have an idea for a law, the first step is to organize all the facts, figures, and arguments for why it is needed and what you hope it will do. Next, you find a representative or senator who will sponsor your bill. You probably will have the best luck with your own representative or senator,

FIGURE 27.5 The National Capitol Building in Washington, D.C., houses both the United States Senate and the House of Representatives. All federal laws are written, debated, and passed in that building.

but someone from another state or district with a special interest in your topic also might be willing to sponsor your bill. A member of the majority party is more likely to be successful with a bill because the majority directs the process and passes most of the legislation in a given session.

If the issue is controversial, the senator or representative usually will send out a "Dear Colleague" letter soliciting support and cosponsorship from other members. After introduction, the bill is sent to the appropriate committee for hearings. There are sixteen standing committees and four select or special committees in the Senate. The House has twenty-three standing and three select committees. These "full" committees rarely hold hearings, however. Instead, bills are sent to one of several hundred subcommittees in each branch. This divides the work load and shares the leadership positions among many individual members. Sometimes the hearing process is very extensive, and it may include field hearings in which ordinary citizens have a chance to express their opinions (figure 27.7)

Contact your legislator to find out if there will be field hearings in your city. If they are scheduled, go even if you don't intend to speak. It is an educational experience and gives you a chance to be involved. Before attending a hearing, find out if others in your community share your position. You will be more effective if you coordinate your presentation. The rules for writing letters to your legislator (table 27.3) generally apply also to testimony at a hearing. Well-organized, factual testimony can make a good impression and have a positive impact both on elected officials and the public.

Almost all of the advance work in getting a bill ready for introduction and in guiding it through the hearings process is

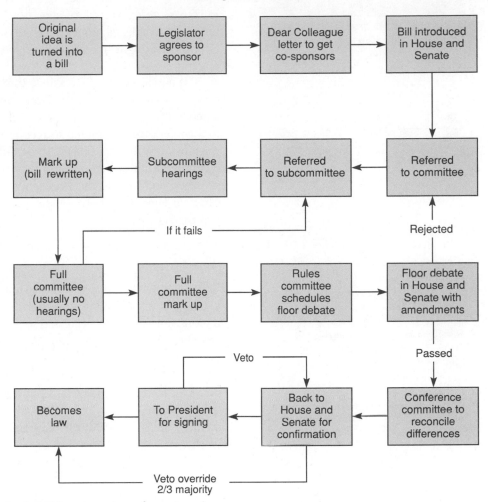

FIGURE 27.6 The path of a bill from inception to becoming a law.

FIGURE 27.7 Citizens line up to speak at a legislative hearing. Speaking about public issues is a right that few people in the world enjoy. By getting involved in the legislative process, you can be informed and have an impact on governmental policy.

done by the legislative staff. These people are key in the success or failure of bills in which you are interested. It is well worth getting to know them and working with them. They often are young, in their twenties or thirties. You could become a legislative aide and gain valuable experience following your higher education.

Few of us actually get involved in the legislative process very often, but you can make your wishes and opinions known by communicating with your elected representatives. Table 27.3 gives some suggestions for how to write effectively. Most representatives get little mail, so a single, well-written letter from a constituent can make a difference. Even on controversial issues, a representative might get only fifty to one hundred letters. Since each congressional district includes about half a million people, the legislator tends to assume that each letter represents the views of five thousand to ten thousand constituents. Your voice has great impact! If you don't have time to write, you can call the

TABLE 27.3 How to write to your elected officials

1. Address your letter properly:
   a. *Your representative*:
      The Honorable _____
      House Office Building
      Washington, D.C. 20515
      Dear Representative _____ ,
   b. *Your senators*:
      The Honorable _____
      Senate Office Building
      Washington, D.C. 20510
      Dear Senator _____ ,
   c. *The president*:
      The President
      The White House
      1600 Pennsylvania Avenue, N.W.
      Washington, D.C. 20500
      Dear Mr. President,

2. Tell who you are and why you are interested in this subject. Be sure to give your return address.

3. Always be courteous and reasonable. You can disagree with a particular position, but be respectful in doing so. You will gain little by being shrill, hostile, or abusive.

4. Be brief. Keep letters to one page or less. Cover only one subject, and come to the point quickly. Trying to cover several issues confuses the subject and dilutes your impact.

5. Write in your own words. It is more important to be authentic than polished. Don't use form letters or stock phrases provided by others. Speak or write from your own personal experiences and interests. Try to show how the issue affects the legislator's own district and constituents.

6. If you are writing about a specific bill, identify it by number (for instance, H.R. 321 or S.123). You can get a free copy of any bill or committee report by writing to the House Document Room, U.S. House of Representatives, Washington, D.C. 20515 or the Senate Document room, U.S. Senate, Washington, D.C. 20510.

7. Ask your legislator to vote a specific way, support a specific amendment, or take a specific action. Otherwise you will get a form response that says: "Thank you for your concern. Of course I support clean air, pure water, apple pie, and motherhood."

8. If you have expert knowledge or specifically relevant experience, share it. But don't try to intimidate, threaten, or dazzle your representative. Don't pretend to have vast political influence or power. Legislators quickly see through artifice and posturing; they are professionals in this field.

9. If possible, include some reference to the legislator's past action on this or related issues. Show that you are aware of his or her past record and are following the issue closely.

10. Follow up with a short note of thanks after a vote on an issue that you support. Show your appreciation by making campaign contributions or working for candidates who support issues important to you.

11. Try to meet your senators and representatives when they come home to campaign, or visit their office in Washington if you are able. If they know who you are personally, you will have more influence when you call or write.

12. Join with others to exert your combined influence. An organization is usually more effective than isolated individuals.

local or Washington office. You probably will not talk directly to your representative or senator, but your opinion will be registered by an aide. Another option is to send a telegram. Western Union has a night rate of only a few dollars for short political opinion messages.

If a bill makes it successfully through the committee process, it goes to the floor of the House and Senate for debate and amendments. If passed by a majority of both houses, the bill goes to a conference committee to reconcile any differences in the versions from the two bodies. If the differences were great, the compromise bill will go back to the respective houses for confirmation and then to the President for signing. The President can veto legislation, but Congress can override that veto by a two-thirds vote.

The Courts

In the past few decades, the judicial branch has been dominated by an activist philosophy that often has made the courts more responsive to environmental and social concerns than the other branches of government. Citizens can seek relief in the courts from unjust laws or actions in two ways. One way is to bring a civil suit, asking for payment of damages that were caused by a private individual, corporation, or governmental agency and that injured you or your property. The other way is to ask for a judgment by the court about the constitutionality of laws passed by Congress or the adequacy and legality of regulations established by an administrative agency. If the court finds a law or regulation to be improper, it can issue an injunction to stop implementation or application of that law or regulation. It also can order an agency to rewrite its regulations and may even direct what the new regulations should be.

There are several problems that make lawsuits difficult for ordinary citizens who want to protect the environment or some other public resource:

1. You have to show that the action or law you oppose is illegal; that may be difficult to do.

2. You have to establish that you have standing in court (a right to be heard). To do so, you must prove that you are directly affected, sometimes a difficult requirement.

3. Suits are expensive; a major suit might cost hundreds of thousands of dollars in legal fees, court fees, witnesses, etc.

4. It may take years before a suit is finally settled, and by that time it may be too late to save the resource.

5. You have to prove that the defendant (an agency, corporation, or individual) is responsible for the harm that you allege. A corporation can admit that they produce a toxic chemical, but you have to show, beyond reasonable doubt, that it was their *specific* chemical that caused the problem.

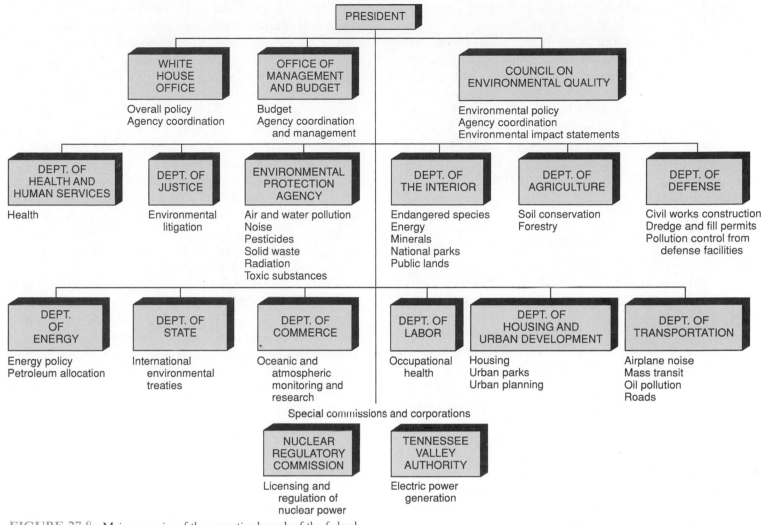

FIGURE 27.8 Major agencies of the executive branch of the federal government with responsibility for resource management and environmental protection.

Source: Data from U.S. General Accounting Office.

In spite of these difficulties, the courts often have been the most successful place for environmental organizations to bring about changes in how we manage our environment. More than one hundred public-interest law firms in the United States specialize in social and environmental issues. Several environmental organizations, such as the Environmental Defense Fund, the National Resources Defense Council, and the Sierra Club Legal Foundation, act primarily or exclusively through litigation (bringing suits in court).

The Executive Branch

Of the roughly 14 million federal employees in the United States, the vast majority work in the executive branch. Many of these civil servants are in the administrative agencies that carry out and enforce the laws passed by the legislature. They monitor, manage,

control, and protect our resources and our environment. Figure 27.8 shows the major agencies and branches of government that have environmental responsibilities. The Environmental Protection Agency (EPA) is often regarded as the main guardian of environmental quality. It has responsibility for regulating air and water pollution, solid and hazardous wastes, toxic substances, noise, radiation, and certain pesticides.

The Department of the Interior manages the national parks, wildlife refuges, wild rivers, historic sites, BLM lands, the Bureau of Reclamation, and the Bureau of Mining. It is by far the largest land manager in the country (chapter 13). The Department of Agriculture administers the national forests, the Agricultural Research Service, the Soil Conservation Service, and the Food Safety Inspection Service. The Department of Health administers the Public Health Service and the Food and Drug Administration. The Department of Labor is responsible for occupational safety and health. The Department of Energy is responsible for nuclear energy and fossil fuels.

These administrative agencies represent a tremendous bureaucracy, which often is ponderous and unresponsive, but then, we have a big country, and running it is a big job. For the most part, our civil service is made up of dedicated, professional people. If you ever have traveled in a less developed country, you will appreciate how honest and open our government is. Our system has been criticized for responding incrementally to problems. It may make improvements more slowly than we would like in some situations, but it also makes mistakes more slowly. If it had the power to make sudden leaps, it might make them in the wrong direction.

Environmental Impact Statements

One of the most useful tools for those concerned about environmental protection is the provision of the National Environmental Policy Act of 1970 requiring all federal agencies to prepare an **Environmental Impact Statement (EIS)** to analyze the effects of any major program that it plans to undertake. As we discussed earlier in this chapter, it often is difficult for citizens to prove they are personally injured by a pollutant or an action of the government. It usually is not difficult, however, to show that an agency has prepared an inadequate EIS or has failed to consider important environmental consequences of their actions.

Intervening has become much easier since the Freedom of Information Act (FOIA) requires federal agencies to make public all minutes of meetings, correspondence, and other official documents. This means that you can find out what an agency considered when preparing an EIS, who they talked to, and why they decided as they did. It often is fairly easy to show that undue consideration was given to commercial interests or that environmental values were overlooked or deliberately ignored. In some cases, simply asking for an EIS will prompt an agency to reconsider a harmful project. In other cases, you may have to ask the courts to order one. Often the delay caused by litigation over an EIS gives time to organize other opposition. Sometimes just generating a discussion of the adverse effects of a project is enough to kill it. Letters to your legislators are considered private correspondence and cannot be made public under the FOIA.

International Environmental Laws

There is no international legislature with authority to pass laws comparable to our Congress, nor are there international agencies with power to regulate resources on a global scale. There is an international court at the Hague in the Netherlands, but it has little power to enforce its decisions. Powerful nations can simply ignore the court, as the United States did when it was found in violation of international law after it mined the harbors of Nicaragua in 1983. Still, there is movement toward environmental protection on a worldwide scale through special covenants, treaties, and multilateral agreements.

In 1982, the Third United Nations Conference on the Law of the Sea produced a comprehensive convention that addressed almost every issue concerning jurisdiction over ocean waters and use of ocean resources. The product of nine years of debate and negotiation, this treaty may be a model for international environmental protection. There already have been positive results from application of the agreements on pollution control, marine mammal protection, navigation safety, export and dumping of toxic wastes, and other aspects of the marine environment.

One of the biggest issues of contention at present is deep-seabed mining, primarily of manganese nodules (chapter 9). The industrial nations assert that resources should be appropriated on a "first come, first served" basis. They argue that those who invest in developing and harvesting the resource deserve to enjoy the profits. The less developed nations claim that these resources form a common heritage of all humankind that should benefit everyone, not just those wealthy enough to afford the technology to harvest them. They see this as the first application of a *new economic order* (chapter 8), in which we share more equally in resource benefits. An International Seabed Authority has been proposed that would carry out mining, distribute revenues, transfer technology to less developed countries, and compensate land-based producers of the same minerals for adverse economic effects. To some critics, this approaches international socialism.

There have been several other examples of successful international conventions and treaties. The Antarctic Treaty of 1961 reserves the Antarctic continent for peaceful scientific research and bans all military activities in the region. Exploitation of Antarctic mineral resources raises the same questions as seabed minerals. The 1979 Convention on Long-Range Transboundary Air Pollution was the first multilateral agreement on air pollution and the first environmental accord involving all the nations of Eastern and Western Europe and North America. The 1989 Accord on Chlorofluorocarbon Emissions (chapter 23) was achieved remarkably quickly to address the threat to the stratospheric ozone shield. It recognizes the need for economic expansion by the less developed countries and requires greater reductions from the bigger polluters than the smaller ones. Table 27.4 lists some factors that affect the success or failure of international environmental protection agreements.

NONGOVERNMENTAL ORGANIZATIONS

The blossoming of environmental concern and activism in the 1960s and 1970s resulted in the formation of many large and small pressure groups, advisory agencies, research groups, "network" associations, political parties, professional societies, charitable organizations, and other groups concerned about environmental quality and resource use (figure 27.9). We lump these groups together as **nongovernmental organizations (NGO)**. They have become a powerful aspect of environmental

TABLE 27.4 Factors affecting international environmental laws

1. *Severity of the problem.* Once a problem is widely acknowledged as critical, it is easier to make progress.
2. *Science.* Good data is needed on the extent of the problem and possible solutions.
3. *Geography.* The extent to which the problem is transboundary and where its effects are felt.
4. *Law.* Whether countries have laws protecting the environment and whether foreigners have access to court and administrative proceedings to enforce those laws.
5. *Domestic interests and pressures.* Who favors and who opposes action on the issue in each country.
6. *Formal institutions and policies.* Whether there is a mechanism in place for cooperative action among the interested countries.
7. *History.* Whether there is a tradition of cooperation or conflict among the countries.
8. *Relative economic strength, military power, and population.* Large disparities in size or strength may hinder agreement unless the stronger country depends on the weaker for the resource.
9. *Outside influences.* Third parties can influence negotiations positively or negatively.
10. *Timing.* It may be easier to reach agreement before various interests are entrenched or at times when other changes in societies or economies are occurring.

Source: From *WORLD RESOURCES, 1987.* Copyright © 1987 by the International Institute for Environment and Development and the World Resources Institute. Reprinted by permission of Harper Collins Publishers Inc.

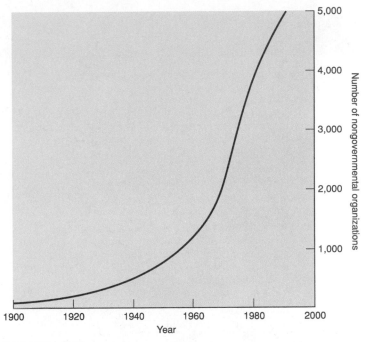

FIGURE 27.9 Growth of international nongovernmental organizations (NGOs). If we add to this number approximately 1,100 groups with governmental links, 1,100 with a more formal structure, 600 religious groups, and 4,500 NGOs with an international orientation but based in a single country, there are more than 12,000 environmentally concerned NGOs in the world.

protection in most of the world. Using the communications media to influence public opinion has become one of the most important tools of the environmental movement (table 27.5).

Diverse Approaches and Emphases

The NGOs use many different approaches for environmental protection. Some are highly professional and combine private individuals with representatives of government agencies on quasi-governmental boards or standing committees with considerable power. Others are at the fringes of society, sometimes literally voices crying in the wilderness. Some work for political change; others specialize in gathering and disseminating information, and still others undertake direct action to protect a specific resource.

In many countries with a parliamentary form of government, "green" parties based on environmental interests have become a political force in recent years. The largest and most powerful green party is *Die Grünen* in Germany, which controls about 10 percent of the seats in the *Bundestag* or parliament. (figure 27.10). It is a revolutionary, socialist, egalitarian, council-style movement, committed to participatory democracy, environmental protection, and a radical transformation of society. A fundamental premise is that ceaseless industrial growth destroys both people and the environment and must, therefore, be stopped. The Greens have formed coalitions with antinuclear weapons protesters, feminists, human rights advocates, and other public interest groups, but have struggled continuously over whether or not to work with the ruling centrist party. Political realists argue that they could be more effective within a larger coalition; idealists vow never to compromise their principles.

The "winner-takes-all" political system in the United States makes it extremely difficult to start a new political party. There are "green committees of correspondents" that work to introduce environmental issues into party politics and to support Green candidates. Campus Greens are strong at many colleges and universities. The four key values they espouse are (1) ecological wisdom, (2) peace and social justice, (3) grassroots democracy, and (4) freedom from violence. In 1990, Alaska became the first state to give a Green Party official standing. Jim Sykes, the Green gubernatorial candidate, received 3.2 percent of the total vote, enough to guarantee the party a spot on future ballots.

Among the most successful organizing techniques used by the Greens are protest marches, demonstrations, civil disobedience, and other participatory public actions and media events. Other environmental organizations also find these techniques effective in attracting attention to a particular problem and bringing about a quick change. Greenpeace, for instance, carries out well-publicized confrontations with whalers, seal hunters, toxic-waste dumpers, and others who threaten very specific and visible resources (figure 27.11). Greenpeace may well be the

TABLE 27.5 Using the media to influence public opinion

Shaping opinion, reaching consensus, electing public officials, and mobilizing action are accomplished primarily through the use of the communications media. To have an impact in public affairs, it is essential to know how to use these resources. The following list contains some suggestions for how to do this:

1. *Assemble a press list.* Learn to write a good press release by studying books from your public library on press relations techniques. Get to know reporters from your local newspapers and TV stations.

2. *Appear on local radio and TV talk shows.* Get experts from local universities and organizations to appear.

3. *Write letters to the editor, feature stories, and news releases.* You may include black and white photographs. Submit them to local newspapers and magazines. Don't overlook weekly community shoppers and other "freebie" newspapers, which usually are looking for newsworthy material.

4. *Try to get editorial support from local newspapers, radio, and TV stations.* Ask them to take a stand supporting your viewpoint. If you are successful, send a copy to your legislator and to other media managers.

5. *Put together a public service announcement, and ask local radio and TV stations to run it (preferably not at 2 A.M.).* Your library or community college may well have audiovisual equipment that you can use. Cable TV stations usually have a public service channel and will help with production.

6. *If there are public figures in your area who have useful expertise, ask them to give a speech or make a statement.* A press conference, especially in a dramatic setting, often is a very effective way of attracting attention.

7. *Find music stars or media personalities to support your position.* Ask them to give a concert or performance, both to raise money for your organization and to attract attention to the issue. They might like to be associated with your cause.

8. *Hold a media event that is photogenic and newsworthy.* Clean up your local river and invite the photographers to accompany you. Picket the corporate offices of a polluter, wearing eye-catching costumes and carrying humorous signs. Don't be violent, abusive, or obnoxious; it will backfire on you. Good humor usually will go farther than threats.

9. *If you hear negative remarks about your issue on TV or radio, ask for free time under the Fairness Doctrine to respond.* Stations need to do a certain amount of public service to justify relicensing and may be happy to accommodate you.

10. *Ask your local TV or newspaper to do a documentary or feature story about your issue or about your organization and what it is trying to do.*

FIGURE 27.10 Delegates to the Green party convention in Germany debate the political, social, economic, and environmental platform. Although the Green's lost almost all their seats in the National Parliment in 1990, they remain a force in local politics.

about fifteen cents on the dollar, far cheaper than any other method of saving this area. In a similar case of quiet diplomacy, Janet Welsh Brown, a senior associate at the Washington-based World Resources Institute, wrote a personal letter to Nicaraguan President Daniel Ortega in August, 1987, questioning the wisdom of logging a unique virgin rain forest in southern Nicaragua. This low-key approach was enough to cancel a twenty-year, $50 million logging contract.

Mainline Environmental Organizations

Among the oldest, largest, and most influential environmental groups in the United States are The National Wildlife Federation, The Audubon Society, Sierra Club, and The Izaak Walton League (chapters 13 and 15). Although each of these groups was militant—even radical—in its formative stages, they now tend to be more staid and conservative. People often join as much for their publications and social aspects as for their stands on environmental issues.

Still, these groups are powerful and important forces in environmental protection. Their mass membership, large professional staff, and long history give them a degree of respectability and influence not found in newer, smaller groups. The Sierra Club, for instance, with some four hundred thousand members and chapters in nearly every state, has a national staff of about four hundred, an annual budget approaching $20 million, and twenty full-time professional lobbyists in Washington. In a 1985 survey that asked congressional staff and officials of government agencies to rate the effectiveness of groups that attempt to influence federal policy on pollution control, the top five were voluntary environmental organizations. In spite of their large budgets and important connections, the American Petroleum Institute, the Chemical Manufacturers Association, and the Edison Electric Institute ranked far behind the environmental groups in terms of influence.

largest environmental organization in the world, with some 2.5 million contributing members.

In contrast to these highly visible groups, others choose to work behind the scenes, but their impact may be equally important. In 1987, Conservation International, an offshoot of the Nature Conservancy, negotiated the purchase of $650,000 of Bolivia's external debt in exchange for preservation of 1.6 million hectares of tropical forest and grassland in the Beni River region of northern Bolivia. The total cost was only $100,000,

FIGURE 27.11 Greenpeace activists try to stop the killing of whales by placing themselves between the whaling ship and its quarry. In 1985, they were narrowly missed by a harpoon fired directly over their heads by a Russian whaler.

Some environmental groups, such as the Environmental Defense Fund (EDF), The Nature Conservancy (TNC), National Resources Defense Council (NRDC), and The Wilderness Society (WS) have limited contact with ordinary members except through their publications. They depend on a professional staff to carry out the goals of the organization through litigation (EDF and NRDC), land acquisition (TNC), or lobbying (WS). Although not often in the public eye, these groups can be very effective because of their unique focus. The Nature Conservancy buys land of high ecological value that is threatened by development. With more than $50 million in cash purchases and $44 million in donated lands, it now owns some nine hundred parcels of land, the largest privately owned system of nature sanctuaries in the world. Altogether, it has saved more than one million hectares (2.47 million acres) of land in its forty-year history.

The NRDC claims to be the most effective of all groups concerned with pollution control because of its emphasis on litigation. An approach that proved to be very effective in the mid-eighties was "citizen suits" that sued violators of pollution discharge regulations for damages to the environment. In landmark decisions during 1985 and 1986, federal judges ordered such companies as Bethlehem Steel and the Gwaltney Meatpacking Company to pay millions of dollars in damages for pollution discharges. Since companies are required by law to report their violations to the EPA, gathering evidence that pollution has occurred is simply a matter of obtaining records through the Freedom of Information Act. In December, 1987, the Supreme Court ruled that damages could be claimed only for direct personal losses, rather than general damage to the environment. This will make "citizen suits" more difficult to prosecute, but will not rule out their effectiveness altogether.

Broadening the Environmental Agenda

The whole environmental movement in the United States tends to be overwhelmingly white, middle-class suburbanites. Few blue-collar workers and even fewer minorities are involved, especially in leadership or professional positions. This is unfortunate in several ways. The movement will probably never be successful until it can put together a broad-based coalition of support. Furthermore, poor people are often most seriously affected by toxic waste dumps, noise, urban blight, and air and water pollution. They should be included in discussions of how to remedy these problems. Unfortunately, most environmental groups have acquired a reputation for caring about plants, animals, and wilderness areas more than about people. Until we can convince everyone that they have a stake in environmental protection, we are not likely to achieve our goals.

Ecotage and Direct Action

A striking contrast to these mainline conservation organizations are the "direct action" groups, such as Earth First!, Sea Shepherd, and a few other groups that form either the "cutting edge" or the "radical fringe" of the environmental movement, depending on your outlook. The main tactics of these groups are passive resistance, civil disobedience, and attention-grabbing actions, such as guerrilla street theater, picketing, protest marches, road blockades, and other demonstrations. Many of these techniques are borrowed from the civil rights movement and Mahatma Gandhi's passive, nonviolent civil disobedience, which had roots in the *stavagrahas* that go back to the beginnings of Indian culture (chapter 12).

Members of Earth First! perched in Douglas firs marked for felling in Oregon, chained themselves to a giant tree-smashing bulldozer in Texas to prevent forest clearing, and blockaded roads being built into wilderness areas (figure 27.12). In some cases,

FIGURE 27.12 A protestor chains herself to a tree in east Texas in an attempt to stop the giant tree-smashing machine (nicknamed Godzilla) in the background from destroying more forest. Civil disobedience actions such as this are effective in attracting media coverage and public attention but may not have much long-term effect unless coupled with other approaches.

the protests can be humorous and lighthearted, as when the Earth First! members dressed in bear costumes to protest grizzly bear management policies in Yellowstone National Park.

In some cases, however, protests can be dangerous and violent. In May, 1987, a worker at the Louisiana-Pacific sawmill in Cloverdale, California, was seriously injured when the sawblade hit a large spike in the redwood log he was sawing. No one has admitted spiking the tree, but it is a technique described in some detail in Earth First!'s 1985 book *Ecodefense: A Field Guide to Monkey Wrenching* by Dave Foreman and Bill Haywood. *Monkey wrenching* is a term for environmentalist sabotage, or **ecotage,** a concept made popular by author Edward Abbey's book *The Monkey Wrench Gang*. Other ecotage tactics frequently discussed in the monthly Earth First! newsletter are "decommissioning" of heavy construction equipment, planting sharp spikes in trails used by off-road vehicles, pulling up survey stakes for unwanted developments, and destroying billboards.

In 1991, Foreman and three others pleaded guilty to conspiracy to sabotage a nuclear power plant in Arizona. The FBI had infiltrated the group and made hundreds of hours of tape recordings. Earth First! supporters claim that FBI agents acted as prevocateurs, trying to instigate dangerous and illegal acts. The threat of harassment and arrest had its intended effect, however, driving many of the principal figures in the radical environmental group into retirement or underground. Foreman, for instance, resigned from Earth First! saying that it was compromised by the infiltration and that it was drifting too much into social issues and away from the fight for pure wilderness. The near-fatal bombing of a car in which two Earth First! activists were riding in 1990 also frightened and angered many activists. Clearly, powerful and ruthless forces in America regard militant environmentalism as a threat.

Sea Shepherd, a radical offshoot from Greenpeace, is determined to stop the killing of marine mammals. Dissatisfied with peaceful protests, they have taken direct action by sinking whaling vessels and destroying machinery at whale processing stations. Greenpeace has criticized this destructive approach, perhaps because of memories of the sinking of its ship *Rainbow Warrior* by the French secret service in New Zealand in 1985. They and other critics of ecotage see these acts as environmental terrorism that gives all environmentalists a bad name. Sea Shepherd's Paul Watson says, "Our aggressive nonviolence is just doing what must be done, according to the dictates of our own conscience." Earth First!'s Dave Foreman (sometimes described as a kamikaze environmentalist) says, "Extremism in defense of Mother Earth is no vice." What do you think? Do ends justify means?

Environmental Philosophy

The debate about why and how we should protect the environment and its resources is not new. It goes back at least to the beginnings of conservationism in the nineteenth century when the altruistic preservationists, led by John Muir, fought bitterly with the utilitarian conservationists, led by Gifford Pinchot (chapter 12).

Then, as now, conservationists were progressives who sought to work within the existing system for the most part, to reform it, rather than tear it down. They were pragmatists who sought compromise and accommodation, rather than confrontation. Their approach to resource management was scientific, professional, activist, and technological. Husbanding and stewardship of resources are key concepts in this view. As humanists and populists, they considered the goals of conservation to be preventing waste and saving resources for prudent, constructive, and efficient human use. Their motto was, "For the greatest good, for the greatest number, for the longest time." Pinchot said, "There may be just as much waste in neglecting the development and use of certain natural resources as there is in their destruction."

In contrast, the preservationists were antimodernists who distrusted technology and civilization. They were visionary, mystical, zealous, and uncompromising in their defense of nature;

egalitarian in regarding humans as just ordinary members of the biological community; and often pantheists, appreciative of nature as a source of beauty, knowledge, and power.

These themes have reemerged recently in a set of principles called **deep ecology**, by Norwegian philosopher Arne Naess. Among his tenents are a belief in voluntary simplicity, being in touch with nature, decentralization, personal freedom, cultural and biological diversity, egalitarianism (for people and other beings), the sacredness of nature, and an obligation to act in defense of the environment. Naesses' followers tend to focus on personal growth in small, alternative communities, separate from the dominant social paradigm. Believing that humans have exceeded the carrying capacity of the biosphere and that there are too many of us and too much technology on the planet, they argue that growth of both population and technology must be stopped, or even reversed, if Earth is to survive. They believe that other life-forms and natural features have a right to exist whether they benefit us or not. Important principles in deep ecology are reinhabitation (learning how to dwell in and care for a specific place), **bioregionalism** (allegiance to a biome or ecosystem rather than a nation) (table 27.6), living frugally, and returning to a simple agrarian (or even pre-agrarian) life-style. They are often very interested in native American cultures, Eastern and occult religions, popular psychology, and other "self-actualizing" new age movements.

Historian Steven Fox has pointed out that environmental groups tend to have predictable life cycles. They start as zealous, revolutionary groups, often led by an innovative, charismatic leader who sees threats to the environment and motivates others to work for change. Groups start as small, informal, experimental organizations trying new techniques and new approaches. As the group experiences successes and gains membership and power, however, it becomes more cautious and less revolutionary. The strains of being "ad hoc," flexible, and democratic begin to wear on the members. They want stability and professionalism as they develop public and political credibility. It becomes important to preserve the power and prestige that the group has accumulated. Soon the group becomes entrenched and bureaucratic. Radical members within the group become impatient with conservatism and pragmatism, and break away to form a new, revolutionary group.

At the base of the controversy between these radical and conservative environmentalists are some deep differences in opinion about society and human nature. Which of these philosophies you ascribe to makes a great deal of difference in how we might attain a sustainable future. If you are thinking of joining an environmental group, you should examine your feelings about which kind of a group you would feel most comfortable with. You might ask yourself the following questions:

1. Is it best to be progressive and reformist or radical and revolutionary; should we work to improve existing structures, or should we tear them down and start over?

TABLE 27.6 Test your bioregional knowledge

1. Trace the water you drink from precipitation to tap.
2. How many days until the moon is full (plus or minus a couple of days)?
3. Describe the soil type around your home.
4. What were the primary subsistence techniques of the culture(s) that lived in the area before you?
5. Name five native edible plants in your bioregion and their season(s) of availability.
6. From what direction do winter storms generally come in your region?
7. Where does your garbage go?
8. How long is the growing season where you live?
9. On what day of the year are the shadows the shortest where you live?
10. Name five trees in your area. Are any of them native? If you can't name names, describe them.
11. Name five resident and any migratory birds in your area.
12. What is the land-use history by humans in your bioregion during the past century?
13. What primary geological event/process influenced the landform where you live?
14. What species have become extinct in your area?
15. What are the major plant associations in your region?
16. From where you are reading this, point north.
17. What spring wildflower is consistently among the first to bloom where you live?
18. What kinds of rocks and minerals are found in your bioregion?
19. Were the stars out last night?
20. Name some beings (nonhuman) that share your place.
21. Do you celebrate the turning of the summer and winter solstice? If so, how do you celebrate?
22. How many people live next door to you? What are their names?
23. How much gasoline do you use a week, on the average?
24. What energy costs you the most money? What kind of energy is it?
25. What developed and potential energy resources are in your area?
26. What plans are there for massive development of energy or mineral resources in your bioregion?
27. What is the largest wilderness area in your bioregion?

Deep Ecology: Living as if Nature Mattered by Bill Devall and George Sessions, copyright 1985 by Gibbs Smith, Publisher; Salt Lake City, UT.

2. Are you anthropocentric or biocentric? Do humans have a special importance in the world, or are we "plain members of a biotic community"?

3. Do you believe in conservation, preservation, or some combination of the two views? Is the purpose of saving resources to preserve them for use by later generations or simply to save them for their own sake?

4. Do you favor professionalism (with representative democracy) or participatory democracy? Are efficiency

FIGURE 27.13 Students cleaning up the beach in Santa Cruz, California, after a tanker collision. Ordinary people can make a contribution to improving the environment. It can be personally rewarding and can foster public spirit.

and effectiveness the measure of success, or is direct participation by everyone more important?

5. Would you prefer integration of environmentalism into the broadly agreed-upon national agenda or do you find yourself fundamentally in opposition to the dominant paradigm?

6. Can technology offer solutions to our problems, or will it inevitably make them worse?

7. Is environmentalism a science, religion, or social movement?

8. Is further growth possible—even necessary—or impossible?

9. Are you pessimistic or hopeful? Are we approaching an Armageddon or a Renaissance?

What Can You Do Personally to Make the World Better?

Regardless of whether we choose to join a large organization or work in a small group, there are things that each of us can do to improve our environment (Box 27.1). The following list offers some suggestions. You probably will come up with others when you give the subject some thought.

1. *Think globally and act locally.* We need to be aware of global conditions, but each of us should also work to improve our own particular place. (figure 27.13)

2. *Continue your education.* Learning shouldn't stop when you leave school, nor should all your information come from "talking heads" on television. Read good books and a daily newspaper; discuss current events with your friends.

3. *Vote.* You can't improve your world by refusing to vote. If you don't like the choices available, work to get candidates on the ballot who represent your interests.

4. *Learn about the ecology of your bioregion.* Develop a sense of place that puts you in contact with the natural features and cycles of your environment.

5. *Think about the consequences of your profession and your life-style.* If they are damaging to other people or to the environment, adjust your behavior accordingly. Try to persuade friends, family, and co-workers to do the same.

6. *Work with others.* Being part of a community of people with similar interests gives you support and increases your effectiveness. If you can't find an appropriate group, start one of your own.

7. *Don't feel responsible for every problem in the world.* Protect your own energy, spirit, and sanity by focusing on a few things that you can do something about.

8. *Take time for the good things in life.* Enjoy art, music, literature, religion, friendship, love, and natural beauty.

The second national Earth Day was celebrated throughout the United States in April, 1990. An estimated two million people gathered in parks and on college campuses to enjoy the spring sunshine, listen to music, play games, see celebrities, and be part of the scene. For some people, this event was more spectacle and fun than a real commitment to change, but most people who attended the lectures, marches, clean-ups, workshops, and other activities associated with Earth Day have a sincere concern about the environment and our global future. The New York Times/CBS News poll regularly asks people to respond to the statement: "Protecting the environment is so important that requirements and standards cannot be too high, and continuing environmental improvements must be made regardless of cost." In 1980, only 45 percent of those polled agreed, but by 1990, that number had risen to 80 percent.

Sustaining this high level of concern and transforming it into a permanent change in public policy, institutions, and behavior is a big challenge. One way we all can have an impact on the environment is through our purchases. After Earth Day, a plethora of products were rushed to the market and advertised as environmentally friendly, biodegradable, recyclable, all natural, made of recycled materials, or environmentally wise. How can we know which are environmentally correct and which are just marketing scams?

Some plastic garbage bags, for instance, are advertised with great fanfare to be biodegradable or photodegradable. They may break down under laboratory conditions but in a landfill , protected from oxygen, light, and water, they probably will last for centuries. They do take up less space than paper, however. Some plastics manufacturers also claim that recycling plastic has less environmental impact than repulping paper, but if there are no programs for recycling plastics, that claim doesn't mean much. Who are we to believe, and what is the correct thing to do? As Kermit the Frog says, "It's not easy being green."

To assist consumers with purchasing decisions, about a dozen certification programs now review and rate the environmental impact of goods on the market. Some are funded and operated by the industries that make the goods they rate. Their objectivity is somewhat suspect. Others are operated by independent agencies. Among those that seem most legitimate are Green Seal, Green Cross, and Good Earthkeeping. They verify product claims of environmental safety and carry out life-cycle cost analyses that measure environmental impacts of product manufacture, use, and disposal. The model for these programs is the Blue Angel seal from Germany, which has a decade of experience in environmental certification. While this is still a very imprecise field, it should provide a great service to consumers once its methodology becomes more firmly established.

Although there are many difficulties in life, there are many joys as well.

9. *Live simply and frugally*. You will be better off and so will your environment. Remember that the best *things* in life are not *things*.

10. *Don't feel guilty, or try to make others feel guilty*. A good rule is that you should give two compliments for every criticism.

11. *Don't be discouraged*. It is important to face facts honestly and to be realistic about the state of the world, but it doesn't help to wallow in despair. Don't dwell on negatives. Do what you can to improve the world, and take pleasure and pride in the small victories and elements of success.

Recreational development is a job not of building roads into lovely country, but of building receptivity into the still unlovely human mind . . . all history consists of successive excursions from a single starting point, to which man returns again and again to organize yet another search for a durable scale of values.
Aldo Leopold

SUMMARY

In this chapter, we have reviewed ways our current choices in resource use and organization of society are determining what our lives and those of our children will be like in the future. We have seen causes for pessimism about the state of our environment but also reasons for optimism about our future if we work together.

There is a growing consensus that population growth must be controlled if we are to stay within the carrying capacity of Earth. Whether economic growth and social development are tolerable depends on your point of view. Development and social justice are tightly linked. Sustainable development is regarded by many as the best solution for nondestructive economic growth and real progress in alleviating poverty, suffering, and oppression. The four bases of sustainable development are scientific information, ethical principles, hope for the future, and sharing of benefits of growth.

Environmental laws are the foundation of most environmental protection. The global nature of resources and pollution make international laws and conventions essential. Every citizen should know how laws are established, administered, and enforced. A variety of nongovernmental organizations, political parties, environmental clubs and societies, professional bodies, and other groups play important roles in environmental protection. Of the varied tactics applied by these groups, direct violent action (ecotage) is the most controversial.

Since the earliest origins of conservation, altruistic preservationists have debated utilitarian conservationists over the purposes and tactics to be used in preserving resources. The latest version of this debate is between "deep ecologists" and political pragmatists. Which group you support depends on your beliefs, temperament, philosophical position, and political persuasion.

There is much that each of us can do to help preserve our environment and conserve resources. Our actions not only will help society in general, but they often yield direct personal benefits as well.

■ Review Questions

1. What are some reasons to be optimistic or pessimistic about the future?
2. What are ecojustice and sustainable development? How might we attain these goals?
3. What is environmental ethics? How does it differ from morals or from general ethical principles?
4. Describe the distinction between moral agents and moral subjects. Are their differences in the rights, duties, and moral obligations of these moral classes?
5. What is moral extensionism? How does it apply to environmental ethics?
6. Describe and discuss the principles for resolving competing ethical claims and ethical dilemmas.
7. Outline the path a proposed law takes as it passes through Congress. At what points can citizens express their opinions and wishes?
8. Give some examples of some international environmental conventions and treaties. Have they been effective?
9. What are ecotage and direct action? Who advocates these confrontational approaches and why?
10. What is deep ecology? What are its principles and how is it applied to our environmental crises?

■ Questions for Critical Thinking

1. Is progress possible? Can technology, understanding, and human enterprise make the world a better place in which to live?
2. What would your life be like if you were living one hundred, one thousand, or ten thousand years ago? Would that be preferable to living today?
3. What kinds of economic and social development do you think are compatible with or necessary for a healthy environment?
4. What arguments would you make for or against a utilitarian ethic?
5. What rights would you ascribe to other species? Is there a moral distinction between humans and other animals? What rights do plants, rocks, and geological features have?
6. Describe the philosophical bases of stewardship and ecofeminism. Are the two terms mutually exclusive?

7. Many Americans do not get involved in politics and do not even vote. Why? What are the consequences?
8. What are the benefits and disadvantages of international environmental laws? How can we encourage international cooperation on environmental issues?
9. Some environmental groups advocate direct, violent action to attack environmental problems; others prefer to work within existing systems to bring about gradual progressive reform. Which tactic do you favor? Why?
10. What can you do personally to make the world a better place? What actions are best done individually and which ones require group efforts?

■ Key Terms

anthropocentric (p. 559)
bill (p. 562)
biocentrism (p. 560)
bioregionalism (p. 571)
common law (p. 562)
deep ecology (p. 571)
ecofeminism (p. 561)
ecojustice (p. 557)
ecotage (p. 570)
environmental ethics (p. 558)
Environmental Impact Statement (EIS) (p. 566)
fidelity (p. 560)
instrumental value (p. 559)
moral agents (p. 559)
moral subjects (p. 559)
nihilists (p. 558)
nongovernmental organization (NGO) (p. 566)
noninterference (p. 560)
nonmalfeasance (p. 560)
regulations (p. 562)
relativists (p. 558)
statutory law (p. 562)
stewardship (p. 560)
universalists (p. 558)
utilitarianism (p. 559)

SUGGESTED READINGS

■ Andruss, V. et al. 1991. *Home! Reinhabiting North America*. Santa Cruz, CA: New Society Publishers. An anthology of essays, poems, and illustrations by some of the leading bioregional philopsophers and environmental activists.

■ Berger, J. 1990. *Environmental Restoration*. Washington, D.C.: Island Press. How Americans are working to renew our damaged environment.

■ Berry, W. *The Unsettling of America: Culture and Agriculture*. New York: Avon Books, 1977. An eloquent and passionate declaration of land mismanagement, the waste of human and natural resources, and the necessity to develop a sense of place and an ethic of treating the land kindly.

■ Bookchin, M. 1990. *Remaking Society: Pathways to a Green Future*. Boston, MA: South End Press. A challenging analysis of the social, moral, economic, and ecological basis of our current condition.

■ Borrelli, P. Winter, 1986. "Epiphany: Religion, Ethics, and the Environment." *The Amicus Journal* 7(3):34. Excellent review of bioethics and the environmental movement.

■ Brown, L., and E. Wolf. "The New World Order." In *State of the World 1991*. Washington, D.C.: Worldwatch Institute, some good suggestions for a sustainable world.

- Chase, S. (ed.) 1991. *Defending the Earth: A Dialog between Murray Bookchin and Dave Foreman*. Boston, MA: South End Press. An excellent presentation of the debate between deep and social ecology.
- Devall, B., and G. Sessions. *Deep Ecology*. Layton, Utah: Gibbs M. Smith, Inc, 1985. An excellent summary of the philosophy and application of deep ecology.
- Durning, A. 1991. "Asking How Much Is Enough." *State of the World*. Washington D.C. World-Watch Institute. How we curb unnecessary consumption and reach a sustainable society.
- *(An) Environmental Agenda for the Future*. Washington, D.C.: Island Press, 1985. Writings by the leaders of conservation organizations in the United States that address the most serious questions about the future of our resources.
- Foreman, D., and B. Haywood. *Ecodefense: A Field Guide to Monkey Wrenching*. 2d ed. Tucson, Arizona: Earth First!, 1987. Expanded, revised, and field-tested hints about direct action!
- Fox, S. *John Muir and His Legacy: The American Conservation Movement*. Boston: Little, Brown and Company, 1981. An interesting and highly readable biography of Muir and the history of environmental organizations.
- Graber, L. *Wilderness as Sacred Space*. Washington, D.C.: Association of American Geographers, 1976. Well-written study of the religious basis of wilderness preservation and the geopiety of wilderness "purists."
- Manes, C. 1990. *Green Rage: Radical Environmentalism and the Unmaking of Civilization*. Boston: Little Brown and Company. A brief history of radical environmentalism by Miss Anthropy of Earth First.
- Marsh, G. *Man and Nature: Or Physical Geography as Modified by Human Action*. New York: Charles Scribner, 1864. (Reprinted in 1965 by Harvard University Press.) Called the "fountainhead of the conservation movement," this pioneering work is an enduring classic.
- Naess, A. 1973. "The Shallow and the Deep: Long-Range Ecology Movement." *Inquiry* 16:95. The original use of the term *deep ecology*.
- Nash, R. *Wilderness and the American Mind*. New Haven, Connecticut: Yale University Press, 1982. An important book for a theoretical understanding of historical attitudes, religious beliefs, and the development of an ethical attitude toward wilderness.
- Nash, R. 1989. *The Rights of Nature*. Madison: The University of Wisconsin Press. A history of environmental ethics that brings us up to Green polotics and radical environmentalisms.
- Passmore, J. *Man's Responsibility for Nature: Ecological Problems and Western Traditions*. New York: Charles Scribner & Sons, 1974. A good discussion of the literary, religious, and philosophical basis of our attitudes toward nature.
- Plant, J. *Healing the Wounds: The Promise of Ecofeminism*. Santa Cruz: New Society Publishers, 1989. An anthology of articles by leading ecofeminists, feminist theologians, and activists.
- Repetto, R. *World Enough and Time: Successful Strategies for Resource Management*. New Haven, Connecticut: Yale University Press, 1986. An excellent review of the prospects and possibilities for sustainable development.
- Rolston, H. *Environmental Ethics*. Philadelphia: Temple University Press, 1988. A pioneering synthesis of our duties to and values of the natural world.
- Russell, D. Fall, 1987. "The Monkeywrenchers." *The Amicus Journal* 9(4):28. A good review of the new radical environmentalists.
- Sagoff, M. 1988. *The Economy of the Earth*. Cambridge: Cambridge University Press. A good summary of environmental philosophy and law.
- Sale, K. *Dwellers in the Land: The Bioregional Vision*. San Francisco: Sierra Club Books, 1985. An excellent introduction to the politics, history, and economy of the bioregional vision.
- Schumacher, E. *Small Is Beautiful: Economics as if People Mattered*. London: Blond & Briggs, 1973. A classic reevaluation of economics, growth, and social values from the father of the appropriate technology movement.
- Seed, J., J. Macy, P. Fleming, and A Naess. 1988. *Thinking Like a Mountain: Toward a Council of All Beings*. Philadelphia: New Society Publishers. Poems, songs, essays, and workshop agendas fostering a biocentric view of nature.
- Shiva, V. *Staying Alive: Women, Ecology, and Development*. London: Zed Books, 1988. Examines the position of women in relation to nature, especially the *Chipko* movement in India.
- Singer, P. 1990. *Animal Liberation*. New York: Avon Books. The most important statement of the animal liberation movement.
- Swift, R. September/October 1987. "What if the Greens Achieved Power?" *Utne Reader* 23:30. Part of a special issue on Green politics and the environment drawn from many different sources in the "alternative press." Good source of information.
- Taylor, P. W. *Respect for Nature*. Princeton, New Jersey: Princeton University Press, 1986. A compelling analysis of human and environmental ethics leading to a biocentric outlook and a sense of respect for nature.
- Thoreau, H. *Walden*. Concorde, Massachusetts: several publishers, 1854. Usually combined with the essay, "Civil Disobedience." Perhaps the most influential nature writing in American history. Regarded by some as the original source of deep ecology.
- Tuan, Yi Fu. *Topophilia: A Study of Environmental Perception, Attitudes, and Values*. New York: Prentice Hall, 1974. A comparison of different cultures in time and place and their attitudes toward self and nature.
- Ward, B., and R. Dubos. *Only One Earth: The Care and Maintenance of a Small Planet*. New York: W.W. Norton and Company, 1972. An unofficial report commissioned by the secretary-general of the United Nations Conference on the Human Environment.

A P P E N D I X A

Periodic Chart
of the Elements

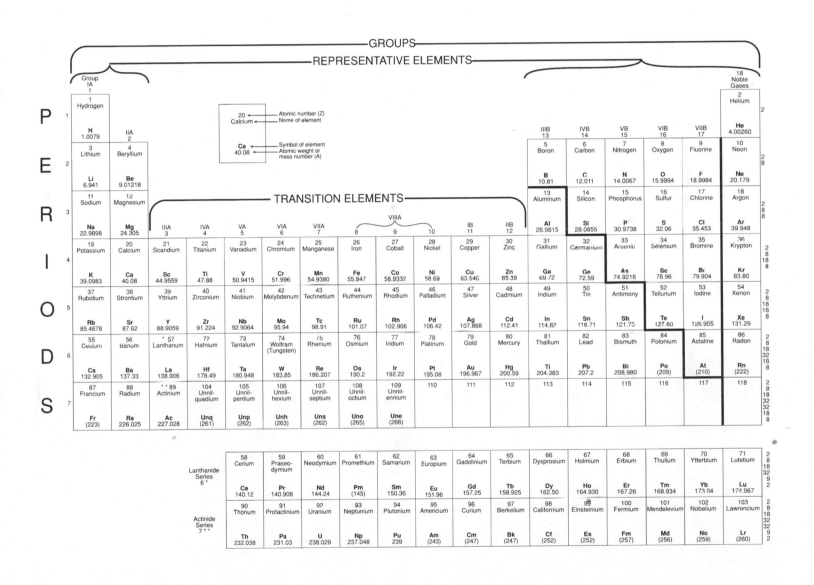

A P P E N D I X *B*

Geologic Time and Formations

Time-rock Units of the Geologic Column	TIME UNITS OF THE GEOLOGIC TIME SCALE Numbers are absolute dates in millions of years before present.			MILESTONES IN THE HISTORY OF LIFE	Relative Lengths of Major Time Divisions In True Scale	
	Eras	Periods	Epochs			
	CENOZOIC	Quaternary	Holocene 0.01	Expansion of hominids	66.4	Cenozoic
			Pleistocene 1.8			
			Pliocene 5.3	Dominance of elephants, horses, large carnivores	245	Mesozoic
		Tertiary	Miocene 23.7	Development of whales, bats, monkeys, horses Coevolution of insects and flowering plants		
			Oligocene 36.6	Grazing animals widespread, grasses abundant	570	Paleozoic
			Eocene 57.8	Primitive horses		
			Paleocene 66.4	Rapid development of mammals		
	MESOZOIC	Cretaceous 66.4 144		Extinction of dinosaurs and ammonites Development of flowering plants		
		Jurassic 208		Climax of dinosaurs, cycads abundant Earliest record of birds		
		Triassic 245		First primitive mammals; conifers and cycads abundant; appearance of dinosaurs; rapid development in reptiles		Precambrian
	PALEOZOIC	Permian 286		Development of conifers, extinction of trilobites Spread of reptiles, including mammal-like reptiles		
		Pennsylvanian (Upper Carboniferous) 320		Earliest primitive reptiles Abundant insects Coal-forming forests widespread		
		Mississippian (Lower Carboniferous) 360		Echinoderms and bryozoa abundant Spreading of fish faunas		
		Devonian 408		Appearance of amphibians Earliest forests		
		Sulurian 438		Earliest record of land plants and animals First jawed fish		
		Ordovician 505		Primitive fishes, the first known vertebrates Diverse communities of marine shelled invertebrates		
		Cambrian 570		Marine invertebrate faunas abundant; trilobites predominant		
	PRECAMBRIAN TIME			Development of first multicellular animals Expansion of simple marine plants Origin of life	4600	Origin of earth

From James C. Brice, et al., *Laboratory Studies in Earth History*, 4th ed. Copyright © 1989 Wm. C. Brown Publishers, Dubuque, Iowa. All Rights Reserved. Reprinted by permission.

A P P E N D I X *C*

Units of Measurement
Metric/English Conversions

LENGTH

1 meter = 39.4 inches = 3.28 feet = 1.09 yard
1 foot = 0.305 meters = 12 inches = 0.33 yard
1 inch = 2.54 centimeters
1 centimeter = 10 millimeter = 0.394 inch
1 millimeter = 0.001 meter = 0.01 centimeter = 0.039 inch
1 fathom = 6 feet = 1.83 meters
1 rod = 16.5 feet = 5 meters
1 chain = 4 rods = 66 feet = 20 meters
1 furlong = 10 chains = 40 rods = 660 feet = 200 meters
1 kilometer = 1,000 meters = 0.621 miles = 0.54 nautical miles
1 mile = 5,280 feet = 8 furlongs = 1.61 kilometers
1 nautical mile = 1.15 mile

AREA

1 square centimeter = 0.155 square inch
1 square foot = 144 square inches = 929 square centimeters
1 square yard = 9 square feet = 0.836 square meters
1 square meter = 10.76 square feet = 1.196 square yards = 1 million square millimeters
1 hectare = 10,000 square meters = 0.01 square kilometers = 2.47 acres
1 acre = 43,560 square feet = 0.405 hectares
1 square kilometer = 100 hectares = 1 million square meters = 0.386 square miles = 247 acres
1 square mile = 640 acres = 2.59 square kilometers

VOLUME

1 cubic centimeter = 1 milliliter = 0.001 liter
1 cubic meter = 1 million cubic centimeters = 1,000 liters
1 cubic meter = 35.3 cubic feet = 1.307 cubic yards = 264 U.S. gallons
1 cubic yard = 27 cubic feet = 0.765 cubic meters = 202 U.S. gallons
1 cubic kilometer = 1 million cubic meters = 0.24 cubic mile = 264 billion gallons
1 cubic mile = 4.166 cubic kilometers
1 liter = 1,000 milliliters = 1.06 quarts = 0.265 U.S. gallons = 0.035 cubic feet
1 U.S. gallon = 4 quarts = 3.79 liters = 231 cubic inches = 0.83 imperial (British) gallons
1 quart = 2 pints = 4 cups = 0.94 liters
1 acre foot = 325,851 U.S. gallons = 1,234,975 liters = 1,234 cubic meters
1 barrel (of oil) = 42 U.S. gallons = 159 liters

MASS

1 microgram = 0.001 milligram = 0.000001 gram
1 gram = 1,000 milligrams = 0.035 ounce
1 kilogram = 1,000 grams = 2.205 pounds
1 pound = 16 ounces = 454 grams
1 short ton = 2,000 pounds = 909 kilograms
1 metric ton = 1,000 kilograms = 2,200 pounds

TEMPERATURE

Celsius to Fahrenheit $°F = (°C \times 1.8) + 32$
Fahrenheit to Celsius $°C = (°F - 32) \div 1.8$

ENERGY AND POWER

1 erg = 1 dyne per square centimeter
1 joule = 10 million ergs
1 calorie = 4.184 joules
1 kilojoule = 1,000 joules = 0.949 British Thermal Units (BTU)
1 megajoule = MJ = 1,000,000 joules
1 kilocalorie = 1,000 calories = 3.97 BTU = 0.00116 kilowatt-hour
1 BTU = 0.293 watt-hour
1 kilowatt-hour = 1,000 watt-hour = 860 kilocalories = 3,400 BTU
1 horsepower = 640 kilocalories
1 quad = 1 quadrillion kilojoules = 2.93 trillion kilowatt-hours

APPENDIX D

Environmental Organizations

This is a sampling of nongovernmental national and international environmental organizations. For the more complete, annually updated Conservation Directory, send $15 to the National Wildlife Federation, 1400 Sixteenth Street NW, Washington, D.C., 20036. The directory includes many professional associations and state or local organizations not listed here. Your local library may have a copy. Most of the organizations publish a magazine or newsletter.

■ African Wildlife Foundation, 1717 Massachusetts Avenue NW, Washington, D.C. 20036 (202–265–8393). Finances and operates wildlife conservation projects in Africa.

■ American Committee for International Conservation, c/o Mike McCloskey, Sierra Club, 330 Pennsylvania Avenue SE, Washington, D.C. 20003 (202–547–1144). An association of nongovernmental organizations concerned with international conservation.

■ American Farmland Trust, 1920 N Street NW, Suite 400, Washington, D.C. 20036 (202–659–5170). Works for preservation of family farms and for soil conservation.

■ American Museum of Natural History, Central Park West at 79th Street, New York, NY 10024 (212–769–5000). Conducts natural history research and publishes educational material.

■ Appalachian Mountain Club, 5 Joy Street, Boston, MA 02108 (617–523–0636). Sponsors trail maintenance, outdoor education, and recreational hikes and climbs. Operates a mountain hut system on the Appalachian trail.

■ Bioregional Institute, 233 Miramar Street, Santa Cruz, CA 95060 (408–425–0264). Works toward the restoration of an ecological harmony with the land. Many bioregional organizations have excellent newsletters.

■ Center for Environmental Education, 1725 DeSales Street NW, Suite 500, Washington, D.C. 20036 (202–429–5609). A nongovernmental, nonprofit organization that sponsors programs for protection of endangered species.

■ Center for Science in the Public Interest, 1501 16th Street NW, Washington, D.C. 20036 (202–332–9110). National consumer advocacy organization that focuses on health and nutrition.

■ Clean Water Action Project, 317 Pennsylvania Avenue SE, Washington, D.C. 20003 (202–547–1196). Works for clean water and for protection of natural resources. Conducts voter education and public awareness projects.

■ Conservation International, 1015 18th Street NW, Suite 1000, Washington, D.C. 20036 (202–429–5660). Buys land or trades foreign debt for land set aside for nature preserves in developing countries.

■ Cousteau Society, Inc., 930 W. 21st Street, Norfolk, VA 23517 (804–627–1144). Produces television films, lectures, books, and research on ocean quality and other resource issues.

■ Ducks Unlimited, Inc., One Waterfowl Way, Long Grove, IL 60047 (312–438–4300). Perpetuates waterfowl by purchasing and protecting wetland habitat.

■ Earth Island Institute, 300 Broadway, Suite 28, San Francisco, CA 94133. A clearinghouse for international information on environmental and resource issues. Founded by David Brower to bring together sources of conservation action and news.

■ Environmental Defense Fund, Inc., 257 Park Avenue South, New York, NY 10010 (212–505–2100). Protects environmental quality and public health through litigation and administrative appeals.

■ Environmental Law Institute, 1616 P Street NW, Suite 200, Washington D.C. 20036 (202–328–5150). Sponsors research and education on environmental law and policy.

■ Food and Agriculture Organization of the United Nations, Via delle Terme di Caracalla, Rome 00100 Italy (Tele: 57971). A special agency of the United Nations established to improve food production, nutrition, and the health of all people.

■ Friends of the Earth, 530 Seventh Street SE, Washington, D.C. 20003 (202–543–4312). Affiliated with groups in thirty-two countries around the world. Works to form public opinion and to influence government policies to protect nature.

■ Fund for Animals, Inc., 200 West 57th Street, New York, NY 10019 (212–246–2096). Advocacy group for humane treatment of all animals.

■ Greater Yellowstone Coalition, P.O. Box 1874, 420 West Mendenhall, Bozeman, MT 59715 (406–586–1593). A small but influential organization dedicated to the preservation and protection of the Greater Yellowstone ecosystem.

■ Greenpeace USA, 1436 U Street NW, Washington, D.C. 20009 (202–462–1177). Worldwide organization that works to halt nuclear weapons testing, to protect marine animals, and to stop pollution and environmental degradation.

■ Humane Society of the United States, 2100 L Street NW, Washington D.C. 20037 (202–452–1100). Dedicated to the protection of both domestic and wild animals.

■ Institute for Food and Development Policy, 1885 Mission Street, San Francisco, CA 94103. An international research and education center working for social justice and environmental protection.

■ Institute for Social Ecology, P.O. Box 89, Plainsfield, VT 85667. Offers courses for credit through Goddard College and workshops on social ecology, ecofeminism, urban design, and ecological planning.

■ International Alliance for Sustainable Agriculture, 1701 University Avenue SE, Minneapolis, MN 55414. An alliance of organic farmers, researchers, consumers, and international organizations dedicated to sustainable agriculture.

■ International Union for Conservation of Nature and Natural Resources (IUCN), Avenue du Nont-Blanc CH-1196 Gland, Switzerland (022–64–71–81). Promotes scientifically based action for the conservation of wild plants, animals, and resources.

■ Izaak Walton League of America, Inc., 1401 Wilson Boulevard, Level B, Arlington, VA 22209 (703–528–1818). Educates the public on land, water, air, wildlife and other conservation issues.

■ Land Institute, Route 3, Salina, KS 67401 (913–823–5376). Carries out research on perennial species, prairie polycultures, and sustainable agriculture. Has training programs and conferences.

■ League of Conservation Voters, 2000 L Street NW, Suite 804, Washington, D.C. 20036 (202–785–8683). A nonpartisan national political campaign committee that strives to elect environmentally responsible public officials. Publishes an annual evaluation of voting records of Congress.

■ National Audubon Society, 950 Third Avenue, New York, NY 10022 (212–832–3200). One of the oldest and largest conservation organizations, Audubon has many educational and recreational programs as well as an active lobbying and litigation staff.

■ National Parks and Conservation Association, 1015 31st Street NW, Washington, D.C. 20007 (202–944–8530). A private nonprofit organization dedicated to preservation, promotion, and improvement of our national parks.

■ National Wildlife Federation, 1400 Sixteenth Street NW, Washington, D.C. 20036 (202–797–6800). Specializes in wildlife conservation but recognizes the importance of habitat and other resources to all living things. More than 5 million members.

■ Natural Resources Defense Council, Inc., 122 East 42nd Street, New York, NY 10168 (212–949–0049). An environmental organization that monitors government agencies and brings legal action to protect the environment.

■ (The) Nature Conservancy, 1815 North Lynn Street, Arlington, VA 22209 (703–841–5300). Works with state and federal agencies to identify ecologically significant natural areas. Manages a system of over 1000 nature sanctuaries nationwide.

■ New Alchemy Institute, Box 432, Woods Hole, MA 02543 (617–563–2655). Conducts research and provides education on sustainable agriculture, aquaculture, and bioshelters.

■ Planet/Drum Foundation, P.O. Box 31251, San Francisco, CA 94131. Promotes wise use of resources and new attitudes toward nature.

■ Population Reference Bureau, 777 14th Street NW, Suite 800, Washington, D.C. (202–639–8040). Gathers, interprets, and publishes information on social, economic, and environmental implications of world population dynamics. Excellent data source.

■ Project Lighthawk, P.O. Box 8136, Santa Fe, NM 87504 (505–982–9656). Uses light aircraft for aerial surveys of forest condition, range management, and other conservation issues. Offers a unique vantage point to environmentalists.

■ Rainforest Action Network, 300 Broadway, Suite 28, San Francisco, CA 94133 (415–398–4404). Shares offices with Earth Island Institute and focuses on actions designed to save rainforests around the world.

■ Resources for the Future, 1616 P Street NW, Washington D.C. 20036 (202–328–5000). Conducts research and provides education about natural resource conservation issues.

■ Rodale Institute, 222 Main Street, Emmaus, PA 18049 (215–967–5171). A leading research institute for organic farming and alternative crops. Publishes magazines, books, and reports on regenerative farming.

■ Save-the-Redwoods League, 114 Sansome Street, Room 605, San Francisco, CA 94104 (415–362–2352). Buys land, plants trees, and works with state and federal agencies to save redwood trees.

■ Scientists' Institute for Public Information, 355 Lexington Avenue, New York, NY 10017 (212–661–9110). Enlists scientists and other experts in public information programs and public policy forums on a variety of environmental issues.

■ Sea Shepherd Conservation Society, P.O. Box 7000-S, Redondo Beach, CA 90277 (213–373–6979). An international marine conservation action program. Carries out field campaigns to call attention to and stop wildlife destruction and resource misuse.

■ Sierra Club, 730 Polk Street, San Francisco, CA 94109 (415–776–2211). Founded in 1892 by John Muir and others to explore, enjoy, and protect the wild places of the earth. Conducts outings, educational programs, volunteer work projects, litigation, political action, and administrative appeals. Has one of the most comprehensive programs of any conservation organization.

■ Student Conservation Association, Inc. Box 550, Charlestown, NH 03603 (603–826–5206). Coordinates environmental internships and volunteer jobs with state and federal agencies and private organizations for students and adults.

■ United Nations Environment Programme, P.O. Box 30552, Nairobi Kenya; and United Nations, Rm. DC2-0803, New York, NY 10017 (212–963–8138). Coordinates global environmental efforts with United Nations agencies, national governments, and nongovernmental organizations.

■ Waldebridge Ecological Centre, Worthyvale Manor Farm, Camelford, Cornwall P132 9TT England (Tele: 0840 212711). Conducts research and education on a variety of environmental issues. Publishes *The Ecologist*, an excellent global environmental journal.

■ Wilderness Society, 1400 I Street NW, 10th Floor, Washington, D.C. 20005 (202–842–3400). Dedicated to preserving wilderness and wildlife in America.

■ World Resources Institute, 1735 New York Avenue NW, Washington D.C. 20006 (202–638–6300). A policy research center that publishes excellent annual reports on world resources.

■ Worldwatch Institute, 1776 Massachusetts Avenue NW, Washington D.C. 20036 (202–452–1999). A nonprofit research organization concerned with global trends and problems. Publishes excellent periodic reports and annual summaries.

■ Zero Population Growth, Inc., 1601 Connecticut Avenue NW, Washington D.C. 20009 (202–332–2200). Advocates of worldwide population stabilization.

APPENDIX *E*

United States Government Agencies

■ Bureau of Land Management, C and 18th Street NW, Washington, D.C. 20240. Administers about half of all public lands, mainly in the western United States. Follows policy of multiple use for maximum public benefit.

■ Bureau of Mines, 2401 E Street NW, Washington, D.C. 20241 (202–634–1004). Oversees mineral production, mine safety, and mine reclamation.

■ Bureau of Reclamation, Washington, D.C. 20240. Builds and operates federal water projects, mostly in the western states with the Army Corps of Engineers.

■ Council on Environmental Quality, 722 Jackson Place NW, Washington, D.C. 20503 (202–395–5750). Advises the president on environmental matters.

■ Department of Agriculture, 14th Street and Independence Avenue SW, Washington, D.C. 20250. Manages national forests and grasslands and oversees farm prices, farm policies, and soil conservation.

■ Department of Health and Human Services: Food and Drug Administration, 5600 Fishers Lane, Rockville, MD 20857 (301–443–1544). Enforces laws requiring that foods and drugs be pure, safe, and wholesome.

■ Department of the Interior, C Street between 18th and 19th, NW, Washington, D.C. 20240 (202–343–1100). Administers national parks, monuments, wildlife refuges, and public lands.

■ Environmental Protection Agency, 401 M Street SW, Washington, D.C. 20460. Enforces clean air and clean water laws. Identifies, regulates, and purifies toxic and hazardous materials.

■ The Forest Service, P.O. Box 96090, Washington, D.C. 20013 (202–447–3957). Administers national forests and grasslands.

■ National Park Service, Interior Building, P.O. Box 37127, Washington, D.C. 20013 (202–343–6843). Administers the national park system.

■ Soil Conservation Service, P.O. Box 2890, Washington, D.C. 20013 (202–447–4543). Provides technical and educational assistance for soil conservation and watershed protection.

■ United States Fish and Wildlife Service, Washington, D.C. 20240. Carries out wildlife research and management. Enforces game, fish, and endangered species laws. Administers the national wildlife refuges.

G L O S S A R Y

A

abiotic Environmental factors that are nonliving components of ecosystems.

absorptive or saprobic nutrition Mode of feeding in which an organism absorbs small, dissolved food molecules through the cell membrane, as decomposer organisms typically do.

acid precipitation The deposition of wet acidic solutions or dry acidic particles from the air; commonly known as acid rain but also includes acid fog, snow, etc.

active solar system A heating system that pumps a heat-absorbing, fluid medium (air, water, or an antifreeze solution) through a relatively small collector.

acute effects A single exposure to a toxin that results in an immediate health crisis.

aerobic respiration The intracellular breakdown of sugar or other organic compounds in the presence of oxygen that releases energy and produces carbon dioxide and water.

aerosols Small particles or droplets suspended in a gas.

aesthetic degradation Undesirable changes in chemical or physical characteristics of the atmosphere, such as noise, odor, and light pollution, that are not life-threatening but that reduce the quality of life.

age-specific death rates Mortality rates for specific age classes.

albedo A description of a surface's reflective properties.

allergens Substances that activate the immune system and cause an allergic response; may not be directly antigenic themselves but may make other materials antigenic.

alpha particles Helium nuclei consisting of two protons and two neutrons.

alpine The high, treeless biogeographic zone of mountains that consists of slopes above the timberline.

altruistic preservation A philosophy of preserving nature for its own sake.

ambient air The air that surrounds us.

amino acid An organic compound containing an amino group and a carboxyl group; amino acids are the units or building blocks that make peptide and protein molecules.

anaerobic respiration The incomplete intracellular breakdown of sugar or other organic compounds in the absence of oxygen that releases some energy and produces organic acids and/or alcohol.

anemia Low levels of hemoglobin due to iron deficiency or lack of red blood cells.

animalia One of the five kingdom classifications; includes organisms that are multicellular, have eukaryotic cells, lack cell walls, are usually motile, and feed by eating other organisms.

annual A plant that lives for a single growing season.

anthracite Hard, dense, rocklike coal that has a glassy texture, does not make coke, and burns with a nonluminous flame.

anthropocentrism The belief that humans hold a special place in nature; being centered primarily on humans and human affairs.

antigens Substances that stimulate the production of and react with specific antibodies.

aquifers Porous, water-bearing layers of sand, gravel, and rock below Earth's surface; reservoirs for groundwater.

arithmetic growth A pattern of growth that increases at a constant amount per unit time, such as 1, 2, 3, 4 or 1, 3, 5, 7.

artesian well The result of a pressurized aquifer intersecting the surface or being penetrated by a pipe or conduit, from which water gushes without being pumped; also called a spring.

asphyxiants Chemicals that exclude oxygen or actively interfere with oxygen uptake and distribution; includes inert chemicals, such as nitrogen gas or halothane, that can displace oxygen and fill enclosed spaces.

asthma A distressing disease characterized by shortness of breath, wheezing, and bronchial muscle spasms.

atom The smallest unit of matter that has the characteristics of an element; consists of three main types of subatomic particles: protons, neutrons, and electrons.

autotroph An organism that synthesizes food molecules from inorganic molecules by using an external energy source, such as light energy.

B

barrier islands Low, narrow, sandy islands that form offshore from a coastline.

BAT The Best Available Technology; the best pollution control technology available; the Clean Water Act effectively negates this category by stipulating that equipment must be economically feasible.

bedrock Unbroken, solid rock overlain by sand, gravel, or soil.

benthos Bottom subcommunity; all of the organisms living on the bottom of a body of water.

beta particles High-energy electrons released by radioactive decay.

bill A piece of legislation introduced in Congress and intended to become law.

bioaccumulation The selective absorption and concentration of molecules by cells.

biocentrism The belief that all creatures have rights and values; being centered on nature rather than humans.

bioregionalism Organization of human activities according to natural geographic or ecological boundaries and associations. This philosophy emphasizes a sense of place and living within the resources of one's local ecosystem.

biodegradable plastics Plastics that can be decomposed by microorganisms.

biogeochemical cycles Movement of matter within or between ecosystems; caused by living organisms, geological forces, or chemical reactions. The cycling of nitrogen, carbon, sulfur, oxygen, phosphorus, and water are examples.

biogeographical area An entire ecosystem and its associated land, water, air, and wildlife resources.

biological or biotic factors Organisms and products of organisms that are part of the environment and potentially affect the life of other organisms.

biological community The populations of plants, animals, and microorganisms living and interacting in a certain area at a given time.

biological oxygen demand (BOD) A standard test for measuring the amount of dissolved oxygen utilized by aquatic microorganisms over a five-day period.

biological resources Earth's organisms.

biomagnification Increase in concentration of certain stable chemicals (e.g., heavy metals or fat-soluble pesticides) in successively higher trophic levels of a food chain or web.

biomass The total mass or weight of all the living organisms in a given population or area.

biomass fuel Organic material produced by plants, animals, or microorganisms that can be burned directly as a heat source or converted into a gaseous or liquid fuel.

biomass pyramid A metaphor or diagram that explains the relationship between the amounts of biomass at different trophic levels.

biome A broad, regional type of ecosystem characterized by distinctive climate and soil conditions and a distinctive kind of biological community adapted to those conditions.

biomolecule An organic molecule produced by a living organism.

biosphere The zone of air, land, and water at the surface of the earth that is occupied by organisms.

biota All organisms in a given area.

biotic Pertaining to life; environmental factors created by living organisms.

biotic potential The maximum reproductive rate of an organism, given unlimited resources and ideal environmental conditions. Compare with environmental resistance.

birth control Any method used to reduce births, including celibacy, delayed marriage, contraception, methods that prevent implantation of fertilized zygotes, and induced abortions.

bituminous coal Soft coal that has high contents of volatile gases and ash.

black lung disease Inflammation and fibrosis caused by accumulation of coal dust in the lungs or airways. *See* respiratory fibrotic agents.

bog An area of waterlogged soil that tends to be peaty; fed mainly by precipitation; low productivity; some bogs are acidic.

boreal forest A broad band of mixed coniferous and deciduous trees that stretches across northern North America (and also Europe and Asia); its northernmost edge, the **taiga**, intergrades with the arctic **tundra**.

BPT The Best Practical Control Technology; the best technology for pollution control available at reasonable cost and operable under normal conditions.

bronchitis An inflammation of bronchial linings that causes persistent cough, copious production of sputum, and involuntary muscle spasms that constrict airways.

C

cancer Invasive, out-of-control cell growth that results in malignant tumors.

captive breeding Raising plants or animals in zoos or other controlled conditions to produce stock for subsequent release into the wild.

carbohydrate An organic compound consisting of a ring or chain of carbon atoms with hydrogen and oxygen attached; examples are sugars, starches, cellulose, and glycogen.

carbon cycle The circulation and reutilization of carbon atoms, especially via the processes of photosynthesis and respiration.

carbon monoxide (CO) Colorless, odorless, nonirritating but highly toxic gas produced by incomplete combustion of fuel, incineration of biomass or solid waste, or partially anaerobic decomposition of organic material.

carbon sink Places of carbon accumulation, such as in large forests (organic compounds) or ocean sediments (calcium carbonate); carbon is thus removed from the carbon cycle for moderately long to very long periods of time.

carbon source Source of carbon that reenters the carbon cycle; cellular respiration and combustion.

carcinogens Substances that cause cancer.

carnivores Organisms that mainly prey upon animals.

carrying capacity The maximum number of individuals of any species that can be supported by a particular ecosystem on a long-term basis.

cash crops Those crops produced for sale, especially luxury crops that bring high prices on foreign markets.

cellular organelle An intracellular structure specialized for a particular function; examples are nucleus, mitochondria, chloroplasts, Golgi apparatus or dictyosomes, lysosomes, endoplasmic reticulum, and others.

cellular respiration The process in which a cell breaks down sugar or other organic compounds to release energy used for cellular work; may be **anaerobic** or aerobic depending on the availability of oxygen.

cellular work Using energy to do useful work, such as active transport, synthesis of biomolecules, and movement, within a living cell.

chain reaction A self-sustaining reaction in which the fission of nuclei produces subatomic particles that cause the fission of other nuclei.

chemical bond The force that holds molecules together.

chemosynthesis Autotrophic synthesis of organic compounds by certain bacteria; uses energy from inorganic compounds.

chloroplasts Chlorophyll-containing organelles in eukaryotic organisms; sites of photosynthesis.

chronic effects Long-lasting results of exposure to a toxin; can be a permanent change caused by a single, acute exposure or a continuous, low-level exposure.

chronic food shortages Long-term undernutrition and malnutrition; usually caused by people's lack of money to buy food or lack of opportunity to grow it themselves.

city A differentiated community with a sufficient population and resource base to allow residents to specialize in arts, crafts, services, and professional occupations.

clearcut Cutting every tree in a given area, regardless of species or size; an appropriate harvest method for some species; can be destructive if not carefully controlled.

climate A description of the long-term pattern of weather in a particular area.

climax community A relatively stable, long-lasting community reached in a successional series; usually determined by climate and soil type.

closed canopy A forest where tree crowns spread over 20 percent of the ground; has the potential for commercial timber harvests.

coal gasification The heating and partial combustion of coal to release volatile gases, such as methane and carbon monoxide; after pollutants are washed out, these gases become efficient, clean-burning fuel.

coal washing Coal technology that involves crushing coal and washing out soluble sulfur compounds with water or other solvents.

Coastal Zone Management Act Legislation of 1972 that gave federal money to thirty seacoast and Great Lakes states for development and restoration projects.

co-composting Microbial decomposition of organic materials in solid waste into useful soil additives and fertilizer; often, extra organic material in the form of sewer sludge, animal manure, leaves, and grass clippings are added to solid waste to speed the process and make the product more useful.

cofactor Nonprotein components needed by enzymes in order to function; often minerals or vitamins.

cogeneration The simultaneous production of electricity and steam or hot water in the same plant.

cold front A moving boundary of cooler air displacing warmer air.

coliform bacteria Bacteria that live in the intestines (including the colon) of humans and other animals; used as a measure of the presence of feces in water or soil.

commensalism A symbiotic relationship in which one member is benefited and the second is neither harmed nor benefited.

common law A body of unwritten principles based on custom and the precedents of previous legal decisions.

community ecology The study of interactions of all populations living in the ecosystem of a given area.

competitive exclusion A principle recognizing that when two or more populations are competing for the same resource, one population will eventually dominate and cause the other population to dwindle, become extinct, or adapt to an ecological niche that reduces or eliminates competition.

compound A molecule made up of two or more kinds of atoms held together by chemical bonds.

condensation The aggregation of water molecules from vapor to liquid or solid when the saturation concentration is exceeded.

condensation nuclei Tiny particles that float in the air and facilitate the condensation process.

conifers Needle-bearing trees that produce seeds in cones.

conservation of matter In any chemical reaction, matter changes form; it is neither created nor destroyed.

Consolidated Metropolitan Statistical Area (CMSA) An area with more than one million people formed by the merger of two or more population centers meeting other specified requirements of the Census Bureau.

consumer An organism that obtains energy and nutrients by feeding on other organisms or their remains. *See also* heterotroph.

consumption The fraction of withdrawn water that is lost in transmission or that is evaporated, absorbed, chemically transformed, or otherwise made unavailable for other purposes as a result of human use.

contour plowing Plowing along hill contours; reduces erosion.

control rods Neutron-absorbing material inserted into spaces between fuel assemblies in nuclear reactors to regulate fission reaction.

conurbation An area with more than one million people that is formed by the merger of two or more population centers and that meets other specified requirements.

convection currents Rising or sinking air currents that stir the atmosphere and transport heat from one area to another. Convection currents also occur in water; *see* spring overturn.

conventional pollutants The seven substances (sulfur dioxide, carbon monoxide, particulates, hydrocarbons, nitrogen oxides, photochemical oxidants, and lead) that make up the largest volume of air quality degradation; identified by the Clean Air Act as the most serious threat of all pollutants to human health and welfare; also called criteria pollutants.

cool deserts Deserts of the Western mountain valleys and Great Basin characterized by cold winters and sagebrush.

coral reefs Prominent oceanic features composed of hard, limy skeletons produced by coral animals; usually formed along edges of shallow, submerged ocean banks or along shelves in warm, shallow, tropical seas.

core The dense, intensely hot mass of molten metal, mostly iron and nickel, thousands of kilometers in diameter at Earth's center.

core region The primary industrial region of a country; usually located around the capital or largest port; has both the greatest population density and the greatest economic activity of the country.

Coriolis effect The influence of friction and drag on air layers near Earth; deflects air currents to the direction of Earth's rotation.

cover crops Plants, such as rye, alfalfa, or clover, that can be planted immediately after harvest to hold and protect the soil.

criteria pollutants *See* conventional air pollutants.

critical factor The single environmental factor closest to a tolerance limit for a given species at a given time. *See* limiting factors.

croplands Lands used to grow crops.

crude birth rate The number of births in a year divided by the midyear population.

crude death rates The number of deaths per thousand persons in a given year; also called crude mortality rates.

crust The cool, lightweight, outermost layer of Earth's surface that floats on the soft, pliable underlying layers; similar to the "skin" on a bowl of warm pudding.

cultural eutrophication An increase in biological productivity and ecosystem succession caused by human activities.

cyclone collectors A gravitational settling device that removes heavy particles from an effluent stream.

D

deciduous plants Trees and shrubs that shed their leaves at the end of the growing season.

decline spiral A catastrophic deterioration of a species, community, or whole ecosystem; accelerates as functions are disrupted or lost in a downward cascade.

decomposers Fungi and bacteria that break complex organic material into smaller molecules.

deep ecology An integrated combination of beliefs, life-style, and activism based on voluntary simplicity, decentralization, personal freedom, the sacredness of nature, and direct action in environmental protection.

degradation (of water resource) Deterioration in water quality due to contamination or pollution; makes water unsuitable for other desirable purposes.

delta Fan-shaped sediment deposit found at the mouth of a river.

demographic transition A pattern of falling death and birth rates in response to improved living conditions; could be reversed in deteriorating conditions.

demography Vital statistics about people: births, marriages, deaths, etc.; the statistical study of human populations relating to growth rate, age structure, geographic distribution, etc., and their effects on social, economic, and environmental conditions.

denitrifying bacteria   Free-living soil bacteria that convert nitrates to gaseous nitrogen and nitrous oxide.

dependency ratio   The number of nonworking members compared to working members for a given population.

desertification   Denuding and degrading a once-fertile land, initiating a desert-producing cycle that feeds on itself and causes long-term changes in soil, climate, and biota of an area.

dew point   The temperature at which condensation occurs for a given concentration of water vapor in the air.

dieback   A sudden population decline; also called a population crash.

discharge   The amount of water that passes a fixed point in a given amount of time; usually expressed as liters or cubic feet of water per second.

disease   A deleterious change in the body's condition in response to destabilizing factors, such as nutrition, chemicals, or biological agents.

dissolved oxygen (DO) content   Amount of oxygen dissolved in a given volume of water at a given temperature and atmospheric pressure; usually expressed in parts per million (ppm).

diversity (species diversity, biological diversity)   The number of species present in a community (species *richness*), as well as the relative *abundance* of each species.

DNA   Deoxyribonucleic acid; the long, double helix molecule in the nucleus of cells that contains the genetic code and directs the development and functioning of all cells.

dominant plants   Those plant species in a community that provide a food base for most of the community; usually take up the most space and have the largest biomass.

drip irrigation   Uses pipe or tubing perforated with very small holes to deliver water one drop at a time directly to the soil around each plant. This conserves water and reduces soil waterlogging and salinization.

dry alkali injection   Spraying dry sodium bicarbonate into flue gas to absorb and neutralize acidic sulfur compounds.

E

ecofeminism   A philosophy that applies feminist critical analysis and perspective to ecological issues. In this view, oppression of both women and nature by patriarchial society are similar and have related destructive effects.

ecojustice   Justice in the social order and integrity in the natural order.

ecological equivalents   Different species that occupy similar ecological niches in similar ecosystems in different parts of the world.

ecological niche   The functional role and position of a species (population) within a community or ecosystem, including what resources it uses, how and when it uses the resources, and how it interacts with other populations.

ecological succession   A progression of communities replacing each other at a given site over a period of time; eventually a long-lasting **climax community** may result. *See also* primary succession and secondary succession.

ecology   The study of environmental factors and how organisms interact with them.

economic development   A rise in real income *per person*; usually associated with new technology that increases productivity or resources.

economic growth   An increase in the total wealth of a nation; if population grows faster than the economy, there may be real economic growth, but the share per person may decline.

ecosystem   A specific biological community and its physical environment interacting in an exchange of matter and energy.

ecosystem restoration   Reinstate an entire community of organisms to as near its natural condition as possible.

ecotage   Direct action (guerrilla warfare) or sabotage in defense of nature; also called monkeywrenching.

edge effect   Occurs when the transitional zone between two communities provides habitat for a different set of organisms or provides habitat that can be used by members of both adjacent communities.

electron   A negatively charged subatomic particle that orbits around the nucleus of an atom.

electron acceptor   An atom that accepts or receives one or more electrons from another atom.

electron transfer   The transfer of an electron from one atom to another.

electrostatic precipitators   The most common particulate controls in power plants; fly ash particles pick up an electrostatic surface charge as they pass between large electrodes in the effluent stream, causing particles to migrate to the collecting plate.

element   A molecule composed of one kind of atom; cannot be broken into simpler units by chemical reactions.

emigration   The movement of members from a population.

emission standards   Regulations for restricting the amounts of air pollutants that can be released from specific point sources.

emphysema   An irreversible, obstructive lung disease in which airways become permanently constricted and alveoli are damaged or destroyed.

endangered species   A species considered to be in imminent danger of extinction.

energy   The capacity to do work (i.e., to change the physical state or motion of an object).

energy cascade   A series of sequential physical or chemical reactions (usually involving electron transfers) that result in stepwise energy degradation as stated in the Second Law of Thermodynamics. Used in metabolic reactions, such as aerobic respiration, to carry out cellular work.

energy efficiency   A measure of energy produced compared to energy consumed.

energy pyramid   A representation of the loss of useful energy at each step in a food chain.

energy recovery   Incineration of solid waste to produce useful energy.

environment   A combination of external biological and nonbiological factors that influence the life of a cell or organism.

environmental ethics   A search for moral values and ethical principles in human relations with the natural world.

Environmental Impact Statement   An analysis, required by provisions in the National Environmental Policy Act of 1970, of the effects of any major program a federal agency plans to undertake; also called an EIS.

environmental resistance   All the limiting factors that tend to reduce population growth rates and set the maximum allowable population size or carrying capacity of an ecosystem.

environmental resources   Anything an organism needs that can be taken from the environment.

environmental science   The study of the complex interactions of human populations with matter and energy resources; it incorporates aspects of the natural and social sciences, business, law, technology, and other fields.

enzymes   Molecules, usually proteins or nucleic acids, that act as catalysts in biochemical reactions.

epidemiology   The study of the distribution and causes of disease and injuries in human populations.

epiphyte   A plant that grows on a substrate other than the soil, such as the surface of another organism.

equilibrium community   Also called a **disclimax community**; a community subject to periodic disruptions, usually by fire, that prevent it from reaching a climax stage.

estuary Enclosed or semi-enclosed body of water that forms where a river enters the ocean and creates an area of mixed fresh water and ocean water.

eukaryotic cell A cell containing a membrane-bounded nucleus and membrane-bounded organelles.

eutectic chemicals Phase-changing chemicals used in heat storage systems to store a large amount of energy in a small volume.

eutrophic Rivers and lakes rich in organisms and organic material (eu = good; trophic = nutrition).

evaporation The process in which a liquid is changed to vapor (gas phase).

evergreens Coniferous trees and broad-leaved plants that retain their leaves year-round.

exhaustible resources Generally considered the Earth's geologic endowment: minerals, nonmineral resources, fossil fuels, and other materials present in fixed amounts in the environment.

existence value The importance we place on just knowing that a particular species or a specific organism exists.

exponential growth Growth at a constant rate of increase per unit of time; can be expressed as a constant fraction or exponent. *See* geometric growth.

external costs Expenses, monetary or otherwise, borne by someone other than the individuals or groups who use a resource.

extinction The irrevocable elimination of species; can be a normal process of the natural world as species out-compete or kill off others or as environmental conditions change.

extirpate To destroy totally; extinction caused by direct human action, such as hunting, trapping, etc.

F

family planning Controlling reproduction; planning the timing of birth and having as many babies as are wanted and can be supported.

famines Acute food shortages characterized by large-scale loss of life, social disruption, and economic chaos.

fauna All of the animals present in a given region.

fecundity The physical ability to reproduce.

fen An area of waterlogged soil that tends to be peaty; fed mainly by upwelling water; low productivity.

feral A domestic animal that has taken up a wild existence.

fermentation (alcoholic) A type of anaerobic respiration that yields carbon dioxide and alcohol; used in commercial fermentation processes, including production of raised bakery dough products and alcoholic beverages.

fertility Measurement of actual number of offspring produced through sexual reproduction; usually described in terms of number of offspring of females since paternity can be difficult to determine.

fibrosis The general name for accumulation of scar tissue in the lung.

fidelity A principle that forbids misleading or deceiving any creature capable of being mislead or deceived. We are to be truthful in our dealings with others.

filters A porous mesh of cotton cloth, spun glass fibers, or asbestos-cellulose that allows air or liquid to pass through but holds back solid particles.

fire-climax community An equilibrium community maintained by periodic fires; examples include grasslands, chapparal shrubland, and some pine forests.

First World The industrialized capitalist or market economy countries of Western Europe, North America, Japan, Australia, and New Zealand.

flora All of the plants present in a given region.

flue-gas scrubbing Treating combustion exhaust gases with chemical agents to remove pollutants. Spraying crushed limestone and water into the exhaust gas stream to remove sulfur is a common scrubbing technique.

fluidized bed combustion High pressure air is forced through a mixture of crushed coal and limestone particles lifting the burning fuel and causing it to move like a boiling fluid. Fresh coal and limestone are added continuously to the top of the combustion bed while ash and slag are drawn off below.

food aid Financial assistance intended to boost less developed countries' standards of living.

food chain A linked feeding series; in an ecosystem, the sequence of organisms through which energy and materials are transferred, in the form of food, from one trophic level to another.

food surpluses Excess food supplies.

food web A complex, interlocking series of individual food chains in an ecosystem.

forest management Scientific planning and administration of forest resources for sustainable harvest, multiple use, regeneration, and maintenance of a healthy biological community.

freezing condensation A process that occurs in the clouds when ice crystals trap water vapor. As the ice crystals become larger and heavier, they begin to fall as rain or snow.

fresh water Water other than seawater; covers only about 2 percent of Earth's surface, including streams, rivers, lakes, ponds, and water associated with several kinds of wetlands.

front The boundary between two air masses of different temperature and density.

frontier An unexploited natural area at the leading edge of human settlement.

frontier mentality The idea that the world has an unlimited supply of resources for human use regardless of the consequences to natural ecosystems and the biosphere.

fuel assembly A bundle of hollow metal rods containing uranium oxide pellets; used to fuel a nuclear reactor.

fuel-switching A change from one fuel to another.

fuelwood Branches, twigs, logs, wood chips, and other wood products harvested for use as fuel.

fugitive emissions Substances that enter the air without going through a smokestack, such as dust from soil erosion, strip mining, rock crushing, construction, and building demolition.

fungi One of the five kingdom classifications; consists of nonphotosynthetic, eukaryotic organisms with cell walls, filamentous bodies, and absorptive nutrition.

G

Gaia hypothesis A theory that the living organisms of the biosphere form a single complex interacting system that creates and maintains a habitable Earth; named after Gaia, the Greek Earth mother goddess.

gamma rays Very short wavelength forms of the electromagnetic spectrum.

garden city A new town with special emphasis on landscaping and rural ambience.

gasohol A mixture of gasoline and ethanol.

gene A unit of heredity; a segment of DNA nucleus of the cell that contains information for the synthesis of a specific protein, such as an enzyme.

gene banks Storage for seed varieties for future breeding experiments.

general fertility rate Representation of population age structure and fecundity; crude birth rate multiplied by the percentage of fecund women (between approximately fifteen and forty-four years of age) by 1,000.

genetic assimilation The disappearance of a species as its genes are diluted through crossbreeding with a closely related species.

geometric growth Growth that follows a geometric pattern of increase, such as 2, 4, 8, 16, etc. *See* exponential growth.

germ plasm Genetic material that may be preserved for future agricultural, commercial, and ecological values (plant seeds or parts or animal eggs, sperm, and embryos).

green revolution Dramatically increased agricultural production brought about by "miracle" strains of grain; usually requires high inputs of water, plant nutrients, and pesticides.

groundwater Water held in gravel deposits or porous rock below Earth's surface; does not include water or crystallization held by chemical bonds in rocks or moisture in upper soil layers.

gully erosion Removal of layers of soil, creating channels or ravines too large to be removed by normal tillage operations.

H

Hadley cells Circulation patterns of atmospheric convection currents as they sink and rise in several intermediate bands.

half-life (of radioisotopes) The length of time required for half the nuclei in a sample to change into another isotope.

hazardous chemicals Dangerous chemicals, including flammables, explosives, irritants, sensitizers, acids, and caustics; may be relatively harmless in diluted concentrations.

hazardous waste Any discarded material containing substances known to be toxic, mutagenic, carcinogenic, or teratogenic to humans or other life-forms; ignitable, corrosive, explosive, or highly reactive alone or with other materials.

health A state of physical and emotional well-being; the absence of disease or ailment.

heat A form of energy transferred from one body to another because of a difference in temperatures.

heat capacity The amount of heat energy that must be added or subtracted to change the temperature of a body; water has a high heat capacity.

heat of vaporization The amount of heat energy required to convert water from a liquid to a gas.

herbivore An organism that eats only plants.

heterotroph An organism that is incapable of synthesizing its own food and, therefore, must feed upon organic compounds produced by other organisms.

homeostasis Maintaining a dynamic, steady state in a living system through opposing, compensating adjustments.

Homestead Act Legislation passed in 1862 allowing any citizen or applicant for citizenship over twenty-one years old and head of a family to acquire 160 acres of public land by living on it and cultivating it for five years.

host organism An organism that provides lodging for a parasite.

hot desert Deserts of the American Southwest and Mexico; characterized by extreme summer heat and cacti.

human ecology The study of the interactions of humans with the environment.

human resources Human wisdom, experience, skill, and enterprise.

humus Sticky, brown, insoluble residue from the bodies of dead plants and animals; gives soil its structure, coating mineral particles and holding them together; serves as a major source of plant nutrients.

hypothyroidism Listlessness and other metabolic symptoms caused by low thyroid hormone levels.

I

igneous rocks Crystalline minerals solidified from molten magma from deep in Earth's interior; basalt, rhyolite, andesite, lava, and granite are examples.

inbreeding depression In a small population, an accumulation of harmful genetic traits (through random mutations and natural selection) that lowers viability and reproductive success of enough individuals to affect the whole population.

industrial timber Trees used for lumber, plywood, veneer, particleboard, chipboard, and paper; also called roundwood.

inertial confinement A nuclear fusion process in which a small pellet of nuclear fuel is bombarded with extremely high-intensity laser light.

infiltration The process of water percolation into the soil and pores and hollows of permeable rocks.

inflammatory response A complex series of interactions between fragments of damaged cells, surrounding tissues, circulating blood cells, and specific antibodies; typical of infections.

informal economy Small-scale family businesses in temporary locations outside the control of normal regulatory agencies.

insolation Incoming solar radiation.

instrumental value Value or worth of objects that satisfy the needs and wants of moral agents. Objects that can be used as a means to some desirable end.

intangible resources Abstract commodities, such as open space, beauty, serenity, genius, information, diversity, and satisfaction.

integrated pest management IPM is an ecologically-based pest control strategy that relies on natural mortality factors, such as natural enemies, weather, cultural control methods, and carefully applied doses of pesticides.

internal costs The expertises (monetary or otherwise) borne by those who use a resource.

internalizing costs Planning so that those who reap the benefits of resource use also bear all the external costs.

interplanting The system of planting two or more crops, either mixed together or in alternating rows, in the same field; protects the soil and makes more efficient use of the land.

interspecific competition In a community, competition for resources between members of *different* species.

intraspecific competition In a community, competition for resources among members of the *same* species.

ionizing radiation High-energy electromagnetic radiation or energetic subatomic particles released by nuclear decay.

ionosphere The lower part of the thermosphere.

ions Atoms that have picked up or lost extra electrons, leaving them with a net negative or positive charge.

irritants Corrosives (strong acids), caustics (alkaline reagents), and other substances that damage biological tissues on contact.

irruptive growth *See* Malthusian growth.

island effects Reductions in species diversity caused by reduction in ecosystem area.

isotopes Different forms of the same element.

J

J curve A growth curve that depicts exponential growth; called a **J** curve because of its shape.

jet streams Powerful winds or currents of air that circulate in shifting flows; similar to oceanic currents in extent and effect on climate.

K

known resources Minerals or other useful environmental materials or services that are identified and partially mapped; may not be environmentally or socially acceptable or economically feasible to exploit.

kwashiorkor A widespread human protein deficiency disease resulting from a starchy diet low in protein and essential amino acids.

L

land rehabilitation A utilitarian program to repair damage and make land useful to humans.

landfills Land disposal sites for solid waste; operators compact refuse and cover it with a layer of dirt to minimize rodent and insect infestation, wind-blown debris, and leaching by rain.

landscape ecology Ecological study that deals with understanding the effects of topography and soils on the development of biological communities.

law of diminishing returns At some stages of economic development adding more workers or more labor will not increase productivity; instead, average output and standard of living decrease with each additional worker.

LD50 A chemical dose lethal to 50 percent of a test population.

less developed countries (LDC) Nonindustrialized nations characterized by low per capita income, high birth and death rates, high population growth rates, and low levels of technological development.

life expectancy The average number of years lived by a group of individuals after reaching a given age; the probable number of years of survival for an individual of a given age.

life span The longest period of life reached by a type of organism.

lignite Soft, brown to black coal of low BTU value with original plant components still discernible.

limiting factors Chemical or physical factors that limit the existence, growth, abundance, or distribution of an organism. The **principle of limiting factors** states that for each physical factor in the environment, there are both minimum and maximum limits (**tolerance limits**) beyond which a particular species cannot survive. The single factor closest to a tolerance limit for a given species at a given time is the **critical factor.**

lipid A nonpolar organic compound that is insoluble in water but soluble in solvents, such as alcohol and ether; includes fats, oils, steroids, phospholipids, and carotenoids.

liquid metal fast breeder A nuclear power plant that converts Uranium 238 to Plutonium 239; thus, it creates more nuclear fuel than it consumes. Because of the extreme heat and density of its core, the breeder uses liquid sodium as a coolant.

logistic growth Growth rates regulated by internal and external factors that establish an equilibrium with environmental resources; species may grow exponentially when resources are unlimited but slowly as the carrying capacity is reached. *See* **S** curve.

longevity The length or duration of life; compare to survivorship.

low-head hydropower Small-scale hydro technology that can extract energy from small headwater dams; causes much less ecological damage.

M

magma Molten rock from deep in Earth's interior; called lava when it spews from volcanic vents.

magnetic confinement A technique for enclosing a nuclear fusion reaction in a powerful magnetic field inside a vacuum chamber.

malignant tumor A mass of cancerous cells that have left their site of origin, migrated through the body, invaded normal tissues, and are growing out of control.

malnourishment A nutritional imbalance caused by lack of specific dietary components or inability to absorb or utilize essential nutrients.

Malthusian growth A population explosion followed by a population crash; also called irruptive growth.

mantle A hot, pliable layer of rock that surrounds Earth's core and underlies the cool, outer crust.

marasmus A widespread human protein deficiency disease caused by a diet low in calories and protein or imbalanced in essential amino acids.

marine Living in or pertaining to the sea.

market equilibrium The dynamic balance between supply and demand under a given set of conditions in a "free" market (one with no monopolies or government interventions).

marsh Wetland without trees; in North America, this type of land is characterized by cattails and rushes.

mass burn Incineration of unsorted solid waste.

matter Something that occupies space and has mass.

megacity *See* megalopolis.

megalopolis Also known as a megacity or supercity; megalopolis indicates an urban area with more than ten million inhabitants.

megawatt (MW) Unit of electrical power equal to one thousand kilowatts or one million watts.

mesosphere The atmospheric layer above the stratosphere and below the thermosphere; the middle layer; temperatures are usually very low.

metabolism All the energy and matter exchanges that occur within a living cell or organism; collectively, the life processes.

metamorphic rock Igneous and sedimentary rocks modified by heat, pressure, and chemical reactions.

micro-hydro generators Small power generators that can be used in low-level rivers to provide economical power for four to six homes, freeing them from dependence on large utilities and foreign energy supplies.

Milankovitch cycles Periodic variations in tilt, eccentricity, and wobble in Earth's orbit; Milutin Milankovitch suggested that it is responsible for cyclic weather changes.

milpa agriculture An ancient farming system in which small patches of tropical forests are cleared and perennial polyculture agriculture practiced and is then followed by many years of fallow to restore the soil; also called swidden agriculture.

mineral A naturally occurring, inorganic, crystalline solid with definite chemical composition and characteristic physical properties.

mitochondria Cellular organelles found in eukaryotic cells; sites of aerobic respiration.

mixed perennial polyculture Growing a mixture of different perennial crop species (where the same plant persists for more than one year) together in the same plot; imitates the diversity of a natural system and is often more stable and more suitable for sustainable agriculture than monoculture of annual plants.

molecule A combination of two or more atoms.

Monera One of the five kingdom classifications; contains prokaryotic cells, those that do not have a membrane-bounded nucleus, chloroplasts, or mitochondria; includes bacteria, including blue-green bacteria.

monoculture agroforestry Intensive planting of a single species; an efficient wood production approach, but one that encourages pests and disease infestations and conflicts with wildlife habitat or recreation uses.

monsoon A seasonal reversal of wind patterns caused by the different heating and cooling rates of the oceans and continents.

montane coniferous forests Coniferous forests of the mountains consisting of belts of different forest communities along an altitudinal gradient.

morbidity Illness or disease.

more developed countries (MDC) Industrialized nations characterized by high per capita incomes, low birth and death rates, low population growth rates, and high levels of industrialization and urbanization.

moral agents Beings capable of making distinctions between right and wrong and acting accordingly. Those whom we hold responsible for their actions.

moral subjects Beings that are not capable of distinguishing between right or wrong or that are not able to act on moral principles and yet are susceptible of being wronged by others. This category assumes some rights or inherent values in moral subjects that gives us duties or obligations towards them.

mortality Death rate in a population; the probability of dying.

mulch Protective ground cover, including manure, wood chips, straw, seaweed, leaves, and other natural products, or synthetic materials, such as heavy paper or plastic, that protect the soil, save water, and prevent weed growth.

multiple use Many uses that occur simultaneously; used in forest management; limited to mutually compatible uses.

mutagens Agents, such as chemicals or radiation, that damage or alter genetic material (DNA) in cells.

mutation A change, either spontaneous or by external factors, in the genetic material of a cell; mutations in the gametes (sex cells) can be inherited by future generations of organisms.

mutualism A symbiotic relationship between individuals of two different species in which both species benefit from the association.

N

NAAQS National Ambient Air Quality Standard; federal standards specifying the maximum allowable levels (averaged over specific time periods) for regulated pollutants in ambient (outdoor) air.

natality The production of new individuals by birth, hatching, germination, or cloning.

natural history The study of where and how organisms carry out their life cycles.

natural increase Crude death rate subtracted from crude birth rate.

natural resources Goods and services supplied by the environment.

natural selection The mechanism for evolutionary change in which environmental pressures cause certain genetic combinations in a population to become more abundant; genetic combinations best adapted for present environmental conditions tend to become predominant.

neo-Malthusian Modern advocates of the Malthusian point of view that human populations will grow to the maximum carrying capacity of the environment or until conditions become intolerable.

net energy yield Total useful energy produced during the lifetime of an entire energy system minus the energy used, lost, or wasted in making useful energy available.

neurotoxins Toxic substances, such as lead or mercury, that specifically poison nerve cells.

neutron A subatomic particle, found in the nucleus of the atom, that has no electromagnetic charge.

new towns Experimental urban environments that seek to combine the best features of the rural village and the modern city.

nihilists Those who believe the world has no meaning or purpose other than a dark, cruel, unceasing struggle for power and existence.

nitrate-forming bacteria Bacteria that convert nitrites into compounds that can be used by green plants to build proteins.

nitrite-forming bacteria Bacteria that combine ammonia with oxygen to form nitrites.

nitrogen cycle The circulation and reutilization of nitrogen in both inorganic and organic phases.

nitrogen-fixing bacteria Bacteria that convert nitrogen from the atmosphere or soil solution into ammonia that can then be converted to plant nutrients by nitrite- and nitrate-forming bacteria.

nitrogen oxides Highly reactive gases formed when nitrogen in fuel or combustion air is heated to over 650°C (1,200°F) in the presence of oxygen or when bacteria in soil or water oxidize nitrogen-containing compounds.

noncriteria pollutants See unconventional air pollutants.

nongovernmental organizations A term referring collectively to pressure and research groups, advisory agencies, political parties, professional societies, and other groups concerned about environmental quality, resource use, and many other issues.

noninterference A principle that requires us to refrain from interfering with the lives or destinies of other moral agents or moral subjects except to satisfy essential needs.

nonmalfeasance Doing no harm to any being that has goods (values, interests, moral values) of their own.

nonpoint sources Scattered, diffuse sources of pollutants, such as runoff from farm fields, golf courses, construction sites, etc.

nonrenewable resources Materials or services from the environment that are not replaced or replenished by natural processes at a rate comparable to our use of the resource; a resource depleted or exhausted by use.

northern coniferous forest See boreal forest.

nuclear fission The radioactive decay process in which isotopes split apart to create two smaller atoms.

nuclear fusion A process in which two smaller atomic nuclei fuse into one larger nucleus and release energy; the source of power in a hydrogen bomb.

nucleic acids Large organic molecules made of nucleotides that function in the transmission of hereditary traits, in protein synthesis, and in control of cellular activities.

nucleus The center of the atom; occupied by protons and neutrons. In cells, the organelle that contain the chromosomes (DNA).

nuées ardentes Deadly, denser-than-air mixtures of hot gases and ash ejected from volcanoes.

numbers pyramid A diagram showing the relative population sizes at each trophic level in an ecosystem; usually corresponds to the biomass pyramid.

O

oceanic islands Islands in the ocean; formed by breaking away from a continental landmass, volcanic action, coral formation, or a combination of sources; support distinctive communities.

ocean shorelines Rocky coasts and sandy beaches along the oceans; support rich, stratified communities.

offset allowances A controversial component of air quality regulations that allows a polluter to avoid installation of control equipment on one source with an "offset" pollution reduction at another source.

old soils Soils from which rainwater has carried away most of the soluble aluminum and iron and has left behind clay and rust-colored oxides.

oligotrophic Condition of rivers and lakes that have clear water and low biological productivity (oligo = little; trophic = nutrition); are usually clear, cold, infertile headwater lakes and streams.

omnivore   An organism that eats both plants and animals.

open canopy   A forest where tree crowns cover less than 20 percent of the ground; also called woodland.

open range   Unfenced, natural grazing lands; includes woodland as well as grassland.

open system   A system that exchanges energy and matter with its environment.

optimum   The most favorable condition in regard to an environmental factor.

orbital   The space or path in which an electron orbits the nucleus of an atom.

organic compounds   Complex molecules organized around skeletons of carbon atoms arranged in rings or chains; includes biomolecules, molecules synthesized by living organisms.

overburden   Overlying layers of noncommercial sediments that must be removed to reach a mineral or coal deposit.

overnutrition   Receiving too many calories.

overshoot   The extent to which a population exceeds the carrying capacity of its environment.

oxygen cycle   The circulation and reutilization of oxygen in the biosphere.

oxygen sag   Oxygen decline downstream from a pollution source that introduces materials with high biological oxygen demands.

ozone   A highly reactive molecule containing three oxygen atoms; a dangerous pollutant in ambient air. In the stratosphere, however, ozone forms an ultraviolet absorbing shield that protects us from mutagenic radiation.

P

Pacific Coast coniferous forests   The several kinds of cool, moist forest ecosystems from Alaska to northern California. *See* temperate rain forest.

parabolic mirrors   Curved mirrors that focus light from a large area onto a single, central point, thereby concentrating solar energy and producing high temperatures.

parasite   An organism that lives in or on another organism, deriving nourishment at the expense of its host, usually without killing it.

parent material   Undecomposed mineral particles and unweathered rock fragments beneath the subsoil; weathering of this layer produces new soil particles for the layers above.

particulate material   Atmospheric aerosols, such as dust, ash, soot, lint, smoke, pollen, spores, algal cells, and other suspended materials; originally applied only to solid particles but now extended to droplets of liquid.

parts per billion (ppb)   Number of parts of a chemical found in one billion parts of a particular gas, liquid, or solid mixture.

parts per million (ppm)   Number of parts of a chemical found in one million parts of a particular gas, liquid, or solid mixture.

parts per trillion (ppt)   Number of parts of a chemical found in one trillion (10^{12}) parts of a particular gas, liquid, or solid mixture.

passive heat absorption   The use of natural materials or absorptive structures without moving parts to gather and hold heat; the simplest and oldest use of solar energy.

pasture   Enclosed domestic meadows or managed grazing lands.

patchiness   Within a larger ecosystem, the presence of smaller areas that differ in some physical conditions and thus support somewhat different communities; a diversity-promoting phenomenon.

pathogen   An organism that produces disease in a host organism, an alteration of one or more metabolic functions in response to the presence of the organism.

peat   Deposits of moist, acidic, semidecayed organic matter.

pellagra   Lassitude, torpor, dermatitis, diarrhea, dementia, and death brought about by a diet deficient in tryptophan and niacin.

peptides   Two or more amino acids linked by a peptide bond.

perennial   A plant that survives and reproduces each year.

perennial species   Plants that grow for more than two years.

permafrost   A permanently frozen layer of soil that underlies the arctic tundra.

permanent retrievable storage   Long-term storage of supertoxic materials in secure buildings, salt mines, or bedrock caverns where they can be inspected and retrieved for repacking or for transfer if a better storage process is developed.

pest   Any organism that reduces the availability, quality, or value of a useful resource.

pesticide   Any chemical that kills, controls, drives away, or modifies the behavior of a pest.

pH   A value that indicates the acidity or alkalinity of a solution on a scale of 0 to 14, based on the proportion of H^+ ions present.

photochemical oxidants   Products of secondary atmospheric reactions. *See* smog.

photodegradable plastics   Plastics that break down when exposed to sunlight or to a specific wavelength of light.

photosynthesis   The biochemical process by which green plants and some bacteria capture light energy and use it to produce chemical bonds. Carbon dioxide and water are consumed while oxygen and simple sugars are produced.

photovoltaic cell   An energy-conversion device that captures solar energy and directly converts it to electrical current.

physical or abiotic factors   Nonliving factors, such as temperature, light, water, minerals, and climate, that influence an organism.

phytoplankton   Microscopic, free-floating, autotrophic organisms that function as producers in aquatic ecosystems.

pioneer species   In primary succession on a terrestrial site, the plants, lichens, and microbes that first colonize the site.

plankton   Primarily microscopic organisms that occupy the upper water layers in both freshwater and marine ecosystems; photosynthetic protists comprise the **phytoplankton**; nonphotosynthetic protists and small invertebrate animals comprise the **zooplankton.**

plankton subcommunity   Primarily microscopic organisms that float freely within the water column, thus occupying the upper levels in a vertically stratified aquatic community. *See* plankton.

Plantae   One of the five kingdom classification categories; includes multicellular, photosynthetic, eukaryotic organisms with cell walls.

plasma   A hot, electrically neutral gas of ions and free electrons.

poachers   Those who hunt wildlife illegally.

point sources   Specific locations of highly concentrated pollution discharge, such as factories, power plants, sewage treatment plants, underground coal mines, and oil wells.

pollution   To make foul, unclean, dirty; any physical, chemical, or biological change that adversely affects the health, survival, or activities of living organisms or that alters the environment in undesirable ways.

polycentric complex   Cities with several urban cores surrounding a once dominant central core.

population   A group of individuals of the same species occupying a given area.

population crash A sudden population decline caused by predation, waste accumulation, or resource depletion; also called a dieback.

population explosion Growth of a population at exponential rates to a size that exceeds environmental carrying capacity; usually followed by a population crash.

population hurdle A need for investment in infrastructure and social services in a rapidly growing population that prevents the capital investment necessary for real economic development.

population momentum A potential for increased population growth as young members reach reproductive age.

power The rate of energy delivery; measured in horsepower or watts.

predation The act of feeding by a predator.

predator An organism that feeds directly on other organisms in order to survive; live-feeders, such as herbivores and carnivores.

prevention of significant deterioration A clause of the Clean Air Act that prevents degradation of existing clean air; opposed by industry as an unnecessary barrier to development.

price elasticity A situation in which supply and demand of a commodity respond to price.

primary pollutants Chemicals released directly into the air in a harmful form.

primary sewage treatment A process that removes solids from sewage before it is discharged or treated further.

primary standards Regulations of the 1970 Clean Air Act; intended to protect human health.

primary succession An ecological succession that begins in an area where no biotic community previously existed.

principle of competitive exclusion A result of natural selection whereby two similar species in a community occupy different ecological niches, thereby reducing competition for food.

producer An organism that synthesizes food molecules from inorganic compounds by using an external energy source; most producers are photosynthetic.

production frontier The maximum output of two competing commodities at different levels of production.

productivity The amount of biological matter or biomass produced in a given area during a given unit of time.

prokaryotic Cells that do not have a membrane-bounded nucleus or membrane-bounded organelles.

promoters Agents that are not carcinogenic but that assist in the progression and spread of tumors; sometimes called cocarcinogens.

pronatalist pressures Influences that encourage people to have children.

proteins Chains of amino acids linked by peptide bonds.

Protista One of the five kingdom classification categories; includes organisms that exist as single eukaryotic cells or as colonies of eukaryotic cells.

proton A positively charged subatomic particle found in the nucleus of an atom.

public trust A doctrine obligating the government to maintain public lands in a natural state as guardians of the public interest.

pull factors (in urbanization) Conditions that draw people from the country into the city.

push factors (in urbanization) Conditions that force people out of the country and into the city.

R

radioactive An unstable isotope that decays spontaneously and releases subatomic particles or units of energy.

radioactive decay A change in the nuclei of radioactive isotopes that spontaneously emit high-energy electromagnetic radiation and/or subatomic particles while gradually changing into another isotope or different element.

radionucleides Isotopes that exhibit radioactive decay.

rain forest A forest with high humidity, constant temperature, and abundant rainfall (generally over 380 cm [150 in.] per year); can be tropical or temperate.

rain shadow Dry area on the downwind side of a mountain.

rangeland Grasslands and open woodlands suitable for livestock grazing.

raprenox A process to reduce nitrogen oxide emissions by mixing combustion gases with isocyanic acid, which converts nitrogen oxides to dimolecular nitrogen gas.

recharge zone Area where water infiltrates into an aquifer.

recycling Reprocessing of discarded materials into new, useful products; not the same as reuse of materials for their original purpose, but the terms are often used interchangeably.

red tide A population explosion or bloom of minute, single-celled marine organisms called dinoflagellates. Billions of these cells can accumulate in protected bays where the toxins they contain can poison other marine life.

reduced tillage systems Systems, such as minimum till, conserv-till, and no-till, that preserve soil, save energy and water, and increase crop yields.

refuse-derived fuel Processing of solid waste to remove metal, glass, and other unburnable materials; organic residue is shredded, formed into pellets, and dried to make fuel for power plants.

regenerative farming A less intensive style of farming that uses little or no inorganic fertilizer, pesticides, water, machinery, and fossil fuel energy; yields may be lower, but so are the input costs of supplies and material.

regulations Rules established by administrative agencies; regulations can be more important than statutory law in the day-to-day management of resources.

relative humidity At any given temperature, a comparison of the actual water content of the air with the amount of water that could be held at saturation.

relativists Those who believe moral principles are always dependent on the particular situation.

renewable resources Resources normally replaced or replenished by natural processes; resources not depleted by moderate use; examples include solar energy, biological resources such as forests and fisheries, biological organisms, and some biogeochemical cycles.

residence time The length of time a component, such as an individual water molecule, spends in a particular compartment or location before it moves on through a particular process or cycle.

resilience The ability of a community or ecosystem to recover from disturbances.

resistance (inertia) The ability of a community to resist being changed by potentially disruptive events.

resource partitioning In a biological community, various populations sharing environmental resources through specialization, thereby reducing direct competition. *See also* **ecological niche.**

respiratory fibrotic agents Special class of irritants, including chemical reagents and particulate materials, that damages the lungs, causing scar tissue formation that lowers respiratory capacity.

rill erosion The removing of thin layers of soil as little rivulets of running water gather and cut small channels in the soil.

risk Probability that something undesirable will happen as a consequence of exposure to a hazard.

risk assessment Evaluation of the short-term and long-term risks associated with a particular activity or hazard; usually compared to benefits in a cost-benefit analysis.

RNA (ribonucleic acid) A nucleic acid used for transcription and translation of the genetic code found on DNA molecules.

rock cycle The process whereby rocks are broken down by chemical and physical forces; sediments are moved by wind, water, and gravity, sedimented and reformed into rock, and then crushed, folded, melted, and recrystallized into new forms.

ruminant animals Cud-chewing animals, such as cattle, sheep, goats, and buffalo, with multichambered stomachs in which cellulose is digested with the aid of bacteria.

runoff The excess of precipitation over evaporation; the main source of surface water and, in broad terms, the water available for human use.

run-of-the-river flow Ordinary river flow not accelerated by dams, flumes, etc. Some small, modern, high-efficiency turbines can generate useful power with run-of-the-river flow or with a current of only a few kilometers per hour.

rural area An area in which most residents depend on agriculture or the harvesting of natural resources for their livelihood.

S

Sagebrush Rebellion A coalition of cattlemen, miners, loggers, developers, farmers, politicians, and others who wanted to see more local control over land management and natural resources.

salinity Amount of dissolved salts (especially sodium chloride) in a given volume of water.

salinization A process in which mineral salts accumulate in the soil, killing plants; occurs when soils in dry climates are irrigated profusely.

saltwater intrusion Movement of saltwater into freshwater aquifers in coastal areas where groundwater is withdrawn faster than it is replenished.

saturation point The maximum concentration of water vapor the air can hold at a given temperature.

scavenger An organism that feeds on the dead bodies of other organisms.

S curve A curve that depicts logistic growth; called an S curve because of its shape.

secondary pollutants Chemicals modified to a hazardous form after entering the air or are formed by chemical reactions as components of the air mix and interact.

secondary recovery technique Pumping pressurized gas, steam, or chemical-containing water into a well to squeeze more oil from a reservoir.

secondary sewage treatment Bacterial decomposition of suspended particulates and dissolved organic compounds that remain after primary sewage treatment.

secondary standards Regulations of the 1970 Clean Air Act intended to protect materials, crops, visibility, climate, and personal comfort.

secondary succession Succession on a site where an existing community has been disrupted.

Second World The industrialized, socialist, centrally planned economy nations of Eastern Europe and the Soviet Union and its allies.

secure landfill A solid waste disposal site lined and capped with an impermeable barrier to prevent leakage or leaching. Drain tiles, sampling wells, and vent systems provide monitoring and pollution control.

sedimentary rock Deposited material that remains in place long enough or is covered with enough material to compact into stone; examples include shale, sandstone, breccia, and conglomerates.

sedimentation The deposition of organic materials or minerals by chemical, physical, or biological processes. Sediments can be transported from their source to their place of deposition by gravity, wind, water, or ice. If subjected to sufficient heat, pressure, or chemical reactions, sediments can solidify into sedimentary rock.

selective cutting Harvesting only mature trees of certain species and size; usually more expensive than clearcutting, but it is less disruptive for wildlife and often better for forest regeneration.

seriously undernourished Those who receive less than 80 percent of their minimum daily caloric requirements; are likely to suffer permanently stunted growth, mental retardation, and other social and developmental disorders.

shantytowns Settlements created when people move onto undeveloped lands and build their own shelter with cheap or discarded materials; some are simply illegal subdivisions where a landowner rents land without city approval; others are land invasions.

sheet erosion Peeling off thin layers of soil from the land surface; accomplished primarily by wind and water.

sinkholes A large surface crater caused by the collapse of an underground channel or cavern; often triggered by groundwater withdrawal.

sludge Semisolid mixture of organic and inorganic materials that settles out of wastewater at a sewage treatment plant.

slums Legal but inadequate multifamily tenements or rooming houses; some are custom built for rent to poor people, others are converted from some other use.

smog The term used to describe the combination of smoke and fog in the stagnant air of London; now often applied to photochemical pollution products or urban air pollution of any kind.

social justice Equitable access to resources and the benefits derived from them; a system that recognizes inalienable rights and adheres to what is fair, honest, and moral.

soil A complex mixture of weathered mineral materials from rocks, partially decomposed organic molecules, and a host of living organisms.

soil horizons Horizontal layers that reveal a soil's history, characteristics, and usefulness.

southern pine forest United States coniferous forest ecosystem characterized by a warm, moist climate.

species A population of morphologically similar organisms that can reproduce sexually among themselves but that cannot produce fertile offspring when mated with other organisms.

species diversity The number and relative abundance of species present in a community.

species recovery plan A plan for restoration of an endangered species through protection, habitat management, captive breeding, disease control, or other techniques that increase populations and encourage survival.

spring overturn Springtime lake phenomenon that occurs when the surface ice melts and the surface water temperature warms to its greatest density at 4°C and then sinks, creating a convection current that displaces nutrient-rich bottom waters.

squatter towns Shantytowns that occupy land without owner's permission; some are highly organized movements in defiance of authorities; others grow gradually.

stability In ecological terms, a dynamic equilibrium among the physical and biological factors in an ecosystem or a community; relative homeostasis.

stable runoff The fraction of water available year round; usually more important than total runoff when determining human uses.

Standard Metropolitan Statistical Area (SMSA) An urbanized region with at least 100,000 inhabitants with strong economic and social ties to a central city of at least 50,000 people.

statutory law   Rules passed by a state or national legislature.

steady-state economy   Characterized by low birth and death rates, use of renewable energy sources, recycling of materials, and emphasis on durability, efficiency, and stability.

stewardship   A philosophy that holds that humans have a unique responsibility to manage, care for, and improve nature.

stomates   The small openings in leaves, herbaceous stems, and fruits through which gases and water vapor pass.

strategic minerals   Materials a country cannot produce itself but which it uses for essential materials or processes.

stratosphere   The zone in the atmosphere extending from the tropopause to about 50 km (30 mi) above Earth's surface; temperatures are stable or rise slightly with altitude; has very little water vapor but is rich in ozone.

streambank erosion   Washing away of soil from banks of established streams, creeks, or rivers, often as a result of the removal of trees and brush along streambanks or cattle damage to the banks.

stress   Physical, chemical, or emotional factors that place a strain on an animal. Plants also experience physiological stress under adverse environmental conditions.

stress-related disease   See stress-shock.

stress-shock   A loose set of physical, psychological, and/or behavioral changes thought to result from the stress of excess competition and extreme closeness to other members of the same species.

strip-farming   Planting different kinds of crops in alternating strips along land contours; when one crop is harvested, the other crop remains to protect the soil and prevent water from running straight down a hill.

strip-mining   Removing surface layers over coal seams using giant, earth-moving equipment; creates a huge open-pit from which coal is scooped by enormous surface-operated machines and transported by trucks; an alternative to deep mines.

subbituminous coal   Banded, black, and fairly soft low-energy coal that has woody layers commonly visible.

sublimation   The process by which water can move between solid and gaseous states without ever becoming liquid.

subsidence   A settling of the ground surface caused by the collapse of porous formations that result from withdrawal of large amounts of groundwater, oil, or other underground materials.

subsoil   A layer of soil beneath the topsoil that has lower organic content and higher concentrations of fine mineral particles; often contains soluble compounds and clay particles carried down by percolating water.

sulfur dioxide   A colorless, corrosive gas directly damaging to both plants and animals.

supercity   See megacity or megalopolis.

supply/demand curve   The relationship between the available supply of a commodity or service and its price.

surface mining   Some minerals are also mined from surface pits. See strip-mining.

surface tension   A condition in which the water surface meets the air and acts like an elastic skin.

survivorship   The percentage of a population reaching a given age or the proportion of the maximum life span of the species reached by any individual.

sustainable agriculture   An ecologically sound, economically viable, socially just, and humane agricultural system. Stewardship, soil conservation, and integrated pest management are essential for sustainability.

sustainable development   Using renewable resources in harmony with ecological systems to produce a rise in real income per person and an improved standard of living for everyone.

sustained yield   Utilization of a renewable resource at a rate that does not impair or damage its ability to be fully renewed on a long-term basis.

swamp   Wetland with trees, such as the extensive swamp forests of the southern United States.

swidden agriculture   See milpa agriculture.

symbiosis   The intimate living together of members of two different species; includes **mutualism, commensalism,** and, in some classifications, **parasitism.**

synergism   The combined effects of two or more factors on a system.

synergistic effects   When an injury caused by exposure to two environmental factors together is greater than the sum of exposure to each factor individually.

systemic   A condition or process that affects the whole body; many metabolic poisons are systemic.

T

taiga   The northernmost edge of the boreal forest, including species-poor woodland and peat deposits; intergrading with the arctic tundra.

tailings   Mining waste left after mechanical or chemical separation of minerals from crushed ore.

technopolis   Also called a vertical city; this model of city development proposes that cities grow vertically instead of horizontally.

tectonic plates   Huge blocks of Earth's crust that slide around slowly, pulling apart to open new ocean basins or crashing ponderously into each other to create new, larger landmasses.

temperate rain forest   The cool, dense, rainy forest of the northern Pacific coast; enshrouded in fog much of the time; dominated by large conifers.

teratogens   Chemicals or other factors that specifically cause abnormalities during embryonic growth and development.

terracing   Shaping the land to create level shelves of earth to hold water and soil; requires extensive hand labor or expensive machinery, but it enables farmers to farm very steep hillsides.

territoriality   An area surrounding an organism's home site or nesting site that the organism actively defends, especially against members of its own species; the territory may be occupied and defended by an individual, a mated pair, or a social group, depending on species.

tertiary sewage treatment   The removal of inorganic minerals and plant nutrients after primary and secondary treatment of sewage.

thermal plume   A plume of hot water discharged into a stream or lake by a heat source, such as a power plant.

thermodynamics   A branch of physics that deals with transfers and conversions of energy.

thermodynamics, first law   Energy can be transformed and transferred, but cannot be destroyed or created.

thermodynamics, second law   With each successive energy transfer or transformation, less energy is available to do work.

thermoplastics   Soft plastics composed of single-chain, unlinked polymers, such as polyethylene, polypropylene, polyvinylchloride, polystyrene, and polyester, that can be remelted and reformed to make useful products.

thermoset polymers   Hard plastics composed of cross-linked molecular networks, such as acrylic, phenolic, or epoxy resins, that cannot be remelted or recycled.

thermosphere   The highest atmospheric zone; a region of hot, dilute gases above the mesosphere extending out to about 1,600 km (1,000 mi) from Earth's surface.

Third World Less developed countries that are not capitalistic and industrialized (First World) or centrally-planned socialist economies (Second World); not intended to be derogatory.

threatened species While still abundant in parts of its territorial range, this species has declined significantly in total numbers and may be on the verge of extinction in certain regions or localities.

tied ridges A series of ridges running at right angles to each other so that runoff is blocked in all directions and water is allowed to soak into the soil.

timberline In mountains, the highest-altitude edge of forest that marks the beginning of the treeless alpine tundra.

tolerance limits *See* limiting factors.

topsoil The first true layer of soil; layer in which organic material is mixed with mineral particles; thickness ranges from a meter or more under virgin prairie to zero in some deserts.

tornado A violent storm characterized by strong swirling winds and updrafts; tornadoes form when a strong cold front pushes under a warm, moist air mass over the land.

total fertility rate The number of children born to an average woman in a population during her entire reproductive life.

toxins Poisonous chemicals that react with specific cellular components to kill cells or to alter growth or development in undesirable ways; often harmful, even in dilute concentrations.

tragedy of the commons Depletion or degradation of a commons, or publicly-held resource, to which everyone has access but no one has ownership or a sense of responsibility.

transitional zone A zone in which populations from two or more adjacent communities meet and overlap.

transpiration The evaporation of water from plant surfaces, especially through stomates.

trauma Injury caused by accident or violence.

trombe wall An interior, heat-absorbing wall; may be water-filled glass tubes that absorb heat rays and let light into interior rooms.

trophic level A step in the movement of energy through an ecosystem; an organism's feeding status in an ecosystem.

tropical rain forest A species-rich biome type; consists of stratified communities of broad-leaved shrubs, trees, epiphytes, lianas, numerous insects, birds, and other animals; abundant rainfall, year-round warm to hot temperatures, and old and nutrient-poor soils.

tropical seasonal forest Semi-evergreen or partly deciduous forests tending toward open woodlands and grassy savannas dotted with scattered, drought-resistant tree species; distinct wet and dry seasons, hot year-round.

tropopause The boundary between the troposhere and the stratosphere.

troposphere The layer of air nearest to Earth's surface; both temperature and pressure usually decrease with increasing altitude.

tsunami Giant seismic sea swells that move rapidly from the center of an earthquake; they can be 10 to 20 meters high when they reach shorelines hundreds or even thousands of kilometers from their source.

tundra Treeless arctic or alpine biome characterized by cold, harsh winters, a short growing season, and potential for frost any month of the year; vegetation includes low-growing perennial plants, mosses, and lichens.

U

unconventional air pollutants Toxic or hazardous substances, such as asbestos, benzene, beryllium, mercury, polychlorinated biphenyls, and vinyl chloride, not listed in the original Clean Air Act because they were not released in large quantities; also called noncriteria pollutants.

undernourished Those who receive less than 90 percent of the minimum dietary intake over a long-term time period; they lack energy for an active, productive life and are more susceptible to infectious diseases.

undiscovered resources Potential supplies of a mineral or other useful material believed to exist based on history, scientific theory, or general knowledge of geology, biology, or geography of an unexplored area.

universalists Those who believe that some fundamental ethical principles are universal and unchanging. In this vision, these principles are valid regardless of the context or situation.

utilitarianism A philosophy that regards an action as right if it produces the greatest good for the greatest number of people.

upwelling Convection currents within a body of water that carry nutrients from bottom sediments toward the surface.

urban area An area in which a majority of the people are not directly dependent on natural resource-based occupations.

urbanization An increasing concentration of the population in cities and a transformation of land use to an urban pattern of organization.

utilitarian conservation Saving resources because of their value to humans; to provide the greatest good for the greatest number for the longest time; a philosophy argued by Theodore Roosevelt and Gifford Pinchot.

V

vertical city *See* technopolis.

vertical stratification The vertical distribution of specific subcommunities within a community.

village A collection of rural households linked by culture, custom, and association with the land.

visible light A portion of the electromagnetic spectrum that includes the wavelengths used for photosynthesis.

vitamins Organic molecules essential for life that we cannot make for ourselves; we must get them from our diet; they act as enzyme cofactors.

volatile organic compounds Organic chemicals that evaporate readily and exist as gases in the air.

W

warm front A long, wedge-shaped boundary caused when a warmer advancing air mass slides over neighboring cooler air parcels.

waste stream The steady flow of varied wastes, from domestic garbage and yard wastes to industrial, commercial, and construction refuse.

water cycle The recycling and reutilization of water on Earth, including atmospheric, surface, and underground phases and biological and nonbiological components.

water droplet coalescence A mechanism of condensation that occurs in clouds too warm for ice crystal formation.

waterlogging Water saturation of soil that fills all air spaces and causes plant roots to die from lack of oxygen; a result of overirrigation.

water table The top layer of the zone of saturation; undulates according to the surface topography and subsurface structure.

weather A description of the physical conditions of the atmosphere (moisture, temperature, pressure, and wind).

weathering Changes in rocks brought about by exposure to air, water, changing temperatures, and reactive chemical agents.

wetlands Ecosystems of several types in which rooted vegetation is surrounded by standing water during part of the year. *See also* swamps, marshes, bogs.

wilderness An area of undeveloped land affected primarily by the forces of nature; an area where humans are visitors who do not remain.

Wilderness Act Legislation of 1964 recognizing that leaving land in its natural state may be the highest and best use of some areas.

wildlife Plants, animals, and microbes that live independently of humans; plants, animals, and microbes that are not domesticated.

wildlife refuges Areas set aside to shelter, feed, and protect wildlife; due to political and economic pressures, refuges often allow hunting, trapping, mineral exploitation, and other activities that threaten wildlife.

wind farms Large numbers of windmills concentrated in a single area; usually owned by a utility or large-scale energy producer.

windbreak Rows of trees or shrubs planted to block wind flow, reduce soil erosion, and protect sensitive crops from high winds.

withdrawal A description of the total amount of water taken from a lake, river, or aquifer.

woodland A forest where tree crowns cover less than 20 percent of the ground; also called open canopy.

work The application of force through a distance; requires energy input.

world conservation strategy A system of maintaining essential ecological processes, preserving genetic diversity, and ensuring that utilization of species and ecosystems is sustainable.

World Ocean The interconnected world seas and oceans.

X

X ray Very short wavelength in the electromagnetic spectrum; can penetrate soft tissue; although it is useful in medical diagnosis, it also damages tissue and causes mutations.

Y

yellowcake The concentrate of 70 to 90 percent uranium oxide extracted from crushed ore.

young soils Soils that haven't weathered much and, therefore, are rich in soluble minerals from parent rocks, such as silicon, iron, and aluminum.

Z

zero population growth (ZPG) The number of births at which people are just replacing themselves; also called the replacement level of fertility.

zone of aeration Upper soil layers that hold both air and water.

zone of leaching The layer of soil just beneath the topsoil where water percolates, removing soluble nutrients that accumulate in the subsoil; may be very different in appearance and composition from the layers above and below it.

zone of saturation Lower soil layers where all spaces are filled with water.

CREDITS

PHOTOGRAPHS

Table of Contents
Page vi: Photo by David Swanlund, Courtesy of Save the Redwoods League; p. vii: © David Wells/The Image Works; p. viii: © Robert and Linda Mitchell; p. xii: © Walter Frerck/Odyssey Production; p. xiv: © Wolfgang Kaehler.

Part Openers
One: Photo by David Swanlund, Courtesy of Save-The-Redwoods League; Two: © David Wells/The Image Works; Three: © Robert and Linda Mitchell; Four: © Walter Frerck/Odyssey Production; Five: © Wolfgang Kaehler.

Chapter 1
Opener: Science VU/Visuals Unlimited; 1.2: © R. Frerck/Odyssey Productions; 1.4: © Visuals Unlimited; 1.5: © Science VU/Visuals Unlimited; 1.6: © Stan W. Elems/Visuals Unlimited.

Chapter 2
Opener: Earth Satellite Corporation/Science Photo Library/Photo Researchers, Inc.; BOX 2.1: Science VU/Visuals Unlimited; 2.3 & 2.4: Earth Satellite Corporation/Science Photo Library/Photo Researchers, Inc.; 2.6: © John Shaw/Tom Stack and Associates; 2.10: Maptec International LTD/Science Photo Library/Photo Researchers, Inc.; 2.13C: © Laurence Pringle, 1972/Photo Researchers, Inc.

Chapter 3
Opener: James Brock; 3.1: © Hungerford; 3.9: James Brock; 3.11: © Tom Myers; 3.17: © J. Brinton/Visuals Unlimited.

Chapter 4
4.6: © M. K. Wicksten/Visuals Unlimited; 4.7: © Robert C. Simpson/Valan Photos; 4.9 & 4.10: © John D. Cunningham/Visuals Unlimited; 4.12: © V. Ahmadjian and J. B. Jacobs/Visuals Unlimited; 4.13: © D. Wilder/Tom Stack and Associates; BOX 4.1: © Tom & Pat Leeson/Photo Researchers, Inc.; BOX 4.2: © Fred McConnaughy, Photo Researchers, Inc.; 4.15: Dr. Carl W. Bollwinkel; 4.16: © J. A. Wilkinson/Valan Photos; 4.17A & B: U.S. Forest Service; 4.18: Daryl Smith.

Chapter 5
Opener: © Stephen S. Krasemann/Photo Researchers, Inc.; 5.1: © Stephen J. Krasemann/Photo Researchers, Inc.; 5.6: © Richard Weymouth Brooks/Photo Researchers, Inc.; 5.7: © John D. Cunningham/Visuals Unlimited; 5.8: © Ron Spomer/Visuals Unlimited; 5.9 & 5.10: © B. L. Heidel/Visuals Unlimited; 5.11: © Joe McDonald/Visuals Unlimited; 5.12: Chippewa Valley Museum; 5.13 & 5.14: © John D. Cunningham/Visuals Unlimited; 5.15: © Glenn Oliver/Visuals Unlimited; 5.16: © Doug Sokell/Visuals Unlimited; 5.17: © Patrick K. Armstrong/Visuals Unlimited; 5.18: © Lucy Jones/Visuals Unlimited; 5.19: © Robert and Linda Mitchell; 5.20: © Francois Gohler/Photo Researchers, Inc.; 5.21: © Len Clifford/Visuals Unlimited; 5.22: © Robert and Linda Mitchell; 5.24: Earth Satellite Corporation/Science Photo Library/Photo Researchers, Inc.; 5.30: Fred L. Rose; 5.31: © Frank M. Hanna/Visuals Unlimited; 5.32: © 1989 Bruce Berg/Visuals Unlimited.

Chapter 6
Opener: © David Wells/The Image Works; 6.1: A. Devaney, Inc. NY; 6.4: © Tom McHugh/Photo Researchers, Inc.

Chapter 7
7.1: AP/Wide World Photos © Sylvan Wittwer/Visuals Unlimited; 7.19: A: © Jim Shaffer; B: © H. Morgan/Photo Researchers Inc.; C: © Bob Coyle; D: © Ray Ellis/Photo Researchers, Inc.; E & F: © Bob Coyle.

Chapter 8
Opener: © Audrey Lang/Valan Photos; 8.1: © Jerry Irwin; 8.4: © Lowell Georgia/Photo Researchers, Inc.; 8.5: © Audrey Lang/Valan Photos; 8.6: © T. Kitchen/Valan Photos; 8.8: © John D. Cunningham/Visuals Unlimited; BOX 8.1: © Frank J. Miller/Photo Researchers, Inc.; BOX 8.2: © Joe McDonald/Visuals Unlimited.

Chapter 9
Opener: © Victor H. Hutchison/Visuals Unlimited; 9.1: © Victor H. Hutchison/Visuals Unlimited; 9.7: U.S.G.S./E. B. Hardin; 9.11: © Joseph Netts/Photo Researchers, Inc.; 9.14: U.S.G.S./A. Keith; 9.15: © Herman Kokojah/Black Star; 9.16: © David Weintraub/Photo Researchers, Inc.

Chapter 10
10.1: © Allan Tannenbaum/Sygma; 10.3: © Omikron/Photo Researchers, Inc.; 10.4: © Biophoto Association/Photo Researchers, Inc.; 10.5: U.S.D.A.; 10.10: © Link/Visuals Unlimited; 10.11: © Victor Englebert/Photo Researchers, Inc.; 10.12: © Chester Higgins, Jr./Photo Researchers, Inc.; BOX 10.1: © Audrey Topping Cohen/Photo Researchers, Inc.

Chapter 11
11.1: © Pam Hickman/Valan Photos; 11.7 & 11.8: Soil Conservation; 11.9: Library of Congress; 11.12: © John Cunningham/Visuals Unlimited; 11.14: © Joe Monroe/Photo Researchers; 11.15: © Wolfgang Kaehler; BOX 11.1: Terry Gips with "International Alliance for Sustainable Agriculture."

LINE ART AND TEXT

Chapter 1
Figure 1.8 *THE LIMITS OF GROWTH: A report for The Club of Rome's Project on the Predicament of Mankind*, by Donella H. Meadows, Dennis L. Meadows, Jórgen Randers, William W. Behrens, III. A Potomac Associates book published by Universe Books, N.Y., 1972. Graphics by Potomac Associates.

Chapter 2
Figure 2.11 From Leland G. Johnson, *Biology*, 2d ed. Copyright © 1987 Wm. C. Brown Publishers, Dubuque, Iowa. All Rights Reserved. Reprinted by permission.

Chapter 3
Figure 3.6 From "Trophic Structure and Productivity of Silver Springs, Florida" by Howard T. Odum, *Ecological Monographs*, 1957, 55–112. Copyright © 1957 Ecological Society of America, Tempe, AZ. Reprinted by permission.

Chapter 5
Figure 5.2 Reprinted by permission of Macmillan Publishing Company from *Communities and Ecosystems*, by Robert H. Whittaker. Copyright © 1975 by Robert H. Whittaker. **Figure 5.3** From Leland G. Johnson, *Biology*, 2d ed. Copyright © 1987 Wm. C. Brown Publishers, Dubuque, Iowa. All Rights Reserved. Reprinted by permission.

Chapter 6
Figure 6.2 From R. L. Brown, et al., "Twenty-two Dimensions of the Population Problem" in *Worldwatch Paper 5*, March 1976. Copyright © 1976 Worldwatch Institute, Washington, DC. Reprinted by permission. **Figure 6.10** From data in A. S. Bodenheimer, *Monographiae Biologicae*, 6:1–276, 1958. Copyright © 1958 Kluwer Academic Publishers, The Netherlands, as appeared in Edward J. Kormandy, *Concepts of Ecology*. Reprinted by permission of Kluwer Academic Publishers, The Netherlands, and Prentice-Hall, Publishers, Inc., Englewood Cliffs, NJ.

Chapter 7
Figure 7.2 "World Population Growth Through History" chart (Washington, D.C.: Population Reference Bureau, Inc., 1985). **Figure 7.3** From *POPULATION, RESOURCES, ENVIRONMENT*, 2/E. by Paul R. Ehrlich and Anne H. Ehrlich.

Copyright © 1970, 1972 by W. H. Freeman and Company. Reprinted by permission. **Figure 7.4** Thomas Merrick, with PRB staff, "World Population in Transition" *Population Bulletin*, Vol. 41, No. 2 (Washington, D.C.: Population Reference Bureau, Inc., January 1988 reprint). **Figure 7.5** From *Illustrated Atlas of the World* © 1985 by Rand McNally, R. L. 89-S-176. **Figure 7.6** From C. Clark, *Population Growth and Land Use*. Copyright © 1967 St. Martin's Press, New York, NY. Reprinted by permission of St. Martin's Press and Macmillan Magazines Ltd., London, England. **Box Figure 7.2** Source: H. Yuan Tien, "China: Demographic Billionaire," *Population Bulletin* Vol. 38, No. 2 (April 1983): table 2; and H. Yuan Tien, "The New Census of China," *Population Today* Vol. 19, No. 1 (January 1991): table 2. **Figure 7.7** From T. McKeown, et al., "An Interpretation of the Modern Rise of Population in Europe" in *Population Studies*, Volume 26/3, November 1972. Copyright © 1972 The Population Investigation Committee of the London School of Economics, London, England. Reprinted by permission. **Figure 7.8** From United Nations Population Division data estimates, 1986, as appeared in *The State of the World's Children 1987*. Reprinted by permission of UNICEF. **Figure 7.10** Carl Haub, "Understanding Population Projections." *Population Bulletin*, Vol. 42, No. 4 (Washington, D.C.: Population Reference Bureau, Inc., December 1987). **Figure 7.15** From T. McKeown, et al., "An Interpretation of the Modern Rise of Population in Europe" in *Population Studies*, Volume 26/3, November 1972. Copyright © 1972 The Population Investigation Committee of the London School of Economics, London, England. Reprinted by permission. **Figure 7.18** From United Nations Population Division as appeared in UNICEF, *The State of the World's Children*. Copyright © 1987 Oxford University Press, Oxford, England. Reprinted by permission of the United Nations, New York, NY and Oxford University Press. **Figure 7.22** *1988 World Population Data Sheet* (Washington, D.C: Population Reference Bureau, Inc., April 1988).

Chapter 8
Figure 8.14 Population Reference Bureau, Inc., Washington, D.C. **Figure 8.15** *Models of Doom: A Critique of "Limits to*

Growth" by H. S. D. Universe Books, New York, 1973; and Chatto & Windus Ltd., London, England. **Figure 8.16** *Models of Doom: A Critique of "Limits to Growth"* by H. S. D. Universe Books, New York, 1973; and Chatto & Windus Ltd., London, England. **Figure 8.17** From R. A. Carpenter and J. A. Dixon, "Ecology Meets Economics" in *Environment*, June 1985. Reprinted with permission of the Helen Dwight Reid Educational Foundation. Published by Heldref Publications, 4000 Albemarle St., N.W., Washington, D.C. 20016. Copyright © 1985.

Chapter 9
Figure 9.2 From Carla W. Montgomery, *Physical Geology*, 2d ed. Copyright © 1990 Wm. C. Brown Publishers, Dubuque, Iowa. All Rights Reserved. Reprinted by permission. **Box Figure 9.1** *Star Tribune*, October 14, 1990.

Chapter 10
Figure 10.2 From Robert Repetto, editor, *The Global Possible*. Copyright © Yale University Press, New Haven, CT. Reprinted by permission. **Figure 10.6** From *World Resources 1990–1991*. Copyright © 1990 by the World Resources Institute. Reprinted by permission of Oxford University Press, Inc.

Chapter 11
Figure 11.6 From *Living in the Environment: An Introduction to Environmental Science*, 5/E, by G. Tyler Miller, Jr. © 1988 by Wadsworth, Inc. Reprinted by permission of the publisher.

Chapter 12
Figure 12.3 From United Food and Agriculture Organization, *Unasylva*, Vol. 28, No. 112–113, 1976. Copyright © 1976 Food and Agriculture Organization of the United Nations, Rome, Italy. Reprinted by permission. **Figure 12.12** From Dregne, *Desertification of Arid Lands*. Copyright © 1984 Harwood Academic Publishers. Reprinted by permission. **Figure 12.14** From *WORLD RESOURCES 1987*. Copyright © 1987 by the International Institute for Environment and Development and the World Resources Institute. Reprinted by permission of Basic Books, a division of HarperCollins Publishers Inc.

Chapter 13

Figure 13.2 *Northern Sun News*, Minneapolis, MN, June 1988. Reprinted by permission. **Figure 13.20** Newsweek, July 28, 1986. Copyright © 1986 Newsweek, Inc. Used by permission.

Chapter 14

Figure 14.9 From Vinzenz Ziswiler, *Extinct and Vanishing Animals*. Copyright © 1967 Springer-Verlag, Heidelberg, Germany. Reprinted by permission. **Figure 14.10** "From *Environment*, 16(10):31–34, 1974. Reprinted with permission of the Helen Dwight Reid Educational Foundation. Published by Heldref Publications, 4000 Albemarle St., N.W., Washington, D.C. 20016. Copyright © 1974." **Figure 14.14** From *World Resources 1990–1991*. Copyright © 1990 by the World Resources Institute. Reprinted by permission of Oxford University Press, Inc.

Chapter 15

Figure 15.13 From R. H. MacArthur and E. O. Wilson, *The Theory of Island Biogeography*. Copyright © Princeton University Press, Princeton, NJ.

Chapter 16

Figure 16.4 From W. G. Kendrew, *Climate*. Copyright © 1930 Oxford University Press, Oxford, England. Reprinted by permission. **Figure 16.17** From Carla W. Montgomery, *Physical Geology*, 2d ed. Copyright © 1990 Wm. C. Brown Publishers, Dubuque, Iowa. All Rights Reserved. Reprinted by permission.

Chapter 18

Figure 18.2 From R. Dorf, *The Energy Factbook*. Copyright © 1981 McGraw-Hill Publishing Company, New York, NY. Reprinted by permission.

Chapter 19

Figure 19.5 With permission of the National Council on Radiation Protection and Measurements. **Box Figure 19.1** From Rolf Lange, et al., "Dose Estimates from the Chernobyl Accident" in *Nuclear Technology*, 82:311–323, September 1988. Copyright © 1988 by the American Nuclear Society, La Grange Park, Illinois. Reprinted by permission of the publisher and author; University of California, Lawrence Livermore National Laboratory, and Department of Energy. **Figure 19.8**

Courtesy of Northern States Power Company, Minneapolis, MN. **Figure 19.9** Courtesy of Northern States Power Company, Minneapolis, MN. **Figure 19.15** From "Managing Canada's Nuclear Fuel Wastes." Reprinted by permission of Atomic Energy of Canada Limited. **Figure 19.20** From C. Flavin, *Worldwatch Report #75, 1987*. Worldwatch Institute, Washington, DC. Reprinted by permission.

Chapter 20

Figure 20.17 Reproduced from *State of the World*, 1986, General Editor, Lester Brown. By permission of W. W. Norton & Company, Inc. Copyright © 1986 by Worldwatch Institute. **Figure 20.20** From R. Dorf, *The Energy Factbook*. Copyright © 1981 McGraw-Hill Publishing Company, New York, NY. Reprinted by permission.

Chapter 21

Figure 21.7 "Reprinted with permission from *The New England Journal of Medicine*, page 1226, 1986." **Figure 21.15** From Paul Slovic, "Perception of Risk" in *Science*, volume 236, April 1987. Copyright 1987 by the American Association for the Advancement of Science, Washington, DC. Reprinted by permission of the publisher and author.

Chapter 22

Figure 22.3 From William U. Chandler, *Investing in Children*, Worldwatch Paper 64, 1985, page 32. Copyright © Worldwatch Institute, Washington, DC. Reprinted by permission. **Figure 22.15** From Organization for Economic Cooperation and Development (OECD), *The State of the Environment*, 1985, page 84. Reprinted by permission.

Chapter 23

Figure 23.20 Courtesy General Motors, Detroit, MI.

Chapter 25

Figure 25.5 From *Prospects of World Urbanization*, 1988 (United Nations publication, Sales No. E.89.XIII.8), p. 28.

Chapter 26

Figure 26.2 Courtesy Town & Country Planning Association, London, England. **Figure 26.5** Used with permission. *AMAZING® Stories*. **Figure 26.8** Reprinted with permission of Sierra Club Books.

ILLUSTRATORS CREDITS

Bowring Cartographics: Figures 5.2, 5.4, 5.5, 7.5, 9.3, 9.5, 9.8, 10.7, 12.3, 12.12, 13.2, 13.8, 13.10, 13.16, 13.20.

Diphrent Strokes, Inc.: Figures 1.8a, 1.8b, 1.8c, 2.1, 2.5, 2.8a-d, 2.11, 2.13a&b, 3.2, 3.5, 3.6, 3.18, 7.7, 7.12, 7.14, 7.15, 7.18, 8.15, 9.2, 9.6, 9.10, 10.6, 12.4, 14.2, 14.14, 15.13, 16.2, 18.2, 18.4, 18.13, 19.4, 19.9, 19.12, 20.5, 21.7, 21.15, 22.3, 22.12, 24.3, 24.8, 24.15, 25.5, 25.7, Box Figures 7.2, 9.1, 16.2b, 17.2, 20.1, 23.2.

Norman Frisch: Figures 5.26, 5.27, 5.29, 9.4, 9.13, 18.11.

Marjorie C. Leggitt: Figures 3.3, 3.4, 4.3, 4.4, 4.8, 4.11, 7.9, 11.4, 11.17, 14.3, 14.5, Box Figure 4.4.

Don Luce: Figures 4.14, 5.23, 5.25, 11.5.

Don Luce/Marjorie C. Leggitt: Figure 3.7.

Laurie O'Keefe: Figures 1.1, 4.1, 7.20a-c, 8.18, 21.4.

Laurie O'Keefe/Don Luce: Figures 3.10, 3.13, 3.16, 3.19, 4.5.

Rolin Graphics: Figures 1.7, 2.2, 2.7, 2.9, 2.12, 3.14, 3.15, 4.2, 5.3, 5.28, 6.2, 6.3, 6.5, 6.6, 6.7, 6.8, 6.9, 6.10, 6.11, 7.2, 7.3, 7.4, 7.6, 7.8, 8.2, 8.3, 8.7, 8.9, 8.10, 8.11, 8.13, 8.14, 8.16, 8.17, 9.9, 9.12, 10.2, 10.8, 11.3, 11.11, 11.13, 11.16, 12.14, 13.6, 13.7, 13.11, 13.13, 13.18, 14.8, 14.9, 14.10, 14.11, 18.5, 18.6.

INDEX

McPhee, John, 379
Madagascar periwinkle, 265
Magma, 154
Magnesium, 153, 174
Magnetic confinement, 386
Magnification, 429
MAGNOX, 373, 376
Maize, 176, 177, 179
Malaria, 419, 420, 421
Malathion, 424
Malignant tumors, 425
Malnourishment, 172
Malnutrition, 187
Malthus, Thomas, 90, 107, 121, 139, 178
Malthusian checks, 107
Malthusian growth, 95
Malthusian strategy, 96
Mammals, 263, 282
Mammoth Hot Springs, 255
Management
    biological resources. *See* Biological resources
        BLM. *See* Bureau of Land Management
        (BLM)
    coal. *See* Coal
    forest. *See* Forests
    IPM. *See* Integrated pest management (IPM)
    land. *See* Land management era
    natural gas. *See* Natural gas
    oil. *See* Oil
    petroleum. *See* Petroleum
    radioactive waste. *See* Radioactive waste
    of rangelands. *See* Rangelands
    topography, 208
    of water. *See* Water
Man and Biosphere (MAB), 293
Man and Nature, 241
Mancozeb, 211
Maneb, 211, 424
Manganese, 157, 174
Mangosteen, 264
Mantle, 153, 154
Mao Zedong, 110, 186
Marasmus, 172
Marbled Murlets, 251
Marine ecosystems, 84–87
Marine Protection Research and Sanctuaries Act
    (1972), 461
Marine symbiosis, 55
Market efficiencies, 138
Market equilibrium, 138
Marketing, of pollution, 144
Marsh, George Perkins, 241
Marshes, 83
Martinique, 166
Marx, Karl, 107, 134
Massachusetts Institute of Technology (MIT), 142,
    331, 378
Mass burn, 498, 499
Mass extinctions, 268
Mass transit, 549–51
Material cycles, 32–41
Material economy, 138
Mather, Stephen, 242
Matter
    conservation of, 26
    in ecosystems, 25–43, 29–41
    and energy of atoms, 16–19
    and energy and life, 15
Mauritania, 404
MDC. *See* More developed countries (MDC)
Meadows, Donnela, 142
Means, and ends, 559

Measles, 421
Measurement, of toxicity, 430–33. *See also* Units
Meat, 176, 178
Mech, David, 93
Mechanical weathering, 156
Media, and public opinion, 568
Medicines, 265, 507
Megacity, 518
Megalopolis, 518
Mekong River, 201
Melons, 178
Mendes, Francisco Chico, 226–27
Mendes, Ilzmar, 227
Mercury, 355, 423
Mesosphere, 327
Mesquite plants, 35
Metabolic degradation, 429–30
Metabolism, 18
Metals, 157–58, 159, 469, 507
    and coal, 354–55
    heavy. *See* Heavy metals
    recycling, 500
Metamorphic rocks, 157
Methane, 23, 327, 331, 404–5
Methionine, 172
Methylethylketone, 423
Methylisobutylketone, 423
Metric/English conversions, 581
Metropolis, 518
Mexico City, 5, 163, 165
Michaels, Patrick, 331
Microbes, 211
Micro-hydro generators, 408
Microlending, 147
Microwave, 20
Mifegyne, 126
Mifepristone, 126
"Migration between the Core and the Periphery,"
    519
Milankovitch, Milutin, 339
Milankovitch cycles, 339
Milk, 176, 178
Mill, John Stuart, 134, 149, 559
Mille Lacs Refuge, 288
Millet, 176
Millirem (mrem), 369
Mills, 160, 161
Milpa agriculture, 225
Minamata disease, 445–46
Mineral Commodity Summaries, 160
Mineralogy, economic, 157–60
Mineral resources, conserving, 160–61
Minerals, 153, 156, 173–75
    nonmetal, 158
    strategic, 158–60
Minerals and Mining Act (1866), 240
Minimills, 160, 161
Minimum till, 209, 212
Minimum wrong, 561
Mining, 161–62, 352–53, 380
Minnesota Mining and Manufacturing Company
    (3M), 11, 504
Mirrors, parabolic, 397
Mississippi River, 201
Misty Fjords, 254
MIT. *See* Massachusetts Institute of Technology
    (MIT)
Mixed perennial polyculture, 225
Mobility, effects of, 115
Models
    and fertility, 118
    World Computer Model, 142

Models of Doom, 142
Mojave National Park, 255
Mollisols, 197
Molybdenum, 174
Momentum, population, 98
Monkey Wrench Gang, The, 570
Monoculture agroforestry, 223
Monsanto, 11
Monsoon, 337
Montane coniferous forests, 74, 75
Monuments, and air pollution, 482–83
Moody, Judith B., 155
Moon, William Least Heat, 64
Moose, 91–93
Moral agents, 559
Moral extensionism, 559–60
Moral subjects, 559
Morbidity, 419
More developed countries (MDC), 108–9, 141,
    176
Morphine, 265
Mortality, 97–98, 419, 442
Mortality rate, 97, 112–13
Mosquito, 420
Mount Pele, 166
Mount St. Helens, 166–67
Mount Vesuvius, 166
Mouse, 101
Moving water, 391, 412
Muir, John, 242, 319, 570
Mulch, 209
Multiple use, 242
Mumford, Lewis, 541
Municipal sewage treatment, 458–60
Murder, 528–29
Murlets, 251
Musca domestica, 94
Muscarine, 431
Muscle power, 344
Mushroom poison, 431
Mus Musculus, 101
Mutagens, 210, 418, 424
Mutualism, 39, 100
Mycorrhizae, 54

n

NAAQS. *See* National Ambient Air Quality
    Standards (NAAQS)
Naess, Arne, 571
Nagasaki, 374
Nagoya Water Research Institute, 477
Nance, R. Damian, 155
Narmada River, 146
NASA. *See* National Aeronautics and Space
    Administration (NASA)
Natality, 96–97, 111–12
Natality rate, 111–12
National Aeronautics and Space Administration
    (NASA), 244, 410
National Ambient Air Quality Standards (NAAQS),
    487, 488, 489
National Center for Atmospheric Research, 331
National Environmental Policy Act, 513
National Forest Multiple Use and Sustained Yield
    Act (1960), 242
National Forest Service, 239
National Institute of Environmental Health
    Sciences, 210
National Institute of Occupational Health and
    Safety, 505, 533
National land-use policy, 243
National Park "Organic" Act (1916), 241

Troposphere, 327
Tryptophane, 172
TSP. *See* Total suspended particulates (TSP)
Tsunami, 165–66
Tumors, 425
Tundra, 76–77
Tuolumne River, 319
Turtle excluder device (TED), 270
TVA. *See* Tennessee Valley Authority (TVA)
Twain, Mark, 338

u

UAE. *See* United Arab Emirates (UAE)
Ultraviolet (UV), 20
  light, 371
  radiation, 470–71
Unconventional pollutants, 472
Undernourished, 171, 187
Undiscovered resources, 136
UNESCO, 293
United Arab Emirates (UAE), 349
United Fruit Company, 183
United Nations, 171, 177, 179, 180, 195, 566
United Nations Childrens Fund (UNICEF), 421, 442
United States
  birth rates, 119
  energy use, 350
  erosion in, 199–200
  Government agencies, 587. *See also entries below*
  land resources, 239–60
  land-use decisions, 241
  and suburbs, 541–43
  urbanization in, 519–20
United States Bureau of Land Management (BLM), 239, 243, 244, 249, 257, 565
United States Bureau of Mines, 137, 160
United States Bureau of Reclamation (BuRec), 243
United States Department of Agriculture, 565
United States Department of Agriculture Forest Service, 244
United States Department of Defense (DOD), 244
United States Department of Energy (DOE), 244, 380–81, 401, 410
United States Department of Health, 565
United States Department of Housing and Urban Development (HUD), 532
United States Department of Labor, 565
United States Department of the Air Force, 244
United States Department of the Army, 244
United States Department of the Interior, 242, 243, 244, 272
United States Department of the Navy, 244
United States Department of Transportation (DOT), 244
United States Fish and Wildlife Service (USFWS), 243, 244, 277, 285
United States Food and Drug Act, 432–33, 436
United States Food and Drug Administration (FDA), 424, 565
United States Forest Service (USFS), 239, 245, 247, 249
United States General Accounting Office (GAO), 200, 253
United States Geological Survey (USGS), 136, 300, 449
United States Government agencies, 587
United States National Academy of Sciences, 232, 264
United States National Cancer Insitute, 265
United States Public Health Service, 486, 565

United Technologies, 410
Units
  of energy, 344, 345
  of measurement, 581
  of radiation, 368, 369
  water, 301
Universalist, 558
University of Illinois, 331
University of Minnesota, 331
University of Sussex, 142
University of Virginia, 331
University of Wisconsin, 331, 436
Unwin, Raymond, 540
Upwelling, 46
Uranium, 355, 380
Urban, 518
Urban area, 517
Urbanization, 516–35
  causes of, 522–23
  degrees of, 517–19
  and government policy, 523
  transportation and city growth, 527–30
  trends and terms, 517–22
  in United States, 519–20
  and urban problems, 524–26
  world, 520–22
Urban reform, 539–44
Urban renewal, 544–45
Urechis caupo, 55
Urethanes, 211
Use, current and multiple, 242, 243
USFS. *See* United States Forest Service (USFS)
USFWS. *See* United States Fish and Wildlife Service (USFWS)
USGS. *See* United States Geological Survey (USGS)
Utilitarian conservation, 242
Utilitarianism, 242, 558–59
Utilization efficiencies, 391–92

v

Valdez Otter Rescue Center, 457
Valine, 172
Vanadium, 355
Van Der Ryc, Sim, 138
Vaporization, heat of, 22
Vascular plants, 263
Vas deferens, 125
Vasectomy, 125
Vaux's Swift, 251
Vegetable oils, 176
Vegetables, 176, 178, 181
Vegetation, 21, 70, 71
Vegetation zones, 221
Velsicol Chemical Company, 508
Vertical city, 544
Vertical stratification, 49–50
Vietnam War, 79, 243
Village, 517, 518
Vinblastine, 265
Vincristine, 265
Vining, Daniel R., Jr., 519
Violet, 266
Vitamins, 174, 175
VOC. *See* Volatile organic compounds (VOC)
Volatile organic compounds (VOC), 468, 471–72
Volcanoes, 166–67
Voltaire, 561

w

Warbler, 51
Warm front, 336

Warmth, 20
Wasatch Plateau, 17
Washington Public Power Supply Consortium (WPPSS), 384
Waste
  cost and safety considerations, 498–99
  disposal methods, 455–60, 496–97
  exporting, 497–98
  and government regulations, 513
  hazardous, 493, 504–13
  organic, 501
  oxygen-demanding, 443–44
  radioactive. *See* Radioactive waste
  and recycling, 500–503
  solid, 493, 494–504
  toxic, 493, 504–13
Waste lagoons, 508–9
Waste stream, 495, 503–4
Water, 202, 344–45
  and air pollution, 480–81
  availability and use, 307–13
  and density, 82
  and drought cycles, 309–11
  fresh, 81–83
  and freshwater shortages, 313–16
  and groundwater cycle, 314–15
  household use, 321
  legislation, 461–62
  major compartments, 304–7
  management and conservation, 320–22
  moving, 391, 412
  pollution. *See* Water pollution
  proposals and projects, 316–19
  quality conditions, 448–55
  quantities used, 311
  supplies, 307–9
  surface, 448–50
  and temperature, 82, 302
  units of measurement, 301
Water cycle, 32–34
Watergate, 243
Waterless toilet, 321
Waterlogging, 202
Water planet, 21–22
Water pollution, 439–63
  categories, 441
  control, 455–61
  defined, 440
  and ecosystem disruption, 441
  and groundwater, 450–52
  and infant mortality, 442
  and infectious agents, 440–43
  legislation, 461–62
  and ocean pollution, 452–55
  and oxygen-demanding wastes, 443–44
  quality conditions, 448–55
  source reduction, 455
  toxic inorganic, 445–46
  types and effects, 440–48
  and urban problems, 524
Water resources, 299–324
Watershed management, 320
Water-soluble vitamins, 174, 175
Water table, 305
Watson, Paul, 570
Watt, James, 345
Wattenberg, Ben, 118
Wave energy, 391, 412
Wavelengths, 20
Waves, 20, 165–66
Wax gourd vegetable, 181
Wealth of Nations, The, 235